WileyPLUS Learning Space

An easy way to help your students **learn**, **collaborate**, and **grow**.

Personalized Experience

Students create their own study guide while they interact with course content and work on learning activities.

Flexible Course Design

Educators can quickly organize learning activities, manage student collaboration, and customize their course—giving them full control over content as well as the amount of interactivity between students.

Clear Path to Action

With visual reports, it's easy for both students and educators to gauge problem areas and act on what's most important.

Instructor Benefits

Assign activities and add your own materials

Guide students through what's important in the interactive e-textbook by easily assigning specific content

Set up and monitor collaborative learning groups

Assess learner engagement

Gain immediate insights to help inform teaching

Student Benefits

- Instantly know what you need to work on
- Create a personal study plan
- Assess progress along the way
- Participate in class discussions
- Remember what you have learned because you have made deeper connections to the content

are dedicated to supporting you from idea to outcome.

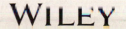

9th Edition

Environment

Peter H. Raven
Missouri Botanical Garden

David M. Hassenzahl
California State University at Chico

Mary Catherine Hager

Nancy Y. Gift
Berea College

Linda R. Berg
former affiliations:
University of Maryland, College Park
St. Petersburg College

WILEY

PUBLISHER & VICE PRESIDENT Petra Recter

EXECUTIVE EDITOR   Ryan Flahive

ASSISTANT EDITOR   Julia Nollen

EDITORIAL ASSISTANT   Kathryn Hancox

MARKETING MANAGER   Suzanne Bochet

SENIOR CONTENT MANAGER   Kevin Holm

TEXT AND COVER DESIGNER   Madelyn Lesure

PHOTO DEPARTMENT MANAGER   Malinda Patelli

SENIOR PHOTO EDITOR   Lisa Gee

SENIOR PRODUCT DESIGNER   Geraldine Osnato

CREATIVE DIRECTOR   Harry Nolan

SENIOR PRODUCTION EDITOR   Sandra Dumas

SENIOR DESIGNER   Madelyn lesure

COVER IMAGE © Alexander B. Murphy

MEETING THE CHALLENGE: © Corbis

YOU CAN MAKE A DIFFERENCE: © Royalty-free/Corbis

FOOD FOR THOUGHT: ©Aboli/iStockphoto

LEARNING OBJECTIVES : © MacieJ Noskowski/iStockphoto

CASE IN POINT: © Corbis Digital Stock/Corbis

This book was typeset in 10/12 Stix at SPi-Global and printed and bound by Quad Graphics/Versailles. The cover was printed by Quad Graphics/Versailles.

Founded in 1807, John Wiley & Sons. Inc. has been a valued source of knowledge and understanding for more than 200 years, helping people around the world meet their needs and fulfill their aspirations. Our company is built on a foundation of principles that include responsibility to the communities we serve and where we live and work. In 2008, we launched a Corporate Citizenship Initiative, a global effort to address the environmental, social, economic, and ethical challenges we face in our business. Among the issues we are addressing are carbon impact, paper specifications and procurement, ethical conduct within our business and among our vendors, and community and charitable support. For more information, please visit our website: www.wiley.com/go/citizenship.

The paper in this book was manufactured by a mill whose forest management programs include sustained yield-harvesting of its timberlands. Sustained yield-harvesting principles ensure that the number of trees cut each year does not exceed the amount of new growth.

This book is printed on acid-free paper. ∞

ISBN-13 (paperback) 978-1118-87582-7

ISBN-13 (AP Edition) 978-1119-05305-7

ISBN-13 (binder-ready version) 978-1118-978443

Printed in the United States of America.

10 9 8 7 6 5 4 3 2 1

TEXT CREDITS: Chapter 2: excerpt from the "Wilderness Essay," a letter written by Wallace Stegner to David Pesonen of the U. of California's Wildland Research Center; Section on successful pollution control efforts: *EPA's 2008 Report on the Environment: Highlights of Trends*. US EPA; Section on natural resources, the environment, and the national income accounts, Main source: Levin, J. "The Economy and the Environment: Revising the National Income Accounts," *IMF Survey* (June 4, 1990). Reprinted by permission of the International Monetary Fund, Washington, D.C.; quote from Bill McKibben from keynote delivered at first meeting of the Association for the Advancement of Sustainability in Higher Education, May 2006; list of 8 principles of deep ecology: Naess, A. and D Rothenberg. *Ecology, Community and Lifestyle*. Cambridge, UK: Cambridge University Press (2001); quote in section on environmental world-views: Seed, J., Macy, J. Fleming P. and Naess, A. *Thinking like a Mountain: Towards a Council of All Beings*. Philadelphia, New Society Publishers (1988); Chapter 12: information on health effects from Chernobyl from Health effects of the Chernobyl accident: an overview, World Health Organization (2006); Chapter 13: Section on Drinking Water Problems from World Health Organization *World Health Report 2008*; Chapter 19: Quote by Ulf Merbold: From Ulf Merbold of the German Aerospace Liaison. Reprinted by permission; Chapter 24: Quote from Bruntland Report: *Our Common Future*. World Commission on Environment and Development. Cary, NC : Oxford University Press (1987). Reprinted by permission of Oxford University; Quote from Kai N. Lee: *Compass and Gyroscope: Integrating Science and Politics for the Environment*. Washington, D.C.: Island Press (1993); 5 recommendations around which the chapter is organized: Brown, L. *Plan B 2.0: Rescuing a Planet under Stress and a Civilization in Trouble*. New York: W.W. Norton (2006); Quote in EnviroNews on Reforestation in Kenya: by Ole Danbolt Mjoes, the Nobel Committee Chair.

Preface

The environmental challenges that today's students will face throughout their lives are characterized by a seeming paradox: They are both increasingly global and increasingly local. Threats to local food production include global climate change. An energy resource can endanger a rare species when it is extracted and global public health after it is used, as when mercury from coal burning reaches the ocean. Improving the conditions of our environment requires that we understand how the choices we make impact air, water, soil, and organisms, as well as their interrelationships. Science is the most appropriate and effective approach to gaining that understanding. It is therefore critical that students learn about the science behind energy, climate change, and other environmental issues that affect them, not only because they will make decisions about energy and climate change but also because they will experience the repercussions if these problems are not dealt with effectively.

The overarching concept of environmental sustainability has never been more important to the field of environmental science than it is today. Sustainability, a central theme of *Environment,* is integrated throughout the text. Yet the more we learn about the environment, the more we realize that interactions among different components of the environment are many and complex. Therefore, a second important theme of *Environment* is environmental systems. Understanding how change to one component affects other processes, places, and organisms is essential to managing existing problems, avoiding future problems, and improving the world in which we live.

From the opening pages, we acquaint students with current environmental issues—issues that have many dimensions and that defy easy solutions. We begin by examining the scientific, historical, ethical, governmental, and economic underpinnings of environmental science. This provides a conceptual foundation for students that they can then bring to bear on the rest of the material in the book. We next explore the basic ecological principles that govern the natural world and consider the many ways in which humans affect the environment. Later chapters examine in detail the effects of human activities, including overpopulation, energy production and consumption, depletion of natural resources, and pollution. Throughout *Environment,* 9th Edition, we relate the topics of a given chapter to food, energy, and climate change, which further reinforces the interactions of environmental systems.

While we avoid unwarranted optimism when presenting these problems, we do not see value in the gloomy predictions of disaster so commonly presented by the media. Instead, we encourage students to take active, positive roles to understand and address the environmental challenges of today and tomorrow.

Raven, Hassenzahl, Hager, Gift, and Berg's *Environment,* 9th Edition, is intended as an introductory text for undergraduate students, both science and nonscience majors. Although relevant to all students, *Environment,* 9th Edition, is particularly appropriate for those majoring in education, journalism, government and politics, and business, as well as the traditional sciences. We assume that our students have little prior knowledge of environmental science. Important ecological concepts and processes are presented in a straightforward, unambiguous manner.

All of the chapters have been painstakingly researched, and extraordinary efforts have been made to obtain the most recent data available. Both instructors and students will benefit from the book's currency because environmental issues and trends are continually changing.

Environment, 9th Edition, integrates important information from many different fields, such as biology, geography, chemistry, geology, physics, economics, sociology, natural resources management, law, and politics. Because environmental science is an interdisciplinary field, this book is appropriate for use in environmental science courses offered by a variety of departments, including (but not limited to) biology, geology, geography, and agriculture.

New to the 9th Edition is an emphasis on food and the environment. This includes a substantially updated chapter on food by new team member Nancy Y. Gift, increased attention to food systems woven throughout the text, and a challenge in each chapter for students to think about food and environment in contexts that affect them. Co-author Mary Catherine Hager likewise brings expertise in ecosystems, population biology, and water ecology. Together, their contributions deepen our ability to present the complex systems that make up our environment in a clear and compelling fashion.

Effective Learning Tools in *Environment*, 9th Edition

A well-developed pedagogical plan that facilitates student mastery of the material has always been a hallmark of *Environment*. The 9th Edition has continued to refine the **learning tools** to help students engage with the key material and apply it to their daily lives. Pedagogical features in this 9th edition include:

Chapter Introductions illustrate certain concepts in the chapter with stories about some of today's most pressing environmental issues.

NEW Food for Thought features in each chapter challenge students to consider how issues in the chapter relate to some aspect of food systems.

 In Your Own Backyard feature provides a critical thinking question at the beginning of each chapter and connects the broad themes of the text to local issues and resources students can investigate.

Learning Objectives at the beginning of each section head indicate in behavioral terms what the student must be able to do to demonstrate mastery of the material in the chapter.

Review Questions at the end of each section give students the opportunity to test their comprehension of the learning objectives.

EnviroNews features provide additional topical material about relevant environmental issues.

EnviroNews on Campus features report on recent campus and student efforts to improve the environment.

 Meeting the Challenge boxes profile environmental success stories.

 You Can Make a Difference boxes suggest specific courses of action or lifestyle changes students can make to improve the environment.

Tables and Graphs, with complete data sources cited in the text, summarize and organize important information.

Marginal Glossaries, located in every chapter, provide handy definitions of the most important terms.

 Case in Point features offer a wide variety of in-depth case studies that address important issues in the field of environmental science.

Energy and Climate Change features relate energy and/or climate change to the topics in each chapter. This feature is easy to locate by its it icon—a compact fluorescent lightbulb superimposed over the sun. This icon is also paired with key graphs and illustrations in each chapter wherever energy and climate change are discussed, and accompanied by a critical thinking question.

Review of Learning Objectives with Selected Key Terms restates the chapter learning objectives and provide a review of the material presented. Boldfaced selected key terms, including marginal glossary terms, are integrated within each summary, enabling students to study vocabulary words in the context of related concepts.

Critical Thinking and Review Questions, many new to this edition, encourage critical thinking and highlight important concepts and applications. At least one question in each chapter provides a systems perspective; another question relates climate and energy to the chapter. Visual questions have been added to each chapter.

Suggested Reading lists for each chapter are available online to provide current references for further learning.

Major Changes in the Ninth Edition

A complete list of all changes and updates to the 9th Edition is too long to fit in the Preface, but several of the more important changes to each chapter follow:

Chapter 1, Introducing Environmental Science and Sustainability, has a new chapter introduction on food systems and sustainability, and a new section on renewable and nonrenewable resources.

In Chapter 2, Environmental Laws, Economics, and Ethics, environmental policies were updated through 2014 and a feature on food and zoning policy was added.

Chapter 3, Ecosystems and Energy, includes an updated discussion of how humans have affected the Antarctic food web.

Chapter 4, Ecosystems and the Physical Environment, features recent examples of deadly and destructive tornadoes and tropical cyclones, and new evidence of how mangroves protect areas against typhoons.

Chapter 5, Ecosystems and Living Organisms, features an expanded section on logistic population growth with added emphasis on implications for sustainable yield and pest control.

Chapter 6, Major Ecosystems of the World, has an added connection between ecosystem productivity across biomes and how that relates to human food available in those biomes.

Chapter 7, Human Health and Environmental Toxicology, features expanded coverage of endemic diseases including ebola, HIV/AIDS, and polio.

Chapter 8, The Human Population, features all-new population data through 2013, and updated discussions of global AIDS, human migration, aging populations, and other pertinent demographic issues.

Chapter 9, The Urban Environment, includes increased emphasis on urban land use and relationship with food production, decreased emphasis on differences between developed/less developed urban centers.

Chapter 10, Energy Consumption, features additional materials on food and energy consumption, and new images of fuel cell vehicles and plug in electric hybrid vehicles.

Chapter 11, Fossil Fuels, now reflects the shifts in domestic energy production in the United States as natural gas imports have declined substantially as hydraulic fracturing has grown, and domestic oil production has increased significantly.

Chapter 12, Renewable Energy and Nuclear Power, includes details on the Fukushima Daiichi nuclear power plant meltdown, and an updated chapter introduction on biofuels.

Chapter 13, Water: A Limited Resource, includes a section on Mississippi River flooding with an extensive evaluation of flood control efforts, offers a new EnviroNews on climate change and crops in Southern Africa, and features current trends in U.S. drought and aquifer depletion, as well as recent data on worldwide availability of clean water and new examples of global water conflicts.

Chapter 14, Soil Resources, now addresses biodynamic farming, and has increased information about micorrhizae

Chapter 15, Mineral Resources, offers a new EnviroNews on conflict minerals and a revised look at the link between industrialization and mineral consumption.

Chapter 16, Biological Resources, includes the most recent numbers of known and endangered species, a substantive update on the status of Earth's biodiversity hotspots, new approaches to combating invasive species, and a new EnviroNews on Campus featuring habitat restoration.

Chapter 17, Land Resources, includes updated National Park information, and discussion of harvesting rainforest fruit in the Food for Thought feature.

Chapter 18, Food Resources, as noted above, is the most substantial update in this edition, with particular focus on urban agriculture practices, green roofs, a discussion of food insecurity and the relationship of policy to hunger, and a new section on Grow Appalachia.

Chapter 19, Air Pollution, features an update of the chapter introduction on southeast Asian air pollution to 2014, and reorganized learning objectives for ozone, acid deposition, and indoor air pollution sections.

Chapter 20, Global Climate Change, features data and figures updated to be consistent with the most recent Intergovernmental Panel on Climate Change reports.

Chapter 21, Water Pollution, includes updated data throughout and features new examples of energy-related water pollution in a revised chapter introduction, as well as a revamped discussion of pharmaceuticals in wastewater.

Chapter 22, Pest Management, includes expanded discussion of Roundup, GM crops, Bt, and pest control in kitchen gardens.

Chapter 23, Solid and Hazardous Wastes, features updated data throughout the chapter, including current statistics on e-waste generation and legislation, and trends in U.S. recycling programs.

Chapter 24, Tomorrow's World, has increased focus on solutions and hopefulness, with additional material on **Wangari Maathai**.

For Students

Different learning styles, different levels of proficiency, different levels of preparation—each of your students is unique. *WileyPLUS Learning Space* empowers them to take advantage of their individual strengths. With *WileyPLUS Learning Space*, students receive timely access to resources that address their demonstrated needs, and get immediate feedback and remediation when needed.

Integrated multimedia resources include:

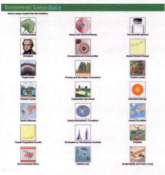

- **Environmental Science Basics** provides a suite of animated concepts and tutorials to give students a solid grounding in key basic environmental concepts. Concepts ranging from global climate change to sustainable agriculture are presented across 21 modules in easy-to-understand language.

For Instructors

WileyPLUS Learning Space empowers you with the tools and resources you need to make your teaching even more effective:

- You can customize your classroom presentation with a wealth of resources and functionality from PowerPoint slides to a database of rich visuals. You can even add your own materials to your *WileyPLUS Learning Space* course.

- With *WileyPLUS Learning Space* you can identify those students who are falling behind and intervene accordingly, without having to wait for them to come to office hours.

- *WileyPLUS Learning Space* simplifies and automates such tasks as student performance assessment, making assignments, scoring student work, keeping grades, and more.

- **Project Activities** relating to the Virtual Field Trips and In Your Own Backyard questions allow instructors to bring learning outside of the classroom and assign critical thinking questions and projects. Students will have the ability to submit completed Project Activities through their *WileyPLUS Learning Space* course.

- **Test Bank** is available on both the instructor companion site and in WileyPLUS *Learning Space*. Containing approximately 60 multiple-choice and essay test items per chapter, this Test Bank offers assessment of both basic understanding and conceptual applications. The Environment, 9th Edition, Test Bank is offered in two formats: MS Word files and a Computerized Test Bank through Respondus. The easy-to-use test-generation program fully supports graphics, print tests, student answer sheets, and answer keys. The software's advanced features allow you to create an exam to your exact specifications.

- **Instructor's Manual,** originally by our co-author David Hassenzahl, is available on both the instructor companion site and in *WileyPLUS Learning Space*. The Instructor's Manual now provides over 90 creative ideas for in-class activities. Also included are answers to all end-of-chapter and review questions prepared by.

- **All Line Illustrations and Photos** from *Environment*, 9th Edition, in jpeg files and PowerPoint format, are available both on the instructor companion site and in *WileyPLUS Learning Space*.

- **PowerPoint Presentations,** are tailored to the topical coverage and learning objectives of *Environment*, 9th Edition. These presentations are designed to convey key text concepts, illustrated by embedded text art. An effort has been made to reduce the number of words on each slide and increase the use of visuals to illustrate concepts. All are available on the instructor companion site and in *WileyPLUS Learning Space*.

- **Animations.** Select text concepts are illustrated using flash animation for student self-study or classroom presentation.

Also Available

Environmental Science: Active Learning Laboratories and Applied Problem Sets, 2rd Edition, by Travis Wagner and Robert Sanford, both of the

University of Southern Maine, is designed to introduce environmental science students to the broad, interdisciplinary field of environmental science. It presents specific labs that use natural and social science concepts and encourages a hands-on approach to evaluating the impacts from the environmental/human interface. The laboratory and homework activities are designed to be low cost and to reflect a sustainable approach in both practice and theory. *Environmental Science: Active Learning Laboratories and Applied Problem Sets*, 2rd Edition, is available as a stand-alone or in a package with *Environment*, 9th Edition. Contact your Wiley representative for more information.

For more Environmental Science Case Studies, visit **customselect.wiley.com** to view the **Environmental Science Regional Case Study Collection** and customize your course materials with a rich collection of local and global examples (http://customselect.wiley.com/collection/escasestudies).

Book Companion Site   (www.wiley.com/college/raven)

For Students

- Flashcards and Animations
- Quantitative and Essay Questions, as well as useful website links

For Instructors

- Visual Library; all images in jpg and PowerPoint formats
- Instructor's Manual; Test Bank; Lecture PowerPoint Presentations

Instructor Resources are password protected.

Wiley may provide complementary instructional aids and supplements or supplement packages to those adopters qualified under our adoption policy. Please contact your sales representative for more information. If as an adopter or potential user you receive supplements you do not need, please return them to your sales representative or send them to: Attn: Wiley Returns Department, Heller Park Center, 360 Mill Road, Edison, NJ 08817.

Acknowledgments

The development and production of *Environment*, 9th Edition, was a process involving interaction and cooperation among the author team and between the authors and many individuals in our home and professional environments. We are keenly aware of the valuable input and support from editors, colleagues, and students. We also owe our families a debt of gratitude for their understanding, support, and encouragement as we struggled through many revisions and deadlines.

THE EDITORIAL ENVIRONMENT

Preparing this book has been an enormous undertaking, but working with the outstanding editorial and production staff at John Wiley & Sons has made it an enjoyable task. We thank our Publisher Petra Recter and Executive Editor Ryan Flahive for their support, enthusiasm, and ideas. We also thank Senior Product Designer Geraldine Osnato for overseeing and coordinating the development of the supplements and media components.

We are grateful to Marketing Manager Suzanne Bochet for her superb marketing and sales efforts. We thank Lisa Gee for her contribution as photo researcher, helping us find the wonderful photographs that enhance the text. We appreciate the artistic expertise of Senior Designer Madelyn Lesure for the interior design and Senior Production Editor Sandra Dumas for her efforts in maintaining a smooth production process. Our colleagues and students have provided us with valuable input and have played an important role in shaping *Environment*, 9th Edition. We thank them and ask for additional comments and suggestions from instructors and students who use this text. You can reach us through our editors at John Wiley & Sons; they will see that we get your comments. Any errors can be corrected in subsequent printings of the book, and more general suggestions can be incorporated into future editions.

THE PROFESSIONAL ENVIRONMENT

The success of *Environment*, 9th Edition, is due largely to the quality of the many professors and specialists who have read the manuscript during various stages of its preparation and provided us with valuable suggestions for improving it. In addition, the reviewers of the first eight editions made important contributions that are still part of this book.

Reviewers and Virtual Focus Group Participants of the 9th Edition

John Aliff, *Georgia Perimeter College*;
Sara Alexander, *Baylor University*;
John Anderson, *Georgia Perimeter College*;
Christine Baumann Feurt, *University of New England*;
Zygmunt Osiecki, *Chatham University*;
Mary Beth Kolozvary, *Siena College*;
Jacob Kagey, *University of Detroit Mercy*;
Howard Whiteman, *Murray State University*;
Patti Howell, *South Georgia College*;
Brian Ernsting, *University of Evansville*;
Robert Bruck, *North Carolina State University*;
Arielle Burlett, *Chatham University*;
Jennifer Cole, *Northeastern University*;
Iver Duedall, *Brevard Community College*;
Elizabeth Johnson, *Post University*;
Julie Kilbride, *Hudson Valley Community College*;
John Krolak, *Georgia Perimeter College*;
Andrew Lapinski, *Reading Area Community College*;
Jody Martin de Camilo, *St. Louis Community College*;
Jessica Mooney, *Chatham University*;
Sherri Morris, *Bradley University*;
Daniel Ratcliff, *Rose State College*;
Katherine Van de Wal, *Community College of Baltimore County–Essex*;
Dennis Walsh, *MassBay Community College*;
Karen Wehn, *Buffalo State College and Erie Community College*;
John Weishampel, *University of Central Florida*;
James Yount, *Brevard Community College*;

Reviewers of Earlier Editions

James Anthony Abbott, *Stetson University*;
David Aborn, *University of Tennessee, Chattanooga*;
Marc Albrecht, *University of Nebraska, Kearney*;
John Aliff, *Georgia Perimeter College*;
Sylvester Allred, *Northern Arizona University*;
Diana Anderson, *Northern Arizona University*;
Abbed Babaei, *Cleveland State University*;
Lynne Bankert, *Erie Community College*;
Dora Barlaz, *The Horace Mann School*;
David Bass, *University of Central Oklahoma*;
Richard Bates, *Rancho Santiago Community College*;
George Bean, *University of Maryland*;
Mark Belk, *Brigham Young University*;
David Belt, *Penn Valley Community College*;
Bruce Bennett, *Community College of Rhode Island, Knight Campus*;
Eliezer Bermúdez, *Indiana State University*;
Patricia Beyer, *University of Pennsylvania, Bloomsburg*;
Rosina Bierbaum, *University of Michigan*;
Rodger Bland, *Central Michigan University*;
David Boose, *Gonzaga University*;
Richard Bowden, *Allegheny College*;
William Bowie, *Caleway Community College*;
Frederic Brenner, *Grove City College*;
Christopher Brown, *University of Kansas*;
John Dryden Burton, *McLennan Community College*;
Kelly Cain, *University of Arizona*;
Stefan Cairns, *Central Missouri State University*;
Catherine Carter, *Georgia Perimeter College*;
Ann Causey, *Prescott College*;
Michelle Cawthorn, *Georgia Southern University*;
Brad S. Chandler, *Palo Alto College*;
Ravi Chandran, *De Vry University*;
Jeff Chaumba, *University of Georgia*;

Robert Chen, *University of Massachusetts, Boston*;
Gary Clambey, *North Dakota State University*;
Lu Anne Clark, *Lansing Community College*;
Robert Clegern, *University of Maryland University College*;
George Cline, *Jacksonville State University*;
Jennifer Rivers Cole, *Northeastern University*;
Harold Cones, *Christopher Newport College*;
Bruce Congdon, *Seattle Pacific University*;
Rebecca Cook, *Lambuth University*;
Ellen J. Crivella, *Ohio State University*;
John H. Crow, *Rutgers State University*;
Nate Currit, *Pennsylvania State University*;
Ruth Darling, *Sierra College*;
Karen DeFries, *Erie Community College*;
Armando de la Cruz, *Mississippi State University*;
Michael L. Denniston, *Georgia Perimeter College*;
Kelli Doehler, *University of Nevada, Las Vegas*;
Lisa Doner, *Plymouth State University*;
Michael L. Draney, *University of Wisconsin, Green Bay*;
JodyLee Duek, *Pima Community College*;
James P. Dunn, *Grand Valley State University*;
Jean Dupon, *Menlo College*;
Laurie Eberhardt, *Valparaiso University*;
Thomas Emmel, *University of Florida*;
Donald Emmeluth, *Montgomery Community College, Fulton*;
Margaret "Tim" Erskine, *Lansing Community College*;
Timothy Farnham, *University of Nevada, Las Vegas*;
Allen Lee Farrand, *Bellevue Community College*;
Huan Feng, *Montclair State University*;
Kate Fish, EarthWays, *St. Louis, Missouri;*
Steve Fleckenstein, *Sullivan County Community College*;

Andy Friedland, *Dartmouth College*;
Carl Friese, *University of Dayton*;
Allan A. Gahr, *Gordon College*;
Bob Galbraith, *Crafton Hills College*;
Carey Gazis, *Central Washington University*;
Megan Gibbons, *Birmingham-Southern College*;
Jeffrey Gordon, *Bowling Green State University*;
Joseph Goy, *Harding University*;
Noah Greenwould, *Center for Biological Diversity*;
Stan Guffey, *University of Tennessee*;
Mark Gustafon, *Texas Lutheran University*;
Elaine J. Hanford, *Collin College*;
John Harley, *Eastern Kentucky University*;
Neil Harriman, *University of Wisconsin, Oshkosh*;
Denny Harris, *University of Kentucky*;
Syed E. Hasan, *University of Missouri*;
David Hassenzahl, *University of Nevada, Las Vegas*;
Bronwen Haugland, *Truckee Meadows Community*;
Michelle M. Hluchy, *Alfred University*;
Leland Holland, *Pasco-Hernando Community College*;
James Horwitz, *Palm Beach Community College*;
Dan Ippolito, *Anderson University*;
Solomon Isiorho, *Indiana University/Purdue University, Fort Wayne*;
Jeff Jaeger, *University of Nevada, Las Vegas*;
John Jahoda, *Bridgewater State College*;
Jan Jenner, *Talladega College*;
Richard A. Jensen, *Hofstra University*;
David Johnson, *Michigan State University*;
Elizabeth Johnson, *Post University*;
Patricia Johnson, *Palm Beach Community College*;
Linda Johnston, *University of Kansas*;
Thomas Justice, *McLennan Community College*;
Karyn Kakiba-Russell, *Mount San Antonio College*;
Colleen Kelley, *Northern Arizona University*;
William E. Kelso, *Louisiana State University*;
Erica Kipp, *Pace University*;
Nancy Knowlton, Scripps Institution of Oceanography, *University of California, San Diego*;
Penny Koines, *University of Maryland*;
Allen Koop, *Grand Valley State University*;
Karl F. Korfmacher, *Denison University*;
Katrina Smith Korfmacher, *Denison University*;
Tim Kusky, *St. Louis University*;
Michael Larsen, *Campbell University*;
Robert Larsson Essex County *Community College*;
Ernesto Lasso de la Vega, *Edison College*;
Ericka Lawson, *Baylor College of Medicine (formerly Columbia College)*;
Norm Leeling, *Grand Valley State University*;
Joe Lefevre, *State University of New York, Oswego*;
James K. Lein, *Ohio University*;
Tammy Liles, *Lexington Community College*;
Patrick Lloyd, *Kingsborough Community College*;
James Luken, *Northern Kentucky University*;

Timothy Lyon, *Ball State University*;
Michael MacCracken, *Climate Change Programs Climate Institute*;
Andy Madison, *Union University*;
Blase Maffia, *University of Miami*;
Jane M. Magill, *Texas A&M University*;
Alberto Mancinelli, *Columbia University*;
Eric Maurer, *University of Cincinnati*;
Julie Maxson, *Metropolitan State University*;
Michael McCarthy, *Eastern Arizona College*;
William Russ McClain, *Davis and Elkins College*;
Linda Matson McMurry, *San Pedro Academy/ Texas Department of Agriculture*;
George Middendorf, *Howard University*;
Chris Migliaccio, *Miami-Dade Community College*;
Gary Miller, *University of North Carolina, Asheville*;
Mark Morgan, *Rutgers University*;
Michael Morgan, *University of Wisconsin, Green Bay*;
James Morris, *University of South Carolina*;
Andrew Neill, *Joliet Junior College*;
Helen Neill, *University of Nevada, Las Vegas*;
Lisa Newton, *Fairfield University*;
Zia Nisani, *Antelope Valley Community College*;
Amy Northrup, *University of Nevada, Las Vegas*;
Diane O'Connell, *Schoolcraft College*;
James T. Oris, *Miami University of Ohio*;
Robert Paoletti, *Kings College*;
Richard Pemble, *Moorhead State University*;
Barry Perlmutter, *College of Southern Nevada*;
Scott Peterson, *Nuclear Energy Institute*;
Thomas E. Pliske, *Florida International University*;
Ervin Poduska, *Kirkwood Community College*;
Sandra Postel, *Global Water Policy Project*;
Lowell Pritchard, *Emory University*;
James Ramsey, *University of Wisconsin, Stevens Point*;
Elizabeth Reeder, *Loyola College*;
James J. Reisa, Environmental Sciences and Toxicology, *National Research Council*;
Maralyn Renner, *College of the Redwoods*;
Howard Richardson, *Central Michigan University*;
Mary M. Riestenberg, *College of Mount St. Joseph*;
Virginia Rivers, *Truckee Meadows Community College*;
W. Guy Rivers, *Lynchburg College*;
Barbara C. Reynolds, *University of North Carolina, Asheville*;
Sandy Rock, *Bellevue Community College*;
Vikki Rodgers, *Babson College*;
Barbra Roller, *Miami Dade Community College, South Campus*;
Richard Rosenberg, *Pennsylvania State University*;
John Ruppert, *Fairleigh Dickinson University*;
Wendy Ryan, *Kutztown University*;
Neil Sabine, *Indiana University East*;
Patricia A. Saunders, *Ashland University*;
Frank Schiavo, *San Jose State University*;

Jeffery Schneider, *State University of New York, Oswego*;
Judith Meyer Schultz, *University of Cincinnati*;
Thomas Schweizer, *Princeton Energy Resources International*;
Sullivan Sealey, *University of Miami*;
Surbhi Sharma, *University of Nevada, Las Vegas*;
Marvin Sigal, *Gaston College*;
Heather Skaza, *University of Nevada, Las Vegas*;
Mark Smith, *Fullerton College*;
Mark Smith, *Victor Valley Community College*;
Val H. Smith, *University of Kansas*;
William J. Smith, Jr., *University of Nevada, Las Vegas*;
Nicholas Smith-Sebasto, *University of Illinois*;
Robert Socolow, *Princeton University*;
Joyce Solochek, *Marquette University*;
Morris Sotonoff, *Chicago State University*;
Krystyna Stave, *University of Nevada, Las Vegas*;
Steven Steel, *Bowling Green State University*;
Ashley Steinhart, *Fort Lewis College*;
Allan Stevens, *Snow College*;
Richard Strange, *University of Tennessee*;
Richard P. Stringer, *Harrisburg Area Community College, Lancaster*;
Andrew Suarez, *University of Illinois*;
Lynda Swander, *Johnson County Community College*;
Karen Swanson, *William Paterson University*;
Jody A. C. Terrell, *Texas Woman's University*;
Kip Thompson, *Ozarks Technical Community College*;
Norm Thompson, *University of Georgia*;
Jerry Towle, *California State University, Fresno*;
Jack Turner, *University of South Carolina, Spartanburg*;
Mike Tveten, *Pima Community College*;
Eileen Van Tassel, *Michigan State University*;
G. Peter van Walsum, *Baylor University*;
Bruce Van Zee, *Sierra College*;
Laine Vignona, *University of Wisconsin, River Falls*;
Matthew Wagner, *Texas A&M University*;
Rachel Walker, *University of Minnesota*;
Phillip Watson, *Ferris State University*;
Alicia Whatley, *Troy State University*;
Jeffrey White, *Indiana University*;
Susan Whitehead, *Becker College*;
John Wielichowski, *Milwaukee Area Technical College*;
Jeffrey Wilcox, *University of North Carolina, Asheville*;
James Willard, *Cleveland State University*;
Ray E. Williams, *Rio Hondo College*;
Akiko Yamane, *California State University, Fresno*;
James Robert Yount, *Brevard Community College*.

About the Authors

Peter H. Raven, one of the world's leading botanists, has dedicated nearly five decades to conservation and biodiversity as president (now emeritus) of the Missouri Botanical Garden and professor of Botany at Washington University in St. Louis, where he has cultivated a world-class institution of horticultural display, education, and research. Described by *Time* magazine as a "Hero for the Planet," Dr. Raven champions research around the world to preserve endangered species and is a leading advocate for conservation and a sustainable environment.

Dr. Raven is a trustee of the National Geographic Society and is a past president of the American Association for the Advancement of Science. He is the recipient of numerous prizes and awards, including the prestigious National Medal of Science in 2001, the highest award for scientific accomplishments in this country; Japan's International Prize for Biology; the Environmental Prize of the Institut de la Vie; the Volvo Environment Prize; the Tyler Prize for Environmental Achievement; and the Sasakawa Environment Prize. He also has held Guggenheim and MacArthur fellowships.

Dr. Raven received his Ph.D. from the University of California, Los Angeles, after completing his undergraduate work at the University of California, Berkeley.

Dr. David M. Hassenzahl, dean of the College of Natural Sciences at the California State University at Chico, is an internationally recognized scholar of sustainability and risk analysis. His leadership, research, teaching, and outreach efforts focus on incorporating scientific information and expertise into public decisions, with particular emphasis on the management, interpretation, and communication of uncertainty. He has dedicated the last three decades to addressing subjects as diverse as climate change, energy, toxic chemicals, nuclear materials, and public health; has presented on these topics on four continents; and has supported decision making in public, private, and not-for-profit contexts.

Photo by Jason A. Halley.
Courtesy of California State University, Chico

Previously, Dr. Hassenzahl served as founding dean of the Falk School of Sustainability at Chatham University, and Chair of the Department of Environmental Studies at the University of Nevada, Las Vegas. Dr. Hassenzahl holds a B.A. in environmental science and paleontology from the University of California at Berkeley and a Ph.D. in science, technology, and environmental policy from Princeton University.

A Senior Fellow of the National Council for Science and the Environment, Dr. Hassenzahl's work has been supported by the National Science Foundation and the National Aeronautics and Space Administration. He has received the Society for Risk Analysis Outstanding Educator Award and the UNLV Foundation Distinguished Teaching Award.

Mary Catherine Hager is a professional science writer and editor specializing in life and earth sciences. She received a double-major B.A. in environmental science and biology from the University of Virginia and an M.S. in zoology from the University of Georgia. Ms. Hager worked as an editor for an environmental consulting firm and as a senior editor for a scientific reference publisher. For more than 20 years, she has written and edited for environmental science, biology, and ecology textbooks primarily targeting college audiences. Additionally, she has published articles in environmental trade magazines and edited federal and state assessments of wetlands conservation issues. Her writing and editing pursuits are a natural outcome of her scientific training and curiosity, coupled with her love of reading and effective communication.

Nancy Y. Gift, Compton Chair of Sustainability at Berea College, in Berea, Kentucky, enjoys teaching in a wide range of sustainability-related courses. She has taught women's nature writers, first-year writing, grant writing, ecology, weed science, sustainable agriculture, history of African-American farmers, and mathematics for the environment; she continually teaches introductory sustainability and environmental science. She has written two books for the public on the role of weeds in healthy lawns, and she is an associate editor for *Agronomy Journal*. She holds a B.A. in biology from Harvard University, an M.S. in crop and soil science from the University of Kentucky, and a Ph.D. in crop science from Cornell University.

Nancy Y. Gift

Dr. Gift currently works with faculty across a variety of disciplines to encourage the infusion of sustainability in a wide range of fields. She won the Buhl Professor of Humanities award at Chatham University for her work there in teaching interdisciplinary courses. She taught environmental biology courses at the University of Chicago. Before committing to life as an academic, she worked at Cornell as a summer forest laborer for Arnot Forest, as an apple post-harvest storage lab technician, as a statistical analyst in veterinary epidemiology, and as an extension educator in sustainable weed management.

Linda R. Berg is an award-winning teacher and textbook author. She received her Ph.D. from the University of Maryland, College Park, after completing her B.S. and M.S. at the same institution. Her recent interests involve the Florida Everglades and conservation biology.

Dr. Berg formerly taught at the University of Maryland, College Park, for almost two decades, followed by St. Petersburg College in Florida for one decade. She taught introductory courses in environmental science, biology, and botany to thousands of students and received numerous teaching and service awards. Dr. Berg is also the recipient of many national and regional awards, including the National Science Teachers Association Award for Innovations in College Science Teaching, the Nation's Capital Area Disabled student Services Award, and the Washington Academy of Sciences Award in University Science Teaching.

During her career as a professional science writer, Dr. Berg has authored or co-authored several leading college science textbooks. Her writing reflects her teaching style and love of science.

Contents

HECTOR GUERRERO / AFP / Getty Images

Introducing Environmental Science and Sustainability

Industrial production of chicken requires many inputs, including feed, heating and cooling, and often antibiotics and hormones to accelerate growth. It also generates waste streams that can lead to air and water pollution if not treated or managed.

One of the best ways to understand our complex relationship with the global environment is to use food as a lens. Culture, price, personal tastes, and availability often determine our food choices. However, we rarely think about how a particular meal comes to our plate, and how its production impacts the environment.

Consider a simple chicken sandwich. To make bread, commercial bakeries rely on wheat from farms that require large amounts of inputs (land, irrigation water, fertilizers, and pesticides) and mechanization (usually diesel-fueled trucks and tractors). Using land for agriculture displaces native plants and animals, excess fertilizers and pesticides enter waterways and flow to the ocean, and burning diesel releases pollutants into the atmosphere. Once wheat is harvested, it is sent to a plant that grinds it into flour. This requires additional energy and produces a stream of organic waste material. The wheat is then shipped to a bakery, which adds water, sugar, yeast, corn syrup, vitamins and minerals, preservatives, oil, and other ingredients—each of which has also been processed and transported. The bread is then packed into plastic bags and delivered to stores and restaurants, hundreds or thousands of miles from where the wheat was grown. Each step uses energy, adds packaging, and generates solid and liquid wastes. Packaging, as well as uneaten or damaged bread, is usually thrown into the garbage and sent to a landfill.

Commercially grown chicken produces even more impact on the environment, since chicken feed grain has to be grown, processed, and delivered to a poultry farm (**photo**). Raising, processing, cooking, packaging, and delivering the chicken all require energy and material inputs and result in pollution and waste. In addition, chickens and other animals are often given antibiotics to make them grow faster. This can lead to antibiotic-resistant diseases, and the antibiotics can disrupt natural ecosystems.

There are more sustainable alternatives to the chicken sandwich. Wheat, grain, and chicken can be grown using methods that greatly reduce impacts on the environment. Foods that are grown near where they are consumed require less energy for transportation. Alternative pest management can reduce demand for pesticides and antibiotics. Bulk packaging and composted food waste can reduce or even eliminate the need for landfills. But even when these practices are adopted, growing and processing food requires land, water, energy, and other inputs.

Humans have developed agriculture over several thousand years. In so doing, we have altered ecosystems, cut down forests, shifted waterways, and driven plants and animals to extinction. At the same time, we have helped some plants and animals—including wheat and chicken, but rats and cockroaches as well—to become dominant species. Our agricultural practices have even contributed to climate change, which in turn forces us to adapt our food production practices. Knowing how something as simple as a choice between two sandwiches can have wide-ranging impacts on the environment is a great point from which to begin to understand the relationship between humans and the environment.

In Your Own Backyard...

Where and by whom is food grown near where you live? Look in your cupboards and refrigerator: Where and by whom are most of the foods you eat grown? How might switching to locally grown foods affect your diet and your food budget?

Human Impacts on the Environment

- Explain how human activities affect global systems.
- Describe the factors that characterize human development and how they impact environment and sustainability.

Earth is remarkably suited for life. Water, important both in the internal composition of organisms and as an external environmental factor affecting life, covers three-fourths of the planet. Earth's temperature is habitable—neither too hot, as on Mercury and Venus, nor too cold, as on Mars and the outer planets. We receive a moderate amount of sunlight—enough to power photosynthesis, which supports almost all the life-forms that inhabit Earth. Our atmosphere bathes the planet in gases and provides essential oxygen and carbon dioxide that organisms require. On land, soil develops from rock and provides support and minerals for plants. Mountains that arise from geologic processes and then erode over vast spans of time affect weather patterns, provide minerals, and store reservoirs of fresh water as ice and snow that melt and flow to lowlands during the warmer months. Lakes and ponds, rivers and streams, wetlands, and groundwater reservoirs provide terrestrial organisms with fresh water.

Earth's abundant natural resources have provided the backdrop for a parade of living things to evolve. Life has existed on Earth for about 3.8 billion years. Although early Earth was inhospitable by modern standards, it provided the raw materials and energy needed for early life-forms to arise and adapt. Some of these early cells evolved over time into simple multicellular organisms—early plants, animals, and fungi. Today, several million species inhabit the planet. A representative sample of Earth's biological diversity includes intestinal bacteria, paramecia, poisonous mushrooms, leafhoppers, prickly pear cacti, seahorses, dogwoods, angelfish, daisies, mosquitoes, pitch pines, polar bears, spider monkeys, and roadrunners (**Figure 1.1**).

About 100,000 years ago—a mere blip in Earth's 4.5-billion-year history—an evolutionary milestone began with the appearance of modern humans in Africa. Large brains and the ability to communicate made our species successful. Over time, our population grew; we expanded our range throughout the planet and increasingly impacted the environment with our presence and our technologies. These technologies have allowed many people in the world lives with access to well-lit and air-conditioned buildings, effective medical treatment, high-speed transportation, and uninterrupted food supplies. This has been particularly true in North America, Western Europe, and Japan; increasingly, many urban residents in China, India, South America, and parts of Africa have similar access to wealth and material goods.

Today, the human species is the most significant agent of environmental change on our planet. Our burgeoning population and increasing use of energy, materials, and land transform natural systems to meet our needs and desires. Our activities consume ever-increasing amounts of Earth's abundant but finite resources—rich topsoil, clean water, and breathable air. This alteration of natual systems eradicates many types of ecosystems and thousands upon thousands of unique species that inhabit them. Evidence continues to accumulate that human-induced climate change alters the natural environment in disruptive ways. Human activities are disrupting global **systems**.

This book introduces the major impacts that humans have on the environment. It considers ways to better manage those impacts, while emphasizing that each possible choice has the potential to cause additional impacts. Most important, it explains the value of minimizing human impact on our planet. Our lives and well-being, as well as those of future generations, depend on our ability to manage Earth's environmental resources effectively.

system A set of components that interact and function as a whole.

INCREASING HUMAN NUMBERS

Figure 1.2, a nighttime satellite photograph of North America, including the United States, Mexico, and Canada, depicts the home of about 470 million people. The tiny specks of light represent cities, with the great metropolitan areas, such as New York along the northeastern seacoast, ablaze with light.

The driver of all other environmental problems, the one that links all others, is the many people who live in the area shown in this picture. According to the United Nations, in 1950 only eight cities in the world had populations larger than 5 million, the largest being New York, with 12.3 million. By 2011 Tokyo, Japan had 17.8 million inhabitants, with 37.2 million inhabitants in the greater Tokyo metropolitan area. The combined population of the world's 10 largest urban agglomerations was over 200 million (see Table 9.1).

In 2011, the human population as a whole passed a significant milestone: 7 billion individuals. Not only is this figure incomprehensibly large, but also our population grew this large in a brief span of time. In 1960, the human population was only 3 billion (**Figure 1.3**). By 1975, there were 4 billion people, and by 1999, there were 6 billion. The 7.2 billion people who currently inhabit our planet consume great quantities of food and water, use a great deal of energy and raw materials, and produce much waste.

Despite family planning efforts in many countries, population growth rates do not change quickly. Several billion people will be

Figure 1.1 **A male greater roadrunner carries a desert spiny lizard it has captured.** Life abounds on Earth, and every organism is linked to many others, including humans. Photographed in New Mexico.

©Deco/Alamy

Figure 1.2   **Satellite view of North America at night.** This image shows most major cities and metropolitan areas in the United States, Mexico, and Canada.

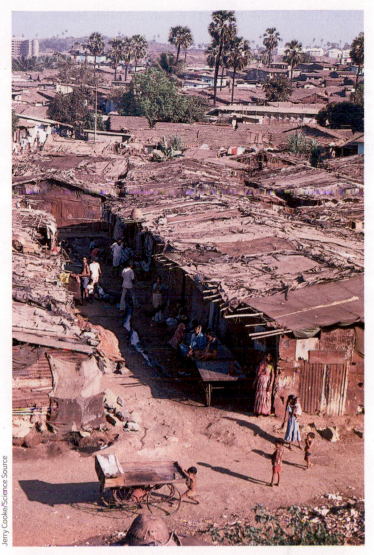
Jerry Cooke/Science Source

Figure 1.4   **Slum in Mumbai (Bombay), India.** Many of the world's people live in extreme poverty. One trend associated with poverty is the increasing movement of poor people from rural to urban areas. As a result, the number of poor people living in or around the fringes of cities is mushrooming.

added to the world in the twenty-first century, so even if we remain concerned about the impacts of a growing population and even if our solutions are effective, the coming decades may be clouded with tragedies. The conditions of life for many people may worsen considerably.

On a global level, nearly one of every two people live in extreme **poverty** (**Figure 1.4**). One measure of poverty is having a per capita income of less than $2 per day, expressed in U.S. dollars adjusted for purchasing power. More than 2.5 billion people—about 40% of the total world population—currently live at this level of poverty. Poverty is associated with low life expectancy, high infant mortality, illiteracy, and inadequate access to health services, safe water, and balanced nutrition. According to the UN Food and

poverty A condition in which people cannot meet their basic needs for adequate food, clothing, shelter, education, or health.

Agricultural Organization, at least 1 billion people (many of them children) lack access to the food needed for healthy, productive lives.

Most demographers (people who study human populations) expect the world population to stabilize before the end of the current century. Worldwide, fertility rates have decereased to a current average of about three children per family, and this average is projected to continue to decline in coming decades. Expert projections for world population at the end of the twenty-first century range from about 8.3 billion to 10.9 billion, depending largely on how fast the fertility rate decreases (see Figure 8.3).

No one knows whether Earth can support so many people indefinitely. Among the tasks we must accomplish is feeding a world population considerably larger than the present one without undermining the natural resources that support us. Our ability to achieve this goal will determine the quality of life for our children and grandchildren.

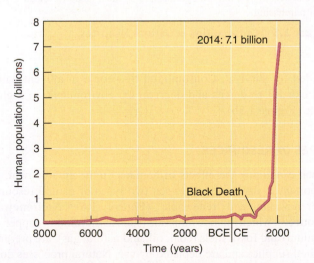

Figure 1.3   **Human population growth.** It took thousands of years for the human population to reach 1 billion (in 1800) but only 130 years to reach 2 billion (1930). It took only 30 years to reach 3 billion (1960), 15 years to reach 4 billion (1975), 12 years to reach 5 billion (1987), 12 years to reach 6 billion (1999), and 13 years to reach 7 billion (2012). (Population Reference Bureau)

DEVELOPMENT, ENVIRONMENT, AND SUSTAINABILITY

Until recently, demographers differentiated countries as highly developed, moderately developed, and less developed. The United States, Canada, Japan, and most of Europe, which represent 19% of the world's population but more than 50% of global economic activity, are **highly developed countries**. Development in this context is based mainly on total wealth of the country. The world's poorest countries, including Bangladesh, Kenya, and Nicaragua, are considered **less developed countries (LDCs).** Cheap, unskilled labor is abundant in LDCs, but capital for investment is scarce. Most LDC economies are agriculturally based, often for only one or a few crops. As a result, crop failure or a lower world market value for that crop is catastrophic to the economy. Hunger, disease, and illiteracy are common in LDCs.

highly developed countries Countries with complex industrial bases, low rates of population growth, and high per capita incomes.

less developed countries (LDCs) Developing countries with a low level of industrialization, a high fertility rate, a high infant mortality rate, and a low per capita income (relative to highly developed countries).

However, recent decades have seen substantial increases in wealth for many urban residents in previously less developed countries, including China, India, Brazil, and Mexico. These countries have substantial *income disparities,* meaning that other urban residents and most of the rural inhabitants of those remain poor, and lack access to cars, electricity, fresh water, and modern medical technology. Consequently, using the total wealth or income of a country may not usefully describe the well-being of people in that country. More appropriate measures can include the percentage of residents who make more than $2 per day, have access to fresh water and electricity, or have access to education.

REVIEW

1. What is one example of a global system?
2. How do the total wealth of a country and income disparity relate to sustainability?

Population, Resources, and the Environment

LEARNING OBJECTIVES

- Differentiate between renewable and nonrenewable resources.
- Explain the impact of population and affluence on consumption.
- Define *ecological footprint.*
- Describe the three most important factors that determine human impact on the environment.

The relationships among population growth, use of natural resources, and environmental degradation are complex. We address the details of resource management and environmental problems in this and later chapters, but for now, let us consider two useful generalizations: (1) The resources essential to an individual's survival are small, but a rapidly increasing population tends to overwhelm and deplete local soils, forests, and other natural resources (**Figure 1.5a**). (2) In highly developed countries, individual resource demands are large, far above what is

(a) A trend towards larger homes leads to greater consumption in a number of ways: more energy to heat and cool, more distance to travel between homes, and more furnishings inside the home.

(b) The "tiny house" movement seeks to simplify individual living, and as a consequence reduces consumption. Tiny house occupants have no place to accumulate material goods, and so must be more selective in their purchasing.

Figure 1.5   **Consumption of natural resources.**

needed for survival. Consumption by people in affluent nations can exhaust resources and degrade the environment on a global scale (**Figure 1.5b**).

TYPES OF RESOURCES

When examining the effects of humans on the environment, it is important to distinguish between two types of natural resources: nonrenewable and renewable (**Figure 1.6**). **Nonrenewable resources**, which include minerals (such as aluminum, copper, and uranium) and fossil fuels (coal, oil, and natural gas), are present in limited supplies and are depleted by use. Natural processes do not replenish nonrenewable resources within a reasonable period on the human time scale. Fossil fuels, for example, take millions of years to form.

In addition to a nation's population, several other factors affect how nonrenewable resources are used, including how efficiently the resource is extracted and processed as well as how much of it

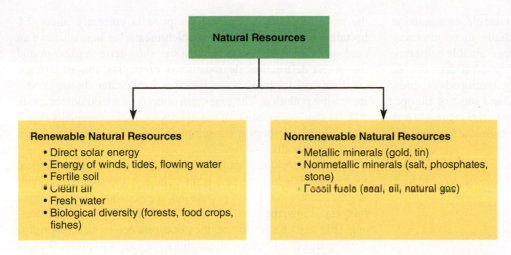

Figure 1.6 Natural resources. Nonrenewable resources are replaced on a geologic time scale, and their supply diminishes with use. Renewable resources can be (but are not always) replaced on a fairly short time scale; as will be explained in later chapters, most renewable resources are derived from the sun's energy.

is required or consumed by different groups. People in the United States, Canada, and other highly developed nations tend to consume most of the world's nonrenewable resources. Nonetheless, Earth has a finite supply of nonrenewable resources that sooner or later will be exhausted. In time, technological advances may provide substitutes for some nonrenewable resources. Slowing the rate of population growth and consumption will buy time to develop such alternatives.

Some examples of **renewable resources** are trees, fishes, fertile agricultural soil, and fresh water. Nature replaces these resources fairly rapidly (on a scale of days to centuries), and they can be used forever as long as they are not overexploited in the short term. In developing countries, forests, fisheries, and agricultural land are particularly important renewable resources because they provide food. Indeed, many people in developing countries are subsistence farmers who harvest just enough food so that they and their families can survive.

Rapid population growth can cause the overexploitation of renewable resources. For example, large numbers of poor people must grow crops on land inappropriate for farming, such as on mountain slopes or in tropical rain forests. Although this practice may provide a short-term solution to the need for food, it does not work in the long term: When these lands are cleared for farming, their agricultural productivity declines rapidly, and severe environmental deterioration occurs. Renewable resources are usually only *potentially* renewable. They must be used in a *sustainable* way—in a manner that gives them time to replace or replenish themselves.

The effects of population growth on natural resources are particularly critical in developing countries. The economic growth of developing countries is often tied to the exploitation of their natural resources, usually for export to highly developed countries. Developing countries are faced with the difficult choice of exploiting natural resources to provide for their expanding populations in the short term (to pay for food or to cover debts) or conserving those resources for future generations.

It is instructive to note that the economic growth and development of the United States, Canada, and other highly developed nations came about through the exploitation and, in some cases, the destruction of resources. Continued economic growth in highly developed countries now relies significantly on the importation of these resources from less developed countries. One

of the reasons economic growth in highly developed countries has been possible is the uneven distribution of both renewable and nonrenewable resources around the world. Many very poor countries—Ethiopia, for example—have only limited fossil-fuel resources.

RESOURCE CONSUMPTION

Consumption is the human use of materials and energy. Consumption, which is both an economic and a social act, provides the consumer with a sense of identity as well as status among peers. Advertisers promote consumption as a way to achieve happiness. Western culture encourages spending and consumption well beyond that which is necessary for survival.

In general, people in highly developed countries are extravagant consumers; their use of resources is greatly out of proportion to their numbers. A single child born in a highly developed country may have a greater impact on the environment and on resource depletion than 12 or more children born in a developing country. Many natural resources are used to provide the automobiles, air conditioners, disposable diapers, cell phones, DVD players, computers, clothes, newspapers, athletic shoes, furniture, boats, and other comforts of life in highly developed countries. Yet such consumer goods represent a small fraction of the total materials and energy required to produce and distribute them. According to the Worldwatch Institute, a private research institution in Washington, D.C., Americans collectively consume almost 10 billion tons of materials every year. The disproportionately large consumption of resources by highly developed countries affects natural resources and the environment as much as or more than the population explosion in the developing world.

Unsustainable Consumption Consumption in a country is unsustainable if the level of demand on its resource base damages the environment or depletes resources to such an extent that future generations will have lower qualities of life. In comparing human impact on the environment in developing and highly developed countries, we see that unsustainable consumption can occur in two ways. First, environmental quality and resource depletion can result from too many people, even if those people consume few resources per person. This is the current situation in many developing nations.

In highly developed countries, unsustainable consumption results when individuals consume substantially more resources than necessary for survival. Both types of unsustainable consumption have the same effect—pollution, environmental degradation, and resource depletion. Many affluent, highly developed countries, including the United States, Canada, Japan, and most of Europe, consume unsustainably: *Highly developed countries represent less than 20% of the world's population, yet they consume significantly more than half of its resources.*

According to the Worldwatch Institute, highly developed countries account for the lion's share of total resources consumed:

- 86% of aluminum used
- 76% of timber harvested
- 68% of energy produced
- 61% of meat eaten
- 42% of the fresh water consumed

These nations also generate 75% of the world's pollution and waste.

ECOLOGICAL FOOTPRINT

Environmental scientists **Mathis Wackernagel** and **William Rees** developed the concept of ecological footprint to help people visualize what they use from the environment. Each person has an **ecological footprint**, an amount of productive land, fresh water, and ocean required on a continuous basis to supply that person with food, wood, energy, water, housing, clothing, transportation, and waste disposal. The *Living Planet Report 2010*, produced by scientists at the Global Footprint Network, World Wildlife Fund, and Zoological Society of London, estimates that since about 1975, the human population has been consuming more of the productive land, water, and other resources than Earth can support (**Figure 1.7**). In 2010, annual consumption was about 50% more than Earth produces. This is an unsustainable consumption rate.

The *Living Planet Report* estimates that Earth has about 11.4 billion hectares (28.2 billion acres) of productive land and water. If we divide this area by the global human population, we see that each person is allotted about 1.6 hectares (4.0 acres). However,

the average global ecological footprint is currently about 2.7 hectares (6.7 acres) per person, which means we humans have an *ecological overshoot*. We can see the short-term results around us—forest destruction, degradation of croplands, loss of biological diversity, declining ocean fisheries, local water shortages, and increasing pollution. The long-term outlook, if we do not seriously address our consumption of natural resources, is potentially disastrous. Either per-person consumption will drop, population will decrease, or both.

In the developing nation of India, the per capita ecological footprint is 0.9 hectare (2.2 acres); India is the world's second-largest country in terms of population, so even though its per capita footprint is low, the country's footprint is high: 986.3 million hectares (**Figure 1.8**). In France, the per capita ecological footprint is 4.9 hectares (12.1 acres); although its per capita footprint is high, France's footprint as a country is 298.1 million hectares, which is lower than India's, because its population is much smaller. In the United States, the world's third-largest country, the per capita ecological footprint is 7.9 hectares (19.5 acres); the U.S. footprint as a country is 2,457 million hectares! If all people in the world had the same lifestyle and level of consumption as the average North American, and assuming no changes in technology, we would need about four additional planets the size of Earth.

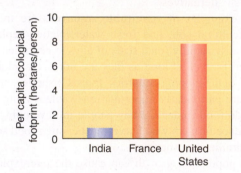

(a) The average ecological footprint of a person living in India, France, or the United States. For example, each Indian requires 0.9 hectare (2.2 acres) of productive land and ocean to meet his or her resource requirements.

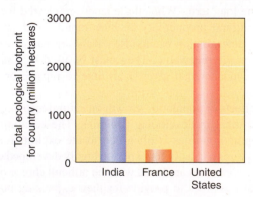

(b) The total ecological footprint for the countries of India, France, and the United States. Note that India, although having a low per capita ecological footprint, has a significantly higher ecological footprint as a country because of its large population. If everyone in the world had the same level of consumption as the average American, it would take the resources and area of five Earths.

Figure 1.8 **Ecological footprints.** (Data from World Wildlife Fund, *Living Planet Report 2012*)

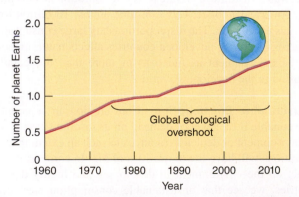

Figure 1.7 **Global ecological overshoot.** Earth's ecological footprint has been increasing over time. By 2010, humans were using the equivalent of 1.5 Earths, a situation that is not sustainable. (Data from World Wildlife Fund, *Living Planet Report 2010*)

As developing countries increase their economic growth and improve their standard of living, more and more people in those nations purchase consumer goods. More new cars are now sold annually in Asia than in North America and Western Europe combined. These new consumers may not consume at the high level of the average consumer in a highly developed country, but their consumption has increasingly adverse effects on the environment. For example, air pollution caused by automotive traffic in urban centers in developing countries is terrible and getting worse every year. Millions of dollars are lost because of air pollution–related health problems in these cities. One of society's challenges is to provide new consumers in developing countries (as well as ourselves) with less polluting, less consuming forms of transportation.

THE *IPAT* MODEL

Generally, when people turn on the tap to brush their teeth in the morning they do not think about where the water comes from or about the environmental consequences of removing it from a river or the ground. Similarly, most North Americans do not think about where the energy comes from when they flip on a light switch or start a car. We do not realize that all the materials in the products we use every day come from Earth, nor do we grasp that these materials eventually are returned to Earth, much of them in sanitary landfills.

Such human impacts on the environment are difficult to assess. They are estimated using the three factors most important in determining environmental impact *(I)*:

1. The number of people *(P)*

2. Affluence, which is a measure of the consumption or amount of resources used per person *(A)*

3. The environmental effects (resources needed and wastes produced) of the technologies used to obtain and consume the resources *(T)*

These factors are related in this way:

$$I = P \times A \times T$$

model A representation of a system; describes the system as it exists and predicts how changes in one part of the system will affect the rest of the system.

In science, a **model** is a formal statement that describes the behavior of a system. The *IPAT* model, which biologist **Paul Ehrlich** and physicist **John Holdren** first proposed in the 1970s, shows the mathematical relationship between environmental impacts and the forces driving them.

For example, to determine the environmental impact of emissions of the greenhouse gas CO_2 from motor vehicles, multiply the population times the number of cars per person (affluence/consumption per person) times the average car's annual CO_2 emissions per year (technological impact). This model demonstrates that although increasing motor vehicle efficiency and developing cleaner technologies will reduce pollution and environmental degradation, a larger reduction will result if population and per capita consumption are also controlled.

The *IPAT* equation, though useful, must be interpreted with care, in part because we often do not understand all the environmental impacts of a particular technology on complex environmental systems. Motor vehicles are linked not only to global warming from CO_2 emissions but also to local air pollution (tailpipe exhaust), water pollution (improper disposal of motor oil and antifreeze), and solid waste (disposal of nonrecyclable automobile parts in sanitary landfills). There are currently more than 600 million motor vehicles on the planet, and the number is rising rapidly.

The three factors in the *IPAT* equation are always changing in relation to one another. Consumption of a particular resource may increase, but technological advances may decrease the environmental impact of the increased consumption. For example, there are more television and computer screens in the average household than there were 20 years ago (increased affluence) and more households (increased population). However, many of the new units are flat screens, which require less materials to produce and less energy to operate (more efficient technology). Thus, consumer trends and choices may affect environmental impact.

Similarly, the average fuel economy of new cars and light trucks (sport-utility vehicles, vans, and pickup trucks) in the United States declined from 22.1 miles per gallon in 1988 to 20.4 miles per gallon in the early 2000s, in part because of the popularity of sport-utility vehicles (SUVs). In addition to being less fuel efficient than cars, SUVs emit more emissions per vehicle mile. More recently, hybrids have helped to increase the average fuel economy (**Figure 1.9**). Such trends and uncertainties make the *IPAT* equation of limited usefulness for long-term predictions.

The *IPAT* equation is valuable because it helps identify what we do not know or understand about consumption and its environmental impact. The National Research Council of the U.S. National Academy of Sciences[1] has identified research areas we must address, including the following: Which kinds of consumption have the greatest destructive impact on the environment? Which groups in society are responsible for the greatest environmental disruption? How can we alter the activities of these environmentally disruptive groups? It will take years to address such

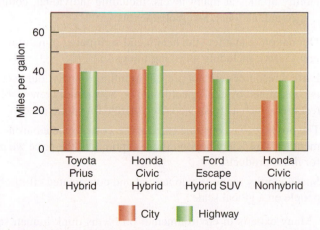

Figure 1.9 **Fuel-efficient hybrids.** Shown are city and highway miles per gallon for three hybrid models available in 2014. (U.S. Department of Energy)

[1] The National Research Council is a private, nonprofit society of distinguished scholars. It was organized by the National Academy of Sciences to advise the U.S. government on complex issues in science and technology.

questions, but the answers should help decision makers in government and business formulate policies to alter consumption patterns in an environmentally responsible way. Our ultimate goal should be to reduce consumption so that our current practices do not compromise the ability of future generations to use and enjoy the riches of our planet.

REVIEW

1. How do renewable resources differ from nonrenewable resources?
2. How are human population growth and affluence related to natural resource depletion?
3. What is an ecological footprint?
4. What does the *IPAT* model demonstrate?

Sustainability

LEARNING OBJECTIVES

- Define *sustainability*.
- Relate Garrett Hardin's description of the tragedy of the commons in medieval Europe to common-pool resources today.
- Briefly describe sustainable development.

One of the most important concepts in this text is **sustainability**. A sustainable world is one in which humans can

sustainability The ability to meet current human economic and social needs without compromising the ability of the environment to support future generations.

have economic development and fair allocation of resources without the environment going into a decline from the stresses imposed by human society on the natural systems (such as fertile soil, water, and air) that maintain life. When the environment is used sustainably, humanity's present needs are met without endangering the welfare of future generations (**Figure 1.10**). Environmental sustainability applies at many levels, including individual, community, regional, national, and global levels.

- Our actions can affect the health and well-being of natural *ecosystems*, including all living things.
- Earth's resources are not present in infinite supply; our access is constrained by ecological limits on how rapidly renewable resources such as fresh water regenerate for future needs.
- The products we consume can impose costs to the environment and to society beyond those captured in the price we pay for those products.
- Sustainability requires a concerted and coordinated effort of people on a global scale.

Many experts in environmental problems think human society is not operating sustainably because of the following human behaviors:

- We extract nonrenewable resources such as fossil fuels as if they were present in unlimited supplies.

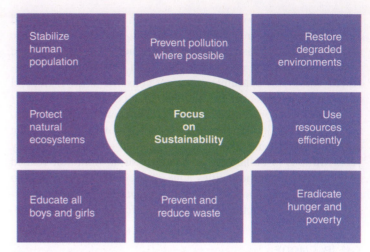

Figure 1.10 **Sustainability.** Sustainability requires a long-term perspective to protect human welfare and natural resource assets, such as the efforts shown here.

- We consume renewable resources such as fresh water and forests faster than natural systems can replenish them (**Figure 1.11**).
- We pollute the environment with toxins as if the capacity of the environment to absorb them is limitless.
- A small fraction of the human population dominates a large percentage of Earth's resources.
- Our numbers continue to grow despite Earth's finite ability to feed us, sustain us, and absorb our wastes.

If left unchecked, these activities could threaten Earth's life-support systems to such a degree that recovery is impossible. If major resources like agricultural land, fisheries, and fresh water are exhausted to the point that they cannot recover quickly, substantial human suffering would result. Thus managing these resources sustainably means more than protecting the environment: sustainability promotes human well-being.

Topham/The Image Works

Figure 1.11 **A logger cuts down the last standing tree on a clear-cut forest slope.** Logging destroys the habitat for forest organisms and increases the rate of soil erosion on steep slopes. Photographed in Canada.

At first glance, the issues may seem simple. Why do we not just reduce consumption, improve technology, and limit population growth? The answer is that various interacting ecological, societal, and economic factors complicate the solutions. Our inadequate understanding of how the environment works and how human choices affect the environment is a major reason that problems of sustainability are difficult to resolve. The effects of many interactions between the environment and humans are unknown or difficult to predict, and we generally do not know if we should take corrective actions before our understanding is more complete.

SUSTAINABILITY AND THE TRAGEDY OF THE COMMONS

Garrett Hardin (1915–2003) was a professor of human ecology at the University of California–Santa Barbara who wrote about human environmental dilemmas. In 1968, he published his classic essay, "The Tragedy of the Commons," in the journal *Science*. He contended that our inability to solve many environmental problems is the result of a struggle between short-term individual welfare and long-term environmental sustainability and societal welfare.

Hardin used the commons to illustrate this struggle. In medieval Europe, the inhabitants of a village shared pastureland, called the commons, and each herder could bring animals onto the commons to graze. If the villagers did not cooperatively manage the commons, each might want to bring more animals onto it. If every herder in the village brought as many animals onto the commons as possible, the plants would be killed from overgrazing, and the entire village would suffer. Thus, an unmanaged commons would inevitably be destroyed by the people who depended on it.

Hardin argued that one of the outcomes of the eventual destruction of the commons would be private ownership of land, because when each individual owned a parcel of land, it was in that individual's best interest to protect the land from overgrazing. A second outcome Hardin considered was government ownership and management of such resources, because the government had the authority to impose rules on users of the resource and thereby protect it.

Hardin's essay has stimulated a great deal of research in the decades since it was published. In general, scholars agree that degradation of the self-governing commons—now called **common-pool resources**—typically is not a problem in closely knit communities. Indeed, sociologist **Bill Freudenberg** has pointed out that medieval commons were successfully managed but became degraded after they were privatized. Economist **Elinor Ostrum** demonstrated that common pool resources can be sustainably managed by communities with shared interests, strong local governance, and community-enforced accountability.

common-pool resources Those parts of our environment available to everyone but for which no single individual has responsibility—the atmosphere and climate, fresh water, forests, wildlife, and ocean fisheries.

As one goes from local to regional to global common-pool resources, the challenges of sustainably managing resources become more complex. In today's world, Hardin's parable has particular relevance at the global level. These modern-day commons are experiencing increasing environmental stress (see, for example, the discussion of climate change in Chapter 20). No individual, jurisdiction, or country owns common-pool resources, and they are susceptible to overuse. Although exploitation may benefit only a few, everyone on Earth must pay for the environmental cost of exploitation.

The world needs effective legal and economic policies to prevent the short-term degradation of common-pool resources and ensure their long-term well-being. We have no quick fixes because solutions to global environmental problems are not as simple or short term as are solutions to some local problems. Most environmental ills are inextricably linked to other persistent problems such as poverty, overpopulation, and social injustice—problems beyond the capacity of a single nation to resolve. The large number of participants who must organize, agree on limits, and enforce rules complicates the creation of global treaties to manage common-pool resources. Cultural and economic differences among participants make finding solutions even more challenging.

Clearly, all people, businesses, and governments must foster a strong sense of **stewardship**—shared responsibility for the sustainable care of our planet. Cooperation and commitment at the international level are essential if we are to alleviate poverty, stabilize the human population, and preserve our environment and its resources for future generations.

stewardship Shared responsibility for the sustainable care of our planet.

GLOBAL PLANS FOR SUSTAINABLE DEVELOPMENT

In 1987, the World Commission on Environment and Development released a groundbreaking report, *Our Common Future* (see Chapter 24). A few years later, in 1992, representatives from most of the world's countries met in Rio de Janeiro, Brazil, for the *UN Conference on Environment and Development*. Countries attending the conference examined environmental problems that are international in scope: pollution and deterioration of the planet's atmosphere and oceans, a decline in the number and kinds of organisms, and destruction of forests.

In addition, the Rio participants adopted *Agenda 21*, an action plan of **sustainable development** in which future economic development, particularly in developing countries, will be reconciled with environmental protection. The goals of sustainable development are achieving improved living conditions for all people while maintaining a healthy environment in which natural resources are not overused and excessive pollution is not generated. Three factors—environmentally sound decisions, economically viable decisions, and socially equitable decisions—are necessary for truly sustainable development. To use sustainability as a guiding principle for environmental management requires that we think about how these three factors interact as parts of a complex and interlinked system (**Figure 1.12**).

sustainable development As defined by the Brundtland report, economic development that meets the needs of the present generation without compromising the ability of future generations to meet their own needs.

A serious application of the principles of environmental sustainability to economic development will require many changes in such fields as population policy, agriculture, industry, economics,

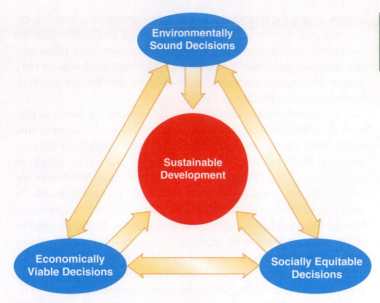

Figure 1.12   Sustainable development, a systems concept. Using sustainable development as an organizing principle for environmental management requires us to recognize that economic development, social justice, and the environment are linked in many and complex ways. We must consider whether economic decisions harm the environment or deplete natural resources, whether resource management decisions are socially equitable, and whether societal decisions impact economic opportunities for current and future generations.

and energy use. In 2000, representatives from 189 countries met at the *UN Millennium Summit* and committed their countries to a global partnership with goals known as the *Millennium Development Goals* (see Chapter 8).

Despite such international summits, we have made only limited progress in improving the quality of life for poor people or in solving the world's most serious environmental problems. Many governments have shifted their attention from serious environmental priorities to other challenges such as terrorism, worsening international tensions, and severe economic problems. Meanwhile, scientific warnings about important environmental problems such as global climate change have increased.

Despite the lack of significant change at the international level, some nations, states, and municipalities have made important environmental progress. Many countries have enacted more stringent air pollution laws, including the phasing out of leaded gasoline. More than 100 countries have created sustainable development commissions. Corporations that promote environmentally responsible business practices have joined to form the World Business Council for Sustainable Development. The World Bank, which makes loans to developing countries, has invested billions of dollars in sustainable development projects around the world.

REVIEW

1. What is sustainability?
2. What is the tragedy of the commons?
3. What are the three foundations of sustainable development?

Environmental Science

LEARNING OBJECTIVES

- Define *environmental science* and briefly describe the role of Earth systems in environmental science.
- Outline the scientific method.

Environmental science encompasses the many interconnected issues involving human population, Earth's natural resources, and environmental pollution. Environmental science combines information from many disciplines, such as biology, geography, chemistry, geology, physics, economics, sociology, demography (the study of populations), cultural anthropology, natural resources management, agriculture, engineering, law, politics, and ethics. **Ecology**, the branch of biology that studies the interrelationships between organisms and their environment, is a basic tool of environmental science.

> **environmental science** The interdisciplinary study of humanity's relationship with other organisms and the nonliving physical environment.

Environmental scientists try to establish general principles about how the natural world functions. They use these principles to develop viable solutions to environmental problems—solutions based as much as possible on scientific knowledge. Environmental problems are generally complex, so our understanding of them is often less complete than we would like it to be. Environmental scientists are often asked to reach a consensus before they fully understand the systems that they study. As a result, they often make recommendations based on probabilities rather than precise answers.

Many of the environmental problems discussed in this book are serious, but environmental science is not simply a "doom-and-gloom" listing of problems coupled with predictions of a bleak future. To the contrary, its focus is, and our focus as individuals and as world citizens should be, on identifying, understanding, and finding better ways to manage the stresses that human activities place on environmental resources and systems.

EARTH SYSTEMS AND ENVIRONMENTAL SCIENCE

One of the most exciting aspects of environmental science and many other fields of science is working out how *systems* that consist of many interacting parts function as a whole. As discussed in the chapter introduction, Earth's climate is a system that in turn is composed of smaller, interdependent systems, such as the atmosphere and ocean; these smaller systems are linked and interact with one another in the overall climate system.

The systems approach provides a broad look at overall processes, as opposed to the details of individual parts or steps. A commuter in city traffic may be quite familiar with the production of CO_2 by a car engine, but that knowledge does not automatically translate into an understanding of the global effect of millions of motor vehicles emitting CO_2. Thus, using a systems perspective helps scientists gain valuable insights that are not always obvious from looking at individual components within the system.

Also, problems arise from *not* thinking about systems. For example, if a company decides to burn waste oil to avoid its leaking into groundwater, pollution shifts from groundwater to the air. A systems perspective would require company executives to think about the trade-offs between the two disposal methods and, more importantly, about alternatives that might avoid generating waste oil in the first place.

Environmental scientists often use models to describe the interactions within and among environmental systems. Many of these models are computer simulations that represent the overall effect of competing factors to describe an environmental system in numerical terms. Models help us understand how the present situation developed from the past or how to predict the future course of events, including the long-term impacts of decisions or choices we make today. Models also generate additional questions about environmental issues. (Appendix III contains an introduction to modeling.)

A natural system consisting of a community of organisms and its physical environment is known as an **ecosystem**. In ecosystems, biological processes (such as photosynthesis) interact with physical and chemical processes to modify the composition of gases in the atmosphere, transfer energy from the sun through living organisms, recycle waste products, and respond to environmental changes with resilience. Natural ecosystems are the foundation for our concept of environmental sustainability.

Ecosystems are organized into larger and larger systems that interact with one another (discussed in Chapter 3). At a global level are Earth systems, which include Earth's climate, atmosphere, land, coastal zones, and the ocean. Environmental scientists use a systems approach to try to understand how human activities are altering global environmental parameters such as temperature, CO_2 concentration in the atmosphere, land cover, changes in nitrogen levels in coastal waters, and declining fisheries in the ocean.

Many aspects of Earth systems are in a steady state or, more accurately, a **dynamic equilibrium**, in which the rate of change in one direction is the same as the rate of change in the opposite direction. *Feedback* occurs when a change in one part of a system leads to a change in another part. Feedback can be negative or positive. In a **negative feedback system**, a change in some condition triggers a response that counteracts, or reverses, the changed condition (**Figure 1.13a**). A negative feedback mechanism works to keep an undisturbed system in dynamic equilibrium. For example, consider fish in a pond. As the number of fish increases, available food decreases and fewer fish survive; thus, the fish population declines.

In a **positive feedback system**, a change in some condition triggers a response that intensifies the changing condition (**Figure 1.13b**); a positive feedback mechanism leads to greater change from the original condition. Positive feedback can be very disruptive to an already disturbed system. For example, melting of polar and glacial ice can lead to greater absorption of solar heat by the exposed land area, which in turn leads to more rapid melting. As you will see throughout this text, many negative and positive feedback mechanisms operate in the natural environment.

SCIENCE AS A PROCESS

One key to the successful solution of any environmental problem is a careful evaluation of conditions, causes, and effects. Science—a system for managing and producing information—is the most effective way to do this evaluation. It is important to understand clearly just what science is, as well as what it is not. Many people think of science as a body of knowledge—a collection of facts about the natural world and a search for relationships among these

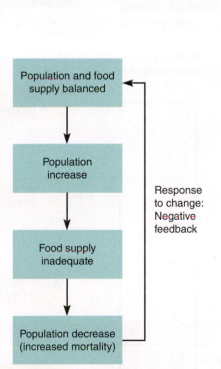

(a) **Negative feedback.** In this simplified example, the initial balance between a population of fish and its food supply is ultimately restored. Thus, in a negative feedback system, the response to change opposes the change.

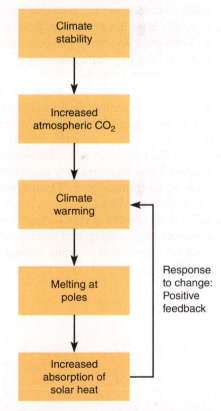

(b) **Positive feedback.** In this simplified example of a positive feedback system, the response to change increases, or amplifies, the deviation from the original point.

Figure 1.13   Feedback systems.

facts. However, science is also a dynamic *process,* a systematic way to investigate the natural world. Scientists seek to describe the apparent complexity of our world with general scientific laws (also called *natural laws*). Scientific laws are then used to make predictions, solve problems, or provide new insights.

Scientists collect objective **data** (singular, *datum*), the information with which science works. Data are collected by observation and experimentation and then analyzed or interpreted. Conclusions are inferred from the available data and are not based on faith, emotion, or intuition.

Scientists publish their findings in scientific journals, and other scientists examine and critique their work. Confirming the validity of new results by *repeatability* is a requirement of science—observations and experiments must produce consistent results when other scientists repeat them. Scrutiny by other scientists reveals any inconsistencies in results or interpretation, and these errors are discussed openly. Thus, science is *self-correcting* over time.

No absolute certainty or universal agreement exists about anything in science. Science is an ongoing enterprise, and generally accepted ideas must be reevaluated in light of newly discovered data. Scientists never claim to know the "final answer" about anything because scientific understanding changes. However, this does not prevent us from using current knowledge in environmental science to make environmental decisions. Science represents the best information, and thus the best opportunity to make informed decisions.

Uncertainty does not mean that scientific conclusions are invalid. Overwhelming evidence links exposure to tobacco smoke and the incidence of lung cancer. We cannot state with absolute certainty which smokers will get lung cancer, but this uncertainty does not mean no correlation exists between smoking and lung cancer. On the basis of the available evidence, we say each individual who smokes has an increased *risk* of developing lung cancer, and we can say with great confidence that far more smokers will develop lung cancer than will nonsmokers.

Importantly, science cannot tell what we *should* do when faced with an environmental challenge, only what result we can expect given different choices. Science can provide better guidance than can religion or political preference on what will happen if we release pesticides that eventually reach the ocean or greenhouse gases that can change the climate. But values, politics, religion, and culture must determine whether we find the loss of waterfowl or changes in the climate worth the benefits we get from spraying pesticides or burning fossil fuels.

The Scientific Method   The established processes scientists use to answer questions or solve problems are collectively called the **scientific method** (**Figure 1.14**). Although the scientific method has many variations, it basically involves five steps:

> **scientific method** The way a scientist approaches a problem by formulating a hypothesis and then testing it by means of an experiment.

1. **Recognize a question or unexplained occurrence in the natural world.** After a problem is recognized, one investigates relevant scientific literature to determine what is already known about it.

2. **Develop a *hypothesis*, or educated guess, to explain the problem.** A good hypothesis makes a prediction that can be tested and possibly disproved. The same factual evidence is often used to formulate several alternative hypotheses; each must be tested.

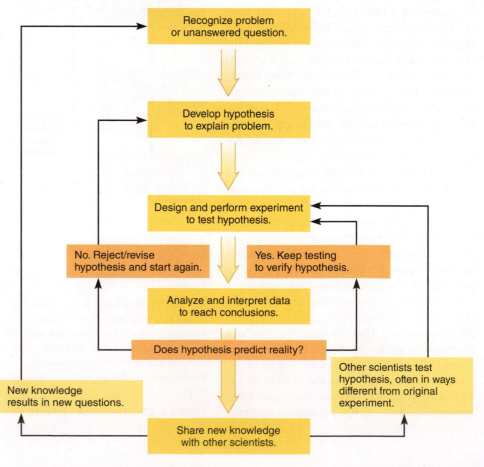

Figure 1.14   Scientific method. The basic steps of the scientific method are shown in yellow. Scientific work rarely proceeds in so straightforward a manner; examples of additional paths are shown in orange.

Recognize problem or unanswered question.

Develop hypothesis to explain problem.

Design and perform experiment to test hypothesis.

No. Reject/revise hypothesis and start again.

Yes. Keep testing to verify hypothesis.

Analyze and interpret data to reach conclusions.

Does hypothesis predict reality?

New knowledge results in new questions.

Other scientists test hypothesis, often in ways different from original experiment.

Share new knowledge with other scientists.

3. **Design and perform an experiment to test the hypothesis.** An experiment involves collecting data by making careful observations and measurements. Much of the creativity in science involves designing experiments that sort out the confusion caused by competing hypotheses. The process never "proves" anything; instead, it disproves or falsifies alternative hypotheses until only the most plausible hypothesis is left.

4. **Analyze and interpret the data to reach a conclusion.** Does the evidence match the prediction stated in the hypothesis—that is, do the data support or refute the hypothesis? Should the hypothesis be modified or rejected based on the observed data?

5. **Share new knowledge.** Publishing articles in scientific journals or books and presenting the information at meetings permits others to understand and critique methods and findings, and repeat the experiment or design new experiments that either verify or refute the work.

Although the scientific method is usually described as a linear sequence of events, science is rarely as straightforward or tidy as the scientific method implies. Good science involves creativity and openness to new ideas in recognizing questions, developing hypotheses, and designing experiments. Scientific knowledge progresses by trial and error. Many creative ideas end up as dead ends, and temporary setbacks or reversals of direction often occur as knowledge progresses. Scientific knowledge often expands haphazardly, with the "big picture" emerging slowly from confusing and sometimes contradictory details.

Scientific work is often incorrectly portrayed in the media as "new facts" that have just come to light. At a later time, additional "new facts" that question the validity of the original study are reported. If one were to read the scientific papers on which such media reports are based, one would find that the scientists made tentative conclusions based on their data. Science progresses from uncertainty to less uncertainty, not from certainty to greater certainty. Science leads to a better understanding of nature over time, despite the fact that science never "proves" anything.

Most often, many factors influence the processes scientists want to study. Each factor that influences a process is a **variable**. Ideally, to evaluate alternative hypotheses about a given variable, we run experiments that hold all other variables constant so that they do not confuse or mislead us. To test a hypothesis about a variable, two forms of the experiment are done in parallel. In an **experimental group**, we alter the chosen variable in a known way. In a **control group**, we do not alter that variable. In all other respects the two groups are the same. We then ask, "What is the difference, if any, between the outcomes for the two groups?" Any difference is the result of the influence of that variable because all other variables remained the same. Much of the challenge of environmental science lies in designing control groups and in successfully isolating a single variable from all other variables.

Theories   Theories explain scientific laws. A **theory** is an integrated explanation of numerous hypotheses, each supported by a large body of observations and experiments and evaluated by the peer review process. A theory condenses and simplifies many data that previously appeared unrelated. A good theory grows as additional information becomes known. It predicts new data and suggests new relationships among a range of natural phenomena.

A theory simplifies and clarifies our understanding of the natural world because it demonstrates relationships among classes of data. Theories are the solid ground of science, the explanations of which we are most sure. This definition contrasts sharply with the general public's use of the word *theory*, implying lack of knowledge or a guess—as in, "I have a theory about the existence of life on other planets." In this book, the word *theory* is always used in its scientific sense, to refer to a broadly conceived, logically coherent, and well-supported explanation.

Absolute truth is not possible in science, only varying degrees of uncertainty. Science is continually evolving as new evidence comes to light, and its conclusions are always provisional or uncertain. It is always possible that the results of future experiments will contradict a prevailing theory, which will then be replaced by a new or modified theory that better explains the scientific laws of the natural world.

Climate Change: Hypotheses and Theory   A dominant theory in environmental science today is that carbon dioxide (CO_2) and other gases released from burning fossil fuels has caused and will continue to cause Earth's climate to change. Clearly, however, the climate is a complex system, with many variables that change over a long period. We cannot run an experiment to test the hypothesis that releasing greenhouse gases over a century causes global temperatures to increase. Understanding climate change requires us to observe what is happening, compare those observations to what existing theory predicts will happen, and adapt our theories based on new observations. The theory of climate change draws on physics, chemistry, oceanography, atmospheric science, astronomy, and other scientific fields.

Many of the components of climate theory can be tested directly (**Figure 1.15**). As one example, scientists have tested the hypothesis that some gases, which we call greenhouse gases, can absorb energy (where energy absorption is a variable). In one such experiment, they filled identical containers with air, varied the concentrations of CO_2 in the containers, and then added heat. They observed temperature increases in the containers with more CO_2. While such experiments do not prove that CO_2 absorbs heat, they demonstrate that it is reasonable to think the hypothesis is correct. Indeed, heat absorption by CO_2 has been understood for over a century.

Climate scientists have combined this experimental evidence about greenhouse gases with our understanding of other parts of the climate, such as natural variability, solar variability, and reflectivity of ice and other surfaces, to establish a theory of climate change. This theory has uncertainties, but it predicts that as greenhouse gas concentrations increase, global atmospheric temperature will increase, sea level will rise, precipitation patterns will change, and glaciers and ice caps will melt (**Figure 1.16**). As we will see in Chapter 20, observations have confirmed this general theory, while allowing us to refine it and make more reliable predictions about the future. Climate theory does not tell us what we should do to avoid climate change, only what we can expect from different possible decisions.

Figure 1.15 **Climate change analysis.** This equipment is part of a long-term project to explore the release of carbon dioxide (a greenhouse gas) from the melting tundra in Alaska.

 Carbon dioxide is released by warming, and contributes to warming. Is this an example of a positive feedback or a negative feedback?

Figure 1.16 **Melting iceberg.** Water streams from an iceberg that was once part of the Ilulisat Kangerlua Glacier, Greenland. While icebergs have broken off of glaciers throughout geologic history, evidence shows that the rate at which glaciers are melting has accelerated rapidly over the past hundred years.

REVIEW

1. What is environmental science? Why is a systems perspective so important in environmental science?

2. What are the steps of the scientific method? Does the scientific process usually follow these steps? Why or why not?

Addressing Environmental Problems

LEARNING OBJECTIVES

- List and briefly describe the five stages in addressing environmental problems.
- Briefly describe the history of the Lake Washington pollution problem of the 1950s and how it was resolved.

We have shown the strengths and limitations of science—what science can and cannot do. Before examining the environmental problems in the remaining chapters of this text, let us consider the elements that contribute to addressing those problems. What is the role of science? Given that we can never achieve complete certainty in science, at what point are scientific conclusions considered certain enough to warrant action? Who makes the decisions, and what are the trade-offs?

ADDRESSING ENVIRONMENTAL PROBLEMS

Viewed simply, there are five stages in addressing an environmental problem (**Figure 1.17**):

1. **Scientific assessment.** The first stage in addressing any environmental problem is scientific assessment, the gathering of information. The problem is defined. Data are then collected, and experiments or simulations are performed.

2. **Risk analysis.** Using the results of a scientific investigation, we can analyze the potential effects of doing nothing or of intervening—what is expected to happen if a particular course of action is followed, including any adverse effects the action might generate. In other words, the risks of one or more remediation (correction or cleanup) options are considered.

3. **Public education and involvement.** Public participation and commitment are an essential part of addressing most environmental problems. The public can be a source of both knowledge and values, and many individuals and groups have a stake in the outcome of decisions. People are generally more willing to work together to solve a problem if they have the opportunity to participate from the start.

4. **Political action.** Affected parties select and implement a course of action. Ideally, science provides information on what *can* be done, but in the political process, opinions often differ about how to interpret evidence when selecting a course of action. Often, what people think *should* be done affects their beliefs about science and scientists.

5. **Long-term evaluation.** The results of any action taken should be carefully monitored, both to see if the environmental problem is being addressed and to improve the initial assessment and modeling of the problem.

These five stages represent an ideal approach to systematically addressing environmental problems. In real life, addressing environmental problems is rarely so neat and tidy, particularly when the problem is of regional or global scale, or when those bearing the costs are not those who stand to benefit from a new policy. Quite often, the public becomes aware of a problem, which triggers discussion regarding remediation before the problem has been clearly identified. Also, we often do not know what scientific information is needed until stage 2, 3, or even 4, which means that to make informed decisions, we often need to ask scientists to develop additional research.

To demonstrate the five steps as they operate in an ideal situation, let us consider a relatively simple environmental problem recognized and addressed in the 1950s—pollution in Lake Washington. This problem, unlike many environmental issues we face today, was relatively easy to diagnose and solve.

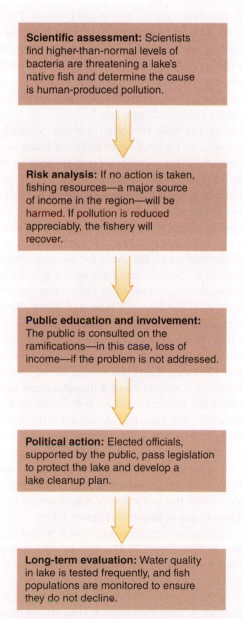

Figure 1.17   Addressing environmental problems. These five steps provide a framework for addressing environmental problems, such as the hypothetical example given. Solving environmental problems rarely proceeds in such a straightforward manner.

CASE IN POINT

LAKE WASHINGTON

Lake Washington is a large, deep freshwater lake on the eastern boundary of Seattle (**Figure 1.18**). During the first part of the twentieth century, the Seattle metropolitan area expanded eastward toward the lake from the shores of Puget Sound, putting Lake Washington under increasingly intense environmental pressures. Between 1941 and 1954, 10 suburban sewage treatment plants began operating around the lake. Each plant treated the raw sewage to break down the organic material within it and released the effluent (treated sewage) into the lake. By the mid-1950s, a great deal of treated sewage had been poured into the lake.

Scientists at the University of Washington were the first to note the effects of this discharge on the lake. Their studies indicated that large masses of cyanobacteria (photosynthetic bacteria) were growing in the lake. Cyanobacteria require a plentiful supply

Figure 1.18 **Lake Washington.** This large freshwater lake forms the eastern boundary of Seattle, Washington.

of nutrients such as nitrogen and phosphorus, and deepwater lakes such as Lake Washington do not usually have many dissolved nutrients. The increase in filamentous cyanobacteria indicated that the quality of water in Lake Washington was diminishing.

In 1955, the Washington Pollution Control Commission, citing the scientists' work, concluded that the treated sewage effluent was raising the levels of dissolved nutrients to the point of serious pollution. The sewage treatment was not eliminating many chemicals, particularly phosphorus, a major component of detergents. Mats of cyanobacteria formed a smelly green scum over the surface of the water. The bacteria that decompose cyanobacteria when they die multiplied explosively, consuming vast quantities of oxygen, until the lake's deeper waters could no longer support oxygen-requiring organisms such as fishes and small invertebrates.

Scientific assessment of an environmental problem verifies that a problem exists and builds a sound set of observations on which to base a solution. Scientists predicted that the lake's decline could be reversed: If the pollution was stopped, the lake would slowly recover. They outlined three steps necessary to save the lake:

1. Comprehensive regional planning by the many suburbs surrounding the lake

2. Complete elimination of sewage discharge into the lake

3. Research to identify the key nutrients causing the cyanobacteria to grow

It is one thing to suggest that treated sewage no longer be added to Lake Washington and quite another to devise an acceptable remediation option. Further treatment of sewage could remove some nutrients, but it might not be practical to remove all of them. The alternative was to dump the sewage somewhere else—but where?

In this case, officials decided to discharge the treated sewage into Puget Sound. In their plan, a ring of sewers built around the lake would collect the treated sewage and treat it further before discharging it into Puget Sound.

The plan to further treat the sewage was formulated to minimize the environmental impact on Puget Sound. It was assumed that the treated effluent would have less of an impact on the greater quantity of water in Puget Sound. Also, phosphate does not control cyanobacterial growth in Puget Sound as it does in Lake Washington. The growth of photosynthetic bacteria and algae in Puget Sound is largely limited by tides, which mix the water and transport the tiny organisms into deeper water, where they cannot get enough light to grow rapidly.

Despite the Washington Pollution Control Commission's conclusions, local sanitation authorities were not convinced that urgent action was necessary. Public action required further education, and scientists played a key role. They wrote articles for the general public that explained what nutrient enrichment is and what problems it causes. The general public's awareness increased as local newspapers published these articles.

Cleaning up the lake presented serious political problems because there was no regional mechanism in place to permit the many local suburbs to act together on matters such as sewage disposal. In late 1957, the state legislature passed a bill permitting a public referendum in the Seattle area regarding the formation of a regional government with six functions: water supply, sewage disposal, garbage disposal, transportation, parks, and planning. The referendum was defeated, apparently because suburban voters felt the plan was an attempt to tax them for the city's expenses. An advisory committee immediately submitted a revised bill limited to sewage disposal to the voters. Over the summer there was widespread discussion of the lake's future, and when the votes were counted, the revised bill passed by a wide margin.

At the time it was passed, the Lake Washington plan was the most ambitious and most expensive pollution-control project in the United States. Every household in the area had to pay additional taxes for construction of a massive trunk sewer to ring the lake, collect the effluent, treat it, and discharge it into Puget Sound. Meanwhile, the lake had deteriorated further. Visibility had declined from 4 m (12 ft) in 1950 to less than 1 m (3 ft) in 1962 because the water was clouded with cyanobacteria. In 1963, the first of the waste treatment plants around the lake began to divert its effluent into the new trunk sewer. One by one, the others diverted theirs, until the last effluent was diverted in 1968. The lake's condition began to improve (**Figure 1.19**).

Water transparency returned to normal within a few years. Cyanobacteria persisted until 1970 but eventually disappeared. By 1975, the lake was back to normal, and today the lake remains clear. Continuing to protect the water quality even as population around the lake grows has required a systems perspective. Rather than just clean up wastewater, more recent efforts have included strategies that reduce the generation of wastes, such as water recycling and efforts to reduce oil and other industrial wastes.

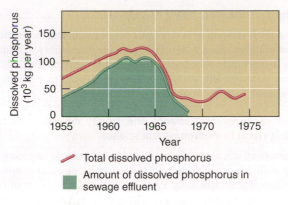

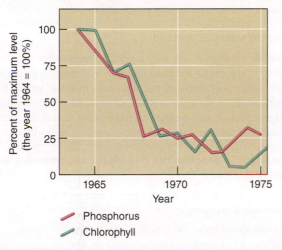

(a) Dissolved phosphorus in Lake Washington from 1955 to 1974. Note that the level of dissolved phosphorus declined in the lake as the phosphorus contributed by sewage effluent (shaded area) declined.

(b) Cyanobacterial growth from 1964 to 1975, during Lake Washington's recovery, as measured indirectly by the amount of chlorophyll, the pigment involved in photosynthesis. Note that as the level of phosphorus dropped in the lake, the number of cyanobacteria (that is, the chlorophyll content) declined.

Figure 1.19   Nutrients in Lake Washington compared with cyano bacterial growth. (From W.T. Edmondson) *The Uses of Ecology: Lake Washington and Beyond.* University of Washington Press (1991).

ENVIRONEWS

Green Roofs

A roof that is completely or partially covered with vegetation and soil is known as a *green roof.* Also called *eco-roofs*, green roofs can provide several environmental benefits. For one thing, the plants and soil are effective insulators, reducing heating costs in winter and cooling costs in summer. The rooftop mini-ecosystem filters pollutants out of rainwater and reduces the amount of stormwater draining into sewers. In urban areas, green roofs provide wildlife habitat, even on the tops of tall buildings. A city with multiple green roofs provides "stepping stones" of habitat that enable migrating birds and insects to pass unharmed through the city. Green roofs can also serve as local food resources (see Chapter 18 opener); some restaurants have begun including rooftop gardens for herbs, vegetables, and even beekeeping. Green roofs can also provide an outdoor refuge for people living or working in the building as they allow urban systems to more closely resemble the natural systems they have replaced.

The largest individual green roof in the United States is on the Ford Motor Company factory in Dearborn, Michigan. Planted with nine varieties of hardy plants that require little or no maintenance, the Ford green roof absorbs excess rainwater. This water is naturally cleaned as it passes through a series of wetlands before entering the Rouge River, allowing Ford to avoid costly water treatment.

REVIEW

1. What are the steps used to solve an environmental problem?
2. What was the Lake Washington pollution problem of the 1950s? How was it addressed?

REVIEW OF LEARNING OBJECTIVES WITH SELECTED KEY TERMS

• **Explain how human activities affect global systems.**

Earth consists of many physical and biological **systems**. Its abundant resources have allowed many forms of life to thrive and evolve. Humans, through our growing population and technology, have exploited these resources to the point that we are putting the environment at risk.

• **Describe the factors that characterize human development and how they impact environment and sustainability.**

Human development is typically characterized by factors associated with wealth, such as access to energy resources and medical technology. Historically, **highly developed countries** have represented less than 20% of the global population but account for more than 50% of resource use. **Less**

developed countries (LDCs) are developing countries with high **poverty** rates, low levels of industrialization, high fertility rates, high infant mortality rates, and very low per capita incomes (relative to highly developed countries). Increasingly, many of the world's countries, such as China and India, have mixed development, with some urban residents owning considerable wealth but other urban and most rural residents living in poverty.

• **Differentiate between renewable and nonrenewable resources.**

Renewable resources are those that nature replaces fairly rapidly (on a scale of days to centuries), and can be used forever as long as they are not overexploited in the short term. **Nonrenewable resources** are present in limited supplies and are depleted by use.

• **Explain the impact of population and affluence on consumption.**

As population increases, people can exceed the capacity of a region to support basic needs for food, shelter, and clean water. When **consumption** by individuals substantially exceeds these basic needs, the resources in a region will be exceeded even more quickly. In either case, consumption that exhausts both **nonrenewable** and **renewable resources** is unsustainable.

• **Define** *ecological footprint*.

An individual's **ecological footprint** is the amount of productive land, fresh water, and ocean required on a continuous basis to supply that person with food, energy, water, housing, material goods, transportation, and waste disposal.

• **Describe the three most important factors that determine human impact on the environment.**

One **model** of environmental impact *(I)* has three factors: the number of people *(P)*; the affluence per person *(A)*, which is a measure of the consumption or amount of resources used per person; and the environmental effect of the technologies used to obtain and consume those resources *(T)*. This model uses an equation to represent the relationship between environmental impacts and the forces that drive them:

$$I = P \times A \times T$$

• **Define** *sustainability*.

Sustainability is the ability to meet current human natural resource needs without compromising the ability of future generations to meet their needs; in other words, it is the ability of humans to manage natural resources indefinitely without the environment going into a decline from the stresses imposed by human society on the natural systems that maintain life.

• **Relate Garrett Hardin's description of the tragedy of the commons in medieval Europe to common-pool resources today.**

Garrett Hardin contended that our inability to solve many environmental problems is the result of a struggle between short-term individual welfare and long-term environmental sustainability and societal welfare. In today's world, Hardin's parable has particular relevance at the global level. **Common-pool resources** are those parts of our environment that are available to everyone but for which no single individual has responsibility—shared resources such as the atmosphere, fresh water, forests, wildlife, and ocean fisheries.

• **Briefly describe** *sustainable development*.

Sustainable development is economic development that meets the needs of the present generation without compromising the ability of future generations to meet their own needs. Three factors—environmentally sound decisions, economically viable decisions, and socially equitable decisions—interact to promote sustainable development.

• **Define** *environmental science* **and briefly describe the role of Earth systems in environmental science.**

Environmental science is the interdisciplinary study of humanity's relationship with other organisms and the nonliving physical environment. Environmental scientists study *systems*; each system is a set of components that interact and function as a whole. A natural system consisting of a community of organisms and its physical environment is an **ecosystem**. Ecosystems are organized into larger and larger systems that interact with one another. At a global level are Earth systems, which include Earth's climate, atmosphere, land, coastal zones, and the ocean.

• **Outline the scientific method.**

The **scientific method** is the way a scientist approaches a problem by formulating a hypothesis and then testing it by means of an experiment. There are many variations of the scientific method, which basically involves these steps: State the problem or unanswered question; develop a **hypothesis**; design and perform an experiment to test the hypothesis; analyze and interpret the **data**; and share the conclusion with others.

• **List and briefly describe the five stages in addressing environmental problems.**

1. Scientific assessment involves gathering information about a potential environmental problem.

2. Risk analysis evaluates the potential effects of intervention.

3. Public education and involvement occur when the results of scientific assessment and risk analysis are placed in the public arena.

4. Political action is the implementation of a particular course of action by elected or appointed officials.

5. Long-term evaluation monitors the effects of the action taken.

• **Briefly describe the history of the Lake Washington pollution problem of the 1950s and how it was resolved.**

Lake Washington exemplifies a successful approach to addressing a relatively simple environmental problem. The pouring of treated sewage into Lake Washington had raised its level of nutrients to the point where the lake supported excessive growth of cyanobacteria. Disposal of the sewage in another way solved the lake's pollution problem.

Critical Thinking and Review Questions

1. Explain why a single child born in the United States can have a greater effect on the environment than 12 or more children born in a developing country.

2. Do you think it is possible for the world to sustain its present population of 7.2 billion indefinitely? Why or why not?

3. Is consumption driven more by population than affluence in highly developed countries? Less developed countries? Explain the difference.

4. In this chapter, we said the current global ecological footprint is 2.7 hectares (6.7 acres) per person. Do you think it will be higher, lower, or the same in 15 years? Explain your answer.

5. How are the concepts of ecological footprint and the *IPAT* model similar? Which concept do you think is easier for people to grasp?

6. Explain the following proverb as it relates to the concept of environmental sustainability: We have not inherited the world from our ancestors; we have borrowed it from our children.

7. Name an additional example of a common-pool resource other than those mentioned in this chapter.

8. Explain why economic well-being, environment, and ethics all contribute to sustainable development.

9. Give an example of an Earth system.

10. Thomas Henry Huxley once wrote, "The great tragedy of science—the slaying of a beautiful hypothesis by an ugly fact." Explain what he meant, based on what you have learned about the nature of science.

11. In the chapter, the term *model* is defined as a formal statement that describes a situation and that can be used to predict the future course of events. On the basis of this definition, is a model the same thing as a hypothesis? Explain your answer.

12. Some people want scientists to give them precise, definitive answers to environmental problems. Explain why this is not possible, and explain its implications for making decisions about climate change.

13. Explain why it might be difficult to make a decision about whether to allow farmers to spray pesticides even if we all agree about the negative health effects of the pesticide.

14. Place the following stages in addressing environmental problems in order and briefly explain each: long-term evaluation, public education and involvement, risk analysis, scientific assessment, political action.

15. What does the term *system* mean in environmental science?

16. In what ways do decisions about energy use and climate change that we make today limit the possibilities available to the next generation? Explain your answer.

17. Examine the graph, which shows an estimate of the discrepancy between the wealth of the world's poorest countries and that of the richest countries.

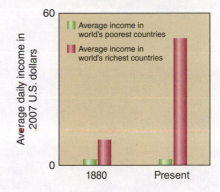

a. How has the distribution of wealth changed from the 1880s to the present? What explains this difference?

b. Based on the trend evident in this graph, predict what the graph might look like in 100 years.

c. Some economists think that our current path of economic growth is unsustainable. Are the data consistent with this idea? Explain your answer.

Food for Thought

For one week, keep track of the food you eat. Where does your food come from? How is it packaged? Did you produce any of it yourself, or do you know the individuals who did? Would you be able to eat only foods grown within 100 kilometers of your house? 500 kilometers? Explain the benefits and challenges of trying to do so.

2

Environmental Laws, Economics, and Ethics

Northern spotted owl. This rare species (*Strix occidentalis caurina*) is found predominantly in old-growth forests in the Pacific Northwest, from northern California to southern British Columbia. Protecting the owl's habitat benefits many other species that reside in the same environment.

During the late 1980s, an environmental controversy began in western Oregon, Washington, and northern California that continues today. At stake were thousands of jobs and the future of large tracts of *old-growth coniferous forest*, along with the existence of organisms that depend on the forest. The northern spotted owl came to symbolize the confrontation because of its dependence on mature forests for habitat (**see photo**).

Old-growth forests have never been logged. Because most of the forests in the United States were logged at one time or another, less than 10% of old-growth forests remain, and this fraction is decreasing. Most old-growth forests in the United States are found in the Pacific Northwest and Alaska.

Old-growth forests provide biological habitats for many species, including the northern spotted owl and 40 other endangered or threatened species. Provisions of the *Endangered Species Act* require the government to protect the habitat of endangered species; under this law, a 1991 court order suspended logging in about 1.2 million hectares (3 million acres) of forest where the owl lives. The timber industry opposed the moratorium, stating that thousands of jobs would be lost if the area were set aside.

The situation was more complex than simply jobs versus the environment, however. From 1977 to 1987, logging in Oregon's national forests increased by over 15%, while automation in the timber industry eliminated an estimated 12,000 jobs. By the 1980s, the rate at which industry was removing trees would have eliminated most remaining old-growth forests by about 2000.

The 1994 **Northwest Forest Plan** provided federal aid to retrain some timber workers for other careers. Hundreds of former loggers were employed to restore watersheds and salmon habitats in the forests they once logged. The plan, along with previous congressional and administrative actions, reserved about 75% of federal timberlands to safeguard watersheds and protect the northern spotted owl and several hundred other species. Logging was resumed on federal forests in Washington, Oregon, and northern California, but at only about one-fifth of the 1980s level.

Timber-cutting interests were not happy with the plan, and they tried unsuccessfully to revoke or revise the laws. Environmental groups were also unhappy; in 1999 they sued the U.S. Forest Service and the Bureau of Land Management, the two agencies overseeing logging of old-growth forests on federal land, maintaining that these agencies had not adequately carried out the provisions of the plan. The judge agreed that the plan required agencies to survey local endangered and threatened species before approving logging plans. The Forest Service conducted these surveys until 2004, when it discontinued the practice, arguing that they were an inconsequential bureaucratic hurdle.

In mid-2008, responding to lawsuits from the timber industry, the Fish and Wildlife Service released a new management plan that would reduce the protected habitat area by 23%. Critics immediately countered that the decision ignored input from agency scientists. In mid-2009, the Northwest Forest Plan was reinstated as the governing policy.

A weakened species becomes vulnerable to other threats as well. In 2013, the Fish and Wildlife Service approved a plan to remove barred owls from northern spotted owl habitats. The barred owl, native to eastern forests, has become established in areas previously dominated by northern spotted owls.

This chapter examines the environmental history of the United States and the roles of government, economics, and ethics in environmental management.

In Your Own Backyard…
Look into a contested environmental issue in your community. Are they described as *environment* versus *economics*? Do you think that's a fair depiction?

A Brief Environmental History of the United States

LEARNING OBJECTIVES

- Briefly outline the environmental history of the United States.
- Describe the environmental contributions of the following people: George Perkins Marsh, Theodore Roosevelt, Gifford Pinchot, John Muir, Aldo Leopold, Wallace Stegner, Rachel Carson, and Paul Ehrlich.
- Distinguish between utilitarian conservationists and biocentric preservationists.

From the establishment of the first permanent English colony at Jamestown, Virginia, in 1607, the first two centuries of U.S. history were a time of widespread environmental exploitation. Land, timber, wildlife, rich soil, clean water, and other resources were cheap and seemingly inexhaustible. Early European settlers did not dream that the bountiful natural resources of North America would one day become scarce. During the 1700s and early 1800s, most Americans had a **frontier attitude**, a desire to conquer and exploit nature as quickly as possible. Concerns about the depletion and degradation of resources occasionally surfaced, but efforts to conserve were seldom made because the vastness of the continent made it seem that we would always have enough resources.

PROTECTING FORESTS

The great forests of the Northeast were leveled within a few generations, and shortly after the Civil War in the 1860s, loggers began deforesting the Midwest at an accelerated rate. Within 40 years, they deforested an area the size of Europe, stripping Minnesota, Michigan, and Wisconsin of virgin forest (**Figure 2.1**). By 1897, the sawmills of Michigan had processed 160 billion board feet of white pine, leaving less than 6 billion board feet standing in the whole state.

Figure 2.1   Logging operations in 1884. This huge logjam occurred on the St. Croix River near Taylors Falls, Minnesota.

During the nineteenth century, many U.S. naturalists began to voice concerns about conserving natural resources. **John James Audubon** (1785–1851) painted lifelike portraits of birds and other animals in their natural surroundings. His paintings, based on detailed field observations, aroused widespread public interest in the wildlife of North America. **Henry David Thoreau** (1817–1862), a prominent U.S. writer, lived for two years on the shore of Walden Pond near Concord, Massachusetts. There he observed nature and contemplated how people could economize and simplify their lives to live in harmony with the natural world. **George Perkins Marsh** (1801–1882) was a farmer, linguist, and diplomat at various times during his life. Today he is most remembered for his book *Man and Nature*, which recognized the interrelatedness of human and environmental systems and provided one of the first discussions of humans as agents of global environmental change. Marsh was widely traveled, and *Man and Nature* was based in part on his observations of environmental damage in areas as geographically separate as the Middle East and his native Vermont.

In 1875, a group of public-minded citizens formed the *American Forestry Association*, with the intent of influencing public opinion against the wholesale destruction of America's forests. Sixteen years later, in 1891, the **General Revision Act** gave the president the authority to establish forest reserves on federally owned land. **Benjamin Harrison** (1833–1901), **Grover Cleveland** (1837–1908), and **Theodore Roosevelt** (1858–1919) used this law to remove 17.4 million hectares (43 million acres) of forest, primarily in the West, from logging.

In 1907 angry northwestern congressmen pushed through a bill rescinding the president's powers to establish forest reserves. Theodore Roosevelt, an important contributor to the conservation movement, responded by designating 21 new national forests that totaled 6.5 million hectares (16 million acres). He then signed into law the bill that prevented him and future presidents from establishing additional forest reserves.

Roosevelt appointed **Gifford Pinchot** (1865–1946) as the first head of the U.S. Forest Service. Both Roosevelt and Pinchot were **utilitarian conservationists** who viewed forests in terms of their usefulness for people—such as in providing jobs. Pinchot supported expanding the nation's forest reserves and managing forests scientifically, such as by harvesting trees only at the rate at which they regrow. Today, national forests are managed for multiple uses, from biological habitats to recreation to timber harvest to cattle grazing.

utilitarian conservationist A person who values natural resources because of their usefulness for practical purposes but uses them sensibly and carefully.

ESTABLISHING AND PROTECTING NATIONAL PARKS AND MONUMENTS

Congress established the world's first national park in 1872 after a party of Montana explorers reported on the natural beauty of the canyon and falls of the Yellowstone River; Yellowstone National Park now includes parts of Idaho, Montana, and Wyoming. In 1890, the *Yosemite National Park Bill* established the Yosemite

Figure 2.2 **President Theodore Roosevelt (left) and John Muir.** This photo was taken on Glacier Point above Yosemite Valley, California.

and Sequoia National Parks in California, largely in response to the efforts of a single man, naturalist and writer **John Muir** (1838–1914) (**Figure 2.2**). Muir, a **biocentric preservationist**, founded the *Sierra Club*, a national conservation organization that is still active on a range of environmental issues.

In 1906, Congress passed the **Antiquities Act**, which authorized the president to set aside certain sites that had scientific, historical, or prehistoric importance, such as the Badlands in South Dakota, as national monuments. By 1916, the U.S. Army was managing 13 national parks and 20 national monuments. (Today the *National Park Service* manages the 58 national parks and 73 national monuments.)

biocentric preservationist A person who believes in protecting nature because all forms of life deserve respect and consideration.

Some environmental battles involving the protection of national parks were lost. John Muir's Sierra Club fought such a battle with the city of San Francisco over its efforts to dam a river and form a reservoir in the Hetch Hetchy Valley (**Figure 2.3**), which lay within Yosemite National Park and was as beautiful as Yosemite Valley. In 1913, Congress voted to approve the dam.

The controversy generated a strong sentiment favoring better protection for national parks, and in 1916 Congress created the National Park Service to manage the national parks and monuments for the enjoyment of the public "without impairment." It was this clause that gave a different outcome to another battle, fought in the 1950s between conservationists and dam builders over the construction of a dam within Dinosaur National Monument. No one could deny that to drown the canyon with 400 feet of water would "impair" it. This victory for conservation established the "use without impairment" clause as the firm backbone of legal protection afforded to U.S. national parks and monuments. Today, discussion is under way to restore Hetch Hetchy, which the state of California estimates would cost as much as $10 billion.

CONSERVATION IN THE MID-TWENTIETH CENTURY

During the Great Depression, the federal government financed many conservation projects to provide jobs for the unemployed. During his administration, **Franklin Roosevelt** (1882–1945) established the Civilian Conservation Corps, which employed more than 175,000 men to plant trees, make paths and roads in national parks and forests, build dams to control flooding, and perform other activities to protect natural resources.

During the droughts of the 1930s, windstorms carried away much of the topsoil in parts of the Great Plains, forcing many farmers to abandon their farms and search for work elsewhere (see Chapter 14). The *American Dust Bowl* alerted the United States to the need for soil conservation, and in 1935, President Roosevelt formed the *Soil Conservation Service*.

Figure 2.3 **Hetch Hetchy Valley in Yosemite.** A view in Hetch Hetchy Valley (left) before and (right) after Congress approved a dam to supply water to San Francisco. (National Archives, WallaceKleck/Terraphotographics/BPS)

Aldo Leopold (1886–1948) was a wildlife biologist and environmental visionary who was influential in the conservation movement of the mid- to late-twentieth century. His textbook *Game Management*, published in 1933, supported the passage of a 1937 act in which new taxes on sporting weapons and ammunition funded wildlife management and research. Leopold also wrote philosophically about humanity's relationship with nature and about the need to conserve wilderness areas. In *A Sand County Almanac*, published in 1949, Leopold argued persuasively for a land ethic and the sacrifices such an ethic requires.

Aldo Leopold had a profound influence on many American thinkers and writers, including **Wallace Stegner** (1909–1993), who penned his famous "Wilderness Essay" in 1962. Stegner's essay, written to a commission conducting a national inventory of wilderness lands, helped create support for passage of the Wilderness Act of 1964. Stegner wrote:

Something will have gone out of us as a people if we ever let the remaining wilderness be destroyed; if we permit the last virgin forests to be turned into comic books and plastic cigarette cases; if we drive the few remaining members of the wild species into zoos or to extinction; if we pollute the last clean air and dirty the last clean streams and push our paved roads through the last of the silence, so that never again will Americans be free in their own country from the noise, the exhausts, the stinks of human and automotive waste. We simply need that wild country available to us, even if we never do more than drive to its edge and look in. For it can be a means of reassuring ourselves of our sanity as creatures, a part of the geography of hope.

During the 1960s, public concern about pollution and resource quality began to increase, in large part because of marine biologist **Rachel Carson** (1907–1964). Carson wrote about interrelationships among living organisms, including humans, and the natural environment (**Figure 2.4**). Her most famous work, *Silent Spring*, was published in 1962. In this work, Carson wrote against the indiscriminate use of pesticides.

Silent Spring heightened public awareness and concern about the dangers of uncontrolled use of DDT and other pesticides, including poisoning of wildlife and contamination of food supplies. Ultimately, this heightened public awareness led to restrictions on the use of certain pesticides. At about this time, the media began to increase its coverage of environmental incidents, such as hundreds of deaths in New York City from air pollution (1963), closed beaches and fish kills in Lake Erie from water pollution (1965), and detergent foam in a creek in Pennsylvania (1966).

In 1968, when the population of Earth was "only" 3.5 billion people (compared to 7.0 billion in 2011), ecologist **Paul Ehrlich** published *The Population Bomb*. In it, he described the unavoidable environmental damage necessary for Earth to support such a huge population, including soil loss, groundwater depletion, and extinctions. Ehrlich's book triggered public debate on the issue of overpopulation and how to address it effectively. As will be discussed later in this chapter, Ehrlich's *environmental worldview* influenced his theory that the world could comfortably hold only around 1 billion people.

Figure 2.4   Rachel Carson. Carson's book *Silent Spring* heralded the beginning of the environmental movement. Photographed in Maine in 1962.

THE ENVIRONMENTAL MOVEMENT OF THE LATE TWENTIETH CENTURY

Until 1970, the voices of **environmentalists** were heard in the United States primarily through societies such as the *Sierra Club* and the *National Wildlife Federation*. There was no generally perceived environmental movement until the spring of 1970, when **Gaylord Nelson**, former senator from Wisconsin, urged Harvard graduate student **Denis Hayes** to organize the first nationally celebrated Earth Day. This event awakened U.S. environmental consciousness to population growth, overuse of resources, and pollution and degradation of the environment. On Earth Day 1970, an estimated 20 million people in the United States demonstrated their support for environmental quality by planting trees, cleaning roadsides and riverbanks, and marching in parades. (**Figure 2.5** presents a list of selected environmental events since Earth Day 1970.)

In the years that followed the first Earth Day, environmental awareness and the belief that individual actions could repair the damage humans were doing to Earth became a pervasive, popular movement. Musicians popularized environmental concerns, and many of the world's religions embraced environmental themes such as protecting endangered species.

By Earth Day 1990, the movement had spread around the world, signaling the rapid growth in environmental consciousness; an estimated 200 million people in 141 nations demonstrated to increase public awareness of the importance of individual efforts. Earth Day continues to be a popular event around the world, and many years include an organizing theme, such as "Think Globally, Act Locally" (1990) and "The Face of Climate Change" (2013, **Figure 2.6**).

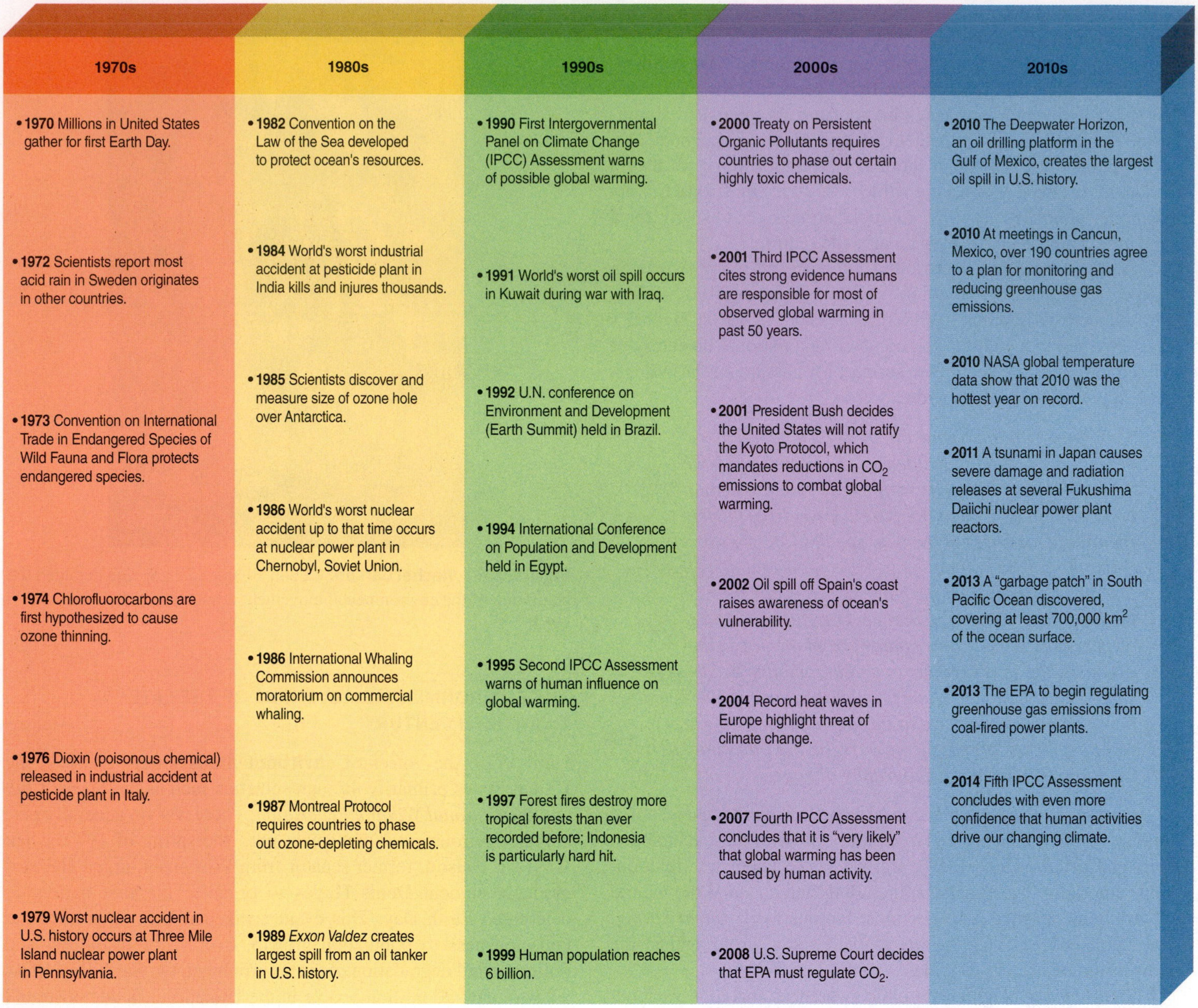

1970s	1980s	1990s	2000s	2010s
• **1970** Millions in United States gather for first Earth Day.	• **1982** Convention on the Law of the Sea developed to protect ocean's resources.	• **1990** First Intergovernmental Panel on Climate Change (IPCC) Assessment warns of possible global warming.	• **2000** Treaty on Persistent Organic Pollutants requires countries to phase out certain highly toxic chemicals.	• **2010** The Deepwater Horizon, an oil drilling platform in the Gulf of Mexico, creates the largest oil spill in U.S. history.
• **1972** Scientists report most acid rain in Sweden originates in other countries.	• **1984** World's worst industrial accident at pesticide plant in India kills and injures thousands.	• **1991** World's worst oil spill occurs in Kuwait during war with Iraq.	• **2001** Third IPCC Assessment cites strong evidence humans are responsible for most of observed global warming in past 50 years.	• **2010** At meetings in Cancun, Mexico, over 190 countries agree to a plan for monitoring and reducing greenhouse gas emissions.
• **1973** Convention on International Trade in Endangered Species of Wild Fauna and Flora protects endangered species.	• **1985** Scientists discover and measure size of ozone hole over Antarctica.	• **1992** U.N. conference on Environment and Development (Earth Summit) held in Brazil.	• **2001** President Bush decides the United States will not ratify the Kyoto Protocol, which mandates reductions in CO_2 emissions to combat global warming.	• **2010** NASA global temperature data show that 2010 was the hottest year on record.
• **1974** Chlorofluorocarbons are first hypothesized to cause ozone thinning.	• **1986** World's worst nuclear accident up to that time occurs at nuclear power plant in Chernobyl, Soviet Union.	• **1994** International Conference on Population and Development held in Egypt.		• **2011** A tsunami in Japan causes severe damage and radiation releases at several Fukushima Daiichi nuclear power plant reactors.
	• **1986** International Whaling Commission announces moratorium on commercial whaling.	• **1995** Second IPCC Assessment warns of human influence on global warming.	• **2002** Oil spill off Spain's coast raises awareness of ocean's vulnerability.	• **2013** A "garbage patch" in South Pacific Ocean discovered, covering at least 700,000 km^2 of the ocean surface.
• **1976** Dioxin (poisonous chemical) released in industrial accident at pesticide plant in Italy.	• **1987** Montreal Protocol requires countries to phase out ozone-depleting chemicals.	• **1997** Forest fires destroy more tropical forests than ever recorded before; Indonesia is particularly hard hit.	• **2004** Record heat waves in Europe highlight threat of climate change.	• **2013** The EPA to begin regulating greenhouse gas emissions from coal-fired power plants.
• **1979** Worst nuclear accident in U.S. history occurs at Three Mile Island nuclear power plant in Pennsylvania.	• **1989** *Exxon Valdez* creates largest spill from an oil tanker in U.S. history.	• **1999** Human population reaches 6 billion.	• **2007** Fourth IPCC Assessment concludes that it is "very likely" that global warming has been caused by human activity.	• **2014** Fifth IPCC Assessment concludes with even more confidence that human activities drive our changing climate.
			• **2008** U.S. Supreme Court decides that EPA must regulate CO_2.	

Figure 2.5   **Selected environmental events, 1970 to the present.**

REVIEW

1. Which occurred first in the U.S. environmental movement: concerns about forest conservation or concerns about pollution?

2. Describe how an individual can influence U.S. environmental history or policy.

3. Explain how the attitudes of utilitarian conservationists toward environmental policy differ from those of biocentric preservationists.

© Octavio Jones/Tampa Bay Times/ZUMAPRESS.com

Figure 2.6   **Earth Day 2013.** Maria Martinez Head Start students, along with faculty and staff, plant a tree as part of Earth Day events in 2014.

U.S. Environmental Legislation

- Explain why the National Environmental Policy Act is the cornerstone of U.S. environmental law.
- Explain how environmental impact statements provide powerful protection of the environment.

By the late 1960s, much of the U.S. public had become increasingly disenchanted with governmental secrecy, and many did not trust industry to work in the public interest. This broad social transformation, which included opposition to the Vietnam War and resistance to racist policies, was reflected in environmental attitudes as well. Galvanized by well-publicized ecological disasters, such as the 1969 oil spill off the coast of Santa Barbara, California, and by overwhelming public support for the Earth Day movement, in 1970 the **Environmental Protection Agency (EPA)** was formed, and the **National Environmental Policy Act (NEPA)** was signed into law. A key provision of NEPA required the federal government to consider the environmental impact of any proposed federal action, such as financing highway or dam construction. NEPA provides the basis for developing detailed **environmental impact statements (EISs)** to accompany every federal recommendation or legislative proposal. These EISs are supposed to help federal officials make informed decisions. Each EIS must include the following:

environmental impact statement (EIS) A document that summarizes potential and expected adverse impacts on the environment associated with a project as well as alternatives to the proposed project; typically mandated by law for public and/or private projects.

1. The nature of the proposal and why it is needed
2. The environmental impacts of the proposal, including short-term and long-term effects and any adverse environmental effects if the proposal is implemented (**Figure 2.7**)
3. Alternatives to lessen the adverse effects of the proposal; providing ways to mitigate the impact of the project

A required step in the EIS process is the solicitation of public comments, which generally provide a broader perspective on the proposal and its likely effects. NEPA established the **Council on Environmental Quality** to monitor the required EISs and report directly to the president. This council had no enforcement powers, and NEPA was originally considered innocuous, generally more a statement of good intentions than a regulatory policy.

During the next few years, environmental activists took people, corporations, and the federal government to court to challenge their EISs and use them to block proposed development. The courts decreed that EISs must be substantial documents that thoroughly analyze the environmental consequences of anticipated projects on soil, water, and organisms. The courts also said that the public must have access to EISs. These rulings put sharp teeth into the law—particularly the provision for public scrutiny, which placed intense pressure on federal agencies to respect EIS findings.

NEPA revolutionized environmental protection in the United States. In addition to overseeing federal highway construction, flood and erosion control, military projects, and many other public works, federal agencies oversee nearly one-third of the land in the United States. Federally owned holdings include extensive fossil-fuel and mineral reserves as well as millions of hectares of public grazing land and public forests, all subject to NEPA. Many states and local governments now require EISs for public (and sometimes private) projects, as do many foreign countries.

NEPA has evolved considerably through four decades of litigation and eight presidential administrations, but the fundamental effect of publicly available information remains. Although almost everyone agrees that NEPA has helped federal agencies reduce adverse environmental impacts of their activities and projects, it has its critics. Environmentalists complain that EISs are sometimes incomplete, include only alternatives that make a predetermined alternative look attractive, or are ignored when decisions are made. Other critics think the EISs delay important projects ("paralysis by analysis") because EISs are too involved, take too long to prepare, and are often the targets of lawsuits.

ENVIRONMENTAL POLICY SINCE 1970

Although legislation to manage many environmental problems existed before 1970, the regulatory system that exists today was largely put into place during the 1970s. Congress has passed many environmental laws that address a wide range of issues, such as endangered species, clean water, clean air, energy conservation, hazardous wastes, and pesticides (**Table 2.1**). These laws have greatly increased federal regulation of pollution, creating a tough interlocking mesh of laws to improve environmental quality.

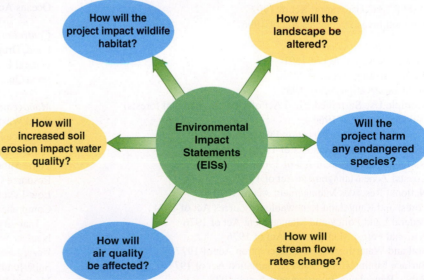

Figure 2.7   Environmental impact statements. These detailed statements help federal agencies and the public consider the environmental impacts of proposed activities. When anticipated impacts are likely to be high, decision makers will feel pressure to seek alternative actions.

Many environmental laws contain provisions that allow private citizens to take violators, whether they are private industries or government-owned facilities, to court for noncompliance. These citizen suits have contributed significantly in the enforcement of environmental legislation.

In the early 1980s, President Reagan attempted to reverse the pro-environmental trend by appointing a pro-business EPA administrator, **Ann Gorsuch**, and leaving many top-level positions in the agency unfilled. Congressional and public backlash led to even more restrictive environmental laws. Gorsuch was replaced with **William Ruckleshaus**, who had been the first EPA administrator and was widely respected for having established the agency's credibility and authority. Ruckleshaus advocated a rational, organized approach to environmental regulation that would avoid the "pendulum" effect of stringent and then relaxed environmental regulation.

Table 2.1   Some Important Federal Environmental Legislation

General
Freedom of Information Act of 1966
National Environmental Policy Act of 1969
National Environmental Education Act of 1990

Conservation of Energy and Renewable Energy Resources
Energy Policy and Conservation Act of 1975
Northwest Power Act of 1980
National Appliance Energy Conservation Act of 1987
Energy Policy Act of 1992
American Recovery and Reinvestment Act of 2008

Conservation of Wildlife
Fish and Wildlife Act of 1956
Anadromous Fish Conservation Act of 1965
Fur Seal Act of 1966
National Wildlife Refuge System Act of 1966
Species Conservation Act of 1966
Marine Mammal Protection Act of 1972
Marine Protection, Research, and Sanctuaries Act of 1972
Endangered Species Act of 1973
Federal Noxious Weed Act of 1974
Magnuson Fishery Conservation and Management Act of 1976
Whale Conservation and Protection Study Act of 1976
Fish and Wildlife Improvement Act of 1978
Fish and Wildlife Conservation Act of 1980
Fur Seal Act Amendments of 1983
Wild Bird Conservation Act of 1992
National Invasive Species Act of 1996

Conservation of Land
General Revision Act of 1891
Taylor Grazing Act of 1934
Soil Conservation Act of 1935
Multiple Use Sustained Yield Act of 1960 (re: national forests)
Wilderness Act of 1964
Land and Water Conservation Fund Act of 1965
Wild and Scenic Rivers Act of 1968
National Trails System Act of 1968
Coastal Zone Management Act of 1972
National Reserves Management Act of 1974
Forest and Rangeland Renewable Resources Act of 1974
Federal Land Policy and Management Act of 1976
National Forest Management Act of 1976
Soil and Water Resources Conservation Act of 1977
Surface Mining Control and Reclamation Act of 1977
Public Rangelands Improvement Act of 1978
Antarctic Conservation Act of 1978
Endangered American Wilderness Act of 1978
Alaska National Interest Lands Act of 1980
Coastal Barrier Resources Act of 1982

Emergency Wetlands Resources Act of 1986
North American Wetlands Conservation Act of 1989
California Desert Protection Act of 1994
Food, Conservation, and Energy Act of 2008 (the latest version of the "farm bill," which has been amended and renamed every 5 years or so since the 1930s)

Air Quality and Noise Control
Noise Control Act of 1965
Clean Air Act of 1970
Quiet Communities Act of 1978
Asbestos Hazard and Emergency Response Act of 1986
Clean Air Act Amendments of 1990

Water Quality and Management
Refuse Act of 1899
Water Resources Research Act of 1964
Water Resources Planning Act of 1965
Clean Water Act of 1972
Ocean Dumping Act of 1972
Safe Drinking Water Act of 1974
National Ocean Pollution Planning Act of 1978
Water Resources Development Act of 1986
Great Lakes Toxic Substance Control Agreement of 1986
Water Quality Act of 1987 (amendment of Clean Water Act)
Ocean Dumping Ban Act of 1988
Oceans Act of 2000

Control of Pesticides
Food, Drug, and Cosmetics Act of 1938
Federal Insecticide, Fungicide, and Rodenticide Act of 1947
Food Quality Protection Act of 1996

Management of Solid and Hazardous Wastes
Solid Waste Disposal Act of 1965
Resource Recovery Act of 1970
Hazardous Materials Transportation Act of 1975
Toxic Substances Control Act of 1976
Resource Conservation and Recovery Act of 1976
Low-Level Radioactive Policy Act of 1980
Comprehensive Environmental Response, Compensation, and Liability ("Superfund") Act of 1980
Nuclear Waste Policy Act of 1982
Hazardous and Solid Waste Amendments of 1984
Superfund Amendments and Reauthorization Act of 1986
Medical Waste Tracking Act of 1988
Marine Plastic Pollution Control Act of 1987
Oil Pollution Act of 1990
Pollution Prevention Act of 1990
State or Regional Solid Waste Plans (RCRA Subtitle D) of 1991

Since 1980, the rate of new major environmental legislation has slowed, but environmental policy continues to evolve through administrative actions and the judicial process. In 1994, President Clinton issued an executive order requiring that environmental justice be considered in all regulatory actions. Environmental policy through the Clinton and George W. Bush administrations was characterized by expanded use of cost–benefit analysis in environmental regulation.

The George W. Bush administration placed higher priority on extracting environmental resources than reducing human impacts on the environment. No major environmental laws were passed between 2000 and 2008, and a number of major land-use regulations—such as preserving habitat for the spotted owl—were relaxed. In addition, in some cases states were barred from passing regulations more stringent than those required by the federal government, most notably the attempt by California and 13 other states to impose automobile efficiency standards on cars and light trucks. One of the first acts of the Obama administration was to reverse this prohibition.

In one significant legal case during the George W. Bush administration, Massachusetts sued the EPA to require that greenhouse gases be regulated under the **Clean Air Act (CAA)**. The EPA made two arguments: (1) that the CAA did not authorize climate change regulations and (2) that the CAA was not an appropriate way to manage climate change since it is a global problem, not amenable to local or even national solutions. The Supreme Court disagreed, concluding that greenhouses gases are pollutants emitted into the air and capable of endangering public health or welfare. It further noted that the EPA claim that it *should not* regulate greenhouse gases was inconsistent with the CAA, since the EPA had acknowledged that greenhouse gases contribute to climate change.

The EPA is usually given the job of translating environmental laws into specific regulations. Before the regulations become official, several rounds of public comments allow affected parties to present their views; the EPA is required to respond to all of these comments. Next, the Office of Management and Budget reviews the new regulations. Some regulations proposed by the EPA must be justified by **cost–benefit analysis**, while others (including those under the Clean Air Act) are barred from considering economic impacts. Implementation and enforcement of environmental regulations often fall to state governments, which must send the EPA detailed plans showing how they intend to achieve regulatory goals and standards. Currently, the EPA oversees thousands of pages of environmental regulations that affect individuals, corporations, local communities, and states.

Environmental laws have not always worked as intended. The Clean Air Act of 1977 required coal-burning power plants to outfit their smokestacks with expensive *scrubbers* to remove sulfur dioxide from their emissions but made an exception for tall smokestacks (**Figure 2.8**). This loophole led to the proliferation of tall stacks that have since produced *acid rain* throughout the Northeast. The Clean Air Act Amendments of 1990, described in Chapter 19, go a long way toward closing this loophole. In addition, many environmental regulations are not enforced properly. According to **Oliver Houck** of the Tulane Law School, environmental laws such as the Clean Water Act have compliance rates of about 50%.

Despite imperfections, U.S. environmental legislation has had overall positive effects. Since 1970:

1. Twenty-three national parks have been established, and the National Wilderness Preservation System now totals more than 43 million hectares (107 million acres).

2. Millions of hectares of land vulnerable to erosion have been withdrawn from farming, reducing soil erosion in the United States by more than 60%.

3. Many previously endangered species are better off now than they were in 1970, and the American alligator, California gray whale, bald eagle, and brown pelican have recovered enough to be removed from the endangered species list. (Dozens of other species, however, such as the manatee, ivory-billed woodpecker, polar bear, and Kemp's sea turtle, have suffered further declines or extinction since 1970.)

4. Energy efficiency and conservation technology have improved markedly for buildings, vehicles, and consumer products.

Although we still have a long way to go, pollution-control efforts have been particularly successful. According to the EPA's most recent *Report on the Environment (2008)*:

1. Both emissions and ambient concentrations of six important air pollutants have dropped by at least 25% since 1970, much more in the case of lead. (Carbon dioxide emissions have continued to rise.)

2. Since 1990, levels of wet sulfate, a major component of acid rain, have dropped by about 33%.

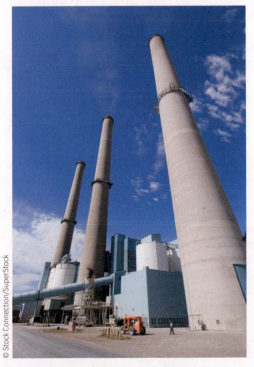

© Stock Connection/SuperStock

Figure 2.8   Tall smokestacks. Such smokestacks, which emit sulfur dioxide from coal-burning power plants, were exempt from the requirement of pollution-control devices under the Clean Air Act of 1977.

3. In 2008, 92% of the U.S. population got its drinking water from community water systems with no violations of EPA standards, up from around 75% in 1993.

4. In 2014, 45% of municipal solid waste was combusted for energy recovery or recovered for composting or recycling, up from 6% in the 1960s. While total municipal solid waste has been increasing, the amount sent to landfills has remained roughly constant since 1980.

5. In 2007, the EPA considered human exposures to contamination to be under control at 93% of listed hazardous waste sites, up from 37% of the sites listed in 2000.

However, not all environmental trends are positive. In some major urban areas, such as Houston and Las Vegas, air quality has begun to deteriorate since about 2000. While technology has reduced environmental impacts of industries and products, population increases have outpaced technological improvements (recall the *IPAT* model from Chapter 1). In addition, despite broad and increasing concern about climate change, no effective U.S. or international policies have been established. Future environmental improvement will require more sophisticated strategies, involving technological and behavior changes.

REVIEW

1. Which law is the cornerstone of U.S. environmental law? Why?

2. In what ways has the U.S. environment improved as the result of regulations?

Economics and the Environment

LEARNING OBJECTIVES

- Explain why economists prefer efficient solutions to environmental problems.
- Describe command and control regulation, incentive-based regulation, and cost-effectiveness analysis.
- Give two reasons why the national income accounts are incomplete estimates of national economic performance.

Economics is the study of how people use their limited resources to try to satisfy unlimited wants. Economists combine theoretical assumptions about individual and institutional behavior with analytical tools that include developing hypotheses, testing models, and analyzing observations and data. They try to understand the consequences of the ways in which people, businesses, and governments allocate their resources.

Economists who work on environmental problems must take a systems perspective. For example, Nobel Prize–winning economist **Amartya Sen** addresses overlapping issues of environment, poverty, nationalism, gender, and governmental structure (**Figure 2.9**). This approach recognizes the complex interactions among the environment, society, health, and well-being.

Economics, especially as applied to public policy, relies on several precepts. First, economics is utilitarian. This means that all goods and services—including those provided by the environment—have value to people and that those values can be converted into some common currency. Thus, in a market economy such as that of the United States, goods and services have dollar values, which are determined by the amount of money someone is willing to pay for them.

This leads to a second precept, the **rational actor model**. Economists assume two things about all individuals. First, individuals have preferences among different goods and services, which economists refer to as **utility**. Second, individuals can and do spend their limited resources (money and time) in a way that provides them the most utility.

Third, in an ideal economy, resources will be allocated efficiently. *Efficiency* is a term economists use to describe multiple individuals getting the greatest amount of goods or services from a limited set of resources. For example, if business plan A will build 9 cars with a given amount of material and number of workers, while business plan B will build 10 equivalent cars with the same resources, business plan B is more efficient and will succeed. To economists, environmental problems arise when market failures occur in one of several ways, principal among which are externalities and inefficiency.

Externalities occur when the producer of a good or service does not have to pay the full costs of production. An example is a blacksmith shop next door to a drip-dry laundry. If soot from the blacksmith shop lands on the laundry, then the laundry owner will have to rewash the laundry and try to pass on the additional cost to its customers. Here, the blacksmith shop is causing a negative externality on the laundry and its customers, who in effect pay part of the cost of the blacksmith operation.

A simple solution—advocated by some economists—is to clearly define rights and ownership. If the blacksmith does not have the right to release soot, then the laundry owner can demand

rational actor model In economics, the assumption that all individuals spend their limited resources in a fashion that maximizes their individual utilities.

utility An economic term referring to the benefit that an individual gets from some good or service. Rational actors try to maximize utility.

externality In economics, the effects (usually negative) of a firm that does not have to pay all the costs associated with its production.

Prakash Singh/Agence FrancePresse/NewsCom

Figure 2.9   Nobel laureate Amartya Sen. Economist Amartya Sen's work includes critiques of traditional assumptions about economic valuation of the environment and recognizes the complex, systemic nature of the environment, society, health, and well-being.

compensation for the added cost of business. The blacksmith then has several options—for example, to close the shop, find a way to control the soot, or pay the laundry owner for the extra cost of business. The efficient solution is the one that costs the least money for everyone. If closing the shop costs $10,000 per year, using soot-control equipment costs $200 per year, and paying to rewash laundry costs $500 per year, the blacksmith should choose to control soot. This same solution will occur even if the black-smith *does* have the right to release soot, since the laundry owner should be willing to pay the blacksmith $200 to use soot-control equipment rather than spending $500 to rewash laundry.

Unfortunately, most environmental externalities are not as clearly defined as the laundry/blacksmith example. Usually, there are many polluters and many affected individuals. A small town may have 5,000 automobiles, each of which generates a different amount of pollution that affects each of 10,000 residents. Damage associated with the releases may depend on time of day as well as where in the town pollution is generated. Even if we could assess those damages, it would be impossible for each driver to compensate each person impacted by the car's emissions. The problem becomes even more extreme for climate change, where billions of individuals generate and are affected by greenhouse gases.

Several economic solutions exist for cases of multiple pollut-ers, all of which rely on the idea that we can identify an efficient or **optimal amount of pollution**. At this optimum, the cost to society of having less pollution is offset by the benefits to society of the activity creating the pollution. To find this optimum requires that we identify the **marginal cost of pollution**, which refers to the cost of a small additional amount of pollution. Determining the marginal cost of pollution involves assessing pollution damage to health, property, agriculture, and aesthetics. Pollution can also reduce **ecosystem services (see Table 5.1)**, which are benefits to humans—including clean water and fresh air—provided by natu-ral systems. Determining the marginal cost of pollution is gener-ally not an easy process.

Similarly, the **marginal cost of abatement** is the cost asso-ciated with reducing (abating) a small additional amount of pol-lution. Untreated wastewater from a pulp mill contains a variety of chemicals and suspended wood fibers. The cost of filtering the fibers is relatively low, requiring mechanical screening. The cost of removing inorganic chemicals, however, may be quite high (see Chapter 21). The societal benefit of filtering is relatively high com-pared to the benefit of removing a small amount of a chemical. If we insist that only perfectly pure water comes out of the pulp mill, the cost of paper will be extremely high.

Figure 2.10 demonstrates how the marginal cost of pollution and the marginal cost of abatement lead to an optimal amount of pollution. The green line represents the marginal cost of abate-ment, and the red line the marginal cost of pollution. When the amount of pollution being generated is high, abatement costs are relatively low. Similarly, when pollution concentration is low, abatement costs are high; in this case, the benefit of additional con-trol is low. Pollution is at an optimal level when marginal costs of pollution and abatement are equal (the point at which the red and green lines intersect). When that is the case, the economic system is efficient, meaning that society would be worse off with either more *or less* pollution. If a pulp mill operating at that optimal level

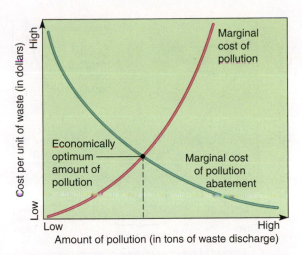

Figure 2.10   Economic optimality and pollution. This figure rep-resents the economically optimal level of pollution in an efficient market. The upward-sloping curve (red line) represents the cost of damage associated with pollution at various levels. As pollution rises, the social cost (in terms of human health and a damaged environment) increases sharply. The downward-sloping curve (green line) represents the cost of reducing pollution to a lower, less damaging level. The intersection of these two curves is the economically optimal point, where a shift in either direction (more or less pollution) will lead to lower total societal benefits.

were to spend an additional $1 on control equipment, the value to society of the reduction in pollution would be less than $1.

It is unusual to find unregulated economic systems at the point of economic optimality. In unregulated markets, those who pollute often must pay only a fraction of the total costs of pollution. Thus, the marginal cost of pollution faced by the polluter (private cost) is substantially lower than the marginal cost of pollution faced by society as a whole. This is represented by **Figure 2.11**, which contains two marginal cost of pollution curves. The marginal cost curve on the left (red line) is the societal marginal cost of pollu-tion, which includes that faced by the polluter. To the right of this is the private marginal cost of pollution curve (blue line), which is the fraction of the overall marginal cost paid by the polluter. With-out some sort of regulation, the amount of pollution generated will be determined by the point at which the private marginal cost of pollution is equal to the marginal cost of abatement (intersection of green and blue lines). Substantially more pollution may be gener-ated at this intersection than is optimal from a societal perspective.

STRATEGIES FOR POLLUTION CONTROL

Economists favor market-based solutions to environmental exter-nality problems. Historically, many environmental regulations have been **command and control** solutions. This means the EPA or other government agency requires that a particular piece of equipment be installed to limit emissions to water, air, or soil. Industries regulated in this fashion often object that they are thereby discouraged from developing lower-cost alternatives that would achieve the same level of pollution control for less money.

Command and control regulations may result in pollu-tion levels that are lower or higher than the economically opti-mum level; for this reason, they are not popular with economists.

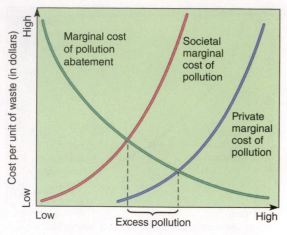

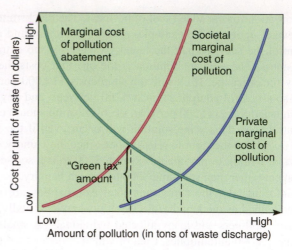

Figure 2.11   **Inefficiency arising from different marginal costs.** This figure represents the marginal costs of pollution faced by society (red line), the private marginal cost of pollution faced by the polluter (blue line), and the marginal cost of abatement (green line). The polluter will generate pollution at the point where the private marginal cost of pollution intersects with the marginal cost of abatement. At this point, substantially more pollution is generated than is optimal from a societal perspective.

Figure 2.12   **The corrective effect of "green taxes."** This figure represents the marginal costs of pollution faced by society (red line), the private marginal cost of pollution faced by the polluter (blue line), and the marginal cost of abatement (green line). A tax is assessed on the polluter equal to the difference between the private marginal cost of pollution and the social marginal cost of pollution. As a result, the entire private marginal cost curve is shifted upward. If the tax is designed well, the cost faced by the polluter will be identical to the marginal societal cost of pollution.

Consequently, most economists, whether they have progressive or conservative views, prefer **incentive-based regulations** over command and control ones. Two major incentive-based approaches that have met with mixed policy success are environmental taxes and *tradable permits*. *Cost-effectiveness analysis* is a cost–benefit-based approach for prioritizing diverse environmental regulations.

A popular incentive-based approach for regulating pollution, particularly in Europe, involves imposing an **emission charge** on polluters. In effect, this charge is a tax on pollution. Economists propose such "green taxes" to correct what they perceive as distortions in the market caused by not including the external cost of driving an automobile or cutting down trees or polluting streams. The purpose of this tax is to force polluters to pay the full cost of pollution. If the tax is set at the correct level, the private marginal cost of pollution will be equivalent to the societal cost of pollution. This is depicted in **Figure 2.12**.

Many European countries have restructured their taxes to take into account environmentally destructive products and activities. Germany increased taxes on gasoline, heating oil, and natural gas while simultaneously lowering its income tax. One result was an increase in carpooling. There were concerns, however, about whether Germany's energy-intensive industries could remain competitive with industries in countries without similar energy taxes. The Netherlands introduced a tax on natural gas, electricity, fuel oil, and heating oil that includes incentives for energy efficiency; income taxes were decreased to offset the tax burden on individuals. Electricity and fuel use has declined as a result. Finland has implemented a carbon dioxide tax. Sweden introduced taxes on carbon and sulfur while simultaneously reducing income taxes. These charges increase the cost of polluting or of overusing natural resources. They also are usually *revenue-neutral* since they are offset by rebates or reductions in other taxes.

If taxes are set at the correct level, users will react to the increase in cost resulting from emission charges by decreasing pollution or decreasing consumption. However, it can be very difficult to identify the economically optimal level of taxation, and taxes on pollution are almost always set too low to have the desired effect on the behavior of people or companies. It is difficult to enact such taxes, especially in the United States, because people object to paying a tax on something they perceive as "free," and they tend to doubt that such taxes will be revenue-neutral.

While environmental taxes are designed to identify and replicate the societal cost of pollution, **tradable permits** rely on identifying the optimal level of pollution. Government sets a cap on pollution and then issues a fixed number of tradable permits (sometimes called **marketable waste-discharge permits**), allowing holders to emit a specified amount of a given pollutant, such as sulfur dioxide. A permit owner can decide whether to generate the pollution or sell the permit, ideally choosing the option that earns the most money. Once the market for permits has been established, industries that can easily reduce emissions will do so and then sell their extra permits to industries that cannot.

Tradable permits can also be designed to reduce emissions over time. The Clean Air Act Amendments of 1990 included a plan to cut acid rain–causing sulfur dioxide emissions by issuing tradable permits for coal-burning electricity utilities. The amount of allowable pollution decreases with each trade. This approach has been effective. Sulfur dioxide emissions are being reduced ahead of schedule at about 50% of the original projected cost.

Cost-effectiveness analysis is an increasingly common regulatory tool. Rather than setting an optimum level of pollution, cost-effectiveness analysis asks, "If we establish this regulation, how much will it cost to achieve some outcome?" Here, the outcome is lives saved or years of life saved. Thus, a regulatory action that costs $5,000 per saved life would be preferable to one that costs $10,000 per saved life. This approach could be used to compare a wide range of interventions, from banning pesticides to requiring catalytic converters on cars.

CRITIQUES OF ENVIRONMENTAL ECONOMICS

Two classes of critiques are leveled at economic approaches to environmental regulation. First, it is difficult to assess the true costs of environmental damage by pollution and pollution abatement. The impacts of pollution on people and nature are extremely uncertain. We often do not know how effective pesticides are at eliminating pests, or whether a pesticide harms humans. If we do not understand the economic benefits of ecosystem services, they may be undervalued. Trying to place a value on quality of life or on damage to natural beauty can be highly controversial (**Figure 2.13**). As you will see in the next few chapters, the web of relationships within the environment is intricate, and environmental systems may be more vulnerable to pollution damage than is initially obvious. When the optimal level of pollution is highly uncertain, too many or too few permits may be issued, and green taxes may be set too high or too low.

Second, we do not all agree that economics is an appropriate decision tool. Economics may not take the risks of unanticipated environmental catastrophes into account and may not consider dynamic changes over time. Utilitarian economics does not account for fairness, so even if society were better off overall, some individuals might lose out. Many people distrust economic approaches to environmental regulation, since they are so strongly associated with industry groups. Economists assign prices based on what people can or do pay for things. This means that the knowledge that a pristine stream or songbird exists will not have much impact in an economic assessment. Additionally, economics dictates that if it is inefficient to save a species, we should not do so. This idea clashes with many religious and ethical beliefs.

NATURAL RESOURCES, THE ENVIRONMENT, AND NATIONAL INCOME ACCOUNTS

Because much of our economic well-being flows from natural, rather than human-made, assets, we should include the use and

Figure 2.13   Agriculture, timber, and scenic beauty at Coruña, Spain. The value of the environment derives from many factors that constitute a complex, interrelated system. These factors include aesthetics, recreation, agriculture, real estate, and ecosystem services. It can be difficult or impossible to accurately assess the monetary value of all possible options for an area such as this one.

misuse of natural resources and the environment in national income accounts. **National income accounts** represent the total income of a nation for a given year. Two measures used in national income accounting are *gross domestic product (GDP)* and *net domestic product (NDP)*. Both GDP and NDP provide estimates of national economic performance used to make important policy decisions.

Unfortunately, current national income accounting practices are misleading and incomplete because they do not incorporate environmental factors. At least two important conceptual problems affect the way national income accounts currently handle the economic use of natural resources and the environment. These problems involve costs and benefits of pollution control and depletion of **natural capital**, which refers to Earth's resources and processes that sustain living organisms (**Figure 2.14**).

Natural Resource Depletion   If a firm produces some product (output) but in the process wears out a portion of its plant and

Figure 2.14   Natural capital and the environment. Goods and services (products) and money (to pay for the products) flow between businesses (production) and consumers (consumption). Economies depend on natural capital to provide sources for raw materials and sinks for waste products.

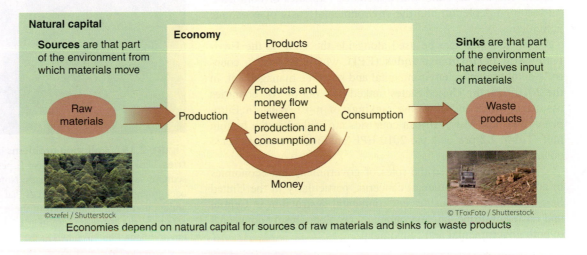

Natural capital

Sources are that part of the environment from which materials move

Economy

Raw materials → Production

Products

Products and money flow between production and consumption

Consumption

Money

Sinks are that part of the environment that receives input of materials

Waste products

©szefei / Shutterstock   © TFoxFoto / Shutterstock

Economies depend on natural capital for sources of raw materials and sinks for waste products

equipment, the firm's output is counted as part of GDP, but the depreciation of capital is subtracted in the calculation of NDP. Thus, NDP is a measure of the net production of the economy after a deduction for used-up capital. In contrast, when an oil company drains oil from an underground field, the value of the oil produced is counted as part of the nation's GDP, but no offsetting deduction to NDP is made to account for the nonrenewable resources used up.

The Cost and Benefits of Pollution Control at the National Level Imagine that a company has the following choices: It can produce $100 million worth of output and, in the process, pollute the local river by dumping its wastes. Alternatively, by properly disposing of its wastes, it can avoid polluting but will only get $90 million of output. Under current national income accounting rules, if the firm chooses to pollute, its contribution to GDP is larger ($100 million rather than $90 million). National income accounts attach no explicit value to a clean river. In an ideal accounting system, the economic cost of environmental degradation (loss of natural capital) would be subtracted in the calculation of a firm's contribution to GDP, and activities that improve the environment—because they provide real economic benefits—would be added to GDP.

Discussing the national income accounting implications of resource depletion and pollution may seem to trivialize these important problems. However, because GDP and related statistics are used regularly in policy analyses, abstract questions of measurement may turn out to have significant effects. Economic development experts have expressed concern that some poor countries, in attempting to raise their GDPs as quickly as possible, overexploit their natural capital and impair the environment. If "hidden" resource and environmental costs were explicitly incorporated into official measures of economic growth, policies that harm the environment might be modified.

Similarly, in industrialized countries, political debates about the environment have sometimes emphasized the impact on GDP of proposed pollution-control measures rather than the impact on overall economic welfare. Better accounting for environmental quality might refocus these debates on the more relevant question of whether, for any given environmental proposal, the benefits (economic and noneconomic) exceed the cost. One option is to replace GDP and NDP with a more comprehensive measure of national income accounting that includes estimates of both natural resource depletion and the environmental costs of economic activities.

One tool that can be used alongside the GDP is the **Environmental Performance Index (EPI)**, which assesses a country's commitment to environmental and resource management. In the 2010 EPI, the United States ranked 61st out of 163 countries assessed, which is below the rankings of most Western European countries. Most African countries are ranked in the bottom half. **Table 2.2** presents the 2010 EPI rankings of some selected countries.

We have examined the roles of government and economics in addressing environmental concerns, particularly in the United States. Now let us consider environmental destruction in Central and Eastern Europe, an environmental problem closely connected to both governmental and economic policies.

Table 2.2 Environmental Performance Index Scores and Rankings for Select Countries

Country	2010 EPI Score	2010 EPI Ranking*
New Zealand	73.4	15
Sweden	86.0	4
Finland	74.7	12
United Kingdom	74.2	14
Canada	66.4	46
Japan	72.5	20
Iceland	93.5	1
Costa Rica	86.4	3
United States	63.5	61
Mexico	67.3	43
China	49.0	121
Sierra Leone	32.1	163
India	48.3	123
Niger	37.6	158

Canada ranks highest among North American countries. The lowest-ranked countries are in sub-Saharan Africa.
*Out of 163 countries
Source: www.yale.edu/epi

CASE IN POINT

ENVIRONMENTAL PROBLEMS IN CENTRAL AND EASTERN EUROPE

The fall of the Soviet Union and communist governments in Central and Eastern Europe during the late 1980s revealed a grim legacy of environmental destruction (**Figure 2.15**). Over the previous decades, the value of natural capital had been ignored. Water was so poisoned from raw sewage and chemicals that it could not be used for industrial purposes, let alone for drinking. Unidentified

© Thomas Hoepker/Magnum Photos

Figure 2.15 **Pollution problems in former communist countries.** This section of a river near Espenhain in Saxony is covered in a scummy layer of chemical wastes. Note the coal-fired power plant in the background, one of many sources that contaminated soils, surface water, and ground water. Thousands of these sites exist in former communist countries, the result of rapid industrialization without regard for the environment. Photographed in the late 1990s.

chemicals leaked out of dumpsites into the surrounding soil and water, while fruits and vegetables were grown nearby in chemical-laced soil. Power plants emitted soot and sulfur dioxide into the air, producing a persistent chemical haze. Buildings and statues eroded, and entire forests died because of air pollution and acid rain. Crop yields fell despite intensive use of chemical pesticides and fertilizers. One of the most polluted areas in the world was the "Black Triangle," which consisted of the bordering regions of the former East Germany, northern Czech Republic, and southwest Poland.

Many Central and Eastern Europeans suffered from asthma, emphysema, chronic bronchitis, and other respiratory diseases as a result of breathing the filthy, acrid air. By the time most Polish children were 10 years old, they suffered from chronic respiratory diseases or heart problems. The levels of cancer, miscarriages, and birth defects were extremely high. Life expectancies are still lower in Eastern Europe than in other industrialized nations; in 2011, the average Eastern European lived to age 71, which was 10 years less than the average Western European.

The economic assumption behind communism was one of high production and economic self-sufficiency, regardless of damages to the environment and natural resources, and so pollution in communist-controlled Europe went largely unchecked. Meeting industrial production quotas always took precedence over environmental concerns, even though production was not carried out for profit. It was acceptable to pollute because clean air, water, and soil were granted no economic value. The governments supported heavy industry—power plants, chemicals, metallurgy, and large machinery—at the expense of the more environmentally benign service industries. As a result, Central and Eastern Europe were overindustrialized with old plants that lacked pollution-abatement equipment.

Communism did little to encourage resource conservation, which can curb pollution. Neither industries nor individuals had incentives to conserve energy because energy subsidies and lack of competition allowed power plants to provide energy at prices far below its actual cost.

More recently, while switching from communism to democracies with market economies, Central and Eastern European governments have faced the overwhelming responsibility of improving the environment. Although no political system in transition will make the environment its first priority, several countries have formulated environmental policies based on the experiences of the United States and Western Europe over the past several decades. However, experts predict it will take decades to clean up the pollution legacy of communism. How much will it cost? The figures are staggering. For example, according to one United Nations estimate, improving the environment in what was formerly East Germany will cost up to $300 billion. This represents the loss of natural capital during the 1960s, 1970s, and 1980s. Eastern Europeans today are paying the costs of their parents' environmental neglect.

The environment of countries in the former Soviet bloc is slowly improving. Some, such as Hungary, Poland, and the Czech Republic, have been relatively successful in moving toward market economies. They generate enough money to invest in environmental cleanup. Latvia, Lithuania, and Slovakia ranked above the United States in the 2010 EPI Assessment. In other countries, such as Romania and Bulgaria, the transition to a market economy has not been as smooth. Economic recovery has been slow in these countries, and severe budgetary problems have forced the environment to take a backseat to political and economic reform. ∎

E N V I R O N E W S
O N   C A M P U S
Greening Higher Education

No longer restricting their environmental education to the classroom, students at many colleges and universities are identifying environmental problems on their campuses, then mapping out and implementing solutions. The tremendous success of many of these efforts has required understanding the campus as a system and demands the cooperation of faculty, administrators, staff, and students. Real-life environmental situations become educational opportunities. Notable examples include the following:

- The 2010 Game Day Challenge, in which 77 colleges and universities diverted over 2,500 tons of waste from landfills during home football game days.

- The Rebel Recycling program at the University of Nevada–Las Vegas, which accepts materials from off campus and collects reusable materials during dorm moveouts. The program was based on the senior thesis of an environmental studies major who continues to run it.

- Participation by hundreds of campuses in the American College and University Presidents' Climate Commitment.

- The reduction of hazardous wastes in chemistry labs across the United States, a result of a student/faculty effort to develop "microscale" experiments at Bowdoin College in Maine.

Overall, student efforts nationally include recycling programs on 80% of campuses and have resulted in a total annual savings of nearly $17 million from 23 campus conservation projects, ranging from transportation initiatives to energy and water conservation, reuse and recycling, and composting. Organizations such as the Campus Ecology program associated with the National Wildlife Federation, the Sierra Student Coalition, the Campus Climate Challenge, and the American Association of Sustainability in Higher Education reflect an increasing level of environmental stewardship found at the nation's institutions of higher learning.

See www.aashe.org

REVIEW

1. What do economists mean by "efficient" regulation?
2. When might command and control regulations be more or less effective than economics-based policies?
3. What does the EPI measure that the GNP does not?

Environmental Justice

- Define *environmental justice* and explain why climate change is an environmental justice issue.

Over the past several decades, there has been increasing recognition that low-income and minority communities in both rural and urban areas face more environmental threats such as pollution and hazardous waste facilities and have fewer environmental amenities such as parks, trees, and clean air. At the same time, members of those communities have less of a voice in planning, including decisions about siting industrial facilities, sanitary landfills, and major transportation routes. Beginning in the late 1970s, environmental sociologist **Robert Bullard** at Clark Atlanta University identified patterns of injustice in Houston. For example, Bullard found that six of Houston's eight incinerators were located on less expensive land in predominantly black neighborhoods.

In addition, residents of low-income communities frequently have less access to other factors that influence health, including quality healthcare, fresh foods, and educational opportunities. The high incidence of asthma in many minority communities is an example of a health condition that may be caused or exacerbated by exposure to environmental pollutants. Few studies have examined how environmental pollutants interact with other socioeconomic factors to cause health problems. Those studies that have been done often fail to conclusively tie exposure to environmental pollutants to the health problems of poor and minority communities.

A 1997 health study of residents in San Francisco's Bayview–Hunters Point area found that hospitalizations for chronic illnesses were almost four times higher than the state average. Bayview–Hunters Point is heavily polluted. It has 700 hazardous waste facilities, 325 underground oil storage tanks, and two Superfund sites. Currently, little scientific evidence shows to what extent a polluted environment is responsible for the disproportionate health problems of poor and minority communities. Progress is being made: The EcoCenter at Heron's Head Park Partnership won a 2010 Achievements in Environmental Justice Award from the EPA for developing an environmental justice education center for the Bayview–Hunters Point neighborhoods (**Figure 2.16**).

ENVIRONMENTAL JUSTICE AND ETHICAL ISSUES

Environmental decisions such as where to locate a hazardous waste landfill have important ethical dimensions. The most basic ethical dilemma centers on the rights of the poor

environmental justice The right of every citizen, regardless of age, race, gender, social class, or other factor, to adequate protection from environmental hazards.

and disenfranchised versus the rights of the rich and powerful. Whose rights should have priority in these decisions? The challenge of **environmental justice** is to find and adopt equitable solutions that respect all groups of people, including those not yet born. Viewed ethically,

Figure 2.16 **Bay View - Hunters Point.** This housing project was among the last parts of the old neighborhood to be replaced.

environmental justice is a fundamental human right. Although we may never completely eliminate environmental injustices of the past, we have a moral imperative to prevent them today so that the negative effects of pollution do not disproportionately affect any segment of society.

In response to these concerns, a growing environmental justice movement has emerged at the grassroots level as a strong motivator for change. Advocates of the environmental justice movement call for special efforts to clean up hazardous sites in low-income neighborhoods, from inner-city streets to rural communities. Many environmental justice groups base their demands on the inherent "rightness" of their position. Other groups want science to give their demands legitimacy. Many advocates cite the need for more research on human diseases that environmental pollutants may influence.

MANDATING ENVIRONMENTAL JUSTICE AT THE FEDERAL LEVEL

In 1994, President Clinton signed an executive order requiring all federal agencies to develop strategies and policies to ensure that their programs do not discriminate against poor and minority communities when decisions are made about where future hazardous facilities are located. The first response to President Clinton's initiative came in 1997 when the Nuclear Regulatory Commission (NRC) rejected a request to build a uranium processing plant near two minority neighborhoods in northern Louisiana. The commission decided that racial considerations were a factor in site selection because the applicant had ruled out all potential sites near predominantly white neighborhoods. This NRC decision sent a message that the U.S. government will protect the rights of vulnerable members of society.

ENVIRONMENTAL JUSTICE AT THE INTERNATIONAL LEVEL

Environmental justice applies to countries as well as to individuals (**Figure 2.17**). Although it is possible to reduce and dispose of

Figure 2.17 International environmental justice. Protestors march for "justicia ambiental," or environmental justice. Photographed in San Salvador.

waste in an environmentally sound manner, industrialized countries have sometimes chosen to send their waste to other countries. (As industrialized nations develop more stringent environmental standards, disposing of hazardous waste at home becomes much more expensive than sending it to a developing nation, where property values and labor costs are lower.) Some waste is exported for legitimate recycling, but other waste is exported strictly for disposal.

The export of solid and hazardous wastes by the United States, Canada, Japan, and the European Union was highly controversial in the 1980s. Developing nations in Africa, Central and South America, and the Pacific Rim of Asia, as well as Eastern and Central European countries, often imported hazardous wastes in return for hard currency. In 1989, the UN Environment Programme developed a treaty, the **Basel Convention**, to restrict the international transport of hazardous waste. Originally, the treaty allowed countries to export hazardous waste only with the prior informed consent of the importing country as well as of any countries that the waste passed through in transit. In 1995, the Basel Convention was amended to ban the export of *any* hazardous waste from industrialized to developing countries.

At the 1996 meeting of the Association for the Advancement of Sustainability in Higher Education, scholar **Bill McKibben** described climate change not only as an environmental justice issue but also as "the greatest social justice issue of all times." He argues that more than any other issue in history, climate change has the most inequitable distribution of benefits (the use of fossil fuels by a small but very wealthy percentage of the world's

population) and risks (the adverse effects of climate change on a large but very poor percentage of the world's population).

Environmental Ethics, Values, and Worldviews

LEARNING OBJECTIVES

- Define *environmental ethics*.
- Define *environmental worldview* and discuss distinguishing aspects of the Western and deep ecology worldviews.

We now shift our attention to the worldviews of different individuals and societies and how those worldviews affect our ability to understand and solve sustainability problems. **Ethics** is the branch of philosophy that is derived through the logical application of human values. These **values** are the principles that an individual or society considers important or worthwhile. Values are not static entities but change as societal, cultural, political, and economic priorities change. Ethics helps us determine which forms of conduct are morally acceptable or unacceptable, right or wrong. Ethics plays a role in whatever types of human activities involve intelligent judgment and voluntary action. When values conflict, ethics helps us choose which value is better, or worthier, than other values.

ENVIRONMENTAL ETHICS

Environmental ethics examines moral values to determine how humans should relate to the natural environment. Environmental ethicists consider such questions as the role we should play in determining the fate of Earth's resources, including other species. Another question is whether we should develop an environmental ethic that is acceptable in the short term for us as individuals but also in the long term for our species and the planet. These are difficult intellectual issues that involve political, economic, societal, and individual trade-offs.

environmental ethics A field of applied ethics that considers the moral basis of environmental responsibility and the appropriate extent of this responsibility.

Environmental ethics considers not only the rights of people living today, both individually and collectively, but also the rights of future generations (**Figure 2.18**). This aspect of environmental ethics is critical because the impacts of today's activities and technologies are changing the environment. In some cases, these impacts may be felt for hundreds or even thousands of years. Addressing issues of environmental ethics puts us in a better position to use science, government policies, and economics to achieve long-term environmental sustainability.

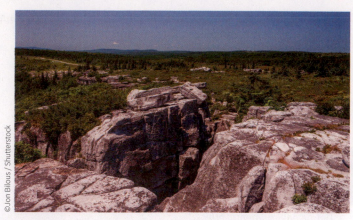

Figure 2.18 **Dolly Sods National Wilderness.** The United States has chosen to protect many areas, such as the Dolly Sods National Wilderness in West Virginia, so that they will be available for enjoyment and use by future generations.

ENVIRONMENTAL WORLDVIEWS

Each of us has a particular **worldview**—a commonly shared perspective based on a collection of our basic values that helps us make sense of the world, understand our place and purpose in it, and determine right and wrong behaviors. These worldviews lead to behaviors and lifestyles that may or may not be compatible with environmental sustainability. Some worldviews share certain fundamental beliefs, whereas others are mutually exclusive. A worldview that is considered ethical in one society may be considered irresponsible or even sacrilegious in another. The following are two extreme, opposing **environmental worldviews**: the Western worldview and the deep ecology worldview. These two worldviews, admittedly broad generalizations, are near opposite ends of a spectrum of worldviews relevant to problems of global sustainability, and each approaches environmental responsibility in a radically different way.

The **Western worldview**, also known as the *expansionist worldview*, is *anthropocentric* (human-centered) and utilitarian. This perspective mirrors the beliefs of the eighteenth-century *frontier attitude*, a desire to conquer and exploit nature as quickly as possible. The Western worldview also advocates the inherent rights of individuals, accumulation of wealth, and unlimited consumption of goods and services to provide material comforts. According to the Western worldview, humans have a primary obligation to humans and are therefore responsible for managing natural resources to benefit human society. Thus, any concerns about the environment are derived from human interests.

environmental worldview A worldview that helps us make sense of how the environment works, our place in the environment, and right and wrong environmental behaviors.

Western worldview An understanding of our place in the world based on human superiority over nature, the unrestricted use of natural resources, and increased economic growth to manage an expanding industrial base to benefit society.

E N V I R O N E W S

Religion and the Environment

A new degree of commitment to environmental issues has grown steadily in many religious faiths and organizations. Many of the world's major religions have collaborated in conferences on religious perspectives and the natural environment. Here are several examples:

- In 2000, more than 1,000 religious leaders met at the Millennium World Peace Summit of Religious and Spiritual Leaders. Protecting the environment was a major topic of discussion.

- In 2001, the U.N. Environment Programme and the Islamic Republic of Iran sponsored an International Seminar on Religion, Culture, and Environment. They considered ways to counter environmental degradation.

- In 2006, the Evangelical Climate Initiative, which includes traditionally conservative evangelicals in the United States, identified global warming as an important religious issue requiring responsible stewardship and attention to future generations.

- The 2010 Baha'i Social and Economic Development Conference focused on the role of religion in dealing with global environmental challenges.

- In 2014, Roman Catholic Pope Francis I identified climate change as a "moral issue."

- On April 3, 2014, evangelical Christians around the United States gathered for a "Day of Prayer and Action" on climate change.

The **deep ecology worldview** is a diverse set of viewpoints that dates from the 1970s and is based on the work of the late **Arne Naess**, a Norwegian philosopher, and others. The principles of deep ecology, as expressed by Arne Naess in *Ecology, Community, and Lifestyle*, include the following:

deep ecology worldview An understanding of our place in the world based on harmony with nature, a spiritual respect for life, and the belief that humans and all other species have an equal worth.

1. All life has intrinsic value. The value of nonhuman life-forms is independent of the usefulness they may have for narrow human purposes.

2. The richness and diversity of life-forms contribute to the flourishing of human and nonhuman life on Earth.

3. Humans have no right to reduce this richness and diversity except to satisfy vital needs.

4. Present human interference with the nonhuman world is excessive, and the situation is rapidly worsening.

5. The flourishing of human life and cultures is compatible with a substantial decrease in the human population. The flourishing of nonhuman life requires such a decrease.

6. Significant improvement of life conditions requires changes in economic, technological, and ideological structures.

7. The ideological change is mainly that of appreciating the quality of life rather than adhering to a high standard of living.

8. Those who subscribe to the foregoing points have an obligation to participate in the attempt to implement the necessary changes.

Compared to the Western worldview, the deep ecology worldview represents a radical shift in how humans relate themselves to the environment. The deep ecology worldview stresses that all forms of life have the right to exist and that humans are not different or separate from other organisms. Humans have an obligation to themselves and to the environment. The deep ecology worldview advocates sharply curbing human population growth. It does not favor returning to a society free of today's technological advances but instead proposes a significant rethinking of our use of current technologies and alternatives. It asks individuals and societies to share an inner spirituality connected to the natural world.

Most people today do not fully embrace either the Western worldview or the deep ecology worldview. The Western worldview emphasizes the importance of humans as the overriding concern in the grand scheme of things. In contrast, the deep ecology worldview is **biocentric** (life-centered) and views humans as one species among others. The planet's natural resources could not support its more than 7 billion humans if each consumed the high level of goods and services sanctioned by the Western worldview. On the other hand, if all humans adhered completely to the tenets of deep ecology, it would entail giving up many of the material comforts and benefits of modern technology.

The world as envisioned by the deep ecology worldview could support only a fraction of the existing human population (recall the discussion of ecological footprints in Chapter 1). These worldviews, although not practical for widespread adoption, are useful to keep in mind as you examine various environmental issues in later chapters. In the meantime, think about your own worldview and discuss it with others. Listen carefully to their worldviews, which will probably be different from your own. Thinking leads to actions, and actions lead to consequences. What are the short-term and long-term consequences of your particular worldview? We must develop and incorporate a long-lasting, environmentally sensitive worldview into our culture if the environment is to be sustainable for us, for other living organisms, and for future generations.

REVIEW

1. What is environmental ethics?
2. What are worldviews? How do Western and deep ecology worldviews differ?

REVIEW OF LEARNING OBJECTIVES WITH SELECTED KEY TERMS

•Briefly outline the environmental history of the United States.

The first two centuries of U.S. history were a time of widespread environmental exploitation. During the 1700s and early 1800s, Americans saw nature primarily as resources to be used. During the nineteenth century, naturalists became concerned about conserving natural resources. The earliest conservation legislation revolved around protecting land—forests, parks, and monuments. By the late twentieth century, environmental awareness had become pervasive, and a suite of environmental laws were in place.

• Describe the environmental contributions of the following people: George Perkins Marsh, Theodore Roosevelt, Gifford Pinchot, John Muir, Aldo Leopold, Wallace Stegner, Rachel Carson, and Paul Ehrlich.

George Perkins Marsh wrote about humans as agents of global environmental change. **Theodore Roosevelt** appointed **Gifford Pinchot** as the first head of the U.S. Forest Service. Pinchot supported expanding the nation's forest reserves and managing forests scientifically. The Yosemite and Sequoia National Parks were established, largely in response to the efforts of naturalist **John Muir**. In *A Sand County Almanac*, **Aldo Leopold** wrote about humanity's relationship with nature. **Wallace Stegner** helped create support for passage of the Wilderness Act of 1964. **Rachel Carson** published *Silent Spring,* alerting the public to the dangers of uncontrolled pesticide use. **Paul Ehrlich** wrote *The Population Bomb,* which raised the public's awareness of the dangers of overpopulation.

• Distinguish between utilitarian conservationists and biocentric preservationists.

A **utilitarian conservationist** is a person who values natural resources because of their usefulness for practical purposes but uses them sensibly and carefully. A **biocentric preservationist** is a person who believes in protecting nature because all forms of life deserve respect and consideration.

• Explain why the National Environmental Policy Act is the cornerstone of U.S. environmental law.

The **National Environmental Policy Act (NEPA)**, passed in 1970, stated that the federal government must consider the environmental impact of a proposed federal action, such as financing a highway or constructing a dam. NEPA established the **Council on Environmental Quality** to monitor required **environmental impact statements (EISs)** and report directly to the president.

• Explain how environmental impact statements provide powerful protection of the environment.

By requiring EISs that are open to public scrutiny, NEPA initiated serious environmental protection in the United States. NEPA allows citizen suits, in which private citizens take violators, whether they are private industries or government-owned facilities, to court for noncompliance.

• Explain why economists prefer efficient solutions to environmental problems.

Economists assume that individuals behave as **rational actors** who seek to maximize **utility**. Groups of individuals behaving according to these assumptions result in economic **efficiency**, the greatest possible total societal benefit. Solutions that are inefficient will spend more on abatement than the pollution costs or will spend less on abatement than pollution reductions are worth.

• **Describe command and control regulation, incentive-based regulation, and cost-effectiveness analysis.**

Governments often use **command and control regulations**, which are pollution-control laws that require specific technologies. **Incentive-based regulations** are pollution-control laws that work by establishing emission targets and providing industries with incentives to reduce emissions. **Cost-effectiveness analysis** is an economic tool used to estimate costs associated with achieving some goal, such as saving a life.

• **Give two reasons why the national income accounts are incomplete estimates of national economic performance.**

National income accounts are a measure of the total income of a nation's goods and services for a given year. An **externality** is a harmful environmental or social cost that is borne by people not directly involved in buying or selling a product. Currently, national income accounting does not include estimates of externalities such as depletion of **natural capital** and the environmental cost of economic activities.

• **Define *environmental justice* and explain why climate change is an environmental justice issue.**

Environmental justice is the right of every citizen, regardless of age, race, gender, social class, or other factor, to adequate protection from environmental hazards. The primary benefit associated with climate change is energy from fossil fuels, which are available to the world's wealthier people, while the risks of climate change are spread across a much larger group of much poorer people.

• **Define *environmental ethics*.**

Environmental ethics is a field of applied **ethics** that considers the moral basis of environmental responsibility and the appropriate extent of this responsibility. Environmental ethicists consider how humans should relate to the natural environment.

• **Define *environmental worldview* and discuss distinguishing aspects of the Western and deep ecology worldviews.**

An **environmental worldview** is a worldview that helps us make sense of how the environment works, our place in the environment, and right and wrong environmental behaviors. The **Western worldview** is an understanding of our place in the world based on human superiority and dominance over nature, the unrestricted use of natural resources, and increased economic growth to manage an expanding industrial base. The **deep ecology worldview** is an understanding of our place in the world based on harmony with nature, a spiritual respect for life, and the belief that humans and all other species have an equal worth.

Critical Thinking and Review Questions

1. Briefly describe each of these aspects of U.S. environmental history: protection of forests, establishment and protection of national parks and monuments, conservation in the mid-20th century, and the environmental movement of the late twentieth century.

2. Describe the environmental contributions of two of the following: George Perkins Marsh, Theodore Roosevelt, Gifford Pinchot, John Muir, Aldo Leopold, Wallace Stegner, Rachel Carson, or Paul Ehrlich.

3. If you were a member of Congress, what legislation would you introduce to deal with each of these problems?

   a. Toxic materials from a major sanitary landfill are polluting some rural drinking-water wells.

   b. Acid rain from a coal-burning power plant in a nearby state is harming the trees in your state. Loggers and foresters are upset.

   c. Oceanfront property is at risk from sea-level rise caused by global warming.

4. Explain why the National Environmental Policy Act is considered the cornerstone of U.S. environmental law. Research some examples of when it has been effective and when it has not.

5. Based on what you have learned in this chapter, do you think the economy is part of the environment, or is the environment part of the economy? Explain your answer.

6. Following the 2010 oil platform collapse and associated oil spill in the Gulf of Mexico, Congress began considering new laws regulating deep-ocean oil drilling. Do you think command and control or incentive-based regulations would be more effective? Explain.

7. Can economic approaches to environmental management adequately account for the complex interactive systems that make up the environment? Why or why not?

8. The Environmental Performance Index has detractors among both environmentalists and industry groups. Based on what you have learned in this chapter, suggest reasons for each group to object to it.

9. Can economic solutions be used to address environmental justice concerns? Explain your answer.

10. Do you agree or disagree with the idea that climate change is an environmental justice issue? Explain your position.

11. Describe how environmental destruction in formerly communist countries relates to natural capital.

12. Determine whether each of these statements reflects the Western worldview, the deep ecology worldview, or both:

   a. Species exist for humans to use.

   b. All organisms, humans included, are interconnected and interdependent.

   c. Humans and nature exist in unity.

   d. Humans are a superior species capable of dominating other organisms.

   e. Humans should protect the environment.

   f. Nature should be used, not preserved.

   g. Economic growth will help Earth manage an expanding human population.

h. Humans have the right to modify the environment to benefit society.

i. All forms of life are intrinsically valuable and have the right to exist.

13. What is this cartoon trying to say? When do you think this cartoon was published?

© Steve Greenberg

Food for Thought

Several generations ago, many people in cities raised edible plants and animals at their homes. Now, however, local zoning laws prohibit livestock and even vegetable gardens in many urban areas. Research zoning laws about growing food in your area, or a nearby town if you live in a rural area. What foods are people allowed to raise at home? What, if any, are the environmental concerns about raising plants and animals for personal consumption in urban areas?

3

Ecosystems and Energy

Stephen J. Krasemann/Science Source

Cordgrass (*Spartina*) in a Chesapeake Bay salt marsh.

A salt marsh in the Chesapeake Bay on the East Coast of the United States contains an assortment of organisms that interact with one another and are interdependent in a variety of ways. This bay is an estuary, a semi-enclosed body of water found where fresh water from a river drains into the ocean. Estuaries, which are complex *systems* under the influence of tides, gradually change from unsalty fresh water to salty ocean water. In the Chesapeake Bay, this change results in freshwater marshes at the head of the bay, brackish (moderately salty) marshes in the middle bay region, and salt marshes on the ocean side of the bay.

A Chesapeake Bay salt marsh consists of flooded meadows of cordgrass (**see photo**). Few other plants are found because high salinity and tidal inundations produce a challenging environment in which only a few plants survive. Both cordgrass and microscopic algae (photosynthetic aquatic organisms) are eaten directly by some animals, and when they die, their remains provide food for other salt marsh inhabitants.

Two major kinds of animal life are noticeably abundant in a salt marsh, insects and birds. Insects, particularly mosquitoes and horseflies, number in the millions. Birds nesting in the salt marsh include seaside sparrows, laughing gulls, and clapper rails. Many other species live in salt marshes. Large numbers of invertebrates, such as shrimps, lobsters, crabs, barnacles, worms, clams, and snails, seek refuge in the water surrounding the cordgrass. Here they eat, hide from predators to avoid being eaten, and reproduce.

Chesapeake Bay marshes are an important nursery for numerous marine fishes—spotted sea trout, Atlantic croaker, striped bass, and bluefish, to name just a few. These fishes typically reproduce in the open ocean, and the young then enter the estuary, where they grow into juveniles.

Almost no amphibians inhabit salt marshes—the salty water dries out their skin—but a few reptiles, such as the northern diamondback terrapin (a semi-aquatic turtle), have adapted. The terrapin spends its time basking in the sun or swimming in the water searching for food—snails, crabs, worms, insects, and fish. Although a variety of snakes live in the dry areas adjacent to salt marshes, only the northern water snake, which preys on fish, is adapted to salty water.

The meadow vole is a small rodent that lives in the salt marsh. Meadow voles are excellent swimmers and scamper about the salt marsh day and night. They eat mainly insects and cordgrass leaves, stems, and roots.

Add to all these visible plants and animals the unseen microscopic world of the salt marsh—countless algae, protozoa, fungi, and bacteria—while considering the environmental challenges these organisms face, and the complexity of a salt marsh ecosystem becomes clear.

In this chapter we begin our study of ecology, which is central to environmental science. Throughout the chapter, which focuses on energy flow, we will encounter many examples from the salt marsh ecosystem.

In Your Own Backyard...

Water quality in the Chesapeake Bay has deteriorated over the years as increasing levels of sediment, sewage, and fertilizer from the land have polluted the water. Identify a body of water near where you live. Does it have similar pollution problems? Why or why not?

What Is Ecology?

- Define *ecology*.
- Distinguish among the following ecological levels: population, community, ecosystem, landscape, and biosphere.

Ernst Haeckel, a 19th-century scientist, developed the concept of **ecology** and named it—*eco* from the Greek word for "house" and *logy* from the Greek word for "study." Thus, *ecology* literally means "the study of one's house." The environment—one's house—consists of two parts, the **biotic** (living) environment, which includes all organisms, and the **abiotic** (nonliving, or physical) surroundings, which include living space, temperature, sunlight, soil, wind, and precipitation (**Figure 3.1**).

ecology The study of systems that include interactions among organisms and between organisms and their abiotic environment.

Ecologists study the vast complex web of relationships among living organisms and their physical environment. Areas of study include, but are not limited to, why organisms are distributed the way they are, why some species are more abundant than others, how the ecological roles of different organisms in their environment vary, and how the interactions between organisms and their environment help to maintain the overall health of our living world.

The focus of ecology is local or global, specific or generalized, depending on what questions the scientist is trying to answer.

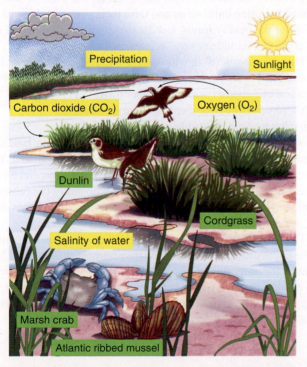

Precipitation

Sunlight

Carbon dioxide (CO₂)

Oxygen (O₂)

Dunlin

Cordgrass

Salinity of water

Marsh crab

Atlantic ribbed mussel

Figure 3.1   Some abiotic and biotic components of a Chesapeake Bay salt marsh. Shown is a mudflat at low tide. Abiotic (nonliving) components are labeled in yellow and biotic (living) components, in green.

One ecologist might determine the temperature or light requirements of a single oak species in a forest, another might study all the organisms that live in that forest, and yet another might examine how nutrients flow between the forest and surrounding areas.

Ecology is the broadest field within the biological sciences, and it is linked to every other biological discipline. The universality of ecology links subjects that are not traditionally part of biology. Geology and earth science are extremely important to ecology, especially when ecologists examine the physical environment of planet Earth. Chemistry and physics are also important; in this chapter, for example, knowledge of chemistry is necessary to understand photosynthesis, and principles of physics illuminate the laws of thermodynamics. Humans are biological organisms, and our activities have a bearing on ecology. Even economics and politics have profound ecological implications, as was discussed in Chapter 2.

How does the field of ecology fit into the organization of the biological world? Ecologists are most interested in the levels of biological organization that include or are above the level of the individual organism (**Figure 3.2**). Individuals of the same **species** occur in **populations**. A population ecologist might study a population of polar bears or a population of marsh grass. Population ecology is discussed in Chapter 5, and human populations in Chapters 8 and 9. Species are considered further in Chapter 16.

species A group of similar organisms whose members freely interbreed with one another in the wild to produce fertile offspring; members of one species generally do not interbreed with other species of organisms.

Populations are organized into **communities**. Ecologists characterize communities by the number and kinds of species that live there, along with their relationships with one another. A community ecologist might study how organisms interact with one another—including feeding relationships (who eats whom)—in an alpine meadow community or in a coral reef community (**Figure 3.3**).

population A group of organisms of the same species that live in the same area at the same time.

community A natural association that consists of all the populations of different species that live and interact within an area at the same time.

Ecosystem is a more inclusive term than *community*. An **ecosystem** includes all the biotic interactions of a community, as well as the interactions between organisms and their abiotic environment. Like other *systems*, an ecosystem consists of multiple interacting and inseparable parts and processes that form a unified whole. An ecosystem—whether terrestrial or aquatic—is a system in which all the biological, physical, and chemical components of an area form a complex, interacting network of energy flow and materials cycling. An ecosystem ecologist might examine how energy, nutrients, organic (carbon-containing) materials, and water affect the organisms living in a desert ecosystem, a forest, or a coastal bay ecosystem.

ecosystem A community and its physical environment.

The ultimate goal of ecosystem ecologists is to understand how ecosystems function. This is not a simple task, but it is important because ecosystem processes collectively regulate the global cycles of water, carbon, nitrogen, phosphorus, and sulfur essential to the survival of humans and all other organisms. As humans increasingly alter ecosystems for their own uses, the

Figure 3.2   Levels of ecological organization. Ecologists study the levels of biological organization from individual organisms to the biosphere.

Figure 3.3   Coral reef community. Coral reef communities have the greatest number of species and are the most complex aquatic community. This close-up of a coral reef in the Caribbean Sea off the coast of Mexico shows a green moray eel, French grunts, and several species of coral. Today many coral reefs worldwide are threatened by global climate change. How can they be protected from warming temperatures?

particular region. Consider a simple **landscape** consisting of a forest ecosystem located adjacent to a pond ecosystem. One connection between these two ecosystems might be great blue herons, which eat fishes, frogs, insects, crustaceans, and snakes along the shallow water of the pond but often build nests and raise their young in the secluded treetops of the nearby forest. Landscapes, then, are based on larger land areas that include several ecosystems.

> **landscape** A region that includes several interacting ecosystems.

The organisms of the **biosphere**—Earth's communities, ecosystems, and landscapes—depend on one another and on the other realms of Earth's physical environment: the atmosphere, hydrosphere, and lithosphere (**Figure 3.4**). The **atmosphere** is the gaseous envelope surrounding Earth; the **hydrosphere** is Earth's supply of water—liquid and frozen, fresh and salty, groundwater and surface water. The **lithosphere** is the soil and rock of Earth's crust. Ecologists who study the biosphere examine global interrelationships among Earth's atmosphere, land, water, and organisms.

> **biosphere** The parts of Earth's atmosphere, ocean, land surface, and soil that contain all living organisms.

The biosphere is filled with life. Where do these organisms get the energy to live? And how do they harness this energy? Let's examine the importance of energy to organisms, which survive only as long as the environment continuously supplies them with energy. We will revisit the importance of energy as it relates to human endeavors in many chapters throughout this text.

natural functioning of ecosystems is changed, and we must determine whether these changes will affect the sustainability of our life-support system.

Landscape ecology is a subdiscipline of ecology that studies ecological processes operating over large areas. Landscape ecologists examine the connections among ecosystems found in a

Review

1. What is ecology?
2. What is the difference between a community and an ecosystem? Between an ecosystem and a landscape?

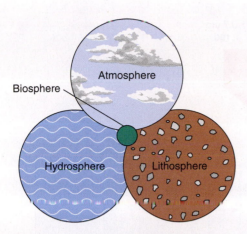

(a) Earth's four realms, represented as intersecting circles, are a system of interrelated parts.

©Jon Arnold Images Ltd./Alamy

(b) In this scene, photographed on Palawan Island, Philippines, the atmosphere contains a cumulus cloud, which indicates warm, moist air. The jagged rocks, formed from volcanic lava flows that have eroded over time, represent the lithosphere. The shallow water represents the hydrosphere. The biosphere includes the green vegetation and humans in the boat as well as the coral reefs visible as darker areas in the water.

Figure 3.4   Earth's four realms.

The Energy of Life

LEARNING OBJECTIVES

- Define *energy* and explain how it is related to work and to heat.
- Use an example to contrast potential energy and kinetic energy.
- Distinguish between open and closed systems.
- State the first and second laws of thermodynamics, and discuss the implications of these laws as they relate to organisms.
- Write summary reactions for photosynthesis and cellular respiration, and contrast these two biological processes.

Energy is the capacity or ability to do work. In organisms, any biological work—such as growing, moving, reproducing, and maintaining and repairing damaged tissues—requires energy. Energy exists in several forms: chemical, radiant, thermal, mechanical,

nuclear, and electrical. *Chemical energy* is energy stored in the bonds of molecules; for example, food contains chemical energy, and organisms use the energy released when chemical bonds are broken and new bonds form. *Radiant energy* is energy, such as radio waves, visible light, and X-rays, that is transmitted as electromagnetic waves (**Figure 3.5**). *Solar energy* is radiant energy from the sun; it includes ultraviolet radiation, visible light, and infrared radiation. *Thermal energy* is **heat** that flows from an object with a higher temperature (the heat source) to an object with a lower temperature (the heat sink). *Mechanical energy* is energy involved in the movement of matter. Some of the matter contained in atomic nuclei can be converted into *nuclear energy*. *Electrical energy* is energy that flows as charged particles. You will encounter these forms of energy throughout the text.

Biologists generally express energy in units of work (*kilojoules, kJ*) or units of heat (*kilocalories, kcal*). One kilocalorie is the energy required to raise the temperature of 1 kg of water by 1°C, equal to 4.184 kJ. The kcal is the unit that nutritionists use to express the energy content of the foods we eat.

Energy can exist as stored energy, called **potential energy**, or as **kinetic energy**, the energy of motion. Think of potential energy as an arrow on a drawn bow, which equals the work the archer did when drawing the bow to its position (**Figure 3.6**). When the string is released, the bow's potential energy is converted to the arrow's kinetic energy of motion. Similarly, the cordgrass that a meadow vole eats has chemical potential energy in the bonds of its molecules. As molecular bonds in the cordgrass are broken by cellular respiration, this energy is converted to kinetic energy and heat as the meadow vole swims in the salt marsh. Thus, energy changes from one form to another.

The study of energy and its transformations is called **thermodynamics**. When considering thermodynamics, scientists use the word *system* to refer to a group of atoms, molecules, or objects being studied.[1] The rest of the universe other than the system is known as the *surroundings*. A **closed system** is self-contained; that is, it does not exchange energy with its surroundings (**Figure 3.7**).[2] Closed systems are very rare in nature. In contrast, an **open system** exchanges energy with its surroundings. This text discusses many kinds of open systems. For example, a city is an open system with an input of energy (as well as food, water, and consumer goods). Outputs from a city system include energy (as well as manufactured goods, sewage, and solid waste). On a global scale, Earth is an open system dependent on a continual supply of energy from the sun.

Regardless of whether a system is open or closed, two laws about energy apply to all things in the universe: the first and second laws of thermodynamics.

---

[1] The meaning of *system* in thermodynamics is different from the meaning of *system* in environmental science (a set of components that interact and function as a whole).

[2] In thermodynamics, an *isolated system* does not exchange energy *or* matter with the surroundings.

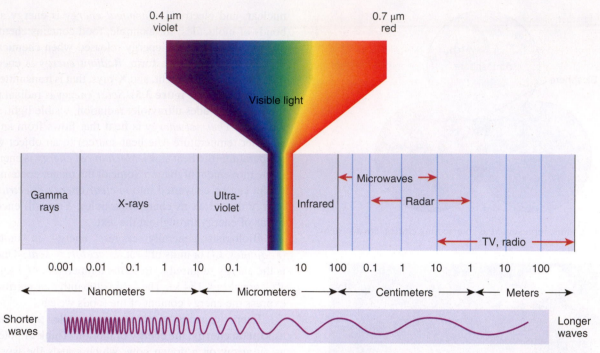

Figure 3.5   The electromagnetic spectrum. The shortest wavelengths are gamma rays, and the longest are TV and radio waves. Visible light occurs between ultraviolet and infrared radiation.

Figure 3.6   Potential and kinetic energy. Potential energy is stored in the drawn bow (shown) and is converted to kinetic energy as the arrow speeds toward its target. Photographed at the 2008 Summer Olympic Games in Beijing, China.

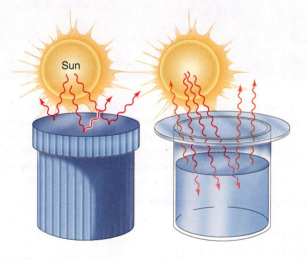

(a) Closed system. Energy is not exchanged between a closed system and its surroundings. A thermos bottle is an approximation of a closed system. Closed systems are rare in nature.

(b) Open system. Energy is exchanged between an open system and its surroundings. Earth is an open system because it receives energy from the sun, and this energy eventually escapes Earth as it dissipates into space.

Figure 3.7   Closed and open systems, with regard to energy.

THE FIRST LAW OF THERMODYNAMICS

first law of thermo-dynamics Energy cannot be created or destroyed, although it can change from one form to another.

According to the **first law of thermodynamics**, an organism may absorb energy from its surroundings, or it may give up some energy into its surroundings, but the total energy content of the organism and its surroundings is always the same. As far as we know, the energy present in the universe at its formation, approximately 15–20 billion years ago, equals the amount of energy present in the universe today. This is all the energy that will ever be present in the universe. Similarly, the energy of any system and its surroundings is constant. A system may absorb energy from its surroundings, or it may give up some energy into its surroundings, but the total energy content of that system and its surroundings is always the same.

The first law of thermodynamics specifies that an organism cannot create the energy it requires to live. Instead, it must capture energy from the environment to use for biological work, a process involving the transformation of energy from one form to another. In photosynthesis, plants absorb the radiant energy of the sun and convert it into the chemical energy contained in the bonds of carbohydrate (sugar) molecules. Similarly, some of that chemical energy may later be transformed by an animal that eats the plant into the mechanical energy of muscle contraction, enabling the animal to walk, run, jump, slither, fly, or swim.

THE SECOND LAW OF THERMODYNAMICS

As each energy transformation occurs, some energy is changed to heat that is released into the cooler surroundings. No other organism can ever reuse this energy for biological work; it is "lost" from the biological point of view. It is not really gone from a thermodynamic point of view because it still exists in the surrounding physical environment. Similarly, the use of food to enable us to walk or run does not destroy the chemical energy once present in the food molecules. After we have performed the task of walking or running, the energy still exists in the surroundings as heat.

second law of thermodynamics When energy is converted from one form to another, some of it is degraded into heat, a less usable form that disperses into the environment.

According to the **second law of thermodynamics**, the amount of usable energy available to do work in the universe decreases over time. The second law of thermodynamics is consistent with the first law; that is, the total amount of energy in the universe is not decreasing with time. However, the total amount of energy in the universe available to do work decreases over time.

Less-usable energy is more diffuse, or disorganized. **Entropy** is a measure of this disorder or randomness; organized, usable energy has low entropy, whereas disorganized energy such as heat has high entropy. Entropy is continuously increasing in the universe in all natural processes, and entropy is not reversible. Another way to explain the second law of thermodynamics, then, is that entropy, or disorder, in a system increases over time.

An implication of the second law of thermodynamics is that no process requiring an energy conversion is ever 100% efficient because much of the energy is dispersed as heat, resulting in an increase in entropy. (*Efficiency* in this context refers to the amount of useful work produced per total energy input.) An automobile engine, which converts the chemical energy of gasoline to mechanical energy, is between 20% and 30% efficient. That is, only 20% to 30% of the original energy stored in the chemical bonds of the gasoline molecules is actually transformed into mechanical energy, or work. In our cells, energy use for metabolism is about 40% efficient, with the remaining energy given to the surroundings as heat.

Organisms are highly organized and at first glance appear to refute the second law of thermodynamics. As organisms grow and develop, they maintain a high level of order and do not become more disorganized. However, organisms maintain their degree of order over time only with the constant input of energy. That is why plants must photosynthesize and why animals must eat food. When relating the second law of thermodynamics to living organisms, the organisms' surroundings must also be considered. Throughout its life, as a plant takes in solar energy and photosynthesizes, it continually breaks down the products of photosynthesis to supply its own energy needs. This breakdown releases heat into the environment. Similarly, as an animal eats and processes food to meet its energy needs, it releases heat into the surroundings. When an organism and its surroundings are taken into account, both laws of thermodynamics are satisfied.

PHOTOSYNTHESIS AND CELLULAR RESPIRATION

Energy is stored in living things as carbon compounds. **Photosynthesis** is the biological process in which light energy from the sun is captured and transformed into the chemical energy of carbohydrate (sugar) molecules (**Figure 3.8**). Photosynthetic pigments such as *chlorophyll*, which gives plants their green color, absorb radiant energy. This energy is used to manufacture the carbohydrate glucose ($C_6H_{12}O_6$) from carbon dioxide (CO_2) and water (H_2O), a process that also releases oxygen (O_2).

Photosynthesis:

$$6CO_2 + 12H_2O + \text{radiant energy} \rightarrow C_6H_{12}O_6 + 6H_2O + 6O_2$$

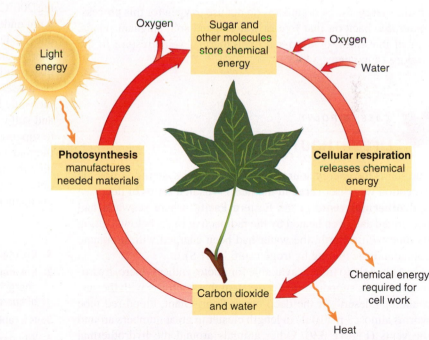

Figure 3.8   Photosynthesis and cellular respiration make up a system. These processes occur continuously in the cells of living organisms. Note that energy flow is *not* cyclic; energy enters living organisms as radiant energy and leaves organisms for the surroundings as heat energy. Given that increasing levels of CO_2 in the atmosphere are linked to climate warming, which process—photosynthesis or cellular respiration—could help reduce warming if it was increased significantly? Explain your answer.

The chemical equation for photosynthesis is read as follows: 6 molecules of carbon dioxide plus 12 molecules of water plus light energy are used to produce 1 molecule of glucose plus 6 molecules of water plus 6 molecules of oxygen. (See Appendix I, Review of Basic Chemistry.)

Plants, some bacteria, and algae perform photosynthesis, a process essential for almost all life. Photosynthesis provides these organisms with a ready supply of energy in carbohydrate molecules, which they use as the need arises. The energy can also be transferred from one organism to another—for example, from plants to the organisms that eat plants. Oxygen, which many organisms require when they break down glucose or similar foods to obtain energy, is a byproduct of photosynthesis.

The chemical energy that plants store in carbohydrates and other molecules is released within the cells of plants, animals, or other organisms through **cellular respiration**. In *aerobic* cellular respiration, molecules such as glucose are broken down in the presence of oxygen and water into carbon dioxide and water, with the release of energy (see Figure 3.8).

Aerobic cellular respiration:

$$C_6H_{12}O_6 + 6O_2 + 6H_2O \rightarrow 6CO_2 + 12H_2O + energy$$

Cellular respiration makes the chemical energy stored in glucose and other food molecules available to the cell for biological work, such as moving around, courting, and growing new cells and tissues. All organisms, including green plants, respire to obtain energy. Some organisms do not use oxygen for this process. *Anaerobic* bacteria that live in waterlogged soil, stagnant ponds, animal intestines, or deep-sea hydrothermal vents respire in the absence of oxygen.

CASE IN POINT

LIFE WITHOUT THE SUN

The sun is the energy source for almost all ecosystems. A notable exception was discovered in the late 1970s in a series of deep-sea **hydrothermal vents** in the Eastern Pacific where seawater had penetrated and been heated by the radioactive rocks below. During its time within Earth, the water had been charged with inorganic compounds, including hydrogen sulfide (H_2S).

Although no light is available for photosynthesis there, hydrothermal vents support a rich ecosystem that contrasts with the surrounding "desert" of the deep-ocean floor. Giant, blood-red tube worms almost 3 m (10 ft) in length cluster in great numbers around the vents (**Figure 3.9**). Other animals around the hydrothermal vents include shrimp, crabs, clams, barnacles, and mussels.

Scientists initially wondered what the ultimate source of energy is for the species in this dark environment. Most deep-sea ecosystems depend on the organic material that drifts down from surface waters; that is, they depend on energy derived from

Figure 3.9 **Hydrothermal vent ecosystem.** Bacteria living in the tissues of these tube worms extract energy from hydrogen sulfide to manufacture organic compounds. These worms lack digestive systems and depend on the organic compounds the bacteria provide, along with materials filtered from the surrounding water. Also visible in the photograph are some crabs (white).

photosynthesis. But hydrothermal vent ecosystems, now known to exist in hundreds of places, are too densely clustered and too productive to depend on chance encounters with organic material from surface waters.

The base of the *food web* in these aquatic oases consists of certain bacteria that survive and multiply in water so hot (exceeding 200°C, or 392°F) that it would not remain in liquid form were it not under such extreme pressure. These bacteria function as producers, but they do not photosynthesize. Instead, they obtain energy and make carbohydrate molecules from inorganic raw materials by carrying out **chemosynthesis**. Chemosynthetic bacteria possess enzymes (organic catalysts) that cause the inorganic molecule hydrogen sulfide to react with oxygen, producing water and sulfur or sulfate. Such chemical reactions provide the energy to support these bacteria and other organisms in deep-ocean hydrothermal vents. Many of the vent animals consume the bacteria directly by filter feeding. Others, such as the giant tube worms, obtain energy from chemosynthetic bacteria that live symbiotically inside their bodies. ■

Review

1. Distinguish among energy, work, and heat.

2. Is water stored behind a dam an example of potential or kinetic energy? What would cause the water to convert to the other form of energy?

3. Is a rabbit an example of a closed system or an open system? Why?

4. When coal is burned in a power plant, only 3% of the energy in the coal is converted into light in a lightbulb. What happens to the other 97% of the energy? Explain your answer using the laws of thermodynamics.

5. Distinguish between photosynthesis and cellular respiration. Which organisms perform each process?

The Flow of Energy through Ecosystems

LEARNING OBJECTIVES

- Define *energy flow*, *trophic level*, and *food web*.
- Summarize how energy flows through a food web, incorporating producers, consumers, and decomposers in your explanation.
- Describe typical pyramids of numbers, biomass, and energy.
- Distinguish between gross primary productivity and net primary productivity, and discuss human impact on the latter.

With the exception of a few ecosystems such as hydrothermal vents, energy enters ecosystems as radiant energy (sunlight), some of which plants trap during photosynthesis. The energy, now in chemical form, is stored in the bonds of organic molecules such as glucose. To obtain energy, animals eat plants, or they eat animals that ate plants. All organisms—plants, animals, and microorganisms—respire to obtain some of the energy in organic molecules. When cellular respiration breaks these molecules apart, the energy becomes available for work such as repairing tissues, producing body heat, or reproducing. As the work is accomplished, the energy escapes the organism and dissipates into the environment as heat (recall the second law of thermodynamics). Ultimately, this

energy flow The passage of energy in a one-way direction through an ecosystem.

heat radiates into space. Once an organism has used energy, that energy becomes unusable for all other organisms. The movement of energy just described is called **energy flow**.

PRODUCERS, CONSUMERS, AND DECOMPOSERS

The organisms of an ecosystem are divided into three categories on the basis of how they obtain nourishment: producers, consumers, and decomposers (**Figure 3.10**). Virtually all ecosystems contain representatives of all three groups, which interact extensively, both directly and indirectly, with one another.

Producers, also called **autotrophs** (Greek *auto*, "self," and *tropho*, "nourishment"), manufacture organic molecules from simple inorganic substances, generally CO_2 and water, usually using the energy of sunlight. In other words, most producers perform the process of photosynthesis. Producers incorporate the chemicals they manufacture into their own bodies, becoming potential food resources for other organisms. Whereas plants are the most significant producers on land, algae and certain types of bacteria are important producers in aquatic environments. In the salt marsh ecosystem discussed in the chapter introduction, cordgrass, algae, and photosynthetic bacteria are important producers.

Animals are **consumers** that use the bodies of other organisms as a source of food energy and bodybuilding materials. Consumers are also called **heterotrophs** (Greek *heter*, "different," and *tropho*, "nourishment"). Consumers that eat producers are **primary consumers** or **herbivores** (plant eaters). Rabbits and deer are examples of primary consumers, as is the marsh periwinkle, a type of snail that feeds on algae in the salt marsh ecosystem.

Secondary consumers eat primary consumers, whereas **tertiary consumers** eat secondary consumers—a mouse eats plants, a snake eats the mouse, a hawk eats the snake. Both secondary and tertiary consumers are flesh-eating **carnivores** that eat other

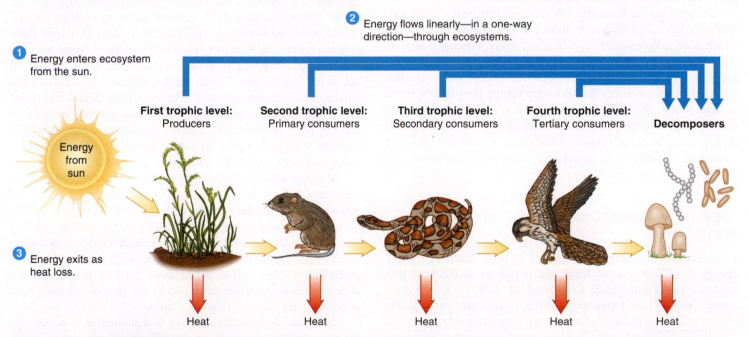

❶ Energy enters ecosystem from the sun.

❷ Energy flows linearly—in a one-way direction—through ecosystems.

❸ Energy exits as heat loss.

First trophic level: Producers

Second trophic level: Primary consumers

Third trophic level: Secondary consumers

Fourth trophic level: Tertiary consumers

Decomposers

Energy from sun

Heat   Heat   Heat   Heat   Heat

Figure 3.10 Energy flow among producers, consumers, and decomposers. In photosynthesis, producers use the energy from sunlight to make organic molecules. Consumers obtain energy when they eat producers or other consumers. Decomposers, such as bacteria and fungi, obtain energy from wastes and dead organic material from producers and consumers. During every energy transaction, some energy is lost to biological systems as it disperses into the environment as heat.

animals. Lions, lizards, and spiders are examples of carnivores, as are the northern diamondback terrapin and the northern water snake in the salt marsh ecosystem. Other consumers, called **omnivores**, eat a variety of organisms, both plant and animal. Bears, pigs, and humans are examples of omnivores; the meadow vole, which eats both insects and cordgrass in the salt marsh ecosystem, is also an omnivore.

Some consumers, called **detritus feeders** or **detritivores**, consume detritus, organic matter that includes animal carcasses, leaf litter, and feces. Detritus feeders, such as snails, crabs, clams, and worms, are especially abundant in aquatic environments, where they burrow in the bottom muck and consume the organic matter that collects there. Marsh crabs are detritus feeders in the salt marsh ecosystem. Earthworms, termites, beetles, snails, and millipedes are terrestrial (land-dwelling) detritus feeders. An earthworm actually eats its way through the soil, digesting much of the organic matter contained there. Detritus feeders work with microbial decomposers to destroy dead organisms and waste products.

Decomposers, also called **saprotrophs** (Greek *sapro*, "rotten," and *tropho*, "nourishment"), are heterotrophs that break down dead organic material and use the decomposition products to supply themselves with energy. They typically release simple inorganic molecules (e.g., CO_2) and mineral salts that producers can reuse. Bacteria and fungi are important decomposers. For example, during the decomposition of dead wood, sugar-metabolizing fungi first invade the wood and consume simple carbohydrates, such as glucose and maltose. When these carbohydrates are exhausted, other fungi, often aided by termites with symbiotic bacteria in their guts, complete the digestion of the wood by breaking down cellulose, the main carbohydrate of wood.

Ecosystems such as the Chesapeake Bay salt marsh contain a variety of producers, consumers, and decomposers, all of which play indispensable roles in ecosystems. Producers provide both food and oxygen for the rest of the community. Consumers maintain a balance between producers and decomposers. Detritus feeders and decomposers are necessary for the long-term survival of any ecosystem because, without them, dead organisms and waste products would accumulate indefinitely. Without microbial decomposers, important elements such as potassium, nitrogen, and phosphorus would remain permanently trapped in dead organisms, unavailable for new generations of organisms.

THE PATH OF ENERGY FLOW: WHO EATS WHOM IN ECOSYSTEMS

In an ecosystem, energy flow occurs in **food chains**, in which energy from food passes from one organism to the next in a sequence (see Figure 3.10). Each level, or "link," in a food chain

trophic level An organism's position in a food chain, which is determined by its feeding relationships.

is a **trophic level** (recall that the Greek *tropho* means "nourishment"). An organism is assigned a trophic level based on the number of energy transfer steps from the source of energy to that level. Producers (organisms that photosynthesize) form the first trophic level, primary consumers (herbivores) the second trophic level, secondary consumers (carnivores) the third trophic level, and so on. At every step in a food chain are decomposers, which respire organic molecules from the carcasses and body wastes of all members of the food chain.

Simple food chains rarely occur in nature because few organisms eat just one kind of organism. More typically, the flow of energy and materials through an ecosystem takes place in accordance with a range of food choices for each organism involved. In an ecosystem of average complexity, numerous alternative pathways are possible. A hawk eating a rabbit is a different energy pathway than a hawk eating a snake. A **food web** is a more realistic **model** of the flow of energy and materials through an ecosystem (**Figure 3.11**). A food web helps us visualize feeding relationships that indicate how a community is organized.

food web A representation of the interlocking food chains that connect all organisms in an ecosystem.

The most important thing to remember about energy flow in ecosystems is that it is linear, or one way. Energy moves along a food chain or food web from one organism to the next as long as it has not been used for biological work. Once an organism has used energy, that energy is lost as heat and is unavailable for any other organism in the ecosystem.

CASE IN POINT

HOW HUMANS HAVE AFFECTED THE ANTARCTIC FOOD WEB

Although the icy waters around Antarctica may seem an inhospitable environment, a complex food web is found there. The base of the food web consists of microscopic, photosynthetic algae present in vast numbers in the well-lit, nutrient-rich water. A huge population of herbivores—tiny, shrimplike **krill**—eat these marine algae (**Figure 3.12a**). Krill, in turn, support a variety of larger animals. A major consumer of krill is the baleen whale, which filters krill out of the frigid water. Baleen whales include blue whales, right whales, and humpback whales (**Figure 3.12b**). Squid and fishes also consume krill in great quantities. These animals, in turn, are eaten by other carnivores: toothed whales such as the sperm whale, elephant seals and leopard seals, king penguins and emperor penguins, and birds such as the albatross and the petrel.

Humans have had an impact on the Antarctic food web, as they have had on most other ecosystems. Before the advent of whaling, baleen whales consumed huge quantities of krill. Until a global ban on hunting large whales was enacted in 1986, whaling steadily reduced the number of large baleen whales in Antarctic waters. As a result of fewer whales eating krill, more krill became available for other krill-eating animals, yet krill populations are thought to have been limited in part by the lack of nutrients once provided in the wastes of baleen whales.

Now that commercial whaling is regulated, it is hoped that the number of large baleen whales will slowly increase, and that appears to be the case for some species. However, the populations of most baleen whales in the Southern Hemisphere are still a fraction of their pre-whaling levels. It is not known whether baleen whales will return to their former position of dominance in terms of krill consumption in the food web. Biologists will continue to

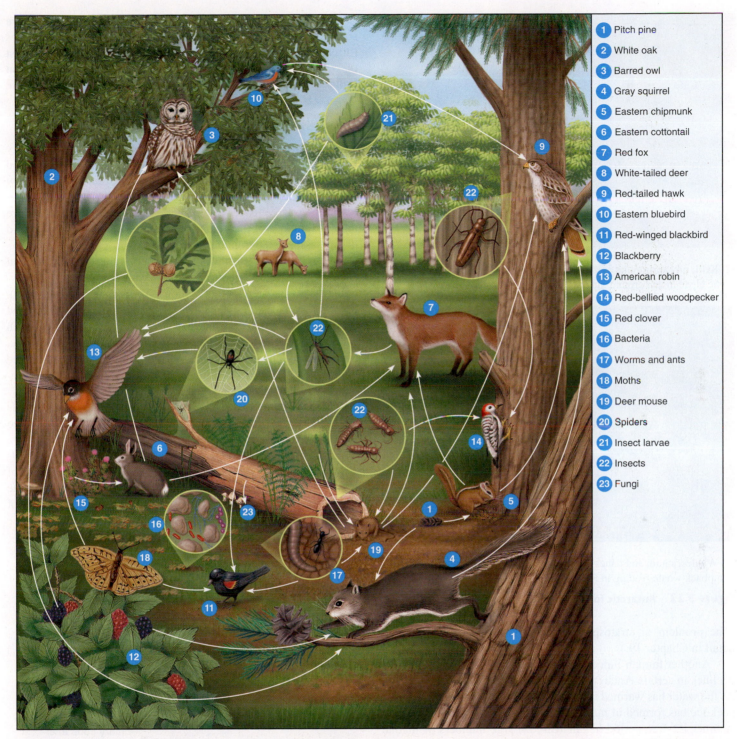

Figure 3.11   **Food web at the edge of an eastern deciduous forest.** This food web is greatly simplified compared to what actually happens in nature. Groups of species are lumped into single categories such as "spiders" and "fungi," other species are not included, and many links in the web are not shown.

monitor changes in the Antarctic food web as the whale populations recover.

Thinning of the ozone layer in the stratospheric region of the atmosphere over Antarctica has the potential to cause long-term effects on the entire Antarctic food web. Ozone thinning allows more of the sun's ultraviolet radiation to penetrate to Earth's surface. Ultraviolet radiation contains more energy than visible light

and can break the chemical bonds of some biologically important molecules, such as deoxyribonucleic acid (DNA). Scientists are concerned that ozone thinning over Antarctica may damage the algae that form the base of the food web in the Southern Ocean. Increased ultraviolet radiation is penetrating the surface waters around Antarctica, and algal productivity has declined, probably as a result of increased exposure to ultraviolet radiation.

(a) Krill, tiny shrimplike animals, live in large swarms and eat photosynthetic algae in and around the pack ice.

(b) Whales, squid, and fishes consume vast numbers of krill. Shown is a humpback whale feeding in the Antarctic.

Figure 3.12 **Antarctic food web.**

(The problem of stratospheric ozone depletion is discussed in detail in Chapter 19.)

Another human-induced change that may be responsible for declines in certain Antarctic populations is global climate change. As the water has warmed in recent decades around Antarctica, less pack ice has formed in many areas during winter months, and the summer ice-free period has increased. Large numbers of marine algae are found in and around the pack ice, providing a critical supply of food for the krill, which reproduce in the area. Years with below-average pack ice cover mean fewer algae, which mean fewer krill reproducing. Also, researchers are finding direct damage to krill from ocean acidification associated with the warming temperatures. Scientists have demonstrated that low krill abundance coincides with unsuccessful breeding seasons for penguins and fur seals, which struggle to find food during warmer winters. Scientists are concerned that climate change will continue to decrease the amount of pack ice and increase ocean acidification, which will reverberate through the food web. (Global climate change, including the effect on Adélie penguins in Antarctica, is discussed in Chapter 20.)

To complicate matters, some commercial fishermen have started to harvest krill to make fishmeal for aquaculture industries (discussed in Chapter 18). Scientists worry that the human harvest of krill may endanger the many marine animals that depend on krill for food. ■

ECOLOGICAL PYRAMIDS

An important feature of energy flow is that most of the energy going from one trophic level to the next in a food chain or food web dissipates into the environment as a result of the second law of thermodynamics. **Ecological pyramids** often graphically represent the relative energy values of each trophic level. There are three main types of pyramids—a pyramid of numbers, a pyramid of biomass, and a pyramid of energy.

A **pyramid of numbers** shows the number of organisms at each trophic level in a given ecosystem, with greater numbers illustrated by a larger area for that section of the pyramid (**Figure 3.13**). In most pyramids of numbers, the organisms at the base of the food chain are the most abundant, and fewer organisms occupy each successive trophic level. In the Antarctic food web, for example, the number of algae is far greater than the number of krill that feed on the algae; likewise, the number of krill is greater than the number of baleen whales, squid, and fishes that feed on krill.

Inverted pyramids of numbers, in which higher trophic levels have *more* organisms than lower trophic levels, are often observed among decomposers, parasites, tree-dwelling herbivorous insects, and similar organisms. One tree may provide food for thousands of leaf-eating insects, for example. Pyramids of numbers are of limited usefulness because they do not indicate the biomass of the organisms at each level, and they do not indicate the amount of energy transferred from one level to another.

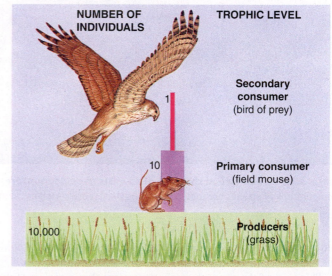

Figure 3.13 **Pyramid of numbers.** This pyramid is for a hypothetical area of temperate grassland; in this example, 10,000 grass plants support 10 mice, which support one bird of prey. Based on the number of organisms found at each trophic level, a pyramid of numbers is not as useful as other ecological pyramids. It provides no information about biomass differences or energy relationships between one trophic level and the next. (Note that decomposers are not shown.)

A **pyramid of biomass** illustrates the total biomass at each successive trophic level in an ecosystem. **Biomass** is a quantitative estimate of the total mass, or amount, of living material; it indicates the amount of fixed energy at a particular time. Biomass units of measure vary: Biomass is represented as total volume, as dry weight, or as live weight. Typically, pyramids of biomass illustrate a progressive reduction of biomass in succeeding trophic levels (**Figure 3.14**). For example, if one assumes about a 90% reduction of biomass for each succeeding trophic level, 10,000 kg of grass should support 1000 kg of grasshoppers, which in turn support 100 kg of toads. The 90% reduction in biomass is used for illustrative purposes only; actual field numbers for biomass reduction in nature vary widely. By this logic, however, the biomass of toad eaters such as snakes could be, at most, only about 10 kg. From this brief exercise, it is apparent that although carnivores do not eat vegetation, a great deal of vegetation is required to support them.

A **pyramid of energy** illustrates the energy content, often expressed as kilocalories per square meter per year, of the biomass of each trophic level in an ecosystem (**Figure 3.15**). These pyramids always have large energy bases and get progressively smaller through succeeding trophic levels. Energy pyramids show that most energy dissipates into the environment when going from one trophic level to the next. Less energy reaches each successive trophic level from the level beneath it because organisms at the lower level use some energy to perform work, and some energy is lost. (Remember, because of the second law of thermodynamics, no biological process is ever 100% efficient.) Energy pyramids explain why there are so few trophic levels: Food webs are short because of the dramatic reduction in energy content at each trophic level. (See "You Can Make a Difference: Vegetarian Diets" in Chapter 18 for a discussion of how the eating habits of humans relate to food chains and trophic levels.)

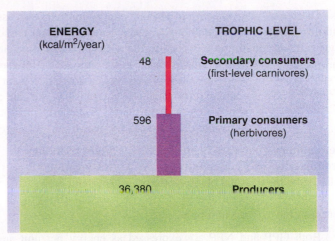

Figure 3.15   **Pyramid of energy.** This pyramid indicates how much energy is present in the biomass at each trophic level in a salt marsh in Georgia and how much is transferred to the next trophic level. Note the substantial loss of usable energy from one trophic level to the next; this loss occurs because of the energy used metabolically and given off as heat. (Note that decomposers are not shown. The 36,380 kcal/m²/year for the producers is gross primary productivity, or GPP, discussed shortly.) (After J. M. Teal. Energy flow in the salt marsh ecosystem of Georgia. *Ecology*, Vol. 43 [1962]).

ECOSYSTEM PRODUCTIVITY

The **gross primary productivity (GPP)** of an ecosystem is the rate at which energy is captured during photosynthesis. (Gross and net primary productivities are referred to as *primary* because plants occupy the first trophic level in food webs.)

> **gross primary productivity (GPP)** The total amount of photosynthetic energy that plants capture and assimilate in a given period.

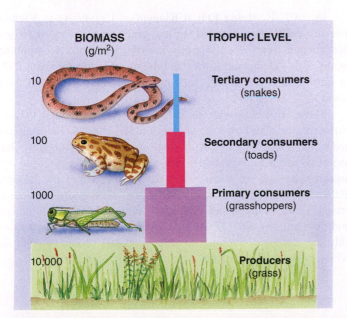

Figure 3.14   **Pyramid of biomass.** This pyramid is for a hypothetical area of temperate grassland. Based on the biomass at each trophic level, pyramids of biomass generally have a pyramid shape with a large base and progressively smaller areas for each succeeding trophic level. (Note that decomposers are not shown.)

Of course, plants respire to provide energy for their own use, and this acts as a drain on photosynthesis. Energy in plant tissues after cellular respiration has occurred is **net primary productivity (NPP)**. That is, NPP is the amount of biomass found in excess of that broken down by a plant's cellular respiration. NPP represents the rate at which this organic matter is actually incorporated into plant tissues for growth.

net primary productivity (NPP) Productivity after respiration losses are subtracted.

Net Primary Productivity	Gross Primary Productivity	Plant Cellular Respiration
(plant growth per unit area per unit time)	= (total photosynthesis per unit area per unit time)	− (per unit area per unit time)

Both GPP and NPP are expressed as energy per unit area per unit time (kilocalories of energy fixed by photosynthesis per square meter per year) or as dry weight (grams of carbon incorporated into tissue per square meter per year).

Only the energy represented by NPP is available as food for an ecosystem's consumers. Consumers use most of this energy for cellular respiration to contract muscles (obtaining food and avoiding predators) and to maintain and repair cells and tissues. Any energy that remains is used for growth and for the production of young, collectively called *secondary productivity*. Any environmental factor that limits an ecosystem's primary productivity—an extended drought, for example—limits secondary productivity by its consumers.

Ecosystems differ strikingly in their productivities (**Figure 3.16**). On land, tropical rainforests have the highest NPP, probably because of their abundant rainfall, warm temperatures, and intense sunlight. Tundra, with its harsh, cold winters, and deserts, with their lack of precipitation, are the least productive terrestrial ecosystems. Wetlands—swamps, marshes, and estuaries that connect terrestrial and aquatic environments—are extremely productive. The most productive aquatic ecosystems are algal beds and coral reefs. Even though the ocean is home to abundant producers, its vast size (large volume) contributes to its low NPP values. A lack of available nutrient minerals in some open ocean regions makes them extremely unproductive, equivalent to aquatic deserts. (Earth's major aquatic and terrestrial ecosystems are discussed in Chapter 6.)

HUMAN IMPACT ON NET PRIMARY PRODUCTIVITY

Humans consume far more of Earth's resources than any other of the millions of animal species. **Peter Vitousek** and colleagues at Stanford University calculated in 1986 how much of the global NPP is appropriated for the human economy and therefore not transferred to other organisms. When both direct and indirect human impacts are accounted for, Vitousek estimated that humans use 32% of the annual NPP of land-based ecosystems. This is a huge amount considering that we humans represent about 0.5% of the total biomass of all consumers on Earth.

In 2001, **Stuart Rojstaczer** and colleagues at Duke University reexamined Vitousek's groundbreaking research. Rojstaczer used contemporary data sets, many of which are satellite-based and more accurate than the data Vitousek analyzed. Rojstaczer's mean value for his conservative estimate of annual land-based NPP appropriation by humans was 32%, like Vitousek's, although Rojstaczer arrived at that number using different calculations.

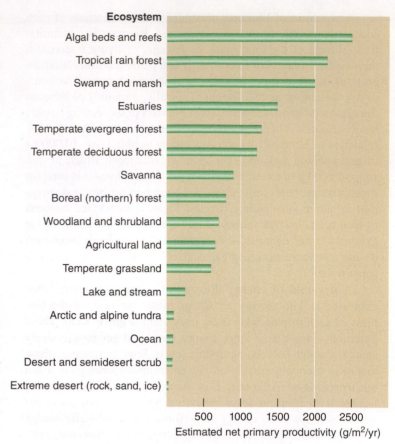

Figure 3.16 **Estimated annual net primary productivities (NPP) for selected ecosystems.** NPP is expressed as grams of dry matter per square meter per year (g/m²/yr). (After R.H. Whittaker, *Communities and Ecosystems*, 2nd edition. New York: Macmillan [1975])

In 2007, **K. Heinz Erb** at Klagenfurt University in Vienna and his colleagues plugged agricultural and forestry statistics that account for 97% of Earth's ice-free land into a computer model. Erb's model indicates that humans appropriate about 25% of Earth's land-based NPP for food, forage (for livestock), and wood.

These studies provide us with estimates, not actual values. However, the take-home message is simple: Human use of global productivity is competing with other species' energy needs. Our use of so much of the world's productivity may contribute to the loss of many species, some potentially useful to humans, through extinction. Human consumption of global NPP could become a serious threat to the planet's ability to support both its nonhuman and human occupants. If we want our planet to operate sustainably, we must share terrestrial photosynthesis products—that is, NPP—with other organisms.

Review

1. What is a food web?
2. How does energy flow through a food web consisting of producers, consumers, and decomposers?
3. What is a pyramid of energy?
4. What is gross primary productivity? Net primary productivity?

ENERGY AND CLIMATE CHANGE

Human Appropriation of NPP

You have seen that humans appropriate a huge amount of Earth's NPP, leaving the remainder for all of Earth's other (nonhuman) organisms. In our efforts against rising CO_2 and global climate change, we may inadvertently be increasing our appropriation of NPP even more. For example, consider the following link between our energy needs and global climate change: increased use of biomass as an energy source. Biomass, such as corn, can be chemically altered to yield alcohol-based fuels for motor vehicles. Biomass is often considered an environmentally friendly source of energy because it does not cause an overall increase of climate-altering CO_2 in the atmosphere. (Growing corn removes CO_2 from the atmosphere, whereas burning fuel made from corn releases CO_2 into the atmosphere; see the figure.) But a closer look at biomass shows it is not so environmentally friendly. If we grow corn and other biomass crops on a large scale for fuel, then we will be taking an even larger share of global NPP, which would leave even less for the planet's organisms. At some point, our use of global NPP

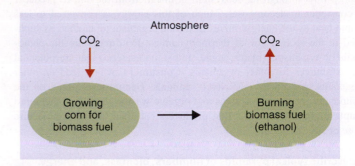

Carbon stored as biomass (corn).

will compromise Earth's systems that sustain us. Indeed, we may have already exceeded the sustainable consumption of NPP.

REVIEW OF LEARNING OBJECTIVES WITH SELECTED KEY TERMS

- **Define *ecology*.**

Ecology is the study of the interactions among organisms and between organisms and their abiotic environment.

- **Distinguish among the following ecological levels: population, community, ecosystem, landscape, and biosphere.**

A **population** is a group of organisms of the same **species** that live in the same area at the same time. A **community** is a natural association that consists of all the populations of different species that live and interact within an area at the same time. An **ecosystem** is a community and its physical environment. A **landscape** is a region that includes several interacting ecosystems. The **biosphere** is the parts of Earth's atmosphere, ocean, land surface, and soil that contain all living organisms.

- **Define *energy* and explain how it is related to work and to heat.**

Energy is the capacity to do work. Energy can be transformed from one form to another but is often measured as **heat**; the unit of heat is the kilocalorie (kcal).

- **Use an example to contrast potential energy and kinetic energy.**

Potential energy is stored energy; **kinetic energy** is energy of motion. Using a bow and arrow as an example, potential energy is stored in the drawn bow and is converted to kinetic energy as the string is released and the arrow speeds toward its target.

- **Distinguish between open and closed systems.**

In thermodynamics, *system* refers to a group of atoms, molecules, or objects being studied; the rest of the universe other than the system is known as the surroundings. A **closed system** is self-contained; that is, it does not exchange energy with its surroundings. In contrast, an **open system** exchanges energy with its surroundings.

- **State the first and second laws of thermodynamics, and discuss the implications of these laws as they relate to organisms.**

According to the **first law of thermodynamics**, energy cannot be created or destroyed, although it can change from one form to another. According to the **second law of thermodynamics**, when energy is converted from one form to another, some of it is degraded into heat, a less usable form that disperses into the environment. The first law explains why organisms cannot produce energy but must continuously capture it from the surroundings. The second law explains why no process requiring energy is ever 100% efficient.

- **Write summary reactions for photosynthesis and cellular respiration, and contrast these two biological processes.**

In photosynthesis:

$$6CO_2 + 12H_2O + \text{radiant energy} \rightarrow C_6H_{12}O_6 + 6H_2O + 6O_2$$

In cellular respiration:

$$C_6H_{12}O_6 + 6O_2 + 6H_2O \rightarrow 6CO_2 + 12H_2O + \text{energy}$$

Plants, algae, and some bacteria capture radiant energy during **photosynthesis** and incorporate some of it into chemical energy contained within carbohydrate molecules. All organisms obtain the energy in carbohydrate and other molecules by **cellular respiration**, in which molecules such as glucose are broken down, releasing energy.

- **Define *energy flow*, *trophic level*, and *food web*.**

Energy flow is the passage of energy in a one-way direction through an ecosystem. A **trophic level** is an organism's position in a **food chain**, which is determined by its feeding relationships. A **food web** is a representation of the interlocking food chains that connect all organisms in an ecosystem.

• **Summarize how energy flows through a food web, incorporating producers, consumers, and decomposers in your explanation.**

Energy flow through an ecosystem is linear, from the sun to producer to consumer to decomposer. Much of this energy is converted to less usable heat as the energy moves from one organism to another, as stipulated in the second law of thermodynamics. **Producers** are the photosynthetic organisms (plants, algae, and some bacteria) that are potential food resources for other organisms. **Consumers**, which feed on other organisms, are almost exclusively animals. **Decomposers** feed on the components of dead organisms and organic wastes, degrading them into simple inorganic materials that producers can then use to manufacture more organic material.

• **Describe typical pyramids of numbers, biomass, and energy.**

A **pyramid of numbers** shows the number of organisms at each successive trophic level. **Biomass** is a quantitative estimate of the total mass, or amount, of living material; it indicates the amount of fixed energy at a particular time. A **pyramid of biomass** illustrates the total biomass at each successive trophic level. **A pyramid of energy** illustrates the energy content of the biomass of each trophic level.

• **Distinguish between gross primary productivity and net primary productivity, and discuss human impact on the latter.**

Gross primary productivity (GPP) is the total amount of photosynthetic energy that plants capture and assimilate in a given period. **Net primary productivity (NPP)** is productivity after respiration losses are subtracted. Scientists have estimated that as much as 32% of the global NPP is appropriated for the human economy. Human use of global productivity is competing with other species' energy needs.

Critical Thinking and Review Questions

1. Draw a food web containing organisms found in a Chesapeake Bay salt marsh.

2. Describe the science of ecology. What are two questions an ecologist might explore in the Chesapeake Bay salt marsh?

3. Which scientist—a population ecologist or a landscape ecologist—would be most likely to study broad-scale environmental issues and land management problems? Explain your answer.

4. What is energy? How are the following forms of energy significant to organisms in ecosystems: (a) radiant energy, (b) mechanical energy, (c) chemical energy?

5. Give two examples of potential energy, and in each case tell how it is converted to kinetic energy.

6. Is this an example of an open system or a closed system? Explain your answer.

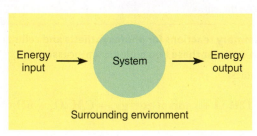

7. How is the first law of thermodynamics related to the movement of an automobile?

8. Give an example of a natural process in which order becomes increasingly disordered.

9. How are photosynthesis and cellular respiration related? Write the overall equations for both processes.

10. Why is the concept of a food web generally preferred over that of a food chain?

11. Could you construct a balanced ecosystem that contained only producers and consumers? Only consumers and decomposers? Only producers and decomposers? Explain the reasons for your answers.

12. Consider a simple ecosystem consisting of a shrub, a worm, a bird, and soil microbes, and identify the producer, primary consumer, secondary consumer, and decomposers. Which of these organisms photosynthesizes? Which carry out cellular respiration? Which give off heat into the surroundings?

13. Suggest a food chain with an inverted pyramid of numbers—that is, greater numbers of organisms at higher rather than at lower trophic levels.

14. Is it possible to have an inverted pyramid of energy? Why or why not?

15. Relate the pyramid of energy to the second law of thermodynamics.

16. What is NPP? Do humans affect the global NPP? If so, how? If not, why?

17. Alcohol fuels produced from corn are often considered a climate-friendly answer to our energy needs. Explain why a large-scale increase in the production of corn to provide fuel could have a negative or even catastrophic impact on ecosystems.

18. How do you interpret this cartoon? What sort of options is the diner considering? Relate your answer to the Food for Thought exercise which is given below.

"I'm trying to eat lower on the food chain."

Edward Koren / The New Yorker Collection/www.cartoonbank.com

Food for Thought

You and a friend order lunch at a restaurant: One of you gets the house salad and the other the specialty burger. Production of the burger required far more land and farming resources than did production of the salad. Explain this difference in terms of ecological pyramids as you identify the likely steps involved in producing your respective meals. Do you consider these steps when making your dining choices?

4

Ecosystems and the Physical Environment

Courtesy Hubbard Brook Experimental Forest

Catchment at Hubbard Brook Experimental Forest. Ecologists who study stream and river systems use catchments (or weirs) to measure the quantity and mineral content of water flowing from a watershed.

Ecosystem processes such as energy flow, the cycling of nutrients, and the effects of natural and human-induced disturbances—for example, air pollution, tree harvesting, and land-use changes—can be studied in both the laboratory and the field. Some ecosystem studies are of particular value because they are long term. Beginning in the late 1950s and continuing to the present, the **Hubbard Brook Experimental Forest (HBEF)**, a 3100-hectare (7750-acre) reserve in the White Mountain National Forest in New Hampshire, has been the site of numerous studies that address the hydrology (e.g., precipitation, surface runoff, and groundwater flow), biology, geology, and chemistry of forests and associated aquatic ecosystems. HBEF is one of the National Science Foundation's 26 long-term ecological research sites.

Experiments at HBEF are often based on field observations. For example, salamander populations surveyed in 1970 have been resurveyed in recent years. Other studies involve controlled experiments. In 1978, scientists added dilute sulfuric acid to a small stream in HBEF to study the effects of acidification. They compared its chemistry and biology to a similar stream that was not

manipulated. This experiment was of practical value because *acid deposition,* a form of air pollution, has acidified numerous lakes and streams in industrialized countries.

Another area of research at HBEF explores the effects of **deforestation**, the clearance of large expanses of forest for agriculture or other uses, on stream ecosystems. When a forest is cleared, the water and minerals that drain into streams increase dramatically. The photograph shows a V-notched dam called a *catchment,* constructed across a stream at the bottom of a valley so scientists can measure water volume and chemical components that flow from the ecosystem. Outflow is generally measured in two separate ecosystems, one of which serves as a control and one of which is experimentally manipulated.

These studies demonstrate that deforestation causes soil erosion and leaching of essential minerals, both of which result in decreased soil fertility. Summer temperatures in streams running through deforested areas are warmer than in shady streams running through control forests. Many stream organisms do not fare well in deforested areas, in part because they are adapted to cooler temperatures.

Detailed ecosystem studies such as those at HBEF also contribute to our practical knowledge about how to maintain water quality, wildlife habitat, and productive forests. **Ecosystem management,** a conservation approach that emphasizes restoring and maintaining the quality of an entire ecosystem rather than the conservation of an individual species, makes use of such knowledge.

This chapter examines the physical components of Earth as a complex system—the interactions and connections of Earth's ecosystems with the abiotic, physical environment on which all life depends. We focus on cycles of matter, solar radiation, the atmosphere, the ocean, weather and climate (including hurricanes), and internal planetary processes (including those that cause earthquakes, tsunamis, and volcanoes).

In Your Own Backyard...

Identify an ecosystem near your home that has been affected by humans and is in need of ecosystem management. What land-use changes have altered this particular ecosystem?

The Cycling of Materials within Ecosystems

LEARNING OBJECTIVES

- Describe the main steps in each of these biogeochemical cycles: carbon, nitrogen, phosphorus, sulfur, and hydrologic.

Chapter 3 described how energy flows in one direction through an ecosystem. In contrast, matter, the material of which organisms are composed, moves through systems in numerous cycles from one part of an ecosystem to another—from one organism to another and from living organisms to the abiotic environment and back again (**Figure 4.1**). These **biogeochemical cycles** involve biological, geologic, and chemical interactions.

Five biogeochemical cycles of matter—carbon, nitrogen, phosphorus, sulfur, and hydrologic (water)—are representative of all biogeochemical cycles. These five cycles are particularly important to organisms because these materials make up the chemical compounds of cells. Carbon, nitrogen, and sulfur are elements that form gaseous compounds, whereas water is a compound that readily evaporates; these four cycles have components that move over long distances in the atmosphere with relative ease. Phosphorus does not form gaseous compounds; as a result, only local cycling occurs easily for that element. Humans are affecting

all these cycles on both local and global scales; we conclude our examination of each cycle with a discussion of these important issues.

THE CARBON CYCLE

Proteins, carbohydrates, and other molecules essential to life contain carbon, so organisms must have carbon available to them. Carbon is found in the atmosphere as a gas, carbon dioxide (CO_2). Carbon is present in the ocean in several forms: dissolved carbon dioxide—that is, carbonate (CO_3^{2-}) and bicarbonate (HCO_3^-)—and dissolved organic carbon from decay processes. Carbon is also present in sedimentary rocks such as *limestone*, which consists primarily of calcium carbonate, $CaCO_3$ (see discussion of the rock cycle in Chapter 15). The global movement of carbon between organisms and the abiotic environment—including the atmosphere, ocean, and sedimentary rock—is known as the **carbon cycle** (**Figure 4.2**).

carbon cycle The global circulation of carbon from the environment to living organisms and back to the environment.

During **photosynthesis**, plants, algae, and certain bacteria remove CO_2 from the air and fix (incorporate) it into chemical compounds such as sugar. Plants use sugar to make other compounds. Thus, photosynthesis incorporates carbon from the abiotic environment into the biological compounds of producers. Those compounds are usually used as fuel for **cellular respiration** by the producer that made them, by a consumer that eats the producer, or by a decomposer that breaks down the remains of the producer or consumer. Thus, cellular respiration returns CO_2 to the atmosphere. A similar carbon cycle occurs in aquatic ecosystems between aquatic organisms and carbon dioxide dissolved in the water.

Sometimes the carbon in biological molecules is not recycled back to the abiotic environment for a long time. A large amount of carbon is stored in the wood of trees, where it may stay for several hundred years or even longer. In addition, millions of years ago, vast coal beds formed from the bodies of ancient trees that did not decay fully before they were buried. Similarly, the organic compounds of unicellular marine organisms probably gave rise to the underground deposits of oil and natural gas that accumulated in the geologic past. Coal, oil, and natural gas, called **fossil fuels** because they formed from the remains of ancient organisms, are vast deposits of carbon compounds, the end products of photosynthesis that occurred millions of years ago (see Chapter 11). The carbon in coal, oil, natural gas, and wood can return to the atmosphere by burning, or **combustion**. In combustion, organic molecules are rapidly oxidized—combined with oxygen—and converted into CO_2 and water, with an accompanying release of heat and light.

The Carbon-Silicate Cycle   On a geologic time scale involving millions of years, the carbon cycle interacts with the silicon cycle in the **carbon-silicate cycle**. The first step of the cycle involves chemical *weathering processes*. Atmospheric CO_2 dissolves in rainwater to form carbonic acid (H_2CO_3), which is a weak acid. As the slightly acidic rainwater moves through the soil, carbonic acid dissociates to form hydrogen ions (H^+) and bicarbonate ions (HCO_3^-). The hydrogen ions enter silicate-rich minerals such as feldspar and change their chemical composition, releasing calcium

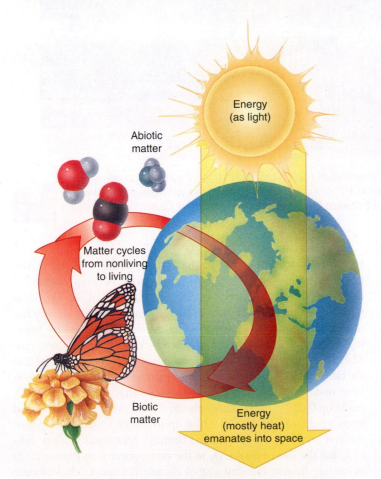

Figure 4.1   Earth's system of energy and matter.   Although energy flows one way through ecosystems, matter continually cycles from the abiotic to the biotic components of ecosystems and back again.

Energy
(as light)

Abiotic
matter

Matter cycles
from nonliving
to living

Biotic
matter

Energy
(mostly heat)
emanates into space

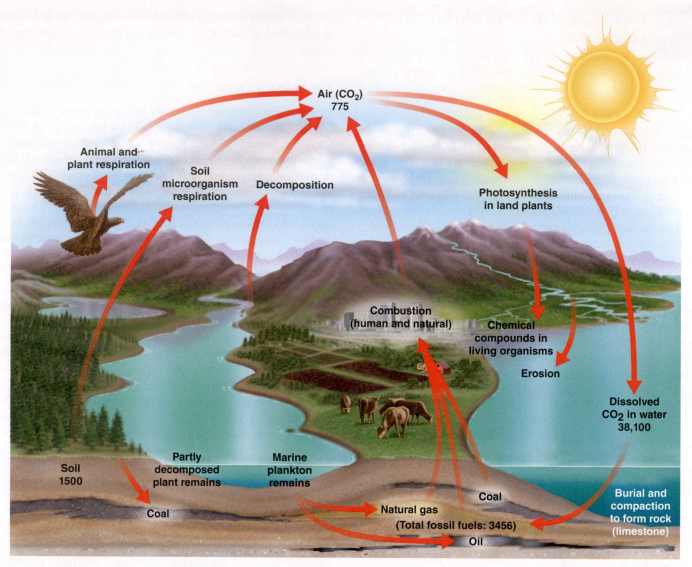

Figure 4.2 **The carbon cycle.** Sedimentary rocks and fossil fuels hold almost all of Earth's estimated 10^{23} g of carbon. The values shown for some of the active pools in the global carbon budget are expressed as 10^{15} g of carbon. For example, the soil contains an estimated 1500×10^{15} g of carbon. How does the carbon cycle influence global climate? (Carbon Dioxide Information Analysis Center, based on data from W.M. Post, Oak Ridge National Laboratory).

ions (Ca^{2+}). The calcium and bicarbonate ions wash into surface waters and eventually reach the ocean.

Microscopic marine organisms incorporate Ca^{2+} and (HCO_3^-) into their shells. When these organisms die, their shells sink to the ocean floor and are covered by sediments, forming calcium carbonate deposits several kilometers thick. The deposits are eventually cemented to form limestone, a sedimentary rock. Earth's crust is active; over millions of years, sedimentary rock on the seafloor may lift to form land surfaces. The summit of Mount Everest, for example, is composed of sedimentary rock. Limestone exposed by the process of geologic uplift slowly erodes as a result of chemical and physical weathering processes, returning CO_2 to the water and atmosphere to participate in the carbon cycle once again.

Alternatively, the geologic process of *subduction* (discussed later in the chapter) buries the carbonate deposits. The increased heat and pressure of deep burial partially melt the sediments, releasing CO_2, which rises in volcanoes and is vented into the atmosphere.

Human-Induced Changes to the Carbon Cycle   Human activities are increasingly disturbing the balance of biogeochemical cycles, including the carbon cycle. Since the end of the 18th century, the advent of the Industrial Revolution, industrial society has used a lot of energy, most of which we have obtained by burning increasing amounts of fossil fuels—coal, oil, and natural gas. This trend, along with a greater combustion of wood as a fuel and the burning of large sections of tropical forests, has shifted carbon from underground deposits to the atmosphere. In the 1700s, CO_2 made up 0.029% of the atmosphere; it now makes up 0.04%, and some scientists project it will be up to 0.06% (double the preindustrial level) by the end of this century. Numerous studies indicate that the increase of CO_2 in the atmosphere (see Figure 20.2) is causing human-induced global climate change. Global climate changes include increasing temperatures, a rise in sea level, altered precipitation patterns, increased wildfires, flooding, drought, heat waves, extinctions of organisms, and agricultural disruption. It has begun to displace people from coastal areas; eventually, millions of people could be displaced.

THE NITROGEN CYCLE

Nitrogen is crucial for all organisms because it is an essential part of biological molecules such as proteins and nucleic acids (for example, DNA). At first glance, a shortage of nitrogen available to organisms would seem impossible. The atmosphere is 78% nitrogen gas (N_2), a two-atom molecule. But each nitrogen molecule is so stable (it has three covalent bonds linking the two atoms) that it does not readily combine with other elements. Atmospheric nitrogen must first be broken apart before the nitrogen atoms can combine with other elements to form proteins and nucleic acids. The **nitrogen cycle**, in which nitrogen

nitrogen cycle The global circulation of nitrogen from the environment to living organisms and back to the environment.

cycles between the abiotic environment and organisms, has five steps: nitrogen fixation, nitrification, assimilation, ammonification, and denitrification (**Figure 4.3**). Bacteria are exclusively involved in all these steps except assimilation.

The first step in the nitrogen cycle, **nitrogen fixation**, is the conversion of gaseous nitrogen to ammonia (NH_3). The process gets its name from the fact that nitrogen is fixed into a form that organisms can use; *fixed* nitrogen refers to nitrogen chemically combined with hydrogen, oxygen, or carbon. Combustion, volcanic action, lightning discharges, and industrial processes, all of which supply enough energy to break apart atmospheric nitrogen, fix considerable nitrogen. Nitrogen-fixing bacteria, including cyanobacteria, carry out biological nitrogen fixation in soil and aquatic environments; they employ the enzyme *nitrogenase* to split atmospheric nitrogen and combine the resulting nitrogen atoms with hydrogen. Nitrogenase functions only in the absence of

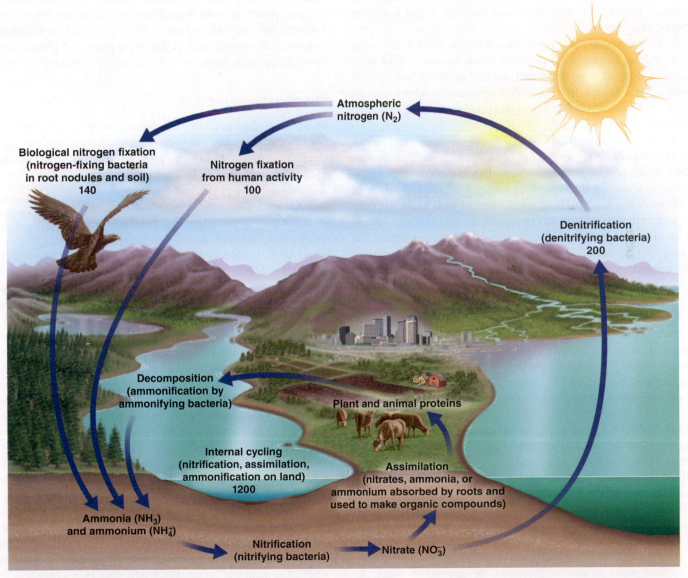

Figure 4.3 **The nitrogen cycle.** The atmosphere holds the largest pool of nitrogen, 4×10^{21} g. The values shown for some of the active pools in the global nitrogen budget are expressed as 10^{12} g of nitrogen per year. For example, each year humans fix an estimated 100×10^{12} g of nitrogen. (Values from W.H. Schlesinger, *Biogeochemistry: An Analysis of Global Change*, 2nd edition. Academic Press, San Diego, 1997, and based on several sources)

oxygen, and the bacteria that use nitrogenase must insulate it from oxygen by some means.

Some nitrogen-fixing bacteria live beneath layers of oxygen-excluding slime on the roots of certain plants. Other important nitrogen-fixing bacteria, *Rhizobium*, live inside special swellings, or nodules, on the roots of legumes such as beans or peas and some woody plants (see the Chapter 5 discussion of mutualism). The relationship between *Rhizobium* and its host plants is mutualistic: The bacteria receive carbohydrates from the plant, and the plant receives nitrogen in a form it can use. In aquatic environments, cyanobacteria perform most nitrogen fixation. Filamentous cyanobacteria have special oxygen-excluding cells that function as the sites of nitrogen fixation.

The conversion of ammonia (NH_3) or ammonium (NH_4^+, formed when water reacts with ammonia) to nitrate (NO_3^-) is **nitrification**. Soil bacteria perform nitrification, a two-step process. First, soil bacteria convert ammonia or ammonium to nitrite (NO_2^-). Then other soil bacteria oxidize nitrite to nitrate. The process of nitrification furnishes these bacteria, called nitrifying bacteria, with energy.

In **assimilation**, plant roots absorb nitrate (NO_3^-), ammonia (NH_3), or ammonium (NO_4^+) and incorporate the nitrogen of these molecules into plant proteins and nucleic acids. When animals consume plant tissues, they assimilate nitrogen by taking in plant nitrogen compounds (amino acids) and converting them to animal compounds (proteins).

Ammonification is the conversion of biological nitrogen compounds into ammonia (NH_3) and ammonium ions (NO_4^+). Ammonification begins when organisms produce nitrogen-containing waste products such as urea (in urine) and uric acid (in the wastes of birds). These substances, as well as the nitrogen compounds that occur in dead organisms, are decomposed, releasing the nitrogen into the abiotic environment as ammonia. The bacteria that perform this process both in the soil and in aquatic environments are called ammonifying bacteria. The ammonia produced by ammonification enters the nitrogen cycle and is once again available for the processes of nitrification and assimilation.

Denitrification is the reduction of nitrate (NO_3^-) to gaseous nitrogen. Denitrifying bacteria reverse the action of nitrogen-fixing and nitrifying bacteria by returning nitrogen to the atmosphere. Denitrifying bacteria prefer to live and grow where there is little or no free oxygen. For example, they are found deep in the soil near the water table, a nearly oxygen-free environment.

Human-Induced Changes to the Nitrogen Cycle From a systems perspective, human activities have disturbed the balance of the global nitrogen cycle. During the 20th century, humans doubled the amount of fixed nitrogen entering the global nitrogen cycle, primarily through the application of fertilizer composed of fixed nitrogen. Several lines of scientific evidence suggest that terrestrial ecosystems enriched with nitrogen have fewer plant species than unenriched terrestrial ecosystems.

Precipitation washes nitrogen fertilizer into rivers, lakes, and coastal areas, where it stimulates the growth of algae. As these algae die, their decomposition by bacteria robs the water of dissolved oxygen, which in turn causes many fishes and other aquatic organisms to die of suffocation. An excess of nitrogen and other nutrients from fertilizer runoff has caused large oxygen-depleted *dead zones* in about 150 coastal areas around the world. (Chapter 21 discusses the dead zone in the Gulf of Mexico.) In addition, nitrates from fertilizer can leach (dissolve and wash down) through the soil and contaminate groundwater. Many people drink groundwater, and nitrate-contaminated groundwater is dangerous to drink, particularly for infants and small children. Nitrate reduces the oxygen-carrying capacity of a child's blood.

Another human activity that affects the nitrogen cycle is the combustion of fossil fuels. When fossil fuels are burned—in automobiles, for example—the high temperatures of combustion convert some atmospheric nitrogen to **nitrogen oxides**, which produce **photochemical smog**, a mixture of air pollutants that injures plant tissues, irritates eyes, and causes respiratory problems. Nitrogen oxides react with water in the atmosphere to form acids that leave the atmosphere as **acid deposition** and cause the pH of surface waters (lakes and streams) and soils to decrease (see Chapter 19). Acid deposition is linked to declining plant and animal populations in aquatic ecosystems and altered soil chemistry on land.

THE PHOSPHORUS CYCLE

Phosphorus does not form compounds in the gaseous phase and does not appreciably enter the atmosphere (except during dust storms). In the **phosphorus cycle**, phosphorus cycles from the land to sediments in the ocean and back to the land (**Figure 4.4**).

phosphorus cycle The global circulation of phosphorus from the environment to living organisms and back to the environment.

As water runs over apatite and other minerals containing phosphorus, it gradually wears away the surface and carries off inorganic phosphate (PO_4^{3-}) molecules. The erosion of phosphorus-containing minerals releases phosphorus into the soil, where plant roots absorb it in the form of inorganic phosphates. Once in cells, phosphates are incorporated into biological molecules such as nucleic acids and ATP (adenosine triphosphate, an organic compound important in energy-transfer reactions in cells). Animals obtain most of their required phosphate from the food they eat, although in some localities drinking water contains a substantial amount of inorganic phosphate. Phosphorus released by decomposers becomes part of the soil's pool of inorganic phosphate for plants to reuse. As in carbon, nitrogen, and other biogeochemical cycles, phosphorus moves through the food web as one organism consumes another.

Phosphorus cycles through aquatic communities in much the same way that it does through terrestrial communities. Dissolved phosphorus enters aquatic communities through absorption and assimilation by algae and plants, which are then consumed by plankton and larger organisms. A variety of fishes and mollusks eat these in turn. Ultimately, decomposers that break down wastes and dead organisms release inorganic phosphorus into the water, where it is available for aquatic producers to use again.

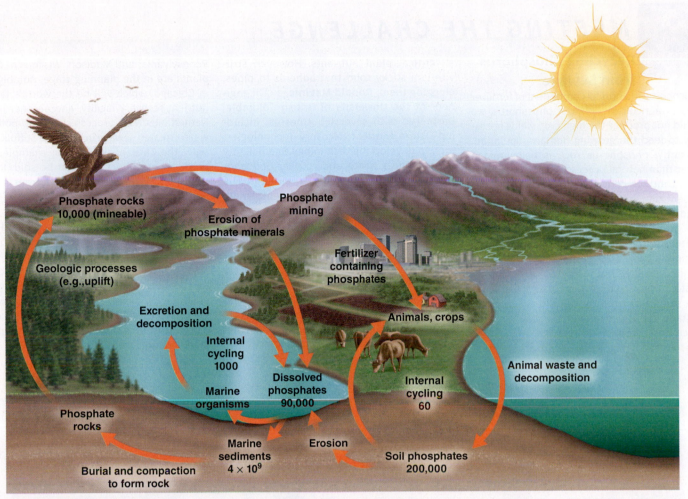

Figure 4.4   The phosphorus cycle. The values shown in the figure for the global phosphorus budget are expressed as 10^{12} g of phosphorus per year. For example, each year an estimated 60×10^{12} g of phosphorus cycles from the soil to terrestrial organisms and back again. (Values from W.H. Schlesinger, *Biogeochemistry: An Analysis of Global Change*, 2nd edition. Academic Press, San Diego, 1997, and based on several sources)

Phosphate can be lost from biological cycles. Streams and rivers carry some phosphate from the land to the ocean, where it can be deposited on the seafloor and remain for millions of years. The geologic process of uplift may someday expose these seafloor sediments as new land surfaces, from which phosphate will once again erode.

A small portion of the phosphate in the aquatic food web finds its way back to the land. Seabirds that eat fishes and other marine animals may defecate on land where they roost. Guano, the manure of seabirds, contains large amounts of phosphate and nitrate. Once on land, these minerals are available for the roots of plants to absorb. The phosphate contained in guano may enter terrestrial food webs in this way, although the amounts involved are quite small.

Human-Induced Changes to the Phosphorus Cycle   Humans affect the phosphorus cycle by accelerating the long-term loss of phosphorus from the land. For example, corn grown in Iowa,

which contains phosphate absorbed from the soil, may fatten cattle in an Illinois feedlot. Part of the phosphate ends up in feedlot wastes, which may eventually wash into the Mississippi River. When people consume beef from the cattle, more of the phosphate ends up in human wastes that are flushed down toilets into sewer systems. Because sewage treatment rarely removes them, phosphates cause water-quality problems in rivers, lakes, and coastal areas. For practical purposes, phosphorus that washes from the land into the ocean is permanently lost from the terrestrial phosphorus cycle (and from further human use), for it remains in the ocean for millions of years. Also, phosphorus is a limiting nutrient to plants and algae in certain aquatic ecosystems. Thus, excess phosphorus from fertilizer or sewage can contribute to enrichment of the water and lead to undesirable changes (recall the discussion of Lake Washington in Chapter 1). "Meeting the Challenge: Recycling Phosphorus from Sewage" examines one approach to conserving phosphorus.

MEETING THE CHALLENGE

Recycling Phosphorus from Sewage

Phosphorus is a critical component of fertilizer used in agriculture. Without phosphorus, we would not produce the high crop yields necessary to feed the growing human population. Currently, phosphorus rock is being mined faster than natural processes can replenish it. If we continue consuming phosphorus at our present rate, it could be depleted early in the 22nd century.

One way to conserve essential phosphate is to recycle it. Much of the phosphate originally mined to produce fertilizer ends up in sewage—both human and animal waste. In sewage treatment plants, a precipitate separates from the solution as bacteria degrade the organic material in sewage. This precipitate, known as *struvite*, consists of phosphorus, magnesium, and ammonium,

all critical plant nutrients. However, struvite is a sticky solid that adheres to pipes, clogging them. **Donald Mavinic**, a civil engineer at the University of British Columbia, partnered with several businesses, including Ostara Nutrient Recovery Technologies, Inc., to develop the technology to efficiently remove the struvite and recover phosphorus during the water treatment process. The phosphorus is sold as Crystal Green®, small, white prills (part a of the figure). For their contributions, Mavinic and his business colleagues received the 2010 Canadian Synergy Award for Innovation.

By late 2014, seven commercial waste water treatment plants were producing Crystal Green in Canada, the United States, and the United Kingdom, including operations in Slough, Berkshire, UK; Saskatoon, Saskatchewan; Portland, Oregon; Suffolk, Virginia; York,

Pennsylvania; and Madison, Wisconsin. Other plants are in the planning stage, notably one in Chicago, which will be the world's largest nutrient recovery facility, and another in the Netherlands. This process cleans up polluted water (treated sewage in York and Suffolk both enter the Chesapeake Bay, for example) *and* produces a product that can be sold. In the York plant, for example, the prills are sold to JR Peters, Inc., a Pennsylvania fertilizer company that blends the prills with other plant nutrients and sells it under the label "Jack's ClassiCote with Crystal Green" (part b of the figure).

This is an example of a financially advantageous solution to a previously serious water pollution problem. Such a system, in which profitable uses are found for another company's wastes, is an example of sustainable manufacturing (see Chapter 15).

(a) **Crystalline prills (small pearl-like granules) of Crystal Green®, a renewable, slow-release fertilizer.** Chemically, the prills consist of 5% nitrogen, 28% phosphorus, and 10% magnesium.

(b) **Fertilizer with Crystal Green®.** The phosphorus in Crystal Green® is slow-release, which means that more gets absorbed by plant roots and less runs off to pollute waterways.

Crystal Green® is a registered trademark of Ostara Nutrient Recovery Technologies, Inc.

Courtesy of JR Peters, Inc.

THE SULFUR CYCLE

sulfur cycle The global circulation of sulfur from the environment to living organisms and back to the environment.

Scientists are still piecing together how the global **sulfur cycle** works. Most sulfur is underground in sedimentary rocks and minerals (e.g., gypsum and anhydrite), which over time erode to release sulfur-containing compounds into the ocean (**Figure 4.5**). Sulfur gases enter the atmosphere from natural sources both in the ocean and on land. Sea spray delivers sulfates (SO_4^{2-}) into the air, as do forest fires and dust storms (desert soils are rich in calcium sulfate, $CaSO_4$). Volcanoes release both hydrogen sulfide (H_2S), a poisonous gas with a smell of rotten eggs, and sulfur oxides (SO_x). Sulfur oxides include sulfur dioxide (SO_2), a choking, acrid gas, and sulfur trioxide (SO_3).

Sulfur gases comprise a minor part of the atmosphere and are not long-lived because atmospheric sulfur compounds are reactive. Hydrogen sulfide reacts with oxygen to form sulfur oxides, and sulfur oxides react with water to form sulfuric acid (H_2SO_4). Although the total amount of sulfur compounds present in the atmosphere at any given time is relatively small, the total annual movement of sulfur to and from the atmosphere is substantial.

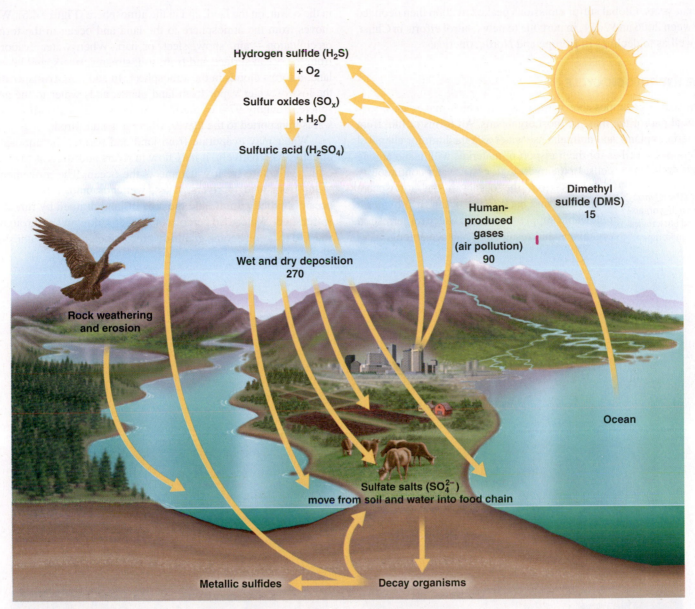

Figure 4.5 **The sulfur cycle.** The largest pool of sulfur on the planet is sedimentary rocks, which contain an estimated 7440×10^{18} g of sulfur. The second-largest pool is the ocean, which contains 1280×10^{18} g of sulfur. The values shown in the figure for the global sulfur budget are expressed in units of 10^{12} g of sulfur per year. For example, the ocean emits an estimated 15×10^{12} g of sulfur per year as the gas dimethyl sulfide. (Values from W.H. Schlesinger, *Biogeochemistry: An Analysis of Global Change*, 2nd edition. Academic Press, San Diego, 1997, and based on several sources)

A tiny fraction of global sulfur is present in living organisms, where it is an essential component of proteins. Plant roots absorb sulfate (SO_4^{2-}) and assimilate it by incorporating the sulfur into plant proteins. Animals assimilate sulfur when they consume plant proteins and convert them to animal proteins. In the ocean, certain marine algae release large amounts of a compound that bacteria convert to dimethyl sulfide, or DMS (CH_3SCH_3). DMS, known for its pungent odor, is released into the atmosphere, where it helps condense water into droplets in clouds and may affect weather and climate. In the atmosphere, DMS is converted to sulfate, most of which is deposited into the ocean.

As in the nitrogen cycle, bacteria drive the sulfur cycle. In freshwater wetlands, tidal flats, and flooded soils, which are oxygen-deficient, certain bacteria convert sulfates to hydrogen sulfide

gas, which is released into the atmosphere, or to metallic sulfides, which are deposited as rock. In the absence of oxygen, other bacteria perform an ancient type of photosynthesis that uses hydrogen sulfide instead of water. Where oxygen is present, the resident bacteria oxidize sulfur compounds to sulfates.

Human-Induced Changes to the Sulfur Cycle   Coal, and to a lesser extent oil, contains sulfur. When these fuels are burned in power plants, factories, and motor vehicles, sulfur dioxide, a major cause of acid deposition, is released into the atmosphere. Sulfur dioxide is also released during the smelting of sulfur-containing ores of such metals as copper, lead, and zinc. Pollution abatement, such as scrubbing smokestack gases to remove sulfur oxides, has reduced the amount of sulfur emissions in some countries in recent years. Global sulfur emissions peaked in 2006 then declined between 2006 and 2011, in part due to new control efforts in China, as well as to declines in Europe and North America.

THE HYDROLOGIC CYCLE

Life would be impossible without water, which makes up a substantial part of the mass of most organisms. All forms of life, from bacteria to plants and animals, use water as a medium for chemical reactions as well as for the transport of materials within and among cells. In the **hydrologic cycle**, water continuously circulates from the ocean to the atmosphere to the land and back to the ocean. It provides a renewable supply of purified water for terrestrial organisms and results in a balance among water

hydrologic cycle The global circulation of water from the environment to living organisms and back to the environment.

in the ocean, on the land, and in the atmosphere (**Figure 4.6**). Water moves from the atmosphere to the land and ocean in the form of precipitation—rain, snow, sleet, or hail. When water evaporates from the ocean surface and from soil, streams, rivers, and lakes on land, it forms clouds in the atmosphere. In addition, **transpiration**, the loss of water vapor from land plants, adds water to the atmosphere. Roughly 97% of the water a plant's roots absorb from the soil is transported to the leaves, where it is transpired.

Water may evaporate from land and reenter the atmosphere directly. Alternatively, it may flow in rivers and streams to coastal estuaries, where fresh water meets the ocean. The movement of water from land to rivers, lakes, wetlands, and, ultimately, the ocean is called **runoff**, and the area of land drained by runoff is a **watershed**. Water also percolates, or seeps, downward through the soil and rock to become **groundwater**, which is fresh water stored

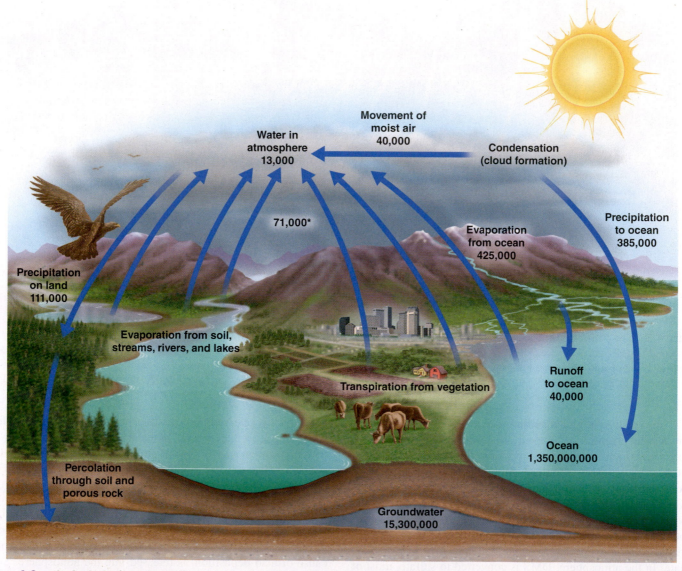

Figure 4.6   **The hydrologic cycle.** Estimated values for pools in the global water budget are expressed as km³, and the values for movements (associated with arrows) are in km³ per year. The starred value (71,000 km³ per year) includes both transpiration from plants and evaporation from soil, streams, rivers, and lakes. (Values from W.H. Schlesinger, *Biogeochemistry: An Analysis of Global Change*, 2nd edition. Academic Press, San Diego, 1997, and based on several sources)

in underground caverns and porous layers of rock. Groundwater may reside in the ground for hundreds to many thousands of years, but eventually it supplies water to the soil, vegetation, streams and rivers, and the ocean.

Regardless of its physical form—solid, liquid, or vapor—or its location, every molecule of water moves through the hydrologic cycle. Tremendous quantities of water are cycled annually between Earth and its atmosphere. The volume of water entering the atmosphere each year is about 389,500 km³ (95,000 mi³). Approximately three-fourths of this water reenters the ocean directly as precipitation; the remainder falls on land.

Human-Induced Changes to the Hydrologic Cycle   Some research suggests that air pollution may weaken the global hydrologic cycle. **Aerosols**—tiny particles of air pollution consisting mostly of sulfates, nitrates, carbon, mineral dusts, and smokestack ash—are produced largely from fossil-fuel combustion and the burning of forests. Once in the atmosphere, aerosols enhance the scattering and absorption of sunlight in the atmosphere and cause clouds to form. Clouds formed in aerosols are less likely to release their precipitation. As a result, scientists think aerosols affect the availability and quality of water in some regions. Climate change caused by CO_2 is also altering the global hydrologic cycle by increasing glacial and polar ice-cap melting and by increasing evaporation in some areas.

We have explored how living things depend on the abiotic environment to supply energy and essential materials (in biogeochemical cycles). Let us now consider five additional aspects of the physical environment that affect organisms: solar radiation, the atmosphere, the ocean, weather and climate, and internal planetary processes.

REVIEW

1. What roles do photosynthesis, cellular respiration, and combustion play in the carbon cycle?

2. What are the five steps of the nitrogen cycle?

3. How does the phosphorus cycle differ from the carbon, nitrogen, and sulfur cycles?

4. What sulfur-containing gases are found in the atmosphere?

Solar Radiation

LEARNING OBJECTIVE

- Summarize the effects of solar energy on Earth's temperature, including the influence of albedos of various surfaces.

The sun makes life on Earth possible. It warms the planet, including the atmosphere, to habitable temperatures. Without the sun's energy, the temperature would approach absolute zero (−273°C) and all water would freeze, even in the ocean. The sun powers the hydrologic cycle, carbon cycle, and other biogeochemical cycles and is the primary determinant of climate. Photosynthetic organisms capture the sun's energy and use it to make the food

molecules required by almost all forms of life. Most of our fuels—wood, oil, coal, and natural gas—represent solar energy captured by photosynthetic organisms. Without the sun, almost all life would cease.

The sun's energy is the product of a massive nuclear fusion reaction and is emitted into space in the form of electromagnetic radiation—especially visible light and infrared and ultraviolet radiation, which are not visible to the human eye (see Figure 3.5). Approximately one-billionth of the total energy released by the sun strikes our atmosphere, and of this tiny trickle of energy, a minute part operates the biosphere.

Clouds and the atmosphere, and, to a lesser extent, surfaces (especially snow, ice, and the ocean) reflect about 30% of the solar radiation (also called *insolation*) that falls on Earth (**Figure 4.7**). Glaciers and ice sheets have high **albedos** and reflect 80% to 90% of the sunlight hitting their surfaces. At the other extreme, asphalt pavement and buildings have low albedos and reflect 10% to 15%, whereas the ocean and forests reflect only about 5%.

> **albedo** The proportional reflectance of solar energy from Earth's surface, commonly expressed as a percentage.

As shown in Figure 4.7, the remaining 70% of the solar radiation that falls on Earth is absorbed and runs the hydrologic cycle, drives winds and ocean currents, powers photosynthesis, and warms the planet. Ultimately, all this energy is lost through the continual radiation of long-wave infrared energy (heat) into space.

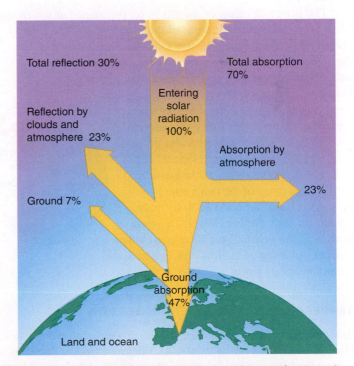

Figure 4.7   Fate of solar radiation that reaches Earth. Most of the sun's energy never reaches Earth. The solar energy that reaches Earth warms the planet's surface, drives the hydrologic cycle and other biogeochemical cycles, produces our climate, and powers almost all life through the process of photosynthesis. (Based on data from K.E. Trenberth, J.T. Fasullo, and J. Kiehl, Earth's Global Energy Budget. *American Meteorological Society,* March 2009.)

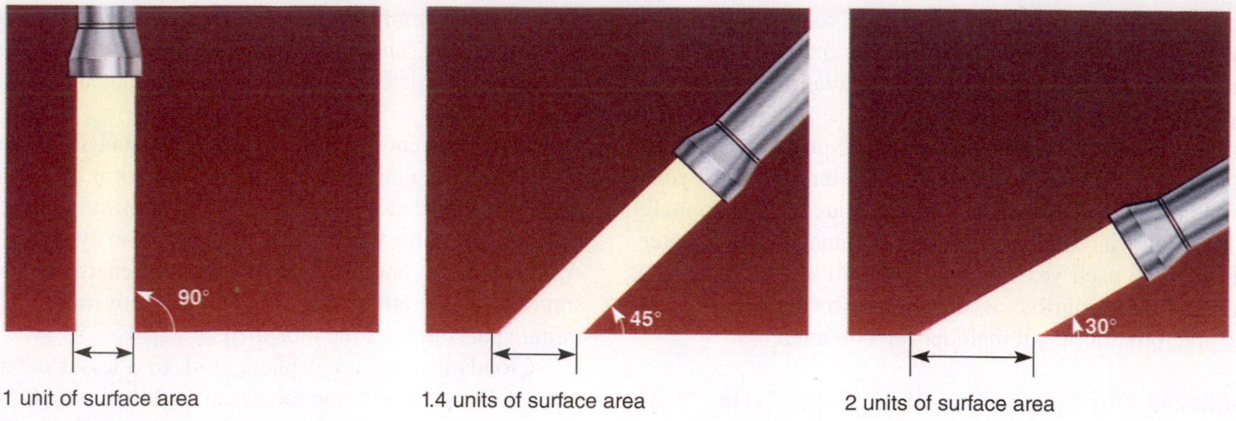

90°

1 unit of surface area

(a) One unit of light is concentrated over 1 unit of surface area.

45°

1.4 units of surface area

(b) One unit of light is dispersed over 1.4 units of surface area.

30°

2 units of surface area

(c) One unit of light is dispersed over 2 units of surface area.

Figure 4.8   Solar intensity and latitude. The angle at which the sun's rays strike Earth varies from one geographic location to another, a result of Earth's spherical shape and its inclination on its axis. Sunlight (represented by the flashlight) that shines vertically near the equator is concentrated on Earth's surface (in a). As one moves toward the poles (see b and c), the light hits the surface more and more obliquely, spreading the same amount of radiation over larger and larger areas.

TEMPERATURE CHANGES WITH LATITUDE

The most significant local variation in Earth's temperature is produced because the sun's energy does not reach all places uniformly. A combination of Earth's roughly spherical shape and the tilt of its axis produces variation in the exposure of the surface to the sun's energy.

The principal effect of the tilt is on the angles at which the sun's rays strike different areas of the planet at any one time (**Figure 4.8**). On average, the sun's rays hit vertically near the equator, making the energy more concentrated and producing higher temperatures. At higher latitudes, the sun's rays hit more obliquely; as a result, their energy is spread over a larger surface area. Also, rays of light entering the atmosphere obliquely near the poles pass through a deeper envelope of air than does light entering near the equator. This causes more of the sun's energy to be scattered and reflected back to space, which in turn further lowers temperatures near the poles. Thus, solar energy that reaches polar regions is less concentrated, and temperatures are lower.

TEMPERATURE CHANGES WITH THE SEASONS

Seasons are determined primarily by Earth's inclination on its axis. Earth's inclination on its axis is 23.5 degrees from a line drawn perpendicular to the orbital plane. During half of the year (March 21 to September 22) the Northern Hemisphere tilts toward the sun, and during the other half (September 22 to March 21) it tilts away from the sun (**Figure 4.9**). The orientation of the Southern Hemisphere is just the opposite at these times. Summer in the Northern Hemisphere corresponds to winter in the Southern Hemisphere.

REVIEW

1. How does the sun affect temperature at different latitudes? Why?
2. What is albedo?

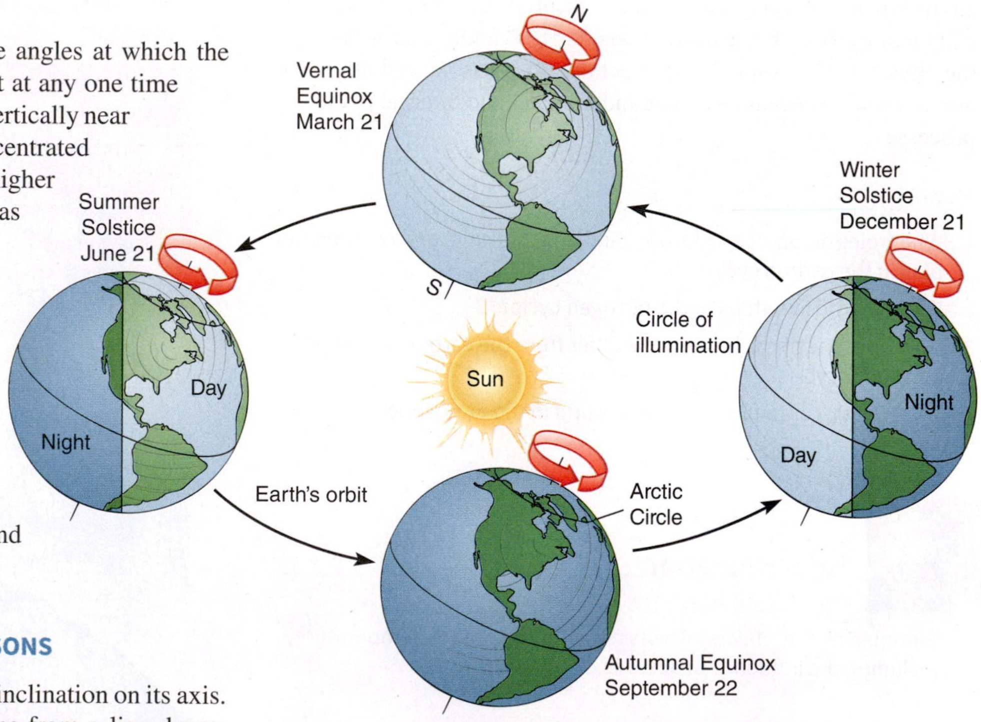

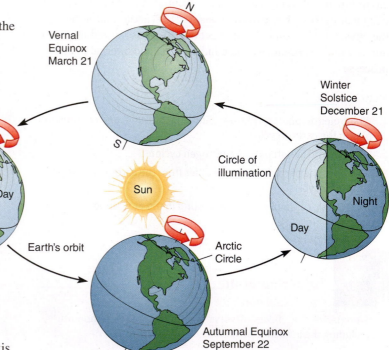

Figure 4.9   Progression of seasons. Earth's inclination on its axis remains the same as it travels around the sun. The sun's rays hit the Northern Hemisphere obliquely during its winter months and more directly during its summer. In the Southern Hemisphere, the sun's rays are oblique during its winter, which corresponds to the Northern Hemisphere's summer. At the equator, the sun's rays are approximately vertical on March 21 and September 22.

The Atmosphere

- Describe the four layers of Earth's atmosphere: troposphere, stratosphere, mesosphere, and thermosphere.
- Discuss the roles of solar energy and the Coriolis effect in producing atmospheric circulation.

The atmosphere is an invisible layer of gases that envelops Earth. Oxygen (21%) and nitrogen (78%) are the predominant gases in the atmosphere, accounting for about 99% of dry air. Other gases, including argon, carbon dioxide, neon, and helium, make up the remaining 1%. In addition, water vapor and trace amounts of various air pollutants, such as methane, ozone, dust particles, microorganisms, and chlorofluorocarbons (CFCs), are present in the air. The atmosphere becomes less dense as it extends outward into space; as a result of gravity, most of the atmosphere's mass is found near Earth's surface (**Figure 4.10**).

The atmosphere performs several ecologically important functions. It protects Earth's surface from most of the sun's ultraviolet radiation and X-rays, as well as from lethal amounts of cosmic rays from space. Without this shielding by the atmosphere, most life would cease to exist. While the atmosphere protects Earth from high-energy radiation, it allows visible light and some infrared radiation to penetrate, and these warm the surface and the lower atmosphere. This interaction between the atmosphere and solar energy is responsible for weather and climate.

Organisms depend on the atmosphere, but they maintain and, in certain instances, modify its composition. Over hundreds of millions of years, photosynthetic organisms converted what was once a carbon dioxide-rich atmosphere to one in which oxygen is the second most common component after nitrogen. A balance between oxygen-producing photosynthesis and oxygen-using respiration maintains the current level of oxygen.

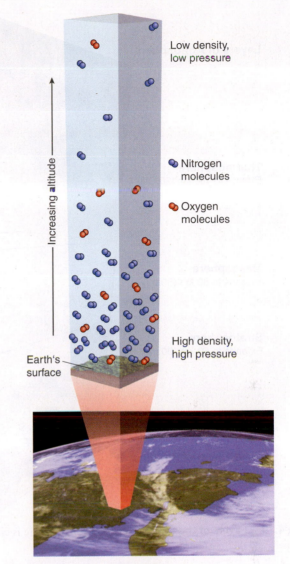

Figure 4.10   Density of the atmosphere. Earth's atmosphere decreases in density and pressure with increasing altitude. The force of gravity is responsible for attracting more air molecules closer to Earth's surface.

LAYERS OF THE ATMOSPHERE

The atmosphere is composed of a series of four concentric layers—the troposphere, stratosphere, mesosphere, and thermosphere (**Figure 4.11**). These layers vary in altitude and temperature with latitude and season. The **troposphere** extends to a height of approximately 12 km (7.5 mi). The temperature of the troposphere decreases with increasing altitude by about 6°C (11°F) for every kilometer. Weather, including turbulent wind, storms, and most clouds, occurs in the troposphere.

troposphere The layer of the atmosphere closest to Earth's surface.

In the next layer of atmosphere, the **stratosphere**, there is a steady wind but no turbulence. Little water is found in the stratosphere, and the temperature is more or less uniform (−45°C to −75°C) in the lower stratosphere; commercial jets fly here. The stratosphere extends from 12 km to 50 km (7.5 mi to 30 mi) above Earth's surface and contains a layer of ozone critical to life because

stratosphere The layer of the atmosphere found directly above the troposphere.

it absorbs much of the sun's damaging ultraviolet radiation. The absorption of ultraviolet radiation by the ozone layer heats the air, and so temperature increases with increasing altitude in the stratosphere.

The **mesosphere**, the layer of atmosphere directly above the stratosphere, extends from 50 km to 80 km (30 mi to 50 mi) above Earth's surface. Temperatures drop steadily in the mesosphere to the lowest in the atmosphere—as low as −138°C.

The **thermosphere** extends from 80 km to 480 km (50 mi to 300 mi) above Earth's surface and is very hot. Gases in the thin air of the thermosphere absorb X-rays and short-wave ultraviolet radiation. This absorption drives the few molecules present to great speeds, raising their temperature in the process to 1000°C or more. The aurora, a colorful display of lights in dark polar skies, is produced when charged particles from the sun hit oxygen or nitrogen molecules in the thermosphere. The thermosphere is important

Layers of Atmosphere

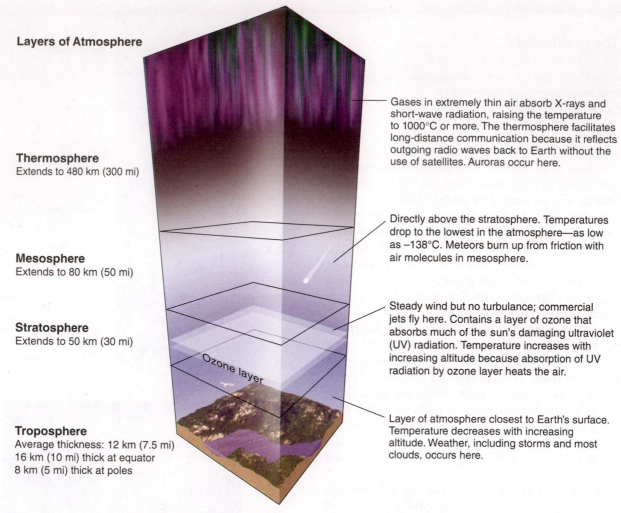

Thermosphere
Extends to 480 km (300 mi)

Gases in extremely thin air absorb X-rays and short-wave radiation, raising the temperature to 1000°C or more. The thermosphere facilitates long-distance communication because it reflects outgoing radio waves back to Earth without the use of satellites. Auroras occur here.

Mesosphere
Extends to 80 km (50 mi)

Directly above the stratosphere. Temperatures drop to the lowest in the atmosphere—as low as −138°C. Meteors burn up from friction with air molecules in mesosphere.

Stratosphere
Extends to 50 km (30 mi)

Ozone layer

Steady wind but no turbulance; commercial jets fly here. Contains a layer of ozone that absorbs much of the sun's damaging ultraviolet (UV) radiation. Temperature increases with increasing altitude because absorption of UV radiation by ozone layer heats the air.

Troposphere
Average thickness: 12 km (7.5 mi)
16 km (10 mi) thick at equator
8 km (5 mi) thick at poles

Layer of atmosphere closest to Earth's surface. Temperature decreases with increasing altitude. Weather, including storms and most clouds, occurs here.

Figure 4.11   Layers of the atmosphere. The troposphere is closest to Earth's surface. The stratosphere is above the troposphere, followed by the mesosphere and thermosphere.

in long-distance communication because it reflects outgoing radio waves back to Earth without the aid of satellites.

ATMOSPHERIC CIRCULATION

In large measure, differences in temperature caused by variations in the amount of solar energy reaching different locations on Earth drive the circulation of the atmosphere. The warm surface near the equator heats the air in contact with it, causing this air to expand and rise in a process known as *convection* (**Figure 4.12**). As the warm air rises, it cools and then sinks again. Much of it recirculates almost immediately to the same areas it has left, but the remainder of the heated air splits and flows in two directions toward the poles. The air chills enough to sink to the surface at about 30 degrees north and south latitudes. This descending air splits and flows over the surface in two directions. Similar upward movements of warm air and its subsequent flow toward the poles occur at higher latitudes farther from the equator. At the poles, the cold polar air sinks and flows toward the lower latitudes, generally beneath the sheets of warm air that simultaneously flow toward the

poles. The constant motion of air transfers heat from the equator toward the poles, and as the air returns, it cools the land over which it passes. This continuous turnover moderates temperatures over Earth's surface.

Surface Winds   In addition to its global circulation patterns, the atmosphere exhibits complex horizontal movements commonly called **winds**. The nature of wind, with its gusts, eddies, and lulls, is difficult to understand or predict. It results in part from differences in atmospheric pressure and from Earth's rotation.

Atmospheric gases have weight and exert a pressure that is, at sea level, about 1013 millibars (14.7 lb per in.²). Air pressure is variable, changing with elevation, temperature, and humidity. Winds tend to blow from areas of high atmospheric pressure to areas of low pressure, and the greater the difference between the high- and low-pressure areas, the stronger the wind.

Earth's rotation influences the direction of wind. Earth rotates from west to east, which causes the east-west movements of surface winds to deflect from their straight-line paths. The moving air swerves to the right of the direction in which it is traveling in the

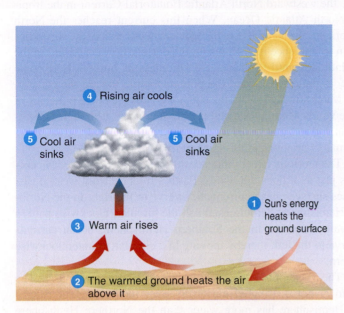

(a) In atmospheric convection, heating of the ground surface heats the air, producing an updraft of less dense, warm air. The convection process ultimately causes air currents that mix warmer and cooler parts of the atmosphere.

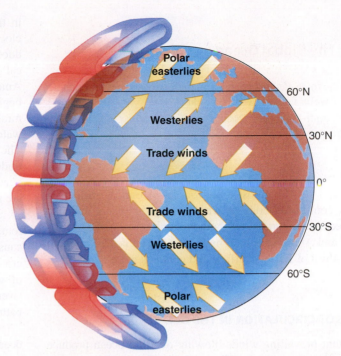

(b) Atmospheric circulation transports heat from the equator to the poles (left side). The heated air rises at the equator, travels toward the poles, and cools in the process so that much of it descends again at around 30 degrees latitude in both hemispheres. At higher latitudes, air circulation is more complex.

Figure 4.12 **Atmospheric circulation.**

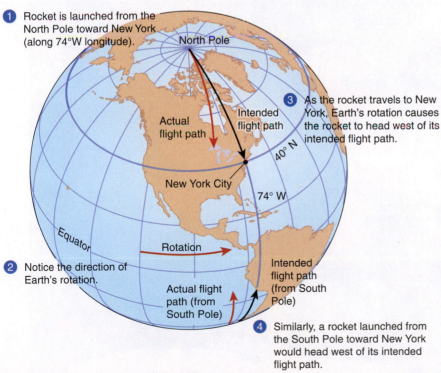

① Rocket is launched from the North Pole toward New York (along 74°W longitude).

③ As the rocket travels to New York, Earth's rotation causes the rocket to head west of its intended flight path.

② Notice the direction of Earth's rotation.

④ Similarly, a rocket launched from the South Pole toward New York would head west of its intended flight path.

Figure 4.13 **Coriolis effect.** Viewed from the North Pole, the Coriolis effect appears to deflect winds and ocean currents to the right. From the South Pole, the deflection appears to be to the left. (Adapted from A.F. Arbogast. *Discovering Physical Geography.* Hoboken, NJ: John Wiley & Sons, Inc., 2007.)

Northern Hemisphere and to the left of the direction in which it is traveling in the Southern Hemisphere (**Figure 4.13**). This tendency is the result of the **Coriolis effect**. The Coriolis effect is greater at higher latitudes and negligible at the equator. Air moving eastward or westward at the equator is not deflected from its path.

The atmosphere has three **prevailing winds** (see **Figure 4.12b**). Prevailing winds that generally blow from the northeast near the North Pole or from the southeast near the South Pole are **polar easterlies**. Winds that generally blow in the midlatitudes from the southwest in the Northern Hemisphere or from the northwest in the Southern Hemisphere are **westerlies**. Tropical winds that generally blow from the northeast in the Northern Hemisphere or from the southeast in the Southern Hemisphere are called **trade winds**.

Coriolis effect The influence of Earth's rotation, which tends to turn fluids (air and water) toward the right in the Northern Hemisphere and toward the left in the Southern Hemisphere.

prevailing winds Major surface winds that blow more or less continually.

REVIEW

1. What is the outermost layer of the atmosphere? Which layer of the atmosphere contains the ozone that absorbs much of the sun's ultraviolet radiation?

2. What basic forces determine the circulation of the atmosphere?

The Global Ocean

- Discuss the roles of solar energy and the Coriolis effect in producing global water flow patterns, including gyres.
- Define *El Niño–Southern Oscillation (ENSO)* and *La Niña*, and describe some of their effects.

The global ocean is a huge body of salt water that surrounds the continents and covers almost three-fourths of Earth's surface. It is a single, continuous body of water, but geographers divide it into four sections separated by the continents: the Pacific, Atlantic, Indian, and Arctic Oceans. The Pacific Ocean is the largest: It covers one-third of Earth's surface and contains more than half of Earth's water.

PATTERNS OF CIRCULATION IN THE OCEAN

The persistent prevailing winds blowing over the ocean produce surface-ocean water **currents** (**Figure 4.14**). The prevailing winds generate **gyres**, large circular ocean currents that often encompass an entire ocean basin. In the North Atlantic Ocean, the tropical trade winds tend to blow toward the west, whereas the westerlies

in the midlatitudes blow toward the east. This helps establish a clockwise gyre in the North Atlantic. That is, the trade winds produce the westward North Atlantic Equatorial Current in the tropical North Atlantic Ocean. When this current reaches the North American continent, it is deflected northward, where the westerlies begin to influence it. As a result, the current flows eastward in the midlatitudes until it reaches the landmass of Europe. Here some water is deflected toward the pole and some toward the equator. The water flowing toward the equator comes under the influence of trade winds again, producing the circular gyre. Although surface-ocean currents and winds tend to move in the same direction, there are many variations to this general rule.

The Coriolis effect influences the paths of surface-ocean currents just as it does the winds. Earth's rotation from west to east causes surface-ocean currents to swerve to the right in the Northern Hemisphere, helping establish the circular, clockwise pattern of water currents. In the Southern Hemisphere, ocean currents swerve to the left, thereby moving in a circular, counterclockwise pattern.

The position of landmasses affects ocean circulation. The ocean is not distributed uniformly over the globe: The Southern Hemisphere has more water than the Northern Hemisphere (**Figure 4.15**). The circumpolar (around the pole) flow of water in the Southern Hemisphere—called the Southern Ocean—is almost unimpeded by landmasses.

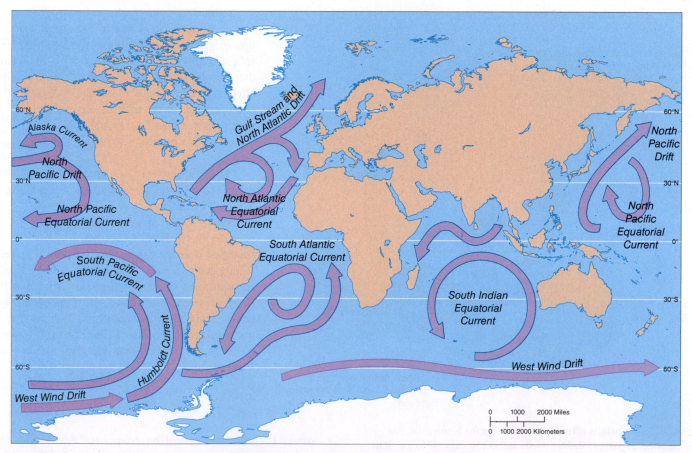

Figure 4.14 **Surface-ocean currents.** Winds largely cause the basic pattern of ocean currents. The main ocean current flow—clockwise in the Northern Hemisphere and counterclockwise in the Southern Hemisphere—results partly from the Coriolis effect.

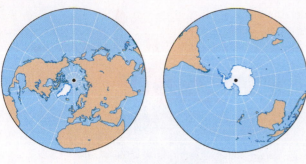

(a) The Northern Hemisphere as viewed from the North Pole.

(b) The Southern Hemisphere as viewed from the South Pole. Ocean currents are freer to flow in a circumpolar manner in the Southern Hemisphere.

Figure 4.15   Ocean and landmasses in the Northern and Southern Hemispheres.

VERTICAL MIXING OF OCEAN WATER

The varying **density** (mass per unit volume) of seawater affects deep-ocean currents. Cold, salty water is denser than warmer, less salty water. (The density of water increases with decreasing temperature down to 4°C.) Colder, salty ocean water sinks and flows under warmer, less salty water, generating currents far below the surface. Deep-ocean currents often travel in different directions and at different speeds than do surface currents, in part because the Coriolis effect is more pronounced at greater depths. **Figure 4.16** shows the circulation of shallow and deep currents—the **ocean conveyor belt**—that moves cold, salty deep-sea water from higher to lower latitudes. Note that the Atlantic Ocean gets its cold, deep water from the Arctic Ocean, whereas the Pacific and Indian Oceans get theirs from the water surrounding Antarctica.

The ocean conveyor belt affects regional and possibly global climate. As the Gulf Stream and North Atlantic Drift (see Figure 4.14) push into the North Atlantic, they deliver an immense amount of heat from the tropics to Europe. As this shallow current transfers its heat to the atmosphere, the water becomes denser and sinks. The deep current flowing southward in the North Atlantic is, on average, 8°C (14°F) cooler than the shallow current flowing northward.

Evidence from seafloor sediments and Greenland ice indicates that the ocean conveyor belt shifts from one equilibrium state to another in a relatively short period (a few years to a few decades). The present ocean conveyor belt reorganized between 11,000 and 12,000 years ago. During this period, heat transfer to the North Atlantic stopped, global temperatures dropped, and both North America and Europe experienced conditions of intense cold. The exact causes and effects of such large shifts in climate are not currently known, but many scientists are concerned that human activities may unintentionally affect the link between the ocean conveyor belt and global climate. The circulation of the conveyor belt in the North Atlantic Ocean could potentially stop as global warming

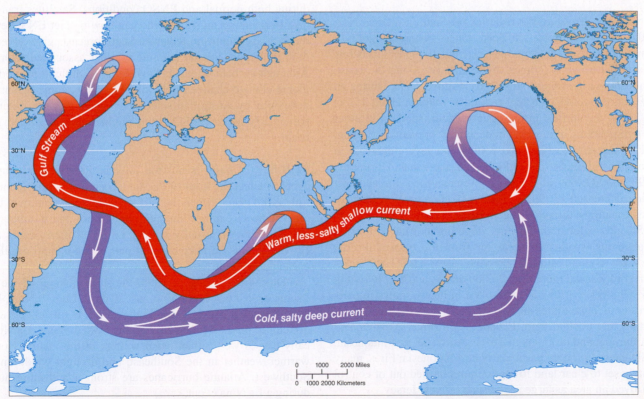

Gulf Stream

Warm, less-salty shallow current

Cold, salty deep current

Figure 4.16   Ocean conveyor belt. This loop consists of both warm, shallow water and cold, deep water. The ocean conveyor belt is responsible for the relatively warm climate in Europe. Changes to the conveyor belt could cause climate changes around the world. How might climate warming trigger a mini ice age in Europe? (After W.S. Broecker, "The Great Ocean Conveyor." *Oceanography*, Vol. 4, 1991.)

causes ice in Greenland to melt. (Melting ice would dilute the cold, salty water so that it doesn't sink.) Some scientists have suggested that this change could lead to a mini ice age in northern Europe. Oceanographers are trying to better understand this phenomenon by measuring the Atlantic Ocean circulation using moored instruments at various depths.

OCEAN INTERACTIONS WITH THE ATMOSPHERE

The ocean and the atmosphere are strongly linked, with wind from the atmosphere affecting the ocean currents and heat from the ocean affecting atmospheric circulation. One of the best examples of the interaction between ocean and atmosphere is the **El Niño–Southern Oscillation** (**ENSO**, or simply **El Niño**) event, which is responsible for much of Earth's interannual (from one year to the next) climate variability.

El Niño–Southern Oscillation (ENSO) A periodic, large-scale warming of surface waters of the tropical eastern Pacific Ocean that affects both ocean and atmospheric circulation patterns.

Normally, westward-blowing trade winds restrict the warmest waters to the western Pacific near Australia. Every three to seven years, however, the trade winds weaken, and the warm mass of water expands eastward to South America, increasing surface temperatures in the eastern Pacific. Ocean currents, which normally flow westward in this area, slow down, stop altogether, or even reverse and go eastward. The name for this phenomenon, **El Niño** (Spanish, "the boy child"), refers to the Christ child because the warming usually reaches the fishing grounds off Peru just before Christmas. El Niños typically last from one to two years.

El Niño can devastate the fisheries off South America. Normally, the colder, nutrient-rich deep water—where remains of dead aquatic organisms decompose—is about 40 m (130 ft) below the surface and **upwells** (comes to the surface) along the coast, partly in response to strong trade winds (**Figure 4.17**). During an ENSO event, the colder, nutrient-rich deep water is about 152 m (500 ft) below the surface in the eastern Pacific, and the warmer surface temperatures and weak trade winds prevent upwelling. The lack of nutrients in the water results in a severe decrease in the populations of anchovies and other marine fishes. During the 1982–1983 ENSO, one of the worst ever recorded, the anchovy population decreased by 99%. However, other species, such as shrimp and scallops, thrive during an ENSO event.

ENSO alters global air currents, directing unusual and sometimes dangerous weather to areas far from the tropical Pacific (**Figure 4.18**). By one estimate, the 1997–1998 ENSO, the strongest on record, caused more than 20,000 deaths and $33 billion in property damages worldwide. It resulted in heavy snows in parts of the western United States and ice storms in eastern Canada; torrential rains that flooded Peru, Ecuador, California, Arizona, and Western Europe; and droughts in Texas, Australia, and Indonesia. An ENSO-caused drought particularly hurt Indonesia: Fires, many deliberately set to clear land for agriculture, burned out of control and destroyed an area as large as the state of New Jersey.

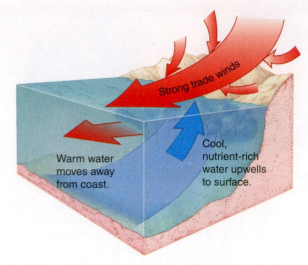

Figure 4.17 **Upwelling.** Coastal upwelling, where deeper waters come to the surface, occurs in the Pacific Ocean along the South American coast. Upwelling provides nutrients for microscopic algae, which in turn support a complex food web. Coastal upwelling weakens considerably during years with El Niño events, temporarily reducing fish populations.

Climate scientists observe and monitor sea-surface temperatures and winds to better understand and predict ENSO events. The **TAO/TRITON array** consists of 70 moored buoys in the tropical Pacific Ocean. These instruments collect oceanic and weather data during both normal and ENSO conditions. Satellites transmit the data to scientists on shore, who can now predict ENSO events as much as six months in advance. Such forecasts give governments time to prepare for the extreme weather associated with ENSO. Although climate scientists have found it difficult to predict how global climate change might affect El Niño, some recent models developed by Australian scientists suggest that the frequency of particularly severe ENSO events might double as climate changes occur.

La Niña El Niño is not the only periodic ocean temperature event to affect the tropical Pacific Ocean. **La Niña** (Spanish, "the little girl") occurs when the surface-water temperature in the eastern Pacific Ocean becomes unusually cool and westbound trade winds become unusually strong. La Niña often occurs after an El Niño event and is considered part of the natural oscillation of ocean temperature.

During the spring of 1998, the surface water of the eastern Pacific cooled 6.7°C (12°F) in just 20 days. Like El Niño, La Niña affects weather patterns around the world, but its effects are more difficult to predict. In the contiguous United States, La Niña typically causes wetter than usual winters in the Pacific Northwest, warmer weather in the Southeast, and drought conditions in the Southwest. Atlantic hurricanes are stronger and more numerous during a La Niña event.

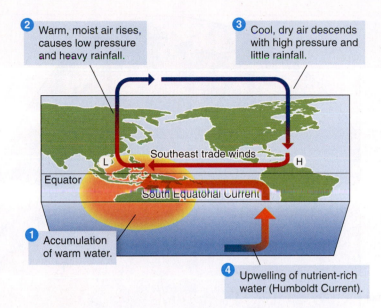

② Warm, moist air rises, causes low pressure and heavy rainfall.

③ Cool, dry air descends with high pressure and little rainfall.

Southeast trade winds

Equator

South Equatorial Current

① Accumulation of warm water.

④ Upwelling of nutrient-rich water (Humboldt Current).

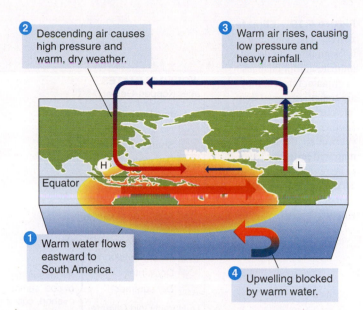

② Descending air causes high pressure and warm, dry weather.

③ Warm air rises, causing low pressure and heavy rainfall.

Equator

① Warm water flows eastward to South America.

④ Upwelling blocked by warm water.

(a) **Normal climate conditions.** ENSO events depend on the relationship of atmospheric circulation to surface-water flow in the Pacific. Normal conditions occur when strong easterly flow pushes warm water into the western Pacific.

(b) **ENSO conditions.** An ENSO event occurs when easterly flow weakens, allowing warm water to collect along the South American coast. Note the relationship between precipitation and the location of pressure systems. During an ENSO event, northern areas of the contiguous United States are typically warmer during winter, whereas southern areas are cooler and wetter.

Figure 4.18 **El Niño—Southern Oscillation** (ENSO). ENSO events drastically alter the climate in many areas remote from the Pacific Ocean.

REVIEW

1. How are the sun's energy, prevailing winds, and surface-ocean currents related?
2. What is the El Niño–Southern Oscillation (ENSO)? What are some of its global effects?

Weather and Climate

LEARNING OBJECTIVE

- Distinguish between weather and climate, and briefly discuss regional precipitation differences.
- Contrast tornadoes and tropical cyclones.

Weather refers to the conditions in the atmosphere at a given place and time; it includes temperature, atmospheric pressure, precipitation, cloudiness, humidity, and wind. Weather changes from one hour to the next and from one day to the next. **Climate** refers to the typical patterns of weather that occur in a place over a period of years.

The two most important factors that determine an area's climate are temperature—both average temperature and temperature extremes—and precipitation—average precipitation, seasonal distribution, and variability. Other climate factors include wind, humidity, fog, and cloud cover. Depending on their layers, altitude, and density, clouds can absorb or reflect sunlight and can retain the planet's outgoing heat; details of the effects clouds have on climate have not yet been resolved. Lightning is an important aspect of climate in some areas because it starts fires.

Day-to-day variations, day-to-night variations, and seasonal variations in climate factors are important dimensions of climate that affect organisms. Latitude, elevation, topography, vegetation, distance from the ocean, and location on a continent or other landmass influence temperature, precipitation, and other aspects of climate. Unlike weather, which changes rapidly, climate usually changes slowly, over hundreds or thousands of years.

Earth has many climates, and because each stays relatively constant for many years, organisms have adapted to them. The many kinds of organisms on Earth are here in part because of the large number of climates—from cold, snow-covered polar climates to tropical climates where it is hot and rains almost every day. A German botanist and climatologist, **Wladimir Köppen**, developed in the early part of the 20th century the most widely used system for classifying climates. **Figure 4.19** shows a world climate map modified from Köppen. Note that there are six climate zones—humid equatorial, dry, humid temperate, humid cold, cold polar, and highland climate—and that each is subdivided into climate types. For example, the three types of humid temperate climates are no dry season, dry winter, and dry summer.

Figure 4.19 **Climate zones.** The boundary lines on this map do not indicate abrupt changes in climate. Instead, they are transition zones from one climate zone to another. In the next 50 years or so, the Köppen map will probably have to be revised to take into account global climate change. (Based on Köppen system.)

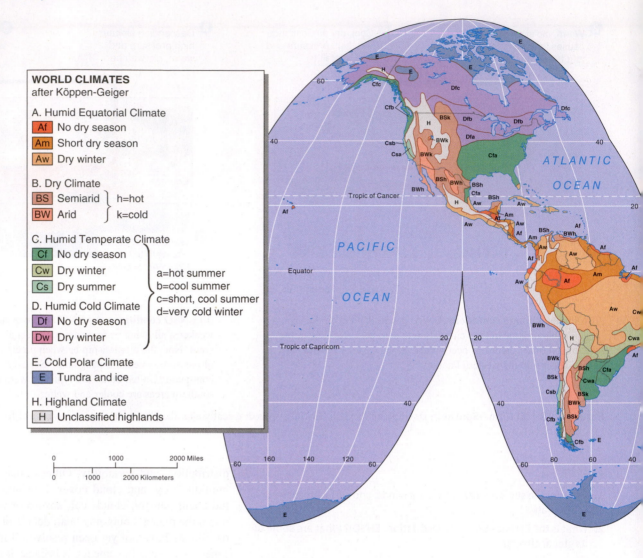

WORLD CLIMATES
after Köppen-Geiger

A. Humid Equatorial Climate
Af   No dry season
Am   Short dry season
Aw   Dry winter

B. Dry Climate
BS   Semiarid   h=hot
BW   Arid   k=cold

C. Humid Temperate Climate
Cf   No dry season
Cw   Dry winter   a=hot summer
Cs   Dry summer   b=cool summer
                        c=short, cool summer
D. Humid Cold Climate   d=very cold winter
Df   No dry season
Dw   Dry winter

E. Cold Polar Climate
E   Tundra and ice

H. Highland Climate
H   Unclassified highlands

PRECIPITATION

Precipitation refers to any form of water, such as rain, snow, sleet, and hail, that falls from the atmosphere. Precipitation varies from one location to another and has a profound effect on the distribution and kinds of organisms present. One of the driest places on Earth is in the Atacama Desert in Chile, where the average annual rainfall is 0.05 cm (0.02 in.). In contrast, Mount Waialeale in Hawaii, Earth's wettest spot, receives an average annual precipitation of 1200 cm (472 in.).

Differences in precipitation depend on several factors. The heavy rainfall of some areas of the tropics results mainly from the equatorial uplift of moisture-laden air. High surface-water temperatures cause the evaporation of vast quantities of water from tropical parts of the ocean, and prevailing winds blow the resulting moist air over landmasses. Heating of the air over a land surface that was warmed by the sun causes moist air to rise. As it rises, the air cools, and its moisture-holding ability decreases (cool air holds less water vapor than warm air). When the air reaches its saturation point—when it cannot hold any additional water vapor—clouds form and water is released as precipitation. The air eventually returns to Earth on both sides of the equator near the Tropics of Cancer and Capricorn (latitudes 23.5 degrees north and

23.5 degrees south, respectively). By then, most of its moisture has precipitated, and the dry air returns to the equator. This dry air makes little biological difference over the ocean, but its lack of moisture over land produces some of the great tropical deserts, such as the Sahara Desert.

Long journeys over landmasses dry the air. Near the windward (the side from which the wind blows) coasts of continents, rainfall is heavy. In temperate areas, continental interiors are usually dry because they are far from the ocean, which replenishes water in the air passing over it.

Mountains, which force air to rise, remove moisture from humid air. The air cools as it gains altitude, clouds form, and precipitation occurs, primarily on the windward slopes of the mountains. As the air mass moves down on the other side of the mountain, it is warmed, thereby lessening the chance of precipitation of any remaining moisture. This situation exists on the West Coast of North America, where precipitation falls on the western slopes of mountains close to the coast. The dry land on the side of the mountains away from the prevailing wind—in this case, east of the mountain range—is a **rain shadow** (**Figure 4.20**).

rain shadow Dry conditions, often on a regional scale, that occur on the leeward side of a mountain barrier; the passage of moist air across the mountains removes most of the moisture.

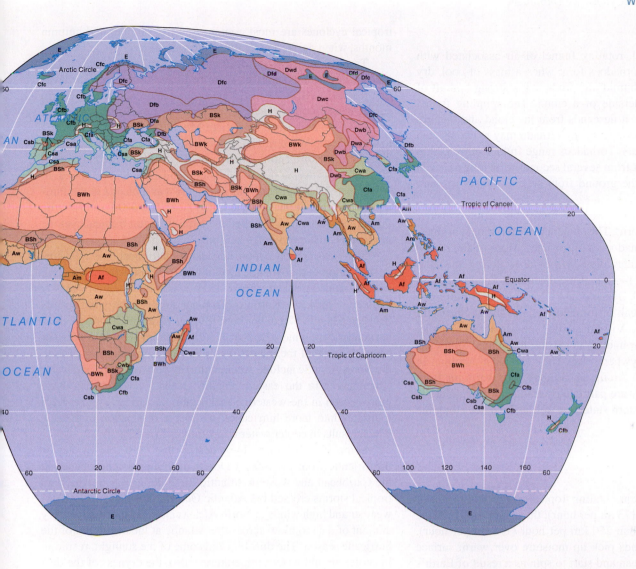

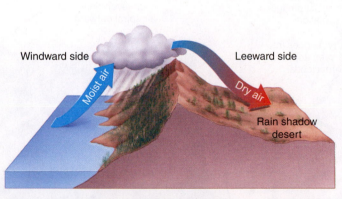

(a) Prevailing winds blow warm, moist air from the windward side. Air cools as it rises, releasing precipitation so that dry air descends on the leeward side.

Earth Satellite Corporation / Science Source

(b) A true-color satellite image of Oregon contrasts the lush vegetation in the moist western region and the more barren arid region to the east of the Cascade Range, which is located in the rain shadow of those mountains.

Figure 4.20   Rain shadow.

TORNADOES

A tornado is a powerful, rotating funnel of air associated with severe thunderstorms. Tornadoes form when a mass of cool, dry air collides with warm, humid air, producing a strong updraft of spinning air on the underside of a cloud. The spinning funnel becomes a tornado when it descends from the cloud and touches the ground. Wind velocity in a strong tornado may reach 480 km per hour (300 mi per hour). Tornadoes range from 1 m to 3.2 km (2 mi) in width. They last from several seconds to as long as seven hours and travel along the ground from several meters to more than 320 km (200 mi).

On a local level, tornadoes have more concentrated energy than any other kind of storm. They can destroy buildings, bridges, and freight trains. Tornadoes kill people: More than 10,000 people in the United States died in tornadoes during the 20th century. In spring of 2011—one of the deadliest tornado years on record with the National Oceanic and Atmospheric Administration (NOAA)—an outbreak of tornadoes across the southeastern states, from Oklahoma to Virginia, killed more than 300 people. Although tornadoes occur in other countries, the United States has more tornadoes than anywhere else. They are most common in the spring months, when Arctic air collides with warm air from the Gulf of Mexico. They are particularly frequent throughout the Great Plains and midwestern states, as well as in states along the Gulf of Mexico coast.

TROPICAL CYCLONES

Tropical cyclones are giant, rotating tropical storms with winds of at least 118 km per hour (73 mi per hour); the most powerful have wind velocities greater than 250 km per hour (155 mi per hour). They form as strong winds pick up moisture over warm surface waters of the tropical ocean and start to spin as a result of Earth's rotation. The spinning causes an upward spiral of massive clouds as air is pulled upward (**Figure 4.21**). Known as *hurricanes* in the Atlantic, *typhoons* in the Pacific, and *cyclones* in the Indian Ocean, tropical cyclones are most common during summer and autumn months, when ocean temperatures are warmest.

Tropical cyclones are destructive when they hit land, not so much from strong winds as from resultant storm surges, waves that rise as much as 7.5 m (25 ft) above the ocean surface. Storm surges cause property damage and loss of life. Some hurricanes produce torrential rains. Hurricane Mitch, the deadliest hurricane to occur in the Western Hemisphere in at least 200 years, hit the Atlantic coast of Central America in 1998 and caused more than 10,000 deaths during the flooding and landslides that followed. In November 2013, Typhoon Haiyan (known as Yolanda in the Philippines) hit Southeast Asia and killed more than 6000 people in the Philippines. Haiyan is believed to be the strongest tropical cyclone in recorded history to make landfall.

Some years produce more hurricanes than others. The 2005 hurricane season in the Atlantic Ocean was the most active on record, with 28 named tropical storms, 15 of which became hurricanes—including Hurricane Katrina. (Tropical storms have wind speeds between 63 and 117 km per hour [39 and 72 mi per hour], whereas hurricanes have wind speeds of 118 km per hour [73 mi per hour] or greater.) Some of the factors that influence hurricane formation in the North Atlantic include precipitation in western Africa and water temperatures in the eastern Pacific Ocean. A wetter-than-usual rainy season in the western Sahel region of Africa (**Figure 4.22**) translates into more hurricanes, as does a dissipation of ENSO, which results in cooler water temperatures in the Pacific.

Some climatologists hypothesize that dust blown across the Atlantic from the Sahara Desert may suppress hurricanes in the Caribbean and western Atlantic. In 2006, a year in which no tropical storms crossed the Atlantic Ocean to form hurricanes, dry weather and high winds in North Africa caused an unusually heavy amount of dust to blow across the Atlantic at the beginning of the hurricane season. The dust blocked some of the sunlight, resulting in cooler air and water temperatures; also, the dryness of the dust and the high winds that blew it across the Atlantic may have suppressed hurricane formation. Similar events are believed to have played a role in a quiet Atlantic hurricane season in 2013.

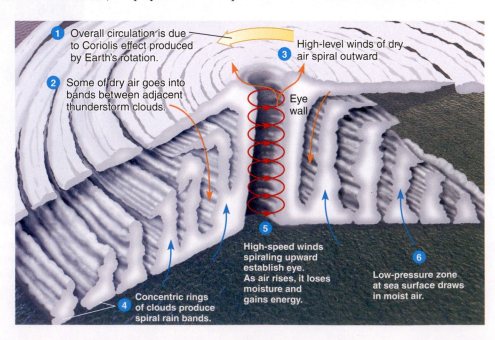

1. Overall circulation is due to Coriolis effect produced by Earth's rotation.
2. Some of dry air goes into bands between adjacent thunderstorm clouds.
3. High-level winds of dry air spiral outward
   Eye wall
4. Concentric rings of clouds produce spiral rain bands.
5. High-speed winds spiraling upward establish eye. As air rises, it loses moisture and gains energy.
6. Low-pressure zone at sea surface draws in moist air.

Figure 4.21   Structure of a cyclone. The numbered processes all take place concurrently. Scientists think that cyclones are becoming stronger as Earth's climate continues to warm.

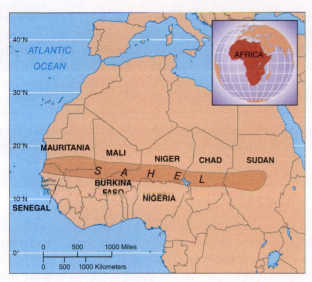

Figure 4.22   The Sahel region in Africa. This area is a transition zone between the Sahara Desert to the north and the moist, tropical rainforest to the south. During the 1970s to 1990s, the Sahel experienced a major drought.

Since 1970, there has been an increase in the intensity of tropical cyclones that have formed. Several factors, such as global atmospheric patterns and vertical wind shear, interact in a complex way to produce tropical cyclone intensity. However, recent evidence suggests that sea-surface temperature is the most important factor, which implies that as the global climate warms, tropical cyclones are becoming more intense (see Chapter 20).

CASE IN POINT

HURRICANE KATRINA

Hurricane Katrina, which hit the north-central Gulf coast along Louisiana, Mississippi, and Alabama in August 2005, was one of the most devastating storms in U.S. history. It produced a storm surge that caused severe damage to the city of New Orleans as well as to other coastal cities and towns in the region. The high waters caused levees and canals to fail, flooding 80% of New Orleans and many nearby neighborhoods.

Most people are aware of the catastrophic loss of life and property damage caused by Katrina. Here we focus on how humans have altered the geography and geology of the New Orleans area in ways that have impaired a balanced natural system and exacerbated the storm damage. This discussion is applicable to many of the world's cities located on river deltas; these cities are also vulnerable to storms and flooding.

The Mississippi River delta formed over millennia from sediments deposited at the mouth of the river. New Orleans is ideally located for industry as well as for sea and river commerce, but the city's development has disrupted delta-building processes. Over the years, engineers have constructed a system of canals to aid navigation and levees to control flooding, since the city is at or below sea level. The canals allowed salt water to intrude and kill the fresh-water marsh vegetation. The levees prevented the

deposition of sediments that remain behind after floodwaters subside. (The sediments are now deposited into the Gulf of Mexico.) Under natural conditions, these sediments replenish and maintain the delta, building up coastal wetlands.

As the city grew, new development took place on wetlands—bayous, waterways, and marshes—that were drained and filled in. Before their destruction, these coastal wetlands provided some protection against flooding from storm surges. We are not implying that had Louisiana's wetlands been intact, New Orleans would not have suffered any damage from a hurricane of Katrina's magnitude. However, had these wetlands been largely unaltered, they would have moderated the damage by absorbing much of the water from the storm surge.

Another reason that Katrina devastated New Orleans is that the city has been subsiding (sinking) for many years, primarily because New Orleans is built on unconsolidated sediment (no bedrock underneath). Many wetlands scientists also attribute this subsidence to the extraction of the area's rich supply of underground natural resources—groundwater, oil, and natural gas. As these resources are removed, the land compacts, lowering the city. New Orleans and nearby coastal areas are subsiding an average of 11 mm each year. At the same time, sea level has been rising an average of 1–2.5 mm per year—with some scientists estimating more dramatic increases—in part because of human-induced changes in climate. ■

REVIEW

1. How do you distinguish between weather and climate? What are the two most important climate factors?

2. Distinguish between tornadoes and tropical cyclones.

Internal Planetary Processes

LEARNING OBJECTIVE

- Define *plate tectonics* and explain its relationship to earthquakes and volcanic eruptions.

Planet Earth consists of layers of different composition and rock strength (**Figure 4.23**). Earth's outermost rigid rock layer (the **lithosphere**) is composed of seven large plates, plus a few smaller ones, that float on the **asthenosphere** (the region of the mantle where rocks become hot and soft). Landmasses are situated on some of these plates. As the plates move horizontally across Earth's surface, the continents change their relative positions. **Plate tectonics** is the study of the dynamics of Earth's lithosphere—that is, the movement of these plates.

Any area where two plates meet—a **plate boundary**—is a site of intense geologic activity (**Figure 4.24**). Three types of plate boundaries—divergent, convergent, and transform—exist; all three types occur both in the ocean and on land. Two plates move apart at a **divergent plate boundary**. When two plates move apart, a ridge of molten rock from the mantle wells up

plate tectonics The study of the processes by which the lithospheric plates move over the asthenosphere.

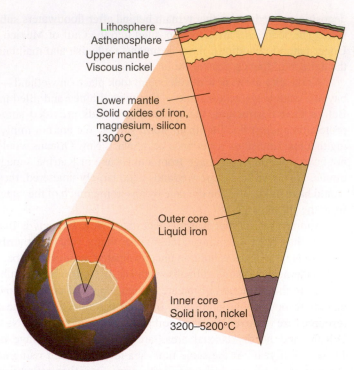

Lithosphere
Asthenosphere
Upper mantle
Viscous nickel

Lower mantle
Solid oxides of iron,
magnesium, silicon
1300°C

Outer core
Liquid iron

Inner core
Solid iron, nickel
3200–5200°C

Figure 4.23 **Earth's internal structure.** A slice through Earth reveals layers of different composition and density.

between them; the ridge continually expands as the plates move farther apart. The Atlantic Ocean is growing as a result of the buildup of lava along the Mid-Atlantic Ridge, where two plates are diverging.

When two plates collide at a **convergent plate boundary**, one of the plates sometimes descends under the other in the process of **subduction**. Convergent collision can also form a mountain range; the Himalayas formed when the plate carrying India converged into the plate carrying Asia. At a **transform plate boundary**, plates move horizontally in opposite but parallel directions. On land, such a boundary is often evident as a long, thin valley due to erosion along the fault line.

Earthquakes and volcanoes are common at plate boundaries. Both the San Francisco area, noted for its earthquakes, and the volcano Mount Saint Helens in Washington State are situated where two plates meet.

EARTHQUAKES

Forces inside Earth sometimes push and stretch rocks in the lithosphere. The rocks absorb this energy for a time, but eventually, as the energy accumulates, the stress is too great and the rocks suddenly shift or break. The energy—released as **seismic waves**, vibrations that spread through the rocks rapidly in all directions—causes one of the most powerful events in nature, an earthquake. Most earthquakes occur along **faults**, fractures where rock moves forward and backward, up and down, or from side to side. Fault zones are often found at plate boundaries.

For example, the Caribbean region is prone to earthquakes due to movements between the North American, South American, and Caribbean plates. Puerto Rico, Jamaica, the Dominican Republic, Martinique, and Guadeloupe have had earthquakes greater than magnitude 7 in the past. In January 2010, an earthquake with a moment magnitude of 7.0 struck in an area approximately 25 km (16 mi) from Port-au-Prince, the capital of Haiti. (Moment magnitude is discussed later in this section.) About 230,000 people were killed, making it one of the deadliest earthquakes on record.

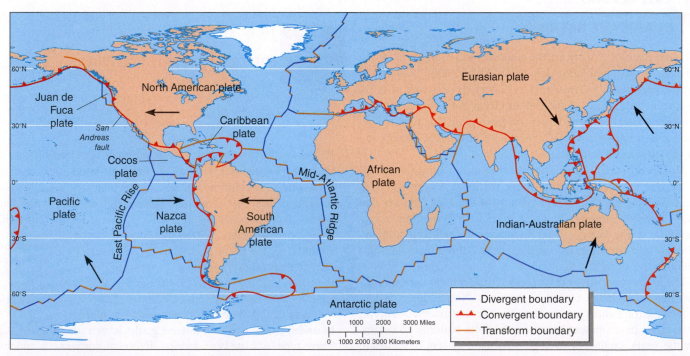

Figure 4.24 **Plates and plate boundary locations.** Seven major independent plates move horizontally across Earth's surface: African, Eurasian, Indian-Australian, Antarctic, Pacific, North American, and South American. Arrows show the directions of plate movements. The three types of plate boundaries are explained in the text.

Most of these people died due to the structural collapse of poorly constructed buildings. About 1 million people whose homes were destroyed became refugees.

The site where an earthquake begins, often far below the surface, is the focus (**Figure 4.25**). Directly above the focus, at Earth's surface, is the earthquake's epicenter. When seismic waves reach the surface, they cause the ground to shake. Buildings and bridges collapse, and roads break. One of the instruments used to measure seismic waves is a seismograph, which helps seismologists (scientists who study earthquakes) determine where an earthquake started, how strong it was, and how long it lasted. In 1935, Charles Richter, a California seismologist, invented the *Richter scale,* a measure of the magnitude of energy released by an earthquake. Each unit on the Richter scale represents about 30 times more released energy than the unit immediately below it. As an example, a magnitude 8 earthquake is 30 times more powerful than a magnitude 7 earthquake and 900 times more powerful than a magnitude 6 earthquake. The Richter scale makes it easy to compare earthquakes, but it tends to underestimate the energy of large quakes.

Although the public is familiar with the Richter scale, seismologists typically do not use it. There are several ways to measure the magnitude of an earthquake, just as there are several ways to measure the size of a person (for example, height, weight, and amount of body fat). Most seismologists use a more accurate scale, the *moment magnitude scale,* to measure earthquakes, especially those larger than magnitude 6.5 on the Richter scale. The moment magnitude scale calculates the total energy that a quake releases.

Seismologists record more than 1 million earthquakes each year. Some of these are major, but most are too small to be felt, equivalent to readings of about 2 on the Richter scale. A magnitude 5 earthquake usually causes property damage. On average, about every five years, a great earthquake occurs with a reading of 8 or higher. Such quakes usually cause massive property destruction and kill large numbers of people. Few people die directly because of the seismic waves; most die because of building collapses or fires started by ruptured gas lines.

Landslides and tsunamis are some of the side effects of earthquakes. A **landslide** is an avalanche of rock, soil, and other debris that slides swiftly down a mountainside. In 2008, a powerful earthquake struck in a mountainous area in Sichuan Province, China. The quake and powerful aftershocks triggered massive landslides and the structural collapse of many buildings. About 70,000 people were killed, and more than 1.5 million people whose homes were destroyed became refugees.

A **tsunami**, a giant sea wave caused by an underwater earthquake, volcanic eruption, or landslide, sweeps through the water at more than 750 km (450 mi) per hour. Although a tsunami may be only about 1 m high in deep-ocean water, it can build to a wall of water 30 m (100 ft) high—as high as a 10-story building—when it comes ashore, often far from where the original earthquake triggered it. Tsunamis have caused thousands of deaths. Colliding tectonic plates in the Indian Ocean triggered tsunamis in 2004 that killed more than 230,000 people in South Asia and Africa. Not only did the tsunamis cause catastrophic loss of life and destruction of property, but also they resulted in widespread environmental damage. Salt water that moved inland as far as 3 km (1.9 mi) polluted soil and groundwater. Oil and gasoline from overturned cars, trucks, and boats contaminated the land and poisoned wildlife. Coral reefs and other offshore habitats were also damaged or destroyed.

In March 2011, an earthquake with a magnitude of 8.9 hit Japan. This earthquake, the strongest ever recorded in Japan, also triggered a disastrous tsunami. Other tsunamis generated by the earthquake hit coastal areas of several Pacific Rim countries, although they caused substantially less damage than in Japan. The death toll in Japan numbered in the thousands and would have been worse except that Japan is one of the best prepared of all nations to face earthquake disasters. Japan lies on a seismically active junction of several tectonic plates. (See Chapter 11 for a discussion of damage to Japan's nuclear power plants from the quake.)

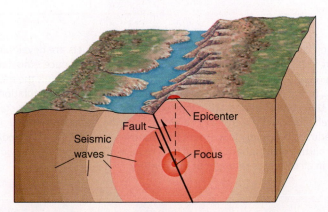

Figure 4.25   Earthquakes. Earthquakes occur when plates along a fault suddenly move in the opposite direction relative to each other. This movement triggers seismic waves that radiate through the crust.

VOLCANOES

The movement of tectonic plates on the hot, soft rock of the asthenosphere causes most volcanic activity. In places, the rock reaches the melting point, forming pockets of molten rock, or **magma**. When one plate slides under or away from an adjacent plate, magma may rise to the surface, often forming volcanoes. Magma that reaches the surface is called **lava**.

Volcanoes occur at three locations: at subduction zones, at spreading plates, and above hot spots. Subduction zones within the Pacific Basin have given rise to hundreds of volcanoes around Asia and the Americas in the region known as the "ring of fire." Plates that spread apart also form volcanoes: Iceland is a volcanic island that formed along the Mid-Atlantic Ridge. The Hawaiian Islands have a volcanic origin, but they did not form at plate boundaries. This chain of volcanic islands is thought to have formed as the Pacific plate moved over a **hot spot**, a rising plume of magma that flowed from deep within Earth's rocky mantle through an opening in the crust (**Figure 4.26**).

The largest volcanic eruption in the 20th century occurred in 1991 when Mount Pinatubo in the Philippines exploded. Despite the evacuation of more than 200,000 people, 338 deaths occurred,

mostly from the collapse of buildings under the thick layer of wet ash that blanketed the area. The volcanic cloud produced when Mount Pinatubo erupted extended upward some 48 km (30 mi). We are used to hearing about human activities affecting climate, but many significant natural phenomena, including volcanoes, also affect global climate. The magma and ash ejected into the atmosphere when Mount Pinatubo erupted blocked much of the sun's warmth and caused a slight cooling of global temperatures for a year or so.

REVIEW

1. What are tectonic plates and plate boundaries?
2. Where are earthquakes and volcanoes commonly located, and why?

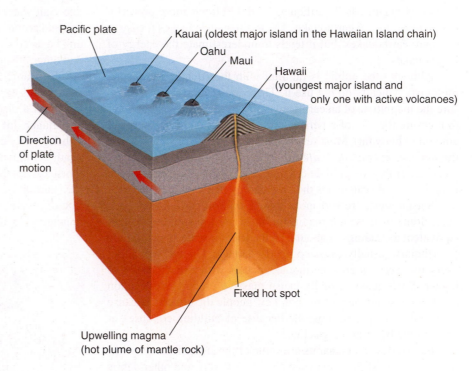

Figure 4.26 Hawaiian hot spot. The hot spot does not move because it is at the top of a rising plume of magma. The islands formed as the Pacific plate moved over the hot spot. Kauai is the oldest Hawaiian island, and Hawaii is the youngest.

ENERGY AND CLIMATE CHANGE

The Carbon Cycle

As we discussed earlier in the chapter, human activities are increasingly disturbing the balance of the carbon cycle. As human demand for fossil energy increases, emissions of the greenhouse gas CO_2 increase. This means carbon that had been captured through photosynthesis over millions of years is being released through combustion (rather than cellular respiration) over a couple of centuries. Plants cannot recapture this carbon through photosynthesis quickly enough to keep the atmospheric CO_2 concentration from increasing. We can only reduce the rate at which the CO_2 concentration increases by reducing emissions.

Many countries traditionally dependent on fossil fuels are developing and implementing policies that could significantly reduce emissions of greenhouse gases. However, so far the countries that emit the most CO_2, including the United States, have not adopted federal regulations; all U.S. control efforts are either voluntary or regulated at the state level. Moreover, about half the global greenhouse gas emissions are produced by rapidly industrializing countries such as China and India. Stabilizing atmospheric CO_2 will thus require major efforts by individuals in all countries throughout this century.

REVIEW OF LEARNING OBJECTIVES WITH SELECTED KEY TERMS

- **Describe the main steps in each of these biogeochemical cycles: carbon, nitrogen, phosphorus, sulfur, and hydrologic.**

1. In the **carbon cycle**, carbon enters plants, algae, and cyanobacteria as carbon dioxide (CO_2), which is incorporated into organic molecules by **photosynthesis**. **Cellular respiration** by plants, by animals that eat plants, and by decomposers returns CO_2 to the atmosphere, making it available for producers again. **Combustion** and weathering also return CO_2 to the atmosphere.

2. In the **nitrogen cycle**, **nitrogen fixation** is the conversion of nitrogen gas to ammonia. **Nitrification** is the conversion of ammonia or ammonium to nitrate. **Assimilation** is the biological conversion of nitrates, ammonia, or ammonium into nitrogen-containing compounds by plants; the conversion of plant proteins into animal proteins is also part of assimilation. **Ammonification** is the conversion of organic nitrogen to ammonia and ammonium ions. **Denitrification** converts nitrate to nitrogen gas.

3. The **phosphorus cycle** has no biologically important gaseous compounds. Phosphorus erodes from rock as inorganic phosphates, and plants absorb it from the soil. Animals obtain phosphorus from their diets, and decomposers release inorganic phosphate into the environment.

4. In the **sulfur cycle**, most sulfur occurs as rocks or as sulfur dissolved in the ocean. Sulfur-containing gases, which include hydrogen sulfide, sulfur oxides, and dimethyl sulfide (DMS), comprise a minor part of the atmosphere and are not long-lived. A tiny fraction of sulfur is present in the proteins of living organisms. Bacteria drive the sulfur cycle.

5. The **hydrologic cycle**, which continuously renews the supply of water essential to life, involves an exchange of water among the land, the atmosphere, and organisms. Water enters the atmosphere by evaporation and **transpiration** and leaves the atmosphere as precipitation. On land, water runs off to lakes, rivers, and the ocean or filters through the ground. **Groundwater** is stored in underground caverns and porous layers of rock.

- **Summarize the effects of solar energy on Earth's temperature, including the influence of albedos of various surfaces.**

Of the solar energy that reaches Earth, 30% is immediately reflected away and the remaining 70% is absorbed. **Albedo** is the proportional reflectance of solar energy from Earth's surface, commonly expressed as a percentage. Glaciers and ice sheets have high albedos, and the ocean and forests have low albedos. Ultimately, all absorbed solar energy is radiated into space as infrared (heat) radiation. A combination of Earth's roughly spherical shape and the tilt of its axis concentrates solar energy at the equator and dilutes solar energy at higher latitudes.

- **Describe the four layers of Earth's atmosphere: troposphere, stratosphere, mesosphere, and thermosphere.**

The **troposphere** is the layer of the atmosphere closest to Earth's surface; weather occurs in the troposphere. The **stratosphere**, found directly above the troposphere, contains a layer of ozone that absorbs much of the sun's damaging ultraviolet radiation. The **mesosphere**, found directly above the stratosphere, has the lowest temperatures in the atmosphere. The **thermosphere** has steadily rising temperatures because the air molecules there absorb high-energy X-rays and short-wave ultraviolet radiation.

- **Discuss the roles of solar energy and the Coriolis effect in producing atmospheric circulation.**

Variations in the amount of solar energy reaching different places on Earth largely drive atmospheric circulation. Atmospheric heat transfer from the equator to the poles produces a movement of warm air toward the poles and a movement of cool air toward the equator, moderating the climate. The atmosphere also exhibits **winds**, complex horizontal movements that result in part from differences in atmospheric pressure and from the Coriolis effect. The **Coriolis effect** is the influence of Earth's rotation, which tends to deflect fluids (air and water) toward the right in the Northern Hemisphere and toward the left in the Southern Hemisphere.

- **Discuss the roles of solar energy and the Coriolis effect in producing global water flow patterns, including gyres.**

Surface-ocean **currents** result largely from prevailing winds, which, in turn, are generated from solar energy. Other factors that contribute to ocean currents include the Coriolis effect, the position of landmasses, and the varying **density** of water. **Gyres** are large, circular ocean current systems that often encompass an entire ocean basin. Deep-ocean currents often travel in different directions and at different speeds than do surface currents. The present circulation of shallow and deep currents, known informally as the **ocean conveyor belt**, affects regional and possibly global climate.

- **Define _El Niño–Southern Oscillation (ENSO)_ and _La Niña_, and describe some of their effects.**

El Niño–Southern Oscillation (**ENSO**) is a periodic, large-scale warming of surface water of the tropical eastern Pacific Ocean that affects both ocean and atmospheric circulation patterns. ENSO results in unusual weather in areas far from the tropical Pacific. During a **La Niña** event, surface water in the eastern Pacific becomes unusually cool.

- **Distinguish between weather and climate, and briefly discuss regional precipitation differences.**

Weather refers to the conditions in the atmosphere at a given place and time, whereas **climate** refers to the typical patterns of weather conditions that occur in a place over a period of years. Temperature and precipitation largely determine an area's climate. Precipitation is greatest where warm air passes over the ocean, absorbing moisture, and is then cooled, such as when mountains force humid air upward.

- **Contrast tornadoes and tropical cyclones.**

A **tornado** is a powerful, rotating funnel of air associated with severe thunderstorms. A **tropical cyclone** is a giant, rotating tropical storm with high winds. Tropical cyclones are called hurricanes in the Atlantic, typhoons in the Pacific, and cyclones in the Indian Ocean.

- **Define _plate tectonics_ and explain its relationship to earthquakes and volcanic eruptions.**

Plate tectonics is the study of the processes by which the lithospheric plates move over the asthenosphere. Earth's lithosphere (outermost rock layer) consists of seven large plates and a few smaller ones. As the plates move horizontally, the continents change their relative positions. **Plate boundaries** are sites of intense geologic activity, such as mountain building, volcanoes, and earthquakes.

Critical Thinking and Review Questions

1. Briefly describe some of the long-term ecological research conducted at Hubbard Brook Experimental Forest (HBEF). What are some of the environmental effects observed in the deforestation study at HBEF?

2. What is a biogeochemical cycle? Why is the cycling of matter essential to the continuance of life?

3. How might deforestation at HBEF alter the biogeochemical cycles involving that ecosystem?

4. Describe how organisms participate in each of these biogeochemical cycles: carbon, nitrogen, phosphorus, and sulfur.

5. What is the basic flow path of the nitrogen cycle?

6. A geologist or physical geographer would describe the phosphorus cycle as a *sedimentary pathway*. What does that mean?

7. Explain why Earth's temperature changes with latitude and with the seasons. Why is it unlikely that the winter Olympics—typically held in February—would take place in the mountains of New Zealand?

8. What are the two lower layers of the atmosphere? Cite at least two differences between them.

9. Describe the general directions of atmospheric circulation.

10. What is a gyre, and how are gyres produced?

11. How does ENSO affect climate on land?

12. What are some of the environmental factors that produce areas of precipitation extremes, such as rain forests and deserts?

13. The system encompassing Earth's global mean surface temperature can be diagrammed as follows: Explain each part of this system.

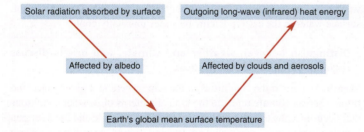

14. How are tornadoes and tropical cyclones alike? How do they differ?

15. Relate the locations of earthquakes and volcanoes to plate tectonics.

16. Examine the following changes that have been identified in the Arctic hydrologic system in the past few decades. Predict the effect of these changes on the salinity in the North Atlantic Ocean.

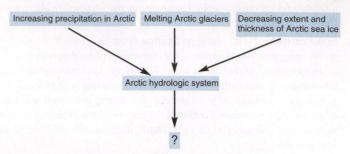

17. What ENSO effects are made fun of in the cartoon below?

By cartoonist Bob Bierman, Copyright Simon Fraser University

Food for Thought

You prepare a small community garden plot and decide to plant peas, tomatoes, and green peppers. How will your green-thumb efforts contribute to any of the biogeochemical cycles, and how will those cycles involve your vegetable patch? Will your choice of plants determine the role your garden plays in any of the cycles?

5

Ecosystems and Living Organisms

Gray Wolf (*Canus lupus*). This Yellowstone wolf, photographed near Canyon Village in 2008, is from the Molly Pack.

Gray wolves originally ranged across North America from northern Mexico to Greenland, but they were trapped, poisoned, and hunted to extinction in most places. By 1960, the only wolves remaining in the lower 48 states were small populations in Minnesota.

Under the provisions of the Endangered Species Act (ESA), wolves were listed as endangered in 1974. Many scientists recommended reintroducing wolves, but this proposal was not acted on for more than two decades. Beginning in 1995, the U.S. Fish and Wildlife Service (FWS) captured a small number of gray wolves in Canada and released them into Yellowstone National Park in Wyoming. The population thrived and has increased to several hundred individuals in the Yellowstone area (see photograph).

Biologists continue to research the effects of the wolf reintroduction on the Yellowstone ecosystem. Wolves prey on elk and occasionally on mule deer, moose, and bison. In some areas of Yellowstone, hunting by wolf packs has helped reduce Yellowstone's elk population, which was at an all-time high before the wolves returned. When elk numbers are not managed properly, they overgraze their habitat, and thousands starve during hard winters.

The reduction and redistribution of Yellowstone's elk population have relieved heavy grazing pressure on aspen, willow, cottonwood, and other plants, particularly at sites where elk cannot see approaching wolves or escape from them rapidly. As a result of a more lush and varied plant composition, herbivores such as beavers and snow hares have increased in number, which in turn supports small predators such as foxes, badgers, and martens. Trout have also increased, as the more extensive willow and cottonwood help protect stream water quality and streambank integrity.

Wolf packs have reduced coyote populations, which has allowed populations of the coyotes' prey, such as ground squirrels, chipmunks, and pronghorn (coyotes prey on the fawns), to increase. Scavengers such as ravens, magpies, bald eagles, wolverines, and bears (both grizzlies and black bears) have benefited from dining on scraps from wolf kills. Researchers will continue to sort out what happens when a top predator such as the wolf is returned to its natural environment.

The reintroduction of wolves remains controversial. Nearby ranchers were against the reintroduction because their livelihood depends on livestock being safe from predators. In response to this concern, ranchers are allowed to kill wolves that attack cattle and sheep, and federal officers can remove any wolf that threatens humans or livestock. The Defenders of Wildlife, an environmental organization, is working with livestock owners to develop nonlethal ways to reduce the chances of wolves attacking livestock.

Today there are more than 5,500 wolves in the continental United States, and the gray wolf has been delisted from federal protection under the ESA. States such as Montana and Idaho now have management plants to sustain their wolf populations.

Regardless of whether an ecosystem is as large as Yellowstone National Park or as small as a roadside drainage ditch, its organisms continually interact with one another and adapt to changes in the environment. This chapter is concerned with making sense of community structure and diversity in a wide variety of populations and communities.

In Your Own Backyard
Search online and find the two largest predators in your state. Is either of them endangered or threatened? Is either of them a threat to livestock or agriculture?

Evolution: How Populations Change over Time

- Define *evolution*.
- Explain the four premises of evolution by natural selection, as proposed by Charles Darwin.
- Identify the major domains and kingdoms of living organisms.

Where did the many species of plants, animals, fungi, and microorganisms in Yellowstone come from? They, like all species living today, are thought to have descended from earlier species by the

evolution Cumulative genetic changes that occur over time in a population of organisms; evolution explains many patterns of distribution and abundance displayed in the natural world.

process of **evolution**. The concept of evolution dates back to the time of Aristotle (384–322 B.C.E), but **Charles Darwin**, a nineteenth-century naturalist, proposed *natural selection*, the mechanism of evolution that the scientific community still accepts today. As you will see, the environment plays a crucial role in Darwin's theory of natural selection as a mechanism of evolution.

Darwin proposed that, from one generation to the next, inherited traits favorable to survival in a given environment would be preserved, whereas unfavorable ones would be eliminated. The result would be **adaptation**, evolutionary modification of a population that improves each individuals' chances of survival and reproductive success in the environment occupied by the population.

NATURAL SELECTION

Darwin proposed the theory of evolution by **natural selection** in his monumental book *The Origin of Species by Means of Natu-*

natural selection The process in which better-adapted individuals—those with a combination of genetic traits better suited to environmental conditions—are more likely to survive and reproduce, increasing their proportion in the population.

ral Selection, published in 1859. Since then, scientists have accumulated an enormous body of observations and experiments that support Darwin's theory. Although biologists still do not agree completely on some aspects of how evolutionary changes occur, the concept of evolution by natural selection is well documented.

As a result of natural selection, the population changes over time; the frequency of favorable traits increases in successive generations, and unfavorable traits decrease or disappear. Evolution by natural selection results from four natural conditions: high reproductive capacity, heritable variation, limits on population growth, and differential reproductive success.

1. **High reproductive capacity.** Each species produces more offspring than will survive to maturity.

2. **Heritable variation.** The individuals in a population exhibit variation. Some traits improve the chances of an individual's survival (a camouflage coat, for example) and reproductive success. The variation necessary for evolution by natural selection must be inherited so that it can be passed to offspring.

3. **Limits on population growth, or a struggle for existence.** Only so much food, water, light, growing space, and so on are available to a population, and organisms compete with one

another for the limited resources available to them. Other limits on population growth include predators and diseases.

4. **Differential reproductive success.** Reproduction is the key to natural selection: The best-adapted individuals reproduce most successfully, whereas less fit individuals die prematurely or produce fewer or less viable offspring. In some cases, enough changes may accumulate over time in geographically separated populations (often with slightly different environments) to produce new species.

Charles Darwin formulated his ideas on natural selection when he was a ship's naturalist on a five-year voyage around the world. During an extended stay in the Galápagos Islands off the coast of Ecuador, he studied the plants and animals of each island, including 14 species of finches. Each finch species was specialized for a particular lifestyle different from those of the other species and different from those of finches on the South American mainland. The finch species, although similar in color and overall size, exhibit remarkable variation in the shape and size of their beaks, which are used to feed on a variety of foods (**Figure 5.1**). Darwin realized that the 14 species of Galápagos finches descended from a single common ancestor—one or a small population of finches that originally colonized the Galápagos from the South American mainland. Over many generations, the surviving finch populations underwent natural selection, making them better adapted to their environments, including feeding on specific food sources. The evolution of these finches continues to be actively studied by researchers.

MODERN SYNTHESIS

One premise on which Darwin based his theory of evolution by natural selection is that individuals transmit traits to the next generation. However, Darwin could not explain *how* this occurs or *why* individuals vary within a population. Beginning in the 1930s and 1940s, biologists combined the principles of genetics with Darwin's theory of natural selection. The result was a unified explanation of evolution known as the **modern synthesis**. In this context, *synthesis* refers to combining parts of several previous theories to form a unified whole. Today the modern synthesis includes our expanding knowledge in genetics, classification, fossils, developmental biology, and ecology.

The modern synthesis explains Darwin's observation of variation among offspring in terms of **mutation**, or changes in the nucleotide base sequence of a gene, or deoxyribonucleic acid (DNA) molecule. Mutations provide the genetic variability on which natural selection acts during evolution. Some new traits may be beneficial, whereas others may be harmful or have no effect at all. As a result of natural selection, beneficial strategies, or traits, persist in a population because such characteristics make the individuals that possess them well suited to thrive and reproduce. In contrast, characteristics that make the individuals that possess them poorly suited to their environment tend to disappear in a population.

The modern synthesis has dominated the thinking and research of many biologists. It has resulted in many new discoveries that validate evolution by natural selection. A vast body of evidence supports evolution, most of which is beyond the scope of this text.

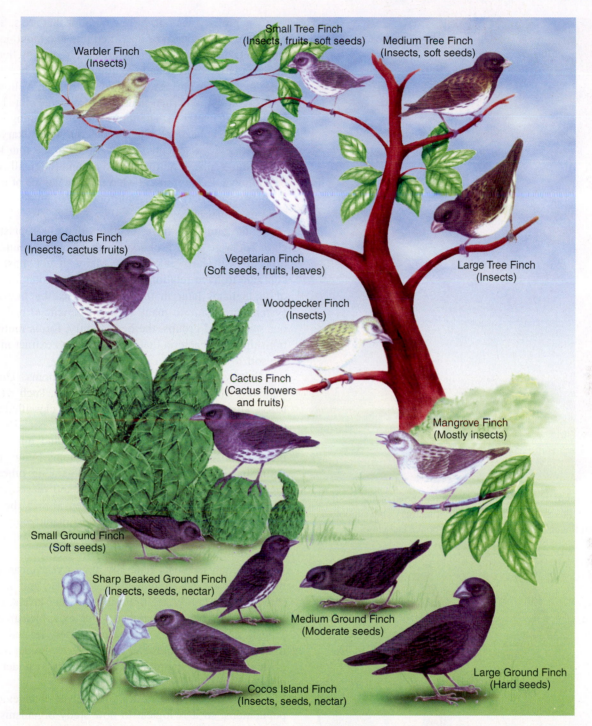

Figure 5.1 **Darwin's finches.** Note the various beak sizes and shapes, which are related to diet. Since 1973 the long-term research of Peter and Rosemary Grant and others has verified and extended Darwin's observations on the evolution of Galápagos finches by natural selection.

This evidence includes observations from biogeography (the study of the geographic locations of organisms), molecular biology, the fossil record, and underlying structural similarities (**Figure 5.2**). In addition, evolutionary hypotheses are tested experimentally.

EVOLUTION OF BIOLOGICAL DIVERSITY: THE ORGANIZATION OF LIFE

Biologists arrange organisms into logical groups to try to make sense of the remarkable diversity of life that has evolved on Earth.

For hundreds of years, biologists regarded organisms as falling into two broad categories—plants and animals. With the development of microscopes, however, it became increasingly obvious that many organisms did not fit well into either the plant kingdom or the animal kingdom. For example, bacteria have a *prokaryotic* cell structure: They lack organelles enclosed by membranes, including a nucleus. This feature, which separates bacteria from all other organisms, is far more fundamental than the differences between plants and animals, which have similar cell structures. Hence, bacteria were neither plants nor animals.

(a) **A pterosaur fossil.** Note the long fourth finger bone on the fossil of this flying reptile. Pterosaur wings consisted of leathery skin supported by the elongated finger bone. The other finger bones had claws.

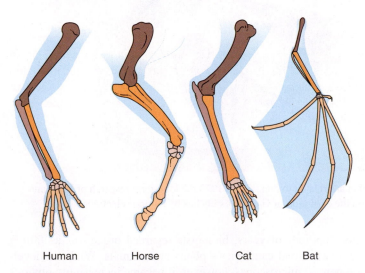

Human   Horse   Cat   Bat

(b) **Similarities among organisms demonstrate how they are related.** The basic underlying similarities among these four limbs indicate that, while proportions of bones have changed in relation to each organism's way of life, the forelimbs have the same basic structure. This structural similarity strongly suggests that these animals shared a common ancestor.

Figure 5.2   **Some evidence for evolution.**

The prokaryotes fall into two groups that are sufficiently distinct from each other to be classified into two kingdoms, **Archaea** and **Bacteria** (**Figure 5.3**). The archaea frequently live in oxygen-deficient environments and are often adapted to harsh conditions; these include hot springs (like Old Faithful, at Yellowstone), salt ponds, and hydrothermal vents (see Case in Point: Life Without the Sun, in Chapter 3). The thousands of remaining kinds of prokaryotes are collectively called bacteria. The eukaryotes, organisms with *eukaryotic* cells, are classified in the four kingdoms of living things, plants, animals, protists, and fungi, all considered to be **Eukarya**. Eukaryotic cells have a high degree of internal organization, containing nuclei, chloroplasts (in photosynthetic cells), and mitochondria.

Some biologists have theorized that chloroplasts and mitochondria were once prokaryotes that began a mutualistic relationship with early eukaryotes. Many, if not most, eukaryotes continue to have relationships with prokaryotes (see discussion of mutualisms later in the chapter). Healthy human digestive systems are home to many types of beneficial bacteria. One of our staple food groups, the bean family, forms mutualistic relationships with bacteria that help the plants extract nitrogen from the soil.

Eukarya are divided into four kingdoms. Unicellular or relatively simple multicellular eukaryotes, such as algae, protozoa, slime molds, and water molds, are classified as members of the kingdom Protista. In addition, three specialized groups of multicellular organisms—fungi, plants, and animals—evolved independently from protists. The kingdoms Fungi, Plantae, and Animalia differ from one another in—among other features—their sources of nutrition. Fungi (molds and yeasts) secrete digestive enzymes into their food and then absorb the predigested nutrients. Plants use radiant energy to manufacture food molecules by photosynthesis. Animals ingest their food, then digest it inside their bodies.

Although scientists generally have a very clear idea of when two species are closely related to each other, and of the relationships within a family, relationships at higher levels, such as kingdom, are still being researched. For example, green algae are protists similar to plants but are not closely related to other protists, such as slime molds and brown algae. As our understanding of gene evolution improves, we get closer to understanding the detailed history of life on Earth.

We have seen that evolution involves changes over time to populations, so it is logical to now study organisms at the population level. Moreover, studying populations of other species provides insights into the biological principles that affect human population changes. (The human population is considered in detail in Chapters 8 and 9.)

REVIEW

1. How do biologists define *evolution?*

2. What are Darwin's four premises of evolution by natural selection?

3. What are the major kingdoms of life? Give an example organism from three kingdoms.

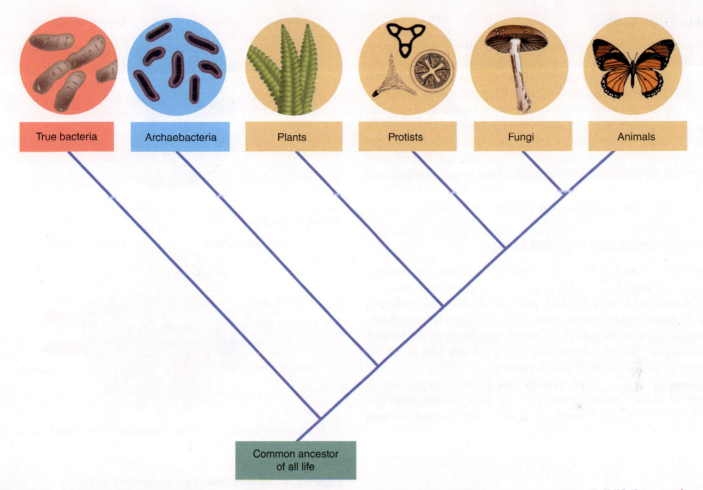

Figure 5.3 **A partial phylogeny of life forms.** To interpret this diagram, start at the bottom with the common ancestor of all life forms, and note the branch points on the way up the main (right-hand) line. The true bacteria were the first to evolve separately from other organisms, followed by the archaebacteria. Eukaryotes (plants, animals, fungi) appeared later in evolutionary history.

Principles of Population Ecology

LEARNING OBJECTIVES

- Define *population,* and explain the four factors that produce changes in population size.
- Use intrinsic rate of increase, exponential population growth, and carrying capacity to explain the differences between J-shaped and S-shaped growth curves.
- Distinguish between density-dependent and density-independent factors that affect population size, and give examples of each.
- Define *survivorship,* and describe type I, type II, and type III survivorship curves.
- Define *metapopulation,* and distinguish between source and sink habitats.

Individuals of a given species that live in the same area at the same time are part of a larger organization—a **population**. Populations exhibit characteristics distinct from those of the individuals within them. Some of the features characteristic of populations but not of individuals are population density, birth and death rates, growth rates, and age structure.

population A group of individuals of the same species that live in the same geographic area at the same time.

Population ecology is the branch of biology that deals with the numbers of a particular species found in an area and how and why those numbers change (or remain fixed) over time. Population ecologists try to determine the population processes common to all populations. They study how a population responds to its environment, such as how individuals in a population compete for food or other resources, and how predation, disease, and other environmental pressures affect the population. A population, whether of bacteria or maples or giraffes, cannot increase indefinitely because of such environmental pressures.

Additional aspects of populations important to environmental science are their reproductive success or failure (that is, extinction) and how populations affect the normal functioning of communities and ecosystems. Scientists in applied disciplines, such as forestry, agronomy (crop science), and wildlife management, must understand population ecology to effectively manage populations of economic importance, such as forest trees, field crops, game animals, and fishes. An understanding of the population dynamics of endangered species plays a key role in efforts to prevent their slide to extinction. Knowing the population dynamics of pest species helps in efforts to prevent their increase to levels that can damage crops.

POPULATION DENSITY

By itself, the size of a population tells us relatively little. Population size is meaningful only when the boundaries of the population are defined. Consider the difference between 1000 mice in 100 hectares (250 acres) and 1000 mice in 1 kitchen. Often, a population is too large to study in its entirety. Such a population is examined by sampling a part of it and then expressing the population in terms of density. Examples include the number of dandelions per square meter of lawn, the number of water fleas per liter of pond water, and the number of cabbage aphids per square centimeter of cabbage leaf. **Population density,** then, is the number of individuals of a species per unit of area or volume at a given time.

HOW DO POPULATIONS CHANGE IN SIZE?

Populations, whether they are sunflowers, eagles, or humans, change over time. On a global scale, this change is due to two factors: the rate at which individuals produce offspring (the birth rate) and the rate at which organisms die (the death rate) (**Figure 5.4a**). In humans, the **birth rate** (b) is usually expressed as the number of births per 1000 people per year, and the **death rate** (d) as the number of deaths per 1000 people per year.

growth rate (r) The rate of change of a population's size, expressed in percent per year.

The **growth rate** (r) of a population is equal to the birth rate (b) minus the death rate (d). Growth rate is also called **natural increase** in human populations.

$$r = b - d$$

As an example, consider a hypothetical human population of 10,000 in which there are 200 births per year (that is, by convention, 20 births per 1000 people) and 100 deaths per year (that is, 10 deaths per 1000 people).

$$r = \underbrace{20/1000}_{b} - \underbrace{10/1000}_{d}$$

$$r = 0.02 - 0.01 = 0.01, \text{ or } 1\% \text{ per year}$$

If individuals in the population are born faster than they die, r is a positive value, and population size increases. If individuals die faster than they are born, r is a negative value, and population size decreases. If r is equal to zero, births and deaths match, and population size is stationary despite continued reproduction and death.

In addition to birth and death rates, **dispersal**, or movement from one region or country to another, is considered when changes in populations on a local scale are examined. There are two types of dispersal: **immigration** (i), in which individuals enter a population and increase its size, and **emigration** (e), in which individuals leave a population and decrease its size. The growth rate of a local population must take into account birth rate (b), death rate (d), immigration (i), and emigration (e) (**Figure 5.4b**). The growth rate equals (the birth rate minus the death rate) plus (immigration minus emigration):

$$r = (b - d) + (i - e)$$

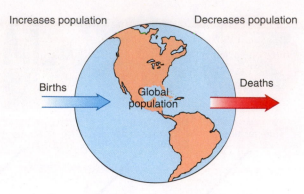

(a) On a global scale, the change in a population is due to the number of births and deaths.

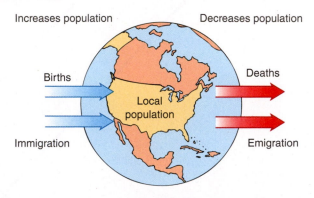

(b) In local populations, such as the population of the United States, the number of births, deaths, immigrants, and emigrants affects population size.

Figure 5.4   Factors that affect population size.

For example, the growth rate of a population of 10,000 that has 100 births (by convention, 10 per 1000), 50 deaths (5 per 1000), 10 immigrants (1 per 1000), and 100 emigrants (10 per 1000) in a given year is calculated as follows:

$$r = \underbrace{(10/1000}_{b} - \underbrace{5/1000)}_{d} + \underbrace{(1/1000}_{i} - \underbrace{10/1000)}_{e}$$

$$r = (0.010 - 0.005) + (0.001 - 0.010)$$

$$r = 0.005 - 0.009 = -0.004, \text{ or } -0.4\% \text{ per year}$$

Is this population increasing or decreasing in size?

MAXIMUM POPULATION GROWTH

The maximum rate that a population could increase under ideal conditions is its **intrinsic rate of increase** (also called *biotic potential*). Different species have different intrinsic rates of increase. Several factors influence a particular species' intrinsic rate of increase. These include the age at which reproduction begins, the fraction of the life span during which an individual can reproduce, the number of reproductive periods per lifetime, and the number of offspring produced during each period of reproduction. These factors, called *life*

intrinsic rate of increase The exponential growth of a population that occurs when resources are not limited.

history characteristics, determine whether a particular species has a large or a small intrinsic rate of increase.

Generally, larger organisms, such as blue whales and elephants, have the smallest intrinsic rates of increase, whereas microorganisms have the greatest intrinsic rates of increase. Under ideal conditions (that is, an environment with unlimited resources), certain bacteria reproduce by dividing in half every 30 minutes. At this rate of growth, a single bacterium would increase to a population of more than 1 million in just 10 hours (**Figure 5.5a**), and the population from a single individual would exceed 1 billion in 15 hours! If you plot the population number versus time, the graph has a J shape characteristic of **exponential population growth** (**Figure 5.5b**). When a population grows exponentially, the larger the population gets, the faster it grows.

> **exponential population growth** The accelerating population growth that occurs when optimal conditions allow a constant reproductive rate over a period of time.

An everyday example of exponential growth is a savings account in which a fixed percentage of interest—say, 2%—is accumulating (i.e., compounding). Assuming you do not deposit or withdraw any money, the amount that your money grows starts increasing slowly and proceeds faster and faster over time, as the balance increases. Exponential population growth works the same way. A small population that is growing exponentially increases at a slow rate initially, but growth proceeds faster and faster as the population increases. Ecologists use the equation at the end of the chapter to predict population levels and growth rates for particular populations.

Regardless of the species, whenever the population is growing at its intrinsic rate of increase, population size plotted versus time gives a curve of the same shape. The only variable is time. A dolphin population will take longer than a bacterial population to reach a certain size (because dolphins do not reproduce as rapidly as bacteria), but both populations will always increase exponentially as long as their growth rates remain constant.

ENVIRONMENTAL RESISTANCE AND CARRYING CAPACITY

Certain populations may exhibit exponential population growth for a short period. Exponential population growth has been experimentally demonstrated in bacterial and protist cultures and in certain insects. However, organisms cannot reproduce indefinitely at their intrinsic rates of increase because the environment sets limits, collectively called **environmental resistance**. Environmental resistance includes such unfavorable environmental conditions as the limited availability of food, water, shelter, and other essential resources (resulting in increased competition), as well as limits imposed by disease and predation.

Using the earlier example, we find that bacteria would never reproduce unchecked for an indefinite period because they would run out of food and living space, and poisonous body wastes would accumulate in their vicinity. With crowding, bacteria would become more susceptible to parasites (high population densities facilitate the spread of infectious organisms such as viruses among individuals) and predators (high population densities increase the likelihood of a predator catching an individual). As the environment deteriorates, their birth rate (b) would decline and their death rate (d) would increase. The environmental conditions might worsen to a point where d would exceed b, and the population would decrease. The number of individuals in a population, then, is controlled by the ability of the environment to support it. As the number of individuals in a population increases, so does environmental resistance, which acts to limit population growth. Environmental resistance is an excellent example of a **negative feedback mechanism**, in which a change in some condition triggers a response that counteracts, or reverses, the changed condition.

Negative feedback acting on a rapidly growing population can eventually reduce the rate of population growth to nearly zero. This leveling out occurs at or near the **carrying capacity (K)**, the limit of the environment's ability to support a population. In nature, the carrying capacity is dynamic and changes in response to environmental changes, such as seasons.

> **carrying capacity (K)** The maximum number of individuals of a given species that a particular environment can support for an indefinite period, assuming there are no changes in the environment.

Time (hours)	Number of bacteria
0	1
0.5	2
1.0	4
1.5	8
2.0	16
2.5	32
3.0	64
3.5	128
4.0	256
4.5	512
5.0	1,024
5.5	2,048
6.0	4,096
6.5	8,192
7.0	16,384
7.5	32,768
8.0	65,536
8.5	131,072
9.0	262,144
9.5	524,288
10.0	1,048,576

(a) When bacteria divide at a constant rate, their numbers increase exponentially. This sample data set assumes a zero death rate, but exponential growth can still occur, albeit more slowly, even if some percentage of bacteria died.

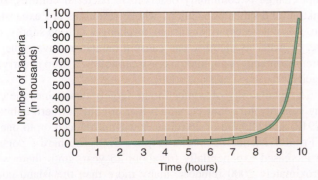

(b) When these data are graphed, the curve of exponential population growth has a characteristic J shape.

Figure 5.5   Exponential population growth.

An extended drought, for example, could decrease the amount of vegetation growing in an area, and this change, in turn, would lower the carrying capacity for deer and other herbivores.

G. F. Gause, a Russian ecologist who conducted experiments during the 1930s, grew a population of a single species, *Paramecium caudatum*, in a test tube. He supplied a limited amount of food (bacteria) daily and replenished the media occasionally to eliminate the buildup of metabolic wastes. Under these conditions, the population of *P. caudatum* increased exponentially at first, but then their growth rate declined to zero, and the population size leveled off. This characteristic population growth pattern theoretically fits most wild organisms, such as ocean fish (**Figure 5.6a**), though students of population ecology can observe many variations from this general pattern. All populations grow best when the number of individuals is high enough for ample reproductive opportunity, but small enough that shortages of resources do not impede growth.

When a population affected by environmental resistance is graphed over a long period (**Figure 5.6b**), the curve has the characteristic S shape of **logistic population growth**. The curve shows an approximate exponential increase initially (note the curve's J shape at the start, when environmental resistance is low), a peak growth rate that occurs when the population is at half the carrying capacity, followed by a leveling out as the carrying capacity of the environment is approached. In logistic population growth, the rate of population growth is proportional to the amount of existing resources, and competition leads to limited population growth. Although logistic population growth is an oversimplification of how most populations change over time, it fits some populations studied in the laboratory, as well as a few studied in nature.

Successful management of either pest species or desirable species requires an understanding of logistic population growth. For example, if we know the carrying capacity of an edible fish, we can calculate the maximum **sustainable harvest** for that species, the number of fish that can be caught each year without reducing the overall population (see Equations at chapter end). This maximum sustainable fishing level will be possible when the fish population is at half the total carrying capacity and achieves its maximum population growth rate. In contrast, in pest control, keeping a pest population well below its maximum growth rate will help minimize the need for ongoing control measures. Additionally, if we work to minimize the carrying capacity for the pest, the necessity for control will be lower. For example, a kitchen that is kept clean of food scraps and dripping water will have a lower carrying capacity for mice and cockroaches. If a city manages feral cat populations by minimizing breeding (through trap-and-neuter programs), the growth rate of the feral cat population will be slower, allowing populations to be kept low more easily. The logistic population growth equation has a wide range of uses in managing both beneficial and pest organisms.

A population rarely stabilizes at *K* (carrying capacity) as shown in Figure 5.6, but may temporarily rise higher than *K*. It will then drop back to, or below, the carrying capacity. Sometimes a population that overshoots *K* will experience a **population crash**, an abrupt decline from high to low population density. Such an

sustainable harvest The number of individuals that can be harvested from a population each year without reducing the total population.

(a) A group of crevalle jack fish.

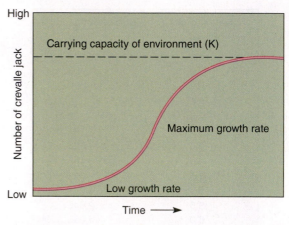

(b) Each habitat, large or small, has a carrying capacity for organisms in it. Populations typically grow slowly when their numbers are small, but growth accelerates (to the maximum growth rate) until environmental resistance (limits to food, breeding sites, habitat, etc.) force growth to slow again, leveling off as the population approaches the carrying capacity (K). Many ocean fish have been overharvested, so their populations are now at the bottom of this curve and have a low population growth rate. Sustainable fishing limits can be set at their highest when the population is at least at half of K.

Figure 5.6   **Logistic Population Growth.**

abrupt change is commonly observed in bacterial cultures, zooplankton, and other populations whose resources are exhausted. Also, *K* is not necessarily constant, as changes in food supply, climate, or water availability can cause a habitat to have richer or poorer resources. A population crash can follow a drop in carrying capacity.

The availability of winter forage largely determines the carrying capacity for reindeer, which live in cold northern habitats. In 1910, a small herd of 26 reindeer was introduced on to one of the Pribilof Islands of Alaska (**Figure 5.7**). The herd's population increased exponentially for about 25 years until there were approximately 2000 reindeer, many more than the island could support, particularly in winter. The reindeer overgrazed the vegetation until the plant life was almost wiped out. Then, in slightly over a decade, as reindeer died from starvation, the number of

(a) A baby reindeer (*Rangifer tarandus*) on one of the Pribilof Islands in the Bering Sea, off the coast of Alaska

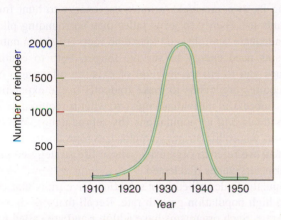

(b) Reindeer experienced rapid population growth followed by a sharp decline when an excess of reindeer damaged the environment. (After V.C. Scheffer. "The Rise and Fall of a Reindeer Herd." Science Month, Vol. 73, 1951)

Figure 5.7   A population crash.

reindeer plunged to 8, about one-third the size of the original introduced population and less than 1% of the population at its peak. Recovery of arctic and subarctic vegetation after overgrazing by reindeer takes 15 to 20 years, and during that time, the carrying capacity for reindeer is greatly reduced.

FACTORS THAT AFFECT POPULATION SIZE

Natural mechanisms that influence population size fall into two categories, density-dependent factors and density-independent factors. These two sets of factors vary in importance from one species to another, and in most cases they probably interact simultaneously to determine the size of a population.

Density-Dependent Factors Certain environmental factors such as predation have a greater influence on a population when its density is greater. For example, if a pond has a dense population of frogs, predators such as waterbirds tend to congregate and eat more of the frogs than if the pond were sparsely populated by

frogs. If a change in population density alters how an environmental factor affects the population, then the environmental factor is called a **density-dependent factor**. As population density increases, density-dependent factors tend to slow population growth by causing an increase in the death rate and/or a decrease in birth rate. Thus, as the population grows closer to K, density-dependent factors have an increasingly negative impact on the growth rate, and population growth approaches zero. Conversely, if population density declines to below K, density-dependent factors cause population growth to increase. Density-dependent factors therefore tend to keep a population at a relatively constant size near the carrying capacity of the environment. (Of course, the environment continually changes, and these changes continually affect the size of the carrying capacity.)

> **density-dependent factor** An environmental factor whose effects on a population change as population density changes.

Predation, disease, and competition are examples of density-dependent factors. As the density of a population increases, predators are more likely to find an individual of a given prey species. When population density is high, the members of a population encounter one another more frequently, and the chance of their transmitting infectious disease organisms increases. As population density increases, so does competition for resources such as living space, food, cover, water, minerals, and sunlight. The opposite effects occur when the density of a population decreases. Predators are less likely to encounter individual prey; parasites are less likely to be transmitted from one host to another; and competition among members of the population for resources such as living space and food declines.

Most studies of density dependence are conducted in laboratory settings where all density-dependent (and density-independent) factors except one are controlled experimentally. Populations in natural settings are exposed to a complex set of variables that continually change, and it is difficult to evaluate the relative effects of a single density-dependent (or density-independent) factor.

Density Dependence and Boom-or-Bust Population Cycles

Lemmings are small rodents that live in the colder parts of the Northern Hemisphere. Lemming populations undergo dramatic increases in their population, followed by crashes over fairly regular time intervals—every three to four years (**Figure 5.8**). This cyclic fluctuation in abundance is often described as a boom-or-bust cycle. Other species such as snowshoe hares and red grouse also undergo cyclic population fluctuations.

The actual causes of such population oscillations are poorly understood, but many hypotheses involve density-dependent factors. The population density of lemming predators, such as weasels, arctic foxes, and jaegers (arctic birds that eat lemmings), may increase in response to the increasing density of prey. As more predators consume the abundant prey, the prey population declines. Another possibility is that a huge prey population overwhelms the food supply. Recent studies on lemming population cycles suggest that lemming populations crash because they eat all the vegetation in the area, not because predators eat them. It appears that prey populations may or may not be controlled by predators, whereas predator populations are most likely held in check by the availability of prey.

(a) The brown lemming (*Lemmus trimucronatus*) lives in the arctic tundra.

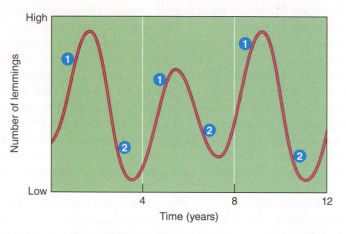

(b) This hypothetical diagram shows the cyclic nature of lemming population oscillations, which are well studied but not well understood. When the population is increasing and high (labeled 1), density-dependent factors are increasingly severe; the population growth rate is dropping, and the population will soon decline. When the population is low and declining (labeled 2), density-dependent factors are relaxed; soon, the population bottoms out and then begins to increase.

Figure 5.8   Lemming population oscillations.

Density-Independent Factors   Density-independent factors are typically abiotic. Random weather events such as a killing frost, severe blizzard, hurricane, or fire may cause extreme and irregular reductions in a population regardless of its size and thus are largely density independent.

density-independent factor An environmental factor that affects the size of a population but is not influenced by changes in population density.

Consider a density-independent factor that influences mosquito populations in arctic environments. These insects produce several generations per summer and achieve high population densities by the end of the season. A shortage of food is not a limiting factor for mosquitoes, nor is there any shortage of ponds in which to breed. Instead, winter puts a stop to the skyrocketing mosquito population. Not a single adult mosquito survives winter, and the entire population grows the following summer from the few eggs and hibernating larvae that survive. The timing and severity of winter weather is a density-independent factor that affects arctic mosquito populations.

REPRODUCTIVE STRATEGIES

Each species has a lifestyle uniquely adapted to its reproductive patterns. Many years pass before a young magnolia tree flowers and produces seeds, whereas within a single season wheat grows from seed, flowers, and dies. A mating pair of black-browed albatrosses produces a single chick every year, whereas a mating pair of gray-headed albatrosses produces a single chick biennially (every other year). Biologists try to understand the adaptive consequences of these various *life history strategies.*

Since evolution involves passing genes on to the next generation, we might think that maximizing reproduction would be the best life strategy. An organism that sexually matures early, has abundant offspring, and whose offspring, in turn, survive and reproduce early and often, might seem to be the most successful organism. All of its energy is invested into reproduction, so it would seem to pass on the most genetic material. An organism with this strategy, however, could not expend any energy toward ensuring its own survival. Animals use energy to hunt for food, and plants use energy to grow taller than surrounding plants (to obtain adequate sunlight) and to extend roots (to obtain nutrients). Organisms need energy to survive longer, both to continue to reproduce and to ensure survival of existing offspring. Nature, then, requires organisms to make tradeoffs in the expenditure of energy. Successful individuals must do what is required to survive as individuals *and* as populations (by reproducing). We refer to organisms whose life strategy emphasizes reproductive success as *r* selected; we refer to organisms whose life strategy emphasizes long-term survival as *K* selected.

Populations described by *r* **selection** have traits that contribute to a high population growth rate. Recall that *r* designates the growth rate. Such organisms have a high *r* and are called *r strategists* or *r-selected species.* Small body size, early maturity, short life span, large broods, and little or no parental care are typical of many *r* strategists, which are usually opportunists found in variable, temporary, or unpredictable environments where the probability of long-term survival is low. Some of the best examples of *r* strategists are mice, insects such as mosquitoes, and common weeds such as the dandelion.

In populations described by *K* **selection**, traits maximize the chance of surviving in an environment where the number of individuals is near the carrying capacity *(K)* of the environment. These organisms, called *K strategists* or *K-selected species,* do not produce large numbers of offspring. They characteristically have long life spans with slow development, late reproduction, large body size, and low reproductive rate. Redwood trees are classified as *K* strategists. Tawny owls, elephants, and other animals that are *K* strategists typically invest in the parental care of their young. Other *K* strategists include long-lived plants, such as the century plant or the California redwood tree. *K* strategists are found in relatively constant or stable environments, where they have a high competitive ability.

SURVIVORSHIP

Ecologists construct **life tables** for plants and animals that show the likelihood of survival for individuals at different times during their lives. Insurance companies originally developed life tables to determine how much policies should cost: Life tables show the relationship between a client's age and the likelihood the client will survive to pay enough insurance premiums to cover the cost of the policy.

Survivorship is the proportion of individuals surviving at each age in a given population. **Figure 5.9** is a graph of the three main

survivorship The probability a given individual in a population will survive to a particular age.

survivorship curves recognized by ecologists. In type I survivorship, as exemplified by humans and elephants, the young (that is, prereproductive individuals) and those at reproductive age have a high probability of living. The probability of survival decreases more rapidly with increasing age, and deaths are concentrated later in life.

In type III survivorship, the probability of death is greatest early in life, and those individuals that avoid early death subsequently have a high probability of survival. In animals, type III survivorship is characteristic of many fish species and oysters. Young oysters have three free-swimming larval stages before adulthood, when they settle down and secrete a shell. These larvae are extremely vulnerable to predation, and few survive to adulthood.

In type II survivorship, intermediate between types I and III, the probability of survival does not change with age. The probability of death is likely across all age groups, resulting in a linear decline in survivorship. This constancy probably results from essentially random events that cause death with little age bias. This relationship between age and survivorship is relatively rare; some lizards have a type II survivorship.

The three survivorship curves are generalizations, and few populations exactly fit one of the three. Some species have one type of survivorship curve early in life and another type as adults. Humans' survivorship curves depend on the public health, war, and socioeconomic conditions in a given region or country.

METAPOPULATIONS

Instead of being distributed as one large population across the landscape, many species occur as local populations in distinct habitat patches. Each local population has its own characteristic birth, death, emigration, and immigration rates. A group of local populations among which individuals occasionally *disperse* (emigrate and immigrate) is called a **metapopulation**. For example, note the three local populations that form the hemlock metapopulation on a mountain landscape in the Great Smoky Mountains (**Figure 5.10**).

metapopulation A set of local populations among which individuals are distributed in distinct habitat patches across a landscape.

The distribution of local populations across the landscape occurs because of local differences in elevation, temperature, amount of precipitation, soil moisture, and availability of soil minerals. Those habitats that increase the likelihood of survival and reproductive success for the individuals living there are called **source habitats**. Source populations generally have greater population densities than populations at less suitable sites, and surplus individuals in the source habitat may disperse to another habitat.

Lower-quality habitats, called **sink habitats**, are areas where the local birth rate is less than the local death rate. Without immigration from other areas, a sink population declines until extinction occurs. If a local population becomes extinct, individuals from a source habitat may recolonize the vacant habitat at a later time. Source and sink habitats, then, are linked to one another by immigration and emigration.

Metapopulations are becoming more common as humans alter the landscape by fragmenting existing habitats to accommodate

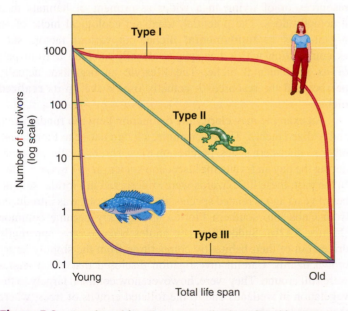

Figure 5.9 Survivorship. These generalized survivorship curves represent the ideal survivorships of species in which death rates are concentrated after reproduction (type I), spread evenly across age groups (type II), and greatest among the young (type III). Although humans, lizards, and fish are shown as examples of each, organisms do have a range of variability; some human populations have relatively high infant mortality, and their survivorship curves may be a hybrid of survivorship curve types.

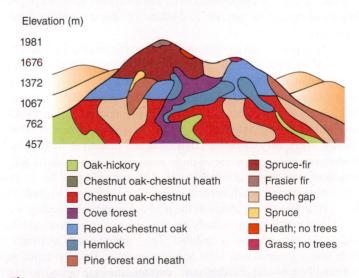

Figure 5.10 Metapopulations. Distribution of vegetation on a west-facing slope in Great Smoky Mountains National Park reveals the mosaic nature of landscapes. How might climate change affect these metapopulations? (Adapted from R.H. Whittaker. Vegetation of the Great Smoky Mountains. *Ecological Monographs*, Vol. 26, 1956)

roads, homes, factories, agricultural fields, and logging. As a result, the concept of metapopulations, particularly as it relates to endangered and threatened species, is an important area of study in conservation biology.

Review

1. What is the effect of each of the following on population size: birth rate, death rate, immigration, and emigration?

2. How do intrinsic rate of increase and carrying capacity produce the J-shaped and S-shaped population growth curves?

3. What are two examples of density-dependent factors that affect population growth? What are two examples of density-independent factors?

4. What are the three main survivorship curves?

5. How does a metapopulation differ from a local population?

Biological Communities

LEARNING OBJECTIVES

- Describe the factors that contribute to an organism's ecological niche.
- Define *competition*, and relate the concepts of competitive exclusion and resource partitioning.
- Define *symbiosis*, and distinguish among mutualism, commensalism, and parasitism.
- Define *predation*, and describe the effects of natural selection on predator-prey relationships.
- Define *keystone species*, and discuss the wolf as a keystone species.

The vast assemblage of organisms that live in populations and metapopulations are organized into communities. The term *community* has a far broader sense in ecology than in everyday speech. For the biologist, a **community** is an association of different populations of organisms that live and interact in the same place at the same time.

The organisms in a community are interdependent in a variety of ways. Species compete with one another for food, water, living space, and other resources. (Used in this context, a *resource* is anything from the environment that meets a particular species' needs.) Some organisms kill and eat other organisms. Some species form intimate associations with one another, whereas other species seem only distantly connected. As discussed in Chapter 3, each organism plays one of three main roles in community life: producer, consumer, or decomposer. The unraveling of the many positive and negative, direct and indirect interactions of organisms living as a community is one of the goals of community ecologists.

Communities vary greatly in size, lack precise boundaries, and are rarely completely isolated. They interact with and influence one another in countless ways, even when the interaction is not readily apparent. Furthermore, communities are nested within one another like Russian nesting dolls—that is, communities exist within communities. A forest is a community, but so is a rotting log in that forest. Insects, plants, and fungi invade a fallen tree as it undergoes a series of decay steps (**Figure 5.11**).

Organisms exist in an abiotic (nonliving) environment that is as essential to their existence as is their biotic (living) environment. Minerals, air, water, and sunlight are just as much a part of a honeybee's environment as the flowers it pollinates and from which it takes nectar. A biological community and its abiotic environment form an **ecosystem** (Chapter 4 considered an ecosystem's abiotic environment). Ecosystems can also include biological communities managed by humans, such as farms.

THE ECOLOGICAL NICHE

Let us consider the way of life of a given species in its community. An ecological description of a species typically includes (1) whether it is a producer, consumer, or decomposer; (2) the kinds of symbiotic associations it forms; (3) whether it is a predator and/or prey; and (4) what species it competes with. Other details are needed, however, to provide a complete picture.

Every organism is thought to have its own role, or **ecological niche**, within the structure and function of an ecosystem. An ecological niche takes into account all aspects of the organism's existence—all the physical, chemical, and biological factors an organism needs to survive, remain healthy, and reproduce. Among other things, the niche includes the local environment in which an organism lives—its **habitat**. An organism's niche also encompasses how the abiotic components of its environment, such as light, temperature, and moisture, interact with and influence it.

> **ecological niche** The totality of an organism's adaptations, its use of resources, and the lifestyle to which it is fitted.

The ecological niche of an organism may be much broader potentially than it is in actuality. Put differently, an organism is potentially capable of using much more of its environment's resources or of living in a wider assortment of habitats than it actually does. The potential, idealized ecological niche of an organism is its **fundamental niche**, but various factors such as competition with other species usually exclude it from part of its fundamental niche. The lifestyle an organism actually pursues and the resources it actually uses make up its **realized niche**.

An example may clarify the distinction between fundamental and realized niches. The green anole, a lizard native to Florida and other southeastern states, perches on trees, shrubs, walls, or fences during the day and waits for insect and spider prey (**Figure 5.12a**). In the past, these little lizards were widespread in Florida. A number of years ago, a related species, the brown anole, was introduced from Cuba into southern Florida and quickly became common (**Figure 5.12b**). Suddenly green anoles became rare, apparently driven out of their habitat by competition from the slightly larger brown lizards. Careful investigation disclosed that green anoles were still around. They were, however, now confined largely to the vegetation in wetlands and to the foliated crowns of trees, where they were less obvious.

The habitat portion of the green anole's fundamental niche includes the trunks and crowns of trees, exterior house walls, and many other locations. Where they became established, brown anoles drove green anoles out from all but wetlands and tree crowns, so the green anole's realized niche became smaller as a result of competition (**Figure 5.12c, d**). Natural communities consist

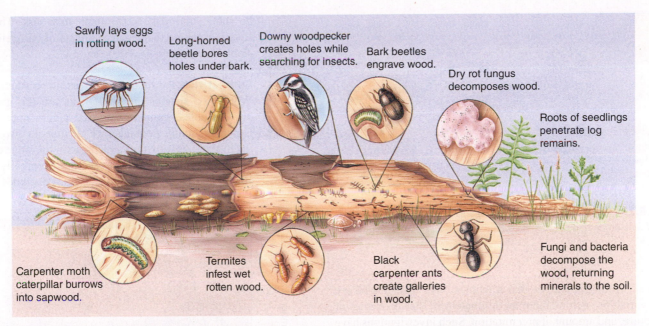

Figure 5.11   **A rotting log community.** Detritivores and decomposers consume a rotting log, which teems with a variety of bacteria, fungi, animals, and plants.

(a) The green anole (*Anolis carolinensis*) is native to Florida.

(b) The brown anole (*Anolis sagrei*) was introduced into Florida.

(c) The fundamental niches of the two lizards initially overlapped. Species 1 is the green anole, and species 2 is the brown anole.

(d) The brown anole outcompeted the green anole, restricting its niche. Note the greatly reduced area of overlap.

Figure 5.12   **Effect of competition on an organism's realized niche.**

of numerous species, many of which compete to some extent, and the interactions among species produce the realized niche of each.

A full description of an organism's niche would involve a long list of necessary and possible resources. In any habitat, only some of these resources tend to be in short supply. A fish living in a lake needs water, but water in the lake is plentiful, not usually limiting to that fish population's ability to increase. Optimal locations for spawning, however, may be limited. A necessary resource that is not abundant enough to allow easy survival or plentiful population growth is considered a **limiting resource.**

limiting resource Any environmental resource that, because it is scarce or at unfavorable levels, restricts the ecological niche of an organism.

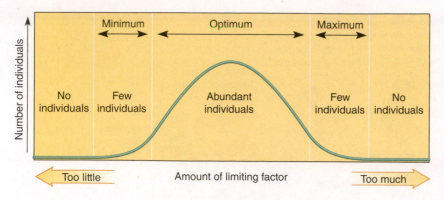

Figure 5.13   Limiting resource. An organism is limited by any environmental resource that exceeds its tolerance or is less than its required minimum.

Most limiting resources that scientists have investigated are simple variables such as the mineral content of soil, extremes of temperature, and amount of precipitation. Such investigations have disclosed that any resource that exceeds an organism's tolerance or is present in quantities smaller than the minimum required limits the occurrence of that organism in an ecosystem (**Figure 5.13**).

No organism exists independently of other organisms. We now consider the three main types of interactions that occur among species in an ecosystem: competition, symbiosis, and predation.

COMPETITION

Competition occurs when two or more individuals attempt to use an essential common resource such as food, water, shelter, living

competition The interaction among organisms that vie for the same resources (such as food or living space) in an ecosystem.

space, or sunlight. Resources are often in limited supply in the environment, and their use by one individual decreases the amount available to others. If a tree in a dense forest grows taller than surrounding trees, it absorbs more of the incoming sunlight, and less sunlight is available for nearby trees. Competition occurs among individuals within a population (**intraspecific competition**) or between species (**interspecific competition; Figure 5.14**).

Ecologists traditionally assumed that competition was the most important determinant of both the number of species found in a community and the size of each population. Today, ecologists recognize that competition is only one of many factors that affect community structure. Furthermore, competition is not always a straightforward, direct interaction. A variety of flowering plants live in a young pine forest and presumably compete with conifers for such resources as soil moisture and soil nutrient minerals. The flowers produce nectar consumed by some insect species that also prey on needle-eating insects, thereby reducing the number of insects feeding on pines. If the flowering plants were removed from the community, would the pines grow faster because they were no longer competing for necessary resources? Or would the increased presence of needle-eating insects caused by fewer omnivorous insects inhibit pine growth?

Few studies have tested the long-term effects on forest species of the removal of one competing species. Long-term effects may

be subtle, indirect, and difficult to assess. They may lower or negate the negative effects of competition for resources.

Competitive Exclusion and Resource Partitioning

When two species are similar, as are the green and brown anoles, their fundamental niches may overlap. However, many ecologists think that no two species indefinitely occupy the same niche in the same community because **competitive exclusion** eventually occurs. In competitive exclusion, one species excludes another

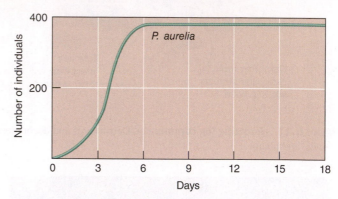

(a) A population of *P. aurelia* grown in a separate culture (in a single-species environment).

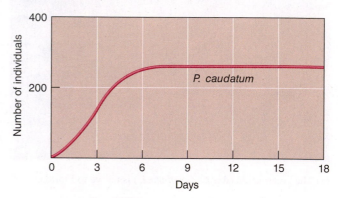

(b) A population of *P. caudatum* grown in a separate culture.

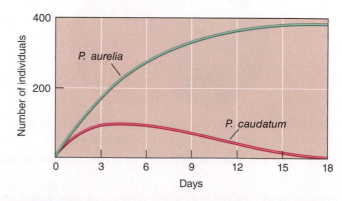

(c) Populations of both species grown in a mixed culture, in competition with each other. *P. aurelia* outcompetes *P. caudatum* and drives it to extinction. (Data for graphs adapted from G.F. Gause, *The Struggle for Existence.* Baltimore: Williams & Wilkins, 1934)

Figure 5.14   Interspecific competition.

from a portion of a niche as a result of competition between species (interspecific competition). Although it is possible for species to compete for some necessary resource without having intense, pervasive competitive interactions, two species with absolutely identical ecological niches cannot coexist. Coexistence *can* occur if the overlap in the two species' niches is reduced. In the lizard example, direct competition between the two species was reduced as the brown anole excluded the green anole from most of its former habitat until the only places open to it were wetland vegetation and tree crowns.

Competition has adverse effects on all species that use a limited resource and may result in competitive exclusion of one or more species. Natural selection should favor those individuals of each species that avoid or at least reduce competition. In **resource partitioning,** coexisting species' realized niches differ from each other in one or more ways. Evidence of resource partitioning in animals is well documented and includes studies in tropical forests of Central and South America that demonstrate little overlap in the diets of fruit-eating birds, primates, and bats that coexist in the same habitat. Although fruits are the primary food for several hundred of these species, the wide variety of fruits available has allowed fruit eaters to specialize, thereby reducing competition.

Resource partitioning may include timing of feeding, location of feeding, nest sites, and other aspects of an organism's ecological niche. **Robert MacArthur**'s study of five North American warbler species is a classic example of resource partitioning (**Figure 5.15**).

SYMBIOSIS

In **symbiosis,** individuals of one species usually live in or on the individuals of another species. At least one of the species—and sometimes both—uses its partner's resources. The partners of a symbiotic relationship, called **symbionts,** may benefit from, be unaffected by, or be harmed by the relationship. Examples of symbiosis occur across all of the domains and kingdoms of life.

symbiosis Any intimate relationship or association between members of two or more species; includes mutualism, commensalism, and parasitism.

Symbiosis is the result of **coevolution,** the interdependent evolution of two interacting species. Flowering plants and their animal pollinators have a symbiotic relationship that is an excellent example of coevolution. Plants are rooted in the ground and lack the mobility that animals have when mating. Many flowering plants rely on animals to help them reproduce. Bees, beetles, birds, bats, and other animals transport the male reproductive units, called pollen grains, from one plant to another, in effect giving plants mobility. How has this come about?

During the millions of years these associations developed, flowering plants evolved in several ways to attract animal pollinators. One of the rewards for the pollinator is food—nectar (a sugary solution) and pollen. Plants often produce food precisely adapted for one type of pollinator. The nectar of bee-pollinated flowers usually contains 30% to 35% sugar, the concentration bees need to make honey. Bees will not visit flowers with lower sugar concentrations in their nectar. Bees also use pollen to make bee-bread, a nutritious mixture of nectar and pollen that bees feed to their larvae.

Plants possess a variety of ways to get the pollinator's attention, most involving colors and scents. Showy petals visually attract the pollinator, much as a neon sign or golden arches attract a hungry person to a restaurant. Insects have a well-developed sense of smell, and many insect-pollinated flowers have strong scents that are also pleasant to humans. A few specialized kinds of flowers have unpleasant odors. The carrion plant produces flowers that mimic the smell of rotting flesh. Carrion flies are attracted to these flowers while they are looking for a place to deposit their eggs, and in the process pollen is transferred.

During the time plants were evolving specialized features to attract pollinators, the animal pollinators coevolved specialized body parts and behaviors to aid pollination and to obtain nectar and pollen as a reward. Coevolution is responsible for the hairy bodies of bumblebees, which catch and hold the sticky pollen for transport from one flower to another. The long, curved beaks of certain birds, which fit into tubular, nectar-bearing flowers, are the result of coevolution (**Figure 5.16**).

(a) Yellow-rumped warbler   (b) Bay-breasted warbler   (c) Cape May warbler   (d) Black-throated green warbler   (e) Blackburnian warbler

Figure 5.15 **Resource partitioning.** In Robert MacArthur's study of five North American warbler species, the warblers' niches initially appeared identical. However, MacArthur determined that individuals of each species spend most of their feeding time in different parts (marked in brown) of the conifer trees they frequent. They also travel differently in the tree canopy, consume different combinations of insects, and nest at slightly different times. (Adapted from R.H. MacArthur, "Population Ecology of Some Warblers of Northeastern Coniferous Forests." *Ecology*, Vol. 39, 1958)

Gallo Images - Peter Chadwick / Getty Images

Figure 5.16   Pollinator coevolution. The southern double-collared Sunbird (*Cinnyrus chellybeus*) has a curved, thin beak, which perfectly fits into the aloe flower. The aloe flower's red color attracts the sunbird.

The thousands, or even millions, of symbiotic associations that result from coevolution fall into three categories: mutualism, commensalism, and parasitism.

Mutualism   In **mutualism**, different species living in close association provide benefits to each other. The interdependent association between nitrogen-fixing bacteria of the genus *Rhizobium* and legumes (plants such as peas, beans, and clover) is an example of mutualism. Nitrogen-fixing bacteria live in nodules in the roots of legumes and supply the plants with all the nitrogen they need. The legumes supply sugar to their bacterial symbionts.

> **mutualism** A symbiotic relationship in which both partners benefit.

Another example of mutualism is the association between reef-building coral animals and microscopic algae. These symbiotic algae, called **zooxanthellae** (pronounced *zoh-zan-thel′ee*), live inside cells of the coral, where they photosynthesize and provide the animal with carbon and nitrogen compounds as well as oxygen. Zooxanthellae have a stimulatory effect on the growth of corals, causing calcium carbonate skeletons to form around their bodies much faster when the algae are present. The corals, in turn, supply their zooxanthellae with waste products such as ammonia, which the algae use to make nitrogen compounds for both partners.

Mycorrhizae (pronounced *my-kor-rye′ zee*) are mutualistic associations between fungi and the roots of about 80% of all plants. The fungus, which grows around and into the root as well as into the surrounding soil, absorbs essential minerals, especially phosphorus, from the soil and provides them to the plant (**Figure 5.17**). In return, the plant provides the fungus with food produced by photosynthesis. Plants grow more vigorously in mycorrhizal relationships and better tolerate *environmental stressors* such as drought and high soil temperatures. Indeed, some plants cannot maintain themselves under natural conditions if the fungi with which they normally form mycorrhizae are not present.

Mycorrhizae are difficult to raise in experimental settings, and most crops lack mycorrhizal associations, putting them at a disadvantage compared with some weeds. Much research remains to be done on how to foster mycorrhizal associations with crop plants; this research could help reduce the need for supplemental fertilizer.

Commensalism   **Commensalism** is an association between two different species in which one benefits and the other is unaffected. One example of commensalism is the relationship between social insects and scavengers that live with the social insects. For example, certain kinds of silverfish move along in permanent association with the marching columns of army ants and share the plentiful food caught in their raids. The army ants derive no apparent benefit or harm from the silverfish.

> **Commensalism** A type of symbiosis in which one organism benefits and the other one is neither harmed nor helped.

Dr. Jeremy Burgess / Science Source

Figure 5.17   Mutualism. Mycorrhizae (*Scleroderma geaster*) densely surround the root of a eucalyptus tree. Eucalyptus is a highly drought-tolerant tree, perhaps because its mycorrhizae help buffer it against environmental stress; mycorrhizae obtain food from the tree.

Another example of commensalism is the relationship between a tropical tree and many **epiphytes**, smaller plants, such as mosses, orchids, bromeliads, and ferns, that live attached to the bark of the tree's branches (**Figure 5.18**). The epiphyte anchors itself to the tree but does not obtain nutrients or water directly from the tree. Its location on the tree lets it obtain adequate light, water (as rainfall dripping down the branches), and required nutrient minerals (washed out of the tree's leaves by rainfall). Thus, the epiphyte benefits from the association, whereas the tree is apparently unaffected. (Epiphytes harm their host if they block sunlight from the host's leaves. When this occurs, the relationship is not considered commensalism.)

Parasitism   In **parasitism**, one organism, the *parasite*, obtains nourishment from another organism, its *host*. Although a parasite may weaken its host, it rarely kills it. Some parasites, such as ticks, live outside the host's body (**Figure 5.19**). Other parasites, such as tapeworms, live within the host. Parasitism is a successful lifestyle: More than 100 parasites live in or on the human species alone.

parasitism A symbiotic relationship in which one organism benefits and the other is adversely affected.

Many parasites do not cause disease. A parasite that causes disease and sometimes the death of a host is known as a **pathogen**. Crown gall disease, caused by a bacterium, occurs in many kinds of plants and results in millions of dollars' worth of damage each year to ornamental and agricultural plants. Crown gall bacteria, which live on detritus (organic debris) in the soil, enter plants through small wounds such as those caused by insects. They cause galls, or tumorlike growths, often at a plant's crown (between the stem and the roots, at or near the soil surface). Although plants seldom die from crown gall disease, they are weakened, grow more slowly, and often succumb to other pathogens.

PREDATION

Predators kill and feed on other organisms. **Predation** includes both animals eating other animals (carnivore-herbivore interactions) and animals eating plants (herbivore-producer interactions). Predation has resulted in the coevolution of predator strategies—more efficient ways to catch prey—and prey strategies—better ways to escape the predator. An efficient predator exerts a strong selective force on its prey, and over time, the prey species may evolve some sort of countermeasure that reduces the probability of being captured. The countermeasure that the prey acquires, in turn, may act as a strong selective force on the predator.

predation The consumption of one species (the prey) by another (the predator).

Adaptations related to predator–prey interactions include predator strategies (pursuit and ambush) and prey strategies (plant defenses and animal defenses). Keep in mind as you read these descriptions that such strategies are not "chosen" by the respective predators or prey. New traits arise randomly in a population as a result of mutation, and the traits may persist under natural selection.

Pursuit and Ambush   A gecko sights a spider and pounces on it. Orcas (killer whales), which hunt in packs, often herd salmon or tuna into a cove so that they are easier to catch (**Figure 5.20**). Any

Figure 5.18   **Commensalism.** Epiphytes are small plants that grow attached to trunks or branches of trees. This orchid (*Phalaenopsis sumatrana*) grows on a tree in a montane rain forest in Malaysia.

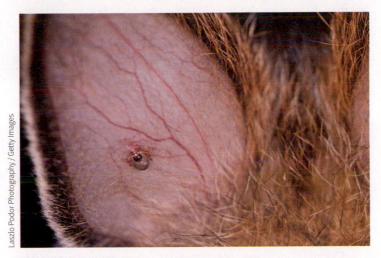

Figure 5.19   **Parasitism.** Tick in a rabbit's ear.

Figure 5.20   **Predation.** Orcas hunting in a pack in Alaska. This cooperative behavior increases hunting success.

trait that increases hunting efficiency, such as the speed of a gecko or the intelligence of orcas, favors predators that pursue their prey. Because these carnivores must process information quickly during the pursuit of prey, their brains are generally larger, relative to body size, than those of the prey they pursue.

Ambush is another effective way to catch prey. The goldenrod spider is the same color as the white or yellow flowers in which it hides. This camouflage prevents unwary insects that visit the flower for nectar from noticing the spider until it is too late. Predators that *attract* prey are particularly effective at ambushing. For example, a diverse group of deep-sea fishes called anglerfish possess rodlike luminescent lures close to their mouths to attract prey.

Plant Defenses against Herbivores   Plants cannot escape predators by fleeing, but they possess adaptations that protect them from being eaten. The presence of spines, thorns, tough leathery leaves, or even thick wax on leaves discourages foraging herbivores from grazing. Other plants produce an array of protective chemicals that are unpalatable or even toxic to herbivores. The active ingredients in such plants as marijuana, opium poppy, tobacco, and peyote cactus may discourage the foraging of herbivores. For example, the nicotine found in tobacco is so effective at killing insects that it is a common ingredient in many commercial insecticides. Milkweeds produce deadly alkaloids and cardiac glycosides, chemicals poisonous to all animals except for a small group of insects.

Defensive Adaptations of Animals   Many animals, such as woodchucks, flee from predators by running to their underground burrows. Others have mechanical defenses, such as the quills of a porcupine and the shell of a pond turtle. Some animals live in groups—a herd of antelope, colony of honeybees, school of anchovies, or flock of pigeons. This social behavior decreases the likelihood of a predator catching one of them unaware; the group has many eyes, ears, and noses watching, listening, and smelling for predators.

Chemical defenses are common among animal prey. The South American poison arrow frog has poison glands in its skin. Its bright **warning coloration** prompts avoidance by experienced predators. Snakes or other animals that have tried to eat a poisonous frog do not repeat their mistake! Some insects have evolved an ability to tolerate milkweed toxins. As a result, they can eat milkweeds without being poisoned, and they accumulate the toxins in their tissues, making themselves toxic to predators (**Figure 5.21**). These insects avoid competition from other herbivorous insects because few other insects can tolerate milkweed toxins. Predators learn to avoid these insects, which are often brightly colored in combinations of red, orange, yellow, white, and black. Not coincidentally, humans use yellow, red, and black in warning signs; we intuitively understand these colors to mean: "Danger." Note that the defense a plant employs against herbivores is actually transformed into a defense herbivores use against predators.

Some animals blend into their surroundings to hide from predators. The animal's behavior often enhances such **cryptic coloration**. The pygmy sea horse resembles gorgonian coral so

Darlyne A. Murawski/National Geographic Creative

Figure 5.21   **Warning coloration.** A milkweed tussock moth caterpillar, toxic from cardiac glycosides in the milkweed it eats, avoids predation by advertising its toxicity with classic warning coloration: orange, black, and white.

closely that the little sea horse was not discovered until 1970, when a zoologist found it in the coral he had placed in an aquarium (**Figure 5.22**). Evolution has preserved and accentuated such camouflage.

KEYSTONE SPECIES

Certain species are more crucial to the maintenance of their community than others. Such species, called **keystone species**, are vital in determining the nature and structure of the entire ecosystem—that is, its species composition and its ecosystem functioning. The fact that other species depend on or are greatly affected by the keystone species is revealed when the keystone species is removed. Keystone species are usually not the most abundant species in the ecosystem.

> **keystone species** A species, often a predator, that exerts a profound influence on a community in excess of that expected by its relative abundance.

Identifying and protecting keystone species are crucial goals of conservation biologists because if a keystone species disappears from an ecosystem, many other organisms in that ecosystem may become more common, rare, or even disappear. One example of a keystone species is a top predator such as the gray wolf, which was discussed in the chapter introduction. Where wolves were hunted to extinction, the populations of deer, elk, and other herbivores increased explosively. As these herbivores overgrazed the vegetation, many plant species that could not tolerate such grazing pressure disappeared. Many smaller animals such as insects were lost from the ecosystem because the plants that they depended on for food were now less abundant. Thus, the disappearance of the wolf resulted in an ecosystem with considerably less biological diversity.

To date, few long-term studies have identified keystone species and determined the nature and magnitude of their effects on the ecosystems they inhabit. Additional studies are urgently

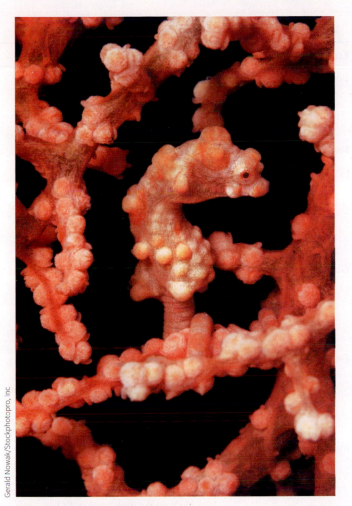

Gerald Nowak/Stockphotopro, Inc

Figure 5.22 **Cryptic coloration.** Pygmy sea horses (*Hippocampus bargibanti*), about the size of small fingernails, are virtually invisible in and around gorgonian coral (*Muricella*).

needed to provide concrete information about the importance of keystone species in conservation biology.

REVIEW

1. What is an ecological niche?
2. What is the principle of competitive exclusion? Of resource partitioning?
3. What is symbiosis? What are the three kinds of symbiosis?
4. Describe how evolution has affected predator–prey relationships.
5. What is a keystone species? Why do we consider the wolf a keystone species?

Species Richness in a Community

LEARNING OBJECTIVES
- Describe factors associated with high species richness.
- Give several examples of ecosystem services.

Species richness varies greatly from one community to another. Tropical rain forests and coral reefs are examples of communities with extremely high species richness. In contrast, geographically isolated islands and mountaintops exhibit low species richness. What determines the number of species in a community? Several factors appear to be significant: the abundance of potential ecological niches, closeness to the margins of adjacent communities, geographic isolation, dominance of one species over others, habitat stress, and geologic history.

> **species richness** The number of species in a community.

Species richness is related to the abundance of potential ecological niches. An already-complex community offers a greater variety of potential ecological niches than does a simple community. For example, in a study of chaparral habitats (shrubby and woody areas) in California, those with structurally complex vegetation provided birds with more kinds of food and hiding places than communities with low structural complexity (**Figure 5.23**).

Species richness is usually greater at the margins of adjacent communities than in their centers. This richness exists because an **ecotone**—a transitional zone where two or more communities meet—contains all or most of the ecological niches of the adjacent communities as well as some niches unique to the ecotone.

Species richness is inversely related to the geographic isolation of a community. Isolated island communities are much less diverse than communities in similar environments found on continents. This lack of diversity is partly the result of the difficulty many species have in reaching and successfully colonizing the island. Sometimes species become locally extinct as a result of random events, and in isolated environments such as islands or mountaintops, extinct species are not readily replaced. Isolated areas are often small and possess fewer potential ecological niches.

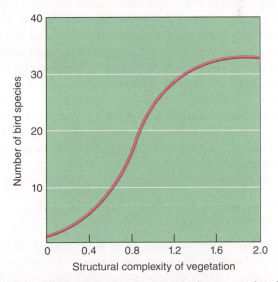

Figure 5.23 **Effect of community complexity on species richness.** As the structural complexity of chaparral vegetation in California increases, the species richness of birds in that habitat increases. (After M.L. Cody and J.M. Diamond, eds., *Ecology and Evolution of Communities.* Harvard University, Cambridge, 1975)

Species richness is reduced when any one species occupies a position of dominance within a community because it may appropriate a disproportionate share of available resources, thus crowding out other species. Ecologist **James H. Brown** and colleagues of the University of New Mexico have studied species competition and diversity in long-term experiments conducted since 1977 in the Chihuahuan Desert of southeastern Arizona. In one experiment, the scientists enclosed their study areas with fencing and then cut holes in the fencing to allow smaller rodents to come and go but to exclude the larger kangaroo rats. The removal of three dominant species, all kangaroo rats, from several plots resulted in an increased diversity of other rodent species. This increase was attributed to less competition for food and to an altered habitat because the abundance of grass species increased dramatically after the removal of the kangaroo rats.

Generally, species richness is inversely related to the environmental stress of a habitat. Only those species capable of tolerating extreme environmental conditions can live in an environmentally stressed community. The species richness of a highly polluted stream is low compared to that of a nearby pristine stream. Similarly, the species richness of high-latitude (farther from the equator) communities exposed to harsh climates is less than that of lower-latitude (closer to the equator) communities with milder climates. Although the equatorial countries of Colombia, Ecuador, and Peru occupy only 2% of Earth's land, they contain 45,000 native plant species. The continental United States and Canada, with a significantly larger land area, possess a total of 19,000 native plant species. Ecuador alone contains more than 1300 native bird species, twice as many as the United States and Canada combined.

Geologic history greatly affects species richness. Tropical rain forests are probably old, stable communities that have undergone few climate changes in Earth's entire history. During this time, myriad species evolved in tropical rain forests, having experienced few or no abrupt climate changes that might have led to their extinction. In contrast, glaciers have repeatedly modified temperate and arctic regions during Earth's history, as the climate alternately cooled and warmed. For example, an area recently vacated by glaciers will have low species richness because few species will have had a chance to enter it and become established. Scientists expect that future climate change will negatively alter species richness.

SPECIES RICHNESS, ECOSYSTEM SERVICES, AND COMMUNITY STABILITY

Ecologists and conservationists have long debated whether the extinction of species threatens the normal functioning and stability of ecosystems. This question is of great practical concern because ecosystems supply human societies with many environmental benefits (**Table 5.1**). Conservationists maintain that ecosystems with greater species richness better supply such **ecosystem services** than ecosystems with lower species richness.

ecosystem services Important environmental benefits that ecosystems provide to people; include clean air to breathe, clean water to drink, and fertile soil in which to grow crops.

Table 5.1 Ecosystem Services

Ecosystem	Services Provided by Ecosystem
Forests	• Purify air and water • Produce and maintain soil • Absorb carbon dioxide (carbon storage) • Provide wildlife habitat • Provide humans with wood and recreation
Freshwater systems (rivers and streams, lakes, and groundwater)	• Moderate water flow and mitigate floods • Dilute and remove pollutants • Provide wildlife habitat • Provide humans with drinking and irrigation water • Provide transportation corridors • Generate electricity • Offer recreation
Grasslands	• Purify air and water • Produce and maintain soil • Absorb carbon dioxide (carbon storage) • Provide wildlife habitat • Provide humans with livestock and recreation
Coasts	• Provide a buffer against storms • Dilute and remove pollutants • Provide wildlife habitat, including food and shelter for young marine species • Provide humans with food, harbors, transportation routes, and recreation
Sustainable agricultural ecosystems	• Produce and maintain soil • Absorb carbon dioxide (carbon storage) • Provide wildlife habitat for birds, insect pollinators, and soil organisms • Provide humans with food and fiber crops

Adapted from p. 527 of *Climate Change Impacts in the United States,* a report of the National Assessment Synthesis Team, U.S. Global Change Research Program, Cambridge University Press [2001].

Traditionally, most ecologists assumed that **community stability**—the absence of change—is a consequence of community complexity. Stability is the result of resistance and resilience. *Resistance* is the ability of a community to withstand environmental *disturbances*, natural or human events that disrupt a community. *Resilience* is the ability of a community to recover quickly to its former state following an environmental disturbance. A community with considerable species richness may function better—that is, have more resistance and resilience—than a community with less species richness. One reason why organic farmers typically grow a wide variety of crop species is that species richness reduces the risk of crop predation and the impact of any environmental disturbance, such as an early frost or hailstorm.

According to this view, the greater the species richness, the less critically important any single species should be. With many possible interactions within the community, any single disturbance is less likely to disrupt enough components of the system to make a significant difference in its functioning. Evidence for this hypothesis includes the fact that destructive outbreaks of pests are more common in cultivated fields, which are low-diversity communities, than in natural communities with greater species richness.

Ongoing studies by **David Tilman** of the University of Minnesota and **John Downing** of the University of Iowa have strengthened the link between species richness and community stability. In their initial study, they established and monitored 207 plots of Minnesota grasslands for seven years. During the study, Minnesota's worst drought in 50 years occurred (1987–1988). The biologists found that those plots with the greatest number of plant species lost less ground cover and recovered faster than did species-poor plots. Further research supported these conclusions and showed a similar effect of species richness on community stability during nondrought years. Similar work at grassland sites in Europe also supports the link between species richness and ecosystem functioning. Sustainable agricultural ecosystems, however, are human-made and therefore inherently different from other ecosystems. Sustainable agriculture is discussed further in Chapter 18.

CASE IN POINT

GARDENS AS ECOSYSTEMS

A suburban or urban garden is a community that supports both humans and wildlife. Of course, vegetables in a garden provide food for both people and insect pests. Spaces between garden plants offer niches where weeds (classic *r*-selected species) can grow, and while some of these weeds compete with garden plants for light, nutrients, and water, other weeds, like purslane (*Portulacca oleracea*), are themselves nutritious food plants for people. A gardener can control weeds with mulch, which makes light a limiting resource for germinating weed seeds. The gardener leaves spaces between vegetable plants to prevent excessive intraspecific competition. Flowers of many herbs and weeds offer food, in the form of pollen and nectar, for adult parasitic wasps, who then provide pest control as an ecosystem service, by laying eggs in caterpillars such as tomato hornworm (*Manduca quinquemaculata*).

While some birds, such as crows, are garden pests, bluebirds, mockingbirds, and other songbirds also help control insect pests through predation. Soil in a garden contains abiotic minerals, as well as bacteria, earthworms, nematodes, fungi, viruses, and insects—each of which may be either a pest or beneficial species from the gardener's perspective. The population density of each species depends on density-dependent factors (available nutrients and water) and density-independent factors (killing frosts). The gardener who monitors pest populations and keeps them well below their maximum growth rates, using natural predators and parasites, will lose less produce.

The gardener relies on mutualisms: between rhizobium bacteria and bean plants to help add nitrogen to the soil, and between honeybees and flowering plants to ensure good harvests of tomatoes and peppers and eggplant. The gardener may impose a type of resource partitioning by planting a wide variety of plants in the garden, from root crops and low-growing greens to bushy tomatoes and tall stalks of corn (**Figure 5.24**). This species richness also helps reduce the effects of environmental stress, such as drought, since different garden plants will demand water at different times in the growing season. ■

Figure 5.24 **A Garden Ecosystem.** A gardener with her harvest of carrots. Note the varied plant heights and the dense growth of a variety of vegetables.

Andy Ryan / Getty Images

REVIEW

1. What are two determinants of species richness? Give an example of each.

2. What are ecosystem services? Describe some ecosystem services a forest provides.

Community Development

LEARNING OBJECTIVE

• Define *ecological succession,* and distinguish between primary and secondary succession.

A community develops gradually, through a sequence of species. The process of community development over time, which involves species in one stage being replaced by different species, is called **ecological succession** or, simply, **succession**. Certain organisms that initially colonize an area are replaced over time by others, which themselves are replaced much later by still others.

The actual mechanisms that underlie succession are not clear. In some cases, an earlier species modifies the environment in some way, thereby making it more suitable for a later species to colonize. Often, succession begins with *r*-selected species, which reproduce quickly. These are later replaced, through competition, with *K*-selected species, which grow more slowly and grow larger.

Ecologists initially thought that succession inevitably led to a stable and persistent community, such as a forest, called a *climax community*. But more recently, the traditional view has fallen out of favor. The apparent stability of a "climax" forest is probably the result of how long trees live relative to the human life span. Mature "climax" communities are not in a state of permanent stability but, rather, in a state of continual disturbance. Over time, a mature forest community changes in species composition and in the relative abundance of each species, despite its overall uniform appearance.

Ecological succession is usually described in terms of the changes in the species composition of the plants growing in an area, although each stage of succession may have its own characteristic kinds of animals and other organisms. The time involved in ecological succession is on the scale of tens, hundreds, or thousands of years, not the millions of years involved in the evolutionary time scale.

PRIMARY SUCCESSION

Primary succession is ecological succession that begins in an environment that has not been inhabited before. No soil exists when primary succession begins. Bare rock surfaces, such as recently formed volcanic lava and rock scraped clean by glaciers, and sand dunes are examples of sites where primary succession may take place (**Figure 5.25**).

primary succession The change in species composition over time in a previously uninhabited environment.

Primary Succession on Bare Rock   Although the details vary from one site to another, on bare rock, lichens are often the most important element in the **pioneer** community—the initial community that develops during primary succession. Lichens secrete acids that help break the rock apart, beginning the process of soil formation. Over time, mosses and drought-resistant ferns may replace the lichen community, followed in turn by tough grasses and herbs. Once enough soil accumulates, grasses and herbs may be replaced by low shrubs, which in turn would be replaced by forest trees in several distinct stages. Primary succession on bare rock from a pioneer community to a forest community often occurs in this sequence: lichens → mosses → grasses → shrubs → trees.

Primary Succession on Sand Dunes   Some lake and ocean shores have extensive sand dunes deposited by wind and water. At first, these dunes are blown about by the wind. The sand dune environment is severe, with high temperatures during the day and low temperatures at night. The sand is also deficient in certain nutrient minerals needed by plants. As a result, few plants can tolerate the environmental conditions of a sand dune.

Henry Cowles developed the concept of succession in the 1880s when he studied succession on sand dunes around the shores

(a) After the glacier's retreat, lichens initially colonize bare gravel, followed by mosses and small shrubs.

(b) At a later time, dwarf trees and shrubs colonize the area (background).

(c)   Still later, spruces dominate the community.

Figure 5.25   Primary succession on glacial moraine. During the past 200 years, glaciers have retreated in Glacier Bay, Alaska. Although these photos were not taken in the same location, they show some of the stages of primary succession on glacial moraine (rocks, gravel, and sand deposited by a glacier). How might climate change be influencing glacial moraine ecosystems?

of Lake Michigan, which has been gradually shrinking since the last ice age. The shrinking lake exposed new sand dunes that displayed a series of stages in the colonization of the land.

Grasses are common pioneer plants on sand dunes around the Great Lakes. As the grasses extend over the surface of a dune, their roots hold it in place, helping to stabilize it. At this point, mat-forming shrubs may invade, further stabilizing the dune. Later, shrubs are replaced by poplars (cottonwoods), which years later are replaced by pines and finally by oaks. Soil fertility remains low, and as a result, other forest trees rarely replace oaks. A summary of how primary succession on sand dunes around the Great Lakes might proceed is in this sequence: grasses → shrubs → poplars (cottonwoods) → pine trees → oak trees.

SECONDARY SUCCESSION

Secondary succession is ecological succession that begins in an environment following destruction of all or part of an earlier community. Clear-cut forests, open areas caused by a forest fire, and abandoned farmland are common examples of sites where secondary succession occurs. Farmers routinely create spaces for secondary succession when they plow land to prepare for crop planting.

secondary succession The change in species composition that takes place after some disturbance destroys the existing vegetation; soil is already present.

During the summer of 1988, wildfires burned approximately one-third of Yellowstone National Park. This natural disaster provided a chance for biologists to study secondary succession in areas that were once forests. After the fires, ash covered the forest floor, and most of the trees, though standing, were charred and dead. Secondary succession in Yellowstone has occurred rapidly. Less than one year later, trout lily and other herbs had sprouted and covered much of the ground. Ten years after the fires, a young forest of knee-high to shoulder-high lodgepole pines dominated the area, and Douglas fir seedlings were also present. Biologists continue to monitor the changes in Yellowstone as secondary succession proceeds.

Old Field Succession   Biologists have extensively studied secondary succession on abandoned farmland. It takes more than 100 years for secondary succession to occur at a single site, but a single researcher can study old field succession in its entirety by observing different sites in the same area that have been undergoing succession for different amounts of time. The biologist may examine county tax records to determine when each field was abandoned.

A predictable succession of communities colonizes abandoned farmland in North Carolina (**Figure 5.26**). The first year after cultivation ceases, an annual plant, crabgrass, dominates. During the second year, horseweed, a larger plant that outgrows crabgrass, is the dominant species. Horseweed does not dominate for more than one year because decaying horseweed roots inhibit

ENVIRONEWS
Update on Honeybee Die-Offs

Bees in the United States are necessary for the pollination of about 90 commercial crops, such as nuts, fruits, and vegetables. Since 2006, many U.S. beekeeping operations have experienced 30% to 90% losses in their bee colonies. The cause of these sudden declines in the adult bee population, known as colony collapse disorder (CCD), was a mystery. No dead bees have been found in or near the colony, so performing autopsies on dead bees has been impossible. Researchers initially investigated many potential causes of CCD, from parasites to bad weather, and some evidence indicated that more than one factor could be interacting.

In 2010 entomologists reported a link between CCD and two parasites, a virus and a fungus. When the virus is present but not the fungus, some bees get sick but the colony as a whole does not collapse. Similarly, when the fungus is present but not the virus, the honeybee population does not collapse. But if the two parasites are present together in the same colony, the result is usually complete devastation. Some pesticides, used in both lawn care and agriculture—to control grubs, which eat grass roots— are also suspected of weakening honeybee navigation and immune systems. Grubs can instead be controlled with nontoxic biological control products, and minimized by cultivating lawns with clover and other low-growing broadleaf plants in combination with grasses.

the growth of young horseweed seedlings. In addition, horseweed does not compete well with other plants that become established in the third year. During the third year after the last cultivation, other weeds—broomsedge, ragweed, and aster—become established. Typically, broomsedge outcompetes aster because broomsedge is drought-tolerant, whereas aster is not.

In years 5 to 15, the dominant plants in an abandoned field are pines such as shortleaf pine and loblolly pine. Through the buildup of soil litter, such as pine needles and branches, pines produce conditions that cause the earlier dominant plants to decline in importance. Over the next century or so, pines give up their dominance to hardwoods such as oaks. The replacement of pines by oaks depends primarily on the environmental changes produced by the pines. The pine litter causes soil changes, such as an increase in water-holding capacity, which young oak seedlings need to become established. In addition, hardwood seedlings are more tolerant of shade than are young pine seedlings. Secondary succession on abandoned farmland in the southeastern United States proceeds in this sequence: crabgrass → horseweed → broomsedge and other weeds → pine trees → hardwood trees.

REVIEW

1. What is ecological succession?
2. How do primary and secondary succession differ?

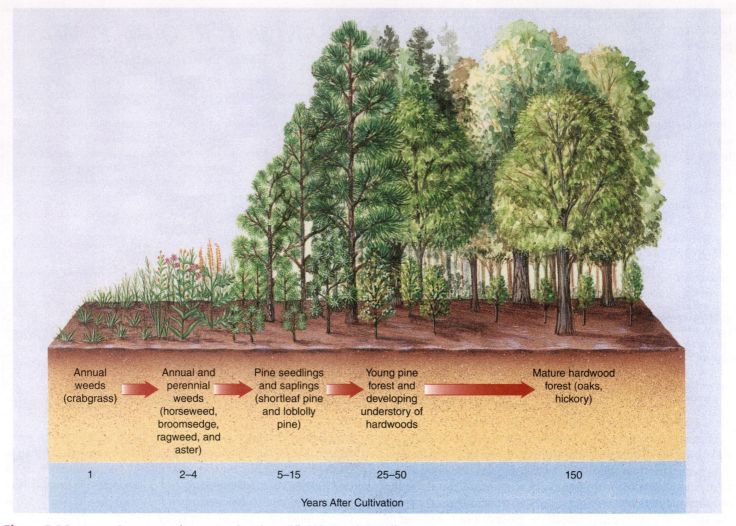

Figure 5.26 **Secondary succession on an abandoned field in North Carolina.**

REVIEW OF LEARNING OBJECTIVES WITH SELECTED KEY TERMS

• **Define** *evolution.*

Evolution is cumulative genetic changes that occur over time in a population of organisms; evolution explains many patterns observed in the natural world.

• **Explain the four premises of evolution by natural selection as proposed by Charles Darwin.**

Natural selection is the process in which better-adapted individuals—those with a combination of genetic traits better suited to environmental conditions—are more likely to survive and reproduce, increasing their proportion in the population. Natural selection, as envisioned by **Charles Darwin**, has four premises:

1. Each species produces more offspring than will survive to maturity.
2. The individuals in a population exhibit heritable variation in their traits.
3. Organisms compete with one another for the resources needed to survive.

4. Those individuals with the most favorable combination of traits are most likely to survive and reproduce, passing their genetic characters on to the next generation.

• **Identify the major groups of living organisms.**

Organisms are classified into six major groups: Archaea, Bacteria, Protista (algae, protozoa, slime molds, and water molds), Fungi (molds and yeasts), Plantae, and Animalia.

• **Define** *population,* **and explain the four factors that produce changes in population size.**

A **population** is a group of individuals of the same species that live in the same geographic area at the same time. **Growth rate** (r) is the rate of change of a population's size, expressed in percent per year. On a global scale (when **dispersal** is not a factor), growth rate (r) is due to **birth rate** (b) and **death rate** (d): $r = b - d$. **Emigration** (e), the number of individuals leaving an area, and **immigration** (i), the number of individuals entering an area, affect a local population's size and growth rate. For a local population (where dispersal is a factor), $r = (b - d) + (i - e)$.

- **Use intrinsic rate of increase, exponential population growth, and carrying capacity to explain the differences between J-shaped and S-shaped growth curves.**

Intrinsic rate of increase is the exponential growth of a population that occurs under ideal conditions. **Exponential population growth** is the accelerating population growth that occurs when optimal conditions allow a constant reproductive rate over a period of time. The **carrying capacity** (K) is the maximum number of individuals of a given species that a particular environment can support for an indefinite period, assuming there are no changes in the environment. Although populations with a constant reproductive rate exhibit exponential population growth for limited periods (the J curve), eventually the growth rate decreases to around zero or becomes negative. The S curve shows an initial lag phase (when the population is small), followed by an exponential phase, followed by a leveling phase as the carrying capacity of the environment is reached.

- **Distinguish between density-dependent and density-independent factors that affect population size, and give examples of each.**

A **density-dependent factor** is an environmental factor whose effects on a population change as population density changes. Predation, disease, and competition are examples. A **density-independent factor** is an environmental factor that affects the size of a population but is not influenced by changes in population density. Hurricanes and fires are examples.

- **Define *survivorship,* and describe type I, type II, and type III survivorship curves.**

Survivorship is the probability that a given individual in a population will survive to a particular age. There are three general types of survivorship curves. In type I survivorship, death is greatest in old age. In type III survivorship, death is greatest among the young. In type II survivorship, death is spread evenly across all age groups.

- **Define *metapopulation,* and distinguish between source and sink habitats.**

Many species exist as a **metapopulation**, a set of local populations among which individuals are distributed in distinct habitat patches across a landscape. **Source habitats** are preferred sites where local reproductive success is greater than local mortality. **Sink habitats** are lower-quality habitats where individuals may suffer death or, if they survive, poor reproductive success.

- **Describe the factors that contribute to an organism's ecological niche.**

An organism's **ecological niche** is the totality of its adaptations, its use of resources, and the lifestyle to which it is fitted. The ecological niche is the organism's place and role in a complex biotic and abiotic system.

- **Define *competition,* and relate the concepts of competitive exclusion and resource partitioning.**

Competition is the interaction among organisms that vie for the same resources (such as food or living space) in an ecosystem. Many ecologists think two species cannot occupy the same niche in the same community for an indefinite period. In **competitive exclusion**, one species excludes another as a result of competition for limited resources. Some species reduce competition by **resource partitioning**, in which they use resources differently.

- **Define *symbiosis,* and distinguish among mutualism, commensalism, and parasitism.**

Symbiosis is any intimate relationship or association between members of two or more species. **Mutualism** is a symbiotic relationship in which both partners benefit. **Commensalism** is a type of symbiosis in which one organism benefits and the other one is neither harmed nor helped. **Parasitism** is a symbiotic relationship in which one organism benefits and the other is adversely affected.

- **Define *predation,* and describe the effects of natural selection on predator–prey relationships.**

Predation is the consumption of one species (the prey) by another (the predator). During **coevolution** between predator and prey, the predator evolves more efficient ways to catch prey, and the prey evolves better ways to escape the predator.

- **Define *keystone species,* and discuss the wolf as a keystone species.**

A **keystone species** is a species, often a predator, that exerts a profound influence on a community in excess of that expected by its relative abundance. One example of a keystone species is the gray wolf. Where wolves were hunted to extinction, the populations of deer, elk, and other herbivores increased. As these herbivores overgrazed the vegetation, many plant species disappeared. Many smaller animals such as insects and trout decreased because the plants they depended on for food were now less abundant. Thus, the disappearance of the wolf resulted in an ecosystem with considerably less biological diversity.

- **Describe factors associated with high species richness.**

Species richness, the number of species in a community, is often great when there are many potential ecological niches, when the area is at the margins of adjacent communities, when the community is not isolated or severely stressed, when one species does not dominate others, and when communities have a long geologic history.

- **Give several examples of ecosystem services.**

Ecosystem services, important environmental benefits that ecosystems provide to people, include clean air to breathe, clean water to drink, and fertile soil in which to grow crops.

- **Define *ecological succession,* and distinguish between primary and secondary succession.**

Ecological succession is the orderly replacement of one community by another. **Primary succession** is the change in species composition over time in a previously uninhabited environment. **Secondary succession** is the change in species composition that takes place after some disturbance destroys the existing vegetation; soil is already present.

Critical Thinking and Review Questions

1. Charles Darwin once said, "It is not the strongest of the species that survive, nor the most intelligent, but the ones most responsive to change." How does this statement relate to the definition of *evolution*?

2. During mating season, male giraffes slam their necks together in fighting bouts to determine which male is stronger and can mate with females. Explain how long necks may have evolved under this scenario, using Darwin's theory of evolution by natural selection.

3. How do the Eukarya and Prokarya differ?

4. How are the following factors related in determining population growth: birth rate, death rate, immigration, and emigration?

5. Draw a graph to represent the long-term growth of a population of bacteria cultured in a test tube containing a nutrient medium that is replenished. Now draw a graph to represent the growth of bacteria in a test tube when the nutrient medium is not replenished. Explain the difference.

6. Which of the following are density-dependent factors: a hurricane, disease, or competition? Why?

7. How do survivorship curves relate to *r* selection and *K* selection in animals?

8. What is the difference between a source habitat and a sink habitat in terms of birth rates and death rates?

9. What is an organism's ecological niche, and why is a realized niche usually narrower, or more restricted, than a fundamental niche?

10. What portion of humans' fundamental niche are we occupying today? Do you think our realized niche has changed over the past 200 years? Why or why not?

11. What is the most likely limiting resource for plants and animals in deserts? How are limiting resources related to competition? Explain your answers.

12. What type of symbiotic relationship—mutualism, commensalism, or predation—do you think exists between the pygmy sea horse and the gorgonian coral pictured in Figure 5.22? Explain your answer.

13. How is predation related to the concept of energy flow through ecosystems (covered in Chapter 3)?

14. Some biologists think that protecting keystone species would help preserve biological diversity in an ecosystem. Explain.

15. Why does species richness vary from one community to another?

16. What kinds of ecosystem services does a forest provide?

17. Describe an example of secondary succession. Begin your description with the specific disturbance that preceded it.

18. Draw a diagram of three concentric circles and label the circles to show the relationships among species, ecosystems, and communities. If you were adding symbiosis, predation, and competition to the simple system you have depicted, in which circle(s) would you place them?

19. Study the graph and determine the overall trend in biological diversity of vertebrate species from 1970 to 2000. Sociobiologist E. O. Wilson says that the five forces responsible for this trend are habitat loss, invasive species, pollution, human population growth, and overconsumption. How do each of these forces relate to energy use and/or climate change?

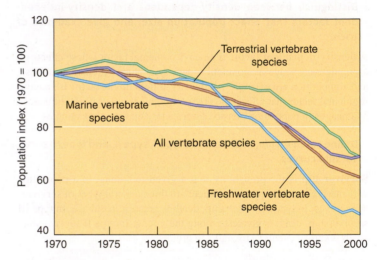

Adapted from Millennium Ecosystem Assessment. *Ecosystems and Human Well-Being: Biodiversity Synthesis*. Washington, D.C.: World Resources Institute [2005]

Food for Thought

Weed Ecology

Weeds, typically pioneer, *r*-selected species, grow quickly and reproduce abundantly.

In agriculture, weeds can interfere with food production. Understanding ecosystems can help farmers manage weeds more effectively. By fostering healthy soils, farmers encourage bacteria, fungi, and insect populations that might consume weed seeds as food. If water is a limiting resource, the farmer might irrigate crops only in the crop row, not allowing weeds between rows to access the water. When we understand weeds as pioneer species, how can we use this knowledge to help us manage weeds in gardens?

Exponential and Logistic Population Growth: Equations

Under exponential population growth, the number of individuals (N) in the population at any single time, t, can be reflected by equation 1:

$$(1) \ N_t = N_o * e^{rt}$$

In this equation N_t represents the population at time t, N_o represents the starting population, e is the base of the natural logarithm (~2.718), r is the natural growth rate of the population, and t is the time since the start of growth.

At any given moment, a student can calculate how much the population will increase in the next time interval. The increase in population can be calculated with equation 2:

$$(2) \ dN/dt = r * N$$

In this equation, dN/dt represents the number of individuals added to the population at any time. In the table of bacterial growth above, r = 1, which means that the population doubles in every time interval.

Although the equation for the logistic growth curve is more complex than the equation for exponential growth, a student can still easily calculate, using equation 3, the predicted growth in population, for any population level, N:

$$(3) \ dN/dt = r * N * (1 - N/K)$$

One way to think about this equation is to note that, as in equation 2, the change in population is proportional to the population level at that moment, N, and to the rate of population growth, r, but is reduced when the population, N, is close to the carrying capacity, K. If N is large, (1-N/K) will be close to zero; if N exceeds K, 1-N/K will become negative, meaning that the population will decrease. But, if N is small, 1-N/K is close to 1, and the population will grow similarly to the exponential growth pattern, at least until the population, N, gets closer to the carrying capacity (K).

6

Major Ecosystems of the World

Burning forest litter and sun flare near Bogong, Victoria, Australia

Peter Walton Photography / Getty Images

A wildfire is any unexpected fire that burns in grass, shrub, and forest areas. Whether started by lightning or humans, wildfires are often a significant environmental force. Those areas most prone to wildfires have wet seasons followed by lengthy dry seasons. Vegetation grows and accumulates during the wet season and then dries out enough during the dry season to burn easily. At the peak of a wildfire season in the American West, where extensive lands are prone to fire, hundreds of new wildfires can break out each day (**see photograph**).

Fires were part of the natural environment long before humans appeared, and plants in many terrestrial ecosystems that we discuss in this chapter have adapted to fire. African savannas, California chaparrals, North American grasslands, and pine forests of the southern United States are some fire-adapted ecosystems. For example, fire helps maintain grasses as the dominant vegetation in grasslands by removing fire-sensitive hardwood trees. Grasses adapted to wildfire have underground stems and buds unaffected by a fire; after fire kills the aerial (aboveground) parts, the untouched underground parts send up new sprouts. Fire-adapted trees such as bur oak and ponderosa pine have

thick, fire-resistant bark; other trees, such as jack pines, depend on fire to open their cones and release the seeds.

Climate change affects rainfall patterns, generally leading to more intense droughts. Partly because of climate change, more frequent and more intense wildfires occur in the western United States as well as other parts of the world. The wildfire season is longer and drier due to an increase in spring and summer temperatures, which causes mountain snows to melt more quickly.

Human interference affects the frequency and intensity of wildfires. If fire is excluded from a fire-adapted ecosystem, organic litter (such as brush and slender trees) accumulates. When a fire does occur, it burns hotter and climbs into the upper canopies of trees. Called *crown fires*, these fires shoot flames high into the air, and wind pushes the flames to nearby trees. Many recent fires in the western United States have been caused in part by decades of fire suppression. Fire suppression efforts continue in many areas because more and more expensive homes have been built in and around fire-adapted ecosystems.

Prescribed burning is an ecological management tool in which the organic litter is deliberately burned under controlled

conditions before it accumulates to dangerous levels. Prescribed burns also suppress fire-sensitive trees, thereby maintaining the natural fire-adapted ecosystem. Fire management experts agree that prescribed burns reduce the damage but do not prevent all fires.

In this chapter, we consider Earth's major ecosystems, both aquatic ecosystems and those on land. We consider how the physical environment, such as presence or absence of wildfires, influences each ecosystem.

In Your Own Backyard...
Many wildfires occur today in places where fire used to be uncommon. Research online to find out how many wildfires occurred in your state in the past year. How many of these were considered severe? Are prescribed fires used to manage any ecosystems in your state?

Earth's Major Biomes

LEARNING OBJECTIVES

- Define *biome* and briefly describe the nine major terrestrial biomes: tundra, boreal forest, temperate rain forest, temperate deciduous forest, temperate grassland, chaparral, desert, savanna, and tropical rain forest.
- Identify one human food in each of the biomes discussed.
- Explain the similarities and the changes in ecosystems observed with increasing elevation and increasing latitude.

Earth has many **climates** based primarily on temperature and precipitation differences. Characteristic organisms have adapted to each climate. A **biome** is quite large in area and encompasses many interacting ecosystems (**Figure 6.1**; also see Figures 4.19 for the world's major climates and Figure 14.8 for the world's major soil types). In terrestrial ecology, a biome is considered the next level of ecological organization above those of community, ecosystem, and landscape.

biome A large, relatively distinct terrestrial region with similar climate, soil, plants, and animals, regardless of where it occurs in the world.

Near the poles, temperature is generally the overriding climate factor (sometimes considered a limiting resource; see Chapter 5), whereas in temperate and tropical regions, precipitation becomes more significant than temperature (**Figure 6.2**). Light is relatively plentiful in most biomes, except in certain environments such as the rainforest floor, or polar regions, where days are short or non-existent in winter. Other abiotic factors to which certain biomes are sensitive include temperature extremes as well as rapid temperature changes, fires, floods, droughts, and strong winds.

We now consider nine major biomes and how humans are affecting them: tundra, boreal forest, temperate rain forest, temperate deciduous forest, temperate grassland, chaparral, desert, savanna, and tropical rain forest.

TUNDRA: COLD BOGGY PLAINS OF THE FAR NORTH

Tundra (also called **arctic tundra**) occurs in the extreme northern latitudes wherever the snow melts seasonally (**Figure 6.3**). The Southern Hemisphere has no equivalent of the arctic tundra because it has no land in the corresponding latitudes. **Alpine tundra** is a similar ecosystem located in the higher elevations of mountains, above the tree line. Alpine tundra can occur at any latitude, even in the tropics.

tundra The treeless biome in the far north that consists of boggy plains covered by lichens and small plants such as mosses; has harsh, very cold winters and extremely short summers.

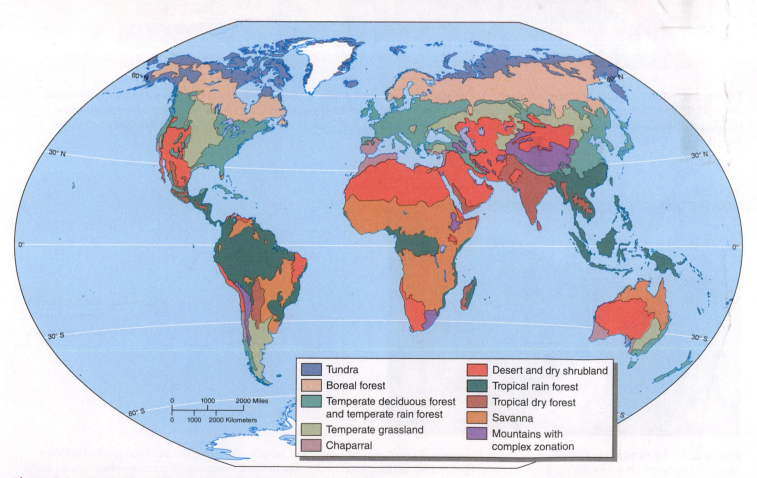

Figure 6.1 **Distribution of the world's terrestrial biomes.** Although sharp boundaries are shown in this highly simplified map, biomes actually blend together at their boundaries. (Based on data from the World Wildlife Fund)

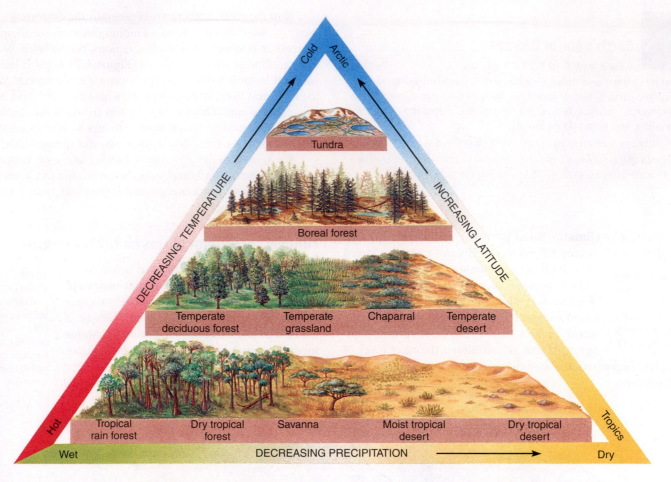

Figure 6.2 **Temperature and precipitation.** The biomes are distributed primarily in accordance with two climate factors, temperature and precipitation. In the higher latitudes, temperature is the more important of the two. In temperate and tropical zones, precipitation is a significant determinant of community composition. (Based on L. Holdridge, *Life Zone Ecology.* Tropical Science Center, San Jose, Costa Rica [1967])

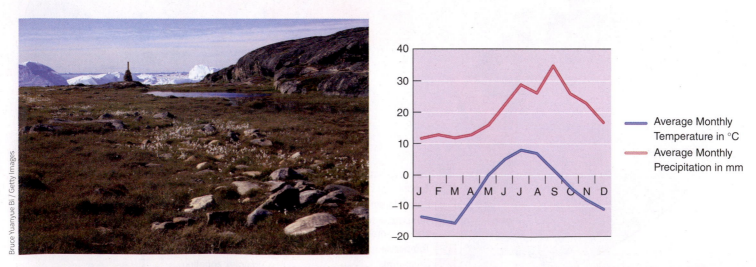

Figure 6.3 **Arctic tundra.** Only small, hardy plants grow in the northernmost biomes that encircles the Arctic Ocean. Photographed in Greenland. Climate graph shows monthly temperature (blue line) and precipitation data (red line) for Ilulissat, Greenland. Climate graphs provide a useful comparison of various climates. During how many months each year is the Greenland tundra at or above freezing? If climate warming continues, how might this climate graph change in 50 years?

Arctic tundra has long, harsh winters and short summers. Although the growing season, with its warmer temperatures, is short (from 50 to 160 days, depending on location), the days are long. Little precipitation (10 to 25 cm, or 4 to 10 in., per year) falls over much of the tundra, most of it during the summer months.

Most tundra soils are geologically young because they formed when glaciers retreated after the last ice age. (Glacier ice, which occupied about 29% of Earth's land during the last ice age, began retreating about 17,000 years ago. Today, glacier ice occupies about 10% of the land.) Tundra soils are usually nutrient-poor and have little organic litter, such as dead leaves and stems, animal droppings, and remains of organisms. Although the soil melts at the surface during the summer, tundra has a layer of **permafrost**, permanently frozen ground that varies in depth and thickness. The thawed upper zone of soil is usually waterlogged during the summer because permafrost interferes with drainage. Permafrost limits the depth that roots penetrate, thereby preventing the establishment of most woody species. The limited precipitation in combination with low temperatures, flat topography (surface features), and the permafrost layer produces a landscape of broad, shallow lakes and ponds, sluggish streams, and bogs.

Tundra has low **species richness** (the number of different species) and low **primary productivity** (the rate at which energy is accumulated; see the section on ecosystem productivity in Chapter 3). Few plant species occur, but individual species often exist in great numbers. Mosses, lichens (such as reindeer moss), grasses, and grasslike sedges dominate tundra. No readily recognizable trees or shrubs grow except in sheltered locations, where dwarf willows, dwarf birches, and other dwarf trees are common. As a rule, tundra plants seldom grow taller than 30 cm (12 in.).

The year-round animal life of the tundra includes lemmings, voles, weasels, arctic foxes, snowshoe hares, ptarmigan, snowy owls, and musk oxen. These animals are adapted to the extreme cold, with coats of thick fur. Lemmings, voles, hares, and musk oxen consume plant material throughout the winter, whereas weasels, foxes, ptarmigan, and owls are efficient predators on the lemmings and other small herbivores. In the summer, caribou migrate north to the tundra to graze on sedges, grasses, and dwarf willow. Dozens of bird species migrate north in summer to nest and feed on abundant insects. Mosquitoes, blackflies, and deerflies occur in great numbers during summer.

Tundra regenerates slowly after it has been disturbed. Even hikers can cause damage. Oil and natural gas exploration and military use have caused long-lasting injury, likely to persist for hundreds of years, to large portions of the arctic tundra (see Chapter 11, Case in Point: The Arctic National Wildlife Refuge). Agriculture is not generally possible within tundra regions, and humans who live on tundra rely primarily on fish and game for their nutrition.

Climate change is beginning to affect the arctic tundra. As the permafrost melts, conifer trees (cone-bearing evergreens) are replacing tundra vegetation. The trees have a lower albedo (reflectivity) than snow, ice, or tundra vegetation, causing additional warming, an example of a *positive feedback mechanism*. In addition, the permafrost line is moving northward, another result of enhanced climate warming. Methane, a greenhouse gas, is trapped in permafrost and releases into the atmosphere when permafrost melts, thereby further enhancing climate change.

BOREAL FORESTS: CONIFER FORESTS OF THE NORTH

Just south of the tundra is the **boreal forest** (also called **taiga**, pronounced tie′guh). The boreal forest stretches across North America and Eurasia, covering approximately 11% of Earth's land (**Figure 6.4**). A biome comparable to the boreal forest is not found in the Southern Hemisphere; as with arctic tundra, there is no land in the corresponding southern latitudes.

boreal forest A region of coniferous forest (such as pine, spruce, and fir) in the Northern Hemisphere; located just south of the tundra.

Winters are extremely cold and severe, although not as harsh as in the tundra. The growing season of the boreal forest is

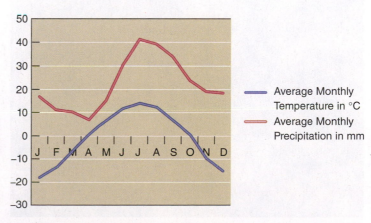

Figure 6.4   Boreal forest. These coniferous forests occur in cold regions of the Northern Hemisphere adjacent to the tundra. Photographed in Autum in Yukon, Canada. Climate graph shows monthly temperature (blue line) and precipitation data (red line) for Whitehorse, Yukon Territory, Canada. How does average monthly precipitation vary with average monthly temperature?

LaraBelova / Getty Images

somewhat longer than that of the tundra. Boreal forest receives little precipitation, perhaps 50 cm (20 in.) per year, and its soil is typically acidic and mineral-poor, with a deep layer of partly decomposed pine and spruce needles at the surface. Permafrost is patchy and, where found, is often deep under the soil. Boreal forest has numerous ponds and lakes in water-filled depressions that were dug by grinding ice sheets during the last ice age.

Black and white spruces, balsam fir, eastern larch, and other conifers dominate the boreal forest, although deciduous trees (trees that shed their leaves in autumn), such as aspen and birch, may form striking stands. Conifers have many drought-resistant adaptations, such as needlelike leaves with a minimal surface area for water loss. Such an adaptation lets conifers withstand the northern winter months when roots cannot absorb water because the ground is frozen.

The animal life of the boreal forest consists of some larger species such as caribou, which migrate from the tundra to the boreal forest for winter; wolves; bears; and moose. Most mammals are medium-sized to small, including rodents, rabbits, and fur-bearing predators such as lynx, sable, and mink. Most species of birds are abundant in the summer but migrate to warmer climates for winter. Wildlife ecologists estimate that one of every three birds in the United States and Canada spends its breeding season in the boreal forests of North America. Insects are abundant, but few amphibians and reptiles occur except in the southern boreal forest.

Most of the boreal forest is not well suited to agriculture because of its short growing season and mineral-poor soil. However, the boreal forest yields lumber, pulpwood for paper products,

animal furs, and other forest products. Currently, boreal forest is the world's primary source of industrial wood and wood fiber, and extensive logging of certain boreal forests has occurred. Mining, drilling for gas and oil, and farming have also contributed to loss of boreal forest. (See Chapter 17 for a discussion of deforestation of this biome.) As in tundra, humans living in the boreal forest biome primarily rely on animal nutrition, though pine nuts, berries from some shrubs, and some wetland plant species, such as cattail, can provide supplements to human diets in this area.

TEMPERATE RAIN FORESTS: LUSH TEMPERATE FORESTS

A coniferous **temperate rain forest** occurs on the northwest coast of North America. Similar vegetation exists in southeastern Australia and in southern South America. Annual precipitation in this biome is high, more than 127 cm (50 in.), and is augmented by condensation of water from dense coastal fogs. The proximity of temperate rain forest to the coastline moderates the temperature so that the seasonal fluctuation is narrow: Winters are mild and summers are cool. Temperate rain forest has relatively nutrient-poor soil, although its organic content may be high. Needles and large fallen branches and trunks accumulate on the ground as litter that takes many years to decay and release nutrient minerals to the soil.

> **temperate rain forest**
> A coniferous biome with cool weather, dense fog, and high precipitation.

The dominant plants in the North American temperate rain forest are large evergreen trees such as western hemlock, Douglas fir, western red cedar, Sitka spruce, and western arborvitae (**Figure 6.5**). Temperate rain forests are rich in epiphytic vegetation (see commensalism, Chapter 5)—smaller plants that grow on the trunks and branches of large trees. Epiphytes in this biome are mainly mosses, club mosses, lichens, and ferns, all of which also carpet the ground. Squirrels, wood rats, mule deer, elk, numerous bird species, and several species of amphibians and reptiles are common temperate rainforest animals. Edible plant species for humans include cattail, fireweed, salmonberry, and stinging nettle; numerous edible mushrooms also are available in this biome.

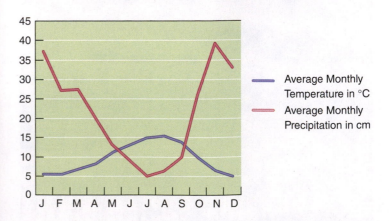

Figure 6.5   **Temperate rain forest.**   This temperate biome has large amounts of precipitation, particularly during winter months.   Photographed in Olympic National Park in Washington State.   Climate graph shows monthly average temperature (blue line) and precipitation data (red line) for Olympic National Park.   What is the range of average monthly temperatures for this biome, and how does it compare to the range of average monthly temperatures in the boreal forest (Figure 6.4)? (data from The Weather Channel)

Temperate rain forest is a rich wood producer, supplying us with lumber and pulpwood. It is also one of the world's most complex ecosystems in terms of species richness. When the logging industry harvests old-growth forest, companies typically replant the area with a **monoculture** (a single species) of trees harvested in 40- to 100-year cycles. The old-growth forest ecosystem, once harvested, never has a chance to regrow through succession. A small fraction of the original old-growth temperate rain forest in Washington, Oregon, and northern California remains untouched. These stable forest ecosystems provide biological habitats for many species, including about 40 endangered and threatened species. The issues surrounding old-growth forests of the Pacific Northwest were explored in the Chapter 2 introduction. Old-growth forest also supports fish habitat in streams and rivers; many Pacific northwest fish species, such as coho and chinook salmon, are extremely valuable and nutritious food sources for humans.

TEMPERATE DECIDUOUS FORESTS: TREES THAT SHED THEIR LEAVES

Hot summers and cold winters are characteristic of the **temperate deciduous forest**, which occurs in temperate areas where precipitation ranges from about 75 to 150 cm (30 to 60 in.) annually. Typically, the soil of a temperate deciduous forest consists of a topsoil rich in organic material and a deep, clay-rich lower layer. As organic materials decay, mineral ions are released. Ions not absorbed by tree roots **leach** (filter) into the clay.

temperate deciduous forest A forest biome that occurs in temperate areas with a moderate amount of precipitation.

The trees of the temperate deciduous forest form a dense canopy that overlies saplings and shrubs. Broad-leaved hardwood trees, such as oak, hickory, maple, and beech, dominate temperate deciduous forests of the northeastern and mideastern United States (**Figure 6.6**). These trees lose their foliage annually. In the southern areas of the temperate deciduous forest, the number of broad-leaved evergreen trees, such as magnolia, increases. Historically, the American chestnut tree was an important food source for both humans and wildlife within the eastern deciduous forest. Other food trees in this biome include nuts (hickory, beech), fruit (pawpaw), as well as sugar maple (syrup) and sassafras (tea). Many understory plants, ranging from berries (blueberry, blackberry) to roots (ginseng, ginger) also act as human food sources in deciduous forest. Humans living in temperate deciduous forest have a high level of dietary diversity available from the surrounding ecosystem.

Temperate deciduous forests originally contained a variety of large mammals, such as puma, wolves, and bison, which are now absent, plus deer, coyote, bears, and many small mammals and birds. In Europe and North America, logging and land clearing for farms, tree plantations, and cities have removed much of the original temperate deciduous forest. Where it has regenerated, temperate deciduous forest is often in a seminatural state modified by recreation, invasive weeds, livestock foraging, and timber harvest. Many forest organisms have successfully become reestablished in these returning forests.

Worldwide, deciduous forests were among the first biomes converted to agricultural use. In Europe and Asia, many soils that

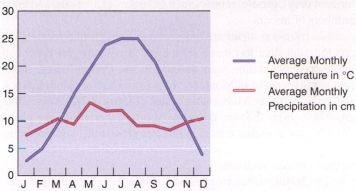

Figure 6.6 **Temperate deciduous forest.** Abundant and diverse deciduous trees produce nuts, fruit, and seeds and support a wide variety of herbivorous insects and birds. Leaves drop to the ground in fall, supporting a host of detritivores in soil. Photographed in Berea College Forest, in Kentucky. Climate graph shows monthly average temperature (blue line) and precipitation data (red line) for Berea, Kentucky. What is the range of average monthly temperatures for this biome, and how does it compare to the range of average monthly temperatures in the boreal forest (Figure 6.4)? (data from The Weather Channel)

originally supported deciduous forests have been cultivated by traditional agricultural methods for thousands of years without a substantial loss in fertility. During the twentieth century, intensive agricultural practices were widely adopted; these, along with overgrazing and deforestation, have contributed to the degradation of some agricultural lands. Most damage to farmland has happened since the end of World War II (see Chapters 17 and 18).

GRASSLANDS: TEMPERATE SEAS OF GRASS

Summers are hot, winters are cold, and rainfall is often uncertain in **temperate grasslands**. Annual precipitation averages 25 to 75 cm (10 to 30 in.). Grassland soil has considerable organic material because the aerial portions of many grasses die off each winter or in wildfires and contribute to the

temperate grasslands A grassland with hot summers, cold winters, and less rainfall than the temperate deciduous forest biome.

organic content of the soil, whereas the roots and rhizomes (underground stems) survive. The roots and rhizomes eventually die and add to the soil's organic material. Many grasses are sod formers—that is, their roots and rhizomes form a thick, continuous underground mat. Although few trees grow except near rivers and streams, grasses grow in great profusion in the deep, rich soil. Periodic wildfires help maintain grasses as the dominant vegetation in grasslands.

Moist temperate grasslands, or *tallgrass prairies*, occur in the United States in parts of Illinois, Iowa, Minnesota, Nebraska, Kansas, and other Midwestern states. Several species of grasses that, under favorable conditions, grow as tall as a person on horseback dominate tallgrass prairies. The land was originally covered with herds of grazing animals, such as pronghorn elk and bison. The principal predators were wolves, although in sparser, drier areas coyotes took their place. Smaller animals included prairie dogs and their predators (foxes, black-footed ferrets, and various birds of prey), grouse, reptiles such as snakes and lizards, and great numbers of insects.

Shortgrass prairies are temperate grasslands that receive less precipitation than the moist temperate grasslands just described but more precipitation than deserts. In the United States, shortgrass prairies occur in Montana, the western half of South Dakota, and parts of other Midwestern states (**Figure 6.7**). Grasses that grow knee-high or lower dominate shortgrass prairies. The plants grow in less abundance than in moister grasslands, and occasionally some bare soil is exposed. Native grasses of shortgrass prairies are drought-resistant.

The North American grassland, particularly the tallgrass prairie, was well suited to agriculture. More than 90% has vanished under the plow, and the remaining prairie is so fragmented that almost nowhere can you see what Native Americans experienced prior to the arrival of European settlers in the Midwest. Today, the tallgrass prairie is considered North America's rarest biome. The rich soils in this biome enabled the North American Midwest, Ukraine, and other moist temperate grasslands to become the breadbaskets of the world, because they provide ideal growing conditions for crops such as corn and wheat, which are also grasses. At the same time, when annual crops replace perennial plant mixtures, seasonally bare ground is vulnerable to erosion. For example, the 1930s Dust Bowl occurred when plowed, bare land remained bare due to drought, and wind blew visible amounts of United States prairie soil as far as Europe. Many ecologists and agricultural scientists are working to understand how grasslands can more sustainably support growth of food crops.

CHAPARRAL: THICKETS OF EVERGREEN SHRUBS AND SMALL TREES

Some hilly temperate environments have mild winters with abundant rainfall combined with hot, dry summers. Such **Mediterranean climates**, as they are called, occur in the area around the Mediterranean Sea as well as in the North American Southwest, southwestern and southern Australia, central Chile, and southwestern South Africa. On mountain slopes of southern California, this Mediterranean-type community is known as **chaparral**. Chaparral soil is thin and often not fertile. Frequent fires occur naturally in this environment, particularly in late summer and autumn.

> **chaparral** A biome with mild, moist winters and hot, dry summers; vegetation is typically small-leaved evergreen shrubs and small trees.

Chaparral vegetation looks strikingly similar in different areas of the world, even though the individual species are not the same. Chaparral usually has a dense growth of evergreen shrubs but may contain short, drought-resistant pine or scrub oak trees that grow 1 to 3 m (3.3 to 9.8 ft) tall (**Figure 6.8**). During the rainy winter season, the environment is lush and green, but the plants lie dormant during the summer. Trees and shrubs often have hard, small, leathery leaves that resist water loss.

Many plants in this biome are fire-adapted and grow best in the months following a fire. Such growth is possible because fire releases nutrient minerals from aerial parts of the plants that burned. Fire does not kill the underground parts and seeds of many plants, and with the new availability of essential nutrient minerals, the plants sprout vigorously during winter rains. Mule deer, wood rats, chipmunks, lizards, and many species of birds are common animals of the chaparral.

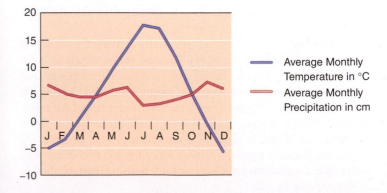

Figure 6.7   Temperate grassland. These bison live on shortgrass prairie in South Dakota. Climate graph shows monthly temperature (blue line) and precipitation data (red line) for the National Bison Range in Montana. How might temperature affect water availability in the summer? (data from The Weather Channel)

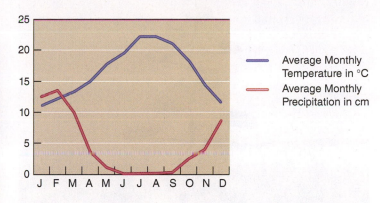

Figure 6.8 **Chaparral.** This chaparral in California is a fire-adapted habitat. Climate graph shows monthly temperature (blue line) and precipitation data (red line) for the Santa Ynez Valley. How do rainfall and temperatures conspire to increase the likelihood of summer wildfires? (data from The Weather Channel)

Chaparral landscapes are not suitable for most agriculture, although some trees (olive, carob) are particularly adapted to growing on chaparral soils. Vineyards may also thrive in chaparral regions, since grapes are vulnerable to fungal diseases, which do not thrive in arid environments. Among livestock, sheep and goats are better suited to this type of landscape and vegetation than cattle, because their hooves do less damage to the fragile soils.

The fires that occur in California chaparral are quite costly to humans when they consume expensive homes built on the hilly chaparral landscape (see Environews). Unfortunately, efforts to prevent the naturally occurring fires sometimes backfire. Removing the chaparral vegetation, whose roots hold the soil in place, causes other problems; witness the mudslides that sometimes occur during winter rains in these areas.

E N V I R O N E W S

Using Goats to Fight Fires

California has about 6000 wildfires each year, and they are becoming increasingly expensive and dangerous to manage because so many people are building homes amid fire-vulnerable chaparral. For one thing, the topography is so steep that firefighters often cannot use mechanized equipment and must be transported to fires by helicopters. Fearing that prescribed burns will get out of control, local governments, such as those of Oakland, Berkeley, Monterey, and Malibu, are increasingly employing a low-tech method to reduce the fuel load—goats.

A herd of 350 goats can denude an entire acre of heavy brush in about a day, but using them entails advance organization and support. First, botanists must walk the terrain to put fences around any small trees and rare or endangered plants; fencing keeps the goats from eating those plants.

Goats are an excellent tool for fire management because they preferentially browse woody shrubs and thick undergrowth—the exact fuel that causes disastrous fires. Fires that occur in areas after goats have browsed there are much easier to contain.

DESERTS: ARID LIFE ZONES

Deserts are dry areas found in both temperate (*cold deserts*) and subtropical regions (*warm deserts*). The low water vapor content of the desert atmosphere results in daily temperature extremes of heat and cold, and a major change in temperature occurs in a single 24-hour period. Deserts vary depending on the amount of precipitation they receive, which is generally less than 25 cm (10 in.) per year. As a result of sparse vegetation, desert soil is low in organic material but is often high in mineral content, particularly the salts sodium chloride (NaCl), calcium carbonate ($CaCO_3$), and calcium sulfate ($CaSO_4$). In some regions, such as areas of Utah and Nevada, the concentration of certain soil minerals reaches levels toxic to many plants. Lithium, an element valuable for use in batteries and medicines, is found in deserts, such as the Atacama Desert in northern Chile.

Plant cover is so sparse in deserts that much of the soil is exposed. Both perennials (plants that live for more than two years) and annuals (plants that complete their life cycles in one growing season) occur in deserts. However, annuals are common only after rainfall. Plants in North American deserts include cacti, yuccas, Joshua trees, and sagebrushes (**Figure 6.9**).

Inhabitants of deserts are strikingly adapted to the demands of their environment. Desert plants tend to have reduced leaves or no leaves, an adaptation that conserves water. In cacti such as the giant saguaro, the stem, which expands accordion-style to store water, carries out photosynthesis; the leaves are modified into spines, which discourage herbivores. Other desert plants shed their leaves for most of the year, growing only during the brief moist season. Many desert plants are provided with spines, thorns, or toxins to resist the heavy grazing pressure often experienced in this food- and water-deficient environment.

Desert animals tend to be small. During the heat of the day, they remain under cover or return to shelter periodically, whereas at night they come out to forage or hunt. In addition to desert-adapted insects, there are a few desert amphibians (frogs

desert A biome in which the lack of precipitation limits plant growth; deserts are found in both temperate and subtropical regions.

Nancy Gift

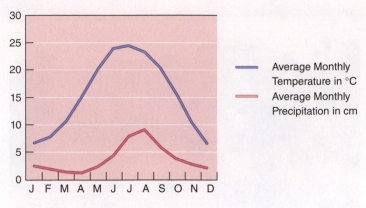

Figure 6.9 **Desert.** This century plant (*Agave americana*) is flowering near Carlsbad Cavern, New Mexico, in May. The plant has succulent leaves with a waxy coating that reduces water loss. Climate graph shows monthly temperature (blue line) and precipitation data (red line) for Carlsbad Canyon. If plants have the highest water needs during flowering and seed set, when might you expect to find flowers in this desert? (data from The Weather Channel)

and toads) and many desert reptiles, such as the desert tortoise, desert iguana, Gila monster, and Mojave rattlesnake. Desert mammals include rodents such as gerbils and jerboas in African and Asian deserts and kangaroo rats in North American deserts. There are also mule deer and jackrabbits in these deserts, oryxes in African deserts, and kangaroos in Australian deserts. Carnivores such as the African fennec fox and some birds of prey, especially owls, live on the rodents and jackrabbits. During the driest months of the year, many desert insects, amphibians, reptiles, and mammals tunnel underground, where they remain inactive.

Some human cultures, such as the Navaho and Pueblo people, have been able to survive on desert agriculture. Agriculture in desert regions is most sustainable when conserved rainfall, rather than groundwater, is used for crops. Some varieties of corn and beans are relatively drought tolerant. Crops such as pearl millet and sorghum use less water, enabling them to be cultivated successfully in some desert regions. Sheep tolerate drought far better than cattle, and can be grazed in some deserts.

Humans have altered North American deserts in several ways. People who drive across the desert in off-road vehicles inflict environmental damage. When the top layer of desert soil is disturbed, erosion occurs more readily, and less vegetation grows to support native animals. Certain cacti and desert tortoises are rare as a result of poaching. Houses, factories, and farms built in desert areas require vast quantities of water, which is imported from distant areas. Increased groundwater consumption by many desert cities has caused groundwater levels to drop. Aquifer depletion in U.S. deserts is particularly critical in southern Arizona and southwestern New Mexico.

SAVANNA: TROPICAL GRASSLANDS

Savanna occurs in areas of low rainfall or seasonal rainfall with prolonged dry periods. The temperatures in tropical savannas vary little throughout the year, and seasons are regulated by precipitation, not by temperature as they are in temperate grasslands. Annual precipitation is 76 to 150 cm (30 to 60 in.). Savanna soil is somewhat low in essential nutrient minerals, in part because it is strongly leached—that is, many nutrient minerals have filtered down out of the topsoil. Savanna soil is often rich in aluminum, which resists leaching, and in some places the aluminum reaches levels toxic to many plants. Although the African savanna is best known, savanna also occurs in South America and northern Australia.

savanna A tropical grassland with widely scattered trees or clumps of trees.

Savanna has wide expanses of grasses interrupted by occasional trees such as acacia, which bristle with thorns that provide protection against herbivores. Both trees and grasses have fire-adapted features, such as extensive underground root systems, that let them survive seasonal droughts as well as periodic fires.

Spectacular herds of hoofed mammals—such as wildebeest, antelope, giraffe, zebra, and elephants—occur in the African savanna (**Figure 6.10**). Large predators, such as lions and hyenas, kill and scavenge the herds. In areas of seasonally varying rainfall, the herds and their predators may migrate annually.

Savannas are rapidly being converted into rangeland for cattle and other domesticated animals, which are replacing the big herds of wild animals. Half of the Cerrado (savanna) in central Brazil has been converted to cropland and pastures since 1970. The problem is more acute in Africa because it has the most

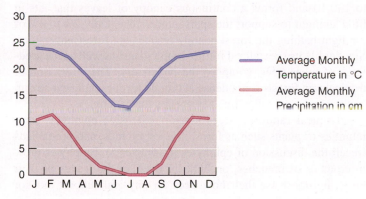

Courtesy Maria Sink Druzel

Figure 6.10 **Savanna.** The greater kudu (*Tragelaphus strepsiceros*) is one of many grazers who inhabit this grassland biome. These kudu were photographed in South Africa. Climate graph shows monthly temperature (blue line) and precipitation data (red line) for Pretoria, South Africa. In what months would savanna be most sensitive to overgrazing? (data from World Weather Online)

rapidly growing human population of any continent. In some places, severe overgrazing (see Chapter 17) and harvesting of trees for firewood have converted savanna to desert, a process called **desertification.**

TROPICAL RAIN FORESTS: LUSH EQUATORIAL FORESTS

tropical rain forest A lush, species-rich forest biome that occurs where the climate is warm and moist throughout the year.

Tropical rain forests occur where temperatures are warm throughout the year and precipitation occurs almost daily. The annual precipitation of a tropical rain forest is typically from 200 to 450 cm (80 to 180 in.). Much of this precipitation comes from locally recycled water that enters the atmosphere by trees' transpiration (loss of water vapor from plants).

Tropical rain forest commonly occurs in areas with ancient, highly weathered, mineral-poor soil. Little organic matter accumulates in such soils because bacteria, fungi, and detritus-feeding ants and termites decompose organic litter quite rapidly. Roots and mycorrhizae quickly absorb nutrient minerals from the decomposing material. Thus, nutrient minerals of tropical rain forests are tied up in the vegetation rather than the soil. Tropical rain forests are found in Central and South America, Africa, and Southeast Asia (see Figure 17.10).

Tropical rain forests are very productive—that is, the plants capture a lot of energy by photosynthesis. Of all the biomes, the tropical rainforest is unexcelled in species richness and variety. A person can travel hundreds of meters without encountering two individuals of the same tree species. The trees of tropical rain forests are typically evergreen flowering plants (**Figure 6.11**). Their roots are often shallow and concentrated near the surface in a mat. The root mat catches and absorbs almost all nutrient minerals released from leaves and litter by decay processes. Swollen bases or braces called *buttresses* hold some trees upright and aid in the extensive distribution of the shallow roots.

Courtesy Brigid Mulloy

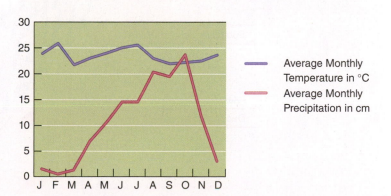

Figure 6.11 **Tropical rainforest.** This Costa Rican rainforest straddles the Continental Divide. Climate graph shows monthly temperature (blue line) and precipitation data (red line) for Santa Rosa National Park, Costa Rica. Many rainforest leaves have a waxy surface. How could this waxy coating be helpful in both wet and dry seasons? (Brigid Mulloy; data from World Weather Online)

A fully developed tropical rain forest has at least three distinct stories, or layers, of vegetation. The topmost story consists of the crowns of occasional, very tall trees, which are exposed to direct sunlight. The middle story reaches a height of 30 to 40 m (100 to 130 ft) and forms a continuous canopy of leaves that lets in little sunlight to support the sparse understory. Only 2% to 3% of the light bathing the forest canopy reaches the forest understory. Smaller plants specialized for life in the shade, as well as the seedlings of taller trees, compose the understory. The vegetation of tropical rain forests is not dense at ground level except near stream banks or where a fallen tree has opened the canopy.

Tropical rainforest trees support extensive epiphytic communities of plants such as ferns, mosses, orchids, and bromeliads (recall the discussion of epiphytes in Chapter 5). Epiphytes grow in crotches of branches, on bark, or even on the leaves of their hosts, but most use their host trees for physical support, not for nourishment.

Lianas (woody tropical vines), some as thick as a human thigh, twist up through the branches of the huge rainforest trees. Once in the canopy, lianas grow from the upper branches of one forest tree to another, connecting the tops of the trees and providing a walkway for many of the canopy's residents. Lianas and herbaceous vines provide nectar and fruit for many tree-dwelling animals.

Not counting bacteria and other soil-dwelling organisms, most tropical rainforest organisms live in the upper canopy. Rainforest animals include the most abundant insects, reptiles, and amphibians on Earth. The birds, often brilliantly colored, are varied, with some, like parrots, specialized to consume fruit and others, like

hummingbirds and sunbirds (Figure 5.16), to consume nectar. Most rainforest mammals, such as sloths and monkeys, live only in the trees and rarely climb to the ground, although some large, ground-dwelling mammals, including elephants, are also found in rain forests.

Although tropical rain forests have an abundance of food species, such as palm fruit, banana, and acai, each grows in low density; plantations of banana or acai are not likely to be sustainable. Biologists know that many rainforest species will become extinct before they are even identified and scientifically described. (See Chapter 17 for more discussion of the ecological impacts of tropical rainforest destruction.)

VERTICAL ZONATION: THE DISTRIBUTION OF VEGETATION ON MOUNTAINS

Hiking up a mountain is similar to traveling toward the North Pole with respect to the major ecosystems encountered (**Figure 6.12**). This elevation-latitude similarity occurs because as one climbs a mountain, the temperature drops just as it does when one travels north. Soils become less rich, because wind and water move dead plant material downhill before it becomes organic matter.

The base of a mountain in Colorado, for example, is covered by deciduous trees, which shed their leaves every autumn. At higher elevations, where the climate is colder and more severe, one might find a coniferous *subalpine forest*, which resembles the northern boreal forest. Higher still, where the climate is quite cold, *alpine tundra* occurs, with vegetation composed of grasses,

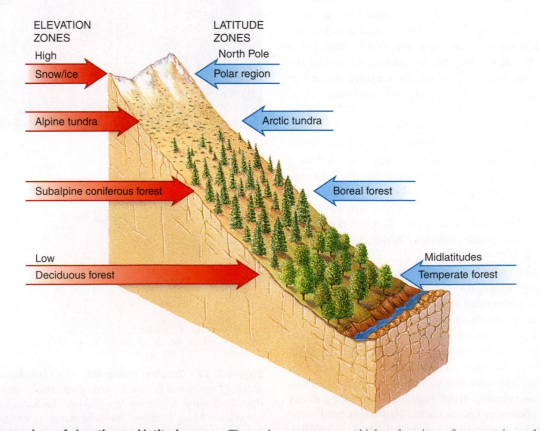

Figure 6.12 **Comparison of elevation and latitude zones.** The cooler temperatures at higher elevations of a mountain produce a series of ecosystems similar to the boreal forest and tundra biomes.

sedges, and small tufted plants; it is called alpine tundra to distinguish it from arctic tundra. The top of the mountain might have a permanent ice or snow cap, similar to that in the polar land areas.

Important environmental differences between high elevations and high latitudes affect the types of organisms found in each place. Alpine tundra typically lacks permafrost and receives more precipitation than arctic tundra. High elevations of temperate mountains do not have the great extremes in day length associated with the changing seasons in biomes at high latitudes. Furthermore, the intensity of solar radiation is greater at high elevations than at high latitudes. At high elevations, the sun's rays pass through less atmosphere, which filters out less ultraviolet (UV) radiation and results in a greater exposure to UV radiation than occurs at high latitudes. Because of these differences, some kinds of agriculture are possible at high altitude that are not viable at high latitudes; many varieties of potatoes, for example, have been grown for generations in small fields in the Peruvian Andes.

Climate change threatens alpine areas. As the climate warms, many plant and animal species are migrating vertically, where it is cooler. However, vertical migration is limited by mountain heights. Some alpine species are now regionally extinct because the climate zone to which they were adapted would be located at altitudes above the mountaintops.

REVIEW

1. What is a biome? What climate and soil factors produce each of the major terrestrial biomes?

2. Which biome adapts most easily to agriculture? Explain your answer.

3. How does vegetation change with increasing elevation and latitude?

Aquatic Ecosystems

LEARNING OBJECTIVES

- Summarize the important environmental factors that affect aquatic ecosystems.
- Briefly describe the eight aquatic ecosystems: flowing-water ecosystem, standing-water ecosystem, freshwater wetlands, estuary, intertidal zone, benthic environment, neritic province, and oceanic province.
- Identify at least one ecosystem service provided by each aquatic ecosystem.

Aquatic life zones differ from terrestrial biomes in almost all respects. Generally speaking, temperature is somewhat less variable in watery environments because the water itself tends to moderate temperature. Water is obviously not an important limiting factor in aquatic ecosystems, though light is frequently a limiting factor.

The most fundamental division in aquatic ecology is probably between freshwater and saltwater environments. **Salinity**, the concentration of dissolved salts such as sodium chloride (NaCl) in a body of water, affects the kinds of organisms present in aquatic ecosystems, as does the amount of dissolved oxygen. Water greatly

interferes with the penetration of light. Floating photosynthetic organisms remain near the water's surface, and vegetation attached to the bottom grows only in shallow water. In addition, low levels of essential nutrient minerals limit the number and distribution of organisms in certain aquatic environments. Other abiotic determinants of species composition in aquatic ecosystems include temperature, pH, and presence or absence of waves and currents.

Aquatic ecosystems contain three main ecological categories of organisms: free-floating plankton, strongly swimming nekton, and bottom-dwelling benthos. **Plankton** are usually small or microscopic organisms that are relatively feeble swimmers. For the most part, they are carried about at the mercy of currents and waves. They cannot swim far horizontally, but some species are capable of large daily vertical migrations and are found at different depths of water depending on the time of day and the season.

Plankton are generally subdivided into two major categories, phytoplankton and zooplankton. **Phytoplankton** are free-floating photosynthetic algae and cyanobacteria that form the base of most aquatic food webs. **Zooplankton** are nonphotosynthetic organisms that include protozoa (animal-like protists), tiny shrimplike crustaceans, and the larval (immature) stages of many animals. In aquatic food webs, zooplankton feed on algae and cyanobacteria and are in turn consumed by newly hatched fish and other small aquatic organisms.

Nekton are larger, more strongly swimming organisms such as fishes, turtles, and whales. **Benthos** are bottom-dwelling organisms that fix themselves to one spot (sponges, oysters, and barnacles), burrow into the sand (worms, clams, and sea cucumbers), or simply walk about on the bottom (crawfish, aquatic insect larvae, and brittle stars).

FRESHWATER ECOSYSTEMS

Freshwater ecosystems include rivers and streams (flowing-water ecosystems), lakes and ponds (standing-water ecosystems), and marshes and swamps (freshwater wetlands). Specific abiotic conditions and characteristic organisms distinguish each freshwater ecosystem. Although freshwater ecosystems occupy a relatively small portion (about 2%) of Earth's surface, they have an important role in the hydrologic cycle: They assist in recycling precipitation that flows as surface runoff to the ocean (see discussion of the hydrologic cycle in Chapter 4). Large bodies of freshwater moderate daily and seasonal temperature fluctuations on nearby land. Freshwater habitats provide homes for many species.

Rivers and Streams: Flowing-Water Ecosystems   The nature

of a **flowing-water ecosystem** changes greatly between its source (where it begins) and its mouth (where it empties into another body of water) (**Figure 6.13**). *Headwater streams*, the small streams that are the sources of a river, are usually shallow, cool, swiftly flowing, and highly oxygenated. Such streams provide high-quality drinking water, though the landscapes rarely support high-density human populations. In contrast, rivers downstream from the headwaters are wider and deeper, cloudy (they contain suspended particulates), not as cool, slower flowing, and less oxygenated. Rivers make

flowing-water ecosystem A freshwater ecosystem such as a river or stream in which the water flows in a current.

Figure 6.13 **Flowing water ecosystem.** In the Canadian Rockies, water flow initiates from snow clouds on mountaintop glaciers (a), proceeds through headwater streams (b) and then to rivers, and finally to the ocean at deltas on the coast. (c) (Photos from Yoho National Park, Kicking Horse Pass, and Victoria, British Columbia)

more reliable water sources, though they also typically require more treatment to provide safe drinking water. Along parts of a river or stream, groundwater wells up through sediments on the bottom. This local input of water moderates the water temperature so that summer temperatures are cooler and winter temperatures are warmer than in adjacent parts of the flowing-water ecosystem.

The various environmental conditions in a river system result in many different habitats. The concept of a river system as a single ecosystem with a gradient in physical features from head-waters to mouth is known as the *river continuum concept*. This gradient results in predictable changes in the organisms inhabiting different parts of the river system. For example, in streams with fast currents, the inhabitants may have adaptations such as hooks or suckers to attach themselves to rocks so that they are not swept away. The larvae of blackflies attach themselves with a suction disk located on the end of their abdomen. Some stream inhabitants, such as immature water-penny beetles, may have flattened bodies to slip under or between rocks. Inhabitants such as fish are stream-lined and muscular enough to swim in the current. Organisms in large, slow-moving streams and rivers do not need such adapta-tions, although they are typically streamlined to lessen resistance when moving through water.

Unlike other freshwater ecosystems, streams and rivers depend on the land for much of their energy. In headwater streams, almost all of the energy input comes from detritus, such as dead leaves carried from the land into streams and rivers by wind or surface runoff. Downstream, rivers contain more producers and are less dependent on detritus as a source of energy than are headwaters.

Human activities have several adverse impacts on rivers and streams. Pollution from industry, agriculture, cities, and lawns alters the physical environment and changes the biotic component downstream from the pollution source. A dam causes water to back up, flooding large areas of land and forming a reservoir, which destroys terrestrial habitat. Below the dam, the once-powerful river is often reduced to a relative trickle, which alters water temperature, sediment transport, and delta replenishment and pre-vents fish migrations. (See Chapter 12 for more discussion of the environmental effects of dams.)

Lakes and Ponds: Standing-Water Ecosystems Zonation characterizes **standing-water ecosystems.** A large lake has three zones: the littoral, limnetic, and profundal zones (**Figure 6.14**). The **littoral zone** is a shallow-water area along the shore of a lake or pond where light reaches the bottom. Emergent vegetation, such as cattails and bur reeds, as well as several deeper-dwelling aquatic plants and algae, live in the littoral zone. The littoral zone is the most productive

standing-water ecosystem A body of fresh water that is surrounded by land and that does not flow; a lake or a pond.

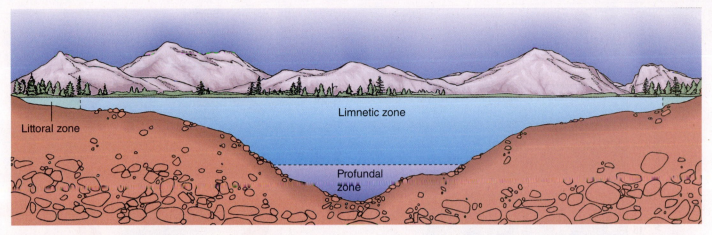

Figure 6.14 Zonation in a large lake. The littoral zone is the shallow-water area around the lake's edge. The limnetic zone is the open, sunlit water away from the shore. The profundal zone, under the limnetic zone, is below where light penetrates.

section of the lake (photosynthesis is greatest here), in part because it receives nutrient inputs from surrounding land that stimulate the growth of plants and algae. Animals of the littoral zone include frogs and their tadpoles; turtles; worms; crayfish and other crustaceans; insect larvae; and many fishes, such as perch, carp, and bass. Surface dwellers such as water striders and whirligig beetles live in still waters. Standing water ecosystems are highly appealing to humans both for the fresh water they provide and for recreational possibilities.

The **limnetic zone** is the open water beyond the littoral zone—that is, away from the shore; it extends down as far as sunlight penetrates to permit photosynthesis. The main organisms of the limnetic zone are microscopic phytoplankton and zooplankton. Larger fishes spend most of their time in the limnetic zone, although they may visit the littoral zone to feed and reproduce. Owing to its depth, less vegetation grows here than in the littoral zone.

The deepest zone, the **profundal zone**, is beneath the limnetic zone of a large lake; smaller lakes and ponds typically lack a profundal zone. Light does not penetrate effectively to this depth, and plants and algae do not live there. Food sinks into the profundal zone from the littoral and limnetic zones. Bacteria decompose dead organisms and other organic material that reach the profundal zone, using up oxygen and liberating the nutrient minerals contained in the organic material. These nutrient minerals are not effectively recycled because there are no producers to absorb and incorporate minerals into the food web. As a result, the profundal zone is both mineral-rich and anaerobic (without oxygen), with few organisms other than anaerobic bacteria occupying it.

Human effects on lakes and ponds include **eutrophication**, which is nutrient enrichment of a body of water with inorganic plant and algal nutrients like nitrates and phosphates. Although eutrophication can be a natural process, human activities often accelerate it. For example, nutrient levels increase due to the runoff of agricultural fertilizers and discharge of treated or untreated sewage. Soil erosion also increases nutrient loads in lakes and ponds. As eutrophication occurs, the number and kinds of aquatic organisms living in the lake change. Eutrophication of lakes is discussed in detail in Chapter 21 (see Figure 21.2).

Thermal Stratification and Turnover in Temperate Lakes.

The marked layering of large temperate lakes caused by how far light penetrates is accentuated by **thermal stratification**, in which the temperature changes sharply with depth. Thermal stratification occurs because the summer sunlight penetrates and warms surface waters, making them less dense. (The density of water is greatest at 4°C. Both above and below this temperature, water is less dense.) In the summer, cool, denser water remains at the lake bottom, separated from the warm, less dense water above by an abrupt temperature transition, the **thermocline** (**Figure 6.15a**). Seasonal distribution of temperature and oxygen (more oxygen dissolves in water at cooler temperatures) affects the distribution of fish in the lake.

In temperate lakes, falling temperatures in autumn cause **fall turnover**, a mixing of the layers of lake water (**Figure 6.15b**). (Such turnovers are not common in the tropics because the seasonal temperatures do not vary much.) As the surface water cools, its density increases, and eventually it displaces the less dense, warmer, mineral-rich water beneath. The warmer water then rises to the surface where it, in turn, cools and sinks. This process of cooling and sinking continues until the lake reaches a uniform temperature throughout.

When winter comes, the surface water may cool below 4°C, its temperature of greatest density, and if it is cold enough, ice forms. Ice forms at 0°C and is less dense than cold water. Thus, ice forms on the surface, and the water on the lake bottom is warmer than the ice on the surface.

In the spring, **spring turnover** occurs as ice melts and the surface water reaches 4°C. Surface water again sinks to the bottom, and bottom water returns to the surface, resulting in a mixing of the layers. As summer arrives, thermal stratification occurs once again. The mixing of deeper, nutrient-rich water with surface, nutrient-poor water during the fall and spring turnovers brings essential nutrient minerals to the surface and oxygenated water to the bottom. The sudden presence of large amounts of essential nutrient minerals in surface waters encourages the growth of large

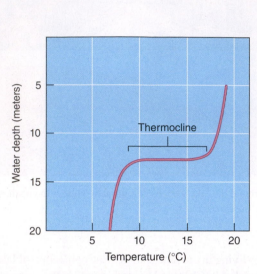

(a) Temperature varies by depth during the summer. An abrupt temperature transition, the thermocline, occurs between the upper warm layer and the bottom cold layer.

Figure 6.15 **Thermal stratification in a temperate lake.**

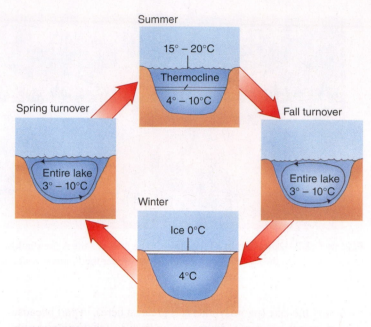

(b) During fall and spring turnovers, a mixing of upper and lower layers of water brings oxygen to the oxygen-depleted depths of the lake and brings nutrient minerals to the mineral-deficient surface water. During summer months, a layer of warm water develops over cooler, deeper water; no turnover occurs to supply oxygen to deeper water, where the oxygen level drops.

algal and cyanobacterial populations, which form temporary **blooms** (population explosions) in the fall and spring. (Harmful algal blooms such as red tides are discussed in Chapter 7.)

Marshes and Swamps: Freshwater Wetlands **Freshwater wetlands** include marshes, in which grasslike plants dominate, and swamps, in which woody trees or shrubs dominate (**Figure 6.16**). Freshwater wetlands include hardwood bottomland forests (lowlands along streams and rivers that are periodically flooded), prairie potholes (small, shallow ponds that formed when glacial ice melted at the end of the last ice age), and peat moss bogs (peat-accumulating wetlands where sphagnum moss dominates). Wetland soils are waterlogged and anaerobic for variable periods. Most wetland soils are rich in accumulated organic materials, in part because anaerobic conditions discourage decomposition.

> **freshwater wetland**
> Land that shallow fresh water covers for at least part of the year and that has a characteristic soil and water-tolerant vegetation.

Wetland plants are highly productive and provide enough food to support a variety of organisms. Wetlands are valued as a wildlife habitat for migratory waterfowl and many other bird species, beaver, otters, muskrats, and game fishes. Wetlands provide natural flood control because they are holding areas for excess water when rivers flood their banks. The floodwater stored in wetlands then drains slowly back into the rivers, providing a steady flow of water throughout the year. Wetlands serve as groundwater recharging areas. One of their most important roles is to help cleanse water by trapping and holding pollutants in the flooded soil. These important environmental functions are called **ecosystem services** (see Table 5.1).

At one time, wetlands were considered wastelands, areas to fill in or drain so that farms, housing developments, and industrial plants could be built. Because they are breeding places for mosquitoes, wetlands were viewed as a menace to public health. Today, the crucial ecosystem services that wetlands provide are widely recognized. Yet despite some legal protection, wetlands are still threatened by agriculture, pollution, dam construction, and urban and suburban development. In many parts of the United States, we continue to lose wetlands (**Figure 6.17**; also see Chapter 17).

Figure 6.16 **Freshwater wetland.** This lake near Salt Springs, Florida, is surrounded by wetlands. Trees and other plants are adapted to life with continually wet roots and low soil oxygen levels.

"WE'RE A PROTECTED SPECIES AND AN ENDANGERED SPECIES, BUT WHEN I SEE A PERSON, I DON'T KNOW IF HE'S PROTECTING US OR ENDANGERING US."

Figure 6.17 **What kinds of common threats are faced by wetlands and by endangered species?**

CASE IN POINT

THE EVERGLADES

The Everglades in the southernmost part of Florida is an expanse of predominantly sawgrass wetlands dotted with small islands of trees. At one time, the "river of grass" drifted south in a slow-moving sheet of fresh water from Lake Okeechobee to Florida Bay (**Figure 6.18**). The Everglades is a haven for wildlife such as alligators, snakes, panthers, otters, raccoons, and thousands of wading birds and birds of prey; the Everglades also provide nurseries for a variety of edible and commercially valuable fish. The region's natural wonders were popularized in an environmental classic, *The Everglades: River of Grass*, written by **Marjory Stoneman Douglas** in 1947. The southernmost part of the Everglades is now protected as a national park.

The Everglades today is about half its original size, and it has many serious environmental problems. Most waterbird populations are down 93% in recent decades, and the area is now home to 50 endangered or threatened species and an increasing number of invasive species. Basically, two problems override all others in

(a) Extent of the original Everglades.

Figure 6.18 **Florida Everglades.**

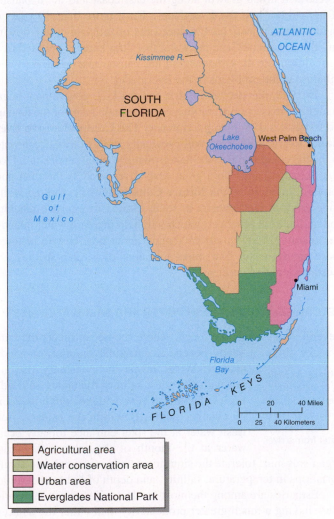

(b) The Everglades today.

- Agricultural area
- Water conservation area
- Urban area
- Everglades National Park

the Everglades today—it receives too little water, and the water it receives is polluted with nutrient minerals from agricultural runoff.

In its original state, water in the Everglades was nutrient-poor. During heavy rains, Lake Okeechobee flooded its banks, creating wetlands that provided biological habitat and helped recharge the Everglades. However, when a hurricane hit the lake in 1928 and many people died, the Army Corps of Engineers built the Hoover Dike along the eastern part of the lake. The Hoover Dike stopped the flooding but prevented the water in Lake Okeechobee from recharging the Everglades. Four canals built by the Everglades Drainage District drained 214,000 hectares (530,000 acres) south of Lake Okeechobee, which was converted to farmland. Fertilizers and pesticides used here make their way to the Everglades, where they alter native plant communities. Phosphorus is particularly harmful because it encourages the growth of nonnative cattails that overrun the native sawgrasses and disrupt the flow of water.

After several tropical storms caused flooding in south Florida in 1947, the Army Corps of Engineers constructed an extensive system of canals, levees, and pump stations to prevent flooding, provide drainage, and supply water to south Florida. These structures diverted excess water to the Atlantic Ocean rather than the Everglades. The resulting lands were drier, which encouraged accelerated urban growth along the east coast and the expansion of agriculture into the Everglades. Thus, more than 80 years of engineering projects have reduced the quantity of water flowing into the Everglades, and the water that does enter is polluted from agricultural runoff. Urbanization has caused loss of habitat, contributing to the problems of the Everglades.

In 1996, the state of Florida and the U.S. government began planning a massive restoration project to undo some of the damage from decades of human interference. **The Comprehensive Everglades Restoration Plan** will eventually supply clean, fresh water to the Everglades and to the fast-growing population of south Florida. Scientists must do a great deal of research as the plan goes forward. In addition, the **Nature Conservancy** is working with a wide array of land owners in south Florida to improve ranching and agricultural practices, which will help increase water quantity and quality. Many questions remain about how best to restore a more natural water flow, repel invasions of foreign species, and reestablish native species. ◼

ESTUARIES: WHERE FRESH WATER AND SALT WATER MEET

Several ecosystems may occur where the ocean meets the land: a rocky shore, a sandy beach, an intertidal mudflat, or a tidal

estuary A coastal body of water, partly surrounded by land, with access to the open ocean and a large supply of fresh water from a river.

estuary. Water levels in an estuary rise and fall with the tides; salinity fluctuates with tidal cycles, the time of year, and precipitation. Salinity changes gradually within the estuary, from fresh water at the river entrance to salty ocean water at the mouth of the estuary. Estuarine organisms must tolerate the significant daily, seasonal, and annual variations in temperature, salinity, and depth of light penetration.

Estuaries are among the most fertile ecosystems in the world, often having a much greater productivity than either the adjacent ocean or the fresh water upriver. This high productivity is brought about by four factors:

1. Nutrients are transported from the land into rivers and creeks that flow into the estuary.
2. Tidal action promotes a rapid circulation of nutrients and helps remove waste products.
3. A high level of light penetrates the shallow water.
4. The presence of many plants provides an extensive photosynthetic carpet and mechanically traps detritus, forming the base of a detritus food web.

Many species, including commercially important fishes and shellfish, spend their larval stages in estuaries among the protective tangle of decaying plants.

Temperate estuaries usually contain **salt marshes**, shallow wetlands dominated by salt-tolerant grasses (see the photograph in the Chapter 3 introduction). Salt marshes have often seemed to be worthless, empty stretches of land to uninformed people. As a result, they have been used as dumps and become severely polluted or have been filled with dredged bottom material to form artificial land for residential and industrial development. A large part of the estuarine environment has been lost in this way, along with many of its ecosystem services, such as biological habitats, sediment and pollution trapping, groundwater supply, and storm buffering. (Salt marshes absorb much of the energy of a storm surge and thereby prevent flood damage elsewhere.) According to the Environmental Protection Agency, estuaries in the United States are in fair to poor condition, with the greatest problems in northeastern and Gulf estuaries.

Mangrove forests, the tropical equivalent of salt marshes, cover perhaps 70% of tropical coastlines (**Figure 6.19**). Like salt marshes, mangrove forests provide valuable ecosystem services. Their interlacing roots are breeding grounds and nurseries for several commercially important fishes and shellfish, such as blue crabs, shrimp, mullet, and spotted sea trout. For example, biologists studied the number of commercially fished yellowtail snapper on coral reefs adjacent to neighboring mangrove-rich and mangrove-poor areas. In one study, they reported that the snapper biomass on reefs

Figure 6.19   Mangrove forest. Red mangroves (*Rhizophora mangle*) have stiltlike roots that support the tree. These roots grow into deeper water as well as into mudflats exposed by low tide. Many animals live in the protection of the root systems of mangrove forests. Photographed in Belize.

near mangrove-rich areas was two times greater than the biomass near mangrove-poor areas. (Recall from Chapter 3 that *biomass* is an estimate of the mass, or amount, of living material.)

Mangrove branches are nesting sites for many species of birds, such as pelicans, herons, egrets, and roseate spoonbills. Mangrove roots stabilize the submerged soil, thereby preventing coastal erosion, and provide a barrier against the ocean during storms such as hurricanes. Mangroves may be even more effective than concrete seawalls in dissipating wave energy and controlling floodwater from tropical storms. Unfortunately, mangroves are under assault from coastal development, unsustainable logging, and aquaculture (see Figure 18.18). Some countries, such as the Philippines, Bangladesh, and Guinea-Bissau, have lost 70% or more of their mangrove forests.

MARINE ECOSYSTEMS

Although lakes and the ocean are comparable in many ways, there are many physical differences. The depths of even the deepest lakes do not approach those of the ocean trenches, with areas that extend more than 6 km (3.6 mi) below the sunlit surface. Tides and currents exert a profound influence on the ocean. Gravitational pulls of both the sun and the moon usually produce two high tides and two low tides each day along the ocean's coastlines, but the height of those tides varies with the season, local topography, and phases of the moon (a full moon causes the highest tides).

Scientists from many countries use a variety of tools to study the ocean's geology, biology, and physical factors. Some drifting buoys sample surface-water temperatures and wind speeds. Other drifting buoys sink to as much as 2 km (1.2 mi), and then return to the surface to provide satellites with deepwater data. Moored buoys, chained to one spot on the ocean floor, measure wave height, current speeds, and carbon dioxide levels. Fiber-optic cables collect data on the ocean floor, such as crustal dynamics. Ships provide a variety of information, from sampling marine life to using sonar to map the ocean floor. Satellites help to track sea ice, oil spills, and algal blooms. Although scientists monitor many aspects of the ocean, much of the marine world remains a mystery. **Rachel Carson**, best known today for *Silent Spring*, her book on the impact of pesticides, wrote lyrically about the mysteries and science of the ocean in her books *The Edge of the Sea, The Sea Around Us,* and *Under the Sea-Wind.*

The immense marine environment is subdivided into several life zones: the intertidal zone, where organisms live along the shore between high and low tides; the benthic environment, where organisms live on or under the seafloor; and the pelagic environment, where organisms live in the water (**Figure 6.20**). The pelagic environment is, in turn, divided into two provinces based on depth: the neritic and oceanic provinces. The ocean comprises the neritic province, consisting of shallow waters close to shore, and the oceanic province, consisting of the rest of the ocean. The oceanic province overlies the ocean floor at depths greater than 200 m.

The Intertidal Zone: Transition Between Land and Ocean

Although the high levels of light and nutrients, together with an abundance of oxygen, make the **intertidal zone** a biologically productive habitat, it is a stressful one. If an intertidal beach is sandy,

intertidal zone The area of shoreline between low and high tides.

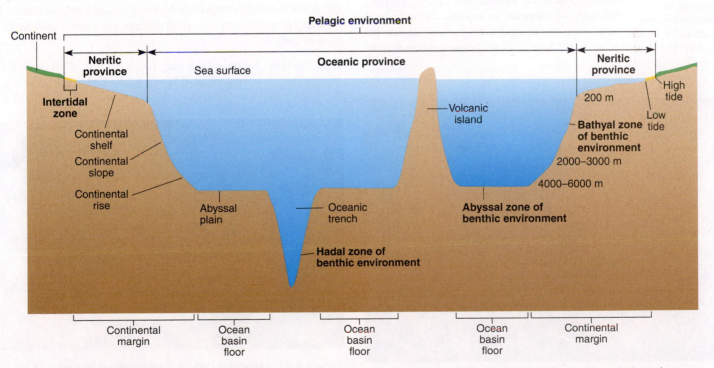

Figure 6.20 Zonation in the ocean. The ocean has three main life zones: the intertidal zone, the benthic environment, and the pelagic environment. The pelagic environment consists of the neritic and oceanic provinces. The neritic province overlies the ocean floor from the shoreline to a depth of 200 m. The oceanic province overlies the ocean floor at depths greater than 200 m. The ocean floor is not a flat expanse; it consists of mountains, valleys, canyons, seamounts, ridges, and trenches. (The slopes of the ocean floor are not typically as steep as shown; they are exaggerated to save space.)

inhabitants must contend with a constantly shifting environment that threatens to engulf them and gives them scant protection against wave action. Consequently, most sand-dwelling organisms, such as mole crabs, are continuous and active burrowers. They usually do not have notable adaptations to survive drying out or exposure because they follow the tides up and down the beach.

A rocky shore provides fine anchorage for seaweeds and marine animals but is exposed to wave action when immersed during high tides and to drying and temperature changes when exposed to air during low tides (**Figure 6.21**). A typical rocky-shore inhabitant has some way of sealing in moisture—perhaps by closing its shell, if it has one—plus a powerful means of anchoring itself to the rocks. Mussels have tough, thread-like anchors secreted by a gland in the foot, and barnacles have special glands that secrete a tightly bonding glue that hardens under water. Rocky-shore intertidal algae, such as rockweed (*Fucus*), usually have thick, gummy coats that dry out slowly when exposed to air, and flexible bodies not easily broken by wave action. Some rocky-shore organisms hide in burrows, under rocks, or in crevices at low tide, and some small crabs run about the splash line.

The intertidal zone, whether sand or rock, takes much of the force of storms such as hurricanes and typhoons. Development of beach areas for housing and recreation is tempting but obviously likely to result in expensive repeated rebuilding efforts.

The Benthic Environment: Seagrass Beds, Kelp Forests, and Coral Reefs

benthic environment
The ocean floor, which extends from the inter-tidal zone to the deep-ocean trenches.

Most of the **benthic environment** consists of sediments (mostly sand and mud) where many animals, such as worms and clams, burrow. Bacteria are common in marine sediments and have even been reported in ocean sediments more than 2.5 km (1.5 mi) below the ocean floor.

The deeper parts of the benthic environment are divided into three zones. They are, from shallowest to deepest, the bathyal, abyssal, and hadal zones. The *bathyal benthic zone* is the benthic environment that extends from a depth of 200 to 4000 m (650 ft to 2.5 mi). The *abyssal benthic zone* extends from a depth of 4000 m to a depth of 6000 m (2.5 to 3.7 mi), whereas the *hadal benthic zone* extends from 6000 m to the bottom of the deepest trenches. (Case in Point: Life Without the Sun in Chapter 3 examines the unusual organisms in the deeper part of the benthic environment.) Here we describe shallow benthic communities that are particularly productive—seagrass beds, kelp forests, and coral reefs.

Seagrass Beds. **Sea grasses** are flowering plants adapted to complete submersion in salty ocean water. They occur only in shallow water, to depths of 10 m (33 ft), where they receive enough light to photosynthesize efficiently. Extensive beds of sea grasses occur in quiet temperate, subtropical, and tropical waters (**Figure 6.22**). Eelgrass is the most widely distributed sea grass along the coasts of North America; the largest eelgrass bed in the world is in Izembek Lagoon on the Alaska Peninsula. The most common sea grasses in the Caribbean Sea are manatee grass and turtle grass.

Sea grasses have a high primary productivity and are ecologically important in shallow marine areas. Their roots and rhizomes help stabilize the sediments, reducing surface erosion. Sea grasses provide food and habitat for many marine organisms, including scallops and crabs. In temperate waters, ducks and geese eat sea grasses, whereas in tropical waters, manatees, green turtles, parrot fish, sturgeon, and sea urchins eat them. These herbivores consume only about 5% of the sea grasses. The remaining 95% eventually enters the detritus food web and is decomposed when the sea grasses die. Decomposing bacteria are, in turn, consumed by a variety of animals such as mud shrimp, lugworms, and mullet (a type of fish).

Kelp Forests. **Kelps**, which may reach lengths of 60 m (200 ft), are the largest brown algae (**Figure 6.23**). Kelps are common in

Figure 6.21   The intertidal zone. A sandy intertidal zone, with one organism, a seaweed with air bladders, well suited to survive until the next tide, and one organism, a Portugese man o' war, trapped onshore by the falling tide

Figure 6.22   Seagrass bed. A bivalve mollusk in a seagrass bed in Almeria, Spain. Sea grasses form underwater meadows that are ecologically important for shelter and food for many organisms. Note the many tiny epiphytic algae and invertebrate animals attached to the seagrass blades.

Figure 6.23 **Kelp forest.** These underwater "forests" support many kinds of aquatic organisms. Photographed off the coast of California.

cooler temperate marine waters of both the Northern and Southern Hemispheres. They are especially abundant in relatively shallow waters (depths of about 25 m, or 82 ft) along rocky coastlines. Kelps are photosynthetic and are the primary food producers for the kelp forest ecosystem. Kelp forests provide habitats for many marine animals. Tube worms, sponges, sea cucumbers, clams, crabs, fishes, and sea otters find refuge in the algal fronds. Some animals eat the fronds, but kelps are mainly consumed in the detritus food web. Bacteria that decompose the kelp remains provide food for sponges, tunicates, worms, clams, and snails. The diversity of life supported by kelp beds almost rivals that found in coral reefs.

Coral Reefs. Coral reefs, built from accumulated layers of calcium carbonate ($CaCO_3$), are found in warm (usually greater than 21°C), shallow seawater. The living portions of coral reefs must grow in shallow waters where light penetrates. Some coral reefs are composed of red coralline algae that require light for photosynthesis. Most coral reefs consist of colonies of millions of tiny coral animals, which require light for the large number of *zooxanthellae*, symbiotic algae that live and photosynthesize in their tissues. In addition to obtaining food from the zooxanthellae that live inside them, coral animals capture food at night with stinging tentacles that paralyze zooplankton and small animals that drift nearby. The waters where coral reefs are found are often poor in nutrients, but other factors are favorable for a high productivity, including the presence of zooxanthellae, a favorable temperature, and year-round sunlight.

Coral reefs grow slowly, as coral organisms build on the calcium carbonate remains of countless organisms before them. On the basis of their structure and underlying geologic features, three kinds of coral reefs exist: fringing reefs, atolls, and barrier reefs (**Figure 6.24**). The most common type of coral reef is the fringing reef. A **fringing reef** is directly attached to the shore of a volcanic island or continent and has no lagoon associated with it.

An **atoll** is a circular coral reef that surrounds a central lagoon of quiet water. An atoll forms on top of the cone of a submerged volcano island. More than 300 atolls are found in the Pacific and Indian Oceans, whereas the Atlantic Ocean, which is geologically different from the other two ocean basins, has few atolls. A lagoon of open water separates a **barrier reef** from the nearby land. The world's largest barrier reef is the Great Barrier Reef, which is nearly 2000 km (more than 1200 mi) in length and up to 100 km (62 mi) across. It extends along the northeastern coast of Australia. The second-largest barrier reef is the Mesoamerican Reef in the Caribbean Sea off the coast of Belize, Mexico, and Honduras.

Coral reef ecosystems are the most diverse of all marine environments. They contain thousands of species of fishes and invertebrates, such as giant clams, snails, sea urchins, sea stars, sponges, flatworms, brittle stars, sea fans, shrimp, and spiny lobsters. The Great Barrier Reef occupies only 0.1% of the ocean's surface, but 8% of the world's fish species live there. Competition is intense, particularly for light and space to grow. According to the National Oceanic and Atmospheric Administration (NOAA), fish dependent on coral reefs provide significant protein for at least one billion people worldwide.

Although coral reefs are often found in shallow, tropical waters, many highly diverse reefs are located in cold water. Cold-water coral reefs, which are found on continental shelves,

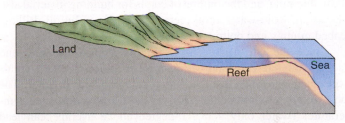

(a) Fringing reef

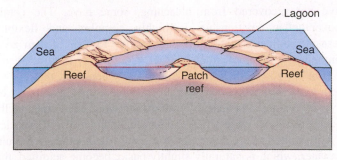

(b) Atoll

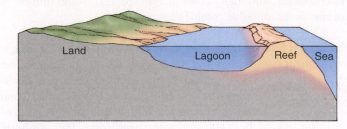

(c) Barrier reef

Figure 6.24 **Types of coral reefs.**

slopes, and seamounts, occur at depths from 50 m to 6 km (164 ft to almost 4 mi). Most of these reefs live in perpetual darkness. Unlike photosynthetic zooxanthellae, these corals rely on capturing prey to supply their food. Like the more familiar shallow-water reefs, cold-water coral reefs are hotbeds of biological diversity. So far, scientists have identified more than 3300 species of cold-water corals.

Coral reefs are ecologically important because they both provide a habitat for many kinds of marine organisms and protect coastlines from shoreline erosion. They provide humans with seafood, pharmaceuticals, and recreational/tourism dollars.

Human Impacts on Coral Reefs.

Although coral formations are important ecosystems, they are being degraded and destroyed. Of 109 countries with large reef formations, 90 are damaging them. According to the latest report of the Global Coral Reef Monitoring Network, most coral reefs (61%) face moderate to severe threats due to local human pressures, though some areas (Hawaii, Australia) have done a better job than others at protecting reefs, with less than 20% of reefs there under threat.

How do we harm coral reefs? In some areas, silt washing downstream from clear-cut inland forests has smothered reefs under a layer of sediment. In addition to pollution from coastal runoff, overfishing, fishing with dynamite or cyanide, disease, and coral bleaching are serious threats. Land reclamation, tourism, oil spills, boat groundings, anchor draggings, hurricane damage, ocean dumping, and the mining of corals for building material also take a toll. Cold-water coral reefs are threatened by oil drilling, seabed mining, and bottom trawling.

Since the late 1980s, corals in the tropical Atlantic and Pacific suffered extensive bleaching, in which stressed corals expel their zooxanthellae, becoming pale or white in color. The most likely environmental stressor is warmer seawater temperatures (water only about 1 degree above average can lead to bleaching) and increased ocean acidity due to higher levels of dissolved carbon dioxide. Although many coral reefs have not recovered from bleaching, some have. The coral-zooxanthellae relationship is highly complex and flexible. Research indicates that corals may lose up to 75% of their zooxanthellae without harming the reef. Corals may hold a "secret reserve" of zooxanthellae, not immediately apparent, that allows them to recover when bleached. In addition, corals take in any of several zooxanthella species, perhaps allowing one zooxanthella species to rescue coral when another abandons it. (Some species of zooxanthellae are more stress-resistant than others.) Should seawater, which is naturally slightly basic, become acidified, the calcium carbonate skeletons of coral animals (and shells of crabs, oysters, clams, and many other marine species) would thin or, in extreme cases, dissolve completely away.

The Pelagic Environment: The Vast Marine System

The **pelagic environment** consists of all of the ocean water, from the shoreline down to the deepest ocean trenches. It is subdivided into several life zones based on depth and degree of light penetration. The upper reaches of the pelagic environment form the **euphotic**

pelagic environment
All ocean water, from the shoreline down to the deepest ocean trenches

zone, which extends from the surface to a maximum depth of 150 m (488 ft) in the clearest ocean water. Sufficient light penetrates the euphotic zone to support photosynthesis. Large numbers of phytoplankton, particularly diatoms in cooler waters and dinoflagellates in warmer waters, produce food by photosynthesis and are the base of marine food webs. The marine system buffers global temperatures against seasonal temperature changes.

The two main divisions of the pelagic environment are the neritic and oceanic provinces.

The Neritic Province.

Organisms that live in the **neritic province,** the water that overlies the continental shelf, are all floaters or swimmers. Zooplankton (including tiny crustaceans; jellyfish; comb jellies; protists such as foraminiferans; and larvae of barnacles, sea urchins, worms, and crabs) feed on phytoplankton in the euphotic zone. Zooplankton are consumed by plankton-eating nekton such as herring, sardines, squid, baleen whales, and manta rays. These, in turn, become prey for carnivorous nekton such as sharks, toothed whales, tuna, and porpoises (**Figure 6.25**). Nekton are mostly confined to the shallower neritic waters (less than 60 m, or 195 ft, deep), near their food.

neritic province The part of the pelagic environment that overlies the ocean floor from the shoreline to a depth of 200 m (650 ft).

The Oceanic Province.

The **oceanic province** is the largest marine environment, comprising about 75% of the ocean's water; the oceanic province is the open ocean that does not overlie the continental shelf. Most of the oceanic province is loosely described as the *deep sea*. (The average depth of the ocean is 4 km, more than 2 mi.) All but the surface waters of the oceanic province have cold temperatures, high hydrostatic pressure, and an absence of sunlight. These environmental conditions are uniform throughout the year.

oceanic province The part of the pelagic environment that overlies the ocean floor at depths greater than 200 m (650 ft).

Most organisms of the deep waters of the oceanic province depend on **marine snow**, organic debris that drifts down into their habitat from the upper, lighted regions of the oceanic province.

Figure 6.25 **Nekton in the neritic province.** A porpoise leaps briefly out of the Pacific Ocean, near the North Island of New Zealand.

© Greg Gilman/Alamy

Figure 6.26 **Hydrothermal vent community.** *Riftia pachyptila* tubeworms and nestled *Bathymodiolus thermophilus* mussels on the East Pacific Rise. This community is on the ocean floor in complete darkness, but produces food using chemosynthesis (like photosynthesis) with heat from a hydrothermal vent.

Organisms of this little-known realm are filter feeders, scavengers, or predators. Many are invertebrates, some of which attain great sizes. The giant squid measures up to 13 m (43 ft) in length, including its tentacles.

Fishes of the deep waters of the oceanic province are strikingly adapted to darkness, high pressure, and scarcity of food. An organism that encounters food infrequently must eat as much as possible when food is available. Adapted to drifting or slow swimming, animals of the oceanic province often have reduced bone and muscle mass. Many of these animals have light-producing organs to locate one another for mating or food capture. The dragon fish has a pocket of red light shining from beneath each eye. Because other species living in the ocean's depths cannot see red light, the dragonfish detects organisms in its surroundings without being seen. (Most animals in the ocean depths are bioluminescent, producing a blue-green light.)

Some deep-sea organisms, however, are producers, using heat from hydrothermal vents as an energy source to fuel a process that chemically parallels photosynthesis. Communities living at hydrothermal vents include clams, mussels, and tubeworms and bacteria living symbiotically within their tissues (**Figure 6.26**). Other organisms, such as barnacles, also live at these vents, in the shelter created by the tubeworms.

National Marine Sanctuaries The United States has set aside **national marine sanctuaries** along the Atlantic, Pacific, and Gulf of Mexico coasts to minimize human impacts and protect unique natural resources and historic sites. These sanctuaries include kelp forests off the coast of California, coral reefs in the Florida Keys, fishing grounds along the continental shelf, and deep submarine canyons, as well as shipwrecks and other sites of historic value (**Figure 6.27**).

The *National Marine Sanctuary Program*, which is part of the National Oceanic and Atmospheric Administration (NOAA), administers the sanctuaries. Like many federal lands, they are managed for multiple purposes, including conservation, recreation, education, mining of some resources, scientific research, and ship salvaging. Commercial fishing is permitted in most of them, although several "no-take" zones exist where all fishing and collecting of biological resources is banned. Several recent studies have shown that these no-take zones for fish promote population increases of individual species as well as species richness (the no-take reserves typically contain 20% to 30% more

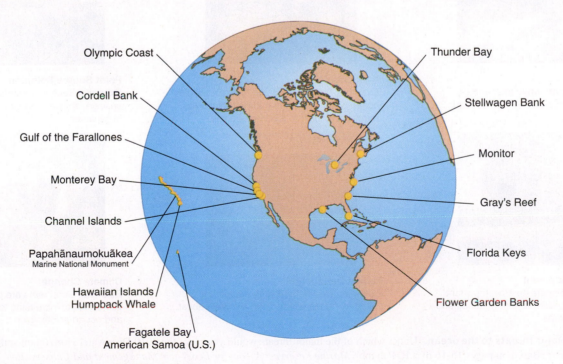

Figure 6.27 **National marine sanctuaries.** (NOAA)

species than nearby fished areas). The reserves have a spillover effect—that is, they boost populations of fish surrounding their borders.

The National Marine Sanctuary system includes 14 sites—13 national marine sanctuaries and one national monument. In 2006, President George W. Bush established the world's largest protected marine area when he designated the northwestern Hawaiian Islands and surrounding waters—an area almost as large as California—as a national monument. Such a designation provides permanent funding to manage and preserve the area. This protected area is home to thousands of species of seabirds, fishes, marine mammals, and coral reef colonies.

Human Impacts on the Ocean The ocean is so vast that it is hard to visualize how human activities could harm it. Such is the case, however (**Figure 6.28**). The development of resorts, cities, industries, and agriculture along coasts alters or destroys many coastal ecosystems, including mangrove forests, salt marshes, seagrass beds, and coral reefs. Coastal and marine ecosystems receive pollution from land, from rivers emptying into the ocean, and from atmospheric contaminants that enter the ocean via precipitation. Disease-causing viruses and bacteria from human sewage contaminate seafood, such as shellfish, and pose an increasing threat to public health. Millions of tons of trash, including plastic, fishing nets, and packaging materials, find their way into coastal and

Nonpoint Source Pollution (runoff from land)
Example: Agricultural runoff (fertilizers, pesticides, and livestock wastes) pollutes water.

Invasive Species
Example: Release of ships' ballast water, which contains foreign crabs, mussels, worms, and fishes.

Overfishing
Example: The populations of many commercial fish species are severely depleted.

Bycatch
Example: Fishermen unintentionally kill dolphins, sea turtles, and seabirds.

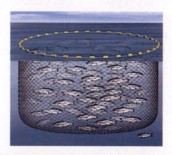

Aquaculture
Example: Produces wastes that can pollute ocean water and harm marine organisms; requires wild fish to feed farmed fish.

Point Source Pollution
Example: Passenger cruise ships dump sewage, shower and sink water, and oily bilge water.

Coastal Development
Example: Developers destroy important coastal habitat, such as salt marshes and mangrove swamps.

Habitat Destruction
Example: Trawl nets (fishing equipment pulled along the ocean floor) destroy habitat.

Climate Change
Example: Coral reefs are particularly vulnerable to increasing temperatures and ocean acidification.

Figure 6.28 **Major threats to the ocean.** Under which of the major threats would you place the 2010 Deepwater Horizon oil spill in the Gulf of Mexico? (Based on pages 15-16 of S.R. Palumbi, *Marine Reserves: A Tool for Ecosystem Management and Conservation*, Pew Oceans Commission [2003])

marine ecosystems. Some of this trash entangles and kills marine organisms. Less visible contaminants of the ocean include fertilizers, pesticides, heavy metals, and synthetic chemicals from agriculture and industry.

Offshore mining and oil drilling pollute the neritic province with oil and other contaminants. Millions of ships dump oily ballast and other wastes overboard in the neritic and oceanic provinces. Fishing is highly mechanized, and new technologies detect and remove every single fish in a targeted area of the ocean. Scallop dredges and shrimp trawls are dragged across the benthic environment, destroying entire communities with a single swipe.

Conservation groups and government agencies have studied these problems for many years and have made numerous recommendations to protect and manage the ocean's resources. For example, from 2002 to 2003, the Pew Oceans Commission, composed of scientists, economists, fishermen, and other experts, verified the seriousness of ocean problems in a series of seven studies.

The 2004 report by the U.S. Commission on Ocean Policy, the first comprehensive review of federal ocean policy in 35 years, recommended improving the ocean and coasts through improved policy and funding initiatives.

The recommended actions will require billions of dollars and years of effort to be realized.

REVIEW

1. What environmental factors are most important in determining the kinds of organisms found in aquatic environments?

2. How do you distinguish between freshwater wetlands and estuaries? Between flowing-water and standing-water ecosystems? Between the neritic and oceanic provinces?

3. List an ecosystem service provided by each of the aquatic ecosystems.

REVIEW OF LEARNING OBJECTIVES WITH SELECTED KEY TERMS

- **Define *biome* and briefly describe the nine major terrestrial biomes: tundra, boreal forest, temperate rain forest, temperate deciduous forest, temperate grassland, chaparral, desert, savanna, and tropical rain forest.**

A **biome** is a large, relatively distinct terrestrial region with similar climate, soil, plants, and animals, regardless of where it occurs in the world. **Tundra** is the treeless biome in the far north that consists of boggy plains covered by lichens and small plants such as mosses; tundra has harsh, very cold winters and extremely short summers. **Boreal forest** is a region of coniferous forest (such as pine, spruce, and fir) in the Northern Hemisphere; it is located south of the tundra. **Temperate rain forest** is a coniferous biome with cool weather, dense fog, and high precipitation. **Temperate deciduous forest** is a forest biome that occurs in temperate areas with a moderate amount of precipitation. **Temperate grassland** is grassland with hot summers, cold winters, and less rainfall than the temperate deciduous forest biome. **Chaparral** is a biome with mild, moist winters and hot, dry summers; vegetation is typically small-leaved evergreen shrubs and small trees. **Desert** is a biome in which the lack of precipitation limits plant growth; deserts are found in both temperate and subtropical regions. **Savanna** is tropical grassland with widely scattered trees or clumps of trees. **Tropical rain forest** is a lush, species-rich forest biome that occurs where the climate is warm and moist throughout the year.

- **Identify at least one human food that grows well in each of the biomes discussed.**

Animal protein, from fishing and hunting, is the primary source of food available in arctic tundra, as well as in boreal forest, though some plants are available in arctic summer. Temperate rain forest fosters edible mushrooms, cattails, and some berries, like salmonberry. Nuts such as hickory, pecan, and chestnut grow in deciduous forest, as well as fruits such as pawpaw. Annual and perennial grasses grow well in prairie biomes. Chaparral can support vineyards and olive orchards. Savanna is best suited for grazing animals, both wild game and livestock. Tropical rain-forest ecosystems support a wide variety of fruit (bananas, acai), edible plants, and game animals.

- **Explain the similarities and the changes in ecosystems observed with increasing elevation and increasing latitude.**

Similar ecosystems are encountered in climbing a mountain (increasing elevation) and traveling to the North Pole (increasing latitude). This elevation-latitude similarity occurs because the temperature drops as one climbs a mountain, just as it does when one travels north.

- **Summarize the important environmental factors that affect aquatic ecosystems.**

In aquatic ecosystems, important environmental factors include **salinity**, amount of dissolved oxygen, and availability of light for photosynthesis.

- **Briefly describe the eight aquatic ecosystems: flowing-water ecosystem, standing-water ecosystem, freshwater wetlands, estuary, intertidal zone, benthic environment, neritic province, and oceanic province.**

A **flowing-water ecosystem** is a freshwater ecosystem such as a river or stream in which the water flows in a current. A **standing-water ecosystem** is a body of fresh water that is surrounded by land and that does not flow (a lake or a pond). **Freshwater wetlands** are lands that shallow fresh water covers for at least part of the year; wetlands have a characteristic soil and water-tolerant vegetation. An **estuary** is a coastal body of water, partly surrounded by land, with access to the open ocean and a large supply of fresh water from a river. Four important marine environments are the intertidal zone, benthic environment, neritic province, and oceanic province. The **intertidal zone** is the area of shoreline between low and high tides. The **benthic environment** is the ocean floor, which extends from the intertidal zone to the deep-ocean trenches. The **pelagic environment** consists of all of the ocean water, from the shoreline down to the deepest ocean trenches. The two main divisions of the pelagic environment are the neritic and oceanic provinces. The **neritic province** is the part of the pelagic environment that overlies the ocean floor from the shoreline to a depth of 200 m. The **oceanic province** is the part of the pelagic environment that overlies the ocean floor at depths greater than 200 m.

• **Identify at least one ecosystem service provided by each aquatic ecosystem.**

Flowing-water ecosystems provide many people with fresh drinking water, and standing water ecosystems provide habitat for aquatic life as well as opportunities for recreation. Wetlands and estuaries both protect land from heavy rainfall events and storm surges. Intertidal zones offer opportunities for recreation, and protect land from hurricanes and typhoons. Benthic environments (seagrass, kelp forests, and coral reefs) are areas rich in biodiversity, which support nurseries of many commercially valuable fish. Larvae of crabs, as well as herring, sardines, and squid, live in the neritic province. Most commercially valuable fish live in the oceanic province. Both neritic and oceanic provinces help buffer landmasses against temperature changes.

Critical Thinking and Review Questions

1. What two climate factors are most important in determining an area's characteristic biome?

2. Offer a possible reason why the tundra has such a low species richness.

3. Describe representative organisms of the forest biomes discussed in the text: boreal forest, temperate deciduous forest, temperate rain forest, and tropical rain forest.

4. Which biome discussed in this chapter is depicted by the information given below? Explain your answer.

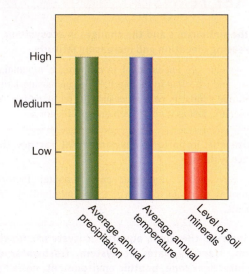

5. In which biome do you live? If your biome does not match a description given in this chapter, how do you explain the discrepancy?

6. Which biomes are best suited for agriculture? Explain why each of the biomes you did not specify is less suitable for agriculture.

7. In biomes less suited for agriculture, what natural food sources are available for humans?

8. Which biome do you think is in greatest immediate danger from human activities? Why?

9. As you walk from the bottom to the top of a mountain, what changes in vegetation would you expect to see?

10. What are two important abiotic factors that affect aquatic ecosystems?

11. Explain the role of freshwater wetlands in water purification.

12. If you were to find yourself on a boat in the Chesapeake Bay, what aquatic ecosystem would you be in? What ecosystem would you be in if you were in the middle of Everglades National Park?

13. Which aquatic ecosystem is often compared to a tropical rain forest? Why?

14. What is the largest marine environment, and what are some of its features?

15. Which of the ocean zones shown would be home to each of the following organisms: lobster, coral, mussel, porpoise, and dragon fish. For those organisms you identified as living in the pelagic environment, are they most likely found in the neritic or oceanic province? Explain your answers.

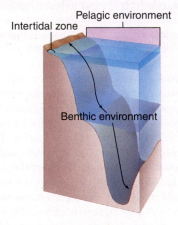

16. What would happen to the organisms in a river with a fast current if a dam was built? Explain your answer.

17. Explain how coral reefs are vulnerable to human activities.

18. Briefly discuss two major human-caused problems in the ocean.

19. Scientists report that in recent years tropical ocean waters have become saltier, whereas polar ocean waters have become less salty. What do you think is a possible explanation for these changes?

Food for Thought

Over the next century, changing temperatures and changes in precipitation are projected to reduce agricultural productivity in some areas and reduce it in others. The figure shows the likely changes in productivity across the globe. Choose a region of interest, and examine how food productivity is likely to change in that region. How might population and trade opportunities change with food productivity?

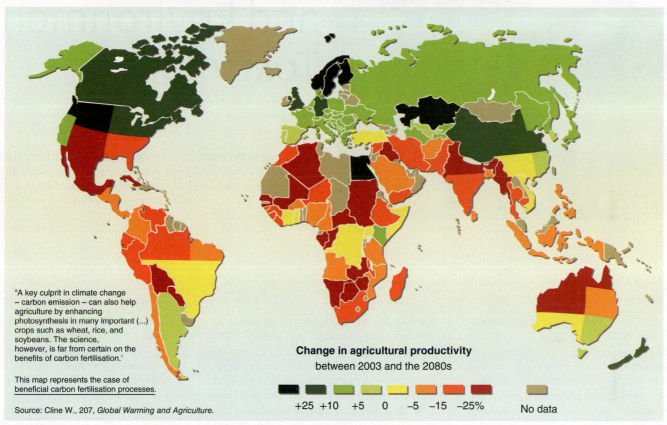

Projected impact of climate change on agricultural yields

"A key culprit in climate change – carbon emission – can also help agriculture by enhancing photosynthesis in many important (...) crops such as wheat, rice, and soybeans. The science, however, is far from certain on the benefits of carbon fertilisation.'

This map represents the case of beneficial carbon fertilisation processes.

Source: Cline W., 207, *Global Warming and Agriculture*.

Change in agricultural productivity
between 2003 and the 2080s

+25 +10 +5 0 −5 −15 −25%   No data

7

Human Health and Environmental Toxicology

Cars line up in the morning at an elementary school. More children are driven to school today than in the past, when most children walked or rode bicycles. Seemingly simple lifestyle choices, such as driving children to school rather than having them walk, can have significant impacts on their health.

© Don Kohlbauer / Zuma Press

Often, when people think about environment and health, they focus on things like pollution and pesticides. Certainly, toxic substances can sicken people, and there's a need to reduce or eliminate them. However, in the United States, pollution is only one of many aspects of the relationship between healthy people and a healthy environment. How individuals and societies interact with the environment has implications for lifelong physical and mental health. Thinking about humans and the environment functioning as a system allows us to understand where and how changes in the system can improve human and environmental health.

Today, for example, many children in suburban America are driven to school, whereas several decades ago they would have walked or bicycled (**see photograph**). Although this change might seem minor, it represents a dramatic change in how children interact with their environment. They are establishing behaviors that will last a lifetime, with long-term health implications. Effects include reduced exercise, increased pollution and accident risks, and the loss of interaction with the natural and physical world.

Consider a typical U.S. suburban school, with a thousand students. Even if only one-third of students are driven to school—the

fraction can be much higher—as many as 300 cars, minivans, SUVs, and pickup trucks may be driving to the school, all at the same time. The large numbers of vehicles present a physical threat to those children (and their parents, siblings, or grandparents) who do walk, especially when drivers are distracted by passengers, other drivers, and cell phones. In addition, hundreds of gasoline-fueled engines idling and accelerating near occupied playgrounds can elevate air pollution levels in areas where children play before each school day begins. Finally, many children who are driven to school on a regular basis miss out on opportunities to experience daily and seasonal changes in weather. Small animals like insects, spiders, and snails go unnoticed, as do the details of trees and grasses.

Children driven to school might otherwise walk a mile or so each day. They miss out on light physical activity that is not made up in other ways. Childhood inactivity is linked to increased adult obesity, which, in turn, is associated with elevated blood pressure, high cholesterol, and a low level of high-density lipoproteins (HDLs, the "good" cholesterol). These conditions harm blood vessels, increasing the risk of stroke, heart disease, and kidney failure. Type 2 diabetes, a

dangerous medical condition, is also related to obesity, as are osteoarthritis and cancers of the colon, gallbladder, prostate, kidney, cervix, ovaries, and uterus.

Over the past century, human life spans and general health have improved markedly. This has come from improved medicine and nutrition, a better understanding of causes and treatments of disease, and reductions in poverty. However, recent studies show that life expectancies have decreased among the poor in the United States, in particular among poor women. Understanding the systemic relationship between human health and the natural, physical, and built environment is critical to a future of improved health.

In this chapter, we explore the factors that underlie health and disease in highly developed and developing countries. In future chapters, we will explore a range of related environment–human health interactions.

In Your Own Backyard...
What changes could you make in your daily routines that would both reduce your use of fossil fuels and improve your health?

Human Health

LEARNING OBJECTIVES

- Contrast health issues in highly developed countries with those in less developed countries.
- Explain the difference between endemic and emerging diseases.

Two indicators of human health in a given country are **life expectancy** (how long people are expected to live) and **infant mortality** (how many infants die before the age of one). These health indicators vividly demonstrate the contrasts in health among different nations. In Japan, the average life expectancy for a woman is 86 and for a man, 79. The **infant mortality rate** in Japan is 2.6, which means that fewer than three infants under age one die per 1,000 live births. In contrast, in the East African country Zambia, the average life expectancy is 42 for women and 41 for men. The infant mortality rate in Zambia is 70, which means that 7% of Zambian infants die before reaching the age of one.

Why is there such a difference between these two countries? A Japanese child receives good healthcare, including vaccinations, and has sufficient nutrition for growth and development. The average Japanese woman gives birth to one or two children and, during her pregnancies, can eat well, rest when necessary, and receive medical attention as needed. As Japanese people age, they may develop the chronic degenerative diseases associated with aging, but they have access to quality healthcare, medications, and rehabilitation services. The average person in Japan spends $550 (in U.S. dollars) on medicines each year.

The picture is very different in Zambia. Many Zambian children receive few immunizations and inadequate nutrition for normal growth and development. In 2007, 15% of Zambian children were underweight (down from 28% in 2001). The average Zambian woman is in poor health due to childhood factors. She may give birth to five or six children. Because she does not receive prenatal care, she has an increased risk of death during childbirth. Moreover, one or more of her children will probably die in infancy. The average Zambian spends about $5 on medicines each year. Those Zambians who survive middle age often die prematurely because they lack treatment for the chronic degenerative diseases associated with aging.

Differences in health and healthcare between highly developed and developing countries highlight the effects of different lifestyles and levels of poverty (**Figure 7.1.**). In developing countries, about 146 million children are underweight, and 3 million of these will die this year. Meanwhile, more than 1 billion people are overweight worldwide (compared to 1 billion who are undernourished), and 300 million people are obese. In the highly developed countries of North America and Europe, about 500,000 people will die this year from obesity-related diseases.

HEALTH ISSUES IN HIGHLY DEVELOPED COUNTRIES

By many measures, the health of people in the United States and other highly developed countries is good. Improved sanitation during the 20th century reduced many diseases, such as typhoid,

Figure 7.1 **Health Clinic in Bangladesh.** This health clinic provides a range of services to Rohingya, a stateless group of Muslims, at the Leda refugee camp.

cholera, and diarrhea, that previously made people ill or killed them. Many childhood diseases, such as measles, polio, and mumps, have been conquered. In 1900, the average life expectancy for Americans was 51 years for women and 48 years for men. Today, the average life expectancy is 80 for women and 75 for men.

In 1900, the three leading causes of death in the United States were pneumonia and influenza, tuberculosis, and gastritis and colitis (diarrhea); these are infectious diseases caused by microorganisms. Currently, the three leading causes of death in the United States are cardiovascular diseases (of the heart and blood vessels), cancer, and chronic obstructive pulmonary disease (of the lungs); these diseases are noninfectious chronic health problems and are associated with aging. A significant fraction of premature deaths in the United States are caused by poor diet, lack of exercise, and smoking. According to a 2008 study, these factors have led to a reduction in life expectancy of 1.3 years for women in 180 of the poorest U.S. counties and for men in 11 of those counties.

Undernutrition is uncommon in the United States, although about one in six of the elderly are undernourished. However, two related types of malnourishment represent a growing health threat. Many Americans consume inadequate amounts of vitamins, fiber, and essential nutrients. In addition, typical American diets include too many total calories, resulting in a significant obesity problem.

Healthcare professionals use the **body mass index (BMI)** to determine whether a person is overweight or obese. To calculate your BMI, multiply your weight by 703, then divide that number by your height in inches squared:

$$BMI = (Weight \times 703) \div Height^2$$

For example, if you are 5 feet tall and weigh 130 pounds: $130 \times 703 = 91{,}390$. Five feet equals 60 inches, and the square of 60 is 3600. Now you can divide: $91{,}390 \div 3600 = 25.4$.

- If your BMI = 18.5 or less, you are considered underweight.
- If your BMI is from 18.5 to 24.9, you are at a healthy weight.
- If your BMI is between 25 and 29.9, you are overweight.
- If your BMI is 30 or more, you are considered obese.

Hierry Falise / Getty_Images

According to these guidelines, a person whose BMI is 26.7 is considered overweight but not obese.

HEALTH ISSUES IN DEVELOPING COUNTRIES

There is good news and bad news regarding the health and well-being of people in developing countries. Many endemic diseases—

endemic disease
A disease that is constantly present, although often varying in prevalence and potency, in a region or population.

those that are constantly present in a region or country—are being managed or eliminated. Gradual improvements in sanitation and drinking-water supplies in moderately developed countries are reducing the incidence of diarrheal diseases such as cholera. Mass immunization programs have eliminated smallpox and reduced the risk of polio, yellow fever, measles, and diphtheria in most countries (see Meeting the Challenge: Global Polio Eradication). Despite these gains, malnutrition, unsafe water, poor sanitation, and air pollution still prevail in many less developed countries (**Figure 7.2**). In addition, research suggests that many unhealthy lifestyles associated with developed countries, including smoking, obesity, and diabetes, are increasingly common in less developed countries. **Table 7.1** contains 10 facts that the World Health Organization considers most important about diseases worldwide.

Although overall life expectancy worldwide has increased to 69 years, four of the very poorest developing countries (Zambia, Zimbabwe, Lesotho, and Afghanistan) still have life expectancies of 45 years or less. HIV/AIDS, which is managed by medical treatments in much of the world, has reduced life expectancies by more than 20 years in the African countries of Botswana, Lesotho, Swaziland, and Zimbabwe. Worldwide, 34 million people were living with HIV/AIDS worldwide in 2011, and 2.5 million of those were newly infected that year.

According to the World Health Organization, about 10 million children under five years of age die each year, most of whom could be saved with inexpensive treatments. Child mortality is particularly serious in Africa, where 14 countries have higher levels

Table 7.1   Ten Facts on the Global Burden of Disease
1. Around 6.6 million children under the age of five died in 2012.
2. Cardiovascular diseases are the leading causes of death in the world.
3. HIV/AIDS is the leading cause of death in Sub-Saharan Africa.
4. Population aging is contributing to the rise in cancer and heart disease.
5. Lung cancer is the most common cause of death from cancer in the world.
6. Complications of pregnancy account for almost 15% of deaths in women of reproductive age worldwide; 99% of these are in developing countries.
7. Mental disorders such as depression are among the 20 leading causes of disability worldwide.
8. Hearing loss, vision problems, and mental disorders are the most common causes of disability.
9. Almost 10% of adults worldwide suffer from diabetes.
10. About 75% of the new infectious diseases affecting humans over the past 10 years were caused by bacteria, viruses, and other pathogens that originated in animals or animal products.

Source: World Health Organization, Global Burden of Disease Study (December 2013).

of child mortality today than they did in 1990. Leading causes of death for children in developing countries include malnutrition, lower respiratory tract infections, diarrheal diseases, and malaria. In sub-Saharan Africa, HIV/AIDS is a significant cause of death for many young children.

EMERGING AND REEMERGING DISEASES

At one time, it was mistakenly thought that all infectious diseases either had been or would soon be conquered. We now know this is not true. **Emerging diseases** are infectious diseases that were not previously found in humans; they typically jump from an animal host to the human species. Acquired immune deficiency syndrome (AIDS) demonstrates why we are concerned about emerging disease: About 36 million people have died from HIV/AIDS since the epidemic began, and 34 million people currently live with the disease (see Chapter 8 introduction). HIV/AIDS is now considered endemic in many places. Epidemiologists think the HIV virus jumped from nonhuman primates to humans as long as 100 years ago, perhaps when humans were exposed to contaminated blood when butchering or eating chimpanzees (**Figure 7.3**). Currently, there is an international network of medical personnel looking for—and trying to prevent the spread of—diseases transmitted from monkeys to humans.

Other emerging diseases include Lyme disease, West Nile virus, Creutzfeld-Jakob disease (the human equivalent of mad cow disease), severe acute respiratory syndrome (SARS), Ebola virus, and monkeypox. In addition, new strains of influenza arise each year, some far more deadly than others. To successfully contain an emerging disease, epidemiologists must recognize the symptoms, identify the disease-causing agent, and inform public health authorities. In turn, public health officials must isolate patients with those symptoms and track down all people who have come into contact with the patients. Meanwhile, medical researchers develop treatment strategies and try to determine and eliminate the origins of the disease.

Robert Caputo/Aurora Photos, Inc.

Figure 7.2   Workers burning trash in Asab, Eritrea. In many developing countries, sorting and burning garbage with little or no protective equipment is common. Workers risk infection, injury, and toxic exposures.

MEETING THE CHALLENGE

Global Polio Eradication

During the 20th century, significant progress was made in eradicating certain infectious diseases caused by viruses. Smallpox, once endemic in many countries, was eliminated worldwide, and polio (more correctly called *poliomyelitis*) was eliminated in North and South America, Europe, Australia, and much of Asia.

Children under the age of five are the main victims of polio, which is spread by contaminated drinking water. The poliovirus attacks the central nervous system (brain and spinal cord), causing paralysis and, in some cases, death. Polio vaccines, first developed in the 1950s, are usually administered orally (by mouth), and a series of four doses is required to provide complete protection to infants.

Supplementary doses are given to children under five years of age.

In 1988, polio was endemic in more than 125 countries. Since 1988, the World Health Organization (WHO) has conducted a vigorous campaign to eliminate polio worldwide. For example, a single case of polio reported in Egypt triggered a campaign to vaccinate at least 1 million children there. India is also making significant progress; after an epidemic in 2002, 100 million children were vaccinated! In 2013, there were fewer than 500 polio cases worldwide, down from about 1,600 in 2008.

In Africa, health experts conduct an annual synchronized campaign to vaccinate children during the peak season for polio. However, in 2003, there was a major setback in efforts to eliminate polio in Africa when the Nigerian state of Kano suspended vaccinations because of concerns about vaccine safety. An outbreak of polio occurred in Nigeria, reinfecting areas that were previously polio-free. Polio also spread from northern Nigeria to other African countries, including Sudan, the Central African Republic, Niger, Chad, Mali, Ivory Coast, Cameroon, Burkina Faso, Guinea, Benin, Botswana, Egypt, and Ethiopia, causing the worst epidemic in years.

In 2004, vaccinations resumed in Kano after health officials found a new supplier of the vaccine. Health officials also conducted a synchronized campaign in nearby countries in an effort to contain the virus in Nigeria. However, the quickness with which polio rebounded after suspension of vaccinations vividly demonstrated that serious efforts at both the national and international levels are needed to make polio the second disease to be globally eradicated.

Reemerging diseases are infectious diseases that existed in the past but, for a variety of reasons, are increasing in incidence or in geographic range. The most serious reemerging disease is tuberculosis. People with HIV/AIDS are more susceptible to tuberculosis because their immune systems are compromised. Thus, the HIV/AIDS pandemic has helped fuel the tuberculosis outbreak. Tuberculosis is associated with poverty, and the increased incidence of urban poverty has also contributed to the emergence of tuberculosis. Other important reemerging diseases include yellow fever, malaria, and dengue fever.

One phenomenon responsible for an increased incidence of tuberculosis and other diseases is the evolution of antibiotic-resistant strains of bacteria. Two factors contribute to *antibiotic-resistant bacteria*. First, antibiotics are often overused. People may take unneeded antibiotics, which accelerates the evolution of resistant bacterial strains. Extensive use of antibiotics in the agricultural industry may contribute to this problem.

Second, in some cases antibiotics are underused—rather than completely wiping out a strain, a few individual bacteria are inadvertently allowed to survive, and they are relatively resistant. This can happen at a societal scale, when a few individuals do not get treated, allowing populations of bacteria to survive. It can also happen at the individual level, when a person fails to complete a course of prescribed antibiotics. Failing to complete a prescription can allow some of the most resistant organisms to survive.

Health experts have determined the main factors involved in the emergence or reemergence of infectious diseases. Some of the most important are these:

- Evolution in the infectious organisms, allowing them to move between animal and human hosts

- Evolution of antibiotic resistance in the infectious organisms

- Urbanization, associated with overcrowding and poor sanitation

- A growing population of elderly people who are more susceptible to infection

Figure 7.3   Bushmeat. In many parts of Africa and elsewhere, protein is scarce, and monkeys and other bushmeat are an essential part of many diets. Hunting and selling bushmeat potentially exposes hunters, vendors, and others to new diseases. Photographed in Gabon.

Lynn Johnson/National Geographic Creative

- Pollution, environmental degradation, and changing weather patterns
- Growth in international travel and commerce
- Poverty and social inequality

CASE IN POINT

INFLUENZA PANDEMICS PAST AND FUTURE

The influenza virus has been a threat to human health for centuries. Every year, one or more new strains of the virus appear and quickly spread around the globe during the flu season, which usually runs from late fall through the winter. In a typical year, anywhere from 5% to 20% of the U.S. population contracts the flu, with symptoms ranging from mild headaches to severe muscle aches, digestive and breathing problems, and high fever. Typically, about 36,000 people in the United States die from the flu each year.

In contrast, a particularly virulent strain during the 1918–1919 flu season may have killed more than 850,000 people in the United States, or nearly 1% of the population of 100 million people. The 1918–1919 flu season corresponded with the end of World War I, a time already characterized by hardships, including hunger and a lack of medicines, as well as a time of unprecedented international travel. The 1918 flu was unusual in that, unlike most strains, it was at least as lethal to healthy people as to the very young, elderly, and infirm.

While individuals can be vaccinated against the flu, successful vaccination requires that researchers predict which influenza strains will appear, successfully prepare enough vaccine for everyone who needs it, and then distribute the vaccine to all those people. In recent years, an increasing fraction of those 65 or older are getting the vaccine (66% in the 2008–2009 flu season, the latest year on record). However, only 33% of those 19 to 49 years old were vaccinated in the same season, which means that a strain similar to that of the 1918–1919 season could be devastating. In addition, such a strain could easily spread around the world via air travel—a scenario envisioned by **Paul Ehrlich** in his 1968 book *The Population Bomb*.

Another recent concern is that of a pandemic of influenza, passing from birds, pigs, or other species to humans. A **pandemic** is a disease that reaches nearly every part of the world and has the potential to infect almost every person. Avian influenza, a strain that appears commonly in birds, tends to be very difficult for humans to contract, since it usually is transferred from bird to human, but not from human to human. It is, however, extremely potent once contracted, with a high fatality rate. A major concern for epidemiologists is the possible evolution of a strain that is easily transferred from human to human—a change that might involve a single genetic mutation. An avian flu pandemic could kill millions, or even billions, of people within a single year.

The swine flu outbreak of 2009 indicates how quickly such a disease can move around the world. The virus, which originated in Mexico, was first identified in April 2009. By early May, it had spread to several dozen countries; by June, it was considered pandemic. The 2009 swine flu strain (referred to as an *H1N1* flu because of its characteristic proteins) has a mixture of human, bird, and swine flu genetic material; H1N1 can be transferred from human to human and between pigs and humans. Two factors prevented the 2009 swine flu from being far more deadly than it was. First, a rapid international response limited its spread, allowing public health officials to prepare to treat the infected. Second, the strain turned out to be less powerful than it could have been. Nonetheless, over 14,000 people worldwide were killed by the 2009 pandemic.

See www.pandemicflu.gov for information on how to be prepared for a flu pandemic. ■

REVIEW

1. What is the average life expectancy for a woman in Japan? A woman in Zambia? Explain the difference.
2. What is the difference between an emerging disease and an endemic disease?

Environmental Pollution and Disease

LEARNING OBJECTIVES

- Summarize the problems associated with chemicals that exhibit persistence, bioaccumulation, and biological magnification in the environment.
- Briefly describe some of the data suggesting that certain chemicals used by humans may also function as endocrine disrupters in animals, including humans.

A 2007 study by researchers at Cornell University suggests that as many as 40% of deaths worldwide are pollution-related. Nonetheless, it is often difficult to establish a direct relationship between environmental pollution and disease (**Figure 7.4**). The relationship is fairly clear for certain pollutants, such as the link between radon and lung cancer (see Chapter 19) or between lead and disorders of the nervous system (see Chapter 21). However, the evidence is less definite for many pollutants, and scientists can only suggest that there is an association between the pollutant and a specific illness. One reason it is so difficult to establish a direct cause and effect is that other factors—such as a person's genetic makeup, diet, level of exercise, and smoking habits—complicate the picture.

Other complications include the fact that certain segments of society are more susceptible than others to adverse health effects from an environmental pollutant. Children are particularly sensitive to pollution, as will be discussed later in this chapter. Other groups at higher risk from pollution include the elderly and people with chronic diseases or compromised immune systems (such as people with HIV/AIDS or those undergoing chemotherapy to treat cancer). Also, people living in poor neighborhoods are often exposed to greater levels of pollution.

Several characteristics of environmental pollutants make them particularly dangerous to both the environment and human health. First, some toxic chemicals persist and accumulate in the

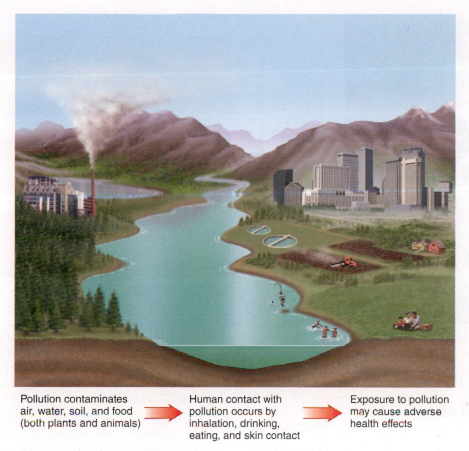

Pollution contaminates air, water, soil, and food (both plants and animals) ➡ Human contact with pollution occurs by inhalation, drinking, eating, and skin contact ➡ Exposure to pollution may cause adverse health effects

Figure 7.4 **Pathways of pollution in the environment.** (EPA)

environment and magnify their concentration in the food web. Second, a diverse group of pollutants affect the body's endocrine system, which produces *hormones* to regulate many aspects of body function.

PERSISTENCE, BIOACCUMULATION, AND BIOLOGICAL MAGNIFICATION OF ENVIRONMENTAL CONTAMINANTS

Some toxic substances exhibit persistence, bioaccumulation, and biological magnification. These substances include certain pesticides (such as DDT, or dichlorodiphenyl-trichloroethane), radioactive isotopes, heavy metals (such as lead and mercury), flame retardants (for example, PBDEs, or polybrominated diphenyl ethers), and industrial chemicals (such as dioxins and PCBs, or polychlorinated biphenyls).

The effects of the pesticide DDT on many bird species first demonstrated the problems with these chemicals. Falcons, pelicans, bald eagles, ospreys, and many other birds are sensitive to traces of DDT in their tissues. A substantial body of scientific evidence indicates that one of the effects of DDT on these birds is that they lay eggs with extremely thin, fragile shells that usually break during incubation, causing the chicks' deaths. After 1972, the year DDT was banned in the United States, the reproductive success of many birds improved (**Figure 7.5**).

Three characteristics of DDT contribute to its impact on birds: persistence, bioaccumulation, and biological magnification. It takes many years for some pesticides to break down into less toxic forms.

The **persistence** of synthetic pesticides is a result of their novel (not found in nature) chemical structures. Natural decomposers such as bacteria have not yet evolved ways to degrade many synthetic pesticides, so they accumulate in the environment and in the food web.

When a pesticide is not metabolized (broken down) or excreted by an organism, it is simply stored, usually in fatty tissues. Over time, the organism may **bioaccumulate**, or *bioconcentrate*, high concentrations of the pesticide.

Organisms at higher levels in food webs tend to have greater concentrations of bioaccumulated pesticide stored in their bodies than those lower in food webs. This increase as the pesticide passes through successive levels of the food web is known as **biological magnification**, or *biological amplification*.

As an example of the biological magnification of persistent pesticides, consider a food chain studied in a Long Island salt marsh that was sprayed with DDT over a period of years for mosquito control: algae and plankton → shrimp → American eel → Atlantic needlefish → ring-billed gull (**Figure 7.6**). The concentration of DDT in water was extremely dilute, on the order of 0.00005 part per million (ppm). The algae and other plankton contained a greater concentration of DDT, 0.04 ppm. Each shrimp grazing on

persistence
A characteristic of certain chemicals that are extremely stable and may take many years to break down into simpler forms through natural processes.

bioaccumulation
The buildup of a persistent toxic substance, such as certain pesticides, in an organism's body, often in fatty tissues.

biological magnification The increased concentration of toxic chemicals, such as PCBs, heavy metals, and certain pesticides, in the tissues of organisms that are at higher levels in food webs.

(a) A bald eagle feeds its chick. Most people know that eagles were on the brink of extinction a few decades ago. However, most people don't understand the scientific research behind the eagles' comeback.

Figure 7.5   Effect of DDT on birds.

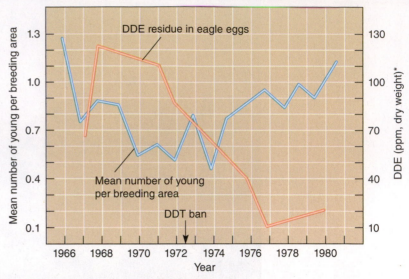

*DDT is converted to DDE in the birds' bodies

(b) A comparison of the number of successful bald eagle offspring with the level of DDT residues in their eggs. DDT was banned in 1972. Note that reproductive success improved after DDT levels decreased. (DDE is a derivative of DDT.) (Reprinted with permission from J. W. Grier, "Ban of DDT and Subsequent Recovery of Reproduction in Bald Eagles." AAAS 1982.)

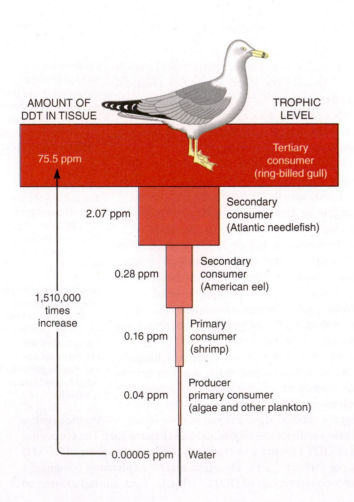

the plankton concentrated the pesticides in its tissues to 0.16 ppm. Eels that ate shrimp laced with pesticide had a pesticide level of 0.28 ppm, and needlefish that ate eels contained 2.07 ppm of DDT. The top carnivores, ring-billed gulls, had a DDT level of 75.5 ppm from eating contaminated fishes. Although this example involves a bird at the top of the food chain, all top carnivores, from fishes to humans, are at risk from biological magnification. Because of this risk, currently approved pesticides in the United States have been tested to ensure they do not persist and accumulate in the environment beyond a level predetermined to be acceptable.

ENDOCRINE DISRUPTERS

Mounting evidence suggests that dozens of industrial and agricultural chemicals—many of which exhibit persistence, bioaccumulation, and biological magnification—are also **endocrine disrupters**. These chemicals, many of which are no longer used in the United States, include

endocrine disrupter
A chemical that mimics or interferes with the actions of the endocrine system in humans and wildlife.

Figure 7.6   Biological magnification of DDT in a Long Island salt marsh. Note how the level of DDT, expressed as parts per million, increased in the tissues of various organisms as DDT moved through the food chain from producers to consumers. The ring-billed gull at the top of the food chain had approximately 1.5 million times more DDT in its tissues than the concentration of DDT in the water. (Based on data from G. M., Woodwell, C. F. Worster, Jr., and P. A. Isaacson. "DDT Residues in an East Coast Sanctuary." *Science* Vol 156 May 12, 1957.)

chlorine-containing industrial compounds known as PCBs and dioxins; the heavy metals lead and mercury; some pesticides such as DDT, kepone, dieldrin, chlordane, and endosulfan; flame retardants (PBDEs); and certain plastics and plastic additives such as phthalates and bisphenol A.

Hormones are chemical messengers produced by organisms in minute quantities to regulate their growth, reproduction, and other important biological functions. Some endocrine disrupters mimic the *estrogens*, a class of female sex hormones, and send false signals to the body that interfere with the normal functioning of the reproductive system. Because both males and females of humans and many other species produce estrogen, endocrine disrupters that mimic estrogen can affect both sexes. Additional endocrine disrupters interfere with the endocrine system by mimicking hormones other than estrogen, such as *androgens* (male hormones such as testosterone) and *thyroid hormones*. Like hormones, endocrine disrupters are active at very low concentrations and, as a result, may cause significant health effects at low doses.

Many endocrine disrupters appear to alter reproductive development in males and females of various animal species. Accumulating evidence indicates that fishes, frogs, birds, reptiles such as turtles and alligators, mammals such as polar bears and otters, and other animals exposed to these environmental pollutants exhibit reproductive disorders and are often left sterile.

A chemical spill in 1980 contaminated Lake Apopka, Florida's third-largest lake, with DDT and other agricultural chemicals with known estrogenic properties. Male alligators living in Lake Apopka in the years following the spill had low levels of testosterone and elevated levels of estrogen. Their reproductive organs were often feminized or abnormally small. The mortality rate for eggs in this lake was extremely high, which reduced the alligator population for many years (**Figure 7.7**).

Humans may also be at risk from endocrine disrupters, as the number of reproductive disorders, infertility cases, and hormonally related cancers (such as testicular cancer and breast cancer) appears to be increasing. More than 60 studies since 1938 have reported that sperm counts in a total of nearly 15,000 men from many nations, including the United States, dropped by more than 50% between 1940 and 1990. Although only one sperm is needed to fertilize an egg, infertility increases as sperm counts decline. Scientists do not know if this apparent decline and environmental factors are linked, however.

Certain *phthalates*—ingredients of cosmetics, fragrances, nail polish, medications, and common plastics used in a variety of food packaging, toys, and household products—are implicated in birth defects and reproductive abnormalities. Animal studies in which rats are fed large doses of phthalates have shown that these compounds can damage fetal development, particularly of reproductive organs. In humans, exposure to certain phthalates may be correlated to the increase since the 1970s in the incidence of *hypospadias*, a birth defect in which the urethral opening (passageway for urine) is on the underside of the penis instead of at its tip. Certain phthalates are also linked to an increased observance of premature breast development in young girls, typically between 6 and 24 months of age. Other phthalates are of negligible concern, according to the National Toxicology Program's Center for the Evaluation of Risks to Human Reproduction.

(a) A young American alligator (*Alligator mississippiensis*) hatches from eggs taken by University of Florida researchers from Lake Apopka, Florida. Many of the young alligators that hatch have abnormalities in their reproductive systems. This young alligator may not leave any offspring.

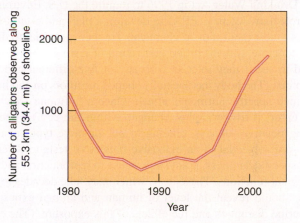

(b) Population of juvenile alligators in Lake Apopka after a chemical spill in 1980. (The chemical spill occurred just prior to the 1980 surveys). (Data from A. R. Woodward, *Florida Fish and Wildlife Conservation Commission.*)

Figure 7.7   **Lake Apopka alligators.**

Bisphenol A (BPA) is a chemical used in the manufacture of many hard plastic polycarbonate products, including baby bottles, toys, and sport-drink containers. Several studies on animals and a smaller number of studies on humans indicate that BPA is an endocrine disrupter. However, it is unclear how much of the BPA in manufactured products is taken up by our bodies and how much of an effect it can have at those concentrations. Consequently, BPA remains a highly contested regulatory issue: The state of California, the Canadian government, and the European Union have all banned its use in baby bottles, while regulators in the United States and Australia have not.

A Centers for Disease Control and Prevention (CDC) report on a diverse sample of the civilian U.S. population showed the presence of 27 different environmental chemicals—heavy metals, nicotine from tobacco smoke, pesticides, and plastics, many of which are thought to be endocrine disrupters. Nearly every chemical was detected in every participant's body at higher levels than expected. Because scientists had never actually measured 24 of the 28 chemicals in humans before, the study could not say if the levels of various environmental chemicals are increasing or decreasing. However, the study provides a good baseline of the U.S. population's exposure to these compounds against which future studies (the CDC plans to continue the tests) can be compared. The study is also a first step toward determining whether any of these chemicals is the hidden cause of modern illnesses or simply a benign byproduct of modern society.

We cannot make definite links between environmental endocrine disrupters and human health problems at this time because of the limited number of human studies. Human exposure to endocrine-disrupting chemicals must be quantified so that we know exactly how much of each of these chemicals affects various communities. Complicating such assessments is the fact that humans are also exposed to natural hormone-mimicking substances in the plants we eat. Soy-based foods such as bean curd and soymilk, for example, contain natural estrogens.

Congress amended the Food Quality Protection Act and the Safe Drinking Water Act in 1996 to require the U.S. Environmental Protection Agency (EPA) to develop a plan and establish priorities to test thousands of chemicals for their potential to disrupt the endocrine system. In the first round of testing, chemicals are tested to see if they interact with any of five different endocrine receptors. (The body has specially shaped receptors on and in cells to which specific hormones attach. Once a hormone attaches to a receptor, it triggers other changes within the cell.) Chemicals testing positive—that is, binding to one or more types of receptors—are subjected to an extensive battery of tests to determine what specific damages, if any, they cause to reproduction and other biological functions. These tests, which may take decades to complete, should reveal the level of human and animal exposure to endocrine disrupters and the effects of this exposure. (The effects of toxic chemicals, including several endocrine disrupters, are discussed throughout the text.)

REVIEW

1. What are the differences among persistence, bioaccumulation, and biological magnification? How are these chemical characteristics interrelated?

2. How did the 1980 chemical spill in Lake Apopka affect alligators?

Determining Health Effects of Environmental Pollution

LEARNING OBJECTIVES

- Define *toxicant,* and distinguish between acute and chronic toxicities.
- Describe how a dose-response curve helps determine the health effects of environmental pollutants.
- Discuss pesticide risks to children.

The human body is exposed to many kinds of chemicals in the environment. Both natural and synthetic chemicals are in the air we breathe, the water we drink, and the food we eat. *All* chemicals, even "safe" chemicals such as sodium chloride (table salt), are toxic if exposure is high enough. A one-year-old child will die from ingesting about two tablespoons of table salt; table salt is also harmful to people with heart or kidney disease.

The study of **toxicants**, or toxic chemicals, is called *toxicology.* It encompasses the effects of toxicants on living organisms as well as ways to prevent or minimize adverse effects, such as developing appropriate handling or exposure guidelines.

> **toxicant** A chemical with adverse human health effects.

The effects of toxicants following exposure can be immediate (acute toxicity) or prolonged (chronic toxicity). **Acute toxicity,** which ranges from dizziness and nausea to death, occurs immediately to within several days following a single exposure. In comparison, **chronic toxicity** generally produces damage to cells (including cancer) or vital organs, such as the kidneys or liver, following a long-term, low-level exposure to chemicals. Toxicologists know far less about chronic toxicity than they do about acute toxicity, in part because the symptoms of chronic toxicity often mimic those of other chronic diseases.

> **acute toxicity** Adverse effects that occur within a short period after exposure to a toxicant.

> **chronic toxicity** Adverse effects that occur some time after exposure to a toxicant or after extended exposure to the toxicant.

We measure toxicity by the extent to which adverse effects are produced by various doses of a toxicant. A **dose** of a toxicant is the amount that enters the body of an exposed organism. The **response** is the type and amount of damage that exposure to a particular dose causes. A dose may cause death (*lethal dose*) or may cause harm but not death (*sublethal dose*). Lethal doses, usually expressed in milligrams of toxicant per kilogram of body weight, vary depending on the organism's age, sex, health, metabolism, genetic makeup, and how the dose was administered (all at once or over a period of time). Lethal doses in humans are known for many toxicants because of records of homicides and accidental poisonings.

IDENTIFYING TOXIC SUBSTANCES

One approach scientists use to determine acute toxicity is to administer various doses of a suspected toxicant to laboratory animals, measure the responses, and use these data to predict the chemical effects on humans. The dose lethal to 50% of a population of test animals is the **lethal dose-50%,** or **LD_{50}.** It is usually reported in milligrams of chemical toxicant per kilogram of body weight. LD_{50} is inversely related to the acute toxicity of a chemical: The smaller the LD_{50}, the more toxic the chemical, and, conversely, the greater the LD_{50}, the less toxic the chemical (**Table 7.2**). The LD_{50} is determined for all new synthetic chemicals—thousands are produced each year—as a way of estimating their toxic potential. It is generally assumed that a chemical with a low LD_{50} for several species of test animals is toxic in humans.

The **effective dose-50%,** or **ED_{50},** is used to evaluate a wide range of biological responses, such as stunted development in the offspring of a pregnant animal, reduced enzyme activity, or onset of hair loss. The ED_{50} causes 50% of a population to exhibit whatever response is under study.

Table 7.2 LD$_{50}$ Values for Selected Chemicals

Chemical	LD$_{50}$ (mg/kg)*
Aspirin	1750
Ethanol	1000
Morphine	500
Caffeine	200
Heroin	150
Lead	20
Cocaine	17.5
Sodium cyanide	10.0
Nicotine	2.0
Strychnine	0.8

*Administered orally to rats.

Table 7.3 Hypothetical Data Set for Animals Exposed to a Chemical

Number of Animals in Test	Number of Animals with Cancer	Dose (mg/kg/day)	Probability of Cancer*
50	0	0.0	0
50	2	5.0	0.04
50	6	10.0	0.12
50	22	20.0	0.44

*The probability of getting cancer at a given dose is the number of animals with cancer at that dose level divided by the total number exposed at that dose level.

One complication of toxicology is that an individual's genes can contribute to that person's response to a specific toxicant. The National Institute of Environmental Health Science has identified several hundred *environmental susceptibility genes*. Subtle differences in these genes affect how the body metabolizes toxicants, making them more or less toxic. Other gene variations allow certain toxicants to bind strongly—or less so—to the genetic molecule DNA. (Generally, when a toxicant binds to our DNA, it is a bad thing.) Genetic variation is one of the most important factors that determine why some people develop lung cancer after years of smoking, but others do not. For example, researchers have identified variations in the P450 gene (which codes for an enzyme) that determine how the body metabolizes some of the cancer-causing chemicals in tobacco smoke.

Traditionally, cancer was the principal disease evaluated in toxicology because many people were concerned about cancer-inducing chemicals in the environment and because cancer is so feared. Environmental contaminants are linked to several other serious diseases, such as birth defects, damage to the immune response, reproductive problems, and damage to the nervous system or other body systems. Although cancer is not the only disease caused or aggravated by toxicants, we focus here on risk assessment as it relates to cancer. Risk assessment for noncancer hazards, such as diseases of the liver, kidneys, or nervous system, is similar to cancer risk assessment.

toxicology The study of the effects of toxic chemicals on human health.

epidemiology The study of the effects of toxic chemicals and diseases on human populations.

Toxicology and **epidemiology** are the two most common methods for determining whether a chemical causes cancer. Toxicologists expose laboratory animals such as rats to varying doses of the chemical and see whether they develop cancer. Epidemiologists look at the historical exposures of humans to the same chemical to see whether exposed groups show increased cancer rates. Each of these methods has advantages and disadvantages but even when used together can provide only rough estimates of a chemical's carcinogenic potential.

Toxicology has the advantage that doses are measured and administered in precise amounts. Usually, two, three, or four groups of animals are exposed to different amounts or doses, including a control group that is not exposed. At the end of the

experiment (about two years for mice and rats), the animals are dissected, and the ratio of animals with tumors to animals without in each group is recorded. This is used to determine a dose-response curve for the chemical. Doses are converted from animal dose to "equivalent human dose" by comparing body weight and metabolism rates. **Table 7.3** depicts a hypothetical toxicological data set.

Several uncertainties limit comparisons of animal laboratory studies to humans. First, the dose levels in the experiments are typically much, much larger than those faced by humans in the environment, so extrapolation from high-dose effects to low-dose effects can be inaccurate. **Extrapolation** is estimating the expected effects at some dose of interest from the effects at known doses. Second, humans and laboratory animals may process the chemicals in different ways. Third, laboratory animals live only about 2 years, while human life expectancy is around 70 to 80 years. Finally, human exposures are much more sporadic, and humans are exposed to a variety of chemicals that may amplify or offset each other.

Epidemiological studies have an advantage in that they look at people who were actually exposed to the chemical. Ideally, a **cohort**, or group of individuals who were exposed to the chemical, is compared to an otherwise similar group who were not exposed. For example, cohorts exposed to benzene in a variety of industrial settings have clearly had higher levels of leukemia than did similar groups who were not exposed.

Epidemiology has several limitations. First, it is usually difficult to reconstruct, or estimate, historical doses. Second, confounding factors may exist—for example, industry workers may also have been exposed to some other chemical that was not recorded. Third, the individuals in an industrial setting—in the benzene case, healthy males between 18 and 60 years of age—may respond differently to the chemical than would others, such as children or pregnant women.

Epidemiology may be more representative than toxicology, but toxicology is usually more precise. Ideally, epidemiological data and toxicological data are combined to provide a clearer picture of the causes of cancer. In the case of benzene, animal tests confirm the epidemiological finding that benzene is a carcinogen but do not improve our understanding of benzene's potency. **Table 7.4** compares the advantages and disadvantages of epidemiology and toxicology.

Table 7.4 Comparison of Advantages and Disadvantages of Toxicological and Epidemiological Studies

Epidemiology	Toxicology	Advantage
Human subjects	Typically animal subjects	Epidemiology
Exposure to multiple chemicals	Exposure to a single chemical	Toxicology
Retrospective (backward-looking)	Prospective (forward-looking)	Toxicology
Arbitrary dose ranges	Specified dose ranges	Toxicology
Estimated doses	Administered doses	Epidemiology
Exposed group genetically diverse	Exposed group genetically homogeneous	
Sample size of 100 to 10,000	Sample size of 10 to 100	Epidemiology
Risk to exposed group near or slightly above background rate	Risk to exposed group substantially above background rate	Toxicology

Source: D. M. Hassenzahl and A. Finkel (2008) "Risk Assessment for Environmental and Occupational Health." In Heggenhougen and Quah, *International Encyclopedia of Public Health.*

Dose-Response Relationships A **dose-response curve** shows the effect of different doses on a population of test organisms (**Figure 7.8**). Scientists first test the effects of high doses and then work their way down to a **threshold** level, the maximum dose with no measurable effect (or, alternatively, the minimum dose with a measurable effect). Toxicologists often assume that doses lower than the threshold level will not have an effect on the organism and are safe.

Many chemicals are essential to humans and animals in small amounts but toxic at higher levels. Vitamin D is a good example: We know that the human body requires vitamin D to properly absorb calcium and phosphorus. Too much or too little vitamin D can cause a variety of symptoms, including digestive complications. Thus the dose-response relationship between vitamin D and health is U-shaped. While the biological effects of chemicals at low concentrations are difficult to assess, some evidence suggests that a variety of chemicals, including some pesticides and trace metals like cadmium, are healthful in small doses but dangerous at higher levels.

This effect, called **hormesis**, occurs when a small exposure improves health, while a larger exposure causes illness. Hormesis is depicted by the blue line in **Figure 7.9**. In this hypothetical data set, animals exposed to relatively low levels of a chemical were less likely to have tumors than those in the control group, while those with higher chemical exposures had clearly elevated levels of tumors. It is possible that there is a threshold (that is, below some concentration, the chemical does not cause cancer) and that the chemical causes tumors at low levels—but not enough to show up in a test of only a few animals. It is also possible that the chemical has a *hormetic* effect and somehow suppresses tumors.

Some evidence suggests that ionizing radiation—the sort of radiation associated with nuclear waste and radon—may have a hormetic effect. However, most radiation experts conclude that while hormesis is possible, it is more likely that radiation causes cancer even at very low doses.

Chemical Mixtures Humans are frequently exposed to various combinations of chemical compounds. Cigarette smoke contains a mixture of chemicals, as does automobile exhaust. The vast

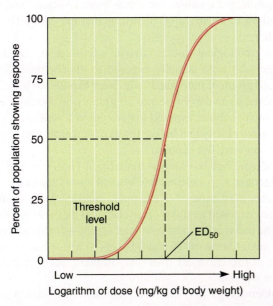

(a) This hypothetical dose-response curve demonstrates two assumptions of classical toxicology: First, the biological response increases as the dose is increased; second, there is a safe dose—a level of the toxicant at which no response occurs. Harmful responses occur only above a certain threshold level.

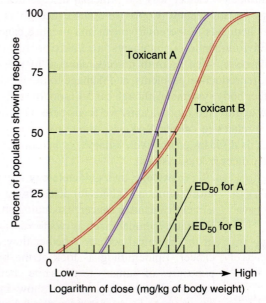

(b) Dose-response curves for two hypothetical toxicants, A and B. In this example, toxicant A has a lower effective dose-50% (ED_{50}) than toxicant B. At lower doses, toxicant B is more toxic than toxicant A.

Figure 7.8 Dose-response curves.

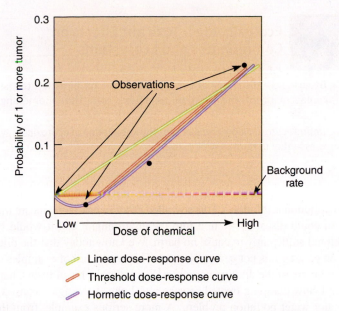

Linear dose-response curve
Threshold dose-response curve
Hormetic dose-response curve

Figure 7.9 **Contrasting dose-response relationships.** Often, only a few observations are available for the effects of a chemical. In such cases, there may be more than one equally plausible explanation for the relationship between the chemical and the expected health effects at low concentrations. In this case, a small amount of the chemical appears to actually reduce the number of tumors; this could be a real effect, or it could be a random effect. Which dose-response curve do you think best explains these observations: linear (green), threshold (orange), or hormetic (blue)?

majority of toxicological studies are performed on single chemicals rather than chemical mixtures, however, and for good reason. Mixtures of chemicals interact in a variety of ways, increasing the level of complexity in risk assessment. Moreover, too many chemical mixtures exist to evaluate them all.

Chemical mixtures interact by additivity, synergy, or antagonism. When a chemical mixture is **additive**, the effect is exactly what one would expect, given the individual effects of each component of the mixture. If a chemical with a toxicity level of 1 is mixed with a different chemical with a toxicity level of 1, the combined effect of exposure to the mixture is 2. A **synergistic** chemical mixture has a greater combined effect than expected; two chemicals, each with a toxicity level of 1, might have a combined toxicity of 3. An **antagonistic** interaction in a chemical mixture results in a smaller combined effect than expected; for example, the combined effect of two chemicals, each with toxicity levels of 1, might be 1.3.

If toxicological studies of chemical mixtures are lacking, how do scientists assign the effects of chemical mixtures? Risk assessors typically assign risk values to mixtures by additivity—adding the known effects of each compound in the mixture. Such an approach usually underestimates but may sometimes overestimate the risk involved, but it is the best approach currently available. The alternative—waiting for years or decades until numerous studies are designed, funded, and completed—is unreasonable.

CHILDREN AND CHEMICAL EXPOSURE

Children are more susceptible to most chemicals than are adults because their bodies are still developing and are not as effective in dealing with toxicants. In addition, because they weigh less than adults, the same dose of a chemical can be more potent. Consider a toxicant with an LD_{50} of 100 mg/kg. A potentially lethal dose for a child who weighs 11.3 kg (25 lb) is $100 \times 11.3 = 1130$ mg, equal to a scant 1/4 teaspoon if the chemical is a liquid. In comparison, the potentially lethal dose for an adult who weighs 68 kg (150 lb) is 6800 mg, or slightly less than 2 teaspoons. Thus, we must protect children from exposure to environmental chemicals because harmful doses are smaller for children than for adults.

In recent years, people have paid increased attention to the health effects of household pesticides on children because it appears that household pesticides are a greater threat to children than to adults. For one thing, children tend to play on floors and lawns, where they are exposed to greater concentrations of pesticide residues. Also, children, especially when very young, are more likely to put items in their mouths.

The EPA estimates that at least 75% of U.S. homes use pesticide products, such as pest strips, bait boxes, bug bombs, flea collars, pesticide pet shampoos, aerosols, liquids, and powders. Several thousand different household pesticides are manufactured, and these contain over 300 active ingredients and more than 2,500 inert ingredients. Each year, poison control centers in the United States receive more than 130,000 reports of exposure and possible poisoning from household pesticides. More than half of these incidents involve children.

Research supports an emerging hypothesis that exposure to pesticides may affect the development of intelligence and motor skills of infants, toddlers, and preschoolers. One study, published in *Environmental Health Perspectives* in 1998, compared two groups of rural Yaqui Indian preschoolers. These two groups, both of which live in northwestern Mexico, shared similar genetic

E N V I R O N E W S

Toxicology Without Animals

There are several reasons why we might want to determine the toxicity of chemicals without using laboratory animals. Regulators are concerned because such studies can take a long time (at least two years for rats and mice) and may still leave questions about the relevance of the findings to human health. Chemical manufacturers are concerned about the cost of such studies as well as the uncertainty in the findings. And many people consider intentionally exposing animals to potentially toxic chemicals unethical.

A recent panel of the National Research Council suggests combining experiments on cells in test tubes with our ever-advancing knowledge of how chemicals interact with each other. This approach allows scientists to determine the *toxicity pathways* by which an individual chemical might (or might not) damage living cells and organs. Several hundred chemicals have already been screened using this technology. Toxicity pathway research is a priority outside the United States as well: By 2013, European cosmetics companies must demonstrate that their products are safe without relying on animal testing.

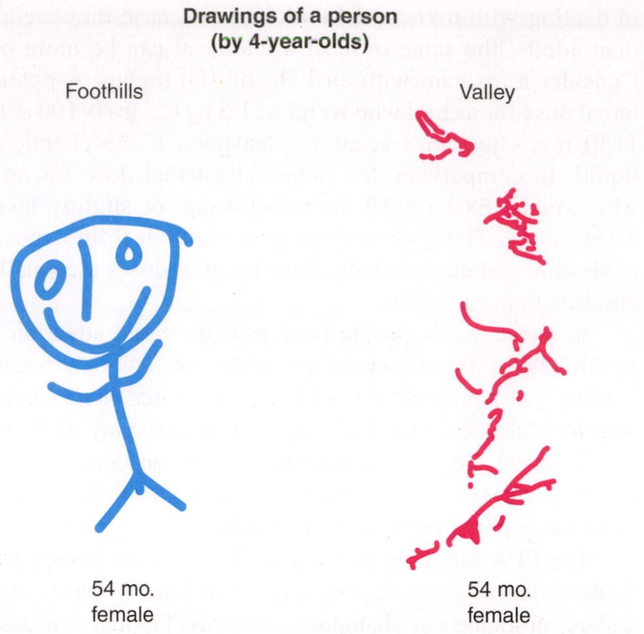

**Drawings of a person
(by 4-year-olds)**

Foothills   Valley

54 mo.   54 mo.
female   female

(a) Most preschoolers who received little pesticide exposure were able to draw a recognizable stick figure.

(b) Most preschoolers who lived in an area where agricultural pesticides were widely used could draw only meaningless lines and circles.

Figure 7.10   Effect of pesticide exposure on preschoolers.
A study in Sonora, Mexico, found that Yaqui Indian preschoolers varied in their motor skills on the basis of the degree of pesticide exposure. The children were asked to draw a person. Two representative pieces of art are shown; both are drawn by 4 1/2-year-old girls.

backgrounds, diets, water mineral contents, cultural patterns, and social behaviors. The main difference between the two groups was their exposure to pesticides: One group lived in a farming community (valley) where pesticides were used frequently (45 times per crop cycle) and the other in an adjacent nonagricultural area (foothills) where pesticides were rarely used. When asked to draw a person, most of the 17 children from the low-pesticide area drew recognizable stick figures, whereas most of the 34 children from the high-pesticide area drew meaningless lines and circles (**Figure 7.10**).

A 2010 study in California by the Center for Health Assessment of Mothers and Children of Salinas found that exposure of pregnant women to pesticides has long-term effects on their children. A 10-year study of 600 women showed a correlation between the mother's pesticide exposure and both delayed mental development and pervasive development disorder in their offspring. The research continues to investigate other potential impacts, and a much larger 10,000-person study has recently begun.

REVIEW

1. What is acute toxicity? chronic toxicity?
2. How are potentially toxic chemicals identified?
3. Why are children particularly susceptible to environmental contaminants such as pesticides?

Ecotoxicology: Toxicant Effects on Communities and Ecosystems

LEARNING OBJECTIVES

- Discuss how discovery of the effects of DDT on birds led to the replacement of the dilution paradigm with the boomerang paradigm.
- Define *ecotoxicology,* and explain why knowledge of ecotoxicology is essential to human well-being.

People used to think—and indeed some still do—that "the solution to pollution is dilution." This so-called *dilution paradigm* meant that you could discard pollution into the environment and it would be diluted sufficiently to cause no harm. We know today that the dilution paradigm is not generally valid. This text is full of examples of the failure of the dilution paradigm. For example, recall from Chapter 1 how dumping treated sewage into Lake Washington caused a major water pollution problem. A more serious example, from the perspective of human well-being, is *Love Canal*, a small community in New York State contaminated by toxic waste dumped into a nearby pit by a chemical company (see Chapter 23). To be fair, the chemical company dumped wastes at a time (1942–1953) when the dilution paradigm was still widely accepted. However, failure of the dilution paradigm brought scant consolation to the people living in Love Canal. They had to abandon their homes, and many have had health problems since leaving the contaminated site.

Today, virtually all environmental scientists have rejected the dilution paradigm in favor of the *boomerang paradigm*: What you throw away can come back and hurt you. The boomerang paradigm was adopted during the latter half of the 20th century after several well-publicized events captured the public's attention. Notable among these events was the discovery that the pesticide DDT was accumulating in birds at the top of the food web. The implications were clear that DDT was an unacceptable threat not only to ecosystem health but also potentially to human health.

As a result of the environmental impacts of DDT and the many other environmental problems that have arisen since DDT, a new scientific field—**ecotoxicology**—was born. Ecotoxicology, also called *environmental toxicology*, is an extension of the field of toxicology, which is human-oriented. However, ecotoxicology is also human-oriented in the sense that humans produce the contaminants that adversely affect the environment.

ecotoxicology The study of contaminants in the biosphere, including their harmful effects on ecosystems.

The scope of ecotoxicology is broad—from molecular interactions in the cells of individual organisms to effects on populations (e.g., local extinctions), communities and ecosystems (e.g., loss of species richness), and the biosphere (e.g., global climate change). Expanding knowledge in ecotoxicology provides many examples of linkages between human health and the health of natural systems.

Ecotoxicology helps policy makers determine the costs and benefits of the many industrial and technological "advances" that affect us and the ecosystems on which we depend. However, most environmental regulations are currently based on data for single species. Scientists have only begun to collect data to determine the environmental status of populations, communities, ecosystems, and higher levels of natural systems. Obtaining this higher-level information is complicated because (1) natural systems are exposed to many **environmental stressors** (changes that tax the environment); (2) natural systems must be evaluated for an extended period to establish important trends; and (3) the results must be clear enough for policy makers and the public to evaluate.

CASE IN POINT

THE OCEAN AND HUMAN HEALTH

The ocean is important to us as a source of both food and natural chemical compounds that could benefit human health. These chemicals have potential uses in many areas, including, among others, novel pharmaceuticals, nutritional supplements, agricultural pesticides, and cosmetics. The discovery and development of beneficial marine compounds is in its infancy. Sponges, corals, mollusks, marine algae, and marine bacteria are some of the ocean's organisms that have the potential to provide these natural chemicals. In addition to serving as a source of food and other products, the ocean absorbs many wastes from human-dominated land areas.

Negative Health Impacts of Marine Microorganisms Marine microorganisms occur naturally in every part of the ocean environment. On average, 1 million bacteria and 10 million viruses occur in each milliliter of seawater! These organisms usually perform their ecological roles (such as decomposition) without causing major problems to humans. However, human activities now have a measurable effect on the ocean, including increases in land-based nutrient runoff and pollution and a small rise in ocean temperatures. These changes are causing an increase in the number and distribution of disease-causing microorganisms, particularly bacteria and viruses that pose a significant health threat to humans. (Recall the discussion of the boomerang paradigm.) For example, humans can become ill—or even die if their immune systems are compromised—from drinking water or eating fish or shellfish (such as mussels or clams) contaminated with disease-causing microorganisms associated with human sewage or livestock wastes. Additional research is needed to study links between human-produced pollution and the increase of disease-causing microorganisms in the ocean system.

Scientists have observed an association in Bangladesh between outbreaks of cholera, caused by a waterborne bacterium, and increases in the surface temperature of coastal waters in the Bay of Bengal. The association is an indirect one. Apparently, the increased temperatures encourage the growth of microscopic plankton, which, in turn, produce ideal growing conditions for the cholera bacterium. Tides carry these contaminated waters into rivers that provide drinking water, increasing the risk of cholera in the local population.

Certain species of harmful algae sometimes grow in large concentrations called *algal blooms*. When some pigmented marine algae experience blooms, their great abundance frequently colors the water orange, red, or brown (**Figure 7.11**). Known as **red tides**, these blooms may cause serious environmental harm and threaten the health of humans and animals. Some of the algal species that form red tides produce toxins that attack the nervous systems of fishes, leading to massive fish kills. Waterbirds such as cormorants suffer and sometimes die when they eat the contaminated fishes. The toxins also work their way up the food web to marine mammals and people. In 1997, more than 100 monk seals, one-third of that endangered species' total population, died from algal toxin poisoning off

Figure 7.11   Red tide. The presence of billions of toxic algae color the water. Photographed in the Gulf of Carpentaria, Australia.

Bill Bachman/Science Source

the West African coast. Humans may also suffer if they consume algal toxins, which often bioaccumulate in shellfish or fishes. Even nontoxic algal species may wreak havoc when they bloom, as they shade aquatic vegetation and upset food web dynamics.

No one knows what triggers red tides, which are becoming more common and more severe, but many scientists think the blame lies with coastal pollution. Wastewater and agricultural runoff to coastal areas contain increasingly larger quantities of nitrogen and phosphorus, two nutrients that stimulate algal growth. Changes in ocean temperatures, such as those attributed to global warming, may also trigger algal blooms. In addition, a possible connection exists between red tide outbreaks in Florida's coastal waters and the arrival of dust clouds from Africa. These dust clouds, which sometimes blow across the Atlantic Ocean, enrich the water with iron and appear to trigger algal blooms.

Because we do not know what causes red tides, no control measures are in place to prevent the blooms or to end them when they occur. However, technologies like satellite monitoring and weather-tracking systems allow better prediction of conditions likely to stimulate blooms. ■

REVIEW

1. What environmental catastrophe was largely responsible for replacement of the dilution paradigm with the boomerang paradigm?

2. What is ecotoxicology?

Decision Making and Uncertainty: Assessment of Risks

LEARNING OBJECTIVES

- Define *risk,* and explain how risk assessment helps determine adverse health effects.
- Explain how risk information can improve environmental decisions.
- Define *ecological risk assessment.*

risk The probability that a particular adverse effect will result from some exposure or condition.

Throughout this chapter, we have discussed various **risks** to human health and the environment. Each of us makes many risk management decisions every day, most of which are based on intuition, habit, and experience. We make these decisions effectively—after all, most of us make it through each day without getting injured or killed.

However, environmental and health decisions often impact many individuals, and the best choices are not always made on an intuitive level. *Risk analysis* is a tool used to organize how we think about complex environmental systems. When we think about risks from a systems perspective, we can decide which of the following it is most effective to do:

- Change our activities to avoid particular risks altogether.

- Limit the extent to which a particular **hazard** can come into contact with us.

- Limit the extent to which the hazard can harm us.

- Provide some sort of offset or compensation if the hazard causes harm.

hazard A condition that has the potential to cause harm.

For example, consider the risk of injury from car accidents. We can redesign cities to allow people to work near where they live. We can establish rules about how fast people can drive or on which side of the street they drive. We can require seatbelts and airbags so that people who are in accidents don't get hurt as badly. Finally, we can purchase insurance to cover expenses, lost time, and suffering. Risk analysis also allows us to think about the tradeoffs between different activities. For example, we could bicycle instead of driving. This would take more time, and we might have a higher risk of an accident, but we would also get some health benefits from the exercise.

RISK INFORMATION AS A DECISION TOOL

Risk management is the process of identifying, assessing, and reducing risks. **Risk assessment** involves using statistical methods to quantify the risks of a particular action so that these risks can be compared and contrasted with other risks. The four steps involved in risk assessment for adverse health effects are summarized in **Figure 7.12**.

Risk is calculated as the probability that some negative effect or event will occur times the consequences (deaths, injuries,

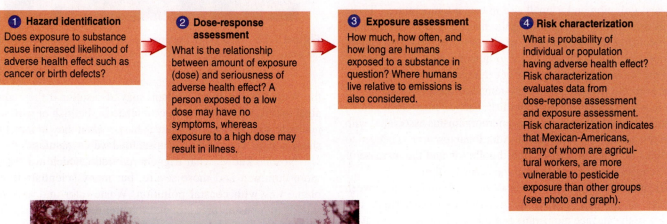

1 Hazard identification
Does exposure to substance cause increased likelihood of adverse health effect such as cancer or birth defects?

2 Dose-response assessment
What is the relationship between amount of exposure (dose) and seriousness of adverse health effect? A person exposed to a low dose may have no symptoms, whereas exposure to a high dose may result in illness.

3 Exposure assessment
How much, how often, and how long are humans exposed to a substance in question? Where humans live relative to emissions is also considered.

4 Risk characterization
What is probability of individual or population having adverse health effect? Risk characterization evaluates data from dose-reponse assessment and exposure assessment. Risk characterization indicates that Mexican-Americans, many of whom are agricultural workers, are more vulnerable to pesticide exposure than other groups (see photo and graph).

Agricultural workers have a greater than average exposure to chemicals such as pesticides. (Sisse Brimberg/National Geographic Image Collection)

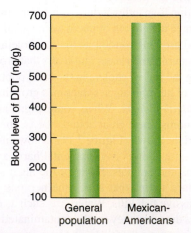

Figure 7.12 **The four steps of a risk assessment for adverse health effects.** (Adapted from *Science and Judgment in Risk Assessment.* Washington, DC, National Academy Press [1994].)

Table 7.5   Probability of Death by Selected Causes for a U.S. Citizen

Cause of Death	One-Year Odds*	Lifetime Odds*
Cardiovascular disease	1 in 300 (3.3×10^{-3})	1 in 4 (2.5×10^{-1})
Cancer, all types	1 in 510 (2.0×10^{-3})	1 in 7 (1.4×10^{-1})
Motor vehicle accidents	1 in 6700 (1.5×10^{-4})	1 in 88 (1.1×10^{-2})
Suicide	1 in 9200 (1.1×10^{-4})	1 in 120 (8.3×10^{-3})
Homicide	1 in 18,000 (5.6×10^{-5})	1 in 240 (4.2×10^{-3})
Killed on the job	1 in 48,000 (2.1×10^{-5})	1 in 620 (1.6×10^{-3})
Drowning in bathtub	1 in 840,000 (1.1×10^{-6})	1 in 11,000 (9.1×10^{-5})
Tornado	1 in 3,000,000 (3.3×10^{-7})	1 in 39,000 (2.6×10^{-5})
Commercial aircraft	1 in 3,100,000 (3.2×10^{-7})	1 in 40,000 (2.5×10^{-5})
Hornet, wasp, or bee sting	1 in 6,100,000 (1.6×10^{-7})	1 in 80,000 (1.3×10^{-5})

*Probability of risk is in parentheses.
Probabilities calculated by L. Berg from multiple sources.

economic losses) if it does occur. Risk is typically reported as a fraction, ranging from 0 (certain not to occur) to 1 (certain to occur). For example, according to the American Cancer Society, in 2002 about 170,000 Americans who smoked died of cancer. This translates into a probability of death of 0.00059 (that is, 5.9×10^{-4}). **Table 7.5** contains probability estimates of death by selected causes in the United States. Note that these are population estimates, are highly uncertain, and most are unequally distributed across the population (e.g., deaths on the job can only occur among the employed, and some jobs are far riskier than others).

Risk assessment is used in several ways for environmental regulation. A regulatory agency may establish a "maximum-risk" standard. For example, the EPA may decide that people should face an additional risk of cancer of no more than 1 in 1 million from trichloroethylene (TCE, a common contaminant) in munici-

cancer potency An estimate of the expected increase in cancer associated with a unit increase in exposure to a chemical.

pal drinking water. If we know the **cancer potency** of TCE and we know how much water the average person drinks, we can calculate the maximum allowable concentration of TCE.

Alternatively, risk managers may be concerned about the expected risk associated with an existing or historical exposure. For example, a company might need to know how much money to set aside to pay medical costs associated with construction workers who were exposed to asbestos. It may be impossible to tell which individuals will contract asbestosis (a lung disease associated with chronic asbestos exposure), but we can estimate the number of disease instances among a large group of individuals.

Many of our decisions about risks have far more to do with our trust in individuals and institutions who manage the risks than with the calculated values of those risks. For example, many people distrust the nuclear industry, which means it doesn't matter how accurately the industry can calculate the risks of a nuclear accident. We worry less about risks that we feel we can control (such as driving or eating) than those outside our control (such as pesticide contamination). We also worry more about things we dread (dying from cancer) than those we don't (bicycling). Effective risk management, then, is not based on calculated risks alone but must also account for intuition, trust, and social conditions.

Cost-Benefit Analysis and Risks   Risk assessment can be an important input to **cost-benefit analysis** (see Chapter 2). In

a cost-benefit analysis, the estimated cost of some regulation to reduce risk is compared with potential benefits associated with that risk reduction. Cost-benefit analysis is an important mechanism to help decision makers formulate environmental legislation, but it is only as good as the data and assumptions on which it is based. Corporate estimates of the cost to control pollution are often many times higher than the actual cost turns out to be. During the debate over phasing out leaded gasoline in 1971, the oil industry predicted a cost of $7 billion per year during the transition, but the actual cost was less than $500 million per year.

Despite the wide range that often occurs between projected and actual costs, the cost portion of cost-benefit analysis is often easier to determine than are the health and environmental benefits. The cost of installing air pollution–control devices at factories is relatively easy to estimate, but how does one put a price tag on the benefits of a reduction in air pollution? What is the value of reducing respiratory problems in children and the elderly, two groups susceptible to air pollution? How much is clean air worth?

Another problem with cost-benefit analysis is that the risk assessments on which such analyses are based are far from perfect. Even the best risk assessments are based on assumptions that, if changed, could substantially alter the estimated risk. Risk assessment is an uncertain science. Cost-benefit analyses and risk assessments are useful in evaluating and addressing environmental problems, but decision makers must recognize the limitations of these methods when developing new government regulations.

The Precautionary Principle   You have probably heard the expression "an ounce of prevention is worth a pound of cure." This statement is the heart of a policy—the **precautionary principle**—advocated by many politicians and environmental activists. According to the precautionary principle, when a new technology or chemical product is suspected of threatening health or the environment, we should undertake precautionary measures even if there is uncertainty about the scope of danger.

precautionary principle The idea that no action should be taken or product introduced when the science is inconclusive but unknown risks may exist.

The precautionary principle may also be applied to existing technologies when new evidence suggests they are more dangerous than originally thought. When observations and experiments suggested that chlorofluorocarbons (CFCs) harm the ozone layer in the stratosphere, the precautionary principle led most countries to phase out these compounds. Studies made after the phase-out supported this step.

To many people, the precautionary principle is common sense, given that science and risk assessment often cannot provide definitive answers to policy makers. The precautionary principle puts the burden of proof on the developers of the new technology or substance, who must demonstrate safety beyond a reasonable doubt.

Although the precautionary principle has been incorporated into decisions in the European Union, the United States, and elsewhere, it has many detractors. Some scientists think that it challenges the role of science and endorses making decisions without the input of science. Advocates of cost-benefit analysis note that the precautionary principle may be extremely expensive or may cause us to hold off on new technologies that are much safer than those already in place.

Some critics contend that the precautionary principle's imprecise definition can reduce trade and limit technological innovations. For example, several European countries made precautionary decisions to ban beef from the United States and Canada because these countries use growth hormones to make cattle grow faster. Europeans contend that the growth hormone might harm humans eating the beef, but the ban, in effect since 1989, is widely viewed as intended to protect their own beef industry.

The rapid expansion of nanotechnology is another area where the precautionary principle might be useful. **Nanotechnology** is the creation of materials and devices on the ultra-small scale of atoms or molecules (**Figure 7.13**). It is increasingly common in a variety of settings, from cosmetics to wastewater treatment to electronics. Because of their very small scale, **nanomaterials** can be environmentally advantageous if they require fewer raw materials than does traditional manufacturing.

However, particles on the nanometer scale (a nanometer is one-billionth of a meter) pose potentially significant but highly uncertain health, safety, and environmental risks. Concerned about this high uncertainty, some policy makers argue that we should take a precautionary approach and avoid using nanotechnologies until they can be shown to be safe. The EPA has decided to regulate nanomaterials that might adversely affect the environment: The burden of proof about product safety will fall on companies that sell nanotechnology.

Fractional Risk Attribution

By one estimate, as many as 150,000 people die each year from health effects associated with climate change, and this rate could reach 300,000 per year by 2030. However, in climate change as well as other environmental cases, it is often difficult to attribute illness, injury, or death to a particular environmental cause. This is the case for first responders (police, firefighters, medical personnel) to the Twin Towers in New York City on September 11, 2001. They were exposed to large amounts of airborne pollutants, and now many of them suffer from cancers and other illnesses. Although we can confidently attribute the increased rate of disease to the exposure, it is very difficult to say which specific cases were caused by the exposure and which would have happened without the exposure. This has created a problem for those first responders, since it is unclear who would be responsible for treating them.

One approach to dealing with this sort of uncertainty is to consider **fractional risk attribution**. This approach allocates the likelihood of each of two or more different possible causes. For example, if there is a doubling of cancers among a group of individuals following some exposure, under fractional risk attribution, we would

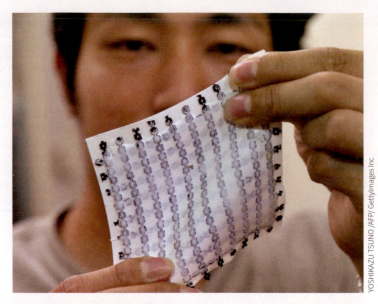

Figure 7.13   Nanotechnology. Japan's Tokyo University researcher Tsuyoshi Sekitani displays an elastic sheet containing carbon nano tubes which conduct electricity (black part) in the silicon rubber (white base) at his laboratory at the University of Tokyo on August 11, 2008.

assign 50% of the "blame" for the cancers to that exposure. We still can't say which cancers were caused by the exposure, but we can say that half of them would not have occurred without it.

This concept could be particularly useful for climate change, where it is difficult to say that any particular weather event was caused by climate change. There is strong evidence that recent record rains, heat, droughts, hurricanes, and other events were caused by the changing climate, but it is impossible to say that any particular death would not have occurred without climate change. We can often estimate the number of various extreme events that would have occurred without climate change, compare this to the number of events that did occur, and use this to estimate the fraction of deaths (as well as property damage, injuries, and illnesses) attributable to climate change.

Risks and Tradeoffs

Some threats to our health, particularly from toxic chemicals in the environment, make big news. Other threats, while larger from a risk assessment perspective, go unnoticed. Media reports of risk events and situations are constrained by the need to entertain as well as inform, and most reporters are not trained in science or risk assessment. Consequently, the risks that get the most attention in the media may not be as detrimental as others that get less coverage.

This does not mean that we should ignore chemicals that humans introduce into the environment. Nor does it mean we should discount stories the news media sometimes sensationalize. These stories serve an important role in getting the regulatory wheels of the government moving to protect us as much as possible from the dangers of our technological and industrialized world. They reflect a distrust in the ability of industry and government to manage risks and thereby suggest opportunities to improve those institutions. Quantifying risks is an important step toward reducing them, but it is as important to understand why and how people respond to risks, independent of their magnitudes.

ECOLOGICAL RISK ASSESSMENT

The EPA and other federal and state environmental monitoring groups are increasingly trying to evaluate ecosystem health using methods developed for assessing risks to human health. Detailed guidelines exist for performing **ecological risk assessment (ERA)**. Although it began as an extension of ecotoxicology, ecological risk assessment covers a much broader set of impacts to ecosystems. It includes an assessment of the exposures of a range of plants and animals to **stressors** (potentially adverse conditions) and the range of impacts associated with those stressors.

Such analyses are difficult because effects may occur on a wide scale, from individual animals or plants in a local area to ecological communities across a large region. Given the hazards and exposure levels of human-induced environmental stressors, ecological effects range from good to bad or from acceptable to unacceptable. Using scientific knowledge in environmental decision making is filled with uncertainty because many ecological effects are incompletely understood or difficult to measure. A real need exists to quantify risk to the environmental system and to develop strategies to cope with the uncertainty.

ERA is increasingly used to manage the diverse threats to **watersheds**. For example, the Yellow River delta area in China includes eight distinct ecosystems (saline seepweed beach, reed marsh, freshwater bodies, forest and meadow, prawn pool and salt pan, paddy fields, dry farmland, and human settlement) (**Figure 7.14**). These face a number of stressors, including agricultural water withdrawal, drought, storm surges, flooding, and oil spills. An ERA for the area concluded that a few areas face the greatest probability of negative effects; managers are now using this information to prioritize risk-reduction activities in the delta.

Courtesy NASA

Figure 7.14 **Yellow River delta.** The Yellow River delta region in China is a highly productive and diverse set of ecosystems that faces multiple human-induced stressors. Ecological risk assessment has guided managers in identifying and dealing with the most vulnerable areas.

REVIEW

1. What is risk assessment?

2. How does cost-benefit analysis differ from the precautionary principle?

3. When is an ecological risk assessment undertaken?

ENERGY AND CLIMATE CHANGE

Chronic Disease and Acute Illness

Although climate change has already begun to cause human deaths and disease, the number of cases is highly uncertain. As with other hazards, climate change–related deaths and illness include both acute and chronic effects. Acute effects include deaths associated with heat waves in Europe in 2006 and the United States in 2010, extreme weather events like Hurricanes Katrina and Ike, wildfires like those in California in 2008, and other natural disasters (see the figure). Chronic effects include an expanded range for tropical diseases and increased airborne allergens. Not all health effects will be negative; for example, milder winters are likely to lead to fewer cases of influenza.

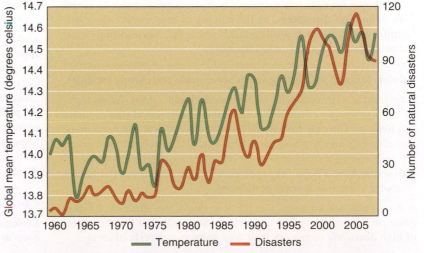

Global mean temperature and heat-related natural catastrophes, 1960–2010. This figure suggests that the number of heat-related natural catastrophes (such as heat waves and droughts) has increased over the past five decades. What is the ratio of heat-related disasters in 2010 to those in 1960? In 1975?

REVIEW OF LEARNING OBJECTIVES WITH SELECTED KEY TERMS

- **Contrast health issues in highly developed countries with those in less developed countries.**

Cardiovascular diseases, cancer, and chronic obstructive pulmonary disease are health problems in the United States and other highly developed nations; many of these diseases are chronic health problems associated with aging and are caused in part by lifestyle choices involving diet, exercise, and smoking. Child mortality is particularly serious in developing countries, where leading causes of death in children include malnutrition, diarrheal diseases, and malaria.

- **Explain the difference between endemic and emerging diseases.**

Endemic diseases are constantly present in a population or region. Some endemic diseases, such as smallpox and polio, have been controlled worldwide, while others remain common. **Emerging diseases** are those that are relatively new to human populations. Because different strains prevail each year, influenza is often thought of as an emerging disease. AIDS was an emerging disease a few decades ago, but is now endemic in many places.

- **Summarize the problems associated with chemicals that exhibit persistence, bioaccumulation, and biological magnification in the environment.**

Chemicals that exhibit **persistence** are extremely stable and may take many years to break down into simpler forms by natural processes. **Bioaccumulation** is the buildup of a persistent toxic substance, such as certain pesticides, in an organism's body, often in fatty tissues. **Biological magnification** is the increased concentration of toxic chemicals, such as PCBs, heavy metals, and certain pesticides, in the tissues of organisms at higher levels in food webs.

- **Briefly describe some of the data suggesting that certain chemicals used by humans may also function as endocrine disrupters in animals, including humans.**

A chemical spill in 1980 contaminated Lake Apopka, in Florida, with DDT and other agricultural chemicals with estrogenic properties. Male alligators living in Lake Apopka in the years following the spill had low levels of testosterone (an androgen) and elevated levels of estrogen. Their reproductive organs were often feminized or abnormally small, and the mortality rate for eggs was extremely high. Humans may also be at risk from **endocrine disrupters**, as the number of reproductive disorders, infertility cases, and hormonally related cancers (such as testicular cancer and breast cancer) appears to be increasing.

- **Define *toxicant,* and distinguish between acute and chronic toxicities.**

A **toxicant** is a chemical with adverse human health effects. **Acute toxicity** refers to adverse effects that occur within a short period after exposure to a toxicant. **Chronic toxicity** refers to adverse effects that occur some time after exposure to a toxicant or after extended exposure to a toxicant.

- **Describe how a dose-response curve helps determine the health effects of environmental pollutants.**

A **dose-response curve** is a graph that shows the effect of different doses on a population of test organisms. Scientists first test the effects of high doses and then work their way down to a **threshold** level.

- **Discuss pesticide risks to children.**

More than half of all reports of exposure and possible poisoning from household pesticides involve children. Some research suggests that exposure to pesticides may affect the development of intelligence and motor skills of infants, toddlers, and preschoolers.

- **Discuss how discovery of the effects of DDT on birds led to the replacement of the dilution paradigm with the boomerang paradigm.**

A paradigm is a generally accepted understanding of how some aspect of the world works. According to the dilution paradigm, "the solution to pollution is dilution." Environmental scientists reject the dilution paradigm in favor of the boomerang paradigm: What you throw away can come back and hurt you. The boomerang paradigm was adopted after the discovery that the pesticide DDT was accumulating in birds at the top of the food web. The implications were that DDT represented an unacceptable threat to ecosystem health and human health.

- **Define *ecotoxicology,* and explain why knowledge of ecotoxicology is essential to human well-being.**

Ecotoxicology is the study of contaminants in the biosphere, including their harmful effects on ecosystems. Ecotoxicology helps policy makers determine the costs and benefits of the many industrial and technological "advances" that affect us and the ecosystems on which we depend.

- **Define *risk,* and explain how risk assessment helps determine adverse health effects.**

Risk is the probability that a particular adverse effect will result from some exposure or condition. **Risk assessment** is the process of estimating those probabilities and consequences. Risk assessments, when properly performed, provide information about the probability and severity of a risk or set of risks.

- **Explain how risk information can improve environmental decisions.**

Risk information can be used to inform several different decision approaches. In **risk management**, risks are assessed and described, and policy choices are chosen to reduce known risks. Risk can serve as an input to **cost-benefit analysis**. Uncertainty associated with a risk assessment for a particular product or material might lead us to apply the **precautionary principle** and thus avoid an uncertain but possibly adverse outcome. **Fractional risk attribution** allows us to allocate responsibility for adverse outcomes based on the likely contributors to those outcomes, even when the cause of individual cases cannot be determined.

- **Define *ecological risk assessment.***

Ecological risk assessment is the process by which the ecological consequences of a range of human-induced **stressors** are estimated.

Critical Thinking and Review Questions

1. What are the three leading causes of death in the United States? How are they related to lifestyle choices?

2. Why are public health researchers concerned about "exporting" health problems associated with developed countries to less developed countries?

3. What accounts for the much lower infant mortality rates in highly developed countries as compared to less developed countries?

4. What was the first viral disease to be globally eradicated? What disease do health officials hope will be eradicated soon?

5. What is a pandemic, and why does the potential for pandemics trouble many public health officials?

6. Distinguish among persistence, bioaccumulation, and biological magnification.

7. How do acute and chronic toxicity differ?

8. What is a dose-response curve? What can it tell us about effects at low doses if experimental information is about high doses?

9. What are the major differences between toxicology and epidemiology? In what ways are they the same? How might decision makers respond when the two approaches lead to different conclusions?

10. Describe the common methods for determining whether a chemical causes cancer.

11. Select one of the two choices to complete the following sentence, and then explain your choice: The absence of certainty about the health effects of an environmental pollutant (is/is not) synonymous with the absence of risk.

12. For each of the following scenarios, suggest whether it would be more appropriate to use cost-benefit analysis, the precautionary principle, or fractional risk attribution. Explain your answers.

    a. Regulating risk associated with mercury from coal-fired power plants

    b. Compensating cancer patients who were exposed to chemicals on the job 20 years earlier

    c. Preparing for climate change–related illnesses in the year 2050

    d. Creating a new policy for nanomaterials in cosmetics and medicines

    e. Preparing for diseases that might transfer from animals to humans in the future

13. How are risk assessments for human health and ecological risk assessments for environmental health alike? How do they differ?

14. Do you expect that a warmer world will improve or worsen human health? Explain your answer.

15. The figure below shows that children's health in the United States is closely linked to both household income and education. Could the environment be one of the factors associated with household income? Household education? Explain.

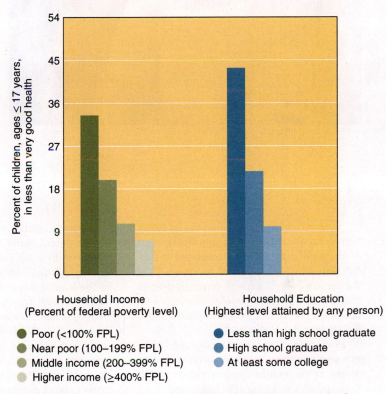

Household income, education, and children's health in the United States.

Food for Thought

In the first half of the 20th century, the U.S. military had to turn away an unexpectedly large number of potential recruits because they were underweight. In response, the U.S. government developed a policy for providing food to millions of Americans. By the end of the 20th century, the military was turning away potential recruits for being overweight. Discuss what this says about the relationship between caloric intake and a healthy diet. Do government-sponsored food programs in your area promote healthful eating? Explain.

8

The Human Population

HIV/AIDS orphans. The orphans of the Chubby Chumbs Fountain of Love orphanage receive their lunch, in Johannesburg, South Africa. Most of these children are AIDS orphans or are HIV positive, or both.

Africa has the most rapidly growing human population of all the continents. Most of Africa's population is concentrated in sub-Saharan Africa (that part of Africa south of the Sahara Desert). Experts currently predict that the sub-Saharan population of 926 million will increase to 1.3 billion by 2025 and is expected to double by 2050.

In spite of these astounding increases, many population experts lowered their estimates of population growth rates in sub-Saharan Africa during the 1990s and early 2000s. Tragically, the estimates were lowered because acquired immune deficiency syndrome (AIDS), caused by the human immune deficiency virus (HIV), is ravaging many countries in the sub-Saharan region. The UN Program on HIV/AIDS estimates that 25 million of the world's 35 million adults and children who are now infected with HIV/AIDS live in sub-Saharan Africa (latest available data). Worldwide, more than 2.3 million people become newly infected with HIV each year, and about 70% of these new infections occur in sub-Saharan Africa. About 15 million children in this region have lost one or both parents to AIDS (see photograph). Many AIDS orphans do not receive adequate education, healthcare, or nutrition.

Despite global progress in expanding the treatment of HIV/AIDS around the world, countries in sub-Saharan Africa are experiencing high death rates from HIV/AIDS, which is the leading cause of death in Africa. The high mortality from HIV/AIDS has caused life expectancies to decline in many African countries. In Swaziland, for example, the average life expectancy declined from a high of about 60 years in the late 1980s to 49 years in 2013. Africa's AIDS crisis has many repercussions beyond a reduced life expectancy. The HIV/AIDS epidemic threatens the economic stability and social support networks of the region and has overwhelmed healthcare systems. Hunger has significantly increased in some countries, owing to the loss of HIV/AIDS-infected agricultural workers; by 2020, the United Nations estimates that AIDS will have killed at least 20% of agricultural workers in southern Africa. Labor shortages are common, and foreign investments have slowed because many investors are wary of the high rate of infection among workers.

The good news is that efforts to improve access to the antiretroviral therapy necessary to treat HIV/AIDS are gaining ground: Globally, the number of people receiving the therapy tripled between 2005 and 2012.

In 2012, approximately 61% of the world's individuals living with HIV/AIDS in low- and middle-income countries—including approximately 68% in sub-Saharan Africa—received treatment. HIV/AIDS deaths and new cases recorded annually have both declined, and the average life expectancy has increased slightly. With these welcome improvements in HIV/AIDS treatment, population experts hope that these sub-Saharan countries can focus on improving family planning programs and controlling birth rates.

The HIV/AIDS epidemic is but one factor affecting the human population. In this chapter, we focus on the current state of the human population.

In Your Own Backyard…

What is the "safe-sex" message regarding HIV transmission? Do you think most students at your school are concerned about HIV transmission? Why or why not?

The Science of Demography

LEARNING OBJECTIVES

- Define *demography* and summarize the history of human population growth.
- Identify Thomas Malthus, relate his ideas on human population growth, and explain why he might or might not be correct.
- Explain why it is impossible to answer precisely how many people Earth can support—that is, Earth's carrying capacity for humans.

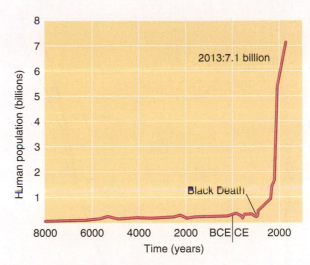

Figure 8.1 **Human population growth.** During the last 1000 years, the human population has been increasing exponentially. Population experts predict that the population will level out during the 21st century, possibly forming the S curve observed in certain other species. (Black Death refers to a devastating disease, probably bubonic plague, that decimated Europe and Asia in the 14th century.) (Population Reference Bureau)

Demography is the science of human population structure and growth. The application of population statistics is called **demographics**.

demography The applied branch of sociology that deals with population statistics and provides information on the populations of various countries or groups of people.

Examine **Figure 8.1**, which shows human population growth. Refer back to Figure 5.5 and compare the two curves. The characteristic J curve of exponential population growth shown in Figure 8.1 reflects the decreasing amount of time it has taken to add each additional billion people to our numbers. It took thousands of years for the human population to reach 1 billion, a milestone that occurred around 1800. It took 130 years to reach 2 billion (in 1930), 30 years to reach 3 billion (in 1960), 15 years to reach 4 billion (in 1975), 12 years to reach 5 billion (in 1987), 12 years to reach 6 billion (in 1999), and 12 years to reach 7 billion (in 2011).

One of the first to recognize that the human population cannot increase indefinitely was **Thomas Malthus**, a British economist (1766–1834). He pointed out that human population growth is not always desirable—a view contrary to the beliefs of his day and to those of many people even today—and that the human population can increase faster than its food supply. The inevitable consequences of population growth, he maintained, are famine, disease, and war. Since Malthus's time, the human population has grown from about 1 billion to 7 billion.

At first glance, it appears that Malthus was wrong. Our population has continued to grow because scientific advances have allowed food production to keep pace with population growth. Malthus might ultimately prove correct because we do not know if our increase in food production is *sustainable*. Have we achieved this increase in food production at the environmental cost of reducing the ability of the land to meet the needs of future populations? The truth is that we still do not know if Malthus was wrong or right.

CURRENT AND FUTURE POPULATION NUMBERS

Our world population, which was 7.1 billion in 2012, increased by about 97 million from 2012 to 2013. This increase was not due to an increase in the birth rate (*b*). In fact, the world birth rate has declined during the past 200 years. Instead, the increase in population was due to a dramatic decrease in the death rate (*d*), which has occurred primarily because greater food production, better medical care, and improvements in water quality and sanitation practices have increased the life expectancies for a great majority of the global population.

Although our numbers continue to increase, the world growth rate (*r*) has declined over the past several years, from a peak of 2.2% per year in the mid-1960s to 1.2% per year in 2013. Population experts at the United Nations and the World Bank have projected that the growth rate will continue to decrease slowly until **zero population growth**—when the birth rate equals the death rate—is attained. If this stabilization occurs, it will be clear that human population follows a steep S curve, rather than a J curve. Experts project that zero population growth will occur toward the end of the 21st century. (You may wish to review *r*, *b*, and *d* as well as J and S curves in Chapter 5.)

The United Nations periodically publishes population projections for the 21st century. In its 2012 revision, it forecast that the human population will be between 8.3 billion (its "low" projection) and 10.9 billion (its "high" projection) in the year 2050, with 9.6 billion thought most likely (its "medium" projection; **Figure 8.2**). These estimates take into account both lower projected fertility levels and higher projected mortality from HIV/AIDS. Lower fertility levels result in an aging population; the percentage of the world population over the age of 65 was about 8% in 2013 and is expected to rise. (Population aging is discussed later in this chapter.) Such population projections are "what-if" exercises: Given certain assumptions about future tendencies in the birth rate, death rate, and migration, an area's population can be calculated for a given number of years into the future.

The main unknown factor in any population growth scenario is Earth's **carrying capacity**. Most published estimates of how many people Earth can support range from 4 billion to 16 billion. These estimates vary widely, depending on what assumptions are made about standard of living, resource production and consumption, technological innovations, and waste generation.

carrying capacity The maximum number of individuals of a population that a particular environment can support for an indefinite period, assuming no changes in the environment.

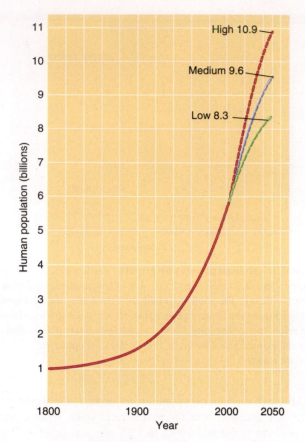

Figure 8.2   **Population projections to 2050.** In 2012, the United Nations revised its projections of the human population in 2050, each based on different fertility rates. The medium projection was 9.6 billion. How will increasing human population numbers from the present to 2050 affect worldwide energy use? (Data from *World Population Prospects*, *The 2012 Revision*, UN Population Division)

If we want all people to have a high level of material well-being equivalent to the lifestyles common in highly developed countries, then Earth will clearly support far fewer humans than if everyone lives just above the subsistence level. Earth's carrying capacity for humans is not decided simply by environmental constraints; human choices and values must be factored into the assessment.

HUMAN MIGRATION

Human migration across international borders is a worldwide phenomenon that has escalated dramatically during the past few decades. People migrate in search of jobs or an improved standard of living; to escape war or persecution for their race, religion, nationality, or political opinions; or to join other family members who have already migrated. The increase in the human population has contributed to the current surge in migration, which population experts predict will continue to increase during the next 30 years or so.

Deteriorating environmental conditions and increased competition for jobs, both brought on by unsustainable population growth, may accelerate international migration. Most international migrants relocate in nearby countries. As a result, each continent has its own characteristic flow of international migrants. In North America, for example, the large number of

unauthorized migrants entering the United States from Mexico and Central America is a controversial political issue. In the European Union, many guest workers from Morocco, the former Yugoslavia, and Turkey unexpectedly became permanent residents, and cultural differences between natives and nonnatives remain an ongoing concern.

Sometimes economic, political, or environmental problems force people to leave their homelands en masse. According to estimates from the UN High Commissioner for Refugees, there were 15.4 million international refugees at the end of 2012 (latest data available). Countries with large numbers of refugees currently include Afghanistan, Somalia, Iraq, the Syrian Arab Republic, and Sudan. Syrians were poised to pass Afghans to become the world's largest refugee group by the end of 2014, with their numbers expected to exceed 4 million. Most refugees flee to neighboring countries, where many are placed in camps, which are often crowded and sometimes unhygienic.

REVIEW

1. Describe human population growth for the past 200 years.
2. Who was Thomas Malthus, and what were his views on human population growth?
3. When determining Earth's carrying capacity for humans, why is it not enough to just consider human numbers? What else must be considered?

Demographics of Countries

LEARNING OBJECTIVES

- Explain how highly developed and developing countries differ in population characteristics such as infant mortality rate, total fertility rate, and age structure.
- Explain how population growth momentum works.

Whereas world population figures illustrate overall trends, they do not describe other important aspects of the human population story, such as population differences from country to country (**Table 8.1**).

Table 8.1   **The World's 10 Most Populous Countries**

Country	2013 Population (in millions)	Population Density (per square kilometer)
China	1357	142
India	1277	388
United States	316	33
Indonesia	249	130
Brazil	196	23
Pakistan	191	230
Nigeria	174	1
Bangladesh	157	1087
Russia	143	8
Japan	127	337

Source: Population Reference Bureau.

As you may recall from Chapter 1, not all countries have the same rate of population increase. Countries can be classified into two groups—highly developed and developing—depending on growth rates, degree of industrialization, and relative prosperity (**Table 8.2**). Note, however, that these categories can be overly simplistic—for example, not taking into account a range of characteristics within countries—yet they help to illustrate differences in population increases.

Nations identified as **highly developed countries** (or *developed countries*), such as the United States, Canada, France, Germany, Sweden, Australia, and Japan, have low rates of population growth and are highly industrialized relative to the rest of the world (**Figure 8.3**). Highly developed countries have the lowest birth rates in the world. Indeed, some countries such as Germany have birth rates just below those needed to sustain their populations and are declining slightly in numbers.

Highly developed countries have low **infant mortality rates**. The infant mortality rate of the United States was 5.9 in 2013, compared with a world rate of 40. Highly developed countries also have longer life expectancies (79 years in the United States versus 70 years worldwide) and high average per capita GNI PPPs ($50,610 in the United States versus $11,690 worldwide). *Per capita GNI PPP* is gross national income (GNI) in purchasing power parity (PPP) divided by midyear population. It indicates the amount of goods and services an average citizen of a particular country could buy in the United States. Perhaps because of high *per capita* GNI PPPs, highly developed countries also have high rates of energy consumption and waste production.

Countries identified as *developing* are grouped into two subcategories, moderately developed and less developed. Mexico, Turkey, Thailand, India, and most South American nations are

> **infant mortality rate**
> The number of infant deaths (under age 1) per 1000 live births.

Table 8.2   Comparison of 2013 Population Data in Developed and Developing Countries

	Developed	Developing	
	(Highly Developed) United States	(Moderately Developed) Venezuela	(Less Developed) Ethiopia
Fertility rate	1.9	2.4	4.8
Projected population change, 2013–2050*	1.3	1.4	2.0
Infant mortality rate	5.9 per 1000	11.6 per 1000	52 per 1000
Life expectancy at birth	79 years	75 years	62 years
Per capita GNI PPP (2012; U.S. $)**	$50,610	$13,120	$1,140
Women using modern contraception	73%	62%	27%

*Includes fertility, mortality, and migration estimates; 2050 population is presented as a multiple of the 2013 population.
**GNI PPP = gross national income in purchasing power parity.
Source: Population Reference Bureau.

LEGEND
- 40+
- 25 – 39
- 15 – 24
- 10 – 14
- <10

Figure 8.3   Birth rates around the world, 2013. Numbers indicate births per 1000 total population. European nations had the fewest births, and African countries had the most. (Population Reference Bureau)

examples of **moderately developed countries**. Their birth rates and infant mortality rates are higher than those of highly developed countries, but they are declining. Moderately developed countries have a medium level of industrialization, and their average per capita GNI PPPs are lower than those of highly developed countries. Examples of **less developed countries** include Bangladesh, Niger, Ethiopia, Laos, and Cambodia. These countries have the highest birth rates, the highest infant mortality rates, the shortest life expectancies, and the lowest average per capita GNI PPPs in the world. These countries also have lower levels of fossil fuel usage and waste production.

Replacement-level fertility is the number of children a couple must produce to "replace" themselves. Replacement-level fertility is usually given as 2.1 children. The number is greater than 2.0 because some infants and children die before they reach reproductive age. Worldwide, the **total fertility rate (TFR)** is currently 2.5, well above the replacement level.

> **total fertility rate (TFR)**
> The average number of children born per woman, given the population's current birth rate.

DEMOGRAPHIC STAGES

In 1945, Princeton demographer **Frank Notestein** recognized four demographic stages, based on his observations of Europe as it became industrialized and urbanized (**Figure 8.4**). During these stages, Europe converted from relatively high to relatively low birth and death rates. To date, all highly developed and moderately developed countries with more advanced economies have gone through this progression, or **demographic transition**. Demographers generally assume that the same demographic transition will occur in less developed countries as they become industrialized.

In the first stage—the **preindustrial stage**—birth and death rates are high, and population grows at a modest rate. Although women have many children, the infant mortality rate is high. Intermittent famines, plagues, and wars also increase the death rate, so the population grows slowly or temporarily declines. If we use Finland to demonstrate the four demographic stages, we can say that Finland was in the first demographic stage from the time of its first human settlements until the late 1700s.

As a result of the improved healthcare and more reliable food and water supplies that accompany the beginning of an industrial

society, the second demographic stage, called the **transitional stage**, has a lowered death rate. The population grows rapidly because the birth rate is still high. Finland in the mid-1800s was in the second demographic stage.

The third demographic stage, the **industrial stage**, is characterized by a decline in birth rate and takes place at some point during the industrialization process. The decline in birth rate slows population growth despite a relatively low death rate. Finland experienced this stage in the early 1900s.

Low birth and death rates characterize the fourth demographic stage, called the **postindustrial stage**. In heavily industrialized countries, people are better educated and more affluent; they tend to desire smaller families, and they take steps to limit family size. The population grows slowly or not at all in the fourth demographic stage. This is the situation in such highly developed countries and groups of countries as the United States, Canada, Australia, Japan, and Europe, including Finland.

Why has the population stabilized in highly developed countries in the fourth demographic stage? The decline in birth rate is associated with an improvement in living standards, although it is not known whether improved socioeconomic conditions have caused a decrease in birth rate or a decrease in birth rate has caused improved socioeconomic conditions. Perhaps both are true. Another reason for the decline in birth rate in highly developed countries is the increased availability of family planning services. Other socioeconomic factors that influence birth rate are increased education, particularly of women, and urbanization of society. We consider these factors in greater detail later in this chapter and in Chapter 9.

We do not know if low birth rates will continue beyond the fourth demographic stage. Low birth rates may be a permanent response to the socioeconomic factors of an industrialized, urbanized society, or they may be a response to the changing roles of women in highly developed countries. Unforeseen changes in the socioeconomic status of women and men in the future may change birth rates. A 2009 study reported in the journal *Nature*, for example, determined that above a certain level of economic development, fertility begins to rise, at least in certain highly developed countries.

The population in many developing countries is approaching stabilization. See **Figure 8.5** and note the general decline in total fertility rates in selected developing countries from the 1960s to

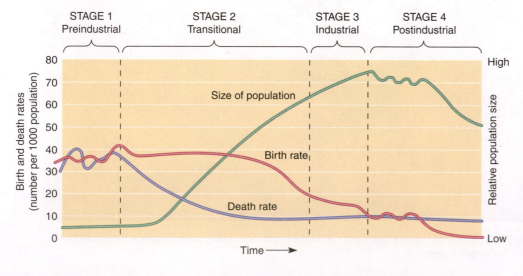

Figure 8.4   Demographic transition.
The demographic transition consists of four demographic stages through which a population progresses as its society becomes industrialized. Note that the death rate declines first, followed by a decline in the birth rate.

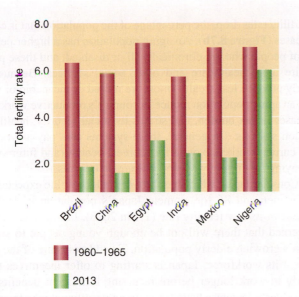

Figure 8.5 **Fertility changes in selected developing countries.** (Population Reference Bureau)

Legend:
1960–1965
2013

of females at each age, from birth to death, are represented in an age structure diagram.

The overall shape of an age structure diagram indicates whether the population is increasing, stable, or shrinking. The age structure diagram of a country with a high growth rate, based on a high fertility rate—for example, Ethiopia or Guatemala—is shaped like a pyramid (**Figure 8.6a**). The probability of future population growth is great because the largest percentage of the population is in the prereproductive age group (0 to 14 years of age). A positive **population growth momentum** exists because when all these children mature, they will become the parents of the next generation, and this group of parents will be larger than the previous group. Even if the fertility rate of such a country has declined to replacement level (that is, couples are having smaller families than their parents did), the population will continue to grow for some time. Population growth momentum, which can be positive or negative, explains how the present age distribution affects the future growth of a population.

In contrast, the more tapered bases of the age structure diagrams of countries with slowly growing, stable, or declining populations indicate that a smaller proportion of the population will become the parents of the next generation (**Figure 8.6b, c**). The age structure diagram of a stable population, one that is neither growing nor shrinking, demonstrates that the numbers of people at prereproductive and reproductive ages are approximately the same. A larger percentage of the population is older—that is, postreproductive—than in a rapidly increasing population. Many countries in Europe have stable populations. In a population shrinking in size, the prereproductive age group is *smaller* than either the reproductive or postreproductive group. Russia, Ukraine, and Germany are examples of countries with slowly shrinking populations.

An estimated 26% of the global human population is under age 15 (**Figure 8.7a**). When these people enter their reproductive

population growth momentum The potential for future increases or decreases in a population based on the present age structure.

2013. The total fertility rate in developing countries decreased from an average of 6.1 children per woman in 1970 to 2.6 in 2013. Although the fertility rates in these countries have declined, most still exceed replacement-level fertility. Consequently, the populations in these countries are still increasing. Even when fertility rates equal replacement-level fertility, population growth will still continue for some time. To understand why this is so, let us examine the age structure of various countries.

AGE STRUCTURE

age structure The number and proportion of people at each age in a population.

To predict the future growth of a population, one must know its **age structure**, or distribution of people by age. The number of males and number of people by age.

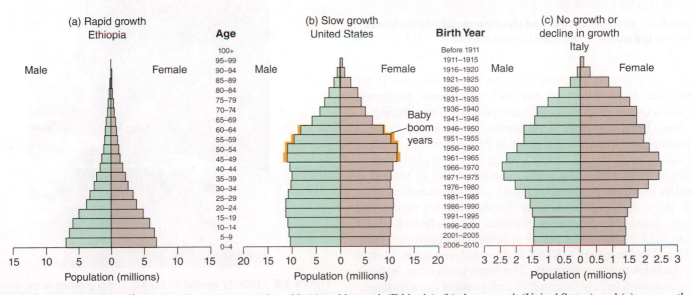

Figure 8.6 **Age structure diagrams.** Shown are countries with (a) rapid growth (Ethiopia), (b) slow growth (United States), and (c) no growth (Italy) or declining population growth. These age structure diagrams indicate that less developed countries such as Ethiopia have a much higher percentage of young people than do highly developed countries. As a result, less developed countries are projected to have greater population growth than highly developed countries. (UN Department of Economic and Social Affairs Population Division)

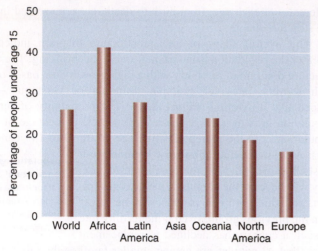

(a) Percentages of the population under age 15 for various regions in 2013. The higher this percentage, the greater the potential for population growth when people in this group reach their reproductive years.

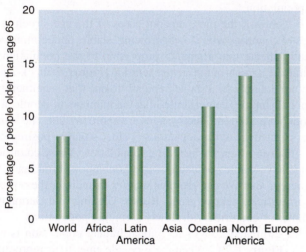

(b) Percentages of the population older than 65 in 2013. Lower fertility rates lead to aging populations.

Figure 8.7 **Prereproductive and elderly populations in various regions of the world.** (Population Reference Bureau)

years, they have the potential to cause a large increase in the growth rate. Even if the birth rate does not increase, the growth rate will increase simply because more people are reproducing.

Since 1950, most of the world population increase has occurred in developing countries—as a result of the younger age structure and the higher-than-replacement-level fertility rates of their populations. In 1950, 67% of the world's population was in developing countries in Africa, Asia (minus Japan), and Latin America. Since 1950, the world's population more than doubled, but most of that growth occurred in developing countries. As a reflection of this fact, in 2013, the number of people in developing countries (including China) had increased to 82.6% of the world population. Most of the population increase during the 21st century will occur in developing countries, largely the result of their younger age structures.

Age Structure: Effects of an Aging Population Declining fertility rates have profound social and economic implications because

as fertility rates drop, the percentage of the population that is elderly increases (**Figure 8.7b**). An aging population has a higher percentage of people who are chronically ill or disabled, and these people require more healthcare and other social services. Because the elderly produce less wealth (most are retired or incapable of working), an aging population reduces a country's productive workforce, increases its tax burden, and strains its social security, health, and pension systems. (In general, these systems are "pay-as-you-go," with current workers paying for retirees' benefits and future workers paying for current workers' retirement benefits.)

Consider Japan, which has the world's longest life expectancy— 80 for men and 86 for women. Japan's population is declining because of its low fertility rate (1.4 in 2013). Japanese leaders are concerned that there will not be enough young people to support Japan's growing elderly population. To address some of the problems in its workforce, Japan is starting to offer incentives to the elderly to work longer before receiving retirement benefits, and some elderly Japanese have taken on nontraditional jobs to expand their working years (**Figure 8.8**).

Russia's fertility rate is low (1.7 in 2013), and the UN Population Division predicts that Russia's population in 2050 will have declined to 132.4 million (its 2013 population was 143.5 million). Like many countries with declining populations, Russia offers incentives to young couples who choose to have children; these include assisting with maternity leave and helping to pay for childcare when the mother returns to the workplace.

Given that the world population continues to increase, many demographers think that providing incentives to encourage women to have more children is not a good approach. After all, if a given

Figure 8.8 **Elderly worker in Japan.** The population in the town of Kamikatsu, Japan, is dominated by the elderly, many of whom seek to stay employed. One thriving industry in the region, run primarily by elderly women, is the collection of decorative leaves that are sold for use in Japanese cuisine.

country raises its birth rate, it is not only increasing the world population but also contributing to its aging population in the future.

Changes in a population's age structure can affect many social problems (such as crime) that at first glance do not appear to be directly related. Sociologists have observed that in an aging population the crime rate may decline. Most crimes involve young adults between the ages of 18 and 24, and in an aging population, these young adults represent a smaller proportion of the population. According to the U.S. Census Bureau, in the late 1990s, the United States experienced a reduction in the rate of violent crimes, and the aging of the U.S. population was cited as at least one factor. (Like all social problems, the causes of crime are complex.)

Thus, aging populations offer a mixed bag of benefits and problems. No country has been faced with an aging population before now, and we do not know how aging populations will function. Despite the uncertainties, most policy analysts think that countries with higher proportions of elderly will probably have to increase the age of retirement (through incentives that encourage people to work after age 65) and decrease benefits for the elderly. Also, because benefits for the elderly will most certainly decline over time, experts recommend that young people begin saving aggressively for their retirements *early* in their careers instead of after their children have grown.

REVIEW

1. What is total fertility rate? Why do you suppose it is much higher in Mozambique than it is in Sweden?
2. What is population growth momentum?

Population and Quality of Life

LEARNING OBJECTIVES

- Relate carrying capacity to agricultural productivity.
- Relate human population to chronic hunger and food insecurity.
- Describe the relationship between economic development and population growth.

As the 21st century proceeds, it will become increasingly difficult to meet the basic needs (a balanced diet, clean water, decent shelter) of all people, especially in countries that have not achieved population stabilization. About 12% of the world's population live in less developed countries. If their rate of population growth continues, many of these countries will double their populations by 2050.

The social, political, and economic problems resulting from continued population growth in less developed countries sometimes result in internal violence and political collapse. Such violent conflict inevitably affects countries that have already achieved stabilized populations and high standards of living. For this reason, population growth is of concern to the entire world community, regardless of where it occurs.

As our numbers increase during the 21st century, environmental degradation, hunger, persistent poverty, economic stagnation, urban deterioration, and health issues will continue to challenge us. Already, the need for food for the increasing number of people living in environmentally fragile arid lands, such as parts of sub-Saharan Africa, has led to overuse of the land for grazing and crop production. When land overuse occurs in combination with an extended drought, formerly productive lands may decline in agricultural productivity—that is, their carrying capacity may decrease. Although it is possible to reclaim such arid lands, the large number of people and their animal herds trying to live off the land make reclamation efforts difficult. (Carrying capacity as it relates to environmental resistance was introduced in Chapter 5.)

No one knows if Earth can sustainably support the 9.6 billion people that is the UN medium projection for 2050 (see Figure 8.2). It is not even clear that Earth can sustainably support the 7.1 billion people we currently have. We cannot quantify the carrying capacity for humans in any meaningful way, in part because our impact on natural resources and the environment involves more than simply population numbers.

To estimate global carrying capacity for humans, we must make certain assumptions about our *quality* of life. Do we assume that everyone in the world should have the same standard of living as average U.S. citizens currently do? If so, then Earth would support a fraction of the humans it could support if everyone in the world had only the barest minimum of food, clothing, and shelter. Also, we do not know if future technologies would completely alter Earth's sustainable population size. We may have already reached or overextended our carrying capacity, and the numerous environmental problems we are currently experiencing may cause the world population increase to come to a halt or even decline precipitously.

POPULATION AND CHRONIC HUNGER

Food security is the condition in which people do not live in hunger or fear of starvation. Many of the world's people—more than 1 billion—do not have food security. These people do not get enough food to thrive; in certain areas of the world, people, especially children, still starve (**Figure 8.9**).

According to the UN Food and Agriculture Organization (FAO), 86 countries are considered low income and food deficient. South Asia and sub-Saharan Africa are the two regions of the world with the greatest **food insecurity**. People with food insecurity live under the threat of starvation. Worldwide, the FAO estimates that as many as 2 billion people face food insecurity intermittently as a result of poverty, drought, or civil strife.

food insecurity The condition in which people live with chronic hunger and malnutrition.

Most of the millions of people who die each year from hunger are not victims of *famine*. Famines are usually attributed to bad weather (droughts or floods), insect outbreaks, armed conflict (which causes a breakdown in social and political institutions), or some other disaster. Famines, which generally receive widespread press coverage, account for between 5% and 10% of the world's hungry people in a given year. The remaining 90% to 95%, which represent people with chronic hunger, receive little attention in the world news.

Chronic hunger, population, poverty, and environmental problems are interrelated, but there is no consensus on the most effective way to stop chronic hunger. Many politicians and economists think the best way to tackle world food problems is to

Figure 8.9 **Hunger.** A young Sudanese boy collapses from hunger at a food distribution center run by a medical charity. Population growth is not the only reason there is not enough food for all the world's peoples. In Sudan, the underlying cause is ongoing civil strife.

Antony Njuguna/Reuters/Landov

economic development
An expansion in a government's economy, viewed by many as the best way to raise the standard of living.

promote the **economic development** of countries that do not produce adequate food supplies for their people. Economic development, which is most effective when associated with stable democratic governments, includes improving roads, extending power grids to isolated rural areas, building schools and medical clinics, and expanding communication networks.

The assumption is that economic development would provide the appropriate technology for the people living in those countries to increase their food production or their food-purchasing ability. Once a country makes these gains, its total fertility rate should decline, helping to lessen the population problem. (Poverty and hunger are revisited in Chapters 18 and 24.)

ECONOMIC EFFECTS OF CONTINUED POPULATION GROWTH

The relationship between economic development and population growth is difficult to evaluate. Population growth affects economic development *and* economic development affects population growth, but the degree to which each affects the other is unclear. Some economists have argued that population growth stimulates economic development and technological innovation. Other economists think a rapidly expanding population hampers developmental efforts. Most major technological advances are now occurring in countries where population growth is low to moderate, an observation that supports the latter point of view.

Are large increases in population a deterrent to economic development? Experts studying the interactions among global problems such as underdevelopment, hunger, poverty, environmental problems, and rapid population growth have concluded that population stabilization alone will not eliminate other world problems. For most of the developing world, however, economic development would profit from slower population growth. Population stabilization will not guarantee higher living standards but will probably promote economic development, which, in turn, will raise the standard of living.

REVIEW

1. How is agricultural productivity related to carrying capacity?
2. How is human population growth related to chronic hunger?
3. How is human population growth related to economic development?

Reducing the Total Fertility Rate

LEARNING OBJECTIVES

- Define *culture* and explain how total fertility rate and cultural values are related.
- Define *gender inequality* and relate the social and economic status of women to total fertility rate.
- Explain how the availability of family planning services affects total fertility rate.

Moving from one place to another used to be a solution for unsustainable population growth, but it is not today. As a species, we have expanded our range throughout Earth, and few habitable areas remain that have the resources to adequately support a major increase in human population. For obvious reasons, increasing the death rate is not an acceptable means of regulating population size. Clearly, the way to control our expanding population is to reduce the number of births. Cultural traditions, women's social and economic status, and family planning all influence the total fertility rate.

CULTURE AND FERTILITY

The values and norms of a society—what is considered right and important and what is expected of a person—are all part of a society's **culture**. A society's culture, which includes its language, beliefs, and spirituality, exerts a powerful influence over individuals by controlling behavior. This control is internalized—that is, it is taken for granted.

culture The ideas and customs of a group of people at a given period; culture, which is passed from generation to generation, evolves over time.

Gender is an important part of culture. Different societies have different gender expectations—that is, varying roles men and women are expected to fill. In parts of Latin America men do the agricultural work, whereas in sub-Saharan Africa this is what women do. With respect to fertility and culture, a couple is expected to have the number of children determined by the cultural traditions of their society.

High total fertility rates are traditional in many cultures. The motivations for having many babies vary from culture to culture, but a major reason for high total fertility rates is high infant and child mortality rates. For a society to endure, it must produce enough children who can survive to reproductive age. If infant and child mortality rates are high, total fertility rates must be high to compensate. Although world infant and child mortality rates are decreasing, culturally embedded fertility levels will take longer to decline. Parents must have enough confidence that the children they already have will survive before they stop having additional babies. Another reason for the lag in fertility decline is cultural: Changing anything traditional, including large family size, usually takes a long time.

Higher total fertility rates in some developing countries are due to the important economic and societal roles of children. In some societies, children usually work in family enterprises such as farming or commerce, contributing to the family's livelihood (**Figure 8.10**). When these children become adults, they provide support for their aging parents.

The International Labour Organization (ILO) estimates that, although child labor has declined in recent years, worldwide, about 144 million children between the ages of 5 and 14 worked in 2012 (the ILO does not count household chores as labor). This estimate represents 11.8% of all children. Almost all of these children live in developing countries. About 38 million child laborers do hazardous work, such as mining and construction, and often suffer from chronic health problems caused by the dangerous, unhealthy workplace conditions to which they are exposed. Children who work long hours do not have childhoods, nor do they receive any education.

In contrast, children in highly developed countries have less value as a source of labor because they attend school and because less human labor is required in an industrialized society. Furthermore, highly developed countries provide many economic and social services for the elderly, so the burden of their care does not fall entirely on their offspring.

Many cultures place a higher value on male children than on female children. In these societies, a woman who bears many sons achieves a high status; the social pressure to have male children keeps the total fertility rate high. For example, the Hindu religion in India traditionally required that parents be buried alongside their son, and having a son still has deep cultural significance in India today.

Religious belief systems are another aspect of culture that affects total fertility rates (TFRs). Several studies in the United States point to differences in TFRs among Catholics, Protestants, and Jews. In general, Catholic women have a higher TFR than either Protestant or Jewish women, and women who do not follow any religion have the lowest TFR of all. However, the observed differences in TFRs may not be the result of religious differences alone. Other variables, such as ethnicity (certain religions are associated with particular ethnic groups) and residence (certain religions are associated with urban or with rural living), complicate any generalizations.

THE SOCIAL AND ECONOMIC STATUS OF WOMEN

Gender inequality exists in most societies, although the extent of the gap between men and women varies in different cultures. Gender disparities include the lower political, social, economic, and health status of women compared to men. For example, more women than men live in poverty, particularly in developing countries. In most countries, women are not guaranteed equality in legal rights, education, employment and earnings, or political participation.

gender inequality The social construct that results in women not having the same rights, opportunities, or privileges as men.

Because sons are more highly valued than daughters, girls are often kept at home to work rather than being sent to school. In most developing countries, a higher percentage of women are illiterate than men (**Figure 8.11**). However, definite progress has been made in recent years in increasing literacy in both women and men and in narrowing the gender gap. Fewer young women and men are illiterate than older women and men within a given country.

Worldwide, some 32 million girls in 2010 were not given the opportunity to receive a primary (elementary school) education, although progress has been made: Nearly twice as many girls were out of school in 1999. As of 2010, however, only one-third of countries worldwide reported gender parity in both primary and secondary education. **Gender parity** in education is the right of every child—boy or girl—to attend school. Fewer women than men attend secondary school (high school). In Africa, no more than 50% of women in any country attend secondary school, and in some countries only 2% to 5% are enrolled.

Laws, customs, and lack of education often limit women to low-skilled, low-paying jobs. In such societies, marriage is usually the only way for a woman to achieve social influence and economic security. Thus, the single most important factor affecting high total fertility rates may be the low status of women in many societies. A significant way to address unsustainable population growth is to improve the social and economic status of women. Next, we will examine how marriage age and educational opportunities, especially for women, affect fertility.

Marriage Age and Fertility   The total fertility rate is affected by the average age at which women marry, which is determined

Figure 8.10   Child at work. This young Mexican boy helps his family's agricultural business by spreading coffee beans to dry. In developing countries, total fertility rates are high partly because children traditionally contribute to the family by working. Older children often have wage-producing jobs that add to the family income.

© Jacques Jangoux/Alamy

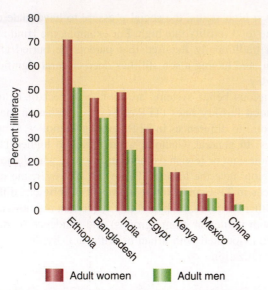

Figure 8.11 **Percent illiteracy of men and women in selected developing countries.** A higher percentage of women than men are illiterate. (From *CIA World Factbook* 2011)

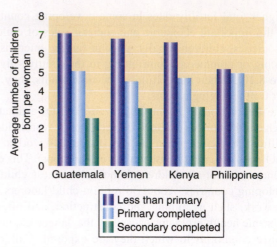

Figure 8.12 **Education and fertility.** The amount of education a woman receives affects the total number of children she has (TFR). The graph shows TFRs for women with different levels of education in several developing countries. (Adapted from E. Murphy, and D. Carr, *Powerful Partners: Adolescent Girls' Education and Delayed Childbearing*. Population Reference Bureau [2007])

by the laws and customs of the society in which they live. Women who marry are more apt to bear children than women who do not marry, and the earlier a woman marries, the more children she is likely to have.

The percentage of women who marry and the average age at marriage vary widely among different societies, but there is generally a correlation between marriage age and total fertility rate. For example, in Pakistan the average age of a woman at marriage is 17, and the TFR is 3.8. In contrast, in Denmark the average age of a woman at marriage is 32, and the TFR is 1.7.

Educational opportunities and fertility In nearly all societies, women with more education tend to marry later and have fewer children. Studies in dozens of countries show a strong correlation between the average amount of education women receive and the total fertility rate (**Figure 8.12**).

Education increases the probability that women will know how to control their fertility, and it provides them with knowledge to improve the health of their families, which results in a decrease in infant and child mortality. A study in Kenya showed that 10.9% of children born to women with no education died by age 5, as compared with 7.2% of children born to women with a primary education and 6.4% of children born to women with a secondary education.

Education increases women's options, providing ways of achieving status besides having babies. Education may have an indirect effect on total fertility rate as well. Children who are educated have a greater chance of improving their living standards, partly because they have more employment opportunities. Parents who recognize this may be more willing to invest in the education of a few children than in the birth of many children whom they cannot afford to educate. The ability of better-educated people to earn more money may be one of the reasons smaller family size is associated with increased family income.

FAMILY PLANNING SERVICES

Socioeconomic factors may encourage people to want smaller families, but reduction in fertility will not become a reality without the availability of **family planning services**. Traditionally, family planning services have focused on maternal and child health, including prenatal care to help prevent infant and maternal death or disability. However, because of gender inequality and cultural constraints, many women cannot ensure their own reproductive health without the support of their partners. Polls of women in developing countries reveal that many who say they do not want additional children still fail to practice any form of birth control. When asked why they do not use birth control, these women frequently respond that their husbands want additional children.

family planning services Services that enable men and women to limit family size, safeguard individual health rights, and improve the quality of life for themselves and their children.

The governments of most countries recognize the importance of educating people about basic maternal and child healthcare (**Figure 8.13**). Developing countries that have had success in significantly lowering total fertility rates credit many of these results to effective family planning programs. Prenatal care and proper birth spacing make women healthier. In turn, healthier women give birth to healthier babies, leading to fewer infant deaths. Globally, the percentage of women using family planning services to limit family size has increased from less than 10% in the 1960s to 56% today. However, the actual *number* of women not using family planning has increased owing to population growth.

Family planning services provide information on reproductive physiology and contraceptives, as well as the actual contraceptive devices, to those who wish to control the number of children they produce or to space their children's births (**Figure 8.14**). Family planning programs are most effective when they are designed with sensitivity to local social and cultural beliefs. Family planning services do not try to force people to limit their family sizes but, rather, attempt to convince people that small families (and the contraceptives that promote small families) are acceptable and desirable.

Contraceptive use is strongly linked to lower total fertility rates. In Canada, the United States, and Northern Europe, where total fertility rates are at replacement levels or lower, the percentage of married women of reproductive age who use contraceptives is greater than 70%. Fertility declines have occurred in developing countries where contraceptives are readily available. For example, research has shown that 90% of the decrease in fertility in 31 developing countries was a direct result of the increased knowledge and availability of contraceptives. Since the 1970s, use of modern contraceptives in East Asia and many areas of Latin America has increased significantly, and these regions have experienced a corresponding decline in birth rate. In areas where contraceptive use has remained low, such as parts of Africa, there was little or no decline in birth rate.

REVIEW

1. How are high total fertility rates related to the economic roles of children in certain cultures?

2. What appears to be the single most important factor affecting high TFRs?

3. How do family planning services influence TFR?

Government Policies and Fertility

LEARNING OBJECTIVES

- Compare how the governments of China and Mexico have tried to slow human population growth.
- Explain the role of population growth momentum in Mexico's population growth.
- Describe at least four of the Millennium Development Goals that came out of the UN Millennium Summit.

Figure 8.13   Family planning. Women in the rural town of Sinnuris, Egypt, learn about family planning and birth control.

The involvement of governments in childbearing and child rearing is well established. Laws determine the minimum age at which people may marry and the amount of compulsory education. Governments may allot portions of their budgets to family planning services, education, healthcare, old-age security, or incentives for smaller or larger family size. The tax structure, including additional charges or allowances based on family size, influences fertility.

In recent years, the governments of at least 78 developing countries in Africa, Asia, Latin America, and the Caribbean have prioritized limiting population growth. They have formulated policies, such as economic rewards and penalties, to achieve this goal. Most countries sponsor family planning projects, many of which are integrated with healthcare, education, economic development, and efforts to improve women's status. The UN Fund for Population Activities supports many of these activities.

CHINA AND MEXICO: CONTRASTING POPULATION GROWTH MEASURES

China, with a mid-2013 population of 1.36 billion, has the largest population in the world, and its population exceeds the combined populations of all highly developed countries. Recognizing that its rate of population growth had to decrease or the quality of life for everyone in China would be compromised, the Chinese government in 1971 began to pursue birth control seriously. It urged couples to marry later, increase spacing between children, and limit the number of children to two.

In 1979, China instigated an aggressive plan to push the nation into the third demographic stage. Announcements were made for incentives to promote later marriages and one-child families (**Figure 8.15**). Local jurisdictions were assigned the task of reaching this goal. A couple who signed a pledge to limit themselves

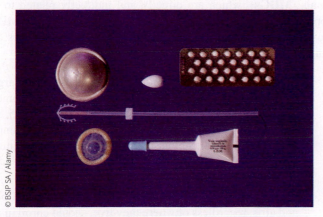

Figure 8.14   Some commonly used contraceptives.

©Johnrochaphoto/Alamy

Figure 8.15 **China's one-child family policy.** A billboard campaign in China promotes the one-child family. Note that the billboard features happy parents with a single female child to counteract the traditional preference for boys.

to a single child might be eligible for such incentives as medical care and schooling for the child, cash bonuses, preferential housing, and retirement funds. Penalties were also instituted, including fines and the surrender of all of these privileges if a second child was born.

China's aggressive plan brought about the most rapid and drastic reduction in fertility in the world, from 5.8 births per woman in 1970 to 2.1 births per woman in 1981. However, it was controversial and unpopular because it compromised individual freedom of choice. In some instances, community and governmental pressures induced women who were pregnant with a second child to get an abortion.

Moreover, the disproportionate number of male versus female babies reported born in recent years (120 boys for every 100 girls in 2000), suggested that many expectant parents aborted female fetuses or killed or abandoned newborn baby girls. In China, sons carry on the family name and traditionally provide old-age security for their parents. Thus, sons are valued more highly than daughters are. Demographers project that by the middle of the 21st century, marriageable males will outnumber marriageable females by 1 million.

China's TFR in 2013 was 1.5. The one-child family policy is now relaxed in rural China, where 47% of all Chinese live. China's recent population control program has relied on education, publicity campaigns, and fewer penalties to achieve its goals. In addition, the Chinese government has been running a "Care for Girls" campaign, and legislation has increased the rights of girls (for example, to inherit). China trains population specialists, and thousands of secondary school teachers integrate population education into the curriculum.

Mexico, with a mid-2013 population of 117.6 million, is the second most populous nation in Latin America. (Brazil, with a population of 195.5 million, is the most populous.) Mexico has a tremendous potential for growth because 30% of its population is

less than 15 years of age. Even with a low birth rate, positive population growth momentum will cause the population to continue to increase in the future because of the large number of young women having babies. However, the approach taken by Mexico contrasts greatly with China's.

Traditionally, the Mexican government supported rapid population growth, but in the late 1960s the government became alarmed at how rapidly the population was increasing. In 1974, the Mexican government instigated several measures to reduce population growth, such as educational reform, family planning, and healthcare. The time was right for such changes, as many Mexican women were already ignoring the decrees of the pro-growth government and the Roman Catholic Church and purchasing contraceptives on the black market. Mexico has had great success in reducing its fertility level, from 6.7 births per woman in 1970 to 2.2 births per woman in 2013.

Mexico's goal includes both population stabilization and balanced regional development. Its urban population makes up 78% of its total population, and most of these people live in Mexico City. Although Mexico is largely urbanized compared with other developing countries, its urban-based industrial economy has been unable to absorb the large number of people in the workforce. (According to the Mexico office of the ILO, about 1.3 million new workers join Mexico's labor force each year.) Unemployment in Mexico is high, and many Mexicans have migrated, both legally and illegally, to the United States.

Mexico's recent efforts at population control include multimedia campaigns. Popular television and radio shows carry family planning messages, such as "Small families live better." Booklets on family planning are distributed, and population education is integrated into the public school curriculum. Social workers receive training in family planning as part of their education.

CASE IN POINT

THE MILLENNIUM DEVELOPMENT GOALS

In 2000, 189 heads of state met at the UN Millennium Summit to address how to meet the needs of the world's impoverished people. This gathering committed their countries to a global partnership with a plan of action known as the **Millennium Development Goals (MDGs)**, which are directly and indirectly related to the population issues discussed in this chapter. Although the MDGs are lofty, each is attainable; each has measurable indicators that can be used to monitor progress.

- **Goal 1: Eradicate extreme hunger and poverty.** Targets include halving the number of people living on an income of less than U.S. $1 (now scaled to $1.25) a day by 2015, and halving the proportion of people who suffer from hunger. Most analysts believe that reaching this goal will require a firm commitment to family planning to curb unsustainable population growth.

- **Goal 2: Achieve universal primary education.** All boys and girls, regardless of where they live, should enroll in and complete a primary education (grades 1 through 5). Limiting family size is essential to meeting this goal because children in large families are less likely to enroll in school and more likely to drop out than children in small families.

- **Goal 3: Promote gender equality and empower women.** *Gender equality* means that men and women have equal power and influence in society. One of the central keys to meeting this goal is that all women have the right to determine if and when they will bear children. Goal 3 is also based on eradicating gender disparity in education.

- **Goal 4: Reduce child mortality.** The target is to reduce by two-thirds the mortality rate of infants and children younger than five years of age. Studies have shown that, given the same amount of healthcare, infants born less than 18 months after an older child are two to four times more likely to die than infants born after 36 months. This goal can be achieved by allowing more time between births; access to family planning and gender equality are keys to longer birth spacing.

- **Goal 5: Improve maternal health.** By 2015, the maternal mortality ratio should be reduced by two-thirds. According to the World Health Organization, access to family planning could prevent 150,000 maternal deaths annually. Family planning improves maternal health by reducing the number of unintended pregnancies and the number of unsafe abortions.

- **Goal 6: Combat HIV/AIDS, malaria, and other diseases.** Reduce the spread of HIV/AIDS, malaria, and other important diseases. For example, the use of condoms indirectly reduces mother-to-infant transmission of HIV/AIDS by preventing unintended pregnancies.

- **Goal 7: Ensure environmental sustainability.** This is a multipart goal that includes reversing the loss of natural resources, halving the number of people without sustainable access to safe drinking water and basic sanitation, and improving the lives of at least 100 million people living in slums (see Chapters 9 and 24). Goal 7 focuses both on relieving the increasing pressure on the environment caused by rising populations in developing countries and addressing the consumptive patterns of people living in highly developed countries.

- **Goal 8: Develop a global partnership for economic development.** The financial aspects of the MDGs are encompassed in this goal: providing measures for debt relief for the external debts in developing countries, creating productive jobs for young people, and providing access to affordable medicines in developing countries.

An unprecedented effort has been extended toward reaching these goals, and many targets have been attained or are in reach.

For example, the number of people living in extreme poverty has indeed been halved, from 47% of the population in 1990 to 22% in 2010, as has the number of people lacking access to safe drinking water. Yet these are massive problems requiring complex solutions. As then UN Secretary-General **Kofi A. Annan** said, "It takes time to train the teachers, nurses, and engineers; to build the roads, schools, and hospitals; to grow the small and large businesses able to create jobs and income needed." These goals should encourage us all to do our part in improving the human condition (see Meeting the Challenge: The Poverty Action Lab).

REVIEW

1. What are the successes and failures of China and Mexico in slowing human population growth?

2. How does population growth momentum affect Mexico's population growth?

3. What are the Millennium Development Goals?

E N V I R O N E W S

The Millennium Villages Project

Sometimes international aid to impoverished people does not reach them, or the aid is piecemeal and provides a temporary, Band-Aid approach that quickly loses its effectiveness after the money is spent. The Millennium Villages Project (MVP) is a different approach to fighting poverty. National and local African governments, the Earth Institute at Columbia University, various aid groups, and other participants work with selected communities in rural Africa to help them achieve the Millennium Development Goals. The MVP has affected 400,000 people—across 10 sub-Saharan countries—who live in 79 villages within 12 groups representing the 12 agro-ecological zones and farming systems in Africa. The villages are all located in areas of food insecurity: At least 20% of all children five years of age or younger are underweight.

The MVP is unique in that it involves a coordinated set of interventions in agriculture, nutrition, health, education, energy, water, and the environment. First, a detailed assessment of current conditions is prepared, both to determine what interventions are most needed and to provide a baseline against which future assessments can monitor the effectiveness of the program. The project involves community-led economic development and is based on the premise that rural African communities can improve their standard of living if they are taught and provided with technologies that simultaneously improve agricultural productivity, health, and education. The program provides assistance for at least five years, during which time the management of improvements is transitioned from the aid groups to the villagers themselves.

MEETING THE CHALLENGE

The Poverty Action Lab

Poverty is a complex, seemingly intractable problem related to many factors, of which economic underdevelopment, lack of education, gender inequality, and degradation of land are just a few. Providing resources to help impoverished people improve their standard of living is critical, but what is the most effective way to spend the money?

In 2003, several economics professors at the Massachusetts Institute of Technology started the **Jameel Poverty Action Lab (J-PAL)** to conduct scientifically rigorous, randomized trials of social and anti-poverty programs. The J-PAL trials, which are controlled experiments, are patterned after the randomized trials conducted by pharmaceutical makers when evaluating the safety of new drugs. The scientific results of trials meant to help the poor assist policy makers from nongovernmental organizations, international organizations, and other groups to formulate more effective programs to reduce poverty.

Dozens of trials are under way or have been completed worldwide. In India, for example, poor elementary schoolchildren who were underachieving in math and literacy were given extra help by tutors who worked with the children in small groups (15–20 students) for two years. The tutors were mostly young, local women who had recently finished secondary school. Students in schools without tutors served as a control. Based on achievement tests, low-achieving students with tutors scored significantly higher than students in control schools, particularly in math. Surprisingly, these students also outscored students in a separate, more expensive program using computer-assisted learning. The low cost of hiring young tutors has enabled other school districts in India to expand the program to their schools.

In Kenya, a randomized trial tested whether providing schoolchildren with free deworming treatments and education on worm prevention improved their overall health, nutrition, and educational advancement. This study determined that school attendance improved by 25% when children were dewormed because they were sick less often. Moreover, the entire community's health benefited because the deworming of schoolchildren reduced the transmission of worm eggs to others in the community.

Researchers concluded that the cost of deworming was significantly less than that of other programs aimed at improving school attendance. J-PAL has been extended to study the effectiveness of social programs in the United States and Europe. In an ongoing study in France, for example, where unemployment is high, particularly among youth, university students with lower socioeconomic backgrounds were randomly placed into either treatment or control groups, with the treatment group receiving mentoring, workshops, and interactions with other students. Researchers hope to determine whether such interventions improve young people's employment prospects.

Achieving Population Stabilization

LEARNING OBJECTIVES

- Explain how individuals can adopt voluntary simplicity to mitigate the effects of population growth.

In this chapter, we have considered how population issues exacerbate global problems, including chronic hunger, poverty, and economic underdevelopment. Population stabilization is critically important if we are to effectively tackle these serious problems.

We have discussed many ways in which developing countries can achieve lower population growth rates. Improving access to public health, increasing the average level of education—especially for women, and providing greater employment for women are all critical to controlling population growth. All depend on community and government involvement.

Highly developed nations must face their own population problems, particularly the environmental costs of unsustainable consumption by affluent people (**Figure 8.16**). Policies should be formulated that support reducing the use of resources, increasing the reuse and recycling of materials, and expunging the throwaway mentality. These policies will show that highly developed countries are serious about the issues of rapid population growth and unsustainable consumption at home as well as abroad.

On a personal level, individuals in highly developed countries should examine their own consumptive practices and consider taking steps to reduce them. Individual efforts to reduce material consumption—sometimes called **voluntary simplicity**—are effective collectively and may influence others to reduce unnecessary consumption. Moving from a lifestyle based on accumulation of wealth and the things money can buy to a lifestyle of voluntary simplicity is called *downshifting*.

voluntary simplicity A way of life that involves wanting and spending less.

JEFF ZELEVANSKY/Reuters/Landov/Landov

Figure 8.16   Warehouse shopping. The United States consumes more goods per person than any other country in the world. The large U.S. population, 316 million in 2013, makes unsustainable consumption an even greater problem. Photographed in Union, New Jersey. How is consumption related to energy use?

Voluntary simplicity is not just living frugally. Instead, it involves desiring less of what money can buy. Each person who scales back his or her consumption benefits the environment (see discussion of ecological footprints in Chapter 1). As a result of needing less money, many people who downshift can afford to work fewer hours. People who advocate voluntary simplicity maintain that their lifestyle gives them fewer "things" but more time—time to do volunteer work, experience the natural world, and enjoy relationships.

Before purchasing a product, an individual striving for simplicity should ask if it is really needed. A material lifestyle—two or three cars in the garage, steaks on the grill, a television in every room, and similar luxury items—may provide short-term satisfaction. Such a lifestyle is not as fulfilling in the long term as learning about the world, interacting with others in meaningful ways, and contributing to the improvement of family and community. In the end, a rich life is measured not by what was owned but by what was done for others.

REVIEW

1. How does voluntary simplicity mitigate the effects of population growth?

ENERGY AND CLIMATE CHANGE

Population

Strong scientific evidence links climate change to increasing CO_2 emissions, most of which are produced by burning fossil fuels: coal, oil, and natural gas. Many factors interact among energy, climate, and population, particularly when considering trends in individual countries. *Per capita CO_2 emissions* in China, for example, are quite low compared to per capita emissions in highly developed countries such as the United States [part (a) of the figure]. But China is industrializing rapidly, and industrialization increases CO_2 emissions. Rapid economic development, combined with the huge number of people in China (more than 1.3 billion), has resulted in its *total annual CO_2 emissions* exceeding total annual emissions in the United States [part (b) of the figure]. China has abundant coal reserves, which have become its energy source of choice as its economy expands. Because burning coal produces more CO_2 *emissions per unit of energy* generated than any other source of fuel, China's CO_2 production per unit of energy is also higher than that in the United States.

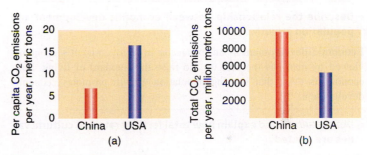

Data from PBL Netherlands Environmental Assessment Agency, *Trends in Global CO_2 Emissions 2013 Report*

REVIEW OF LEARNING OBJECTIVES WITH SELECTED KEY TERMS

- **Define *demography* and summarize the history of human population growth.**

Demography is the applied branch of sociology that deals with population statistics and provides information on the populations of various countries or groups of people. Although it took thousands of years for the human population to reach 1 billion, it took only 130 years to reach 2 billion (in 1930), 30 years to reach 3 billion (in 1960), 15 years to reach 4 billion (in 1975), 12 years to reach 5 billion (in 1987), 12 years to reach 6 billion (in 1999), and 12 years to reach 7 billion (in 2011). World population continues to increase, but the world growth rate (r) has declined since the mid-1960s.

- **Identify Thomas Malthus, relate his ideas on human population growth, and explain why he might or might not be correct.**

Thomas Malthus, a 19th-century British economist, pointed out that the human population could increase faster than its food supply, resulting in famine, disease, and war. Some people think Malthus was wrong, in part because scientific advances have allowed food production to keep pace with population growth. Malthus may ultimately be correct because we do not know if our increase in food production is sustainable.

- **Explain why it is impossible to answer precisely how many people Earth can support—that is, Earth's carrying capacity for humans.**

Carrying capacity is the maximum number of individuals of a given species that a particular environment can support for an indefinite period, assuming there are no changes in the environment. Estimates of how many people Earth can support vary depending on what assumptions are made about standard of living, resource consumption, technological innovations, and waste generation. If we want all people to have a high level of material well-being equivalent to the lifestyles common in highly developed countries, then Earth will clearly support far fewer humans than if everyone lives just above the subsistence level.

- **Explain how highly developed and developing countries differ in population characteristics such as infant mortality rate, total fertility rate, and age structure.**

Infant mortality rate is the number of infant deaths (under age 1) per 1000 live births. **Total fertility rate (TFR)** is the average number of children born to each woman. **Age structure** is the number and proportion of people at each age in a population. **Highly developed countries** have the lowest infant mortality rates, lowest total fertility rates, and oldest age structure. Developing countries have the highest infant mortality rates, highest total fertility rates, and youngest age structure.

- **Explain how population growth momentum works.**

Population growth momentum is the potential for future increases or decreases in a population based on the present age structure. A country can have replacement-level fertility and still experience population growth if the largest percentage of the population is in the prereproductive years.

- **Relate carrying capacity to agricultural productivity.**

When land overuse occurs, often in combination with an extended drought, formerly productive lands may decline in agricultural productivity—that is, their carrying capacity may decrease.

- **Relate human population to chronic hunger and food insecurity.**

The rapid increase in population exacerbates many human problems, including hunger. **Food insecurity** is the condition in which people live with chronic hunger and malnutrition. The countries with the greatest food shortages have some of the highest **TFRs.**

- **Describe the relationship between economic development and population growth.**

Economic development is an expansion in a government's economy, viewed by many as the best way to raise the standard of living. Most economists think that slowing population growth promotes economic development.

- **Define** *culture* **and explain how total fertility rate and cultural values are related.**

Culture comprises the ideas and customs of a group of people at a given period; culture, which is passed from generation to generation, evolves over time. The relationship between TFR and culture is complex. Four factors are primarily responsible for high TFRs: high infant and child mortality rates, the important economic and societal roles of children in some cultures, the low status of women in many societies, and a lack of health and family planning services. Culture influences all of these factors.

- **Define** *gender inequality* **and relate the social and economic status of women to total fertility rate.**

Gender inequality is the social construct that results in women not having the same rights, opportunities, or privileges as men. The single-most important factor affecting high TFRs is the low status of women in many societies.

- **Explain how the availability of family planning services affects total fertility rate.**

Family planning services are services that enable men and women to limit family size, safeguard individual health rights, and improve the quality of life for themselves and their children. TFRs tend to decrease where family planning services are available.

- **Compare how the governments of China and Mexico have tried to slow human population growth.**

In 1979, China began a coercive one-child family policy to reduce TFR. The program is now relaxed in rural China; education and publicity campaigns are used today. In 1974, the Mexican government instigated several measures to reduce population growth, such as educational reform, family planning, and healthcare. Mexico's recent efforts at population control include multimedia campaigns.

- **Explain the role of population growth momentum in Mexico's population growth.**

Even with lower birth rates, Mexico's positive population growth momentum will cause its population to increase because of the large number of young women having babies.

- **Describe at least four of the Millennium Development Goals that came out of the UN Millennium Summit.**

The UN Millennium Summit formed a global partnership with a plan of action, known as the **Millennium Development Goals (MDGs).** The goals are to eradicate extreme hunger and poverty; achieve universal primary education; promote gender equality and empower women; reduce child mortality; improve maternal health; combat HIV/AIDS, malaria, and other diseases; ensure environmental sustainability; and develop a global partnership for economic development.

- **Explain how individuals can adopt voluntary simplicity to mitigate the effects of population growth.**

Voluntary simplicity is a way of life that involves wanting and spending less. Each person who scales back unnecessary consumption lessens the effects of population growth.

Critical Thinking and Review Questions

1. What are some of the factors that have contributed to the huge increase in the incidence of AIDS in sub-Saharan Africa? What are some indications that the trend is reversing?

2. What is demography?

3. What was the human population when you were born? When your parents were born?

4. Some people suggest that Malthus's ideas about human population growth outrunning the environment's ability to support it could happen on a global level. How might this happen in countries considered developing? In highly developed countries?

5. Why is it so difficult to apply the biological concept of carrying capacity directly to the human population?

6. What are total fertility rate and infant mortality rate? Which group of countries has the highest total fertility rates? Which group has the highest infant mortality rates?

7. Which population below is more likely to have a positive population growth momentum? Which is more likely to have a negative population growth momentum? Explain your answers.

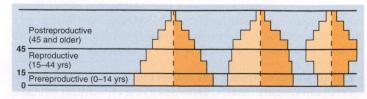

8. How is carrying capacity of the Earth system related to land degradation and agricultural productivity?

9. What is food insecurity? What regions of the world have the greatest food insecurity?

10. How are population growth and economic development related?

11. How do cultural values affect fertility rate? Give examples.

12. What is voluntary simplicity? How is it related to resource consumption?

13. What is family planning? What considerations must family planning programs take into account if they are to reduce fertility rates?

14. China has the world's second largest economy and includes many highly industrialized regions, yet it is typically classified as a developing nation in large part because of its low per capita GNI PPP. What sorts of challenges described in this chapter does China likely face?

15. Why does Mexico have a positive population growth momentum?

16. What are four Millennium Development Goals that the nations of the world have aspired to? Do you consider them attainable? Why or why not?

17. This photograph shows Nigerian students in a classroom. Why do you think the school has many more boys than girls? How is fertility rate related to educational opportunities for women?

©INTERFOTO/Alamy

18. Discuss this statement: The current human population crisis causes or exacerbates all environmental problems, including energy issues and climate change.

World Population Data Sheet Assignment

Use the Population Reference Bureau's *World Population Data Sheet 2014* from www.prb.org to answer the following questions.

1. What is the estimated world population in billions?

2. What is the current birth rate for the world?

3. Which continent has the highest birth rate, and what is this rate?

4. Which country or countries has/have the highest birth rate, and what is this rate?

5. Which continent has the lowest birth rate, and what is this rate?

6. Which country or countries has/have the lowest birth rate, and what is this rate?

7. What is the current death rate for the world?

8. Which country or countries has/have the highest death rate, and what is this rate?

9. How many countries do not have a positive natural increase (i.e., they are either stable or decreasing in population)? In which regions are they found?

10. Which country or countries has/have the highest infant mortality rate, and what is this rate?

11. Which country or countries has/have the lowest infant mortality rate, and what is this rate?

12. Which country or countries has/have the highest total fertility rate, and what is this rate?

13. Which country or countries has/have the lowest total fertility rate, and what is this rate?

14. Which country or countries has/have the highest percentage of the population under age 15, and what is this percentage?

15. Which country or countries has/have the highest percentage of the population over age 65, and what is this percentage?

16. Which country or countries has/have the longest life expectancy, and what is this figure?

17. Which country or countries has/have the shortest life expectancy, and what is this figure?

18. Which country or countries has/have the highest percentage of the population living in urban areas, and what is this percentage?

19. Which country or countries has/have the highest per capita GNI PPP, and what is this figure?

20. Which country or countries has/have the lowest per-capita GNI PPP, and what is this figure?

Food for Thought

Visit the Millennium Villages website and research the food topics addressed by the many past and ongoing projects in the participating villages. What are the primary concerns tackled in efforts to end hunger and provide a sustainable food supply in the Millennium Villages? How might these problems differ from challenges faced in providing a steady and healthy food supply to the hungry in your community?

9

The Urban Environment

Las Vegas, Nevada. Las Vegas functions as a complex system composed of smaller interrelated systems, including the human population, water, food, transportation, and climate systems. (Dennis Flaherty/Science Source)

People are often surprised to learn that Las Vegas originally attracted people for its water. Las Vegas was the site of a spring that Southern Paiutes used for centuries. In the 1600s, travelers on the Santa Fe Trail began stopping at Las Vegas. The first permanent settlement began at the start of the 20th century when the railroad identified Las Vegas as a water stop. Las Vegas grew slowly until the Hoover Dam was built across the Colorado River and provided Las Vegas with an ample water supply. Around 1935, the city began to grow rapidly. With 1.8 million inhabitants today, Las Vegas and the surrounding valley face many challenges and represent an example in which solving environmental problems requires a *systems perspective* (**see photo**).

Water use in Las Vegas is an important environmental issue. Nevada competes with other western states for Colorado River water. Within Nevada, urban areas such as Las Vegas compete for water with agricultural areas. Only about 10 cm (4 in.) of precipitation falls in Las Vegas each year, but it often falls so quickly that flooding is a problem. Since paved ground is impervious to water, water flows along streets and sidewalks rather than soaking into the ground. Runoff carries pollutants—oil from roads, animal wastes from yards, and fertilizers from golf courses—into Lake Mead, from which Las Vegas gets most of its drinking water. This polluted runoff increases the need for water treatment.

Las Vegas, like most urban areas, has a wide array of fast food, restaurants, and grocery stores. However, fresh food is expensive, and vegetable gardening requires consistent access to land, which many urban residents lack. One solution is community gardens, in which a plot of land in a central location is cooperatively managed by a group of people. Vegas Roots is a community gardening organization that supports both food production and community connectedness. Urban agriculture helps improve food access and food quality for urban residents.

Changes in the transportation system impact the water system: The increase in the number of paved roads to handle growing transportation needs in Las Vegas increases the amount of impervious surfaces. Air quality is worsening as more cars and trucks are driven. In addition, road construction contributes air pollution as dust blows from cleared surfaces. Inversion layers (Chapter 19) trap these pollutants, which can accumulate in the atmosphere for days.

Climate change also threatens Las Vegas. Climate models predict higher temperatures and less snowfall in the southwestern United States over the next few decades. Higher temperatures will increase the demand for air conditioning—a major energy use—at the same time that water availability is decreasing.

In this chapter we examine urban populations and trends so that we can better understand the dynamic, densely populated systems known as *cities*. On a global scale, people are increasingly concentrating in cities, and demographers project that cities will be home to two-thirds of all humans by 2050. Many cities will not be able to provide their rapidly expanding populations with basic services, such as clean water, adequate housing, and new schools. Effective planning for the continued rapid growth of cities will require understanding the system interactions among population growth, regional climate, food access, transportation, water use, and energy demand.

> ### In Your Own Backyard…
> *What is the population size of your community? In terms of environmental issues, how does your community compare to Las Vegas?*

Population and Urbanization

LEARNING OBJECTIVES

- Define *urbanization* and describe trends in the distribution of people in rural and urban areas.
- Distinguish between megacities and urban agglomerations.
- Describe some of the problems associated with the rapid growth rates of large urban areas.

Three urban revolutions have transformed human society. During the first urban revolution, from about 8000 to 2000 B.C.E., people moved into cities for the first time; this urban revolution coincided with the advent of agriculture. The second urban revolution occurred from around 1700 to 1950 and coincided with the Industrial Revolution. Modern cities developed and gained prominence as commerce replaced farming as the main way to make a living. As cities became wealthier, urban migration increased, with more people flocking to cities to take advantage of opportunities that the city-based economy provided. Thus, cities experienced dramatic increases in population.

The third urban revolution, which is underway today, differs from the prior two in that the greatest population growth is taking place in cities worldwide, especially in Asia and the global south. Also, the current rate of urban migration makes the earlier two revolutions pale in comparison.

In 2008, the human population reached a milestone: For the first time in history, half of the world's population now lives in urban areas (**Figure 9.1**). The scale of urban growth is impressive: In 1950, less than 30% of the human population lived in cities.

The geographic distribution of people in rural areas, towns, and cities significantly influences the social, environmental, and economic aspects of population growth. When Europeans first settled in North America, the majority of the population consisted of farmers in rural areas. Today, approximately 17% of the people in the United States live in rural areas and 83% live in cities. **Urbanization** involves the movement of people from rural to urban areas, as well as the transformation of rural areas into urban areas.

urbanization The process in which people increasingly move from rural areas to densely populated cities.

How many people does it take to make an urban area or city? The answer varies from country to country; it can be anything from 100 homes clustered in one place to a population of 50,000 residents. A population of 250 qualifies as a city in Denmark, whereas in Greece, a city has a population of 10,000 or more. According to the U.S. Bureau of the Census, a location with 2,500 or more people qualifies as an urban area.

One important distinction between rural and urban areas is not how many people live there but how people make a living there. Most people residing in rural areas have occupations that involve managing natural resources—such as fishing, logging, and farming. In urban areas, most people have jobs that are not directly connected with natural resources.

Cities have grown at the expense of rural populations for several reasons. With the increased mechanization of farms, fewer farmers support an increased number of people, and fewer people have direct access to farmland. Poor people often move to cities because of **land tenure** issues. In many countries, a few wealthy people own most of the land, to which poor farmers are denied access. Women and ethnic minorities sometimes do not have legal rights to own land. Also, in times of war, land ownership is threatened by political and armed conflict. Cities have traditionally provided more types of jobs because cities are the sites of industry, infrastructure development, educational and cultural opportunities, and technological advancements.

CHARACTERISTICS OF THE URBAN POPULATION

Every city is unique in terms of its size, climate, culture, and economic development (**Figure 9.2**). One of the basic characteristics of city populations is their far greater heterogeneity with respect to race, ethnicity, religion, and socioeconomic status than populations in rural areas. People living in urban areas are generally younger than those in the surrounding countryside. The young age structure of cities is due not to a higher birth rate but to the influx of many young adults from rural areas.

Urban and rural areas often have different proportions of men and women. In many countries in Africa, men migrate to the city in search of employment, whereas women tend to care for the farm and children. In countries where farming and land ownership are not viable options for women, cities tend to attract young women rather than young men. Women in some rural areas have little chance of employment after they graduate from high school, so they move to urban areas.

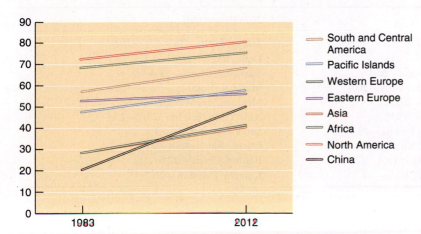

Legend:
- South and Central America
- Pacific Islands
- Western Europe
- Eastern Europe
- Asia
- Africa
- North America
- China

Figure 9.1   The worldwide shift from rural to urban areas, 1983–2012. In most regions, more than 50% of populations live in urban areas. Asia and Africa are still primarily rural, with urban populations increasing yearly. Notice the high rate of rural-to-urban migration in China, whose population recently reached the milestone of 50% urban. (Data collated from the World Bank)

Figure 9.2   Vancouver, Canada. Vancouver is Canada's most important Pacific coast port city. Noted for its lovely parks and gardens, Vancouver has mild weather, with a wet winter season. The greater Vancouver area has a population of 2.3 million, including a large immigrant population, particularly from China, India, the Philippines, Taiwan, and North Korea.

URBANIZATION TRENDS

Urbanization is a worldwide phenomenon. According to the Population Reference Bureau (PRB), just over half of the world population currently lives in urban areas, and the proportion is still rising at over a percent each year. (The PRB considers towns with populations of 2,000 or more as urban.) The percentage of people living in cities compared with rural settings is currently greater in highly developed countries than in developing countries.

In 2012, according to the World Bank, urban inhabitants made up 75% of the total population of countries in Western Europe, North America, and the Middle East (Figure 9.1), but only 40% of the total population in Africa and Southeast Asia. As of 2012, exactly half of China's population lives in urban areas, a proportion that is still increasing.

Although proportionately more people still live in rural settings in Africa and many Southeast Asian countries, urbanization

is increasing rapidly there. For example, the urban population of Indonesia has increased from 24% of that country's 1983 population to 51% in 2012. In 1975, 4 of the world's 10 largest cities were in Latin American or Southeast Asian countries: Mexico City, São Paulo, Buenos Aires, and Kolkata (Calcutta). In 2011, 8 of the world's 10 largest cities were in Latin American or Southeast Asian countries: Delhi, Mexico City, Shanghai, São Paulo, Mumbai (Bombay), Beijing, Dhaka, and Kolkata (Calcutta) (**Table 9.1**).

According to the United Nations, over 21% of the world population in 2012 lived in cities with more than 1 million inhabitants; this proportion has risen by about 1% every three years. The number and size of **megacities** has also increased. In many places, separate urban areas have merged into **urban agglomerations** (**Figure 9.3**). An example is the Tokyo–Yokohama–Osaka–Kobe agglomeration in Japan, which is home to about 50 million people. Other top megacities include Seoul, Jakarta, Delhi, Manila, Shanghai, and Karachi. New York, Mexico City, and São Paolo are 3 of the top 10 megacities in the Western Hemisphere.

About half of the world's 100 largest cities are found in Asia. Mumbai (Bombay) is representative of Asia's rapid urban growth. Its population increased from 7.1 million in 1975, when it was the world's fifteenth-largest city, to 19.7 million in 2007, when it was the seventh-largest. By 2025, Mumbai's population is projected to be 26.6 million, which will make it the fourth-largest city in the world.

Urbanization is increasing in Europe and North America, too, but at a much slower rate. The urban population of the United States, for example, rose from 74% of the total population in 1983 to 83% in 2012. In the United States, most of the migration to cities occurred during the past 150 years, when an increased need for industrial labor coincided with a decreased need for agricultural labor. The growth of U.S. cities over such a long period was typically slow enough to allow important city services such as water

megacities Cities with more than 10 million inhabitants.

urban agglomeration An urbanized core region that consists of several adjacent cities or megacities and their surrounding developed suburbs.

Table 9.1 The World's 10 Largest Cities		
1975	*2011*	*2025 (Projected)*
Tokyo, Japan—26.6*	Tokyo, Japan—37.2	Tokyo, Japan—38.7
New York–Newark, USA—15.9	Delhi, India—22.7	Delhi, India—32.9
Mexico City, Mexico—10.7	Mexico City, Mexico—20.4	Shanghai, China—28.4
Osaka–Kobe, Japan—9.8	New York–Newark, USA—20.4	Mumbai (Bombay), India—26.6
São Paulo, Brazil—9.6	Shanghai, China—20.2	Mexico City, Mexico—24.6
Los Angeles, USA—8.9	São Paulo, Brazil—19.9	New York–Newark, USA—23.6
Buenos Aires, Argentina—8.7	Mumbai (Bombay), India—19.7	São Paulo, Brazil—23.2
Paris, France—8.6	Beijing, China—15.6	Dhaka, Bangladesh—22.9
Kolkata (Calcutta), India—7.9	Dhaka, Bangladesh—15.4	Beijing, China—22.6
Moskva (Moscow), Russia—7.6	Kolkata (Calcutta), India—14.4	Karachi, Pakistan—20.2

*Population in millions.
Source: "Urban Agglomerations 2011," U.N. Population Division Department of Economic and Social Affairs.

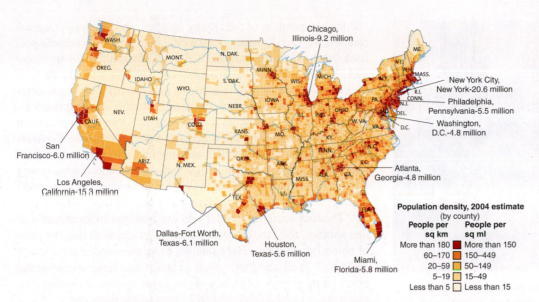

Figure 9.3   **U.S. urban agglomerations.** This map shows estimated population density by county in the contiguous United States. The 10 largest U.S. urban agglomerations are identified along with their 2014 populations. (Map adapted from page 87 in deBlij, H.J. and R.M. Downs, *College Atlas of the World*, National Geographic Society, 2007; Ten largest urban agglomerations in the United States from Demographia World Urban Areas, 10th edition, March 2014)

Population density, 2004 estimate (by county)

People per sq km		People per sq ml
More than 180		More than 150
60–170		150–449
20–59		50–149
5–19		15–49
Less than 5		Less than 15

purification, sewage treatment, education, and adequate housing to keep pace with the influx of people from rural areas. Urban infrastructure in China has sometimes preceded population migrations, with government preemptively erecting housing and other necessary services for workers needed for new factories.

In contrast, some urban areas have grown with less planning, frequently resulting in high levels of urban homelessness or substandard, makeshift housing. For example, in India, 80% of sewage flows directly into waterways, resulting in high pollution levels of surface and groundwater. Urban areas with insufficient infrastructure leave people with a host of problems: homelessness; food insecurity; exceptionally high urban unemployment; increased urban violence; environmental degradation; increasing water and air pollution; and inadequate or nonexistent water, sewage, and waste disposal. Rapid urban growth strains school, medical, and transportation systems.

Substandard housing (slums and squatter settlements) is a critical issue in developing countries (**Figure 9.4**). Squatters illegally occupy the land they build on and cannot obtain city services such as clean water, sewage treatment, garbage collection, paved roads, or police and fire protection. Many squatter settlements are built on land unsuitable for housing, such as high hills where mudslides are a danger during periods of rainfall. Squatters have an insecure existence because they always face the risk of eviction. According to the United Nations, squatter settlements house about one-third of the entire urban population in most developing countries. More than 1 billion people live in areas classified as slums.

Figure 9.4   **Squatter settlement.** The substandard, poor-quality housing in this section of Durban, South Africa, is common in many cities. These houses are built on a steep hill out of whatever materials their owners can salvage and patch together. Squatter settlements like this one often lack running water, sanitation, and electricity.

Peter Chadwick/Science Source

E N V I R O N E W S

Cities in Less Developed Countries

Here are some facts about cities in less developed countries, according to the UN Human Settlements Program:

- One of every four urban households lives in poverty.

- Many children—about 5.8%—die before reaching five years of age.

- About two-thirds of all cities in developing countries do not treat their sewage (wastewater).

- From one-third to one-half of solid waste (garbage) is not collected.

- Almost one-third of all cities have areas that the police consider dangerous.

- Three-fourths of all less developed countries have national laws that recognize the right to obtain adequate housing, although these laws are often not enforced. One-fourth have laws that prohibit women from owning property or obtaining mortgages.

- Buses are the most common form of transportation in cities in the developing world.

Every country has city-dwelling people who lack shelter. Urban scholars estimate that the United States has a total of roughly 600,000 homeless people on any given night, one-third of whom are without shelter (**Figure 9.5**). In Kolkata (Calcutta), India, city officials estimate that 70,000 homeless people sleep in the streets each night, though other organizations estimate this figure to be much higher.

Paul Bradbury/Getty Images

Figure 9.5   Homelessness. Homeless people in the United States often shelter with a combination of a few personal possessions, such as a sleeping bag, and found objects, such as cardboard. In the United States, 35% of the homeless are families with children, 30% have suffered domestic abuse, and nearly a quarter (23%) are military veterans.

REVIEW

1. What is urbanization? What parts of the world have the fastest current rates of urbanization?
2. What is a megacity? An urban agglomeration?
3. What are some of the problems encountered when large urban areas experience rapid growth?

The City as an Ecosystem

LEARNING OBJECTIVES

- Explain how cities can be analyzed from an ecosystem perspective.
- Describe brownfields and food deserts.
- Distinguish between urban heat islands and dust domes.
- Define *compact development*.

As we discussed in the chapter introduction on Las Vegas, cities are complex systems. Many urban sociologists use an ecosystem approach to better understand how cities function and how they change over time. Recall from Chapter 3 that an **ecosystem** is an interacting system that encompasses a biological community and its nonliving, physical environment. Studying urban areas as **urban ecosystems** helps us to better understand how ecosystem services, including the cycling of matter and flow of energy, connect the urban population and its surrounding environment.

urban ecosystem
A heterogeneous, dynamic urban area studied in the context of a broader ecological system.

Urban ecology uses the methods of both natural science and the social sciences to study urban processes, trends, and patterns. Urban ecologists study these processes, trends, and patterns in the context of four variables: population, organization, environment, and technology. They use the acronym POET to refer to these variables.

Population refers to the number of people; the factors that change this number (births, deaths, immigration, and emigration);

and the composition of the city by age, sex, and ethnicity. *Organization* is the social structure of the city, including its economic policies, method of government, and social hierarchy. *Environment* includes both the natural environment, such as whether the city is situated near a river or in a desert, and the city's physical infrastructure, including its roads, bridges, and buildings (**Figure 9.6**). Environment also includes changes in the natural environment caused by humans—air and water pollution, for example. *Technology* refers to human inventions that directly affect the urban environment. Examples of technology include aqueducts, which carry water long distances to cities in arid environments, and air conditioning, which allows people to live in comfort in hot, humid cities.

The four variables (POET) do not function independently of one another; they are interrelated and interact much like the parts of natural ecosystems. For example, energy is needed for all parts of POET. Well-designed urban systems can have a substantial effect on the energy efficiency of the entire system, including how much energy is required to keep it functioning properly.

PHOENIX, ARIZONA: LONG-TERM STUDY OF AN URBAN ECOSYSTEM

The National Science Foundation has 26 Long-Term Ecological Research (LTER) sites where extensive data are gathered on various ecosystems, such as deserts, mountains, lakes, and forests. Two LTER locations—the urban settings of Baltimore and Phoenix—challenge the conventional approach to ecology. With the majority of Americans now living in cities, researchers faced a void when considering the effects of humans on their urban environment. As with all LTER projects, the efforts in Baltimore and Phoenix involve assessing ecosystem health over a long period. The urban research focuses on the ecological effects of human settlement rather than the interactions among the humans themselves.

Many research questions about cities as ecological systems are the same as those for other LTER sites, such as changes in plant and animal populations and the effects of major disturbances such as fire, drought, and hurricanes. For the urban programs, the approach is more complicated because the flow of water, energy, and resources into and out of the sites is linked to the flow of money and the human population (**Figure 9.7**). In some cases political power is connected to better environmental quality of specific (wealthy) neighborhoods.

Researchers are entering new territory to answer broad questions about ecological systems in urban settings. For example, how does urbanization modify the hydrologic cycle of desert ecosystems? How does local climate affect ecosystem services in the urban environment, and what are people's perceptions of the ecosystem services associated with climate? How will the restoration of more natural habitats in the urban environment increase biological diversity? The knowledge gained in urban ecology could increase public awareness and eventually influence policy decisions.

ENVIRONMENTAL PROBLEMS ASSOCIATED WITH URBAN AREAS

The concept of the city as an ecosystem would be incomplete without considering the effects the city has on its natural environment. Growing urban areas affect land use patterns and destroy or fragment

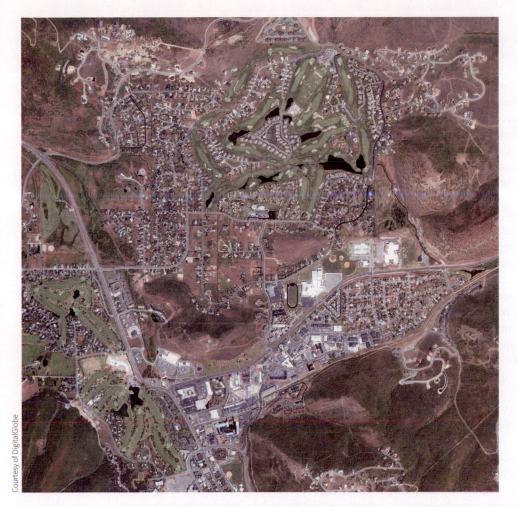

Courtesy of DigitalGlobe

(a) Full view of Park City, Utah. All other views are enlarged and cropped from this shot.

Courtesy of DigitalGlobe

(b) Mixed use (commercial, industrial, and residential).

Courtesy of DigitalGlobe

(c) Single-family residential.

Courtesy of DigitalGlobe

(d) Undeveloped (natural) open space.

Figure 9.6   **Satellite photographs of land-use patterns.** Urban land-use patterns exhibit different configurations. Note the proportions of paved surfaces in each of the landscapes.

Natural capital (inputs)

Energy (fuel)
Clean water
Clean air
Food
Raw and refined materials for construction and industry
Business and consumer products

Products and wastes (outputs)

Waste heat, greenhouse gases
Wastewater, water pollution
Air pollution
Solid waste
Goods, services

Figure 9.7   **The city as a dynamic system.** Like natural ecosystems, cities are open systems. The human population in an urban environment requires inputs from surrounding rural areas and produces outputs that flow into surrounding areas. Not shown in this figure is the internal cycling of materials and energy within the urban system. How might climate change affect the inputs and outputs of the urban system?

wildlife habitat by suburban development that encroaches into former forest, wetlands, desert, or agricultural land in rural areas. For example, large portions of Chicago, Boston, and New Orleans are former wetlands. When parks in urban areas are created for recreation, the original ecosystems often do not return. For example, Chicago has a large area of forest preserve fringing the city, but the ecosystems in those spaces were previously prairie, not forest.

Most cities have multiple blocks of abandoned **brownfields**. Meanwhile, suburban developments continue to expand outward, swallowing natural areas and farmland. Brownfields are available for reuse, but reuse is complicated because many have environmental contaminants that must be cleaned up before redevelopment can proceed. (Superfund sites, discussed in Chapter 23, are contaminated with high levels of hazardous wastes and are not considered brownfields.)

brownfield An urban area of abandoned, vacant factories, warehouses, and/or residential sites whose redevelopment is hindered due to possible contamination.

Brownfields represent an important potential land resource. Pittsburgh, Pennsylvania, is best known for its redevelopment of brownfields that were once steel mills and meatpacking centers. Residential, recreational, and commercial sites now occupy several of these former brownfields (**Figure 9.8**).

Although we often imagine cities being full of necessary conveniences, such as food vendors, some urban areas lack these features. Poorer neighborhoods often lack grocery stores and farmers' markets, and instead have a relative abundance of gas stations, convenience stores, and fast-food restaurants. Neighborhoods where residents lack access to nutritionally high-quality food are called **food deserts**. Residents of such neighborhoods may have higher rates of nutrition-related disease: obesity, diabetes, and heart disease.

food desert A neighborhood in which low-quality, processed foods are far easier to acquire than nutritionally dense, fresh foods.

Cities affect water flow by covering the rainfall-absorbing soil with buildings and paved roads. Storm systems are built to handle the runoff from rainfall, which is polluted with organic wastes (garbage, animal droppings, and such), motor oil, lawn pesticides and fertilizers, and heavy metals (see Figure 21.10). In most cities

across the United States, urban runoff is cleaned up before being discharged into nearby waterways. In many cities, however, high levels of precipitation can overwhelm the treatment plant and result in the release of untreated urban runoff. When this occurs, the polluted runoff contaminates bodies of water far beyond the boundaries of the city. Many cities are built on river deltas or coastal areas, making them vulnerable to flooding as ocean levels rise and storm intensities increase with climate change.

Most workers in U.S. cities have to commute dozens of miles through traffic-congested streets, from suburbs where they live to downtown areas where they work. Automobiles are necessary to accomplish everyday chores because public transportation services suburbs inadequately. This heavy dependence on motor vehicles as our primary means of transportation increases air pollution and causes other environmental problems.

The high density of automobiles, factories, and commercial enterprises in urban areas causes a buildup of airborne emissions, including particulate matter (dust), sulfur oxides, carbon oxides, nitrogen oxides, and volatile organic compounds. Urban areas in nations without air pollution legislation have the worst air pollution in the world. In Mexico City, the air is so polluted that schoolchildren are not permitted to play outside during much of the school year. Although progress has been made in reducing air pollution through national legislation in many countries, the atmosphere in many of their cities still contains higher pollutant levels than health or legal standards dictate.

Urban Heat Islands Streets, rooftops, and parking lots in areas of high population density absorb solar radiation during the day and radiate heat into the atmosphere at night. Heat released by human activities such as fuel combustion is also highly concentrated in cities. The air in urban areas is therefore warmer than the air in the surrounding suburban and rural areas and is known as an **urban heat island** (**Figure 9.9**).

urban heat island Local heat buildup in an area of high population density.

Urban heat islands affect local air currents and weather conditions, particularly by increasing the number of thunderstorms over the city (or downwind from it) during summer months. The uplift of warm air over the city produces a low-pressure cell that draws in cooler air from the surroundings. As the heated air rises, it cools, causing water vapor to condense into clouds and producing thunderstorms. The intensity of urban heat islands can be greatly reduced with green roofs (see Chapter 18 opening photo), which use plants to help moderate temperatures and absorb excess water on building roofs.

The local air circulation patterns of urban heat islands contribute to the buildup of pollutants, especially particulate matter, in the form of **dust domes** over cities (**Figure 9.10a**). Pollutants concentrate in a dust dome because convection (i.e., the vertical motion of warmer air) lifts pollutants into the air, where they remain because of somewhat stable air masses

dust dome A dome of heated air that surrounds an urban area and contains a lot of air pollution.

produced by the urban heat island. If wind speeds increase, the dust dome moves downwind from the city, and the polluted air spreads over rural areas (**Figure 9.10b**). Dust domes can be noticeable to travelers arriving in urban areas, especially when dust is combined with brown smog, which forms when nitrous oxides from vehicles, industry, and lawnmowers react with sunlight.

Figure 9.8 **Brownfield redevelopment in Pittsburgh.** Washington's Landing (formerly Herr's Island) was originally the site of sawmills, stockyards, soap works, and the Pennsylvania Railroad. Most industries had left by the 1970s; brownfield redevelopment began in the 1980s. Today, Washington's Landing features housing, commerce, and recreation, as well as convenient rail-trail access for bike commuters to downtown Pittsburgh.

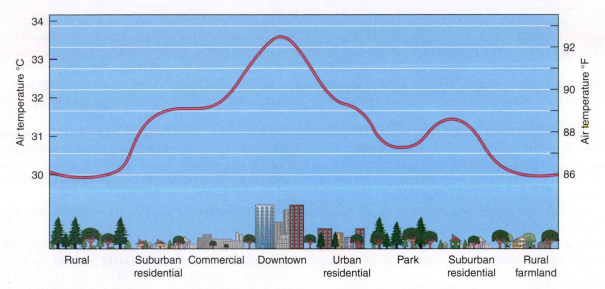

Figure 9.9   Urban heat island. This figure shows how temperatures might vary on a summer afternoon. The city stands out as a heat island against the surrounding rural areas. Note the impact of the park area on local temperature. The spikes in temperature over buildings can be greatly reduced or eliminated with green roofs.

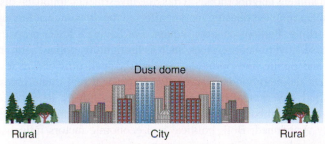

(a) A dust dome of pollutants forms over a city when the air is somewhat calm and stable.

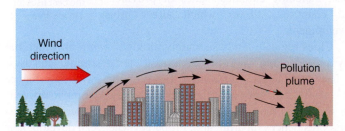

(b) When wind speeds increase, the pollutants move downwind from the city.

Figure 9.10   Dust dome.

Noise Pollution   Sound is called **noise pollution** when it becomes loud or disagreeable, particularly when it results in physiological or psychological harm. Most of the city noise that travels in the atmosphere is of human origin. Vehicles, from trains to trucks to powerboats, produce a great deal of noise. Power lawn mowers, jets flying overhead, leaf blowers, chain saws, jackhammers, cars with stereos booming, and heavy traffic are just a few examples of the outside noise that assails our ears. Indoors, dishwashers, trash compactors, washing machines, televisions, and stereos add to the din.

Prolonged exposure to noise damages hearing. In addition to hearing loss, noise increases the heart rate, dilates the pupils, and causes muscle tension. Evidence exists that prolonged exposure to high levels of noise causes a permanent constriction of blood vessels, which can increase blood pressure, thereby contributing to heart disease. Other physiological effects associated with noise pollution include migraine headaches, nausea, dizziness, and gastric ulcers. Noise pollution also causes psychological stress.

Obviously, producing less noise can reduce noise pollution. Noise can be reduced in a variety of ways, from restricting the use of sirens and horns on busy city streets to engineering motorcycles, vacuum cleaners, jackhammers, and other noisy devices so that they produce less noise. The engineering approach is technologically feasible but is often avoided because consumers associate loud noise with greater power. Putting up shields between the noise producer and the hearer can also help control noise pollution. One example of a sound shield is the noise barriers erected along heavily traveled highways. Plants, such as street trees, also help absorb sound.

ENVIRONMENTAL BENEFITS OF URBANIZATION

Urbanization has the potential to provide tangible environmental benefits that may outweigh the negative aspects. A well-planned city actually benefits the environment by reducing travel-related pollution and preserving rural areas. A solution to urban growth is **compact development**, which uses land efficiently. Dependence on motor vehicles and their associated pollution is reduced as people walk, cycle, or take public transit such as buses or light rails to work and to shopping areas. Because compact development requires fewer parking lots and highways, more room is available for parks, open space, housing, and businesses. Compact development makes a city more livable, and more people want to live there.

compact development
The design of cities in which tall, multiple-unit residential buildings are close to shopping and jobs, and all are connected by public transportation.

Portland, Oregon, provides an example of compact development. Although Portland grapples with many issues, the city government has developed effective land-use policies that dictate where and how growth will occur. The city looks inward to brownfields rather than outward to the suburbs for new development sites. Although the automobile is still the primary means of transportation in Portland, the city's public transportation system is an important part of its regional master plan. Public transportation incorporates light-rail lines, bus routes (many of which have buses arriving every 15 minutes), bicycle lanes, and walkways as alternatives to the automobile. Employers are encouraged to provide bus passes to their employees instead of paying for parking. The emphasis on public transportation has encouraged commercial and residential growth along light-rail and bus stops instead of in suburbs. An urban farm collective helps preserve land available for agricultural production, which helps reduce the need for long-distance food transport.

REVIEW

1. How can a city system be analyzed from an ecosystem perspective?
2. What are brownfields?
3. What types of businesses are present, and which are generally missing, in food deserts?
3. Why are cities associated with urban heat islands and dust domes?
4. What is compact development?

Urban Land-Use Planning

LEARNING OBJECTIVES

- Discuss the use of zoning in land-use planning.
- Relate how a city's transportation infrastructure affects urban development.
- Define *suburban sprawl* and discuss a problem caused or exacerbated by sprawl.

Land use in many cities is based on economic concerns. Taxes pay for the city's infrastructure—its roads, schools, water treatment plants, prisons, and garbage trucks. The city's center—the *central business district*—typically has the highest taxes. Surrounding residential properties have lower taxes than the central business district, but the taxes are still high. Thus, many residential buildings near the central business district are high-rises filled with small apartments or condominiums; the collective tax for the property is high, but individual taxes are more modest. Circling the residential properties—and farther from the city's center—are land-intensive businesses that require lower taxes or that may be supported by taxes: golf courses, cemeteries, water treatment plants, sanitary landfills, and such. Parks and other open spaces are interspersed among the various land uses. People living in the suburbs, often far from the central business district, pay less in taxes but pay higher transportation costs.

High taxes near a city's central business district mean that only more affluent people can afford to live in cities; yet most cities in North America also have poor neighborhoods with few housing options, little or no green space, often inferior schools, and fewer public services. The reasons are complex. As cities became more industrialized, the more affluent citizens fled to the suburbs to avoid the noise and pollution, leaving the poor in the inner cities. Although poor people cannot afford to pay high taxes, they also cannot afford to pay high transportation costs. By living in high-density housing, residents share the tax burden with many other families, mitigating the overall cost to individuals. **Gentrification**, the movement of wealthier people back to older, run-down homes that have been renovated, sometimes displaces the urban poor, who can no longer afford the property taxes or rent to live in a gentrified neighborhood.

Social scientists have examined factors that influence urban development. For example, **David Harvey**, an English geographer, did a detailed analysis of Baltimore, Maryland, in the 1970s. He divided Baltimore's real estate into various areas based on income and ethnicity. Harvey demonstrated that financial institutions and government agencies did not deal consistently with different neighborhoods. For example, banks were less likely to lend money to the inner-city poor for housing than to those living in more affluent neighborhoods. Thus, the investment capitalists' discrimination in the housing market affected the dynamics of buying and selling. Perhaps as a consequence of having a lower property tax base, government services, such as firefighting, police monitoring, and public transportation may not be well provided in less affluent neighborhoods. Harvey concluded that real estate investors and government programs and services largely determined whether a neighborhood remained viable or decayed and was eventually abandoned. Both political and economic factors influence **land-use planning**. Cities do not exist as separate entities; they are part of larger political organizations, including counties, states or provinces, and countries, all of which affect urban development.

> **land-use planning**
> The process of deciding the best uses for land in a given area.

Cities regulate land use mainly through *zoning*, in which the city is divided into **use zones**, areas restricted to specific land uses, such as commercial, residential, farm, or industrial (see Figure 9.6). These categories are often subdivided. For example, residential use zones may designate single-family residences or multi-family residences (apartments). Property owners can develop their properties as long as they meet the zoning ordinances in which the property is located. Often, these rules are very specific, regulating building height, how the building is situated on the property, and what the property can be used for.

Zoning has largely resulted in separate industrial parks, shopping centers, and residential districts. Zoning laws in many urban areas prohibit environmentally friendly behaviors, such as vegetable gardening, drying laundry outside, cultivating a prairie as front lawn, or keeping chickens.

TRANSPORTATION AND URBAN DEVELOPMENT

Transportation and land use are inextricably linked because as cities grow, they expand along public transportation routes. The kinds of transportation available at a particular period in history affect a

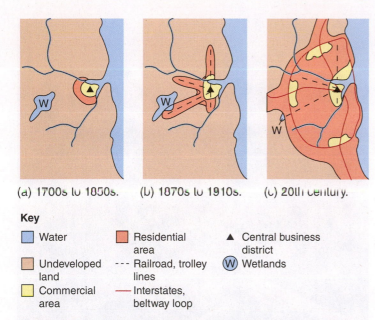

(a) 1700s to 1850s.   (b) 1870s to 1910s.   (c) 20th century.

Key

- 🟦 Water
- 🟧 Undeveloped land
- 🟨 Commercial area
- 🟥 Residential area
- --- Railroad, trolley lines
- — Interstates, beltway loop
- ▲ Central business district
- Ⓦ Wetlands

Figure 9.11   **The relationship between transportation and urban spatial form.** Cities grow outward along transportation routes. During the 20th century, the automobile dramatically expanded suburbanization along roads and highways built to accommodate automobiles. (Adapted from Kaplan, Wheeler, and Holloway. *Urban Geography*. Hoboken, NJ: John Wiley & Sons, Inc., 2004)

city's spatial structure. **Figure 9.11** shows the expansion of a hypothetical city along the eastern coast of North America. From the 1700s to the 1850s, transportation in the city was limited to walking, horse-drawn carriages, and ships (Figure 9.11a). Public transportation infrastructure is expensive, but financial benefits overall are roughly twice the financial benefits of road infrastructure for cars.

Technological advances enabled fixed transportation routes (railroads and electric street trolleys) to spread out of the city from the central business district from the 1870s to 1910s (Figure 9.11b). People could move beyond the city limits quickly and inexpensively. Real estate developers began building housing tracts in areas that were at one time inaccessible. The first suburbs clustered around railway stations, causing the city's outward growth to assume a shape reminiscent of a sea star. Still, the city was relatively contained; as an example, the average commute was only about 1.5 miles in the 1920s.

Cars and trucks forever changed the city, increasing its spatial scale. With the advent of the automobile era during the 20th century and government support of that era through road building, roads expanded the access to more distant development between the "arms" of prior metropolitan development, filling in the wild spaces (Figure 9.11c). Construction of the interstate highway system and outer-city beltway "loops," beginning in the 1950s, encouraged development even farther from the city's central business district. Today, many people live in suburbs far from their place of employment, and daily commutes of 20 or more miles each way are commonplace.

As cities continue to grow, effectively moving people from one place to another within the city system becomes more challenging. Highways create pollution and traffic congestion, whereas light-rail systems take years to build and are extremely expensive.

Scott Dalton/The New York Times/Redux Pictures

Figure 9.12   **Bus rapid transit.** In this system, located in Bogotá, Colombia, people move to and from buses on a moving conveyor belt. They wait for their buses in covered stations (visible in the background). The buses travel quickly because automobiles are prohibited in bus lanes.

Some cities have developed *bus rapid transit systems* to handle large numbers of people quickly. The buses have lanes devoted exclusively to bus transit (**Figure 9.12**); without such lanes, bus travel is unappealing, since the bus can get stuck in the same city traffic as a car, but the bus rider also has to share space with other people, wait through multiple stops, and plan well enough to arrive at work on time in spite of sometimes unpredictable bus schedules.

SUBURBAN SPRAWL

Urbanization and its accompanying **suburban sprawl** affect land use and have generated additional concerns (**Figure 9.13**). Prior to World War II, jobs and homes were concentrated in cities, but during the 1940s and 1950s, jobs and homes began to move from urban centers to the suburbs. New housing and industrial and office parks were built on rural land surrounding the city, along with a suburban infrastructure that included new roads, shopping centers, schools, and the like.

> **suburban sprawl**
> A patchwork of vacant and developed tracts around the edges of cities; contains a low population density.

Further development extended the edges of the suburbs, cutting deeper and deeper into the surrounding rural land and causing environmental problems such as loss of wetlands, loss of biological habitat, air pollution, and water pollution. Meanwhile, people who remained in the city and older suburbs found themselves the victims of declining property values, reduced school funding, and increasing isolation from suburban jobs. This pattern of land use, which has intensified in recent years, has increased the economic disparity between older neighborhoods and newer suburbs. In the past few decades, the rate of land development around most U.S. cities has exceeded that of population growth. Detroit, in particular, has an acute problem with vacant city buildings and lots, none of which support the tax base necessary for urban infrastructure.

Regional planning involving government and business leaders, environmentalists, inner-city advocates, suburbanites, and farmers is urgently needed to determine where new development should take place and where it should not. Also, most metropolitan

JG Photography/Alamy

Figure 9.13   **Suburban sprawl.** Aerial view of suburban sprawl outside Philadelphia, Pennsylvania. These homes are not located within walking distance of bus stops, stores, and restaurants, and most lack vegetable gardens.

areas need to make more efficient use of land that has already been developed, including central urban areas. The Greater Atlanta area is an excellent example of the rapid spread of urban sprawl. During the 1990s and early 2000s, tract houses, strip malls, business parks, and access roads replaced an average of 500 acres of surrounding farms and forests each week. The Greater Atlanta area extends more than 175 km (110 mi) across, almost twice the distance it was in 1990. Atlanta now has 1750 km² (700 mi²) of sprawl.

At the same time, Atlanta has become home to a thriving inner-city garden community, which uses a number of vacant lots for vegetable gardens both for food banks and for other residents. Inner-city gardens help ease the problem of food insecurity (**Figure 9.14**, and see Chapter 18), which is exacerbated as urban development swallows nearby farmland.

U.S. voters have grown increasingly concerned about the unrestricted growth of suburban sprawl. At least 15 states now have comprehensive, statewide growth-management laws. Maryland, for example, has a smart growth plan that protects open space in highly developed areas while promoting growth in areas that could be helped by growth. **Smart growth** is an urban planning and transportation strategy that mixes land uses (commercial, manufacturing, entertainment, gardens, and a range of housing types).

Smart growth incorporates compact development; creates communities in which it is easy to walk from one place to another; and preserves open space, farmlands, and important environmental areas. Because people live near jobs and shopping, the need for a continually expanding highway system is lessened. Smart growth involves taking a long-range look at the "big picture" instead of developing individual tracts of land as they become available. Arlington, Virginia, and Minneapolis–St. Paul, Minnesota, are two examples of cities that have incorporated smart growth policies.

Mike Harrington/Getty Images

Figure 9.14   **Urban gardening.** Vegetables grown on roofs or in vacant urban lots might not provide a high proportion of calories, but they can contribute greatly to nutritional quality.

Although smart growth redirects development efforts to reduce traffic congestion and improve environmental quality, it is not the final answer to population growth in an urban area. **Al Bartlett** at the University of Colorado has pointed out that even development efforts promoting smart growth become overwhelmed as the population continues to grow.

REVIEW

1. How does zoning help regulate urban land use?

2. How are transportation and land use linked?

3. What is suburban sprawl?

Making Cities More Sustainable

LEARNING OBJECTIVES

• List at least five characteristics of an ideal sustainable city.

• Explain how city planners have incorporated environmental sustainability into the design of Curitiba, Brazil.

Most environmental scientists think that, on balance, increased urbanization is better for the environment than having the same number of people living in rural and suburban areas spread across the landscape. However, the challenge is to make these cities more sustainable and more livable through better design. At the same time, rapid urban growth is overburdening the existing infrastructure in many cities, leading to both a lack of urban services and to environmental degradation.

sustainable city A city with a livable environment, a strong economy, and a social and cultural sense of community; sustainable cities enhance the well-being of current and future generations of urban dwellers.

Imagine being commissioned to design a **sustainable city** for 100,000 people. What features should it have? To design a sustainable city, planners must imagine the city as a system. Changes in one variable, such as increasing population density, always impact all other parts of the city system. Greater population density, for example, means a greater demand for housing, food access, public transit and/or parking lots, and energy use for heating and cooling. A mayor or city council cannot simply make these changes. According to social scientists, the most effective urban, regional, and national governments are democratic and participatory, in which local citizens are encouraged to work together to address local problems.

Cities have the potential to produce low levels of energy consumption, resource use, and wastes. A sustainable city should be designed to reduce energy consumption by using energy and other resources efficiently, with codes that require buildings, motor vehicles, and appliances to be energy-efficient (see Meeting the Challenge: Green Architecture). A sustainable city would make use of solar and other forms of renewable energy as much as possible. Fresh produce would be grown within the city, and even street trees could help feed the city's residents healthily.

MEETING THE CHALLENGE

Green Architecture

The Adam Joseph Lewis Center for Environmental Studies at **Oberlin College** near Cleveland, Ohio, is one of the early examples of sustainability in the built environment. Dedicated in 2000, the building is an award-winning example of **green architecture**, which encompasses environmental considerations such as energy conservation, improved indoor air quality, water conservation, and recycled or reused building materials. The Lewis Center is designed to work like a natural ecosystem such as a forest or pond—that is, it obtains its energy from sunlight and produces few wastes that cannot be recycled or reused.

An advanced-design, earth-coupled heat pump, which extracts heat from 24 geothermal wells in the ground, supplies some of the energy for heating and cooling the building. On the roof, an array of photovoltaic (PV) cells collects the sun's energy to meet most of the building's electricity needs. On sunny days, the PV cells produce more electricity than the building uses, so the excess is sold to the Ohio power grid. On cloudy winter days, the building buys the extra energy it needs from the grid.

Energy is saved by the triple-paned windows, which allow sunlight to enter the building but insulate against heat loss during winter and heat gain during summer. Classrooms in the building have motion sensors that detect when people are not in the room, triggering the energy-efficient lighting fixtures and ventilation systems to shut down.

The building's wastewater from sinks and toilets is cleansed and recycled by a *living machine*. This organic system consists of a series of tanks that contain bacteria, aquatic organisms, and plants. These organisms interact to remove organic wastes, nutrients such as nitrogen and phosphorus, and disease-causing organisms from the water as it moves through the tanks. After it has been cleaned, the water is stored and reused in the building's toilets.

The acoustic panels in the auditorium are made from agricultural straw wastes, and the

Courtesy of Oberlin College

Green architecture. The Adam Joseph Lewis Center for Environmental Studies at Oberlin College is an excellent example of ecological design. It uses about 20% of the energy that a new building typically uses. Technical and design innovations may make buildings possible that produce, within the buildings themselves, all of the energy needed for full building function.

MEETING THE CHALLENGE (CONTINUED)

computer table in the resource center is constructed from a bowling alley lane from the old campus bowling alley. All new wood used in the building came from nearby forests that are certified to produce sustainably harvested wood. All the steel frames used in the building are composed of recycled steel. The carpets in the classrooms are installed in squares that are easy to remove. As the carpet wears out, it is not disposed of in a landfill. Instead, the squares are

returned to the manufacturer, where they are recycled into "new" carpet.

The landscape surrounding the building is carefully designed to highlight a variety of natural and agricultural ecosystems. Because most of Oberlin College's students come from urban areas, these landscaped grounds provide an outdoor classroom that exposes students to the natural environment. Much of the area on which Oberlin College was built originally consisted of marshes and deciduous

forests. A wetland that provides a haven for frogs, insects such as dragonflies, and songbirds has been constructed along one side of the building. Native forest trees are planted along another side of the building. Orchards of apple and pear trees, gardens, and a greenhouse are also part of the landscape. Near the orchard is a cistern that collects precipitation runoff from the roof for reuse. The city of Oberlin partners with the college to extend these landscape features beyond the campus.

A sustainable city would reduce pollution and wastes by reusing and recycling materials in the waste stream. Much of the municipal solid waste—paper, plastics, aluminum cans, and such—would be recycled, thereby reducing consumption of virgin materials. Yard wastes would be composted and used to enrich the soil in public places. Sewage (wastewater) treatment would involve the use of living plants in large tanks or, even better, in marshes that could also provide wildlife habitat.

A sustainable city should be designed with large areas of green space that provide habitat for wildlife, thereby supporting biological diversity. The green space would also provide areas of recreation for the city's inhabitants; people are more likely to engage in recreation if local green space exists than if it does not. Humans are a part of the natural world; urban dwellers often are able to forget this fact. Urban living can lead to both a skewed perception of the importance of the natural world to our survival and the misperception that engineering and technology can solve all environmental issues. Having parks and other open spaces within a city encourages residents to spend more time outside, which promotes awareness of the natural world. In addition, substantial evidence indicates that interaction with open space is linked to health; the beneficial effects of spending time in green space are both physical and psychological. Green space is also an ideal allocation of land near rivers or waterfronts, which may be vulnerable to flooding and less suitable for development.

A sustainable city should be people-centered, not automobile-centered (**Figure 9.15**). People would move about the city by walking or bicycling or, for longer distances, mass transit. The use of automobiles would be limited, perhaps by closing certain streets to motor vehicles, and public transport would be readily available, clean, and inexpensive. Restricting automobile use would also reduce fossil-fuel consumption and motor vehicle–generated air pollution.

The people in this sustainable city would grow some of their food—for example, in rooftop gardens, window boxes, greenhouses, and community gardens. Rooftop gardens are particularly popular because they cannot be vandalized. Currently, urban farmers supply about 15% of the food consumed in cities worldwide, and that number is projected to increase. Berlin, Germany, has more than 80,000 urban farmers. In Accra, Ghana, 90% of fresh produce consumed is grown within the city itself. Abandoned

Figure 9.15   Barcelona, Spain, a people-centered city. In the heart of Barcelona, the Ramblas pedestrian walkway provides a people-centered place. Barcelona has other features of sustainable cities, such as green space, urban agriculture, and public transport.

lots and brownfields would be cleaned up and used efficiently and safely, thereby reducing the spread of the city into nearby rural areas—farms, wetlands, and forests.

No city exists that incorporates all the sustainable features we have just discussed. However, many architects, such as German **Albert Speer Jr.,** are beginning to emphasize sustainable building and city planning. Many cities around the world provide unique examples of specific sustainability initiatives. Portland, Oregon, and Pittsburgh, Pennsylvania (discussed earlier in the chapter), are examples of North American cities that exhibit some features of sustainability. Many cities in developing nations are also making progress. One of these is Curitiba, Brazil.

CASE IN POINT

CURITIBA, BRAZIL

Curitiba, a city of 3.1 million people in Brazil, provides a good example of compact development. Curitiba's city officials and planners have had notable successes in public transportation,

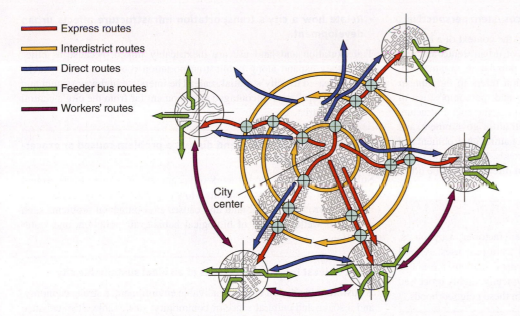

Express routes
Interdistrict routes
Direct routes
Feeder bus routes
Workers' routes

City center

Figure 9.16   Curitiba, Brazil. Curitiba's bus network, arranged like the spokes of a wheel, has concentrated development along the bus lines, saving much of the surrounding countryside from development. (Courtesy of Karl Guide)

traffic management, land-use planning, waste reduction and recycling, and community livability.

The city developed an inexpensive, efficient mass transit system that uses clean, modern buses that run in high-speed bus lanes (**Figure 9.16**). High-density development was largely restricted to areas along the bus lines, encouraging population growth where public transportation was already available. About 2 million people use Curitiba's mass transportation system each day. Curitiba's population has more than tripled since 1975, yet traffic has declined. Curitiba has less traffic congestion and significantly cleaner air, both of which are major goals of compact development. Instead of streets crowded with vehicular traffic, the center of Curitiba is a *calcadao*, or "big sidewalk," that consists of 49 downtown blocks of pedestrian walkways connected to bus stations, parks, and bicycle paths.

Curitiba was the first city in Brazil to use a special low-polluting fuel that contains a mixture of diesel fuel, alcohol, and soybean extract. In addition to burning cleanly, this fuel provides economic benefits for people in rural areas who grow the soybeans and sugar cane (used to make the alcohol). It is estimated that for each million liters of alcohol produced, 50 new jobs are created.

Over several decades, Curitiba purchased and converted flood-prone properties along rivers in the city to a series of interconnected parks crisscrossed by bicycle paths. This move reduced flood damage and increased the per capita amount of green space from 0.5 m^2 in 1950 to 50 m^2 today, a significant accomplishment considering Curitiba's rapid population growth during the same period.

Another example of Curitiba's creativity is its labor-intensive garbage purchase program, in which poor people exchange filled garbage bags for bus tokens, surplus food (eggs, butter, rice, and beans), or school notebooks. This program encourages garbage pickup from the unplanned squatter settlements (where garbage trucks cannot drive) that surround the city. Curitiba supplies more services to these unplanned settlements than most cities do. It tries to provide water, sewer, and bus service to them; the bus service allows the settlers to seek employment in the city.

These changes did not happen overnight. Like Curitiba, most cities can be carefully reshaped over several decades to make better use of space and to reduce dependence on motor vehicles. City planners and local and regional governments are increasingly adopting measures to provide the benefits of sustainable development in the future.

REVIEW

1. What features does a sustainable city possess?

2. Why is Curitiba, Brazil, a model of sustainability?

REVIEW OF LEARNING OBJECTIVES WITH SELECTED KEY TERMS

- **Define *urbanization* and describe trends in the distribution of people in rural and urban areas.**

Urbanization is the process in which people increasingly move from rural areas to densely populated cities. As a nation develops economically, the proportion of the population living in cities increases. In developing nations, most people live in rural settings, but their rates of urbanization are rapidly increasing.

- **Distinguish between megacities and urban agglomerations.**

Megacities are cities with more than 10 million inhabitants. In some places, separate urban areas have merged into an **urban agglomeration,** which is an urbanized core region that consists of several adjacent cities or megacities and their surrounding developed suburbs.

- **Describe some of the problems associated with the rapid growth rates of large urban areas.**

Rapid urban growth often outstrips the capacity of cities to provide basic services. Challenges include substandard housing; poverty; high unemployment; pollution; and inadequate or nonexistent water, sewage, and waste disposal. Rapid urban growth also strains school, medical, and transportation systems. Cities in less developed nations are generally faced with more serious challenges than are cities in highly developed countries.

- **Explain how cities are analyzed from an ecosystem perspective.**

An **urban ecosystem** is an urban area studied in the context of a broader ecological system. **Urban ecology** is the study of urban trends and patterns in the context of four interrelated variables: population, organization, environment, and technology (POET). Population refers to the number of people; the factors that change this number; and the composition of the city by age, sex, and ethnicity. Organization is the social structure of the city, including its economic policies, method of government, and social hierarchy. Environment includes both the natural environment and the city's physical infrastructure, including its roads, bridges, and buildings. Technology refers to human inventions that directly affect the urban environment.

- **Describe brownfields and food deserts.**

A **brownfield** is an urban area of abandoned, vacant factories, warehouses, and residential sites that may be contaminated from past uses. A **food desert** is an area where grocery stores and farmers' markets are not as abundant as fast food and convenience stores. Typically, highly processed foods are easier to purchase than fresh produce in these neighborhoods.

- **Distinguish between urban heat islands and dust domes.**

An **urban heat island** is local heat buildup in an area of high population density. A **dust dome** is a dome of heated air that surrounds an urban area and contains a lot of air pollution.

- **Define** *compact development.*

Compact development is the design of cities in which tall, multiple-unit residential buildings are close to shopping and jobs, and all are connected by public transportation.

- **Discuss the use of zoning in land-use planning.**

Land-use planning is the process of deciding the best uses for land in a given area. The main way that cities regulate land use is by zoning, in which the city is divided into **use zones**, areas restricted to specific land uses, such as commercial, residential, or industrial. Zoning has largely resulted in separation of industrial parks, shopping centers, apartment districts, and other land uses.

- **Relate how a city's transportation infrastructure affects urban development.**

Transportation and land use are inextricably linked, because as cities grow, they expand along public transportation routes. Cars and trucks have increased the city's spatial scale: The interstate highway system and beltway "loops" have encouraged development far from the city's central business district.

- **Define** *suburban sprawl* **and discuss a problem caused or exacerbated by sprawl.**

Suburban sprawl is a patchwork of vacant and developed tracts around the edges of cities; sprawl contains a low population density. Sprawl cuts into the surrounding rural land and causes environmental problems such as loss of wetlands, loss of biological habitat, air pollution, and water pollution.

- **List at least five characteristics of an ideal sustainable city.**

A **sustainable city** is a city with a livable environment, a strong economy, and a social and cultural sense of community; sustainable cities enhance the well-being of current and future generations of urban dwellers. A sustainable city has clear, cohesive urban policies that enable the government infrastructure to manage it effectively. A sustainable city uses energy and other resources efficiently and makes use of renewable energy as much as possible. A sustainable city reduces pollution and wastes by reusing and recycling materials. A sustainable city has large areas of green space. A sustainable city is people-centered, not automobile-centered.

- **Explain how city planners have incorporated environmental sustainability into the design of Curitiba, Brazil.**

Curitiba developed an inexpensive, efficient mass transit system that uses clean, modern buses that run in high-speed bus lanes. High-density development was largely restricted to areas along the bus lines. Curitiba uses a special low-polluting fuel that contains a mixture of diesel fuel, alcohol, and soybean extract. Over several decades, Curitiba purchased and converted flood-prone properties along rivers in the city to a series of interconnected parks crisscrossed by bicycle paths. This move reduced flood damage and increased the per capita amount of green space.

Critical Thinking and Review Questions

1. Which countries are the most urbanized? The least urbanized? What is the urbanization trend today in largely rural nations?

2. Generally, the higher a country's level of urbanization, the lower its level of poverty. Suggest a possible explanation for this observation.

3. What is an urban agglomeration? Give an example.

4. Why are the U.S. urban agglomerations shown in Figure 9.3 each considered a functional system?

5. What are some of the problems brought on by rapid urbanization?

6. Suggest a reason why many squatter settlements are built on floodplains or steep slopes.

7. What are the four variables that urban ecologists study?

8. What is a brownfield? Why is it challenging to redevelop brownfields?

9. What is an urban heat island? A dust dome?

10. How can land-use planning promote compact development?

11. Why is good governance so important in increasing sustainability in cities?

12. Examine the cartoon. How does the cartoon represent urban food availability?

13. How has transportation affected the spatial structure of cities?

14. How has Curitiba, Brazil, been designed to incorporate sustainability?

15. The following graph shows projected CO_2 emissions in India based on the level of urbanization that occurs in the 21st century. Why do you think a higher level of urbanization will result in greater CO_2 emissions? How might India reduce its CO_2 emissions without reducing its energy consumption (which is necessary as India develops economically)?

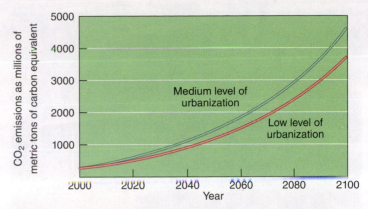

(Data from L. Jiang in L. Jiang, M.H. Young, and K. Hardee, "Population, Urbanization, and the Environment," *World Watch*, September/October 2008.)

Food for Thought

Urban Agriculture

Urban agriculture is beneficial for adults, and children many of whom lack access to fresh produce and who can benefit from the physical exercise of gardening. Fish can be cultivated in small-scale aquaponics facilities, as can be observed in Milwaukee, Wisconsin, as part of **Will Allen**'s urban agriculture initiative, Growing Power.

Urban agriculture in sub-Saharan Africa is more commonly practiced, in part because urban residents have more recently moved from agricultural practice on rural farms. In Dar es Salaam, Tanzania, for example, coffee, bananas, and other fruit trees are cultivated close to many homes. In Zambia, cities may be home to livestock as well as crops, resulting in some problems (disposal of dead animals and of animal wastes) and some solutions, such as improved nutritional status of children. Research laws about urban agriculture in your area. What kinds of activities are permitted, and what kinds of activities are restricted?

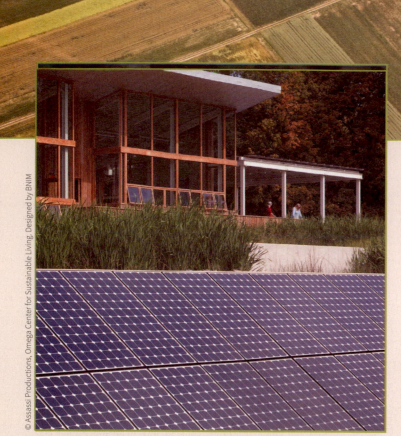

© Assassi Productions, Omega Center for Sustainable Living. Designed by BNIM

10

Energy Consumption

Living Building at the Omega Center for Sustainable Living.
Designed by BNIM, the 2011 American Institute of Architects Firm of
the Year awardee, this building operates with no energy beyond that
produced on site. The greenery in this image is both decorative and a part
of the wastewater treatment system.

Most energy use can be assigned to one of three categories: transportation, industry (including agriculture), and buildings. In highly developed countries, energy use is divided roughly evenly among these three. And while technological improvements can reduce the amount of energy consumed by any of the three categories, recent developments show that it is possible to design commercial and residential buildings that require no energy beyond that provided by the sun.

Architects and engineers constantly push the limits of energy-saving practices and technologies. The Omega Center for Sustainable Living in New York represents the promise of green design (**see photo**). Completed in 2009, the building has now operated for over five years with no net energy use. A system of natural filters and a designed wetland process the wastewater not only for the building but for the rest of the Omega campus as well. The building is the first ever to be certified both LEED Platinum and a Living Building.

LEED (Leadership in Energy and Environmental Design) certification was first developed in 1998 by the U.S. Green Building Council and has changed the way thousands of buildings are designed. A typical LEED-certified building uses far less energy than a traditional design and provides better living and working conditions for the building's users. In order to qualify for different certifications (from basic to Platinum), a building must have sustainable design features in such categories as energy, materials, and indoor air quality.

Although LEED certification is given when a building is completed, a new goal for many designers is to demonstrate that buildings perform as intended over time. One approach is the Living Building Challenge™, whose requirements include that all energy be generated on site from renewable resources, all wastewater be treated on site using natural processes, and no hazardous materials or chemicals be used in construction or operations. The Living Building certification is not awarded until the building has operated successfully for at least a year. So far, about a half dozen facilities have been recognized under this new standard, and a number of additional projects are in the design or construction phases.

Builders need not obtain certification in order to have green design features. Many of the lessons learned by designers trying to meet LEED or Living Building standards are now common practice in the industry. But having standards—and constantly expanding and improving those standards—helps architects understand what works and what the next opportunities are.

A challenge of sustainable design is that it can have much higher upfront costs than traditional building. However, in some cases, improved designs can be less expensive, while in others, the increase in design and construction costs is more than offset by reduced energy and water treatment costs over the life of the building. Harder to quantify are savings from the improved health of occupants, reduced pollution from burning fossil fuels, and reduced reliance on imported energy.

In this chapter, we examine how we use energy and how we can minimize energy use through conservation and efficiency. We look at two useful forms of energy—electricity and hydrogen—that are as environmentally friendly or harmful as the resources used to generate them. In the two following chapters, we consider nonrenewable (fossil fuel and nuclear) and renewable (solar, wind, hydroelectric, geothermal, and tidal) energy resources.

In Your Own Backyard…
What, if any, energy standards are required for new buildings on your campus?

Energy Consumption and Policy

Many energy sources have been described recently as clean, including coal, solar, and nuclear power. However, no energy source is clean; rather, energy sources vary in their advantages and disadvantages. Before we explore in detail the environmental effects associated with fossil fuels, nuclear, hydropower, wind, solar, and other energy resources, we will take a broad look at energy sources, consumption, and policy.

Everything humans do requires energy. We use energy to move and build things as well as to heat, cool, and illuminate our living and work spaces. We use energy to plant, water, harvest, process, ship, and store food. Energy is required to capture energy—to drill for and pump oil, to mine coal and uranium, to build solar panels, and to install wind turbines.

Just a few hundred years ago, almost all the energy used by people was derived from agriculture (including wood, dung, and peat), wind, or water. Energy sources were local, and activities were limited by the amount of useful energy that could be extracted from them. Only a small amount of useful energy can be extracted from a bucket full of wood, as compared to the amount that can be extracted from the same bucket full of gasoline.

A few thousand years ago, the discovery of fire, the domestication of animals, and later the invention of windmills and sailing ships significantly increased people's ability to manipulate their environments, but these advances were minor in comparison to the energy we can capture from fossil fuels; nuclear; and large-scale use of hydropower, solar, wind, and biomass. Electricity and, increasingly, hydrogen allow us to concentrate large amounts of useful energy from a wide range of sources. However, the

concentration of energy has also led to a concentration of wastes associated with energy, including heat and a range of pollutants.

Advantages of an energy source include how concentrated it is, availability, safety, and versatility. Disadvantages include hazard potential, environmental damage, and cost. Crude oil, for example, is a versatile energy source. It is easy to transport and can be made into a variety of different fuels, including diesel, jet fuel, and gasoline. We can store and use gasoline in personal automobiles, although if spilled or ignited, it can cause serious injuries and environmental damage. Similarly, nuclear materials are extremely hazardous to handle. They are not versatile—we use them exclusively to generate electricity. However, when used in well designed and managed reactors, nuclear materials can cause less environmental damage than does coal, the world's main source of electricity.

While crude oil and uranium ore are found in only a few parts of the world, all parts of the world can access solar power. In turn, though, solar has disadvantages, including that it is subject to both seasonal and daily variability, and capturing it requires equipment that can be expensive and environmentally harmful to produce. **Table 10.1** lists some of the advantages and disadvantages of major energy sources; further details on each are found in Chapters 11 and 12.

ENERGY CONSUMPTION

Conspicuous differences in per person energy consumption are found from country to country (**Figure 10.1**). Inhabitants of wealthier countries typically consume much more energy per person than those in poorer countries. Although less than 20% of the world's population lived in highly developed countries in 2014, they used 60% of the commercial energy consumed worldwide.

A comparison of energy requirements for food production illustrates energy consumption differences between countries. Farmers in many countries rely on their own physical energy or the

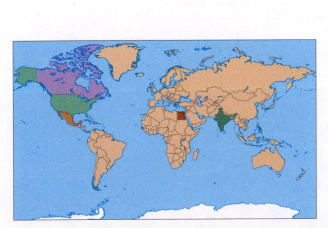

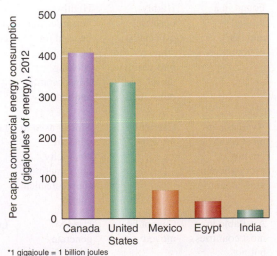

*1 gigajoule = 1 billion joules

Figure 10.1   Annual per capita commercial energy consumption in selected countries, 2012. Energy consumption per person in highly developed countries is much greater than it is in developing countries. (The map is color-coded with the bar graph.) (Compiled from U.S. Department of Energy)

Table 10.1   Advantages and Disadvantages of Several Major Energy Sources

Source	Geographic Distribution	Portability	Versatility	Worst-Case Event	Day-to-day Pollution (Not Climate Change)	Climate Change Potential	Scale	Reliability
Nuclear fission	Uranium found in a limited number of places	Fuel can be moved, but must be used in a fixed location	Used to generate electricity	Reactor failure and release unlikely, but could cause thousands of deaths and long-term contamination	Typically low	Low after construction	Large power plants only	Can run all the time
Solar photovoltaic	Widely available	Limited	Used to generate electricity	Low risk	Low	Very low	Flexible	Daily and seasonal variability
Hydropower	Found in a limited number of places	Cannot be moved	Mostly used to generate electricity, but sometimes for mechanical energy	Dam collapse rare, but could cause thousands of deaths	Low, but permanent disruption to upstream and downstream ecosystems	Low after construction	Flexible but depends on location	Can run all the time
Natural gas	Found in a limited number of places	Can be piped or trucked; often condensed	Can be used for heating, cooking, transportation, and industry	Natural gas plant or pipeline explosion unlikely, but could cause hundreds of deaths	Lowest of the fossil fuels; can burn cleanly	High	Flexible	Can run all the time
Coal	Found in a limited number of places	Fuel can be moved, but must be used in a fixed location	Used to generate electricity, for heating, and in industry	Power plant failure could cause some deaths	Difficult to burn cleanly; releases sulfur, nitrogen, and soot to air, land, and water	Highest	Flexible	Can run all the time
Oil	Found in a limited number of countries	Highly portable, especially when refined into gasoline, diesel, and other fuels	Highly versatile; can be used for heating, cooking, transportation, and industry	Refinery accident could cause some deaths	Refining can be dirty, and burning gasoline, diesel, and other fuels releases pollutants	High	Very flexible	Can run all the time
Wind	Available in most countries, but not everywhere in those countries	Cannot be moved	Mostly used to generate electricity, but sometimes for mechanical energy	Low risk	Low	Low	Flexible	Seasonal and unpredictable variability
Geothermal	Available in most countries, but not everywhere in those countries	Cannot be moved	Used to generate electricity, occasionally for heating	Low risk	Low	Low	Usually mid to large scale	Can run all the time

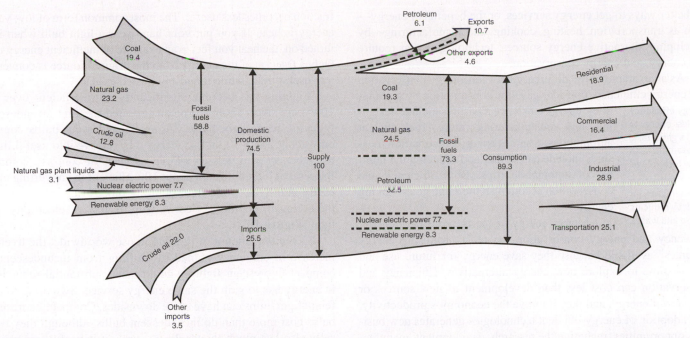

Figure 10.2   Energy consumption in the United States. This diagram provides a range of information about energy supply and use in the United States. The flows on the left indicate the types of energy supply, including those from domestic production (about 75% of supply) and imports (about 25% of supply). Where the energy is used is shown on the right, split among residential, commercial, industrial, and transportation end uses. All numbers are reported as percent of total supply. (Adapted from the U.S. Department of Energy)

energy of animals to plow and tend fields. In contrast, agriculture in the United States involves energy-consuming machines, such as tractors, automatic loaders, and combines. Industrialized agriculture also relies on energy-intensive fertilizers and pesticides. The larger energy input is one reason the agricultural productivity of highly developed countries is greater than that of developing countries.

World energy consumption has increased every year since 1982.[1] From 2009 to 2010, for example, energy consumption increased worldwide by about 4.5%. The increase is not, however, evenly distributed around the world. China more than doubled its energy consumption over the past decade. Similarly, India now uses 70% more energy than it did in 2004. By contrast, energy use in the United States, Germany, and Japan is about the same as it was a decade ago.

A goal of many developing countries is to improve their standard of living. One way to achieve this goal is through economic development, a process usually accompanied by a rise in per capita energy consumption. In addition, as discussed in Chapter 8, the human population continues to increase, and most of this growth occurs in developing countries.

In 2011, the most recent year for which we have accurate numbers, total world energy consumption was almost 540 billion **gigajoules** (GJ). The United States consumed 103 billion GJ, or 19% of the world total (down from 22% just three years earlier). **Figure 10.2** summarizes U.S. energy supply and use in 2011, with sources on the left and usage on the right. In 2011, China

consumed 116 billion GJ (21% of world total) and Kenya, 0.23 billion GJ (0.04%). In 2010, China surpassed the United States in total energy consumption (see Chapter 20).

However, China's population is much larger than that of the United States, so its per capita energy use—the amount used per person—is considerably smaller. The United States has one of the world's highest per capita energy uses, at about 56 GJ per person per year. That is about 6 times as much energy per person as is used in China and about 60 times as much as is used by the average Kenyan.

REVIEW

1. How does the concentration of energy in a source affect how people can use energy?

2. How do the United States, China, and Kenya compare in total energy consumption? In per capita energy consumption?

Energy Efficiency and Conservation

LEARNING OBJECTIVES

*   Explain the relationship between energy services and efficiency.
*   Describe some advantages and disadvantages of conserving energy.

Human energy demand will continue to increase, if only because the human population is growing. In addition, energy consumption continues to increase as developing countries raise their standard of living. However, as energy becomes more expensive, we can look

---

[1] Unless noted otherwise, all energy facts cited in this chapter were obtained from the Energy Information Administration (EIA), the statistical agency of the U.S. Department of Energy (DOE), and are the most recent information available.

for better ways to get **energy services**, or the benefits of energy—such as transportation, heating, cooking, and manufacturing—by developing alternative energy sources, technologies that require less energy, and energy conservation.

As an example of the difference between energy conservation and energy efficiency, consider gasoline consumption by automobiles. **Energy efficiency** measures include designing and manufacturing more fuel-efficient automobiles, whereas **energy conservation** measures include carpooling and reducing the number of automobile trips. Both efficiency and conservation accomplish the same goal—saving energy.

energy efficiency
Using less energy to accomplish a given task, as, for example, with new technology.

energy conservation
Using less energy, as, for example, by reducing energy use and waste.

Many energy experts consider energy efficiency and energy conservation the most promising energy "sources" available because they save energy for future use and buy us time to explore new energy alternatives. Efficiency and conservation can cost less than development of new sources or supplies of energy, and they improve the economy's productivity. The adoption of energy-efficient technologies generates new business opportunities, including the research, development, manufacture, and marketing of those technologies. Many technologies and practices are already known but are slow to be adopted due to both habit and relatively low energy prices.

In addition to economic benefits and energy resource savings, important environmental benefits result from greater energy efficiency and conservation. Using more energy-efficient appliances could cut our CO_2 emissions by millions of tons each year, slowing global climate change. Energy conservation and energy efficiency reduce air pollution, acid precipitation, and other environmental damage related to energy production and consumption.

ENERGY EFFICIENCY

Energy efficiency is a measure of the amount of available energy in a source that is transformed into useful work. In principle, energy efficiency ranges from 0 to 100%. That is, we can transform none, some, or all of the energy in a given source into energy services that are useful to us. For example, burning natural gas for household cooking has an efficiency of close to 100%—almost all the energy contained in the natural gas can be converted into heat in the stove or oven. In contrast, burning natural gas to generate electricity has a maximum efficiency of about 60%. This means that we would need almost twice as much natural gas to generate electricity to cook at home as we would if we burned the gas at the same home.

In practice, of course, choosing which sort of oven to use is more complicated. We have to provide the energy at the place where it is needed, which means piping natural gas to a home or power plant and installing power lines to deliver electricity. As we discuss below, electricity can be generated using a variety of resources in addition to natural gas. We usually need to make such decisions as a home is built or a replacement appliance is purchased, which means committing to the selected energy source for a decade or more.

Recall from Chapter 3 the second law of thermodynamics: Whenever energy is converted from one form to another, some fraction becomes less useful. The most common form of low-value energy is heat. If you put your hand near a light bulb when it is turned on, the heat you feel is an example of inefficient energy use: Only a fraction of the energy from the original source is converted into useful light, while the rest is wasted as heat.

Lighting has become significantly more efficient over the 100 years since Thomas Edison invented the first incandescent (glowing wire) light bulb. Current incandescent bulbs convert only about 2% to 3% of the energy they receive into useful light; the other 97% to 98% is wasted as heat. In contrast, compact fluorescent light bulbs are about 10% efficient in converting electricity into light, while light-emitting diodes (less common, but increasing in popularity) convert up to 20% of electricity into light (**Figure 10.3**).

Given the amount of lighting in use worldwide, the fivefold increase in efficiency gained by shifting from incandescent to compact fluorescent lighting could mean substantial reductions in energy use to gain the same energy service. Switching to different light bulbs can have some downsides. Compact fluorescent bulbs cost more than do incandescent bulbs, although they typically also last much longer. In this case, it is useful to consider the **payback time** necessary to recover the cost of investing in a different light bulb through energy savings and longer-lived light bulbs. For some people, another disadvantage of compact fluorescents is that the light they put out is not as pleasant as that

Figure 10.3 **Thermal images of incandescent, light emitting diode, and compact fluorescent light bulbs.** Thermal imaging provides a measure of the heat loss, and thus the efficiency, of different light bulbs. Here, white is hottest and blue is coolest. The incandescent bulb (top) wastes more energy as heat than do the light emitting diode (left) and compact fluorescent (right) light bulbs. How can shifting from one sort of light bulb to another affect climate change?

Tyrone Turner/National Geographic Image Creative

from an incandescent bulb. However, compact fluorescent light bulbs today provide light of much higher quality than did those a decade ago, and the technology continues to improve. The United States, Australia, and other countries have passed laws restricting sales of incandescent bulbs. (See You Can Make a Difference: Saving Energy at Home for additional energy-saving suggestions for the home.)

The development of more efficient lighting, appliances, automobiles, buildings, and industrial processes has helped reduce energy consumption in highly developed countries, and developing countries have been able to adopt more efficient technologies rather than older, higher-energy alternatives. **Table 10.2** shows that **energy intensity**, a measure of energy use per dollar of GDP, has decreased in most countries over the past three decades. For example, new condensing furnaces require approximately 30% less fuel than conventional gas furnaces. "Superinsulated" buildings use 70% to 90% less heat than do buildings insulated by standard methods (**Figure 10.4**).

The **National Appliance Energy Conservation Act (NAECA)** sets national efficiency standards for refrigerators, freezers, washing machines, clothes dryers, dishwashers, room air conditioners, and ranges/ovens (including microwaves). By one estimate, these standards had decreased electricity use in the United States by 7% by 2010, a reduction equivalent to removing 51 coal-fired power plants from service.

The NAECA requires appliance manufacturers to provide Energy Guide labels on all new appliances. These yellow labels provide estimates of annual operating costs and efficiency levels. Consumers who use this information to buy energy-efficient appliances save hundreds of dollars on utility bills.

Energy Savings in Commercial Buildings   Energy costs often account for 30% of a company's operating budget. Unlike cars, which are traded in every few years, buildings are usually used for 50 or 100 years; thus, a company housed in an older building normally does not have the benefits of new, energy-saving technologies. It makes good economic sense for these businesses to

invest in energy improvements, which often pay for themselves in a few years (**Table 10.3**). The California Energy Commission estimates that "high-performing buildings," which use about 20% less energy, cost $3 to $5 more per square foot to build but save as much as $67 per square foot over the life of the building!

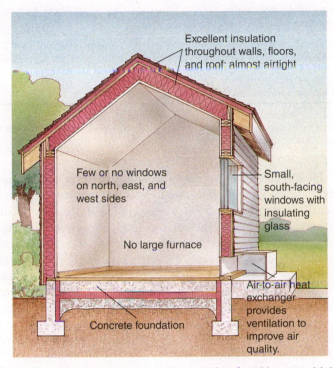

(a) Some of the characteristics of a superinsulated home, which is so well insulated and airtight it does not require a furnace in winter. Heat from the inhabitants, lightbulbs, stove, and other appliances provide almost all the necessary heat.

(b) A superinsulated office building in Toronto, Canada, has south-facing windows with insulating glass. The building is so well insulated it uses no furnace.

Figure 10.4   Superinsulated buildings.

Table 10.2   Comparison of 1980, 2006, and 2012 Energy Intensities for Selected Countries

Country	Energy Intensity*		
	1980	2006	2012
Kenya	4,473	3,393	2,695
India	7,870	7,477	5,860
Japan	7,834	6,492	5,313
Mexico	6,052	6,116	5,209
France	8,684	6,596	5,385
China	37,279	13,780	10,872
United States	15,135	8,841	7,329
Canada	18,701	13,097	10,661

*In Btu per 2000 U.S. dollars of GDP.
Source: Energy Information Administration.

Table 10.3   Energy-Efficiency Upgrades in Selected Commercial Buildings

Project	Energy Payback Time*	Unexpected Benefits Attributed to Project**
Energy-efficient lighting (post office in Nevada)	6 years	6% increase in mail-sorting productivity
Energy-efficient (metal-halide) lighting (aircraft assembly plant in Washington)	2 years	Up to 20% better quality control
Energy-efficient lighting (drafting area of utility company in Pennsylvania)	About 4 years	25% lower absenteeism; 12% increase in drawing productivity
Energy-efficient lighting and air conditioning (office building in Wisconsin)	0 years (paid for by utility rebates); energy savings estimated at 40%	16% increase in worker productivity
Energy-saving daylighting, passive solar heating, heat recovery system (bank in Amsterdam)	3 months	15% lower absenteeism

*How long it takes for energy savings to cover the cost of the project.
**Lighting quality as well as lighting efficiency is improved, resulting in greater worker comfort.
Source: Rocky Mountain Institute.

Energy-efficient upgrades of existing buildings can provide annual energy savings of as much as $2 per square foot. In one successful approach, energy-services companies make detailed assessments of how businesses can improve their energy efficiency. In developing its proposal, an energy-services company guarantees a certain amount of energy savings. It provides the funding to accomplish the improvements, which may be as simple as fine-tuning existing heating, ventilation, and air-conditioning systems or as major as replacing all existing windows and lights. The reduction in utility costs is used to pay the energy-services company, but once the bill is paid, the business benefits from all additional energy savings.

Buildings that produce as much or more energy than they use are called **zero-net-energy** buildings. The Omega building in the chapter introduction is one example. Around the country, individuals are turning their homes into zero-net-energy buildings by upgrading insulation, reducing their demand for energy services, and adding passive and active solar technologies (see Chapter 12). Although zero-net-energy buildings remain rare and often do not have payback times of less than a few decades, those experimenting with them now are providing knowledge and experience for future designers and builders.

Some utilities give customers energy-efficient compact fluorescent light bulbs, air conditioners, or other appliances. They then charge slightly higher rates or a small leasing fee, but the greater efficiency results in savings for both the utility company and the consumer. The utility company makes more money from selling less electricity because it does not have to invest in additional power generation to meet increased demand. The consumer saves because the efficient light bulbs or appliances use less energy, which more than offsets the higher rates.

According to the American Council for an Energy-Efficient Economy, U.S. electric power plants are themselves an important target for improved energy efficiency. Much heat is lost during the generation of electricity. If all this wasted energy were harnessed—for example, by *cogeneration*—it could be used productively, thereby conserving energy. Another way to increase energy efficiency would be to improve our electric grids because about 10% of electricity is lost during transmission. To accomplish this, some energy experts envision that future electricity will be generated far from population centers, converted to supercooled hydrogen, and transported through underground superconducting pipelines. The technology to build such conduits has not yet been developed.

Electric Power Companies and Energy Efficiency   Changes in the regulations governing electric utilities have allowed utilities to make more money by generating less electricity. Such programs provide incentives to save energy and thereby reduce power plant emissions that contribute to environmental problems.

Traditionally, to meet future power needs, electric utilities planned to build new power plants or purchase additional power from alternative sources. Now they often avoid these massive expenses by **demand-side management**, in which they help electricity consumers save energy. Some utilities support energy conservation and efficiency by offering cash awards to consumers who install energy-efficient technologies. When voters in California decided to close the nuclear power plant at Rancho Seco, the Sacramento Municipal Utility District paid customers to buy more efficient refrigerators and to plant trees to shade their houses, thereby lowering air-conditioning costs. These efforts helped the utility company reduce the demand for electricity.

Energy Efficiency in Transportation   Transportation is another sector in which efficiency improvements have been made, but far more could be done. The main energy service provided by automobiles is quickly and comfortably moving people and their things (such as groceries and equipment) from one place to another. However, much of the energy contained in a gallon of gas is wasted as heat. Automobile engines are powered by a series of tiny, controlled explosions, and they operate at high temperatures. Only a fraction of that energy is used to turn the engine, and since engines only operate at peak efficiency within a narrow range of speeds, energy is wasted as the car accelerates, slows, and shifts gears. Additional energy losses occur from braking (brake pads can get very hot in stop-and-go traffic) and from friction when tires contact the road surface. In addition, the automobile itself weighs far more than do the passengers, meaning that most of the energy that is not wasted as heat is used to move metal, plastic, and rubber from place to place!

YOU CAN MAKE A DIFFERENCE

Saving Energy at Home

The average household spends about $1,600 each year on utility bills. This cost can be reduced considerably by investments in energy-efficient technologies. When buying a new home, a smart consumer should demand energy efficiency. Although a more energy-efficient house might cost more, depending on the technologies employed, the improvements usually pay for themselves in two or three years. Any time spent in the home after the payback period means substantial energy savings. Energy efficiency has become an essential element of design codes nationwide and will almost certainly be an important part of future home designs.

Some energy-saving improvements, such as thicker wall insulation, are easier to install while the home is being built. Other improvements can be made in older homes to enhance energy efficiency and, as a result, reduce the cost of heating the homes. Examples include installing thicker attic insulation, installing storm windows and doors, caulking cracks around windows and doors, replacing inefficient appliances and furnaces, and adding heat pumps.

Many of the same improvements provide energy savings when a home is air-conditioned. Additional cooling efficiency is achieved by insulating the air conditioner ducts, especially in the attic; buying an energy-efficient air conditioner; and shading the south and west sides of a house with deciduous trees. Window shades and awnings on south- and west-facing windows reduce the heat a building gains from its environment. Ceiling fans can supplement air conditioners by making a room feel comfortable at a higher thermostat setting. Make sure your ceiling fan is set to draw warm air toward the ceiling in the summer, and reverse this setting in the winter.

Other energy savings in the home include the following (see figure):

- Replace incandescent bulbs with more energy-efficient light bulbs.
- Install a programmable thermostat, which cuts heating and air-conditioning costs up to 33%.
- Lower the temperature setting on water heaters to 140°F (with a dishwasher) or 120°F (without).
- Install low-flow shower heads and faucet aerators to reduce the amount of hot water used.
- Eliminate energy "vampires," or appliances that draw electricity even when they're not in use. (A study by researchers at Cornell University suggests that American households waste $3 billion each year on electricity vampires).

How does a homeowner learn which improvements will result in the most substantial energy savings? In addition to reading the many articles on energy efficiency in newspapers and magazines, a good way to learn about your home is to have a comprehensive energy audit done. Most local utility companies will send an energy expert to your home to perform an audit for little or no charge. The audit will determine the total energy consumed and where thermal losses are occurring (through the ceiling, floors, walls, or windows). On the basis of this assessment, the energy expert will then make recommendations about how to reduce your heating and cooling bills.

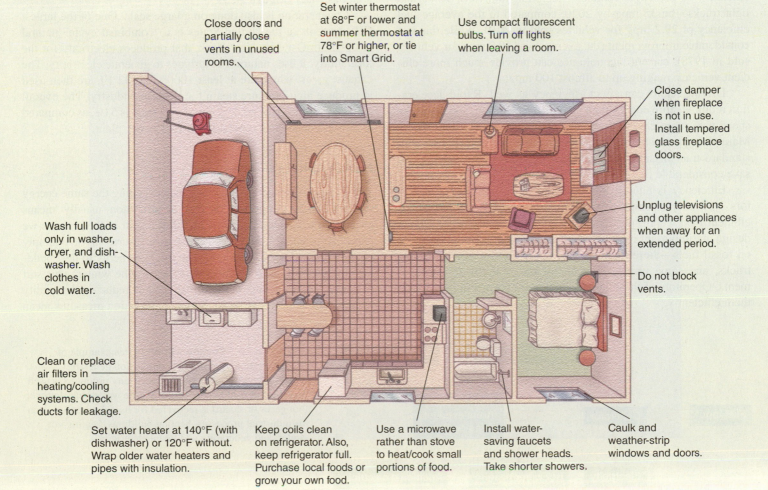

Close doors and partially close vents in unused rooms.

Set winter thermostat at 68°F or lower and summer thermostat at 78°F or higher, or tie into Smart Grid.

Use compact fluorescent bulbs. Turn off lights when leaving a room.

Close damper when fireplace is not in use. Install tempered glass fireplace doors.

Unplug televisions and other appliances when away for an extended period.

Do not block vents.

Wash full loads only in washer, dryer, and dishwasher. Wash clothes in cold water.

Clean or replace air filters in heating/cooling systems. Check ducts for leakage.

Set water heater at 140°F (with dishwasher) or 120°F without. Wrap older water heaters and pipes with insulation.

Keep coils clean on refrigerator. Also, keep refrigerator full. Purchase local foods or grow your own food.

Use a microwave rather than stove to heat/cook small portions of food.

Install water-saving faucets and shower heads. Take shorter showers.

Caulk and weather-strip windows and doors.

Knowing where energy is lost provides opportunities for improvement. Considerable energy can be saved by properly inflating tires, since less friction is generated. Using lighter materials for car bodies, frames, and even engines can reduce weight considerably. Plastics are much lighter than metal but lack the strength, durability, and heat resistance necessary for many important parts. However, advances include the development of ceramic engine parts (ceramics can tolerate high temperatures) and Kevlar structural materials (Kevlar is very strong). Both ceramics and Kevlar weigh less than the metals they replace. Finally, airflow can be a significant source of energy loss through friction, and most car designs include extensive testing to minimize wind resistance.

Modern vehicle designs increase efficiency in other places. For example, the engine of a Toyota Prius gasoline-electric hybrid does not always run, even when the car is in motion. This allows it to run at or near-peak efficiency. Regenerative braking recaptures some of the energy otherwise wasted as the vehicle slows. Electric vehicles operate at much lower temperatures than do gasoline engines; while the electricity used to power electric vehicles primarily comes from burning fossil fuels, it can be done more efficiently at large power plants than under the hood of a car.

A trend that has reduced energy efficiency is the popularity of minivans, SUVs, and light trucks, all of which have higher average gas usage than sedan-type automobiles. In 2007, the United States adopted new standards that will require that the average fuel efficiency for passenger vehicles—including minivans, SUVs, and light trucks—be 35 mpg by 2020, compared to the average fuel efficiency of 29.2 mpg for vehicles sold in 2010. While this is a considerable improvement (the average was 19.9 mpg for vehicles sold in 1978), current technologies can provide much more efficient vehicles, ranging up to almost 100 mpg.

Driving habits can affect efficiency as well. Rapid acceleration requires more energy than smooth acceleration, and changing speed uses more energy than does maintaining a constant speed. Maintaining a constant speed, choosing the right gear (in cars with standard transmission), and minimizing stop-and-go driving can save considerable fuel.

Efficiency is equally important in other transportation sectors. New aircraft, ships, trucks, and trains are all much more fuel-efficient than were older models. Lighter parts and better engine design are important innovations. Since most transportation relies on fossil fuels—diesel, natural gas, or jet fuel for aircraft, ships, trucks, and some trains—rises in oil prices motivate improvements. Operators of these large vehicles are trained in how to use them efficiently.

Energy Efficiency in Industry   Industries are constantly trying to find ways to improve efficiency. For example, technological improvements in the papermaking industry make it possible to use less energy to manufacture paper today than was used just a few years ago. The energy savings from such improvements in efficiency translate into greater profits for the companies employing them.

One energy technology that has matured over the past two decades is **cogeneration**, the production of two useful forms of energy from the same fuel. Cogeneration, or *combined heat and power* (CHP), makes perfect sense from a systems perspective. It involves the generation of electricity through some thermal process (often natural gas); the residual low-temperature steam left over after electricity production is used for building or industrial heating. In CHP, the overall conversion efficiency (that is, the ratio of useful energy produced to fuel energy used) is high because some of what is usually waste heat is used.

Cogeneration can be cost effective on a small scale. Modular CHP systems enable hospitals, factories, college and university campuses, and other businesses to harness steam that is otherwise wasted. In a typical CHP system, electricity is produced in a traditional manner—that is, some type of fuel provides heat to form steam from water. Normally, the steam used to turn the electricity-generating turbine is cooled before being pumped back to the boiler for reheating. In cogeneration, after the steam is used to turn the turbine, it supplies energy to heat buildings, cook food, or operate machinery before it is cooled and pumped back to the boiler as water (**Figure 10.5**).

Cogeneration is also done on a large scale. One of the largest CHP systems in the United States is a "combined cycle" natural gas plant in Oswego, New York, that produces electricity for the local utility. It uses natural gas turbines to generate electricity. The exhaust gases, which are at least 1000°C (1832°F), are then used to produce high-pressure steam for a nearby industry. The overall efficiency of the Oswego cogeneration system is 54%, as compared to 33% efficiency in a typical fossil-fuel power plant.

ENERGY CONSERVATION

While improving energy efficiency means getting the same energy services while using less energy, conservation usually means changes in behaviors, practices, and thus a shift in the services we expect from energy. Some conservation measures involve simply doing with less. For example, we can use less energy to heat buildings by lowering indoor temperatures during the winter. This sort of conservation measure is often maligned as reducing the quality of life, but this need not be so. Some conservation measures mean

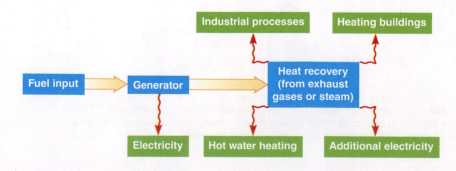

Figure 10.5   Cogeneration. In this example of a cogeneration system, fuel combustion generates electricity in a generator. The electricity produced is used in-house or sold to a local utility. The waste heat (leftover hot gases or steam) is recovered for useful purposes, such as industrial processes, heating buildings, heating hot water, and generating additional electricity.

shifting from one set of energy services to another, equivalent set. Others can even improve quality of life. Keep in mind as well that quality of life is a subjective idea—some people like driving big, powerful cars and drive long distances for fun; others prefer smaller vehicles and would rather avoid driving whenever possible.

Transportation in highly developed countries provides an opportunity for conservation that improves the quality of life. Individuals often commute two or more hours every day in heavy traffic. In many cases, changing commutes would be difficult, but there are a number of ways in which commutes could be reduced while saving time and not detracting from the workday. Examples include early or late start times, fewer but longer workdays, working from home, and satellite offices. In addition, for shorter commutes, walking or bicycling can both improve health and reduce energy use. Recall from the Chapter 7 introduction that increased reliance on cars and trucks has had a measurable negative effect on public health in the United States.

Conservation opportunities abound in agriculture as well. A 2012 study found that as much as 4% of U.S. energy is used to produce food that is wasted. Purchasing and managing food more carefully, then, has great conservation potential. Likewise, locally produced and dried foods require less energy to transport than imports and prepared foods. Producing meat requires far more energy per calorie consumed than does producing vegetables, fruits, and grains.

Similar conservation opportunities can be found in homes and buildings—for example, not heating unused rooms and adjusting thermostats at night and when the building is empty. Lighting can be automated to turn off when a room is not in use. Some businesses even encourage seasonally appropriate clothing so that buildings can be kept cooler in winter and warmer in summer. In addition to saving energy and money, occupants are more comfortable as they go to and from work.

One way to encourage energy conservation is to eliminate **subsidies**, which keep energy prices artificially low because the government thinks they are beneficial to the economy. When prices reflect the true costs of energy, including the environmental costs incurred by its production, transport, and use, energy is used more efficiently. Gasoline prices in the United States do not reflect the true cost of gasoline; even with the recent large increases in gasoline costs, gasoline remains much less expensive in the United States than in other countries. During 2008, Western Europeans paid about twice as much for gasoline as did U.S. citizens (**Table 10.4**). Lower gasoline prices encourage greater consumption. Over the next few years, a more realistic price for gasoline—including a tax to offset externalities (Chapter 2)—could be introduced to encourage people to buy fuel-efficient automobiles, carpool, and use public transportation (**Table 10.5**).

subsidy Government support (such as public financing or tax breaks) of business or institution to promote that group's activity.

REVIEW

1. How can the same energy services be provided with less energy for lighting? For transportation?

2. When does conservation reduce quality of life? Improve quality of life?

Table 10.4 A Comparison of Gasoline Prices in Selected Countries (Including Taxes)

	Regular Gasoline Price (Dollars per Gallon)		
Country	2001	2008	2013
United States	$1.51	$3.37	3.82
Canada	1.78	4.34	4.76
Mexico	2.46	3.04	3.22
Turkey	3.05	9.35	9.89
France	3.58	7.70	7.76

Source: Energy Information Administration.

Table 10.5 A Comparison of Energy Input for Different Kinds of Transportation

Method of Transportation	Energy Input (in BTUs)* per Person, per Mile
Automobile (driver only)	6530
Rail	3534
Carpool	2230
Vanpool	1094
Bus	939

*BTU stands for *British thermal unit*, an energy unit equivalent to 252 calories or 1054 joules.

Electricity, Hydrogen, and Energy Storage

LEARNING OBJECTIVES

- Explain why electricity is a flexible form of energy.
- Provide examples of the role hydrogen could play as a future fuel source.
- Describe the advantages and disadvantages of storing energy.

Hydrogen and electricity are versatile secondary forms of energy; that is, they are not found in nature in usable amounts, but they can be generated from any energy sources. Consequently, they can be as clean or dirty as the fuels used to produce them. Electricity has been in use for over a century and will probably play an increasing role in the future. Hydrogen is likely to be a major fuel in the future, although it is not likely to see widespread use for at least a decade. We often want to store energy when demand does not match up with production.

ELECTRICITY

Electricity, in the context of usable energy, is the flow of electrons in a wire. It continues to be a versatile form of energy, as it can easily be converted into light, heat, or motion. It can be released in very small, precise amounts in delicate electronic equipment or with enough power to move a freight train at high speeds.

Electricity can be generated from almost any energy source. In most cases, electricity is produced when an energy source spins a **turbine**. For example, coal can be burned to turn water into high-temperature steam that drives a turbine, or water can fall from a dam through a turbine. The turbine turns a **generator**, in which a bundle of wires spins around a magnet or a magnet spins inside a bundle of wires (**Figure 10.6**). The spinning causes electrons to move in the wire; this flow of electrons is electricity. Electricity can also come from stored chemical energy. A battery contains chemicals; when a wire is connected to the two ends (poles) of the battery, electrons flow from one pole to the other.

Electrons flowing through a wire in a single direction is **direct current** (DC), whereas electrons moving back and forth very quickly within the wire is **alternating current** (AC). The electrical current can be used to produce heat or power incandescent light bulbs. It can be used to power a variety of electronics, including fluorescent light bulbs, televisions, and computers. Finally, it can be used to drive electric **motors**, which are essentially generators operated in reverse. In a motor, an electrical current is applied to a bundle of wires around a magnet, making the magnet spin. This spinning can be put to other purposes—such as turning the drive shaft of a car or the compressor of a refrigerator or air conditioner.

While early electrical systems were local, with one or two points of production serving a small set of customers, electrical power lines now carry energy from a variety of power plants and other sources to users, who may be hundreds of miles away from the point of generation (see Environews: Smart Grids). This has both the advantage and disadvantage that the environmental impacts associated with an energy source are experienced far from the people who use the energy. For example, the energy from coal burned in rural Arizona provides power to homes, businesses, and industries in Los Angeles. The resulting pollution occurs away from a concentrated human population, but it can be harmful to people, plants, and animals near, and downwind of, where the coal is burned.

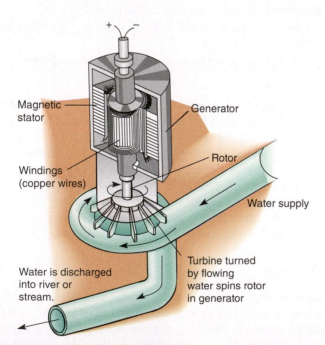

Figure 10.6 **Turbine-driven generator.** Electricity can be produced when a moving gas or liquid (in this case, water) spins a turbine. The turbine is connected to a rotor, around which are windings of copper wire. The wires spin inside a magnetic stator, which doesn't move. This causes electrons to flow in the wire—an electrical current.

HYDROGEN AND FUEL CELLS

Hydrogen is a common element—water molecules contain two hydrogen atoms and one oxygen atom. While water contains little available chemical energy, a hydrogen molecule with two hydrogen atoms (H_2) contains large amounts of available energy. H_2, which is a gas at room temperature, will explode when combined with the plentiful O_2 in the atmosphere, releasing energy and forming water. When chilled to $-253°C$ ($-423°F$), H_2 becomes a liquid and thus takes up much less space than H_2 gas.

Hydrogen has both advantages and disadvantages as an energy source. One advantage is that it has very high energy density, comparable to that of gasoline or liquefied natural gas (LNG) (see Chapter 11). Thus, unlike coal and nuclear energy, hydrogen could substitute for gasoline in automobiles and other forms of transportation. Another advantage is that hydrogen can be produced from any electrical source. **Electrolysis** is the process of using electricity to separate water into O_2 and H_2, which can be separately captured and stored (**Figure 10.7**). Finally, when H_2 is burned with O_2, the two products are available energy and water—no greenhouse gases and no other pollutants except for relatively small amounts of oxides of nitrogen (see Chapter 19). Currently, most of the H_2 in the United States is generated through re-forming natural gas. This is much less expensive than electrolysis but generates considerable amounts of CO_2.

Unfortunately, H_2 has several disadvantages. Its extreme volatility means that it has to be stored, handled, and transported carefully. (Of course, the same is true of gasoline, and H_2 has the advantage of being lighter than air, while spilled gasoline spreads on the ground.) In addition, the process of converting water into

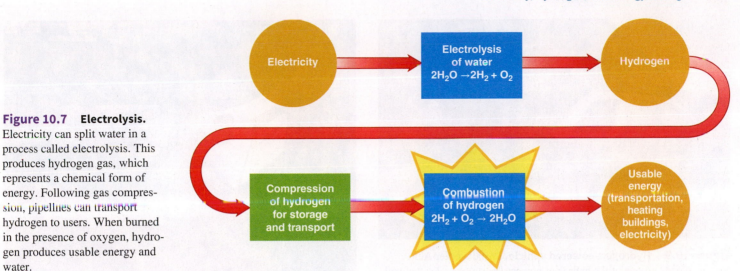

Figure 10.7 **Electrolysis.**
Electricity can split water in a process called electrolysis. This produces hydrogen gas, which represents a chemical form of energy. Following gas compression, pipelines can transport hydrogen to users. When burned in the presence of oxygen, hydrogen produces usable energy and water.

hydrogen is inefficient, so only a fraction of the energy from electricity can be captured when the hydrogen is burned. Since our main sources of electricity continue to be coal, natural gas, hydropower, and nuclear energy, producing hydrogen from this electricity will still create all the environmental problems associated with those sources.

The most promising way to use hydrogen is in a **fuel cell**. A fuel cell is an electrochemical cell similar to a battery (**Figure 10.8**). Unlike batteries, which store a fixed amount of energy, fuel cells produce power as long as they are supplied with fuel. Fuel cell reactants (hydrogen and oxygen) are supplied from external reservoirs, whereas battery reactants are contained within the battery. When hydrogen and oxygen react in a fuel cell, water forms and energy is produced as an electric current. Fuel cells work best when pure oxygen is combined with hydrogen, but in many cases, air (which you will recall is about 18% oxygen) is used instead.

Fuel cells can provide energy for a range of uses. Banks of fuel cells provide electricity for buildings or factories. Some

companies are looking at smaller applications, producing small fuel cells that would replace batteries in cell phones and laptop computers. And, as we discuss further in the next section, fuel cells can be used to power vehicles.

CASE IN POINT

HYDROGEN AND ELECTRICITY FOR TRANSPORTATION

Since transportation accounts for about 70% of the oil used worldwide, most transportation, whether by air, land, or water, relies on fossil fuels. While it may become possible to capture CO_2 from stationary power plants, it is much harder to do so with moving vehicles. Consequently, shifting to alternative power sources for vehicles is a goal in many countries.

The technology to use hydrogen as a substitute for gasoline in vehicles, while improving, is not yet commonly available. **Hydrogen fuel cell vehicles (HFCVs)** are now being used in some industrial and government fleets. Hydrogen vehicles for use on public roads currently can travel no more than about 350 km (220 mi) on a single tank of fuel, whereas 500 km (310 mi) is the typical minimum distance for commercial vehicles. The car pictured in **Figure 10.9** can travel about 480 km (300 mi) on a single load of hydrogen. Honda produced the first commercial HFCV in 2008; several other manufacturers have followed, but the number in production remains low.

Having such vehicles is not particularly useful if H_2 is not readily available. A national hydrogen infrastructure resembling the current infrastructure for gasoline and diesel distribution would have to be developed. At the end of 2013, there were over one hundred hydrogen fueling stations in the United States, about one-third of which were in California. Compare this to the hundreds or thousands of gas stations in any major city.

Iceland, which has no domestic fossil-fuel resources, plans to build the world's first fleet of fuel cell buses, obtaining its hydrogen fuel by using existing geothermal and hydroelectric resources. This suggests another major advantage of fuel cells—independence

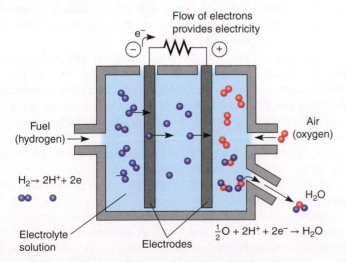

Flow of electrons provides electricity

Fuel (hydrogen)

Air (oxygen)

$H_2 \rightarrow 2H^+ + 2e$

H_2O

$\frac{1}{2}O + 2H^+ + 2e^- \rightarrow H_2O$

Electrolyte solution

Electrodes

Figure 10.8 **Cross section of a hydrogen fuel cell.** Hydrogen is usually produced through electrolysis. If the energy source of the electrolysis is renewable, the chemical energy in the hydrogen can be considered renewable.

Orange County Register /ZUMAPRESS.com

Figure 10.9 **Hydrogen-powered vehicle.** Postal worker Aldo Vasquez loads up one of the hydrogen-powered postal vehicles that will begin testing at the Irvine Post Office on Sand Canyon.

CHROMORANGE/Klaus Willig/picture alliance/Newscom

Figure 10.10 **Plug-in electric hybrid vehicle.** Pure electric vehicles and plug-in electric hybrid vehicles (PEHVs), like the one pictured here, can be charged from regular wall outlets at home. Electric vehicles allow flexibility in fuel sources, since any electrical source can be used. Through pure electric vehicles and PEHVs, energy from wind, nuclear power, hydropower, and solar could become significant for passenger vehicle transportation, which is currently dominated by gasoline.

from energy imports. This independence could come at an environmental cost if we significantly increase coal electrical generation to produce hydrogen.

Several technologies are competing with HFCVs to be the future power source for cars and trucks, and we should probably expect some combination of these to dominate roads within the next several decades. Hybrid vehicles that run on gasoline and electricity are common in the United States (see Chapter 20), but new alternatives include **pure electric vehicles (PEVs)** and **plug-in hybrid electric vehicles (PHEVs)**, which run on electricity and another fuel (**Figure 10.10**). PEVs and PHEVs can be charged through regular wall outlets when demand for electricity is low. The downside of PEVs is that they typically have very short ranges before they must be discharged and so are not as versatile as PHEVs. Either allows us to continue using cars but power them with carbon-free energy sources like wind, solar, or nuclear. A recent Electric Power Research Institute study suggests that 40% of the vehicles on the roads in 2030 could be PHEVs.

Finally, **flexible fuel vehicles (FFVs)**, which have been around as long as the internal combustion engine, are increasingly attractive. Rather than running only on gasoline, FFVs could run on gasoline, ethanol, mixed fuel, or compressed natural gas. FFVs could also be combined with hybrid or PHEV technologies. This would allow consumers to choose among fuels. ■

ENERGY STORAGE

A challenge common to many energy resources is that they are not available when we want or need them. Energy from a few resources, including gasoline, hydroelectric generation, and natural gas, can be easily increased or decreased in response to changing demand. Not all sources are so convenient, however. Some, such as solar and wind, are both intermittent—that is, they are abundant sometimes but sparse or nonexistent at other times—and unpredictable. Other energy sources, especially large coal or nuclear plants, are most effectively operated at a constant power output. Neither intermittent availability nor constant production fits with energy demand (there

are exceptions—summertime sun in the desert, for example, is most intense when air-conditioning demand peaks).

One solution to the mismatch between production and demand is to store unused energy until it is needed, and there are many technologies available to do so. The downside of all energy storage technologies is that they are less than 100% efficient. Every time we convert energy from one form to another, we lose some, and often much, of the useful energy. We can, for example, use batteries to charge other batteries, but with each conversion, less energy would be available.

Energy storage options include a range of energy forms. As just discussed, hydrogen can be used to store energy as chemical energy. Other forms of energy storage can include potential energy, thermal energy (heat), electrochemical energy, kinetic energy, and superconducting magnets.

One example that uses potential energy is **pumped hydroelectric storage**, in which water is pumped up into the basin behind a dam, where it is then available at a later time to generate electricity (see Chapter 12). Pumped hydroelectric storage is common in the United States and has been in use for nearly a century. Pumped hydroelectric storage can be as much as 80% efficient—that is, about 80% of the energy used to pump the water uphill can later be converted back to electricity in a generator.

Potential energy is also the principle behind **compressed air energy storage** (CAES). When energy is abundant, it can be used to compress air in large containers or even in natural or artificially created geological formations, such as caverns left empty after mining operations. When demand increases, the air can be released through a small opening that drives a turbine to turn a generator, producing electricity. CAES plants are operating in Germany and Alabama, and several new plants are expected to come on line in

Figure 10.11   Sulfur-sodium battery for large-scale energy storage. Banks of batteries like these can be used to store large amounts of energy produced when demand is low, to be released as demand increases or when other supplies are reduced.

Courtesy American Electric Power

the next few years, including the Iowa Stored Energy Park, which couples with a wind farm with CAES.

Storing thermal energy is commonly done at large buildings in the summer. At night, when electricity demand is low, water or another liquid is chilled. During the day, this chilled water is used to cool air in the building. The same principle is used, although less commonly, in the winter: When electricity demand is low, liquid is heated, and this hot liquid is used to warm the building during the day.

Batteries are the most common forms of electrochemical energy storage. Batteries can range in size and capacity from those used in watches and hearing aids to lead-acid batteries found in cars to huge banks of sulfur-sodium cells used to store energy at power plants (**Figure 10.11**). The advantage of batteries is that they are typically stable and can hold a charge for an extended period. However, they can be expensive to build and maintain, and because they contain concentrated hazardous chemicals, disposal or recycling of used batteries can create environmental problems (see Chapter 23). Hydrogen is useful for a variety of reasons, including that it can store energy produced from a variety of sources.

In **kinetic energy storage** systems, an electric motor accelerates a *flywheel* (a heavy, spinning object) to a high rotational speed. Later, the energy can be recaptured by operating the motor as a generator that converts the kinetic energy into electrical energy as it slows the spin. The disadvantage of this system is that friction eventually dissipates much of the kinetic energy into wasted heat. Some experimental flywheels that use magnetic bearings and are kept in a vacuum spin as fast as 50,000 revolutions per minute.

Superconducting magnetic energy storage (SMES) takes advantage of a phenomenon that occurs at extremely low temperatures. In these conditions, certain materials called *superconductors* can contain an electrical current almost indefinitely. A disadvantage of SMES systems is that they must be kept below

the temperature of liquid nitrogen (−196°C, or −321°F). Only a very few SMES systems have been built and installed for commercial use. They have mostly been used in industry or to stabilize power grids. However, large-scale SMES projects may be used for energy storage in the future.

REVIEW

1. What are some ways that electricity can be used?
2. What are some uses for hydrogen fuel cells?
3. What are some pros and cons of storing energy as hydrogen? In batteries? In flywheels?

Energy Policy

LEARNING OBJECTIVE

- Discuss how policy decisions shape energy resource and technology development.

Energy policy has always been a major issue in the United States—as in every country. However, over the past several years, energy policy has become increasingly important in political debate. The need for a coherent energy policy is driven by several factors. First, energy is a major driving force in any economy. Second, energy supply is limited, and maintaining supply and developing alternatives require long-term strategy and investment. Third, energy purchases, production, and consumption have substantial environmental downsides that are difficult to manage without governmental intervention.

An effective energy policy for any country has four central objectives. These objectives are best achieved when energy production and consumption are considered from a systems perspective. Since what people want is not energy per se but the services we get from energy, an effective policy should consider how to maintain or increase our quality of life with less energy. Policy should also balance the short-term goal of maintaining secure sources for the energy we use now and the long-term goal of having better—that is, less expensive, less environmentally damaging, and more secure—energy resources in the future.

OBJECTIVE 1: INCREASE ENERGY EFFICIENCY AND CONSERVATION

Over the past several decades, the efficiency of many technologies has improved. But despite gains, global energy use continues to increase. Improvement must occur on many fronts, from individuals conserving heating oil by weatherproofing their homes, to groups of commuters conserving gasoline by carpooling, to corporations developing more energy-efficient products.

Often, it is difficult for individuals or companies to invest in purchases that reduce energy use in the long run if those investments are costly in the short run. Similarly, companies may not recover costs of research that leads to more efficient technologies. In other cases, individuals may not know about energy costs or available opportunities to reduce those costs. One objective of an effective energy policy, then, is to support or provide long-term investment in conservation and efficiency while also informing citizens about true energy costs and alternatives.

An energy policy that was effective, although controversial, was the nationwide 55-mile-per-hour speed limit, which was the law in the United States from 1974 to 1995. Fuel consumption increases approximately 50% if a car is driven at 75 mph rather than at 55 mph. However, many Americans objected that the 55-mph speed limit increased their travel times and that lost time was more expensive to businesses than were energy gains. By contrast, most cities and states have planning policies that promote large, unimpeded roads and freeways, a policy choice that encourages vehicle use.

OBJECTIVE 2: SECURE FUTURE FOSSIL-FUEL ENERGY SUPPLIES

A comprehensive national energy strategy could include the environmentally sound and responsible development of domestically produced fossil fuels, especially natural gas. Three concerns exist about this element of a national energy strategy: security, economics, and the environment. Oil is an internationally traded commodity, and much of it comes from a small number of countries. Countries (including the United States) that import large quantities of oil are rightfully concerned about secure sources. Oil is a global commodity, so increasing oil production in any country increases the global supply but has a limited impact on that country's own supply.

Everyone, environmentalists included, recognizes the need for a dependable energy supply. Securing a future supply of fossil fuels, whether domestic or foreign, is a temporary strategy because fossil fuels are nonrenewable resources that will eventually be depleted, regardless of how efficient our use or how much we conserve. In the United States, maintaining a secure energy supply for the short term should be balanced by developing alternative energy sources for the long term.

OBJECTIVE 3: DEVELOP ALTERNATIVE ENERGY SOURCES

An effective long-term energy strategy will focus on identifying, researching, and developing inexpensive, environmentally benign, and widely available energy resources. An example of how this can work is nuclear power: Without enormous investments by the United States and other countries over the past half century, there would be no commercial nuclear power now. Much smaller investments in solar, wind, and other technologies have meant that those cleaner energy resources have not developed as quickly as has nuclear. A gasoline or carbon tax could finance programs to achieve a sustainable energy future.

One recent policy change in the United States makes it easier for individual electricity consumers to both sell energy back to the grid and buy it from the grid. Early private adopters of rooftop solar panels had to keep their units independent from the grid, since it was difficult to ensure that the quality of electricity produced was the same as that coming from the grid. Modern solar panel designs, however, convert electricity to a form compatible with the grid. An increasing number of consumers are now able to watch their power meters run backward as they supply electricity to the grid!

OBJECTIVE 4: MEET PREVIOUS OBJECTIVES WITHOUT FURTHER DAMAGE TO THE ENVIRONMENT

We must weigh the environmental costs of using a particular energy source against its benefits when it is considered as a practical component of an energy policy. Any country can require that its domestic supplies of fossil fuels be developed with as much attention to the environment as possible. This may make those fuels less competitive on the global market, but it would bring the benefit of a better quality of life for the citizens of that country.

One option in the United States is a 5-cent tax on each barrel of domestically produced oil or ton of domestically produced coal to establish a reclamation fund for some of the environmental damage caused by drilling or mining, production, and refining. This could be accompanied by a similar tariff on oil and coal imported from countries that do not have similar requirements.

REVIEW

1. What are the four objectives described in this section?
2. How do policy decisions affect the development of energy technologies?

Carbon Sequestration and Electricity

Sequestering carbon, or removing it from the waste stream of coal, oil, and natural gas before it gets into the atmosphere, is much easier at a large, stationary power plant than in tens of millions of small, often mobile, engines. Consequently, electric vehicles represent an opportunity to mitigate climate change while still using fossil fuels (see the figure). When gasoline is burned in a car, lawnmower, or chainsaw, it is converted into water, CO_2, and other chemicals.

The technology to remove CO_2 from the exhaust gases is possible but is years away from being developed. Even then, it is not clear that it could be done without enormous reductions in efficiency. In contrast, while removing CO_2 from exhaust streams at power plants is not simple, the technology to do so has been developed. Increasing our reliance on electricity will almost certainly be necessary if we want to both avoid dramatic climate change and continue to burn coal, oil, and natural gas.

Passenger vehicles in the United States. The number of passenger vehicles in the United States has increased from about 74 million in 1960, when average fuel efficiency was 17 miles per gallon, to 270 million in 2010, when average fuel efficiency was 29.2 miles per gallon. If the number of passenger vehicles continues to increase at the same rate until 2020, and the goal of 35 miles per gallon average fuel efficiency is met, will the total gasoline use in 2020 be higher or lower than that used in 2010? (Federal Bureau of Transportation Statistics)

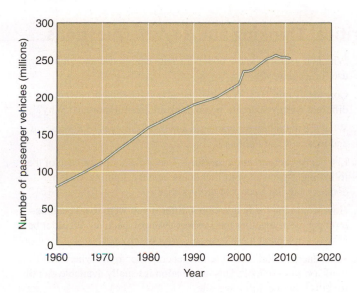

REVIEW OF LEARNING OBJECTIVES WITH SELECTED KEY TERMS

• **Explain the importance of the concentration of energy in a source.**

The concentration of energy in a source determines the amount of useful energy that can be extracted from a given volume or mass of fuel. Concentrated fuels are typically easier to transport and use than are less concentrated fuels. Modern transportation, industry, agriculture, and buildings rely on highly concentrated fuels like coal and gasoline.

• **Describe global energy use.**

Energy use is unevenly distributed around the world, with just 20% of the world's population using 60% of all commercial energy. Global energy use has increased every year for several decades. Some highly developed countries have managed to keep their energy use constant over the past decade. Energy use in China and India is growing rapidly, although per capita energy use in those countries remains below that of highly developed countries like the United States.

• **Explain the relationship between energy services and efficiency.**

Energy services are the benefits we get from using energy. They include transportation, industry, agriculture, commercial, and household energy use. **Energy efficiency** is a measure of the percentage of an energy source that we can transform into energy services. More efficient practices and technologies allow us to get more energy services from a given amount of an energy resource.

• **Describe some advantages and disadvantages of conserving energy.**

Energy conservation saves money, reduces pollution, and reduces our reliance on foreign energy sources. Some forms of conservation, such as maintaining a lower indoor temperature in winter, are seen by some as reducing the quality of life. Other conservation measures, such as switching from driving to walking, can have both benefits (health, cost savings) and downsides (longer commute time).

• **Explain why electricity is a flexible form of energy.**

Electricity can be produced from any energy resource. It is relatively easy to transport and can be used in very small or very large amounts. Finally, it can be put to a range of uses, from heating and lighting to electronics to transportation.

• **Provide examples of the role hydrogen could play as a future fuel source.**

Hydrogen can be generated from a variety of energy sources and can be used for transportation as either a combustible fuel or in a fuel cell.

Consequently, hydrogen has the potential to serve as a major transportation energy source if we transition away from fossil fuels. Hydrogen can also be used to store energy.

• Describe the advantages and disadvantages of storing energy.

Energy resources are not always available when we need them. Because of this mismatch between the timing of energy production and consumption, it can be useful to store energy when it is produced and release it as demand increases. Most energy storage technologies, however, are inefficient—that is, each time energy is stored and released, less of it is available for use. Energy storage technologies take advantage of potential,

thermal, chemical, electrochemical, and kinetic energy, as well as superconducting magnets.

• Discuss how policy decisions shape energy resource and technology development.

Energy policy decisions can encourage or discourage particular energy resources, both through subsidizing the development of some sources but not others and through funding research to develop future energy sources. Policy decisions can influence which technologies get used. Minimum vehicle fuel-efficiency standards promote conservation. Planning that focuses on building large, unimpeded roadways promotes car ownership.

Critical Thinking and Review Questions

1. Why is it useful to know both per capita energy use and total energy use for a country?

2. Consider Table 10.1. Which of the advantages and disadvantages of different energy sources do you think are most important? Why?

3. What are the relative advantages and disadvantages of fluorescent and incandescent lighting?

4. List energy conservation measures you could adopt for each of the following aspects of your life: doing laundry, lighting, bathing, cooking, buying a car, and driving a car.

5. How would the list in question 4 differ if you were to take a long-term perspective rather than a short-term perspective? Would it be easy to implement these changes? Why or why not?

6. What energy sources are used to generate electricity where you live, work, or go to school? This information is usually available on utility bills or at the utility website.

7. When are electricity and hydrogen "climate-neutral"? Can electricity or hydrogen produced using energy from coal be climate-neutral? Why or why not?

8. What forms of energy storage do you use? What are the benefits of these forms of energy storage?

9. What is the national energy policy of the United States? Does it include subsidies? Does it promote efficiency and/or conservation?

10. The German Federal Ministry for Economic Development and Cooperation uses gasoline policies to classify countries into four

categories: (1) high subsidies, (2) subsidies, (3) taxation, (4) high taxation. The November 2008 retail cost to consumers of regular gasoline in representative countries is found in the figure below. How do you think these prices impact conservation and efficiency for passenger vehicles in those countries?

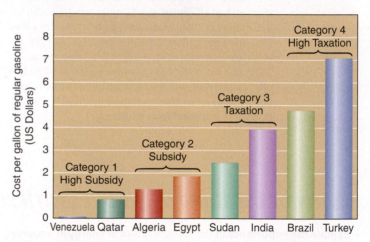

Cost per gallon of regular gasoline in select countries, November 2008. (German Federal Ministry for Economic Development and Cooperation)

Food for Thought

For several days, keep track of places that the foods you eat come from. Make a log of the distance foods have to travel to get to your plate, bowl, cup, or bag. Try to find out how some of the foods are harvested, transported, processed, and cooked. What sorts of energy sources are necessary to get those foods to you? Would you need to change your diet in order to reduce the amount of energy necessary to produce it? In what ways?

11

Fossil Fuels

Ann Heisenfelt/EPA/Photoshot

Deepwater Horizon oil spill in the Gulf of Mexico.

The public health impacts of fossil fuel extraction are often difficult to assess. When the Deepwater Horizon oil-drilling platform exploded in April 2010, over 4 million barrels of crude oil spilled into the Gulf of Mexico (see photo). The explosion itself killed 11 workers, and oil spread across the ocean floor and the coast of the southern United States, killing wildlife and causing extensive ecological damage. Traditional risk analysis is being used to estimate losses directly caused by the oil spill.

The effect on residents of coastal Louisiana, Mississippi, and Florida is less direct. The immediate impact was the closing of beaches and fisheries due to contamination. This resulted in lost income, empty hotels, and idled boats as well as the loss of many service jobs. Daily life for thousands was disrupted and uncertain.

British Petroleum has established a fund to compensate financial losses. However, while compensation replaces lost income for individuals and families, it shifts the economy of a community. For example, boat, dock, and fishing equipment and repairs are no longer needed. Further, we know that the stress associated with disruption and uncertainty can lead to depression, alcoholism, suicide, and other mental health issues. The

long-term impact of this oil spill could be huge, but much of it will be difficult to assess.

Drilling for natural gas can have similarly uncertain, disruptive, and difficult-to-measure public health implications. As we will see in this chapter, in 2007 geologists identified a huge store of natural gas in the northeastern United States, trapped in a formation called the Marcellus Shale. Shale deposits are found as far as 6,000 feet below Earth's surface. Shale gas drilling operations generate substantial wastewater, which must be managed carefully to avoid contaminating the environment, including agricultural and residential water resources.

Although the effects of possible wastewater spills can be estimated, the broader impacts affect physical infrastructure and local communities. Fleets of trucks move drilling equipment and wastewater, increasing traffic and stressing roads and bridges designed for far less traffic. Drilling operations—which can be loud and odorous—are often located near homes and farms. Organic farmers can lose certification when drilling operations come too close.

Extracting mineral and fossil-fuel resources has always had widespread effects. These include economic benefits to communities, including jobs and taxes,

and the resulting resources are used around the world. Extraction can also cause major health and environmental impacts to individuals and communities, many of whom do not share in the benefits. In this chapter, we explore how fossil fuels were formed, how we extract and use them, and some of the associated environmental impacts. Keep in mind that understanding the effects—good and bad—of fossil fuels requires that we think not just about direct benefits and costs but also about broader impacts to environmental and societal systems.

In Your Own Backyard…
What energy resources are generated, refined, or stored near you, and how would a spill affect your community?

Fossil Fuels

LEARNING OBJECTIVES

- Define *fossil fuel* and distinguish among coal, oil, and natural gas.
- Describe the processes that formed coal, oil, and natural gas.
- Relate fossil fuels to the carbon cycle.

fossil fuels Combustible deposits in Earth's crust, composed of the remnants (fossils) of prehistoric organisms that existed millions of years ago. Coal, oil (petroleum), and natural gas are the three types of fossil fuel.

Fossil fuels—coal, oil, and natural gas—supply over 80% of the energy used in North America. A fossil fuel is composed of the remnants of organisms, compressed in an oxygen-free environment. Fossil fuels resulted from photosynthesis (Chapter 3) that captured solar energy millions of years ago.

Fossil fuels are nonrenewable resources—Earth's crust has a finite, or limited, supply of them, and that supply is depleted by use. Although natural processes are still forming fossil fuels, they are forming far too slowly (on a scale of millions of years) to replace the fossil-fuel reserves we are using. Fossil-fuel formation does not keep pace with use, and as fossil fuels are used up, we will have to switch to other forms of energy.

Fossil-fuel resources are concentrated in only a few countries. Africa and South America—with the exception of a few countries—have few fossil-fuel resources. Most of those countries must purchase fossil fuels from other regions or develop alternative energy resources. Either of these approaches requires substantial financial capital—another resource generally lacking in developing countries. Thus, worldwide access to energy is a significant equity issue.

HOW FOSSIL FUELS FORMED

Three hundred million years ago, the climate of much of Earth was mild and warm, and atmospheric carbon dioxide levels were higher. Vast swamps were filled with plant species that have long since become extinct. Many of these plants—horsetails, ferns, and club mosses—were as large as trees (**Figure 11.1**).

In most environments, decomposers such as bacteria and fungi rapidly decompose plants after their death. However, as some ancient swamp plants died, they were covered by water. Their watery grave prevented the plants from decomposing much; wood-rotting fungi cannot act on plant material where oxygen is absent, and anaerobic bacteria, which thrive in oxygen-deficient environments, do not decompose wood rapidly. Over time, more and more dead plants piled up. As a result of periodic changes in sea level, sediment (mineral particles deposited by gravity) accumulated, forming layers that covered the plant material. Aeons passed, and the heat and pressure that accompanied burial converted the nondecomposed plant material into a carbon-rich rock called **coal** and converted the layers of sediment into sedimentary rock. Much later, geologic upheavals raised these layers so that they were nearer Earth's surface.

coal A black, combustible solid composed mainly of carbon, water, and trace elements found in Earth's crust; formed from ancient plants that lived millions of years ago.

Figure 11.1   A Carboniferous forest of Midwestern North America 350 million years ago

© Walter Myers / Stocktrek Images, Inc.

Oil formed when large numbers of microscopic aquatic organisms died and settled in the sediments. As these organisms accumulated, their decomposition depleted the small amount of oxygen present in the sediments. The resultant oxygen-deficient environment prevented further decomposition. Over time, the dead remains were covered and buried deeper in the sediments. Although we do not know the basic chemical reactions that produce oil, the heat and pressure caused by burial presumably aided in the conversion of these remains to the mixture of hydrocarbons (molecules containing carbon and hydrogen) known as oil.

oil A thick, yellow to black, flammable liquid hydrocarbon mixture found in Earth's crust; formed from the remains of ancient microscopic aquatic organisms.

Natural gas, composed primarily of the simplest hydrocarbon, methane, formed in essentially the same way as oil, only at higher temperatures, typically greater than 100°C. Over millions of years, as the remains of organisms were converted to oil or natural gas, the sediments covering them were transformed into sedimentary rock, including sandstone and shale.

natural gas A mixture of energy-rich gaseous hydrocarbons (primarily methane) that occurs, often with oil deposits, in Earth's crust.

FOSSIL FUELS, THE CARBON CYCLE, AND CLIMATE

Burning of fossil fuels represents the completion of the carbon cycle, part of a natural system. Normally, solar energy and carbon dioxide are captured through photosynthesis, stored for weeks or years, then consumed and released. In the case of fossil fuels, however, the energy and carbon accumulated over millions of years but are now being released in just a few hundred years. The carbon dioxide (CO_2) concentration of the atmosphere is rapidly increasing from levels associated with a relatively cool climate to levels associated with a substantially warmer climate.

The equilibrium among CO_2 in the atmosphere, CO_2 dissolved in the ocean, and CO_2 in organic matter changes over long periods—thousands or millions of years. Over the past century, however, we have released so much CO_2 into the atmosphere through consumption of fossil fuels that Earth's CO_2 equilibrium has been disrupted. Chapter 20 explores the consequences of increased CO_2.

REVIEW

1. What are fossil fuels?
2. How are coal, oil, and natural gas formed?
3. What is the relationship between burning fossil fuels and atmospheric carbon dioxide?

Coal

LEARNING OBJECTIVE

- Distinguish between surface mining and subsurface mining.
- Summarize the environmental problems associated with using coal.
- Define *resource recovery* and *fluidized-bed combustion*.

Although coal was used as a fuel for centuries, not until the 18th century did it begin to replace wood as the dominant fuel in the Western world. Since then, coal has had a significant impact on human history. It was coal that powered the steam engine and supplied the energy for the Industrial Revolution, which began in the mid-18th century. Today utility companies use coal to produce electricity, and heavy industries use coal for steel production. Coal consumption has surged in recent years, particularly in the rapidly growing economies of China and India, both of which have large coal reserves.

Coal is found in different grades, largely as a result of the varying amounts of heat and pressure it was exposed to during formation. Coal exposed to high heat and pressure during its formation is drier, is more compact (and therefore harder), and has a higher heating value (that is, a higher **energy density**). Lignite, subbituminous coal, bituminous coal, and anthracite are the four most common grades of coal (**Table 11.1**).

energy density The amount of energy contained within a given volume of an energy source.

Lignite is a soft coal, brown or brown-black in color, with a soft, woody texture. It is moist and produces little heat compared with other types of coal. Lignite is often used to fuel electric power plants. Sizable deposits are found in the western states, and the largest producer of lignite in the United States is North Dakota.

Subbituminous coal is a grade of coal intermediate between lignite and bituminous. Like lignite, subbituminous coal has a relatively low heat value and sulfur content. Many coal-fired electric power plants in the United States burn subbituminous coal because its sulfur content is low. It is found primarily in Alaska and a few western states, such as Montana and Wyoming.

Bituminous coal, the most common type, is also called soft coal, even though it is harder than lignite and subbituminous coal. Bituminous coal is dull to bright black with dull bands. Much bituminous coal contains sulfur, which causes severe environmental problems when the coal is burned in the absence of pollution-control equipment. Nevertheless, electric power plants use bituminous coal extensively because it produces a lot of heat. In the United States, bituminous coal deposits are found in the Appalachian region, near the Great Lakes, in the Mississippi Valley, and in central Texas.

The highest grade of coal, anthracite or hard coal, was exposed to extremely high temperatures during its formation. It is a dark, brilliant black and burns most cleanly—it produces the fewest pollutants per unit of heat released—of all the types of coal because it is not contaminated by large amounts of sulfur. Anthracite has the highest heat-producing capacity of any grade of coal. Anthracite deposits in the United States are largely depleted; most of the remaining deposits are east of the Mississippi River, particularly in Pennsylvania. Coal is usually found in seams, underground layers that vary from 2.5 cm (1 in.) to more than 30 m (100 ft) in thickness. Geologists think that most, if not all, major coal deposits are now identified. Scientists working with coal are therefore concerned less about finding new deposits than about the safety and environmental problems associated with coal, such as the burning coal seam shown in **Figure 11.2**.

COAL RESERVES

Coal, the most abundant fossil fuel, is found primarily in the Northern Hemisphere (**Figure 11.3**). The largest coal deposits are in the United States, Russia, China, Australia, India, Germany, and South Africa. The United States has about 20% of the world's coal supply. According to the World Resources Institute, known world coal reserves could last more than 200 years at the current rate of consumption. In addition, coal resources currently too expensive to develop have the potential to provide enough coal to last for 1,000 or more years (at current consumption rates). For example, some coal deposits are buried more than 5,000 feet inside Earth's

Table 11.1   A Comparison of Different Kinds of Coal

Type of Coal	Color	Water Content (%)	Relative Sulfur Content	Carbon Content (%)	Average Heat Value (BTU/pound)	2012 Cost at Mine for 2000 lb of Coal ($)
Lignite	Dark brown	45	Medium	30	6,000	21.53
Subbituminous coal	Dull black	20–30	Low	40	9,000	13.71
Bituminous coal	Black	5–15	High	50–70	13,000	54.25
Anthracite	Black	4	Low	90	14,000	60.35

Sources: EIA, U.S. Department of Energy, and USGS.

Elleringmann/Laif/Aurora Photos

Figure 11.2 **Burning coal seam.** One problem with mining coal near the surface is that it is highly flammable—a spark can set an entire seam burning, and in some cases it is impossible to extinguish. This burning seam in Black Dragon Mountain in Mongolia releases huge amounts of pollutants into the atmosphere.

crust. Drilling a shaft that deep would cost considerably more than the current price of coal would justify.

COAL MINING

The two basic types of coal mines are surface and subsurface (underground) mines. The type of mine chosen depends on surface contours and on the location of the coal bed relative to the surface. If the coal bed is within 30 m (100 ft) or so of the surface, **surface**

surface mining The extraction of mineral and energy resources near Earth's surface by first removing the soil, subsoil, and overlying rock strata (i.e., the overburden).

mining is usually done (**Figure 11.4**). In one type of surface mining, **strip mining**, a trench is dug to extract the coal, which is scraped out of the ground and loaded into railroad cars or trucks. Then a new trench is dug parallel to the old one, and the *overburden* from the new trench is put into the old trench, creating a *spoil bank*, a

hill of loose rock. Digging the trenches involves using bulldozers, giant power shovels, and wheel excavators to remove the ground covering the coal seam. Surface mining is used to obtain approximately 60% of the coal mined in the United States.

When the coal is deeper in the ground or runs deep into the ground from an outcrop on a hillside, it is mined underground.

subsurface mining The extraction of mineral and energy resources from deep underground deposits.

Subsurface mining accounts for approximately 40% of the coal mined in the United States.

Surface mining has several advantages over subsurface mining: It is usually less expensive and safer for miners, and it generally allows a

more complete removal of coal from the ground. However, surface mining disrupts the land much more extensively than subsurface mining and has the potential to cause several serious environmental problems.

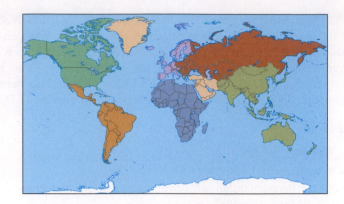

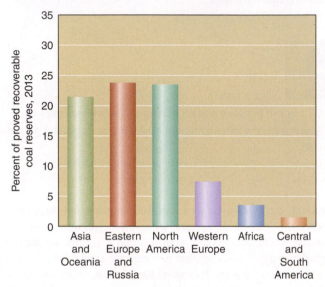

Figure 11.3 **Distribution of coal deposits.** Data are presented as percentages of the 2013 estimated recoverable reserves—that is, of coal known to exist that can be recovered under present economic conditions with existing technologies. The majority of the world's coal deposits are located in the Northern Hemisphere. (The map is color-coded with the bar graph.)

SAFETY PROBLEMS ASSOCIATED WITH COAL

Although we usually focus on the environmental problems caused by mining and burning coal, significant human safety and health risks also occur in the mining process. According to the Department of Energy (DOE), during the 20th century more than 90,000 American coal miners died in mining accidents, although the number of deaths per year declined significantly in the latter part of the century. Miners have an increased risk of cancer and black lung disease, a condition in which the lungs are coated with inhaled coal dust, severely restricting the exchange of oxygen between the lungs and blood. It is estimated that these diseases are responsible for the deaths of at least 2,000 miners in the United States each year.

ENVIRONMENTAL IMPACTS OF THE MINING PROCESS

Coal mining, especially surface mining, has substantial effects on the environment. Prior to passage of the 1977 **Surface Mining Control and Reclamation Act (SMCRA)**, abandoned surface coal mines were usually left as large open pits or trenches. *Highwalls*, cliffs of excavated rock, some more than 30 m (100 ft)

© Jim West/Age Fotostock

Figure 11.4 **Surface coal mining.** The overlying vegetation, soil, and rock are stripped away, and then the coal is extracted out of the ground. Photographed near Douglas, Wyoming.

high, were left exposed. Acid and toxic mineral drainage from such mines, along with the removal of topsoil, which was buried or washed away by erosion, prevented most plants from naturally recolonizing the land. Streams were polluted with sediment and

acid mine drainage Pollution caused when sulfuric acid and dissolved materials such as lead, arsenic, and cadmium wash from coal and metal mines into nearby lakes and streams.

acid mine drainage, produced when rainwater seeps through iron sulfide minerals exposed in mine wastes. Dangerous landslides occurred on hills unstable from the lack of vegetation.

We can restore surface-mined land to prevent such degradation and to make the land productive for other purposes, although restoration is expensive and technically challenging. Among other things, SMCRA requires coal companies to restore areas that have been surface-mined since 1977; requires permits and inspections of active coal mining operations and reclamation sites; prohibits coal mining in sensitive areas such as national parks, wildlife refuges, wild and scenic rivers, and sites listed on the National Register of Historic Places; and stipulates that surface-mined land abandoned prior to 1977 (more than 0.4 million hectares, or 1 million acres) should gradually be restored, using money from a tax that coal companies pay on currently mined coal.

According to the U.S. Office of Surface Mining, more than $1.5 billion has been spent reclaiming the most dangerous abandoned-mine lands; approximately two-thirds of this money was spent in four states—Pennsylvania, Kentucky, West Virginia, and Wyoming. However, so many abandoned mines exist that it is doubtful they can all be restored.

One of the most land-destructive types of surface mining is **mountaintop removal**. A dragline, a huge shovel with a 20-story-high arm, takes enormous chunks out of a mountain, eventually removing the entire mountaintop to reach the coal located below. According to Environmental Media Services, mountaintop removal has leveled between 15% and 25% of the mountaintops in southern West Virginia. The valleys and streams between the mountains are gone as well, filled with debris from the mountaintops. Mountaintop removal is also occurring in parts of Kentucky, Pennsylvania, Tennessee, and Virginia. SMCRA specifically exempts mountaintop removal. In 1977, when SMCRA was passed, existing

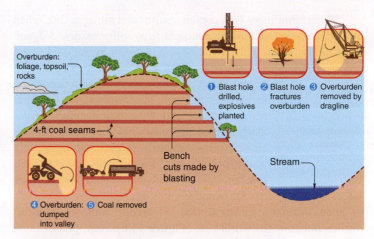

(a) Cross section of a typical mountain before mining.

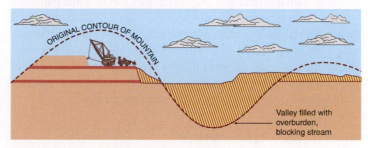

(b) Cross section after mountaintop has been removed.

Figure 11.5 **Removing coal by levelling a mountain.** When coal is removed from mountains, the geography of a region can change. The top of a mountain above a coal seam is scraped off and dumped into the adjacent valley. Next, the coal seam is removed. If another seam lies lower in the mountain, the process is repeated. Streams are buried, lakes are created and destroyed, and large amounts of sediment are washed downstream as bare earth is eroded. (Adapted from E. Reece, "Death of a Mountain," *Harper's Magazine*, April 2005).

technology for mountaintop removal could take out the top seam of coal along a mountain ridge. Today, mountaintop removal takes out as many as 16 seams of coal, from the ridge to the base of the mountain (**Figure 11.5**). In 2014, a federal appeals court upheld earlier court decisions that allow the EPA to limit mountaintop removal because disposing of mountaintop waste rock in valleys violates the Clean Water Act (the disposal buries streams).

ENVIRONMENTAL IMPACTS OF BURNING COAL

Burning coal can affect air and water quality, and impacts range from local (e.g., sooty fallout) to global (climate change and ocean acidification). Burning coal generally contributes more air pollutants (including CO_2) than does burning either oil or natural gas to generate the same amount of useful energy. Coal often contains mercury that is released into the atmosphere during combustion. This mercury moves readily from the atmosphere to water and land, where it accumulates and harms humans as well as wildlife. (See Chapter 19 for a discussion of the human health hazards of elevated levels of mercury.) In the United States, coal-burning electric power plants currently produce one-third of all airborne mercury emissions.

Much bituminous coal contains sulfur and nitrogen that, when burned, are released into the atmosphere as sulfur oxides (SO_2 and SO_3) and nitrogen oxides (NO, NO_2, and N_2O). Sulfur oxides and nitrogen oxides (NO and NO_2) form acids when they react with water. These reactions result in **acid deposition** including *acid rain*. The combustion of coal is responsible for acid deposition, which is particularly prevalent downwind from coal-burning electric power plants. Normal rain is slightly acidic (pH 5.6), but in some areas acid precipitation has a pH of 2.1, equivalent to that of lemon juice. Acidification of lakes and streams has resulted in the decline of aquatic animal populations and is linked to some of the forest decline documented worldwide (**Figure 11.6**). Acid precipitation and forest decline are discussed in greater detail in Chapter 19.

acid deposition A type of air pollution in which acid falls from the atmosphere to the surface as precipitation (acid precipitation) or as dry acid particles.

Although it is relatively easy to identify and measure pollutants such as sulfur oxides in the atmosphere, it is more difficult to trace their exact origins. Air currents transport and disperse air pollutants, which are often altered as they react chemically with other pollutants in the air. Even so, some nations clearly suffer the damage of acid deposition caused by air pollutants produced in other countries, and as a result, acid deposition is an international issue.

MAKING COAL A CLEANER FUEL

We often hear the term *clean coal*. However, at its cleanest, coal still has many environmental downsides. Scrubbers, or desulfurization systems, can remove sulfur from a power plant's exhaust.

Frederica Georgia / Science Source

Figure 11.6   **Trees killed by acid fog, Clingmans Dome, Great Smoky Mountains National Park, Tennessee.**

As polluted air passes through a scrubber, chemicals in the scrubber react with the pollution and cause it to precipitate (settle) out. Modern scrubbers can remove 98% of the sulfur and 99% of the particulate matter in smokestacks. Desulfurization systems are expensive; they cost about $50 to $80 per installed kilowatt, or about 10% to 15% of the construction costs of a coal-fired electric power plant.

In lime scrubbers, a chemical spray of water and lime neutralizes acidic gases such as sulfur dioxide, which remain behind as a calcium sulfate sludge that becomes a disposal problem (see Figure 19.9d). A large power plant may produce enough sludge annually to cover 2.6 km² (1 mi²) of land 0.3 m (1 ft) deep. Although many power plants currently dispose of the sludge in landfills, some have found markets for the material. In **resource recovery**, the sludge is treated as a marketable product rather than as a polluted emission. Some utilities sell calcium sulfate from scrubber sludge to wallboard manufacturers. (Wallboard is traditionally manufactured from gypsum, a mineral composed of calcium sulfate.) Other companies use fly ash, the ash from chimney flues, to make a lightweight concrete that could substitute for wood in the building industry. Some farmers apply calcium sulfate sludge as a soil conditioner. Plants grow better because calcium sulfate neutralizes acids in some soils and increases the water-holding capacity of the soil.

resource recovery The process of removing any material—sulfur or metals, for example—from polluted emissions or solid waste and selling it as a marketable product.

The Clean Air Act Amendments of 1990 required the nation's 111 dirtiest coal-burning power plants to cut sulfur dioxide emissions. Compliance resulted in a total annual decrease of 3.8 million metric tons nationwide, a reduction of about 25% of the total amount of sulfur dioxide emitted in the United States in earlier years. In the second phase of the Clean Air Act Amendments, more than 200 additional power plants made SO_2 cuts by the year 2000, resulting in a total annual decrease of 10 million metric tons nationwide.

A nationwide cap on SO_2 emissions from coal-burning power plants was imposed after 2000. Utilities also cut nitrogen oxide emissions by 2.6 million tons per year, out of 7.2 million tons per year total. Several technologies burn coal in ways that minimize sulfur oxide and nitrogen oxide releases but do not reduce CO_2 emissions.

Fluidized-bed combustion mixes crushed coal with particles of limestone in a strong air current during combustion (**Figure 11.7**). Fluidized-bed combustion takes place at a lower temperature than regular coal burning, and fewer nitrogen oxides are produced. (Higher temperatures cause atmospheric nitrogen and oxygen to combine, forming nitrogen oxides.) Because the sulfur in coal reacts with the calcium in limestone to form calcium sulfate, which then precipitates out, sulfur is removed from the coal during the burning process, so scrubbers are not needed to remove it after combustion.

fluidized-bed combustion A clean-coal technology in which crushed coal is mixed with limestone to neutralize the acidic sulfur compounds produced during combustion.

Several large power plants in the United States now use fluidized-bed combustion. The Clean Air Act Amendments of 1990 provide incentives for utility companies to convert to cleaner technologies, such as fluidized-bed combustion. The cost of

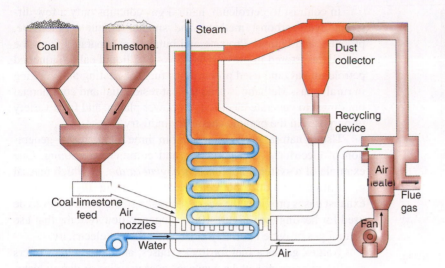

Figure 11.7   Fluidized-bed com-bustion of coal.
Crushed coal and limestone are suspended in air. As the coal burns, limestone neutralizes most of the sulfur dioxide in the coal. The heat generated during combustion converts water to steam, which powers various industrial processes.

installing fluidized-bed combustion compares favorably with the cost of installing desulfurization systems.

Fluidized-bed combustion is more efficient than traditional coal burning—that is, it produces more heat from a given amount of coal—and therefore reduces CO_2 emissions per unit of electricity produced. If improvements of this technology were developed and adopted widely by coal-burning power plants, fluidized-bed combustion could significantly reduce the amount of CO_2 released into the atmosphere. Pressurized fluidized-bed combustion is being developed as a way to reduce CO_2 as well as nitrogen and sulfur oxides. By operating fluidized-bed combustion under high pressure, complete combustion of coal occurs at low temperatures. Sulfur emissions are removed as calcium sulfate, and few nitrogen oxides form because of the low temperatures. Pressurized fluidized-bed combustion is more expensive than regular fluidized-bed combustion because it requires a costly pressurized vessel.

COAL AND CARBON DIOXIDE EMISSIONS

In 2014, the U.S. Supreme Court ruled that the EPA can regulate carbon dioxide emissions from coal power plants. While it will take some time for these regulations to take effect, it will probably mean the closure of older (and dirtier) plants. New power plants may also be built with equipment that captures and stores, or *sequesters*, carbon released when coal is burned. Currently,

carbon capture and storage (CCS) The removal of carbon from fossil-fuel combustion and storage of the carbon, usually underground.

carbon capture and storage (CCS) technology is largely untested. There is one experimental installation at a plant in Germany, about one-twentieth the size of a typical coal power plant, which could serve as a model for future installations.

No country currently requires CCS at new coal power plants. However, it is possible to build a plant with equipment that would make it relatively easy to add CCS equipment in the future. Installing CCS equipment is expensive initially, but much less expensive than retrofitting in the future if CCS becomes a requirement. In Chapter 20, we explore in more detail opportunities for reducing carbon released from burning coal.

REVIEW

1. Which type of coal mining—surface or subsurface mining—is more land-intensive?

2. What are acid mine drainage and acid deposition?

3. What are the environmental benefits of resource recovery? Of fluidized-bed combustion?

Oil and Natural Gas

LEARNING OBJECTIVES

• Define *structural trap* and give two examples.

• Explain what is meant by *peak oil* and why it might concern us.

• Discuss the environmental problems of using oil and natural gas.

• Summarize the continuing controversy surrounding oil drilling in the Arctic National Wildlife Refuge.

Although coal was the most important energy source in the United States during the early 1900s, oil and natural gas became increasingly important, particularly after the 1930s. This change occurred largely because oil and natural gas are more versatile, easier to transport, and cleaner burning than coal. In 2009, oil supplied approximately 37.3% and natural gas 24.7% of the energy used in the United States. In comparison, other U.S. energy sources included coal (20.9%), nuclear power (8.8%), and renewables (8.3%). Globally, in 2012, oil provided 33.1% of the world's energy and natural gas, 23.9%. Other major energy sources included coal (29.9%), renewables (8.6%), and nuclear power (4.5%) (**Figure 11.8**).

Petroleum, or crude oil, is a liquid composed of hundreds of hydrocarbon compounds. During petroleum refining, the compounds are separated into different products—such as gases, gasoline, heating oil, diesel oil, and asphalt—based on their different boiling points (**Figure 11.9**). Oil is also used to produce petrochemicals, compounds in such diverse products as fertilizers, plastics, paints, pesticides, medicines, and synthetic fibers.

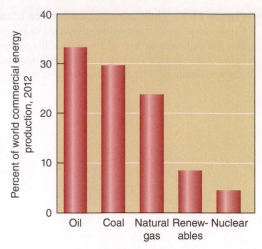

Figure 11.8   World commercial energy sources, 2012. Note the overwhelming importance of oil, coal, and natural gas as commercial energy sources. Alternatives include geothermal, solar, wind, and wood.

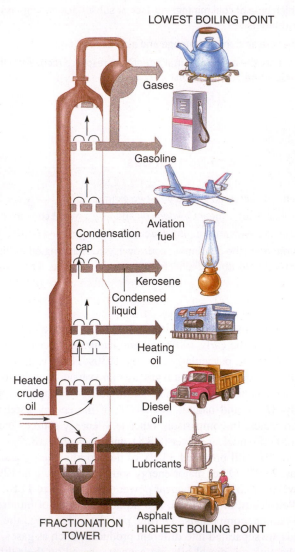

Figure 11.9   Petroleum refining. Crude oil is separated into a variety of products based on their different boiling points. After being heated, they are separated in a fractionation tower, which is about 30 m (100 ft) tall. The lower the boiling point, the higher the compounds rise in the tower.

In contrast to petroleum, natural gas contains only a few different hydrocarbons: methane and smaller amounts of ethane, propane, and butane. Propane and butane are separated from the natural gas, stored in pressurized tanks as a liquid called liquefied petroleum gas, and used primarily as fuel for heating and cooking in rural areas. Methane is used to heat residential and commercial buildings, to generate electricity in power plants, and for a variety of purposes in the organic chemistry industry.

Use of natural gas is increasing in three main areas—generation of electricity, transportation, and commercial cooling. One example of a systems approach is *cogeneration*, in which natural gas is used to produce both electricity and steam; the heat of the exhaust gases provides the energy to make steam from water to be used for heating (see Figure 10.6). Cogeneration systems that use natural gas provide relatively clean and efficient electricity.

Natural gas as a fuel for trucks, buses, and automobiles offers significant environmental advantages over gasoline or diesel: Natural gas vehicles emit up to 93% fewer hydrocarbons, 90% less carbon monoxide, 90% fewer toxic emissions, and almost no soot. Engines that use natural gas are essentially the same as those that burn gasoline. As a fuel, natural gas can be cheaper than gasoline: Individuals can install equipment to compress natural gas in their homes. Many of the compressed natural gas vehicles currently in use in the United States are fleet vehicles, including natural gas–powered transit buses in many cities and taxis in Seattle and Washington, D.C.

Natural gas efficiently fuels residential and commercial air-cooling systems. One example is the use of natural gas in a desiccant-based (air-drying) cooling system, which is ideal for supermarkets, where humidity control is as important as temperature control. Restaurants are also important users of natural gas–powered desiccant-based cooling systems.

The main disadvantage of natural gas is that deposits are often located far from where the energy is used. Because it is a gas and is less dense than a liquid, natural gas costs four times more to transport through pipelines than crude oil. To transport natural gas over long distances, it is first compressed to form liquefied natural gas (LNG), then carried on specially constructed refrigerated ships (**Figure 11.10**).

Through 2007, the United States was importing increasing amounts of LNG, with a peak of around 2,800 million cubic meters per month in late 2007. The industry was concerned that the four facilities for processing imports would be insufficient to meet future projections. However, with the increase of hydraulic fracturing, the need for imported LNG diminished substantially; by 2014, imports were down to 250 million cubic meters per month, about the same rate as in 2001.

EXPLORATION FOR OIL AND NATURAL GAS

Geologic exploration is continually under way in search of new oil and natural gas deposits, usually found together under one or more layers of rock. (Oil and natural gas tend to migrate upward until they reach an impermeable rock layer.) Oil and natural gas deposits are usually discovered indirectly by the detection of **structural traps** (**Figure 11.11**).

Plate tectonic movements sometimes cause the upward folding of sedimentary rock strata (layers); the strata that arch upward can include

structural traps
Underground geologic structures that tend to trap any oil or natural gas if it is present.

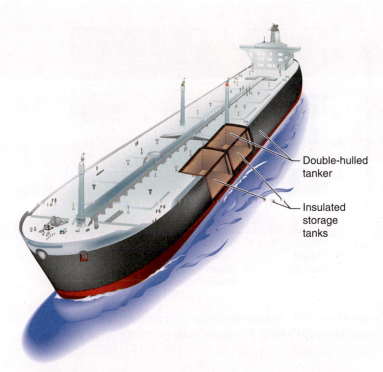

(a) Liquified natural gas is shipped in tankers that carry it in large, cylindrical tanks.

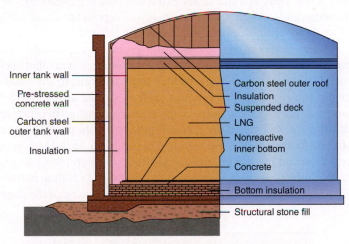

(b) On land, liquefied natural gas is stored in double-walled, insulated tanks. Major LNG accidents are unusual: As LNG heats, it forms a vapor that can burn at a high temperature, but the vapor becomes lighter than air and disperses at about −105°C (−160°F).

Figure 11.10   Liquefied natural gas (LNG) ship and tank. Because natural gas liquefies at very low temperatures, it is transported and stored below −150°C (−260°F).

both porous and impermeable rock. If impermeable layers overlie porous layers, any oil or natural gas present from a source rock such as sandstone may work its way up through the porous rock to accumulate under the impermeable layer.

Many important oil and natural gas deposits (e.g., oil deposits known to exist in the Gulf of Mexico) are found in association with salt domes, underground columns of salt. Salt domes develop when extensive salt deposits form at Earth's surface because of the evaporation of water. All surface water contains dissolved salts. The salts dissolved in ocean water are so concentrated they can

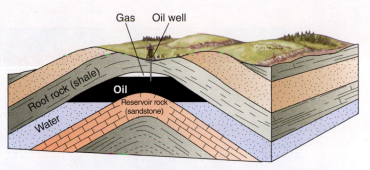

Figure 11.11   Structural trap. The most important of several kinds of structural traps is shown. These traps form when sedimentary rock strata buckle, or fold upward. Oil and natural gas seep through porous reservoir rock such as sandstone and collect under nonporous layers such as a roof of shale. Natural gas accumulates on top of the oil, which in turn floats on groundwater.

be tasted, but even fresh water contains some dissolved material. If a body of water lacks a passage to the ocean, as an inland lake often does, the salt concentration in the water gradually increases. (The Great Salt Lake in North America is an example of a salty inland body of water that formed in this way. Although three rivers empty into the Great Salt Lake, water escapes from the lake only by evaporation, accounting for its high salinity—four times higher than that of ocean water.)

If such a lake were to dry up, a massive salt deposit would remain. Layers of sediment may eventually cover such deposits and convert to sedimentary rock after millions of years. The rock layers settle, and the salt deposit, which is less dense than rock, rises in a column—a salt dome. The ascending salt dome, together with the rock layers that buckle over it, provides a trap for oil or natural gas.

Geologists use a variety of techniques to identify structural traps that might contain oil or natural gas. One method is to drill test holes in the surface and obtain rock samples. Another method is to produce an explosion at the surface and measure the echoes of sound waves that bounce off rock layers under the surface. These data are interpreted to determine whether structural traps are present. However, many structural traps do not contain oil or natural gas.

Three-dimensional seismology produces maps of oil field area and depth, enabling geologists to have a higher rate of success when drilling. Another new technology that improves oil recovery is horizontal drilling. Traditional oil wells are vertical and cannot veer off to follow the contours of underground formations that contain oil. Wells dug with horizontal drilling follow contours, and they generally yield three to five times as much oil as vertical wells.

Even with the new technologies, searching for oil and natural gas is expensive. It costs millions of dollars for the basic geologic analyses to find structural traps. And once oil or natural gas is located, drilling and operating the wells cost additional millions. However, oil companies easily recover these costs once the oil and natural gas begin to flow.

RESERVES OF OIL AND NATURAL GAS

Although oil and natural gas deposits exist on every continent, their distribution is uneven, and a large portion of total oil deposits are clustered relatively close together. Enormous oil fields

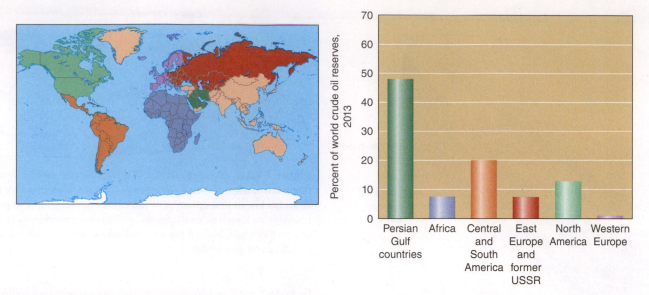

Figure 11.12   Distribution of oil deposits. The Persian Gulf region contains almost 50% of known oil deposits in a relatively small area as compared to other regions of the world. Data are presented as regional percentages of the 2013 world estimate of crude oil reserves. (The map is color-coded with the bar graph.)

containing more than half of the world's total estimated reserves are situated in the Persian Gulf region, which includes Iran, Iraq, Kuwait, Oman, Qatar, Saudi Arabia, Syria, the United Arab Emirates, and Yemen (**Figure 11.12**). In addition, major oil fields are known to exist in Venezuela, Mexico, Russia, Kazakhstan, Libya, and the United States (in Alaska and the Gulf of Mexico).

More than 40% of the world's proved recoverable reserves of natural gas are located in two countries, Russia and Iran (**Figure 11.13**). The United States has more deposits of natural gas than Western Europe, and use of natural gas is more common in North America than in Western Europe. Canada and the United States also extract coal bed methane, a form of natural gas associated with coal deposits.

It is unlikely that major new conventional oil fields will be discovered in the continental United States. However, oil in shale

deposits represents an expanded resource (see below for details about shale oil extraction). In the United States, shale oil, which only a few years ago was rarely extracted, now represents 35% of the domestic supply. Globally, shale oil represents about 10% of total supply.

Large conventional oil deposits probably exist under the continental shelves, the relatively flat underwater areas that surround continents, and in deepwater areas adjacent to the continental shelves. Despite problems such as storms at sea and the potential for major oil spills, many countries engage in offshore drilling for this oil. New technologies, such as platforms the size of football fields, enable oil companies to drill down several thousand feet for oil, making seafloor oil fields once considered inaccessible open for tapping. As many as 18 billion barrels (756 billion gallons) of oil and natural gas may exist in the deep

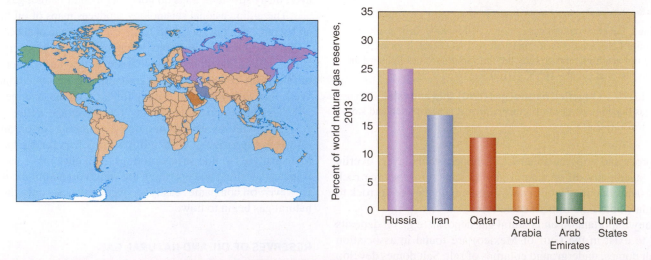

Figure 11.13   Six countries with the greatest natural gas deposits. Data are presented as percentages of the 2013 world estimate of natural gas reserves. Note that Russia and Iran together possess about 42% of the world's natural gas deposits. (The map is color-coded with the bar graph. Qatar and the United Arab Emirates, which are not shown on the map, are tiny countries east of Saudi Arabia.)

water of the Gulf of Mexico, just off the continental shelf, from Texas to Alabama. Continental shelves off the coasts of western Africa and Brazil are also promising. The oil industry is currently developing remote-controlled robots that can install and maintain underwater equipment and pipelines. Environmentalists generally oppose opening the outer continental shelves for oil and natural gas exploration because of the threat a major oil spill would pose to marine and coastal environments. Coastal industries, including fishing and tourism, also oppose oil and natural gas exploration in these areas.

How Long Will Oil and Natural Gas Supplies Last?

It is difficult to project when the world will run out of oil and natural gas, but by some estimates even with oil from shale deposits, the peak level of oil production may have already passed, meaning that global resources are in decline. In contrast, shale deposits could mean many more decades of ample, inexpensive natural gas.

We do not know how many additional oil and natural gas reserves will be discovered, nor do we know if or when technological breakthroughs will allow us to extract more fuel from each deposit. Just a decade ago oil and natural gas in shale deposits were rarely extracted. It is possible that another resource of this magnitude or greater is yet to be discovered. The answer to how long these fuels will last also depends on whether world consumption of oil and natural gas increases, remains the same, or decreases. Economic factors influence oil and natural gas availability and consumption. As reserves are exhausted, prices increase, which drives down consumption and stimulates greater energy efficiency, the search for additional deposits, and the use of alternative energy sources.

Despite adequate oil supplies for the near future, in the long term we will need other resources. Gasoline has been readily available and inexpensive in the United States for most of the past century. An exception was the brief disruption caused when the Organization of Petroleum Exporting Countries (OPEC) limited global oil supply in the early 1970s (**Figure 11.14**). During the next several decades, however, the remaining oil will become more difficult and expensive to obtain. Most experts think we will begin to have serious problems with oil supplies sometime during the 21st century.

Some experts think that global oil production has already reached **peak oil**, the point at which the oil is being withdrawn at the highest possible rate. About 80% of current production comes from oil fields discovered before 1973, and most of these fields have started to decline in production. These analysts say the world must move quickly to develop alternative energy sources because the global demand for energy will only continue to increase even as production declines.

peak oil Also known as "Hubbert's Peak," after the U.S. geologist who first developed the concept; it is the point at which global oil production has reached a maximum rate.

Industry analysts are generally more optimistic. They think that improving technology will allow us to extract more oil out of old oil fields, and are optimistic about undiscovered oil shale resources. New technologies may help us obtain oil from fields formerly unreachable (such as beneath deep-ocean waters). Improved technology may allow us to produce oil from natural gas, coal, and synfuels (discussed later in the chapter). Even so, the most optimistic predictions are for peak oil to occur late this century.

Figure 11.14   Long lines at a filling station, 1973. In 1973 the United States had a short-lived but traumatic gasoline shortage, resulting from an embargo imposed by OPEC. Currently, while gasoline prices can vary greatly, it is always available at some price. How long this ready availability will continue is uncertain; lines like this could be commonplace in the future.

Natural gas is more plentiful than oil. Experts estimate that readily recoverable reserves of natural gas, if converted into a liquid fuel, would be equivalent to between 500 billion and 770 billion barrels of crude oil, enough to keep production rising for at least 10 years after conventional supplies of petroleum have begun to decline. However, if the global use of natural gas continues to increase as it has in recent years, then its supply will not last as long as current projections predict.

Large amounts of natural gas have recently been discovered in shale formations around the world. Removing shale gas can be more expensive and environmentally disruptive than the resources we have been accessing, but shale gas extraction is taking place in an increasing number of places around the country (See Case in Point: The Marcellus Shale).

CASE IN POINT

THE MARCELLUS SHALE

Estimated known reserves of natural gas in the United States have increased in the past few years, from 192 trillion cubic feet (TCF) in 2001 to 2431 TCF in 2013, the most recent year on record. Most of this is due to the discovery of natural gas in large shale deposits

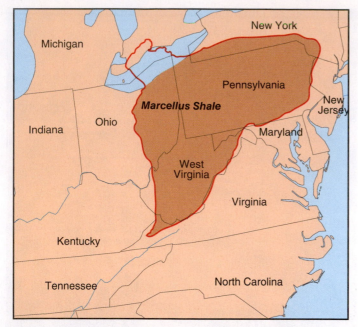

Figure 11.15   Marcellus Shale. The Marcellus Shale formation underlies major sections of 6 states, and parts of several others. The formation ranges from 10 to 120 meters thick and may contain as much as 400 trillion cubic feet of natural gas. How will increased natural gas production in the United States affect our energy security? Our contribution to climate change?

Source: US Bureau of Land Management.

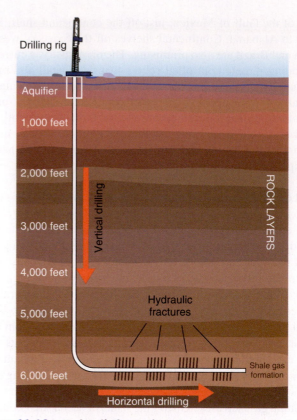

Figure 11.16   Hydraulic fracturing. In order to extract gas from shale, a deep vertical well is drilled, with a horizontal hole bored into the rock layer of interest. Water is then pumped into the well. The pressure breaks natural gas and other hydrocarbons free from the shale; the gas is then extracted from the wellhead at the top. The well must be carefully lined where it crosses an aquifer (thin blue line near top) to prevent contamination.

around the country. For example, the *Marcellus Shale* is found under major parts of six states: Maryland, New York, Pennsylvania, Ohio, Virginia, and West Virginia, and small parts of several others (**Figure 11.15**). Major shale gas deposits have been found beneath more than half of the 48 continental United States. The Marcellus Shale may contain as much as 400 TCF of natural gas; many see this resource as an opportunity to reduce our reliance on imported energy resources and shift away from more environmentally damaging coal.

Shale gas is much more difficult to extract than is gas found above oil in sandstone deposits, where natural gas often vents to the atmosphere if not contained. In contrast, gas must be freed from shale, which means that after a well is drilled, water is pumped down and used to break the shale apart (**Figure 11.16**).

Hydraulic fracturing
Extraction of natural gas that is tightly bound in shale deposits by applying chemicals and water under high pressure.

This process, called **hydraulic fracturing** (sometimes referred to as *fracking*), uses high-pressure water to open or widen gaps that allow natural gas flow. Hydraulic fracturing produces wastewater with high concentrations of salts and hydrocarbons; the wastewater can also include toxic metals and radioactive uranium. While technologies are being developed to recycle the wastewater in place, the standard way to dispose of it now is to pump it back into the ground. In many places, the water must be trucked hundreds of kilometers to suitable disposal sites.

Shale gas deposits are also often found much deeper than are the resources we have already accessed. Coupled with the challenge of extracting gas from shale, this means that drilling, fracturing, and pumping are more intense, and wells must be located closer

together than most current natural gas wells. Consequently, shale gas removal can be environmentally and socially disruptive, as shale gas removal wells are dug near communities, farms, and sensitive natural areas. Even when operated correctly, shale gas extraction can be noisy and odorous; it can also increase traffic and adversely impact roads and bridges. Spills or leaks of hydraulic fracturing liquid can contaminate groundwater or surface water.

As interest in removing gas from the Marcellus Shale increases, state and local governments have introduced a variety of policies and practices. New York and Pennsylvania offer an interesting contrast: While the two states have roughly equivalent amounts of gas that could be removed, New York has been very cautious, with a moratorium on new drilling at the time this book goes to press. In contrast, while some municipalities in Pennsylvania have restricted drilling, there were over 1,000 wells at the end of 2010, and over 6,000 by early 2014. In some places, communities work with drilling companies to plan infrastructure development and traffic management in anticipation of shale gas extraction, while in others, the interactions are much more adversarial. ■

GLOBAL OIL DEMAND AND SUPPLY

One difficult aspect of the oil market is that the world's major oil producers are not its major oil consumers. In 2012, North America and Western Europe consumed 41.8% of the world's total petroleum (down from 50.0% in 2006), yet these same countries produced only 24.4% of the world's crude oil. In contrast, the Persian Gulf region consumed 8.6% of the world's petroleum but produced 30.8% of the world's crude oil. The United States currently imports about one-third of its oil, down from two-thirds just a decade ago.

The imbalance between oil consumers and oil producers will probably worsen in the future because the Persian Gulf region has much higher proven reserves than other countries. At current rates of production, North America's oil reserves will run out decades before those of the Persian Gulf nations, which have 65% of the known world oil reserves and may produce oil at current rates for perhaps a century. This dependence of the United States and other countries on Middle Eastern oil has potential international security implications as well as economic impacts.

Extracting U.S. oil resources more quickly is often suggested as a way to reduce reliance on foreign oil. For example, significant reserves occur off the coast of the United States and in the Arctic National Wildlife Refuge. However, in global markets domestic oil is indistinguishable from foreign oil. So while extraction of domestic oil may increase short-term supplies, it cannot have much impact on the percentage of petroleum that the United States buys from abroad.

ENVIRONMENTAL IMPACTS OF OIL AND NATURAL GAS

Two sets of environmental problems are associated with the use of oil and natural gas: the problems that result from burning the fuels (combustion) and the problems involved in obtaining them (production and transport). We have already mentioned the CO_2 emissions that are a direct result of the combustion of fossil fuels. As with coal, the burning of oil and natural gas produces CO_2. Every gallon of gasoline you burn in your automobile releases an estimated 9 kg (20 lb) of CO_2 into the atmosphere. As CO_2 accumulates in the atmosphere, it insulates the planet, preventing heat from radiating back into space. The global climate is warming more rapidly now than it did during any of the warming periods following the ice ages, and the environmental impact of rapid global climate change could cause substantial human suffering in the future.

Another negative environmental impact of burning oil is acid deposition. Although burning oil does not produce appreciable amounts of sulfur oxides, it does produce nitrogen oxides, mainly through gasoline combustion in automobiles, which contributes approximately half the nitrogen oxides released into the atmosphere. (Coal combustion is responsible for the other half.) Nitrogen oxides contribute to acid deposition and, along with unburned gasoline vapors, the formation of photochemical smog. Poorly tuned engines and diesel-burning vehicles also contribute particulate matter—small particles that are inhaled and cause lung damage and disease (see Chapter 19).

The burning of natural gas, on the other hand, does not pollute the atmosphere as much as the burning of oil. Natural gas is a relatively clean, efficient source of energy that contains almost no sulfur, a contributor to acid deposition. In addition, natural gas produces far less CO_2, fewer hydrocarbons, and almost no particulate matter, as compared to oil and coal.

One of the concerns in oil and natural gas production is the environmental damage that may occur during their transport, often over long distances by pipelines or ocean tankers. A serious spill along the route creates an environmental crisis, particularly in aquatic ecosystems. One of the worst oil spills in Europe's history occurred in 2002 when the oil tanker *Prestige* broke up off the coast of Spain, contaminating hundreds of kilometers of coastline and bringing the large fishing industry there to a halt.

Two extremely damaging oil spills have taken place off the coasts of the United States—in the Gulf of Mexico and in Prince William Sound. Wars and civil unrest have also led to large oil spills, including the largest—during the 1991 Persian Gulf War.

Deepwater Horizon Oil Spill.   On April 22, 2010, the Deepwater Horizon, a drilling platform in the Gulf of Mexico, exploded. As the platform collapsed, its equipment detached from the Macondo oil well, 1,500 m (about 6,000 ft) below the surface and 90 km (50 mi) from the southern coast of the United States. Massive quantities of oil began to gush from the well, spreading across the seafloor and up to the surface. British Petroleum (the company that owned the platform) used a variety of strategies to try to seal the wellhead. After many unsuccessful attempts, the flow of oil was finally stopped in mid-July of 2010.

Between the explosion and July 15, when flow was completely stopped, around 5 million barrels (210 million gallons) of oil gushed from the well. Lighter than water, most of this oil rose to the surface, where wind and currents spread it widely. At its maximum, the oil slick covered an area of nearly 75,000 km^2 (29,000 mi^2). Oil reached the southern coast of Mississippi by early May and the Louisiana coast shortly thereafter (**Figure 11.17a**). Several techniques were used to remove the oil from the ocean surface and shores, including deploying a fleet of over 6,000 ships and boats to scoop up or divert the oil, setting fires to burn oil, and applying dispersants that act like dish soap to prevent the oil from concentrating.

Deepwater platforms are at least 300 m (about 1000 ft) above the ocean surface. There are at least 127 deepwater oil fields in the Gulf of Mexico. In 1990, only about 20,000 barrels of oil were being produced each day from these deepwater fields; by 2007, this had reached nearly 1 million barrels per day. A series of decisions by the U.S. government exempted much of the drilling in these areas from the National Environmental Protection Act (see Chapter 2).

Deepwater oil drilling has a much greater potential for oil spills than does shallow-water drilling, because of both the difficulty in building and maintaining the connection to the well and, as was learned from the Deepwater Horizon event, the extreme challenge of plugging a spill so far below the ocean surface. A blow-out-prevention device, which was designed to stop flow in case of an emergency, failed, as did multiple attempts to cap, bypass, cut off, and bury the wellhead.

The spill killed thousands of plants, plankton, invertebrates, mammals, fishes, and birds (**Figure 11.17b**). Shrimping, a major industry off the coasts of Louisiana and Mississippi, was restricted, although some illegal shrimping continued. Impacted areas included marine sanctuaries and protected wetlands. Tourism was also disrupted: Potential visitors canceled their trips either because the beaches were covered with oil or because they were uncertain about beach contamination.

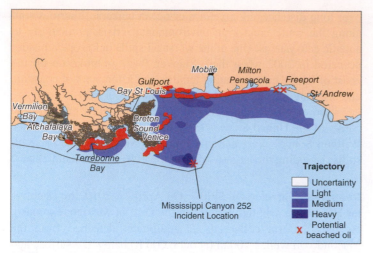

(a) The extent of the spill across the southern coast of the United States on June 29, 2011, when its impact on the shore was greatest.

Trajectory
☐ Uncertainty
☐ Light
☐ Medium
☐ Heavy
✕ Potential beached oil

Mississippi Canyon 252 Incident Location

Carolyn Cole/Photoshot

(b) Birds were among the many organisms that were harmed by the spill.

Figure 11.17   **Deepwater Horizon oil spill.**

CHRIS WILKINS/AFP/Getty Images

(a) An aerial view of the massive oil slick at the southwest end of Prince William Sound.

© Natalie Fobes/Corbis

(b) Workers spray hot water from a crane on the oily cliffs of Eleanor Island, Alaska, USA, after the Exxon Valdez oil spill, 1989.

(c) The extent of the spill (black arrows). Water currents caused the spill to spread rapidly over hundreds of kilometers, killing countless animals, such as sea otters and birds. The Arctic National Wildlife Refuge is located in the northeastern part of Alaska, close to the Trans-Alaska Pipeline, which begins at Prudhoe Bay and extends south to Valdez. The National Petroleum Reserve–Alaska is also shown.

Figure 11.18   *Exxon Valdez* **oil spill, 1989.**

Exxon Valdez **Oil Spill.**   In 1989, the supertanker *Exxon Valdez* hit Bligh Reef and spilled 260,000 barrels (10.9 million gallons) of crude oil into Prince William Sound along the coast of Alaska, creating the largest oil spill from a tanker in U.S. history. As it spread, the oil eventually covered thousands of square kilometers of water and contaminated hundreds of kilometers of shoreline (**Figure 11.18**). According to the U.S. Fish and Wildlife Service and the Alaska Department of Environmental Conservation, more than 30,000 birds (sea ducks, loons, cormorants, bald eagles, and other species) and between 3,500 and 5,500 sea otters died as a result of the spill. The area's killer whale and harbor seal populations declined, and salmon migration was disrupted. Throughout the area, there was no fishing season that year.

Within hours of the spill, scientists began to arrive on the scene to advise both the Exxon Corporation and the government on the best way to try to contain and clean up the spill. But it took much longer for any real action to occur. Eventually, nearly 12,000 workers took part in the cleanup; their activities included mechanized steam cleaning and rinsing, which killed additional shoreline organisms such as barnacles, clams, mussels, eelgrass, and rockweed.

In late 1989, Exxon declared the cleanup "complete." But it left behind, among other problems, contaminated shorelines, particularly rocky coasts, marshes, and mudflats; continued damage to some species of birds (such as the common loon and harlequin duck), fishes (such as murrelet and rockfish), and mammals (such as the harbor seal); and a reduced commercial salmon catch.

One positive outcome of the disaster was passage of the **Oil Pollution Act** of 1990. This legislation establishes liability for damages to natural resources resulting from a catastrophic oil spill, including a trust fund that pays to clean up spills when the responsible party cannot; a tax on oil provides money for the trust fund. The Oil Pollution Act requires double hulls on all oil tankers that enter U.S. waters by 2015. Had the *Exxon Valdez* possessed a double hull, the disaster might not have occurred because only the outer hull might have broken.

The Largest Global Oil Spill   The world's most massive oil spill occurred in 1991 during the Persian Gulf War, when about 6 million barrels (250 million gallons) of crude oil—more than 20 times the amount of the *Exxon Valdez* spill—were deliberately dumped into the Persian Gulf. Many oil wells were set on fire, and lakes of oil spilled into the desert around the burning oil wells. Cleanup efforts along the coastline and in the desert were initially hampered by the war. In 2001, Kuwait began a massive remediation project to clean up its oil-contaminated desert. Progress is slow, and it may take a century or more for the area to completely recover.

CASE IN POINT

THE ARCTIC NATIONAL WILDLIFE REFUGE

The proposed opening of the Arctic National Wildlife Refuge to oil exploration has been a major environment-versus-economy conflict off and on since 1980. On one side are those who seek to protect rare and fragile natural environments; on the other side are those whose higher priority is the development of major U.S. oil supplies.

The refuge, called America's Serengeti, is home to many animal species, including polar bears, arctic foxes, peregrine falcons, musk oxen, Dall sheep, wolverines, and snow geese. It is the calving area for a large migrating herd of caribou: The Porcupine caribou herd contains more than 150,000 head. Dominant plants in this coastal plain of tundra include mosses, lichens, sedges, grasses, dwarf shrubs, and small herbs. Under a thin upper layer of soil is the permafrost layer, which contains permanently frozen water. Although it is biologically rich, the tundra is an extremely fragile ecosystem, in part because of its harsh climate. The organisms living here have adapted to their environment, but any additional stress has the potential to harm or even kill them. Thus, arctic organisms are particularly vulnerable to human activities.

History of the Arctic National Wildlife Refuge   In 1960, Congress declared a section of northeastern Alaska protected because of its distinctive wildlife. In 1980, Congress expanded this wilderness area to form the Arctic National Wildlife Refuge. The Department of the Interior was given permission to determine the potential for oil discoveries in the area, but exploration and development could proceed only with congressional approval.

Pressure to open the refuge to oil development subsided for about five years following the Alaskan oil spill, when public sentiments were strongly against oil companies. In the mid-1990s, however, pro-development interests became more vocal, partly because in 1994, for the first time in its history, the United States imported more than half the oil it used. Although the Department of the Interior concluded that oil drilling in the refuge would harm the area's ecosystem, both the Senate and the House of Representatives passed measures to allow it. (President Clinton vetoed the bill.)

In 2001, President George W. Bush announced his support for opening the refuge to oil drilling, but after contentious debate, the Senate voted in 2005 against doing so. Developing Alaskan oil resources, as well as resources off the shores of coastal states, was again a prioritiy of the Republican platform in the 2008 and to a lesser extent the 2012 presidential elections.

Support for and Opposition to Oil Exploration in the Refuge   Supporters cite economic considerations as the main reason for drilling for oil in the refuge. The United States spends a large proportion of its energy budget to purchase foreign oil. Development of domestic oil could, for a decade or so, improve the balance of trade and make us less dependent on foreign countries for our oil.

The oil companies are eager to develop this particular site because it is near Prudhoe Bay, where large oil deposits are already being tapped. (To date, Prudhoe Bay has produced about 14 billion barrels of crude oil.) Prudhoe Bay has a sprawling industrial complex to support oil production, including roads, pipelines, gravel pads, and storage tanks. The Prudhoe Bay oil deposits peaked in production in 1985 and have declined in productivity since then. As a result, the oil industry is looking for sites that can use the infrastructure already in place.

Conservationists think oil exploration poses permanent threats to the delicate balance of nature in the Alaskan wilderness, in exchange for a temporary oil supply. Further, they point out that using domestic oil is a short-term fix and will in the long run lead to greater dependence on foreign oil. They prefer investing in renewable energy sources and energy conservation—permanent solutions to the energy problem.

Studies, such as one by the U.S. Fish and Wildlife Service, document considerable habitat damage and declining numbers of wolves and bears in the Prudhoe Bay area. (Top predators are usually more susceptible to environmental disruption than are organisms occupying lower positions in a food web.) Because it is not financially practical to restore developed areas in the Arctic to their natural states, development in the Arctic causes permanent changes in the natural environment. ■

REVIEW

1. What are two examples of structural traps?
2. Why do estimates of peak oil vary?
3. What are three environmental problems associated with using oil and natural gas as energy resources?
4. What is the controversy surrounding the Arctic National Wildlife Refuge?

Synfuels and Other Potential Fossil-Fuel Resources

LEARNING OBJECTIVES

- Define *synfuel* and distinguish among tar sands, oil shales, gas hydrates, liquid coal, and coal gas.
- Briefly consider the environmental implications of using synfuels.

synfuel A liquid or gaseous fuel synthesized from coal or other naturally occurring resources and used in place of oil or natural gas.

Synfuels (short for synthetic fuels), materials similar in chemical composition to oil or natural gas, have long been considered possible future sources of fossil fuels. Synfuels include tar sands, oil shales, gas hydrates, liquefied coal, and coal gas. Although more expensive to produce than oil and natural gas, synfuels are becoming economically competitive as fuel prices rise and technology improves.

Tar sands, or oil sands, are underground sand deposits permeated with **bitumen**, a thick, asphalt-like oil. The bitumen in tar sands deep in the ground cannot be pumped out unless it is heated underground with steam to make it more fluid. Once bitumen is obtained from tar sands, it must be refined like crude oil. World tar sand reserves are estimated to contain half again as much fuel as world oil reserves. Major tar sands are found in Venezuela and in Alberta, Canada, where an estimated 300 billion barrels of oil occur in tar sands. Canadian mines are currently producing almost 300 million barrels of oil a year from oil sands.

Pioneers in the American West discovered "oily rocks" when their rock hearths caught fire and burned. Oil shales are sedimentary rocks containing a mixture of hydrocarbons known collectively as **kerogen**. Oil shales are crushed and heated to yield their oil, and the kerogen is refined after it is mined. Only recently has removing oil from shale become cost effective. Large oil shale deposits are located in Australia, Estonia, Brazil, Sweden, the United States, and China. Wyoming, Utah, and Colorado have the largest deposits in the United States. Like tar sands, oil shale reserves may contain half again as much fuel as world oil reserves.

Gas hydrates, also called methane hydrates, are reserves of ice-encrusted natural gas located deep underground in porous rock. Massive deposits are found in the arctic tundra, deep under the permafrost, and in the deep-ocean sediments of the continental slope and ocean floor. Until recently, the U.S. oil industry was not particularly interested in extracting natural gas from gas hydrates because of the expense involved. Several oil companies are currently developing methods to extract gas hydrates. Countries with lots of gas hydrates (e.g., Russia) or with few conventional fossil-fuel deposits (e.g., Japan) have established national gas hydrates programs.

A nonalcohol liquid fuel similar to oil can be produced from coal. It is less polluting than solid coal, but not as clean as oil. **Coal liquefaction** was developed before World War II, but its expense prevented it from replacing gasoline production. Technological improvements have lowered the cost of coal liquefaction. While coal liquefaction is still not cost-competitive with gasoline, there is a major push in the United States to increase production in the near future.

Another synfuel is a gaseous product of coal. **Coal gas** has been produced since the 19th century. As a matter of fact, it was a major fuel used for lighting in American homes until oil and natural gas replaced it during the 20th century. **Coal gasification** is production of the combustible gas methane from coal by reacting it with air and steam (**Figure 11.19**). Several demonstration power plants that convert coal into gas have been constructed in the United States. One advantage of coal gas over solid coal is that coal gas burns almost as cleanly as natural gas. Scrubbers are not needed when coal gas is burned because sulfur is removed during coal gasification. Like other synfuels, coal gas is currently more expensive to produce than fossil fuels.

ENVIRONMENTAL IMPACTS OF SYNFUELS

Although synfuels are promising energy sources, they have many of the same undesirable effects as fossil fuels. Their combustion releases enormous quantities of CO_2 and other pollutants into the atmosphere, thereby contributing to global warming and air

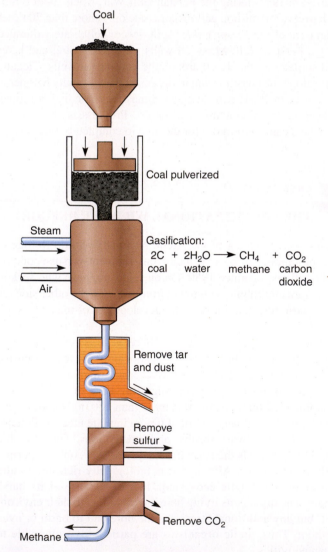

Coal

Coal pulverized

Steam

Air

Gasification:
$$2C + 2H_2O \longrightarrow CH_4 + CO_2$$
coal   water   methane   carbon dioxide

Remove tar and dust

Remove sulfur

Remove CO_2

Methane

Figure 11.19 **Coal gasification.** Shown is one method of coal gasification, in which the combustible gas methane is generated from coal. To follow the steps in coal gasification, start at the top of the figure and work your way down. (Adapted from R. A. Hinrichs.)

pollution. Some synfuels, such as coal gas, require large amounts of water during production and are of limited usefulness in arid areas, where water shortages are already commonplace. Also, enormous areas of land would have to be surface-mined to recover the fuel in tar sands and oil shales.

REVIEW OF LEARNING OBJECTIVES WITH SELECTED KEY TERMS

- **Define *fossil fuel* and distinguish among coal, oil, and natural gas.**

Fossil fuels are combustible deposits in Earth's crust, composed of the remnants of prehistoric organisms that existed millions of years ago. Fossil fuels are nonrenewable resources; Earth has a finite supply of fossil fuels that are depleted by use. **Coal** is a black combustible solid formed from the remains of ancient plants that lived millions of years ago. **Oil** is a thick, yellow to black, flammable liquid hydrocarbon mixture. **Natural gas** is a mixture of gaseous hydrocarbons (primarily methane) that often occurs with oil deposits.

- **Describe the processes that formed coal, oil, and natural gas.**

Coal was formed when partially decomposed plant material was exposed to heat and pressure for aeons, forcing out water and concentrating energy in chemical bonds. Oil and natural gas formed when countless microscopic aquatic organisms died and settled in oxygen-deficient sediments.

- **Relate fossil fuels to the carbon cycle.**

The energy in fossil fuels was captured through photosynthesis. Carbon was captured over a long period and has been stored in fossil fuels for millions of years. Over the past century, much of that carbon has been released into the atmosphere through combustion, a trend that is increasing worldwide.

- **Distinguish between surface mining and subsurface mining.**

Surface mining is the extraction of mineral and energy resources near Earth's surface by first removing the soil, subsoil, and overlying rock strata. **Subsurface mining** is the extraction of mineral and energy resources from deep underground deposits. In the United States, surface mining accounts for 60% of the coal mined, and subsurface mining accounts for 40%.

- **Summarize the environmental problems associated with using coal.**

Surface mining destroys existing vegetation and topsoil. As with all fossil fuels, the combustion of coal produces several pollutants—in particular, large amounts of the greenhouse gas carbon dioxide. **Acid mine drainage** and **acid deposition** are two significant types of pollution associated with coal mining and combustion. Burning coal releases CO_2 to the atmosphere, contributing to global climate disruption and ocean acidification.

- **Define *resource recovery* and *fluidized-bed combustion*.**

Resource recovery is the process of removing any material from polluted emissions or solid waste and selling it as a marketable product. **Fluidized-bed combustion** is a technology in which crushed coal is mixed with limestone to neutralize the acidic sulfur compounds produced during combustion.

- **Define *structural trap* and give two examples.**

Structural traps are underground geologic structures that tend to trap any oil or natural gas present. Structural traps include upward foldings of rock strata and salt domes (underground columns of salt).

- **Explain what is meant by *peak oil* and why it might concern us.**

Peak oil is the point at which global oil production has reached a maximum rate. Once that peak is passed, less and less oil will be removed each year. By some estimates, we are just about at that peak; by others, peak oil remains decades away.

- **Discuss the environmental problems of using oil and natural gas.**

Oil exploration and extraction are a threat to environmentally sensitive areas. Oil spills can occur during extraction or transportation, creating environmental crises. Removing natural gas from shale using **hydraulic fracturing** creates large amounts of contaminated wastewater, and creates noise and dust; moving equipment and water also damages infrastructure. CO_2 emissions released when oil and natural gas are burned contribute to global climate warming. Production of nitrogen oxides when oil is burned contributes to acid deposition.

- **Summarize the continuing controversy surrounding oil drilling in the Arctic National Wildlife Refuge.**

Supporters of drilling in the Arctic National Wildlife Refuge say that development of domestic oil would improve the balance of trade and make us less dependent on foreign countries for our oil. Conservationists think oil exploration poses permanent threats to the delicate balance of nature in the Alaskan wilderness, in exchange for a temporary (and probably relatively small) oil supply.

- **Define *synfuel* and distinguish among tar sands, oil shales, gas hydrates, liquid coal, and coal gas.**

Synfuel is a liquid or gaseous fuel that is synthesized from coal and other naturally occurring resources and used in place of oil or natural gas. Tar sands are underground sand deposits permeated with **bitumen**, a thick, asphalt-like oil. Oil shales are sedimentary rocks containing a mixture of hydrocarbons known collectively as **kerogen**. **Gas hydrates** are reserves of ice-encrusted natural gas located deep underground in porous rock. Coal liquid is a liquid fuel similar to oil, produced from coal by the process of **coal liquefaction**. Another synfuel, **coal gas**, is a gaseous product of coal.

- **Briefly consider the environmental implications of using synfuels.**

Synfuels have many of the same undesirable effects as fossil fuels. Their combustion releases enormous quantities of CO_2 and other pollutants into the atmosphere, thereby contributing to global warming and air pollution. Some synfuels, such as coal gas, require large amounts of water during production and are of limited usefulness in arid areas. Enormously large areas of land would have to be surfacemined to recover the fuel in tar sands and oil shales.

Critical Thinking and Review Questions

1. Describe some advantages and disadvantages of the three types of fossil fuels.

2. The Industrial Revolution may have been concentrated in the Northern Hemisphere because coal is located there. What is the relationship between coal and the Industrial Revolution?

3. Which contributes the most to climate change per unit of useful energy: coal, oil, or natural gas?

4. Few countries in Africa have significant amounts of coal, oil, or natural gas resources. What does this suggest about opportunities for economic development in those countries?

5. Which of the following do you think could most effectively reduce the use of oil in the United States: reducing subsidies on fossil fuels, changing the design of cities, or requiring vehicles to be more efficient? Explain.

6. What impact does drilling for oil in the United States have on short-term U.S. energy supplies? Long-term supplies?

7. How does extraction of natural gas from shale differ from extraction from sandstone?

8. On the basis of what you have learned about coal, oil, and natural gas, which fossil-fuel resources (if any) do you think the United States should exploit during the next 20 years? Explain your rationale.

9. Explain why the U.S. Department of Energy has described coal as a "true measure of the energy strength of the United States." Is this also true of China? India? Why or why not?

10. Which of the negative environmental impacts associated with fossil fuels is most serious? Why?

11. Which major consumer of oil is most vulnerable to disruption in the event of another energy crisis: electric power generation, motor vehicles, heating and air conditioning, or industry? Why?

12. What are the implications of peak oil on future global energy supplies?

13. What is the relationship between fossil fuels and greenhouse gases?

14. Do you think we should permit more oil drilling off the coast of the United States? Why or why not?

15. Some environmental analysts think that much of the conflict in Iraq has been related in part to gaining control over the supply of Iraqi oil. Do you think this is plausible? Explain why or why not.

16. What are the five kinds of synfuels? Why are they not being used more extensively?

17. Distinguish among fluidized-bed combustion, coal liquefaction, and coal gasification.

18. Fossil fuels are nonrenewable resources. Why is this a problem from a systems perspective? (*Hint*: Irreversible change disrupts system stability.)

19. The figure below shows projected annual energy use for the United States, India, and China. How much is each country's energy use expected to increase between 2015 and 2035? What impact will this have on total global carbon dioxide emissions if all of the increase comes from fossil fuels?

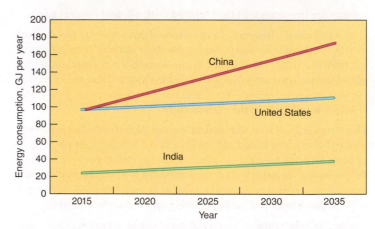

Projected energy use: China, United States, and India.

Food for Thought

In other chapters, we think about the role of energy resources in planting, irrigating, harvesting, processing, and transporting food. But fossil fuels play another important role in agriculture: crude oil is the basis of many of the synthetic pesticides and fertilizers used to produce food on a large scale. Research the pesticides and fertilizers used in food production either where you live or for the foods you eat. Why are they used? What are the alternatives? Which are produced using petroleum as a base? What are the other sources of pesticides and fertilizers?

12

Renewable Energy and Nuclear Power

Array of solar panels in rural Kenya. The panels shown here convert solar energy into electricity, which can then be stored in a battery and used to run lights and small appliances.

As countries around the world seek to improve their standards of living, one option they have is to emulate the fossil-fuel-intensive practices in countries like the United States. But from a systems perspective, it makes better sense for them to ask: "What services do we want from energy, and what combination of characteristics—good and bad—should we be concerned about?" This systems perspective has led researchers to think about appropriate energy sources and uses at different locations around the globe.

In the University of California at Berkeley's Renewable and Appropriate Energy Lab (RAEL), Professor **Daniel M. Kammen** and his students research this topic as well as develop and promote energy solutions. They have found, for example, a thriving market for small-scale solar panels in Kenya (**see photograph**). Several vendors sell units that include an inexpensive solar panel and a rechargeable battery. The battery charges during the day and can be used to run small appliances—radio, television, lights. Such units require neither fossil-fuel-powered generators nor expensive electrical power lines. From a systems perspective, small-scale solar meets the needs of rural Kenyans.

A 2013 RAEL project focuses on energy alternatives in Sarawak, Malaysia. The report finds that it would cost less to produce energy using hybrid energy systems that include small-scale hydroelectricity, conversion of rice husks into biogas (similar to natural gas), and batteries to store electricity. They concluded that such systems would work better than current reliance on diesel generators and would be more reliable and less environmentally disruptive than large-scale hydroelectric dams.

In developing countries, which tend not to have energy-intensive economies, selecting appropriate energy technologies may help avoid some of the downfalls of the energy-intensive lifestyles typical in highly developed countries. Rather than building large coal-fired power plants, Kenyans can adopt small-scale solar electric generators. Elsewhere, small-scale windmills, micro-hydro (small-scale hydroelectric power generation), and other technologies may avoid the need to build unwieldy, inefficient energy infrastructure. And the lesson can be learned in reverse as well: The technology that has made the solar option affordable in Kenya can be repackaged for use in the United States and other countries with fossil-fuel-dependent economies.

Like fossil fuels, which trapped solar energy millions of years ago, many renewable sources originate as solar energy. Winds are driven largely by differential heating of Earth's surface. Hydropower relies on the hydrologic cycle, in which evaporation at low altitudes leads to precipitation at higher altitudes. Biomass comes from growing plants.

Geothermal power, by contrast, is more closely related to nuclear energy: Heat in Earth's interior comes close enough to the surface that it can be used to generate steam.

In this chapter, we explore the range of renewable energy resources available, as well as nuclear power.

In Your Own Backyard
Does your school produce any of its own energy from renewable resources?

Direct Solar Energy

LEARNING OBJECTIVES

- Distinguish between active and passive solar heating and how each is used.
- Contrast the advantages and disadvantages of solar thermal electric generation and photovoltaic panels for converting solar energy into electricity.

The sun produces a tremendous amount of energy, of which only a small portion reaches Earth. Solar energy differs from fossil and nuclear fuels in that it is perpetually available. Solar energy is dispersed over Earth's entire surface rather than concentrated in highly localized areas, as are coal, oil, and uranium deposits. Consequently, to make solar energy useful, we must collect it.

Solar radiation varies in intensity depending on the latitude, season of the year, time of day, and cloud cover. Areas at lower latitudes—closer to the equator—receive more solar radiation annually than do latitudes closer to the North and South Poles. More solar radiation is received during summer than during winter because the sun is directly overhead in the summer and lower on the horizon in the winter. Solar radiation is more intense when the sun is high in the sky (noon) than when it is low in the sky (dawn or dusk). Clouds both scatter incident light and absorb some of the sun's energy, thereby reducing its intensity. The southwestern United States, with its lack of cloud cover and lower latitude, receives the greatest amount of solar radiation annually, whereas the Northeast receives the least (**Figure 12.1**).

Technology for using solar energy directly has been around for millennia. Traditional adobe homes in the southwestern United States, for example, are designed to remain relatively cool in summer and warm in winter (**Figure 12.2**). However, many modern buildings are designed with gas or electric heating and

Figure 12.2 **Adobe home in the southwestern United States.** Adobe has been used for centuries in the southwestern United States, and similar materials can be found in hot, arid regions around the world. Thick walls and carefully placed windows allow the temperature of the building's interior to remain relatively stable through the day and across the seasons.

air conditioning, without attention to the potential advantages of direct solar heating (or shading to avoid heat in summer). Many new building designs rediscover energy-saving benefits of traditional designs.

One example of new technologies to capture solar energy directly is the solar cookers being used by an estimated 1 million people in the rural areas of Africa, Central America, India, and

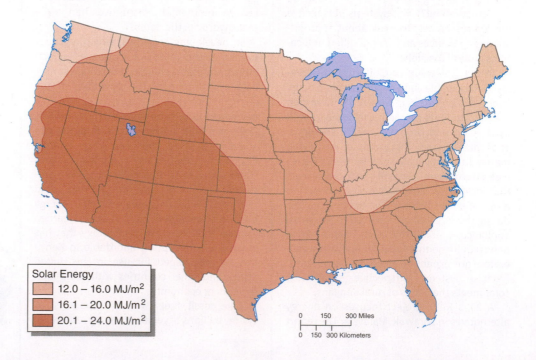

Figure 12.1 **Solar energy distribution over the United States.** This map shows the average daily total of solar energy (on an annual basis) received on a solar collector that tilts to compensate for latitude. The units are in megajoules per square meter. The Southwest is the best area in the United States for year-round solar energy collection. (U.S. Department of Energy)

Solar Energy
- 12.0 – 16.0 MJ/m^2
- 16.1 – 20.0 MJ/m^2
- 20.1 – 24.0 MJ/m^2

0   150   300 Miles

0   150 300 Kilometers

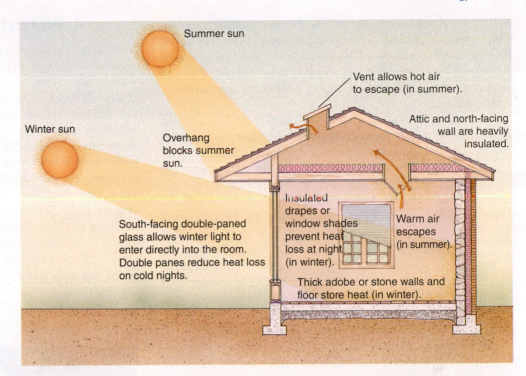

Figure 12.3 Passive solar heating designs. How does passive solar heating reduce climate change?

(a) Several passive designs are incorporated into this home.

China. Recent designs for solar cookers transmit solar light into the cooker; a glass cover prevents transmission of the infrared wavelengths (heat) that would normally escape from the cooker. Pots containing food are placed inside a box on a black metal plate. Solar cookers can reach temperatures of 180°C (just over 350°F) and can boil, bake, simmer, and sauté foods. In average sunlight, a person can cook a full meal in two to four hours.

HEATING BUILDINGS AND WATER

You have probably noticed that the air inside a car sitting in the sun with its windows rolled up becomes much hotter than the surrounding air. Similarly, the air inside a greenhouse remains warmer than the outside air during cold months. This kind of warming occurs partly because the material—such as glass—that envelops the air inside the enclosure is transparent to visible light but impenetrable to heat. Visible light from the sun penetrates the glass and warms the surfaces of objects inside, which, in turn, give off **infrared radiation**—invisible waves of heat. Heat does not escape because infrared radiation cannot penetrate glass, and the air within the glass grows continuously warmer.

passive solar heating
A system of putting the sun's energy to use without requiring mechanical devices to distribute the collected heat.

In **passive solar heating**, solar energy heats buildings without the need for pumps or fans to distribute the heat. Certain design features are incorporated into a passive solar heating system to warm buildings in winter and help them remain cool in summer (**Figure 12.3**). In the Northern Hemisphere, large south-facing windows receive more total sunlight during the day than windows facing in other directions. The sunlight entering through the windows provides heat that is then stored in floors and walls made of concrete or stone, or in containers of water. This stored heat is transmitted throughout

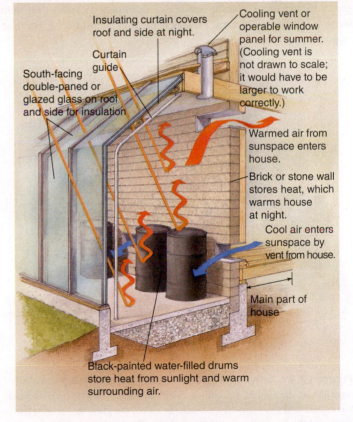

(b) A solar sunspace can be added to existing homes.

the building naturally by convection, the circulation that occurs because warm air rises and cooler air sinks.

Buildings with passive solar heating systems must be well insulated so that accumulated heat does not escape. Depending on the building's design and location, passive heating saves as much as 50% of heating costs.[1] One reason that more buildings are not constructed with passive solar heating features is that they cost a bit more to build than do traditional designs. Often, builders cannot recover costs of features if energy savings cannot offset upfront costs within a few years, even if the building is expected to last for decades.

In **active solar heating**, a series of collection devices mounted on a roof or in a field are used to gather solar energy. The most common collection device is a panel or plate of black metal, which absorbs the sun's energy (**Figure 12.4**). Active solar heating is used primarily for heating water, either for household use or for swimming pools. The heat absorbed by the solar collector is transferred to a liquid inside the panel, which is then pumped to the heat exchanger, where the heat is transferred to water that will be stored in the hot water tank. Because approximately 8% of the energy consumed in the United States goes toward heating water, active solar heating has the potential to supply a significant amount of the nation's energy demand.

The use of active solar energy for space heating, which currently is not as common as for heating water, may become more important when diminishing supplies of fossil fuels force their prices higher. Active solar heating of buildings, now costlier than more conventional forms of space heating, would then become more competitive.

> **active solar heating** A system of putting the sun's energy to use in which a series of collectors absorb the solar energy, and pumps or fans distribute the collected heat.

SOLAR THERMAL ELECTRICITY GENERATION

Systems that concentrate solar energy to heat fluids have long been used for buildings and industrial processes. In **solar thermal electric generation**, electricity is produced in several different ways. One approach is to collect incident sunlight and concentrate it, using mirrors or lenses, to heat a working fluid to high temperatures. In one such design, trough-shaped mirrors, guided by computers, track the sun for optimum efficiency, focus sunlight on oil-filled pipes, and heat the oil to 390°C (735°F) (**Figure 12.5**). The hot oil is circulated to a water storage system and used to change water into superheated steam, which turns a turbine to generate electricity. Alternatively, the heat is used to power a **Stirling engine**. A fluid in a cylinder expands, driving a piston that turns a shaft, providing mechanical energy or producing electricity.

> **solar thermal electricity generation** A means of producing electricity in which the sun's energy is concentrated by mirrors or lenses to either heat a fluid-filled pipe or drive a Stirling engine.

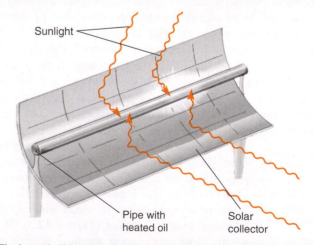

(a) A solar thermal plant in California uses troughs to focus sunlight on a fluid-filled tube.

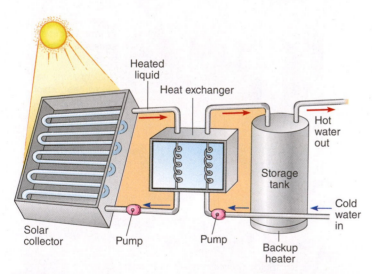

Figure 12.4   Active solar water heating. Solar collectors are mounted on the roof of a building. Each solar panel is a box with a black metal base and glass covering. Sunlight enters the glass and warms the pipes and the liquid flowing through them. The hot liquid (red arrows) heats water, which is further heated to required temperatures by a backup heater that uses electricity or natural gas. Solar domestic water heating can provide a family's hot water needs year-round.

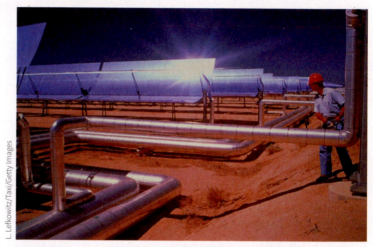

(b) The heated oil is pumped to a water tank, where it generates steam used to produce electricity. For simplicity, arrows show sunlight converging on several points; sunlight actually converges on the pipe throughout its length.

Figure 12.5   Solar thermal electric generation.

---

[1] Unless otherwise noted, all energy facts cited in this chapter were obtained from the Energy Information Administration (EIA), the statistical agency of the U.S. Department of Energy (DOE).

Most electricity in the United States is fed into a grid, a network of cables that carry electricity where it is needed. One disadvantage is that electricity produced from solar sources can only be produced during daylight hours. However, in areas where electricity demand is highest during the day, the fact that solar systems work only during the day is not a serious disadvantage. The world's largest solar thermal system is currently operating in the Mojave Desert in southern California.

Solar thermal energy systems are inherently more efficient than other solar technologies because they concentrate the sun's energy. With improved engineering, manufacturing, and construction methods, solar thermal energy may become cost-competitive with fossil fuels. Solar thermal plants produce electricity at a cost of $0.05 to $0.13 per kilowatt-hour (kWh); the lower end of this range is competitive with coal-fired plants. (**Table 12.1** compares the cost of generating electricity from a variety of sources). This cost should drop over the next decade as the technology matures. In addition, the environmental benefits of solar thermal plants are significant because they do not produce air pollution or contribute to acid rain or global climate change.

PHOTOVOLTAICS

Photovoltaics (PV), which include, in particular, **solar cells**, currently provide more than 21,000 MW of electricity worldwide.

photovoltaics a method of converting sunlight to electricity using layers of materials that either readily give up or absorb electrons.

solar cell A wafer or thin film of solid-state materials, such as silicon or gallium arsenide, that is treated with certain metals so that it generates electricity—that is, a flow of electrons—when it absorbs solar energy.

This is about the same as 21 large nuclear power plants but accounts for only about 0.7% of global electricity. Photovoltaic materials convert sunlight directly into electricity (**Figure 12.6**); they are usually arranged on large panels that absorb sunlight, even on cloudy or rainy days.

While manufacturing PVs creates pollution, while in operation they generate electricity with no pollution and minimal maintenance. They are used on any scale, from small, portable modules to large, multimegawatt power plants (**Figure 12.7**). Our current PV solar cell technology, though used to power satellites, drone aircraft, highway signals, watches, and calculators,

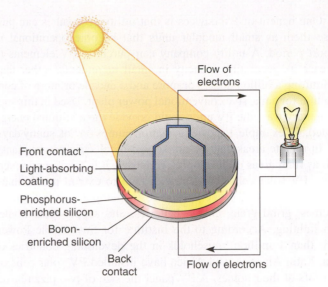

Figure 12.6 **Photovoltaic cells.** Photovoltaic (PV) cells contain silicon and other materials. Sunlight excites electrons, which are ejected from silicon atoms. Useful electricity is generated when the ejected electrons flow out of the PV cell through a wire.

Figure 12.7 **Photovoltaic panels.** These panels are cost effective because they serve two purposes: They generate electricity and shade a parking lot.

has a few limitations that prevent the cells' widespread use to generate electricity. Photovoltaic solar cells are only about 15% to 18% efficient at converting solar energy to electricity (although experimental cells reach 40%), and the number of solar panels needed for large-scale use would require a great deal of land. At current efficiencies, several thousand acres of solar panels would be required to absorb enough solar energy to produce the electricity generated by a single large conventional power plant.

Table 12.1   Generating Costs of Electric Power, 2010

Energy Source	Generating Costs (cents per kilowatt hour)*
Hydropower	5–11
Biomass	6–9
Geothermal	5–10
Wind	4–6
Solar thermal	5–13
Photovoltaics (PV)	15–25
Natural gas	4–5
Coal	4–6
Nuclear power	2–14

*Electricity production and consumption are measured in kilowatt-hours (kWh). As an example, one 50-watt light bulb that is on for 20 hours uses one kilowatt-hour of electricity ($50 \times 20 = 1000$ watt-hours = 1kWh).
Source: Energy Information Agency.

One benefit of PV devices is that utility companies can purchase them as small modular units that become operational in a short period. A utility company can purchase PV elements to increase its generating capacity in small increments, rather than committing a billion dollars or more and a decade or more of construction for a massive conventional power plant. Used in this supplementary way, the PV units could provide the additional energy needed, for example, to power irrigation pumps on hot, sunny days.

In remote areas that are not served by electric power plants, such as rural areas of developing countries, it is more economical to use PV solar cells for electricity than to extend power lines. Photovoltaics are the energy choice to pump water, refrigerate vaccines, grind grain, charge batteries, and supply rural homes with lighting. According to the Institute for Sustainable Power, more than 1 million households in the developing countries of Asia, Latin America, and Africa have installed PV solar cells on the roofs of their homes. A PV panel the size of two pizza boxes can supply a rural household with enough electricity for five lights, a radio, and a television.

The cost of manufacturing PV modules has steadily declined over the past four decades, from an average factory price of almost $90 per watt in 1975 to about $0.90 per watt in 2012. The cost of producing electricity from PVs remains a bit higher than that for other technologies like natural gas ($0.03 to $0.09 per kilowatt-hour), but can be competitive with hydroelectric power ($0.07 per kilowatt-hour).

Future technological progress will make PVs increasingly economically competitive with electricity produced by conventional energy sources. The production of thin-film solar cells, which are much cheaper to manufacture, has decreased costs. Thin-films can be produced as flexible sheets that are incorporated into building materials, such as tiles, window glass, and roofing shingles. More than 120,000 Japanese homes have installed PV solar energy roofing in the past few years, and California has committed to a million solar roofs by 2018. Other promising technological advances are dye-sensitized solar cells and nano-scale PV technologies.

REVIEW

1. What is active solar energy? Passive solar energy?

2. What are the advantages of producing electricity by solar thermal energy? By photovoltaic (PV) solar cells?

Indirect Solar Energy

LEARNING OBJECTIVES

- Define *biomass*, explain why it is an example of indirect solar energy, and outline how it is used as a source of energy.
- Describe the locations that can make optimum use of wind energy and of hydropower, and compare the potential of wind energy and hydropower.

biomass Plant material, including undigested fiber in animal waste, used as fuel.

Combustion of **biomass**—wood and other organic matter—is an example of indirect solar energy because green plants, which use solar energy for photosynthesis, store the energy in

biomass. Windmills, or wind turbines, extract **wind energy** to generate mechanical energy or electricity. The damming of rivers and streams to generate electricity is a type of **hydropower**—the energy of flowing water. Hydropower exists because solar energy drives the hydrologic cycle (see Chapter 4).

wind energy Electric or mechanical energy obtained from surface air currents caused by solar warming of air.

hydropower A form of renewable energy that relies on flowing or falling water to generate mechanical energy or electricity.

ENERGY FROM BIOMASS

Biomass, one of the oldest fuels known to humans, consists of such materials as fast-growing plant and algal crops, crop wastes, sawdust and wood chips, animal wastes, and wood (**Figure 12.8**). Biomass contains chemical energy that comes from the sun's radiant energy, which photosynthetic organisms use to form organic molecules. Biomass is a renewable form of energy when used no faster than it is produced; deforestation and desertification can result when biomass is overused (see Chapter 17). Biomass cannot replace fossil fuels. The entire photosynthesis production of the continental United States amounts to only half of our current energy use—and that would mean devoting it to no other uses, including food, paper, and construction materials.

Biomass fuel, which can be a solid, liquid, or gas, is burned to release its energy. Solid biomass such as wood is burned directly to obtain energy. Biomass—particularly firewood, charcoal (wood that is heated in an oxygen-free environment to concentrate its energy and drive off water), animal dung (primarily undigested plant fiber), and peat (partly decayed plant matter found in bogs and swamps)—supplies a substantial portion of the world's energy. At least half of the human population relies on biomass as their main source of energy. In developing countries, wood is the primary fuel for cooking. Biomass accounts for about 4% of total U.S. energy production. Biomass in the form of low-cost residues from sawmills, paper mills, and agricultural industries is burned in power plants to generate about 7.6 gigawatts (GW) of electricity.

It is possible to convert biomass, particularly animal wastes, into **biogas**. Biogas, which is usually composed of a mixture of gases (mostly methane), is stored and transported like natural gas.

Robert E. Ford/Terraphotographics/Biological Photo Service

Figure 12.8   Biomass. Firewood is the major energy source for much of the developing world. Is the use of firewood always carbon-neutral? Why or why not?

It is a clean fuel—its combustion produces fewer pollutants than either coal or biomass. In India and China, several million family-sized biogas digesters use microbial decomposition of household and agricultural wastes to produce biogas for cooking and lighting. When biogas conversion is complete, the solid remains are removed from the digester and used as fertilizer. Although the technology for biogas digesters is relatively simple, the conditions inside the digester, such as the moisture level and pH, must be carefully monitored if the bacteria are to produce biogas at an optimum level.

Biomass can be converted to liquid fuels, especially **methanol** (methyl alcohol) and **ethanol** (ethyl alcohol), which can be used in internal combustion engines. In many parts of the world, automotive fuels must contain 10% or more ethanol. Biodiesel, made from plant or animal oils, is becoming more popular as an alternative fuel for diesel engines in trucks, farm equipment, and boats. The oil is often refined from waste oil produced at restaurants (e.g., the oil used to make french fries); biodiesel burns much cleaner than diesel fuel.

Although some U.S. energy companies convert sugarcane, corn, or wood crops to alcohol, others are interested in the commercial conversion of agricultural and municipal wastes into ethanol. Costs are high but declining, and an increasing number of waste-to-ethanol conversion systems are coming on line. Several companies are currently building plants that convert biomass (using cornstalks, rice straw, the fibrous residues from processed sugarcane, and sewage sludge) to ethanol. Currently, the profitability of ethanol is possible only because of government subsidies that reduce ethanol's cost. Companies planning waste-to-ethanol processes acknowledge they need government subsidies initially but think they eventually will compete with gasoline without government help.

A major challenge for liquid biomass fuels is that it can take large amounts of energy to produce them. In addition, they are not always "climate-neutral," since more CO_2 may be produced than captured. However, recent research suggests that using switchgrass for ethanol in the Great Plains may actually store more carbon in the ground (as roots and rhizomes) each year than is released from the ethanol. Switchgrass has the added benefit that, when covering large areas, it can reduce erosion and provide habitat for threatened prairie plant and animal species.

Advantages of Biomass Use Biomass is an attractive source of energy because it reduces dependence on fossil fuels and often uses wastes, thereby reducing our waste-disposal problem. Biomass is usually burned to produce energy, so the pollution problems of fossil-fuel combustion, particularly carbon dioxide emissions, are not completely absent in biomass combustion. However, the low levels of sulfur and ash produced by biomass combustion compare favorably with the levels produced when bituminous coal is burned. It is possible to offset the CO_2 released into the atmosphere from biomass combustion by increasing tree planting. As trees photosynthesize, they absorb atmospheric CO_2 and lock it up in organic molecules that make up the body of the tree, thereby providing a carbon "sink." Thus, if biomass is regenerated to replace the biomass used, no net CO_2 is contributed to the atmosphere.

Disadvantages of Biomass Use Use of biomass, especially from plants, poses several problems. For one thing, biomass production requires land, water, and energy. Because use of agricultural land for energy crops competes with the growing of food crops, shifting the balance toward energy production might decrease food production, contributing to higher food prices. In addition, the energy used to produce biomass often requires some use of fossil fuels.

At least half the world's population relies on biomass as its main source of energy. Unfortunately, in many areas people burn wood faster than they replant trees. Intensive use of wood for energy has resulted in severe damage to the environment, including soil erosion, deforestation and desertification, air pollution (especially when burned indoors), and degradation of water supplies.

Crop residues, a category of biomass that includes cornstalks, wheat stalks, and wood wastes at paper mills and sawmills, are increasingly being used for energy. At first glance, it may seem that crop residues, which normally remain in the soil after harvest, would be a good source of energy if they were collected and burned. After all, they are waste materials that will eventually decompose. However, the systems perspective reminds us that crop residues left in and on the ground prevent erosion by holding the soil in place, and their decomposition enriches the soil by making the minerals originally in plant residues available for new plant growth. If all crop residues were removed from the ground, the soil would eventually be depleted of minerals, and its future productivity would decline. Forest residues, which remain in the soil after trees are harvested, fill similar ecological roles.

WIND ENERGY

Wind energy capacity worldwide has increased by 20% to 45% in each of the last 10 years—it is the world's fastest-growing source of energy. Wind, caused by heating of Earth's surface, is an indirect form of solar energy. Radiant energy from the sun is transformed into mechanical energy—the movement of air molecules. Wind is sporadic over much of Earth's surface, varying in direction and magnitude. Like direct solar energy, wind power is a highly dispersed form of energy. Harnessing wind energy to generate electricity has great potential, and wind is increasingly important in supplying our energy needs.

New wind turbines can be huge—100 m tall—and have long blades designed to harness wind energy efficiently (**Figure 12.9**). As turbines have become larger and more efficient, costs for wind power have declined rapidly—from $0.40 per kilowatt-hour in 1980 to $0.04 to $0.06 per kilowatt-hour in 2012. Wind power is cost-competitive with many forms of conventional energy. Advances, such as turbines that use variable-speed operation, may make wind energy an important global source of electricity during the first half of the twenty-first century.

Harnessing wind energy is most profitable in rural areas that receive fairly continual winds, such as islands, coastal areas, mountain passes, and grasslands. The United States currently has about 60 GW of installed capacity, and Germany has about 30 GW of installed capacity. However, this represents about 35% of Germany's total electricity use but only about 8% of U.S. electricity use.

The world's largest concentration of wind turbines is currently located in the Tehachapi Pass at the southern end of the Sierra Nevada in California. In the continental United States, some of

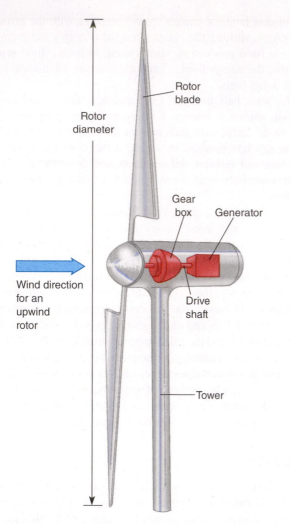

(b) Each of these wind turbines atop Buffalo Ridge in Minnesota generates enough electricity for 250 homes. Buffalo Ridge, which extends from South Dakota across Minnesota into Iowa, is an ideal site for wind farms.

(a) This basic wind turbine design has a horizontal axis (horizontal refers to the orientation of the drive shaft). Airflow causes the turbine's blades to turn 15 to 60 revolutions per minute (rpm). As the blades turn, gears within the turbine spin the drive shaft. This spinning powers the generator, which sends electricity through underground cables to a nearby utility. Wind turbine technology is advancing rapidly, and many changes in design are anticipated. (The tower is not drawn to scale and is much taller than depicted.)

Figure 12.9   Harvesting wind energy.

the best locations for large-scale electricity generation from wind energy are on the Great Plains. The 10 states with the greatest wind energy potential, according to the American Wind Energy Association, are North Dakota, Texas, Kansas, South Dakota, Montana, Nebraska, Wyoming, Oklahoma, Minnesota, and Iowa. In fact, if we developed the wind energy in North Dakota, Texas, and Kansas to their full potential, we could supply more than enough electricity to meet the current needs of the entire United States! U.S. wind power projects are underway in these and many other states. Currently, wind energy is captured and placed into regional electricity grids, but deploying wind energy on a national scale (e.g., wind energy produced in Texas and used in New York City) requires the development of new technologies for storing and distributing energy.

The use of wind power does not cause major environmental problems, although reported bird and bat kills represent one concern. The California Energy Commission estimated that several hundred birds, many of them raptors (birds of prey), turned up dead in the vicinity of the 7,000 turbines at Altamont Pass in California during a two-year study; most had collided with the turbines. Studies later determined that Altamont Pass is a major bird migration pathway. Technical "fixes," such as painted blades and antiperching devices to discourage raptors from roosting on the towers, were implemented at Altamont Pass. Other sites have been required to shut down operations during peak migratory periods. Developers of future wind farm sites currently conduct voluntary wildlife studies and try to locate sites away from bird and bat routes.

Wind produces no waste and is a clean source of energy. It produces no emissions of sulfur dioxide, carbon dioxide, or nitrogen oxides. Every kilowatt-hour of electricity generated by wind power rather than fossil fuels prevents 0.5 to 1 kg (1 to 2 lb) of the greenhouse gas CO_2 from entering the atmosphere. The biggest constraints on wind are cost and public resistance. Wind research and production have not been subsidized to the same extent as nuclear energy and biomass; nonetheless, as costs of other energy sources increase, wind is becoming increasingly competitive. The Not in my backyard (NIMBY) effect (see later in the chapter) plays a mixed role: Some people consider wind farms to be attractive, while others think they are noisy and visually upleasant blights on the landscape.

HYDROPOWER

Hydropower is the world's main renewable source of electrical generation, producing about the same amount of electricity as do the world's nuclear power plants. The sun's energy drives the hydrologic cycle, which encompasses precipitation, evaporation from land and water, transpiration from plants, and drainage and runoff (see Figure 4.6). As water flows from higher elevations back to sea level, we can harness its energy. The potential energy of water held back by a dam is converted to kinetic energy as the water falls down a penstock, where it turns turbines to generate electricity (**Figure 12.10**). Hydropower is more efficient than any other energy source in producing electricity—about 90% of available hydropower energy is converted into electricity. **Table 12.2** summarizes the reasons for and problems with dams.

Hydropower generates approximately 19% of the world's electricity, making it the most widely used form of solar energy. The 10 countries with the greatest hydroelectric production are, in decreasing order, Canada, the United States, Brazil, China, Russia, Norway, Japan, India, Sweden, and France. In the United States, approximately 2,200 hydropower plants produce between 8% and 12% of its electricity, making it the country's leading renewable energy source. Highly developed countries have already built dams at most of their potential sites, but this is not the case in many developing nations. Particularly in undeveloped, unexploited parts of Africa and South America, hydropower represents a great potential source of electricity.

Although most sites for traditional hydropower plants are already in use in the United States, new technological innovations show promise for expanding our hydropower capacity. About 97% of existing U.S. dams currently do not generate electricity, because traditional hydropower technology is suited only for large

Table 12.2	Advantages and Disadvantages of Dams
Reasons to Build Dams	*Problems with Dams*
Electrical power	Ecological disruption downstream
Mechanical power	
Irrigation	• Sediment stopped in dam
Navigation	• Water source diverted
Flood control	• Fish migration halted at dam
Commercial fishing	Ecological disruption in reservoir
Recreation	
• Fishing	• Habitat flooded
• Swimming	• Sediment buildup
• Boating	• Pollution if toxic materials are submerged
	Displacement of people
	Loss of cultural resources
	Catastrophic failure
	Disease
	Seismicity
	Evaporation from reservoir

dams with rapidly flowing water and large flow capacities. The new technologies can produce electricity at smaller dams. Several companies now manufacture turbines that harness electricity from large, slow-moving rivers or from streams with small flow capacities. As these new technologies improve, they have the potential to increase the amount of electricity generated by hydropower without the construction of a single new dam.

Impacts of Dams   One problem associated with hydropower is that building a dam changes the natural flow of a river. A dam causes water to back up, flooding large areas of land and forming a reservoir, which destroys plant and animal habitats. Native fishes are particularly susceptible to dams because the original river ecosystem is so altered. The migration of spawning fish is also disrupted (see the discussion of the Columbia River in Chapter 13). Below the dam, the once-powerful river is reduced to a relative trickle. The natural beauty of the countryside is affected, and certain forms of wilderness recreation are made impossible or less enjoyable, although the dams permit water sports in the reservoir.

At least 200 large dams around the world are associated with reservoir-induced seismicity—earthquakes that occur during and after the filling of a large reservoir behind a dam. The larger the reservoir and the faster it is filled, the greater the intensity of seismic activity. An area does not have to be seismically active to have earthquakes induced by reservoirs.

In arid regions, the creation of a reservoir results in greater evaporation of water because the reservoir has a large and slow moving surface area as compared to the original stream or river. As a result, serious water loss and increased salinity of the remaining water may occur.

If a dam breaks, people and property downstream may be endangered. In addition, waterborne diseases such as schistosomiasis may spread throughout the local population. **Schistosomiasis**, a tropical disease caused by a parasitic worm, damages the liver, urinary tract, nervous system, and lungs. As much as half the population of Egypt suffers from this disease, largely as a result of the

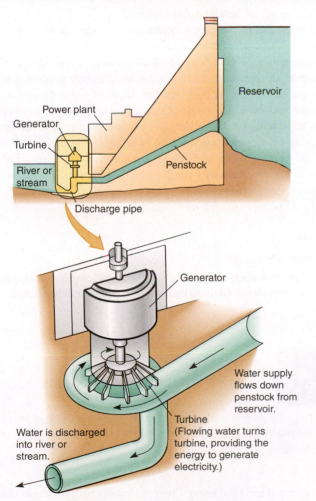

Figure 12.10   Hydroelectric power. A controlled flow of water released down the penstock turns a turbine, which generates electricity.

Aswan Dam, built on the Nile River in 1902 to control flooding but used since 1960 to provide electric power. (The large reservoir behind the dam provides a habitat for the worm, which spends part of its life cycle in the water. The worms infect humans when they bathe, swim, or walk barefoot along water banks or when they drink infected water.)

The environmental and social impacts of a dam may not be acceptable to the people living in a particular area. Laws prevent or restrict the building of dams in certain locations. In the United States, the Wild and Scenic Rivers Act prevents the hydroelectric development of certain rivers, although the number of rivers protected by this law is less than 1% of the nation's total river systems. Other countries, such as Norway and Sweden, have similar laws.

Dams cost a great deal to build but are relatively inexpensive to operate. A dam has a limited life span, usually 50 to 200 years, because over time the reservoir fills in with silt until it cannot hold enough water to generate electricity. This trapped silt, which is rich in nutrients, is prevented from enriching agricultural lands downstream. The gradual depletion of agricultural productivity downstream from the Aswan Dam in Egypt is well documented. Egypt now relies on heavy applications of chemical fertilizers to maintain the fertility of the Nile River valley and its delta.

CASE IN POINT

THE THREE GORGES DAM

For thousands of years, China wanted to dam the Yangze River. Historically, people living in the river basin have faced drought and flood years, often severe enough to kill many through famine or drowning. In the last century, the additional advantage of hydropower made damming the Yangze even more compelling. Consequently, in the 1990s China began work on the Three Gorges Dam (TGD), so named because the 632-km (412-mi)-long reservoir will flood three upstream gorges. In 2003, the reservoir began to fill, and the dam was completed and in full operation at the end of 2008.

The TGD meets several goals. First, it is producing 18 GW of electrical power—equivalent to 18 nuclear power plants or large coal power plants. Given the severe air quality problems facing China (see Chapters 19 and 20) and its current reliance on imported energy, this is a great advantage. Second, agricultural productivity downstream can be optimized. (In 1998, a tenth of China's grain supply was destroyed by flooding that would have been controlled by the dam.) In addition, the new reservoir is used for transportation—large ships can go far upstream in the reservoir—as well as commercial fishing and recreation.

However, the TGD also represents the range of problems associated with building dams. At least 1.5 million people have been displaced, often with inadequate compensation. The dam endangers several species, including the rare Yangtze river dolphin, although this animal may already have been driven to extinction by overfishing and pollution. The reservoir may become highly polluted from industrial areas upstream and contaminated sites in the flooded area. Historical and cultural treasures, including temples, ancient hanging coffins, and massive canyon wall writings,

have been submerged. Also, even as agriculture is improved downstream, thousands of hectares of arable lands upstream have been flooded and thereby removed from production.

China initially had difficulty finding investors, since the uncertainty about design, construction, and effectiveness was so great. Waterborne diseases, including malaria and schistosomiasis, are likely to increase. As people moved away from the reservoir site, entire cities were razed and removed; it is likely that toxic materials as well as human and animal wastes remain, which could contaminate the newly created reservoir. The rate of sedimentation, or buildup of silt behind the dam, is not well understood. Too much sedimentation would undermine all the major reasons for the dam: irrigation, flood control, and hydropower production. ■

OTHER USES OF INDIRECT SOLAR ENERGY

A few other forms of indirect solar energy may become important in the future. Ocean waves are produced by winds, which are caused by the sun; wave energy is therefore considered an indirect form of solar energy. Like other types of flowing water, wave power has the potential to turn a turbine, thereby generating electricity. Norway, Great Britain, Japan, and several other countries are investigating the production of electricity from ocean waves. Two commercial wave power stations are currently operational, off the coasts of Scotland and Portugal.

In the future, we may generate power using ocean temperature gradients, the differences in temperature at various ocean depths. As much as a 24°C difference exists between warm surface water and cold, deeper ocean water. Ocean temperature gradients, which are greatest in the tropics, are the result of solar energy warming the surface of the ocean. **Ocean thermal energy conversion (OTEC)** would take advantage of this temperature difference to produce electricity or to cool buildings. The first commercial OTEC plant is under construction at the Natural Energy Laboratory of Hawaii Authority on the island of Hawaii. As technology improves and costs of other energy sources go up, ocean waves and OTEC may become more viable.

REVIEW

1. What is biomass?
2. What are the advantages and disadvantages of using wind to produce electricity? Of using hydropower to produce electricity?

Other Renewable Energy Sources

LEARNING OBJECTIVE

- Describe geothermal energy and tidal energy, the two forms of renewable energy that are not direct or indirect results of solar energy.

Geothermal and tidal energy are renewable energy sources that are not direct or indirect results of solar energy. **Geothermal energy** is the naturally occurring heat within Earth. **Tidal energy**, caused by changes in water level between high and low tides, is exploited to generate electricity on a limited scale.

GEOTHERMAL ENERGY

Geothermal energy, the natural heat within Earth, arises from the ancient heat within Earth's core, from friction where continental plates slide over one another, and from the decay of radioactive elements. The amount of geothermal energy is enormous. Scientists estimate that just 1% of the heat contained in the uppermost 10 km of Earth's crust is equivalent to 500 times the energy contained in all of Earth's oil and natural gas resources. Nonetheless, because it is difficult to extract, geothermal is not likely to compete with wind, hydropower, or solar energy.

geothermal energy The use of energy from Earth's interior for either space heating or the generation of electricity.

Geothermal energy is typically associated with volcanism. Areas of geologically recent volcanism contain large underground heat reservoirs. As groundwater in these areas travels downward and is heated, it becomes buoyant and rises until it is trapped by an impermeable layer in Earth's crust, forming a **hydrothermal reservoir**. Hydrothermal reservoirs contain hot water and possibly steam, depending on the temperature and pressure of the fluid. Some of the hot water or steam may escape to the surface, creating hot springs or geysers. Hot springs have been used for thousands of years for bathing, cooking, and heating buildings.

Hydrothermal reservoirs are tapped by drilling wells similar to those used for extracting oil and natural gas. One option is to heat water, which can then be used to heat buildings. Alternatively, the heated fluid can be brought to the surface, and the resulting steam is expanded through a turbine to spin a generator, creating electricity (**Figure 12.11**).

The United States, with about a 2.5-GW capacity, is the world's largest producer of geothermal electricity. Electric power is currently produced at geothermal fields in Alaska, California, Nevada, Utah, and Hawaii. The world's largest geothermal power plant is The Geysers, a geothermal field in northern California that provides electricity for 1.7 million homes. Other important producers of geothermal energy include the Philippines, Italy, Japan, Mexico, Indonesia, and Iceland. Total geothermal electric capacity is currently about 8 GW.

Iceland, a country with minimal fossil-fuel or nuclear energy resources, is geographically situated to optimize the use of geothermal energy. Located on the Mid-Atlantic Ridge, a boundary between two continental plates, Iceland is an island of intense volcanic activity and consequently has considerable geothermal resources. Iceland uses geothermal energy to generate electricity and to heat two-thirds of its homes. In addition, most of the fruits and vegetables required by the people of Iceland are grown in geothermally heated greenhouses.

Geothermal energy is considered environmentally benign compared to conventional fossil-fuel-based energy technologies because it emits only a fraction of the air pollutants. The most common environmental hazard associated with geothermal energy is the emission of hydrogen sulfide (H_2S) gas, which

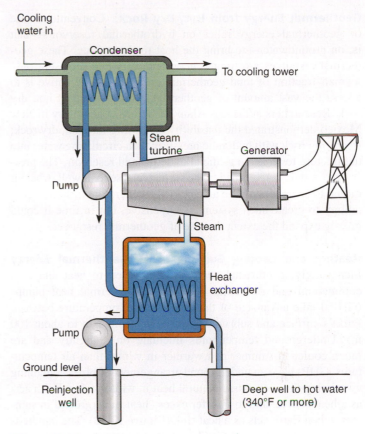

Figure 12.11   Geothermal energy. Shown is one design for a geothermal power plant. Pressurized hot water pumped from underground generates steam in a heat exchanger. The steam turns a turbine and generates electricity. After its use, the steam is condensed and recirculated. The cooler but still pressurized water is reinjected into the ground, reheated, and used again.

comes from the large amounts of dissolved minerals and salts found in the steam or hot water. Some geothermal reservoirs contain H_2S in quantities that require mitigation to meet air quality standards. Air pollution-control methods are highly effective but do increase the energy cost. A lesser concern associated with geothermal energy is that the surrounding land may subside, or sink, as the water from hot springs and their connecting underground reservoirs is removed. Although experience has shown this is not a problem at most geothermal fields, it has occurred at a few of them.

Geothermal Power and Water Use   The water used to transfer heat from geothermal wells to the surface is not inexhaustible. Some geothermal applications recirculate all the water back into the underground reservoir, ensuring decades of heat extraction from a given reservoir. Others consume a portion of the water, eventually depleting water in the underground reservoir. A good example is The Geysers geothermal field, where after nearly 40 years of production, about half the water is depleted but about 95% of the heat remains in the rock.

Geothermal Energy from Hot, Dry Rock Conventional use of geothermal energy relies on hydrothermal reservoirs—that is, on groundwater—to bring the heat to the surface. These geothermal resources are limited geographically and represent only a small fraction of total geothermal energy. An alternative is to access the vast amount of geothermal energy stored in hot, dry rock. Researchers at the Los Alamos National Laboratory in New Mexico demonstrated the feasibility of drilling into hot, dry rock; fracturing rock with hydraulic pressure; then circulating water into the fracture to make an artificial underground reservoir. The pressurized water was returned to the surface in a second well, where it turned to steam and drove an electricity-generating turbine. Technology to create such systems is expensive, but in time it could greatly expand the extent and use of geothermal resources.

Heating and Cooling Buildings with Geothermal Energy Increasingly, geothermal energy is employed to heat and cool commercial and residential buildings. Geothermal heat pumps (GHPs) take advantage of the difference in temperature between Earth's surface and subsurface (at depths from 1 m to about 100 m). Underground temperatures fluctuate only slightly and are much cooler in summer and warmer in winter than air temperatures. GHPs have an underground arrangement of pipes containing circulating fluids to extract natural heat in winter, when Earth acts as a heat source, and to transfer excess heat underground in summer, when Earth acts as a heat sink (**Figure 12.12**). The hundreds of feet of pipe form a ground loop that feeds into a heat pump, which directs the flow of heated or cooled air. Geothermal heating systems can be modified to provide supplemental hot water.

High installation costs currently limit the number of GHPs. However, with the growth of green architecture and rising fuel costs, commercial and residential use of GHPs is on the rise. Benefits include low operating costs, which may be half those of conventional systems, and high efficiency. The Environmental Protection Agency (EPA) estimates that GHPs are the most efficient heating system available, two to three times more efficient than other heating methods, and produce the lowest carbon dioxide emissions.

TIDAL ENERGY

Tides, the alternate rising and falling of the surface waters of the ocean and seas that generally occur twice each day, are the result of the gravitational pull of the moon and the sun. Normally, the difference in water level between high and low tides is about 0.5 m (1–2 ft). Certain coastal regions with narrow bays have extremely large differences in water level between high and low tides. The Bay of Fundy in Nova Scotia has the largest tides in the world, with a difference of up to 16 m (53 ft) between high and low tides.

Water at high tide contains enormous amounts of potential energy as compared to low tide. This **tidal energy** can be captured (with a dam across a bay or a turbine that works much like a wind turbine) and converted into electricity. Tidal power plants currently operate in France, Russia, China, and Canada. However, total global production is only a few MW and is not expected to increase much in the near future.

> **tidal energy** A form of renewable energy that relies on the ebb and flow of the tides to generate electricity.

REVIEW

1. What are the pros and cons of using geothermal energy to produce electricity? Of using tidal power?

Nuclear Power

LEARNING OBJECTIVES

• Distinguish between nuclear energy and chemical energy.
• Describe the nuclear fuel cycle, including the process of enrichment.
• Define *nuclear reactor* and describe a typical nuclear power reactor.

The remainder of this chapter explores **nuclear energy** and its role in electricity generation. In 1905, **Albert Einstein** first hypothesized that mass and energy are related in his now-famous equation $E = mc^2$, in which energy (E) is equal to mass (m) times the speed of light (c) squared. This suggested that nuclear reactions, in which matter is converted to energy, had the potential to release a vast amount of heat.

> **nuclear energy** The energy released by nuclear fission or fusion.

As a way to obtain energy, nuclear processes are fundamentally different from the combustion that produces energy from biomass and fossil fuels. Combustion is a chemical reaction. In ordinary chemical reactions, atoms of one element do not change into atoms of another element, nor does any of their mass (matter) change into energy. The energy released in combustion and other chemical reactions comes from changes in the chemical bonds that hold the atoms together. Chemical bonds are associations between

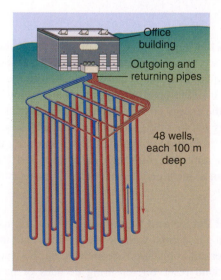

Figure 12.12 Geothermal heat pump. This diagram is based on the geothermal heat pump used at the Philip Merrill Environmental Center in Annapolis, Maryland. The pipes running through the building and underground are a closed loop. In summer, which is shown in the figure, outgoing warm water from the building (red pipes) returns from the ground cooled (blue pipes). In winter, the opposite happens: Outgoing cool water is warmer when it returns to the building. The ground's temperature is a constant 12.2°C (54°F) year-round.

electrons, and ordinary chemical reactions involve the rearrangement of electrons (see Appendix I).

fission The splitting of an atomic nucleus into two smaller fragments, accompanied by the release of a large amount of energy.

In contrast, nuclear energy involves changes in the *nuclei* of atoms; small amounts of matter from the nucleus are converted into large amounts of energy. Two different nuclear reactions release energy: fission and fusion. In **fission**, the process nuclear power plants use, larger atoms of certain elements are split into two smaller atoms of different elements. In **fusion**, the process that powers the sun and other stars, two smaller atoms are combined to make one larger atom of a different element. In both fission and fusion, the mass of the end product(s) is less than the mass of the starting material(s) because a small quantity of the starting material(s) is converted into energy.

fusion The joining of two lightweight atomic nuclei into a single, heavier nucleus, accompanied by the release of a large amount of energy.

Nuclear reactions produce 100,000 times more energy per atom than is available from a chemical bond between two atoms. In nuclear bombs, energy from many atomic fissions is released all at once, producing a tremendous surge of energy that destroys everything in its vicinity. When nuclear energy is used to generate electricity, the nuclear reaction is controlled to produce smaller amounts of energy in the form of heat, which is then converted to electricity.

All atoms are composed of positively charged protons, negatively charged electrons, and electrically neutral neutrons. Protons and neutrons, which have approximately the same mass, are clustered in the center of the atom, making up its nucleus. Electrons, which possess little mass in comparison with protons and neutrons, orbit around the nucleus in distinct regions. Electrically neutral atoms possess identical numbers of positively charged protons and negatively charged electrons.

The **atomic mass** of an element is equal to the sum of protons and neutrons in the nucleus. Each element has a characteristic **atomic number**, or number of protons per atom. In contrast, the number of neutrons in each atom of a given element may vary, resulting in atoms of one element with different atomic masses. Forms of a single element that differ in atomic mass are known as **isotopes**. For example, normal hydrogen, the lightest element, contains one proton and no neutrons in the nucleus of each atom. The two isotopes of hydrogen are **deuterium**, which contains one proton and one neutron per nucleus, and **tritium**, which contains one proton and two neutrons per nucleus. Many isotopes are stable, but some are unstable; the unstable ones, called **radioisotopes**, are radioactive because they spontaneously emit **radiation**, a form of energy consisting of particles. The only radioisotope of hydrogen is tritium.

radioactive decay The emission of energetic particles or rays from unstable atomic nuclei; includes positively charged alpha particles, negatively charged beta particles, and high-energy, electromagnetic gamma rays.

As a radioactive element emits radiation, its nucleus changes into the nucleus of a different, more stable element. This process is called **radioactive decay**. For example, the radioactive nucleus of one isotope of uranium, U-235, decays over time into lead (Pb-207). (Uranium-235 is an isotope of uranium with an atomic mass of 235.) Each radioisotope has its own characteristic rate of decay. The time required for one-half of the total amount of a radioactive substance to change into a different material is called its **radioactive half-life**. The half-lives of different radioisotopes vary enormously (**Table 12.3**). The half-life of iodine (I-232) is only 2.4 hours, the half-life of tritium is 12.3 years, and the half-life of an isotope of uranium (U-234) is 250,000 years.

Uranium ore, the mineral fuel used in conventional nuclear power plants, is a nonrenewable resource present in limited amounts in sedimentary rock in Earth's crust. The steps involved in producing the uranium fuel used in nuclear power plants, from mining to disposal, are collectively called the **nuclear fuel cycle** (**Figure 12.13**).

nuclear fuel cycle The processes involved in producing the fuel used in nuclear reactors and in disposing of radioactive (nuclear) wastes.

Substantial deposits of uranium are found in Australia (20.4% of known world reserves), Kazakhstan (18.2%), the United States (10.6%), Canada (9.9%), and South Africa (8.9%). In the United States, uranium is found in Wyoming, Texas, Colorado, New Mexico, and Utah. Uranium ore contains three isotopes: U-238 (99.28%), U-235 (0.71%), and U-234 (less than 0.01%). Because U-235, the isotope used in conventional fission reactions, is such a minor part (less than 1%) of uranium ore, uranium must be refined after mining to increase the concentration of U-235 to at least 3%. This refining process, called **enrichment**, is energy-intensive.

enrichment The process by which uranium ore is refined after mining to increase the concentration of fissionable U-235.

After enrichment, the uranium fuel used in a **nuclear reactor** is processed into small pellets of uranium dioxide; each pellet contains the energy equivalent of a ton of coal (**Figure 12.14a**). The pellets are placed in **fuel rods**, closed tubes often as long as 4 m (about 13 ft). The fuel rods are then grouped into square **fuel assemblies**, generally of 200 rods each (**Figure 12.14b**). A typical nuclear reactor contains 150 to 250 fuel assemblies.

nuclear reactor A device that initiates and maintains a controlled nuclear fission chain reaction to produce energy for electricity.

In nuclear fission, U-235 is bombarded with neutrons (**Figure 12.15**). When the nucleus of an atom of U-235 is struck by and absorbs a neutron, it becomes unstable and splits into two

Table 12.3 Some Common Radioactive Isotopes Associated with the Fission of Uranium

Radioisotope	Half-Life (years)
Iodine-131	0.02 (8.1 days)
Xenon-133	0.04 (15.3) days
Cerium-144	0.8
Ruthenium-106	1.0
Krypton-85	10.4
Tritium	12.3
Strontium-90	28
Cesium-137	30
Radium-226	1,600
Plutonium-240	6,600
Plutonium-239	24,400
Neptunium-237	2,130,000

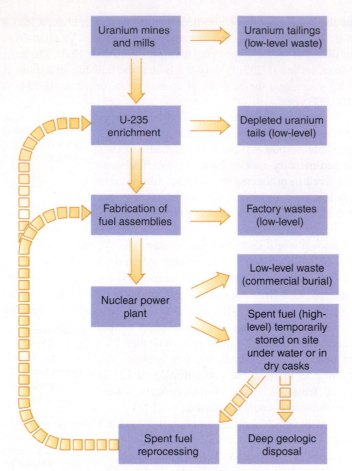

Figure 12.13 Nuclear fuel cycle, from mining to disposal. The left side of the figure shows how mined uranium becomes fuel for nuclear power plants. The right side shows radioactive wastes that must be handled and disposed of during the cycle. Broken lines indicate steps that are not currently occurring. In 1976 the United States stopped reprocessing (reusing spent fuel) for economic and political reasons. Currently, Japan, France, Russia, the United Kingdom, and India are the only countries reprocessing spent fuel for reuse as nuclear fuel. Deep geologic disposal of spent fuel is currently under study in several countries, including the United States.

smaller atoms, each approximately half the size of the original uranium atom. In the fission process, two or three neutrons are also ejected from the uranium atom. They collide with other U-235 atoms, generating a chain reaction as those atoms are split and more neutrons are released to collide with additional U-235 atoms.

The fission of U-235 releases an enormous amount of heat, used to transform water into steam. The steam, in turn, is used to generate electricity. Production of electricity is possible because the fission reaction is controlled. Recall that nuclear bombs make use of uncontrolled fission reactions. If the control mechanism in a nuclear power plant were to fail, a bomblike nuclear explosion could not take place because nuclear fuel only has 3% to 5% U-235, whereas bomb-grade material contains at least 20%—and usually about 85% to 90%—U-235. In the highly unlikely event of an uncontrolled fission reaction, an immense amount of heat could be generated. However, the reactor vessel and massive concrete containment building are designed to contain the heat, along with the attendant radioactivity.

(a) Uranium dioxide pellets, held in a gloved hand, contain about 3% uranium-235, the fission fuel in a nuclear reactor. Each pellet contains the energy equivalent of 1 ton of coal.

(b) The uranium pellets are loaded into long fuel rods, which are grouped into square fuel assemblies.

Figure 12.14 Uranium fuel.

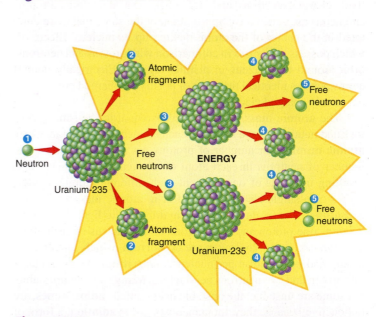

Figure 12.15 Nuclear fission. Starting at the left side of the figure, neutron bombardment (1) of a uranium-235 (U-235) nucleus causes it to split into two smaller radioactive atomic fragments (2) and several free neutrons (3). The free neutrons bombard nearby U-235 nuclei, causing them to split (4) and release still more free neutrons (5) in a chain reaction. Many different pairs of radioactive atomic fragments are produced during the fission of U-235.

ELECTRICITY FROM CONVENTIONAL NUCLEAR FISSION

About two-thirds of U.S. nuclear power plants have pressurized water reactors. These have four main parts: the reactor core, the steam generator, the turbine, and the condenser (**Figure 12.16**).

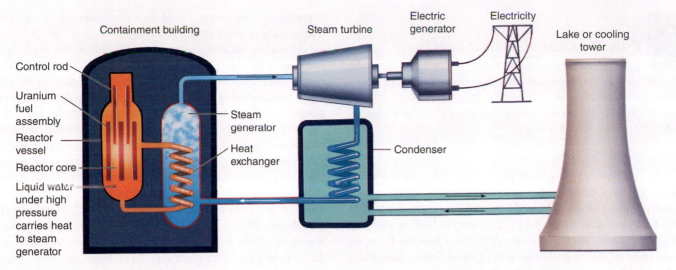

Control rod
Uranium fuel assembly
Reactor vessel
Reactor core
Liquid water under high pressure carries heat to steam generator
Containment building
Steam generator
Heat exchanger
Steam turbine
Electric generator
Electricity
Condenser
Lake or cooling tower

Figure 12.16   Pressurized water reactor. Fission of uranium-235 that occurs in the reactor vessel produces heat, used to produce steam in the steam generator. The steam drives a turbine to generate electricity. The steam then leaves the turbine and is pumped through a condenser before returning to the steam generator. Pumping hot water from the condenser to a lake or massive cooling tower controls excess heat. After it is cooled, the water is pumped back to the condenser. Approximately two-thirds of all nuclear power plants in the United States are of this type.

Fission occurs in the **reactor core,** and the heat produced by nuclear fission is used to produce steam from liquid water in the **steam generator.** The **turbine** uses the steam to drive a generator that produces electricity, and the **condenser** cools the steam, converting it back to a liquid.

The **reactor core** contains the fuel assemblies. Above each fuel assembly is a **control rod** made of a special metal alloy that absorbs neutrons. The plant operator signals the control rod to move either up out of or down into the fuel assembly. If the control rod is out of the fuel assembly, free neutrons collide with uranium atoms in the fuel rods, and fission takes place. If the control rod is completely lowered into the fuel assembly, it absorbs the free neutrons, and fission of uranium no longer occurs. By controlling the placement of the control rods, the plant operator produces the exact amount of fission required.

BREEDER REACTORS AND MIXED OXIDE FUEL (MOX) FOR NUCLEAR FISSION

Uranium ore is mostly U-238, which is not fissionable and is a waste product of conventional nuclear fission. In **breeder nuclear fission,** however, U-238 is converted to plutonium, Pu-239, a fissionable isotope not found in nature. Some of the neutrons emitted in breeder nuclear fission produce additional plutonium from U-238. A breeder reactor thus makes more fissionable fuel than it uses. The fuel is then **reprocessed** to concentrate the Pu-239 for use as fuel.

breeder nuclear fission A type of nuclear fission in which nonfissionable U-238 is converted into fissionable Pu-239.

Because it can use U-238, plutonium-based breeder fission can generate much larger quantities of energy from uranium ore than nuclear fission using U-235. When one adds the uranium reserves in the ground to existing radioactive waste stockpiles, breeder fission has the potential to supply the entire country with electric energy for several centuries.

Although breeder fission sounds promising, it presents concerns about both safety and weapons proliferation. Breeder fission reactors use liquid sodium rather than water as a coolant. Sodium is a highly reactive metal that reacts explosively with water and burns spontaneously in air at the high temperatures maintained in breeder reactors. Loss of this coolant would result in an uncontrolled reaction that would almost certainly rip open the **containment building**, releasing radioactive materials into the atmosphere.

In addition, because it is chemically different from uranium, plutonium is much easier to process into a weapons-grade material. As additional countries develop nuclear power, breeder reactors will allow them to secretly produce bomb-quality plutonium more easily than by processing uranium. The plutonium used for India's first nuclear test bomb in 1974 came from its "civilian" breeder reactor development program. The United States performed the first breeder reactor experiments but abandoned its breeder reactor development program in 1977, during the administration of President Carter, a decision supported by every subsequent U.S. president.

More common, particularly in Europe, are reactors that use **mixed oxide fuel (MOX)**. For MOX reactors, **spent fuel** from standard uranium-based reactors is reprocessed, and the extracted plutonium and U-235 are used as fuel. This option can also be used to mix some of the plutonium from nuclear weapons with uranium and use it to generate electricity. MOX is now used in about 30 European reactors. Through the "Megatons to Megawatts" program, about 17,000 warheads from former Soviet stockpiles were burned as MOX in reactors in the United States between 2011 and 2013.

mixed oxide fuel (MOX) A reactor fuel that contains a combination of uranium oxide and plutonium oxide. The plutonium can come from reprocessed spent fuel from other plutonium stockpiles, including dismantled weapons.

spent fuel The used fuel elements that were irradiated in a nuclear reactor.

REVIEW

1. What is the difference between chemical energy and nuclear energy?

2. What is the nuclear fuel cycle?

3. How does a nuclear reactor produce electricity?

Pros and Cons of Nuclear Energy

L E A R N I N G   O B J E C T I V E S

- Discuss the pros and cons of electric power produced by nuclear energy.
- Describe the nuclear power plant accidents at Three Mile Island, Chernobyl, and Fukushima Daiichi.

One of the reasons nuclear energy proponents argue for the widespread adoption of nuclear energy is that it can have fewer routine environmental impacts than do fossil fuels, particularly coal (**Table 12.4**; see also Table 10.1). The combustion of coal, an extremely dirty fuel, to generate electricity is responsible for more than one-third of the air pollution in the United States. According to the Energy Information Administration (EIA), coal and natural gas power plants are responsible for 67% of the nation's sulfur dioxide emissions, 23% of nitrogen oxide emissions, and 40% of human-made carbon dioxide emissions.

Nuclear power emits no emissions directly into the atmosphere and is a carbon-free source of electricity. At the very least, proponents say we should use nuclear-generated power until other carbon-free energy technologies, such as wind and solar power, become more efficient and cost-competitive. Opponents argue that the underlying problems with fossil fuels and nuclear power are similar: Large-scale technologies have unpredictable but potentially large-scale consequences.

Nuclear power cannot have much direct effect on our dependence on crude oil, since only 3% of the electricity produced in the United States comes from oil. However, while gasoline is currently the major energy source for transportation worldwide, it is difficult to capture the carbon released from gasoline engines. If the world decides to limit greenhouse gas emissions, which seems likely, and if gasoline prices continue to rise, electric vehicles will become a significant portion of the world's automobile fleet. If that happens, nuclear power could then contribute much more to the transportation sector than it does now.

However, nuclear energy generates radioactive waste, such as spent fuel. Nuclear power plants also produce other radioactive wastes, such as radioactive coolant fluids and gases in the reactor. Spent fuel and other radioactive wastes are highly radioactive and dangerous. These wastes present substantial health and environmental hazards and therefore require special storage and disposal measures.

Nuclear power is also not entirely climate-neutral, since several steps involved in producing it, from mining to processing to disposal, currently require substantial amounts of gasoline and diesel. This means that nuclear energy indirectly contributes to the greenhouse effect—about 2 to 6 g of carbon are produced per kilowatt-hour, about two orders of magnitude lower than carbon produced from fossil-fuel electricity. In addition, replacing only 10% of the current U.S. use of fossil fuels would require doubling the number of nuclear power plants, which is a long-term proposition.

While advocates present many reasons why we don't have more nuclear power, a major hurdle is cost. Unfortunately, it is difficult to get a good sense of the true cost of nuclear power—or most energy sources, for that matter—because of various government programs and incentives, including *subsidies*. Currently, it is generally more expensive to produce electricity from nuclear power than from coal, hydropower, or natural gas. Few places remain to build large dams in the United States, though, so new nuclear power is more promising than new hydropower. Coal and natural gas have been heavily subsidized over the past few decades, and both emit substantial amounts of carbon dioxide. If the indirect costs of subsidies and the carbon dioxide externality are added in, nuclear energy appears much more competitive.

In the 1940s and 1950s, the U.S. government provided substantial amounts of funding for nuclear power research—far more than for other non–fossil-fuel energy sources. It does not, however, make sense to factor this into the cost of current nuclear power, because it is what economists call a *sunk cost*—it was spent

---

Table 12.4   Comparison of Environmental Impacts of 1000-MW Coal and Nuclear Power Plants*

Impact	Coal	Nuclear (Conventional Fission)
Land use	17,000 acres	1900 acres
Daily fuel requirement	9000 tons/day	3 kg/day
Availability of fuel, based on present economics	A few hundred years	100 years, maybe longer (much longer with breeder fission)
Air pollution	Moderate to severe, depending on pollution controls	Low**
Climate change risk (carbon dioxide emissions)	Severe	Relatively small**
Radioactive emissions, routine	1 curie	28,000 curies
Water pollution	Often severe at mines	Potentially severe at nuclear waste disposal sites
Risk from catastrophic accidents	Short-term local risk	Long-term risk over large areas
Link to nuclear weapons	No	Yes
Annual occupational deaths	0.5 to 5	0.1 to 1
Certainty about risks	Well known	Highly uncertain

*Impacts include extraction, processing, transportation, and conversion. Assumes coal is strip-mined. (A 1000-MW utility, as a 60% load factor, produces enough electricity for a city of 1 million people.)

**While nuclear electricity generation does not generate air pollution and carbon dioxide directly, many of the steps (mining, construction, and waste disposal, for example) require fossil fuels.

in the past and has no impact on the cost of new nuclear power. The U.S. government is legally committed to take possession of all high-level nuclear waste from commercial nuclear power plants. However, fees paid by the companies producing that waste are sufficient to cover anticipated removal and disposal costs.

Proponents of nuclear energy usually point to France, which obtains 78% of its electricity from nuclear energy because electricity generated by French nuclear power plants is 27% less expensive than that generated from coal. The lower cost of nuclear power is partly explained because France uses the same design for all its nuclear power plants. Nuclear energy proponents believe that the United States could achieve the same economic efficiency. Opponents of nuclear power dismiss the example of France because the French government, like the governments of other countries that use nuclear power, heavily subsidizes the nuclear industry.

Overall, the cost of producing electricity using nuclear power is uncertain. Estimates range from 2 to 14 cents per kilowatt-hour for electricity from new nuclear power plants, compared to 4 to 6 cents per kilowatt-hour from new coal power plants. The uncertainty in the estimates arises from determining how much licensing, construction, fuel, and waste disposal will cost, as well as how to factor subsidies into the equation.

SAFETY ISSUES AT NUCLEAR POWER PLANTS

Although conventional nuclear power plants cannot explode like atomic bombs, accidents can happen in which dangerous levels of radiation are released into the environment and result in human casualties. At high temperatures, the metal encasing the uranium fuel can melt, releasing radiation; this is known as a **meltdown**. Also, the water used in a nuclear reactor to transfer heat can boil away during an accident, contaminating the atmosphere with radioactivity.

The probability that a major accident will occur is considered low by the nuclear industry, but public perception of the risk is high for several reasons. Nuclear power risks are involuntary and potentially catastrophic. In addition, many people are distrustful of the nuclear industry. The consequences of an accident are drastic and life-threatening, both immediately and long after the accident has occurred. We now consider three major accidents, one in the United States (Three Mile Island), a second at Chernobyl in Ukraine, and a third at Fukushima Daiichi in Japan.

Three Mile Island   The most serious nuclear reactor accident in the United States occurred in 1979 at the Three Mile Island power plant in Pennsylvania, the result of human and design errors. A 50% meltdown of the reactor core took place. Had there been a complete meltdown of the fuel assembly, dangerous radioactivity would have been emitted into the surrounding countryside. Fortunately, the containment building kept almost all the radioactivity released by the core material from escaping. Although a small amount of radiation entered the environment, there were no substantial environmental damages and no known human casualties. A study conducted within a 10-mile radius around the plant 10 years after the accident concluded that cancer rates were in the normal range and that no association could be made between cancer rates and radiation emissions from the accident. Numerous

other studies have failed to link abnormal health problems (other than increased stress) to the accident.

Three Mile Island elevated public apprehension about nuclear power. It took 12 years and $1 billion to repair and reopen the other reactor at Three Mile Island. In the aftermath of the accident, public wariness prompted construction delays and cancellations of several new nuclear power plants across the United States.

The accident at Three Mile Island led to a stronger culture of safety practices and training in the nuclear industry. New safety regulations were put in place, including more frequent safety inspections, new risk assessments, and improved emergency and evacuation plans for nuclear power plants and surrounding communities. The meaning of the accident is still debated. Nuclear power proponents argue that since there was no radiation release, it is a success story. Opponents counter that luck was the only reason that there was not a more serious result and that, in any case, the public had been promised that events like that at Three Mile Island would never occur.

Chernobyl   One of the two worst nuclear power plant accidents took place in April 1986 at the Chernobyl plant, located in the former Soviet Union in what is now Ukraine. Explosions ripped apart a nuclear reactor, killing two workers and expelling large quantities of radioactive material into the atmosphere. The effects of this accident were not confined to the area immediately surrounding the power plant: Significant amounts of radioisotopes quickly spread across large portions of Europe. The Chernobyl accident affected and will continue to affect many nations.

The first task faced after the accident was to contain the fire that had broken out after the explosion and prevent it from spreading to other reactors at the power plant. Local firefighters, six of whom later died from exposure to the high levels of radiation, battled courageously to contain the fire. According to the World Health Organization (WHO), by the end of July 1986, 28 firefighters and plant personnel had died from radiation exposure. In addition, 116,000 people who lived within a 30-km (18.5-mi) radius around the plant were quickly evacuated and resettled.

Next, the radioactivity at the power station was cleaned up and contained so that it would not spread. Even in protective gear, workers could stay in the area for only a few minutes at a time. The damaged reactor building was encased in concrete, and the surrounding countryside was partially decontaminated. Highly radioactive soil was removed, and buildings and roads were scrubbed down to remove the radioactive dust that settled on them. A new containment structure is now being built around the plant to prevent intrusion and support further decontamination efforts.

Although cleanup outside the Chernobyl plant boundaries is finished, the people in Ukraine face many long-term problems. Farmland and forest contamination led to reduced agricultural production. Inhabitants in parts of Ukraine still cannot drink the water or consume local produce. The Ukrainian Scientific Center for Radiation Medicine has continually monitored the health of thousands of Chernobyl patients.

A 2003 report by the Ukrainian Intelligence Agency found that a range of factors contributed to the accident. These included violations of safety rules, flawed design, inferior construction, and operator errors. The main design flaw was the lack of a containment building and instability at low power. This type of reactor is not used commercially in either North America or Western Europe because nuclear engineers consider it too unsafe, and a Chernobyl-type accident could not occur at most reactors currently in use. Russia and Lithuania still have reactors like the one in Chernobyl in operation, although they have made safety improvements in them since the Chernobyl explosion.

Human error also contributed significantly. Many operators lacked scientific or technical training, and they made several major mistakes in response to the initial problem. As a result of the disaster at Chernobyl, the Soviet government developed a retraining program for operators at all its nuclear power plants, and safety features were added to existing reactors. Nonetheless, nuclear power plants require many operators, and human error will remain a concern.

One of the disquieting consequences of Chernobyl was the unpredictable course taken by spreading radiation (**Figure 12.17**). Chernobyl's radiation cloud unevenly dumped radioactive fallout over some areas of Europe and Asia, leaving other areas relatively untouched and making it difficult to plan emergency responses for a possible future nuclear accident.

As the health effects of the accident at Chernobyl are monitored over the years, the death toll appears less substantial than was originally expected, but still considerable. In the decade after the accident, epidemiologists observed increased incidence of leukemia in infants and of thyroid cancer and immune abnormalities in children exposed to radioactive iodine from the Chernobyl fallout. Nearly 400,000 adults and more than 1 million children have received government aid for related health problems. Many of those who were exposed to radiation from Chernobyl suffered from debilitating mental stress, anxiety, and depression, the economic and social cost of which are difficult to assess.

Breast cancer, stomach cancer, and other organ cancers are thought to be the major causes of radiation-related deaths, but these are so common in the general population that it is difficult to tell which were caused by radiation from Chernobyl. Thirty years after the accident, the number of cancer deaths is highly disputed, with estimates ranging from 4000 to 20,000. Estimates vary because of uncertainty about both exposure amounts and the effects of small doses of radiation. The most recent Biological Effects of Ionizing Radiation (BEIR VII) expert panel concluded that it is appropriate to assume that even minute doses of **ionizing radiation** are harmful. However, if there is a threshold, or lower level of exposure below which radiation does not cause cancer, that assumption would substantially overestimate the number of cancer-related deaths.

Nuclear reactor designers, engineers, and regulators have learned a great deal from the accident and subsequent cleanup and entombment at Chernobyl. Doctors have learned more about effective treatment of people exposed to massive doses of radiation. Ecologists have learned both about how nuclear materials move through the food chain and, more generally, about how natural systems change when large numbers of humans abruptly move away from an area.

Fukushima Daiichi   On March 11, 2011, a magnitude 9.0 earthquake struck Japan, causing considerable damage across the country. An ensuing tsunami, however, caused much more long-term harm, including severe damage to the Fukushima Daiichi nuclear power station. The plant is located on the northeastern coast of

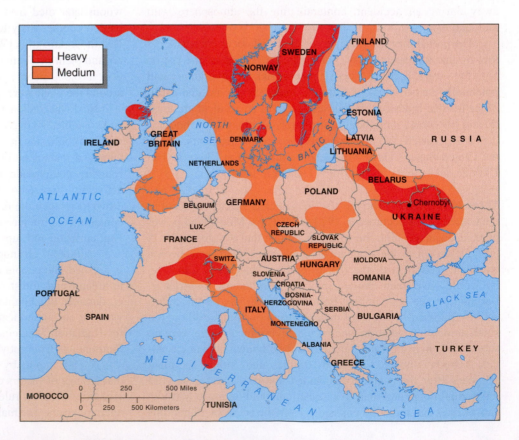

Figure 12.17 **Radioactive fallout from Chernobyl.** Parts of Ukraine, Belarus, Lithuania, Sweden, Norway, France, Italy, and Switzerland received heavy radioactive fallout from the accident.

Japan; the Fukushima Prefecture was one of the areas hit hardest by the tsunami.

The tsunami disrupted both the normal power supply and the backup systems that pump cooling water at the six reactors at Fukishima Daiichi. Without cooling water, the cores experienced meltdowns. At least one explosion occurred; this was not a nuclear explosion, but one caused by a buildup of hydrogen in an over-heated vessel (**Figure 12.18**). It is likely that the entire facility will be not only inoperable but unsafe to humans for decades or centuries.

Considerable amounts of radioactivity were released into the environment, mostly as contaminated cooling water that either leaked into groundwater or was intentionally released into the ocean. To avoid human exposure, the Japanese government evacuated everyone within 20 kilometers, and recommended that everyone within 30 kilometers evacuate as well. Eventually, over 400,000 people were displaced, many of them permanently.

According to a February 2013 World Health Organization (WHO) report, favorable wind conditions during the 2011 disaster dissipated the radiation, leaving only a few hot spots that have slightly higher risks for cancers in children and thyroid cancer in women. However, evacuations in the immediate aftermath of the event led to about 1,600 deaths due to untreated illnesses, as well as physical and psychological fatigue, especially among the elderly.

In addition, the economic effects have been substantial. Beyond the cost of evacuations—a burden made greater by the thousands of Japanese left homeless by the tsunami—farmers and fishing fleets in the surrounding area have been idled. Testing continues to reveal fishes with high radiation contamination, leading South Korea to ban Japanese fish and seafood. The nuclear power plant itself is permanently closed, at a cost of hundreds of millions of dollars. Other harm includes loss of life (several workers were killed trying to stabilize the reactors, and others were exposed to potentially fatal doses of radiation). As with Chernobyl, much of the human costs are associated with dislocation, and stress associated with uncertainty about possible exposures.

Studies since the accident suggest that the facility should have been better designed and prepared for an accident of this magnitude, both to reduce the effects and better respond when it did occur. Managing cooling water and other radiation releases continues to be a challenge at the site, and the extent of radiation contamination on land and in the ocean remain uncertain.

PUBLIC AND EXPERT ATTITUDES TOWARD NUCLEAR ENERGY

An ongoing challenge to both new nuclear power facilities and a permanent repository for high-level wastes is that advocates and opponents are engaged in a vicious cycle of distrust. The NIMBY, or "not in my backyard," response commonly follows a proposal for a nuclear power plant or waste facility. NIMBY prevails in nuclear energy issues partly because, despite expert assurances that a site will be safe, no one can guarantee complete safety, with no possibility of an accident.

The NIMBY response of opponents has parallels in the "irrational public" belief held by many pro-nuclear advocates and experts, as well as the DAD ("decide, announce, defend") approach taken by industry and government. Historical exposures from nuclear testing, misrepresentations of nuclear power safety that became known in the 1970s, and the "impossible" event at Three Mile Island have generated substantial public distrust of both the nuclear industry and governmental regulators. This distrust can be amplified by interactions with experts, who often have little understanding of the nature of public resistance or training in effectively communicating risks. Experts who encounter this distrust misinterpret it as irrationality or ignorance.

Unfortunately, the driving forces behind NIMBY, DAD, and distrust are difficult to resolve. Some nuclear power proponents, believing that the public is uninformed, react with campaigns explaining how "safe" a proposal is. The public feels insulted,

Figure 12.18 Damaged Fukushima Daiichi Nuclear Power Plant. This image, captured by an unmanned drone on March 24, 2011, shows the severely damaged Unit 3 (left) and Unit 4 (right). Note the proximity to the ocean.

ABACA / Newscom

which leads to further distrust and dismissal of expert arguments, which the experts again misinterpret as irrationality and antiscientific attitudes.

This vicious cycle persists between nuclear advocates who chronically misunderstand the nature of public opposition and the resistance of many people to anything nuclear. Breaking it will be essential for a nuclear renaissance and a successful nuclear waste management plan.

REVIEW

1. In generating electricity, how do the environmental effects of coal combustion and conventional nuclear fission compare?

2. What were the effects of the nuclear power plant accidents at Three Mile Island in Pennsylvania, Chernobyl in Ukraine, and Fukushima Daiichi in Japan?

Radioactive Wastes

LEARNING OBJECTIVES

- Distinguish between low-level and high-level radioactive wastes.
- Explain the pros and cons of on-site storage and deep geologic disposal.

Radioactive wastes are classified as either low-level or high-level. Produced by nuclear power plants, university research labs, nuclear medicine departments in hospitals, and industries, **low-level radioactive wastes** include glassware, tools, paper, water, and other materials contaminated by radioactivity. The **Low-Level Radioactive Waste Policy Act,** passed in 1980, specified that all states are responsible for the waste they generate, and it encouraged states to develop facilities to handle low-level radioactive wastes by 1996. Three sites currently accept waste for the entire country—Washington State, South Carolina, and Utah. New technologies to compact low-level radioactive waste have dramatically reduced the volume being disposed of in these nuclear dumps.

low-level radioactive wastes Radioactive solids, liquids, or gases that give off small amounts of ionizing radiation.

The primary **high-level radioactive wastes** produced during nuclear fission are spent fuel rods and assemblies. High-level radioactive wastes are also generated during the reprocessing of spent fuel. Produced by nuclear power plants and nuclear weapons facilities, high-level radioactive wastes are among the most dangerous human-made hazardous wastes.

high-level radioactive wastes Radioactive solids, liquids, or gases that give off large amounts of ionizing radiation.

Fuel rods absorb neutrons, thereby forming radioisotopes, and can be used for only about three years before becoming highly radioactive spent fuel. In 2010, the United States had about 370,000 cubic meters of spent fuel stored temporarily at 65 nuclear power plants around the country. Worldwide, about 15,000 metric tons of spent fuel are produced each year. As the radioisotopes in spent fuel decay, they produce considerable heat, are extremely toxic to organisms, and remain radioactive for thousands of years. Their dangerous level of radioactivity requires special handling. Secure storage of these materials must be guaranteed for thousands of years, until they decay sufficiently to be safe.

It is possible that nuclear power generation will increase substantially in the upcoming decades. In the long term, most waste experts conclude that deep geologic burial is the best solution.

For permanent disposal, the National Research Council has concluded that high-level radioactive wastes should be stored deep underground. An appropriate storage site must have geologic stability and little or no water flowing nearby, which might cause containers to corrode and transport the waste away. Currently, the United Sates has no designated waste storage location (see Case In Point: Yucca Mountain). Meanwhile, radioactive wastes continue to accumulate. Most commercially operated nuclear power plants store their spent fuel in huge indoor pools of water on-site. Storage in water is necessary for at least a few years because some of the materials with short half-lives decay. However, water-based storage is not appropriate for long periods.

Increasingly, nuclear wastes are being transferred from on-site wet storage to **dry cask storage,** which consists of large cylinders of concrete and steel that hold 10 or more metric tons of high-level waste. More than two dozen commercially operated plants have built dry casks, and many others have committed to do the same, due to uncertainty about the future of underground storage (**Figure 12.19**). There are two reasons why power plant owners object to dry cask storage. First, such facilities are extremely expensive to build. Second, once waste is transferred to dry cask storage, it is much safer than when it was stored in water. Consequently, the argument for shipping it to permanent underground storage becomes much less compelling.

Worldwide, most countries have committed to eventually having deep geologic disposal. However, to date no country has opened a permanent site. France continues to reprocess its fuel and has two potential sites under investigation. Neither Japan nor Canada currently has identified candidate sites.

WASTES WITH SHORT HALF-LIVES AND HIGH-LEVEL LIQUID WASTES

Some radioactive wastes are produced directly from the fission reaction. Uranium-235, the reactor fuel, may split in several different ways, forming smaller atoms, many of which are radioactive.

Figure 12.19 **Dry cask storage for spent fuel.** Each cask, designed to last at least 40 years, is monitored and will be replaced if leakage occurs.

Most of these, including krypton-85 (half-life, 10.4 years), strontium-90 (half-life, 28 years), and cesium-137 (half-life, 30 years), have relatively short-term radioactivity. In 300 to 600 years, they will have decayed to the point where they are safe.

The safe storage of fission products with relatively short half-lives is of concern because fission produces larger amounts of these materials than of the materials with extremely long half-lives. Health concerns exist because many of the shorter-lived fission products mimic essential nutrients and tend to concentrate in the body, where they continue to decay, with harmful effects (see the discussion of bioaccumulation in Chapter 7). One of the common fission products, strontium-90, is chemically similar to calcium. If strontium-90 was accidentally released into the environment from improperly stored radioactive waste, it could become incorporated into human and animal bones and teeth in place of calcium. In like manner, cesium-137 replaces potassium in the body and accumulates in muscle tissue, and iodine-131 concentrates in the thyroid gland.

High-level liquid wastes are dangerously unstable and difficult to monitor. For that reason, they are converted to solid form before they are stored. The U.S. government wants to store high-level liquid wastes in enormous glass or ceramic logs that are contained in stainless steel canisters. The glass or ceramic contains boron, which absorbs neutrons and so will prevent the logs from exploding. Solidifying liquid waste into solid glass or ceramic logs, known as **vitrification**, is an established practice in Europe. The United States began producing glass logs in 1996, although it does not yet have a permanent disposal site for these hazardous logs. High-level liquid wastes are temporarily stored in large underground tanks in New York, Washington State, Idaho, and South Carolina.

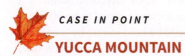

CASE IN POINT

YUCCA MOUNTAIN

In 1954, before the first commercial power plant opened, the U.S. government decided to be responsible for all high-level waste associated with electricity production. In 1982, passage of the **Nuclear Waste Policy Act** required the first site to be operational by 1998. It also legally required the U.S. government to take ownership of nuclear wastes. Since 1999, radioactive wastes from the manufacture of nuclear weapons have been stored permanently in deep underground salt beds at the Waste Isolation Pilot Project (WIPP) near Carlsbad, New Mexico. However, the deadline for completion of an operational high-level radioactive civilian waste repository was postponed from 1998 to 2010 to 2017—and probably will be postponed again.

In a 1987 amendment to the Nuclear Waste Policy Act, Congress identified **Yucca Mountain** in Nevada as the only candidate being formally considered as a permanent underground storage site for high-level radioactive wastes from commercially operated power plants (**Figure 12.20**). Yucca Mountain could take the 42,000-plus tons of spent fuel produced in the United States to date plus spent fuel that will be produced until about 2025. At that time, Yucca Mountain would be full, and a new geologic depository will be needed.

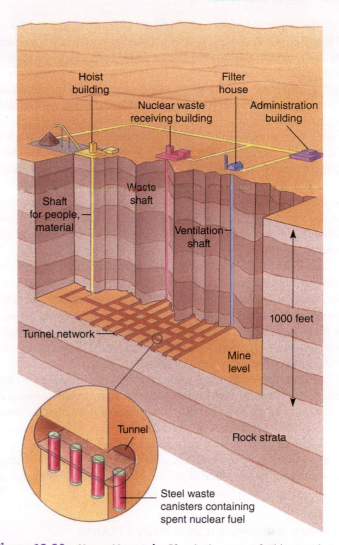

Figure 12.20 **Yucca Mountain.** If and when spent fuel is stored at Yucca Mountain, it will be in a huge complex of interconnected tunnels located in dense volcanic rock 300 m (1000 ft) beneath the mountain crest. Canisters containing high-level radioactive waste may be stored in the tunnels.

The U.S. Department of Energy (DOE) has spent billions of dollars conducting feasibility studies on Yucca Mountain's geology. Results indicate that the site is safe, at least from volcanic eruptions and earthquakes. However, the suitability of Yucca Mountain has been mired in scientific, management, cost, public opposition, and scheduling controversies, and Nevadans oppose the selection of their state for a radioactive waste site. In 2002 Congress finally approved the choice of Yucca Mountain as the U.S. nuclear waste repository, despite the state of Nevada's opposition. In 2004, U.S. federal courts decided that any permanent burial site must meet standards of the EPA for the next 1 million years—an increase from the previous 10,000-year standard. Guaranteeing safety over a million-year period stretches the credibility of any scientific assessment. In 2008, the DOE formally submitted its application for a license to operate the facility, but under the Obama administration, the application has been withdrawn. For some time, the nuclear industry has been weighing whether to file a lawsuit to require that the process go forward.

The Yucca Mountain site, some 145 km (90 mi) northwest of Las Vegas, is controversial in part because it is near a volcano (its last eruption may have occurred 20,000 years ago) and active earthquake fault lines. The possibility of a volcanic eruption at Yucca Mountain is currently considered remote (1 chance in 10,000 during the next 10,000 years). Concerns that earthquakes might disturb the site and raise the water table, resulting in radioactive contamination of air and groundwater, were examined when a magnitude 5.6 earthquake occurred in 1992 about 20 km (about 12 mi) from Yucca Mountain. Scientists were already monitoring water table elevation, and they measured a 1-m change caused by the earthquake. The water table is some 800 m (2625 ft) beneath the mountain crest, so water elevation changes due to earthquake activity are not considered a serious problem by most (but not all) experts.

Transporting high-level wastes from nuclear reactors and weapons sites is a major concern of opponents of the Yucca Mountain site. The typical shipment would travel an average of 2,300 miles, and 43 states would have these materials passing through on their way to Yucca Mountain. Eight states—Illinois, Indiana, Iowa, Kansas, Missouri, Nebraska, Utah, and Wyoming—would contain major transportation corridors to Yucca Mountain.

Around the world, more than two dozen countries plan to store their spent uranium fuel in deep underground deposits. Many of these countries are having similar challenges of public opposition to proposed sites. France, long considered a nuclear power success story, has been unable to select a permanent site. In Sweden, apparent agreement on a site came only after the country agreed to discontinue nuclear power—that decision has since been revisited as energy demand and climate change concerns increase. ■

DECOMMISSIONING NUCLEAR POWER PLANTS

Nuclear power plants are initially licensed to operate for 40 years, although the Nuclear Regulatory Commission (NRC) has granted permission for 49 reactors to extend their operating licenses for an additional 20 years. As nuclear power plants age, certain critical sections, such as the reactor vessel, become brittle or corroded. Consequently, extensions are granted when operators demonstrate that safety systems and physical structures can be maintained for that long. At the end of their operational usefulness, nuclear power plants are not simply abandoned or demolished because many parts have become contaminated with radioactivity.

Three options exist when a nuclear power plant is closed: storage, entombment, and decommissioning. If an old plant is put into storage, the utility company guards it for 50 to 100 years, while some of the radioactive materials decay. This decrease in radioactivity makes it safer to dismantle the plant later, although accidental leaks during the storage period are still a concern.

Most experts do not consider **entombment**, permanently encasing the entire power plant in concrete, a viable option because the tomb would have to remain intact for at least 1000 years. Accidental leaks would likely occur during that time, and we cannot guarantee that future generations would inspect and maintain the "tomb."

decommission To dismantle an old nuclear power plant after it closes.

The third option for the retirement of a nuclear power plant is to **decommission** the plant immediately after it closes. The workers who dismantle the plant must wear protective clothing. Some portions of the plant are too radioactive for workers to safely dismantle, although advances in robotics may make it feasible to tear down these sections. As the plant is torn down, small sections of it are transported to a permanent low-level radioactive waste storage site.

According to the International Atomic Energy Agency, worldwide, 110 nuclear power plants were permanently retired as of 2010 (23 of these in the United States), and many nuclear power plants are nearing retirement age. Also as of 2010, 145 operational plants around the world were 30 years old or older (the oldest two were 47 years old). In the United States, nuclear power plant operators must establish a decommissioning fund at the time a facility is built.

REVIEW

1. What is low-level radioactive waste, and how is it disposed of? What is high-level radioactive waste, and how is it currently stored?
2. What are some of the advantages of storing high-level radioactive waste at Yucca Mountain? What are some of the disadvantages?

The Future of Nuclear Power

LEARNING OBJECTIVE

• Briefly describe how future nuclear power reactors might differ from current plants.

In an effort to promote nuclear energy, nuclear and utility executives have developed a plan that addresses the safety and economic issues associated with nuclear power. They envision building a series of "new generation" nuclear reactors designed to be 10 times safer than current reactors. Costs could be held in line by standardizing nuclear power plants rather than custom-building each one. When France standardized its plant designs, costs were lowered considerably. The plan calls for improving building schedules and streamlining the regulatory process.

The NRC has certified four new advanced reactor designs and is reviewing three others. Nuclear experts consider the new fission reactors safer than the nuclear reactors that are currently operating in the United States. The new generation of nuclear power plants, though smaller, simpler in design, less expensive to build, and safer to operate, will still produce high-level radioactive wastes and have a potential link to nuclear weapons, but they will produce little or no greenhouse gases.

The International Atomic Energy Agency classifies reactors as large if they produce more than 700 MW of electricity. The earliest commercial nuclear reactors were quite small, but by the end of 2012, 305 of the world's 439 reactors were large; the remaining 134 were small or medium reactors (SMRs). It is likely that in the future SMRs will make up a larger proportion of the world's nuclear inventory.

SMRs have several advantages. First, new technologies, including the Pebble Bed Modular Reactor being developed in South Africa and pressurized water reactors being developed in the Republic of Korea and Argentina, are easily scalable—that is, they can be made bigger or smaller as needed. Historically, most nuclear reactors were designed in one size—large. Second, in many places, due to energy demand and transmission infrastructure, smaller reactors are more appropriate. Any possible nuclear renaissance will probably consist of many smaller facilities rather than a few large ones.

Fusion, the atomic reaction that powers the stars, may one day be an important source of energy. In fusion, two lighter atomic nuclei are brought together under conditions of high heat and pressure so that they combine, producing a larger nucleus. Fusion makes the energy produced by burning fossil fuels seem trifling: 30 mL (1oz) of fusion fuel has the energy equivalent of 270,000 L (70,000 gal) of gasoline.

Isotopes of hydrogen are the fuel for fusion. In one type of fusion reaction, the nuclei of deuterium and tritium combine to form helium, releasing huge amounts of thermal energy (heat) in the process—heat that would be converted to electricity. Deuterium, or heavy hydrogen, is present in water and is relatively easy to separate from normal hydrogen. Tritium, which is radioactive,

is not found in nature; this human-made hydrogen isotope is formed during the fusion reaction by bombarding another element, lithium, with neutrons.

Unfortunately, many technological difficulties have been encountered in efforts to stage a controlled fusion reaction. It takes phenomenally high temperatures (millions of degrees) to make atoms fuse. To date, the best fusion experiments generate about one-third of the energy used to heat the fuel.

Another challenge is confining the fuel. At extremely high temperatures, a gas separates into negative electrons and positive nuclei. This superheated, ionized gas, called **plasma**, has a tendency to expand. Confinement of the plasma is necessary so that the nuclei are close enough to one another to fuse, but a regular container does not work because as soon as the nuclei hit the container walls, they lose so much energy they cannot fuse.

REVIEW

1. How will future nuclear power reactors differ from current plants?

REVIEW OF LEARNING OBJECTIVES WITH SELECTED KEY TERMS

- **Distinguish between active and passive solar heating and describe how each is used.**

Passive solar heating is a system of putting the sun's energy to use without requiring mechanical devices (pumps or fans) to distribute the collected heat. Currently, about 7% of new homes built in the United States have passive solar features. **Active solar heating** is a system of putting the sun's energy to use in which a series of collectors absorb the solar energy, and pumps or fans distribute the collected heat. Active solar heating is used for heating water and, to a lesser extent, space heating.

- **Contrast the advantages and disadvantages of solar thermal electric generation and photovoltaics for converting solar energy into electricity.**

Solar thermal electric generation is a means of producing electricity in which the sun's energy is concentrated by mirrors or lenses onto a fluid-filled pipe; the heated fluid is used to generate electricity. Solar thermal plants are not yet cost-competitive with traditional fuels, but they are more efficient than other direct solar technologies, and they do not produce air pollution or contribute to acid rain or global climate change. **Photovoltaics (PV)** include **solar cells**, wafers or thin-films of solid-state materials, such as silicon or gallium arsenide, that are treated with certain metals so that they generate electricity—that is, a flow of electrons—when they absorb solar energy. PV devices generate electricity with no pollution and minimal maintenance, but most are only about 10% to 15% efficient at converting solar energy to electricity.

- **Define** *biomass,* **explain why it is an example of indirect solar energy, and outline how it is used as a source of energy.**

Biomass consists of plant material that is used as fuel. Biomass is an example of indirect solar energy because it includes organic materials produced by photosynthesis. Biomass is burned directly to produce heat or electricity or converted to solid (charcoal), gas (**biogas**), or liquid (**methanol** and **ethanol**) fuels. Biomass is already being used for energy on a large scale, particularly in developing nations. India and China have several million biogas digesters that produce biogas from household and agricultural wastes.

- **Describe the locations that can make optimum use of wind energy and of hydropower, and compare the potential of wind energy and hydropower.**

Wind energy is electric energy obtained from surface air currents caused by the solar warming of air. Harvesting wind energy to generate electricity has great potential because it is currently the most cost-competitive of all forms of solar energy. Harnessing wind energy is most profitable in areas with fairly continual winds, such as islands, coastal areas, mountain passes, and grasslands. **Hydropower** relies on flowing or falling water to generate electricity. The damming of rivers and streams to generate electricity is the major form of hydropower. Currently, hydropower produces about 19% of the world's electricity. Environmental and social problems associated with hydropower include ecological destruction upstream and downstream, increased evaporation of water, disease and pollution, displacement of people, and inundation of farmland.

- **Describe geothermal energy and tidal energy, the two forms of renewable energy that are not direct or indirect results of solar energy.**

Geothermal energy is the use of energy from Earth's interior for either space heating or the generation of electricity. Geothermal energy can be obtained from hydrothermal reservoirs of heated water near Earth's surface. The established technology for extracting geothermal energy from heated areas of Earth's crust involves drilling wells and bringing the steam or hot water to the surface. **Tidal energy** is a form of renewable energy that relies on the ebb and flow of the tides to generate electricity; it is currently used on a very limited scale.

- **Distinguish between nuclear energy and chemical energy.**

In ordinary chemical reactions, the atoms of one element do not change into atoms of another element, nor does any of their mass (matter) change into energy. In contrast, **nuclear energy** is the energy released by nuclear **fission** or **fusion**. In nuclear energy small amounts of matter from atomic nuclei are converted into large amounts of energy.

- **Describe the nuclear fuel cycle, including the process of enrichment.**

The **nuclear fuel cycle** includes the processes involved in producing the fuel used in nuclear reactors and in disposing of radioactive wastes (or nuclear wastes). **Enrichment**, which is part of the nuclear fuel cycle, is the process by which uranium ore is refined after mining to increase the concentration of fissionable U-235.

- **Define *nuclear reactor* and describe a typical nuclear power reactor.**

A **nuclear reactor** is a device that initiates and maintains a controlled nuclear fission chain reaction to produce energy for electricity. A typical reactor contains a **reactor core**, where fission occurs; a **steam generator**; a **turbine**; and a **condenser**. The reactor core contains **fuel rods** filled with pellets of uranium dioxide. Above each **fuel assembly** is a **control rod** that is moved into or out of the fuel assembly, thereby producing the amount of fission required. The fission of U-235 releases heat that converts water to steam, used to generate electricity. Safety features include a **containment building** built of steel-reinforced concrete. Conventional nuclear fission uses U-235, which makes up about 3% to 5% of uranium after enrichment. **Spent fuel** is the used fuel elements that were irradiated in a nuclear reactor.

- **Discuss the pros and cons of electric power produced by nuclear energy.**

Nuclear power can serve as an alternative to electricity generation from coal and natural gas, both of which contribute greenhouse gases to the atmosphere. One reason proponents of nuclear energy argue for its widespread adoption is that it has less of an environmental impact than fossil fuels, particularly coal. Nuclear energy emits few pollutants into the atmosphere and provides power without producing carbon dioxide. However, it generates highly radioactive waste such as spent fuel; every country that uses nuclear power is seeking a permanent waste-disposal site. Safety remains a concern at nuclear power plants.

- **Describe the nuclear power plant accidents at Three Mile Island, Chernobyl, and Fukushima Daiichi.**

The most serious nuclear reactor accident in the United States occurred in 1979 at the Three Mile Island power plant in Pennsylvania. A partial meltdown of the reactor core took place, although the containment building kept almost all the radioactivity from escaping. In 1986 a nuclear reactor accident occurred in Chernobyl in the former Soviet Union (now Ukraine). One or two explosions ripped apart a nuclear reactor and expelled large quantities of radioactive material into the atmosphere, resulting in widespread environmental pollution as well as serious local contamination. At the Fukushima Daiichi nuclear power plant in Japan, three reactors melted down following inundation by a tsunami in March 2011. Over 400,000 people were evacuated, and the impacts of the accident and resulting radiation will continue to have impacts for decades to come.

- **Distinguish between low-level and high-level radioactive wastes.**

Low-level radioactive wastes are radioactive solids, liquids, or gases that give off small amounts of ionizing radiation. **High-level radioactive wastes** are solids, liquids, or gases that initially give off large amounts of ionizing radiation.

- **Explain the pros and cons of on-site storage and deep geologic disposal.**

On-site storage in liquid is necessary for a short period of time, as spent fuel rods cool. However, spent fuel rods cannot be safely stored in liquid for long periods. The National Research Council suggests that deep geologic storage is the safest and most secure long-term solution for high-level nuclear wastes. Until a storage facility is licensed and constructed, on-site **dry cask storage** in large, expensive steel and concrete cylinders will be increasingly common.

- **Briefly describe how future nuclear power reactors might differ from current plants.**

The next generation of nuclear power plants is expected to be smaller, simpler, safer, and more cost-effective than the reactors currently in use around the world. The NRC has already approved four new designs, and more are proposed. Nuclear fusion as a source of energy is many years from becoming a reality.

Critical Thinking and Review Questions

1. Explain the following statement: Unlike fossil fuels, solar energy is not resource-limited but is technology-limited.

2. Biomass is considered an example of indirect solar energy because it is the result of photosynthesis. Given that plants are the organisms that photosynthesize, why are animal wastes considered biomass?

3. One advantage of the various forms of renewable energy, such as solar, thermal, and wind energy, is that they cause no net increase in atmospheric carbon dioxide. Is this true for biomass? Why or why not?

4. Some energy experts refer to the Great Plains states as "the Saudi Arabia of wind power." Explain what the reference means.

5. Japan wishes to make use of solar power, but it does not have extensive tracts of land for building large solar power plants. Which solar technology do you think is best suited to Japan's needs? Why?

6. Why is the potential for future energy from wind energy in the United States greater than the potential from hydropower?

7. How does nuclear energy differ from chemical energy?

8. What are the main steps in the nuclear fuel cycle? Is it a true cycle? Explain your answer.

9. How is breeder nuclear fission different from conventional nuclear fission?

10. What is spent fuel?

11. Should the air pollution (including greenhouse gases) produced from fossil fuels used to mine and refine uranium, build nuclear power plants, and transport nuclear wastes be attributed to nuclear energy? Why or why not?

12. How does the disposal of radioactive wastes pose technical problems? Political problems?

13. What are the main arguments for and against the United States developing additional nuclear power plants to provide us with electricity over the next several decades? Which perspective do you find most convincing?

14. The graph to the right depicts the percent growth of global electricity produced by different energy sources from 2004 to 2012. If these growth rates continue, what percent could wind and solar provide in 2020? 2050?

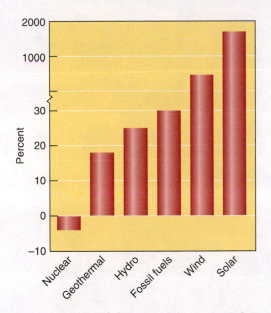

Average Annual Growth in Global Electricity, by Fuel, from 2004 to 2011.

Food for Thought

The use of large amounts of corn to produce methanol that powers motor vehicles is controversial in the United States. One of the major issues is concern that corn can be used either as food for people and livestock or as a biomass-based energy source. Research this issue. Is there evidence that using corn for fuel increases the cost of corn for food? How much of the U.S. corn crop is used for each? How does this concern change if we think about turning corn oil that has already been used for cooking into biodiesel?

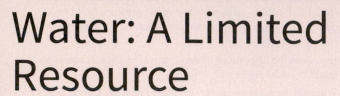

13

Water: A Limited Resource

Domestic water in Lagos, Nigeria. One woman in this photo from Lagos, Nigeria, is filling a bucket with water from a communal tank. In many less less-developed countries, access to safe, clean water for household use is limited and expensive. Poor residents of large towns often pay more for water than do the wealthy, spend a much larger fraction of their time acquiring it, and use part of their similarly limited fuel supply boiling it.

Stuart Franklin/Magnum Photos

Around the world, about 765 million people live without adequate access to water—many have fewer than 10 L (about 2.6 gal) of clean water per day. Worldwide, water is often limited in quality, quantity, or both. This is difficult to imagine for most people in the United States, Canada, and other developed nations, for nearly all of whom water is ample and clean. The typical American uses about 340 L (90 gal) each day, and often waters plants and washes cars with water that is safe enough to drink.

The UN Development Programme (UNDP) estimates that inhabitants of the slums in Lagos, Nigeria (**see photograph**), pay 5 to 10 times as much for water as do those in wealthier neighborhoods. In some places, the poor spend as much as 20% of their income on water. Contrast this with the 0.15% spent by one of the authors of this book.

A systems perspective helps us understand why water is both unsafe and unavailable in much of the world. Consider a typical small Canadian town. A public or private utility company purchases water, often from a single source. The utility company transports and cleans the water, which it then distributes through an infrastructure of pumps and pipes. This complex system requires employees; computerized information processing; coordination with government, other utilities, and consumers; energy to pump the water; and both energy and chemicals to clean it. Total cost is large, but coordination makes the cost per person relatively low. Consumers have a reliable source of water at a predictable cost. Different parts of the system are managed by experts, who are held accountable.

Contrast this with the slums of Lagos. The same system components exist: Water must be cleaned, transported, bought, and sold, all of which require people, energy, and organization. However, there is little money to develop and maintain the infrastructure, so water from pipes is sporadic, if available at all. When the water does run, it might be contaminated from contact with human or animal waste. An alternative is to buy water from a vendor's tanks or jugs. But freelance water dealers are unpredictable, the quality of the water is unknown, and prices vary from day to day. A consumer might boil the water to kill off biological contaminants, but this requires energy, which is also in limited supply.

As part of its report, the UNDP proposed that access to enough safe water—at least 20 L (5.2 gal) per day—should be considered a basic human right. Great progress has been made in many nations toward providing access to a reliable water supply, but Nigeria lags behind. Proposals that might alleviate some of the problems in Lagos include:

- Make water available at a low cost to poor residents, with a price that goes up with higher usage.
- Provide public financing for water infrastructure development.
- Include water access as part of broader poverty-reduction programs.
- Hold water providers accountable for consistency and safety.

While enough water exists in the world for all people, problems of distribution and quality assurance, increasing global population, and water supply disruption due to climate change make universal access to water an issue that will face us for decades to come.

In Your Own Backyard…
What is your main source for drinking water?

The Importance of Water

LEARNING OBJECTIVES

- Describe the structure of a water molecule and explain how hydrogen bonds form between adjacent water molecules.
- Describe surface water and groundwater.

The view of planet Earth from outer space reveals that it is different from other planets in the solar system. Earth is a predominantly blue planet because of the water that covers three-fourths of its surface. Water has a tremendous effect on our planet: It helps shape the continents, it moderates our climate, and it allows organisms to survive.

Life on Earth would be impossible without water. All life-forms, from unicellular bacteria to multicellular plants and animals, contain water. Humans are approximately 70% water by body weight. We depend on water for our survival as well as for our convenience: We drink it; cook with it; travel on it; wash with it (**Figure 13.1**); and use an enormous amount of it for agriculture, manufacturing, mining, energy production, and waste disposal.

Although Earth has plenty of water, about 97% of it is salty and not available for use by most terrestrial organisms. Fresh water is distributed unevenly, resulting in serious regional water supply problems. Conflicts often arise because water used for one purpose is not available for others. Even regions with readily available fresh water have problems maintaining the quality and quantity of water.

Worldwide, freshwater use is increasing. This is in part because the human population is expanding and in part because individuals are using more water. Humans now use well over 50% of Earth's accessible, renewable fresh water. A growing number of countries experience water shortages as population growth, human activities, and climate change place increasing demands on a limited water supply.

PROPERTIES OF WATER

Water is composed of molecules of H_2O, each consisting of two atoms of hydrogen and one atom of oxygen. Water exists in any of three forms: solid (ice), liquid, and vapor (water vapor or steam). Water molecules are **polar**—that is, one end of the molecule has a positive electrical charge, and the other end has a negative charge (**Figure 13.2**). The negative (oxygen) end of one water molecule is attracted to the positive (hydrogen) end of another water molecule, forming a **hydrogen bond** between the two molecules. Hydrogen bonds are the basis for many of water's physical properties, including its high melting/freezing point (0°C, 32°F) and high boiling point (100°C, 212°F). Because most of Earth has a temperature between 0°C and 100°C, most water exists in the liquid form that organisms need.

Water absorbs a great deal of solar heat without its temperature rising substantially. This high *heat capacity* allows the ocean to moderate climate, particularly along coastal areas, and the ocean does not experience the wide temperature fluctuations common on land.

Water must absorb a lot of heat before it **vaporizes**, or changes from a liquid to a vapor. When it does evaporate, it carries the heat, called *heat of vaporization*, with it into the atmosphere. Thus, evaporating water has a cooling effect. That is why your body is cooled when perspiration evaporates from your skin.

Water is sometimes called the "universal solvent." While this is an exaggeration, many materials do dissolve in water. In nature, water is never completely pure because it contains dissolved gases

(a) Each water molecule consists of two hydrogen atoms and one oxygen atom. Water molecules are polar, with positively and negatively charged areas.

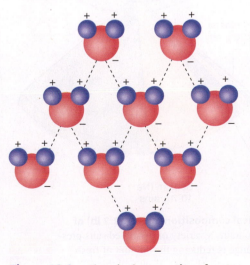

(b) The polarity causes hydrogen bonds (represented by dashed lines) to form between the positive areas of one water molecule and the negative areas of others. Each water molecule forms up to four hydrogen bonds with other water molecules.

Figure 13.2   Chemical properties of water.

Jodi Cobb/ National Geographic Creative

Figure 13.1   Young brick workers in India bathe with water from an irrigation pipe.

from the atmosphere and dissolved mineral salts from the land. Seawater contains a variety of dissolved salts, including sodium chloride, magnesium chloride, magnesium sulfate, calcium sulfate, and potassium chloride (**Figure 13.3**). Water's ability as a solvent has a major drawback: Many pollutants dissolve in water.

In general, water expands when heated and contracts when cold. As water cools, it contracts and becomes denser until it reaches 4°C (39°F), the temperature at which it is densest. When the temperature of water falls below 4°C, it becomes less dense. Ice (at 0°C) floats on the denser, slightly warmer, liquid water. Water freezes from the top down rather than from the bottom up, and aquatic organisms survive beneath the frozen surface.

THE HYDROLOGIC CYCLE AND OUR SUPPLY OF FRESH WATER

In the **hydrologic cycle**, water continuously circulates through the environment, from the ocean to the atmosphere to the land and back to the ocean (see Figure 4.6). The result is a balance among water in the ocean, on the land, and in the atmosphere. The hydrologic cycle continually renews the supply of fresh water on land, which is essential to terrestrial organisms.

Approximately 98% of Earth's water is in the ocean and contains a high amount of dissolved salts (**Figure 13.4**). Seawater

is too salty for human consumption and for most other uses. For example, if you watered your garden with seawater, your plants would die. Most fresh water is unavailable for easy consumption because it is frozen as polar or glacial ice or is in the atmosphere or soil. Lakes, creeks, streams, rivers, and groundwater account for only a small portion—about 0.53%—of Earth's fresh water.

Surface water is water found on Earth's surface in streams and rivers; lakes, ponds, and reservoirs; and **wetlands**, areas of land covered with water for at least part of the year. The **runoff** of precipitation from the land replenishes surface waters and is considered a renewable, though finite, resource. A **drainage basin**, or **watershed**, is the area of land drained by a single river or stream. Watersheds range in size from less than 1 km² for a small stream to a huge portion of the continent for a major river system such as the Mississippi River. **Table 13.1** lists the world's 10 largest watersheds.

Earth contains underground formations that collect and store water. This water originates as precipitation that seeps into the soil and finds its way down through cracks and spaces in sand, gravel, or rock until it is stopped by

surface water Precipitation that remains on the surface of the land and does not seep down through the soil.

runoff The movement of fresh water from precipitation (including snowmelt) to rivers, lakes, wetlands, and, ultimately, the ocean.

drainage basin A land area that delivers water into a stream or river system.

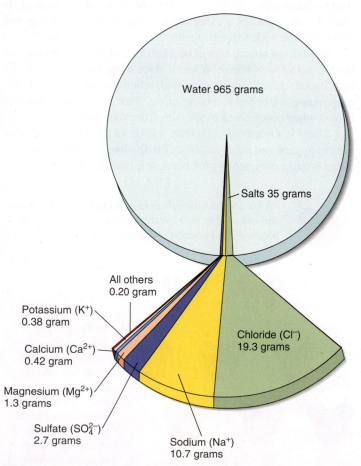

Figure 13.3 **Chemical composition of 1 kg (2.2 lb) of seawater.** Seawater contains a variety of dissolved salts present as ions. Climate change is reducing the amount of fresh water contained in ice caps and glaciers. How will this affect the composition of seawater?

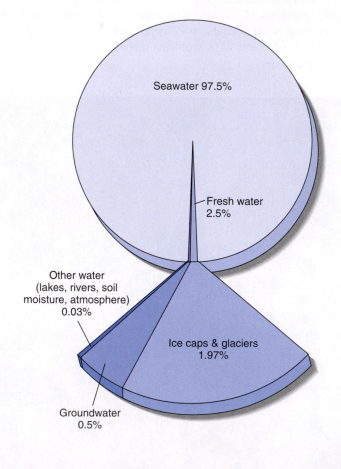

Figure 13.4 **Distribution of water.** Although three-fourths of Earth's surface is covered with water, substantially less than 1% is available for humans. Most water is salty, frozen, or inaccessible in the soil and atmosphere.

Table 13.1   **The World's 10 Largest Watersheds**

Watershed	Region	Area of Watershed (thousand km²)
Amazon	South America	6145
Congo	Africa	3731
Nile	Africa	3255
Mississippi	North America	3202
Ob	Asia	2972
Paraná	South America	2583
Yenisey	Asia	2554
Lena	Asia	2307
Niger	Africa	2262
Yangtze	Asia	1722

Source: Water Resources of the World, World Resources Institute (2010).

an impenetrable layer; there it accumulates as **groundwater**. Groundwater flows through permeable sediments or rocks slowly, typically covering distances of several millimeters to a few meters per day. Eventually, it is discharged into rivers, wetlands, springs, or the ocean. Thus, surface water and groundwater are interrelated parts of the hydrologic cycle.

Aquifers are underground reservoirs that are either unconfined or confined (**Figure 13.5**). In **unconfined aquifers**, the layers of rock above are porous and allow surface water directly above them to seep downward, replacing the aquifer contents. The upper limit of an unconfined aquifer, below which the ground is saturated with water, is the **water table**. The water table varies in depth depending on the amount of precipitation occurring in an area. In deserts, the water table is generally far below the surface. In contrast, lakes, streams, and wetlands occur where the water table intersects with the surface. A well goes dry when the water table has dropped below the depth of the well.

groundwater The supply of fresh water under Earth's surface that is stored in underground aquifers.

aquifers Underground caverns and porous layers of sand, gravel, or rock in which groundwater is stored.

water table The upper surface of the saturated zone of groundwater.

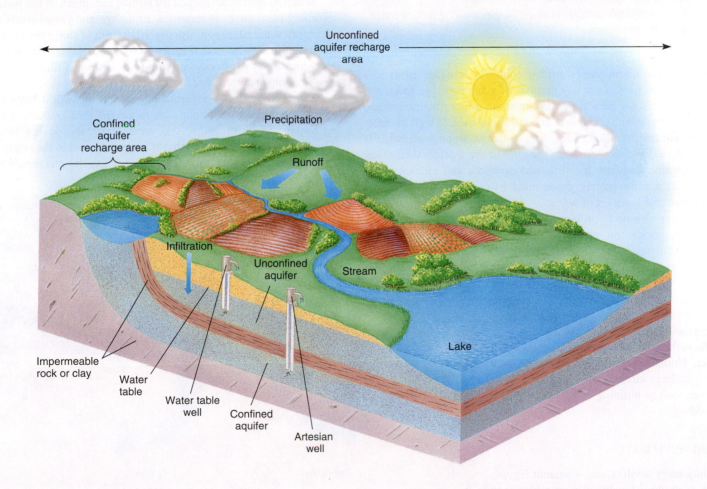

Figure 13.5 **Groundwater.** Excess surface water seeps downward through soil and porous rock layers until it reaches impermeable rock or clay. An unconfined aquifer has groundwater recharged by surface water directly above it. In a confined aquifer, groundwater is stored between two impermeable layers and is often under pressure. Artesian wells, which produce water from confined aquifers, often do not require pumping because of this pressure.

A **confined aquifer**, or **artesian aquifer**, is a groundwater storage area between impermeable layers of rock. The water in a confined aquifer is trapped and often under pressure. Its recharge area (the land from which water percolates to replace groundwater) may be hundreds of kilometers away.

Most groundwater is considered a nonrenewable resource because it has taken hundreds or even thousands of years to accumulate, and usually only a small portion of it is replaced each year by percolation of precipitation. The recharge of confined aquifers is particularly slow.

REVIEW

1. How do hydrogen bonds form between adjacent water molecules?
2. What is surface water? Groundwater?

Water Use and Resource Problems

LEARNING OBJECTIVES

- Describe the role of irrigation in world water consumption.
- Define *floodplain* and explain and evaluate flood control efforts, using the Mississippi River basin as an example.
- Relate some of the problems caused by overdrawing surface water and aquifer depletion.

Water consumption varies among countries, ranging from several liters per person per day in areas of acute shortage to several hundred liters per person per day in some more affluent nations. This encompasses agricultural and industrial uses as well as direct individual consumption. The greatest user of water worldwide is agriculture. Irrigation accounts for 70% of the world's total water consumption, industry for 19%, and domestic and municipal use for 11%, a ratio largely determined by countries that use a lot of water.

Water resource problems fall into three categories: too much, too little, and poor quality/contamination. (Chapter 21 addresses the third category.) We cannot prevent floods and droughts because they are part of natural climate variations. However, human activities often exacerbate their seriousness, and climate change affects their frequency, severity, and location. Humans often court disaster when they make environmentally unsound decisions, such as building in an area prone to flooding.

TOO MUCH WATER

Many early civilizations—ancient Egypt, for example—developed near rivers that periodically spilled over, inundating the surrounding land with water. When the water receded, a thin layer of sediment rich in organic matter remained and enriched the soil. These civilizations flourished partly because of their agricultural productivity, which was the result of floods replenishing nutrients in the soil.

Flooding results from system interactions among human activities and natural phenomena. Modern floods are more disastrous in terms of property loss than those of the past because humans often remove water-absorbing plant cover from the soil and construct buildings on **floodplains**. These activities increase the likelihood of both floods and flood damage.

floodplain The area bordering a river channel that has the potential to flood.

Forests, particularly on hillsides and mountains, trap and absorb precipitation to provide nearby lowlands with some protection from floods. When woodlands are cut down, particularly if they are clear-cut, the altered area cannot hold water nearly as well. Heavy rainfall then results in rapid runoff from the exposed, barren hillsides, which causes soil erosion and puts lowland areas at extreme risk of flooding.

When a natural area—an area undisturbed by humans—is inundated with heavy precipitation, the plant-protected soil absorbs much of the excess water. Water not absorbed runs off into a river channel, which may then spill over its banks onto the floodplain. Because rivers meander, the flow is slowed, and the swollen waters rarely cause significant damage to the surrounding area.

When an area is developed for human use, much of the water-absorbing plant cover is removed. Buildings and paved roads do not absorb water, so runoff, usually in the form of storm sewer runoff, is significantly greater (**Figure 13.6**). People who build homes or businesses on the floodplain of a river will most likely experience flooding at some point.

Increasingly, local governments around the world have put zoning restrictions on floodplains to curtail development. During floods in California in 2006, many expensive levees failed (**Figure 13.7a**). Rather than rebuild levees adjacent to the rivers to try to prevent floods (**Figure 13.7b**), river experts recommended

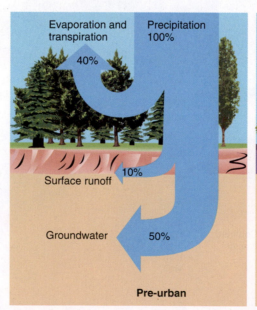
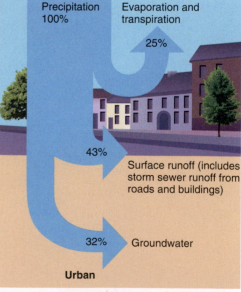

(a) The fate of precipitation in Ontario, Canada, before urbanization.

(b) After Ontario was developed, surface runoff increased substantially, from 10% to 43%.

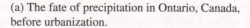

Figure 13.6 **How development changes the natural flow of water.**

that California try a different approach in which rivers are allowed to occupy part of the floodplain during a flood (**Figure 13.7c**). Smaller levees are built some distance from the river's edge. The new approach is less expensive, results in less damage during floods, and provides some of the natural benefits of floods, such as improved habitat for waterfowl and other wildlife and replenishment of the soil in the floodplain.

(a) A levee broke and water surged onto the floodplain in Merced, California, in 2006.

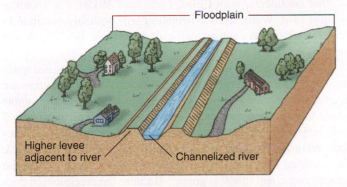

Floodplain

Higher levee adjacent to river — Channelized river

(b) To control flooding, many rivers are channelized (straightened and deepened), with high levees adjacent to the river. This flood-control method is expensive and may or may not prevent floods.

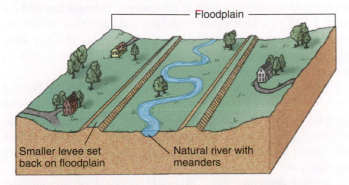

Floodplain

Smaller levee set back on floodplain — Natural river with meanders

(c) Scientists now recommend letting rivers meander naturally through much of their floodplains, with smaller levees set back some distance from the rivers. The floodplains absorb much of the river's water, forming a buffer between the river and developed areas. When a flooding river spills over its banks, it creates a wetland that is an important wildlife habitat.

Figure 13.7   Flood management.

CASE IN POINT

THE FLOODS OF 2011

Major flooding occurs periodically in the central United States. The Mississippi River, which spans 31 states and two Canadian provinces, floods frequently. A 1993 flood engulfed farms and towns in nine midwestern states, spreading over 9.3 million hectares (23 million acres) and causing property damage in excess of $12 billion. Major Mississippi River flooding also took place in 2001 and 2008.

In late spring of 2011, unusually heavy rains in the upper reaches of the floodplain, coupled with abundant snowmelt, triggered near-record flooding that moved through the Mississippi River drainage basin for weeks. According to the U.S. Census Bureau, flooding in Illinois, Missouri, Kentucky, Arkansas, Tennessee, Mississippi, and Louisiana affected more than 43,000 people and caused approximately $2.8 billion in damage to more than 21,000 homes and businesses and half a million hectares (1.2 million acres) of farmland (**Figure 13.8**).

The elaborate flood-control system along the Mississippi—constructed and controlled by the Army Corps of Engineers—is the world's largest, with its levees, floodways, spillways, pumps, and reservoirs, and it is at the center of much controversy. Though many levee failures did occur in 2011, the system prevented more catastrophic flooding. A first-ever response to a flooding event, floodwaters were deliberately released and diverted through three control structures, a floodway in Missouri and two spillways in Louisiana, to successfully prevent flooding of urban areas such as Cairo, Illinois, and Baton Rouge and New Orleans, Louisiana. Rural areas instead received the brunt of the floodwaters (**Figure 13.9**).

Not everyone considers the extreme alteration and control of the Mississippi River to be desirable. Ecologists promote removing control structures where possible, to allow broader, more natural flooding rather than the more violent flows associated with levee breaches. Natural floodplains better absorb periodic flooding, provide more wildlife habitat, and offer pollution control by absorbing floodwater rather than channeling it down the river. Also, in the Mississippi River basin, natural flooding would distribute across farmland valuable nutrients that are now channeled out into the Gulf of Mexico.

For the past hundred years or so, people in the Midwest have drained wetlands to produce farmland or land on which to build homes, not recognizing the ability of wetlands to moderate floods. Levees created a false sense of security that allowed farms and communities to operate along floodplains.

Solutions to prevent flooding along the Mississippi River remain elusive. Following the 1993 flooding, the U.S. Flood Plain Management Task Force issued a report concluding that people should use floodplains more wisely and rely less on levees to control flooding. Restoring floodplains to their natural condition, moving some towns to higher ground, and promoting wetland development would all further this recommendation. However, the political will is lacking to promote this approach in the face of business and residential demands for maintaining more structured flood protection. In spite of the tremendously high costs involved, estimated at $2 billion, the Corps of Engineers is rebuilding and repairing levees damaged in the 2011 floods. ■

AP Photo/The Merced Sun-Star, Jack Bland

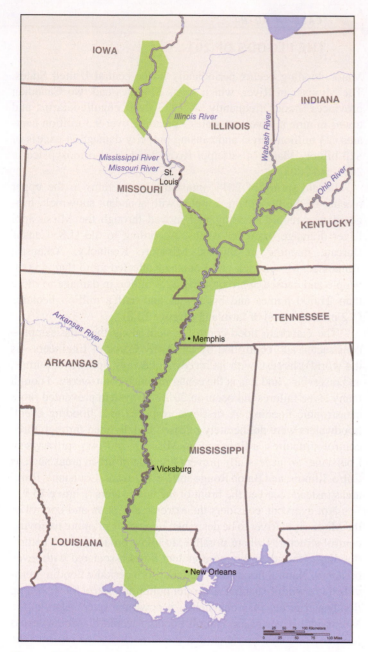

Figure 13.8 **Flooded stretches along the Mississippi River, May 9–13, 2011.** (Adapted from U.S. Census Bureau)

Figure 13.9 **An aerial view of a farm near Rock Port, Missouri submerged in Missouri River floodwaters June 24, 2011.**

Irrigation, which increases the agricultural productivity of arid and semiarid lands, has become increasingly important worldwide in efforts to produce enough food for burgeoning populations **(Figure 13.10)**. Since 1955, the amount of irrigated land has more than tripled; Asia has more agricultural land under irrigation than do other continents, with China, India, and Pakistan accounting for most of it. Water use for irrigation will probably continue to increase in the 21st century, particularly in Asia, but at a slower rate than in the last half of the 20th century.

Population growth in arid and semiarid regions intensifies water shortage. More people need food, so additional water resources are diverted for irrigation. Also, the immediate need for food prompts people to remove natural plant cover to grow crops on marginal lands subject to frequent drought and subsequent crop losses. Livestock overgraze the small amount of plant cover in natural pastures. The resulting bare soil cannot absorb the water as well when the rains do come, and runoff is greater. Because the precipitation does not replenish the soil, crop productivity is poor and people are forced to cultivate food crops on additional marginal land.

Removing too much fresh water from a river or lake can have disastrous consequences in local ecosystems. Humans can remove perhaps 30% of a river's flow without greatly affecting the natural ecosystem. In some places, however, considerably more is withdrawn for human use. In the arid American Southwest, 70% or more of surface water is often removed.

When surface water is overdrawn, wetlands dry up. Natural wetlands play many roles, such as serving as a breeding ground for many species of birds and other animals. Estuaries, where rivers empty into seawater, become saltier when surface waters are overdrawn, and this change in salinity reduces the productivity associated with estuaries.

Aquifer depletion from excessive removal of groundwater lowers the water table. Prolonged aquifer depletion can empty an aquifer, eliminating it as a water resource. In addition, aquifer depletion from porous sediments causes **subsidence,** or sinking, of the land above it. Some areas of the San Joaquin Valley in California have sunk almost 10 m (33 ft) in the past 50 years.

aquifer depletion The removal of groundwater more rapidly than it can be recharged by precipitation or melting snow.

TOO LITTLE WATER

Arid lands, or deserts, are fragile ecosystems in which plant growth is limited by lack of precipitation. **Semiarid lands** receive more precipitation than deserts but are subject to frequent and prolonged droughts. For example, a multiyear drought that began in 2000 was still affecting the arid and semiarid lands of the western United States in the spring of 2014. A large part of central and southern California was considered under exceptional drought, and municipal water supplies were dwindling rapidly. No one knows how long the drought will last, but previous droughts in Arizona have lasted longer than two decades. Many scientists think that the current drought is associated with climate change and may represent a long-term shift in water resources in the western United States.

Figure 13.10 **Agricultural use of water.** Center-pivot irrigation produces massive green circles. Each circle is the result of a long irrigation pipe that extends along the radius from the circle's center to its edge and slowly rotates, spraying the crop. Photographed in Texas.

The limestone bedrock of Florida erodes as groundwater moves through it, sometimes causing a **sinkhole**, a large surface cavity or depression where an underground cave roof has collapsed. Sinkholes occur more frequently when droughts or excessive pumping of water causes a lowering of the water table.

saltwater intrusion The movement of seawater into a freshwater aquifer located near the coast; caused by aquifer depletion.

Saltwater intrusion occurs along coastal areas when groundwater is depleted faster than it recharges (**Figure 13.11**). Saltwater intrusion is also occurring in low-lying parts of the world due to sea-level rise associated with global climate change. Well water in such areas eventually becomes too salty for human consumption or other uses. Once it occurs, saltwater intrusion is difficult to reverse.

REVIEW

1. Which human activity is responsible for 70% of global water consumption?

2. How might allowing more natural flooding along the Mississippi River prove beneficial?

3. What are some of the problems associated with overdrawing surface water? With aquifer depletion?

Water Problems in the United States and Canada

LEARNING OBJECTIVE

• Relate the background for the following U.S. water problems: Mono Lake, the Colorado River basin, drought in Texas, and the Ogallala Aquifer.

Compared with many countries, the United States has a plentiful supply of fresh water. Despite this overall abundance, many U.S. regions have severe water shortages because of geographic and seasonal variations. (**Figure 13.12** shows the average annual precipitation in North America.)

SURFACE WATER

The increased use of U.S. surface water for agriculture, industry, and personal consumption since the 1960s has caused many water problems. The populations of many U.S. regions have grown during this period, with particularly rapid growth in the Southwest and Florida. If water consumption in these and other areas continues to increase, the availability of surface waters could become a serious regional problem, even in places that have never experienced water shortages.

Water problems are particularly severe in the American West and Southwest. Much of this large region is arid or semiarid and

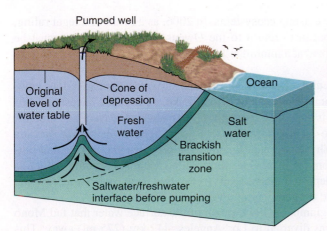

(a) Normally, fresh groundwater overlies salty groundwater. The well supplies fresh water as long as pumping is not excessive.

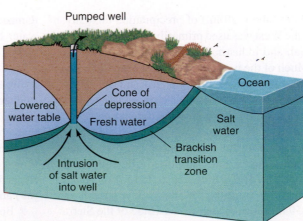

(b) However, the removal of large amounts of fresh groundwater causes the brackish transition zone to migrate. The well draws up salty groundwater unfit to drink.

Figure 13.11 **Saltwater intrusion.**

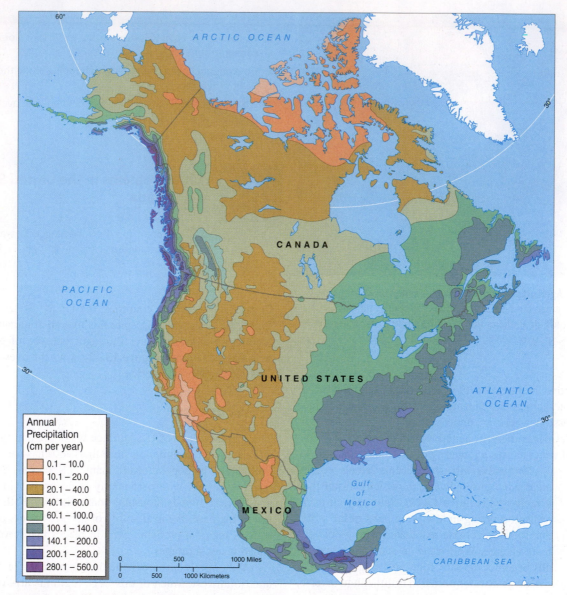

Annual
Precipitation
(cm per year)

	0.1 – 10.0
	10.1 – 20.0
	20.1 – 40.0
	40.1 – 60.0
	60.1 – 100.0
	100.1 – 140.0
	140.1 – 200.0
	200.1 – 280.0
	280.1 – 560.0

Figure 13.12 **Average annual precipitation in North America.** Climate change is expected to reduce precipitation in the western United States. Will this impact wetter areas or areas that are already relatively dry?

receives less than 50 cm (about 20 in.) of precipitation annually. Historically, water in the West was used primarily for irrigation, but municipal, commercial, and industrial uses now compete heavily for available water. Much of the water used in the West and Southwest originates as snow in the Rocky Mountains and the Sierra Nevada; climate change appears to be causing reduced snowfall—and thus less total water available for a growing population.

The development of new sources of water continues to meet expanding water needs in the West and Southwest. Water is diverted from distant sources and transported via **aqueducts** (large conduits) (**Figure 13.13**). As long ago as 1913, Los Angeles started bringing in water from the Owens Valley, an area of California 400 km (250 mi) north, along the east side of the Sierra Nevada. Dams and water-holding basins were established to ensure year-round supply. Now, however, the closest, most practical water sources are already tapped, and experiences such as that at Mono Lake suggest that removing water can dramatically

damage existing ecosystems. In 2006, as a result of a legal ruling, water began to return to the Owens River, which now must be maintained at a minimum level.

Mono Lake   Removing too much surface water has serious environmental repercussions. Mono Lake, a salty lake in eastern California, is a striking example of this practice. Rivers and streams that are largely formed from snowmelt in the Sierra Nevada replenish Mono Lake. Evaporation provides the only natural outflow from the lake. Over time, Mono Lake is becoming saltier as rivers deposit dissolved salts (recall that fresh water contains some salt) and evaporation removes water, but not salt.

Beginning in 1941, much of the surface water that fed Mono Lake was diverted to Los Angeles, 442 km (275 mi) away. This change in water flow led to changes throughout the Mono Lake ecosystem. As the water level dropped (about 14 m, or 46 ft), increased salinity adversely affected brine shrimp and alkali fly

Figure 13.13   An aqueduct in California's Mojave Desert. Water is carried almost 400 km to Los Angeles.

populations. This, in turn, affected more than 80 species of water-birds that feed on the shrimp and flies. Dust storms from the exposed lakebed began to pose a health hazard and to cause violations of federal air pollution standards.

A court order halted water diversions from Mono Lake in 1989, and in 1994 the state of California worked out an agreement on Mono Lake water rights between the Los Angeles water authority and environmental groups. Less water is being diverted to Los Angeles, and Mono Lake will return to about 72% of its original volume by around 2015. The National Audubon Society expects hundreds of thousands of migratory and nesting birds to return to the lake's shores to nest.

reclaimed water Treated wastewater that is reused in some way, such as for irrigation, manufacturing processes that require water for cooling, wetland restoration, or groundwater recharge.

The city of Los Angeles is using state funds to develop water conservation and **reclaimed water** projects to replace water supplies from Mono Lake. By 2015, California's reclaimed water projects are expected to produce enough water to make up for the water lost from Mono Lake.

The Colorado River Basin   One of the most serious water supply problems in the United States is in the Colorado River basin. The river's headwaters are formed from snowmelt in Colorado, Utah, and Wyoming, and its major tributaries—collectively called the upper Colorado—extend throughout these states. The lower Colorado River runs through part of Arizona and then along the border between Arizona and both Nevada and California before crossing into Mexico and emptying into the Gulf of California.

The Colorado River provides water for more than 30 million people, including the cities of Denver, Las Vegas, Albuquerque, Phoenix, Los Angeles, and San Diego, with plans in Utah to divert Colorado River water to Salt Lake City. It irrigates 1.4 million hectares (3.5 million acres) of fruit, vegetable, and field crops worth $1.5 billion per year. The Colorado River has 49 dams, 11 of which produce electricity by hydropower. More than 30 Native American tribes live along the Colorado River and claim rights to some of its water. The river provides $1.25 billion per year in revenues from almost 30 million people who use it for recreation.

The most important of all the treaties regulating use of Colorado River water is the 1922 **Colorado River Compact**. It stipulates an annual allotment of 7.5 million **acre-feet** of water to the lower Colorado (California, Nevada, Arizona, and New Mexico) and the remainder to the upper Colorado (Colorado, Utah, and Wyoming). However, the Colorado River Compact overestimated the average annual flow of the Colorado River, which at the time was thought to be 15 million acre-feet. This overallocation was enshrined in the multistate agreement.

acre-foot The amount of water needed to cover an acre of land 1 foot deep. An acre-foot is equal to 326,000 gallons and is enough to supply eight typical U.S. residents for one year.

Mexico receives a share of the Colorado River, as stipulated by a 1944 treaty. Consequently, the Colorado River water is often completely consumed before it can reach the Pacific Ocean, causing serious problems for the ecosystem and inhabitants of the Colorado River delta (**Figure 13.14**). To compound the problem, as more and more water is used, the lower Colorado becomes increasingly salty as it flows toward Mexico; in places, the Colorado River is saltier than the ocean.

In 2003 California agreed to withdraw no more water from the Colorado River than the Colorado River Compact permits. To make up the difference, farmers in California's Imperial Valley

Figure 13.14   Colorado River delta at the Gulf of California, Mexico. As a result of diversion for irrigation and other uses in the United States, the Colorado River usually dries up before reaching the Gulf of California in Mexico.

agreed to sell some water they would normally use for irrigation to thirsty cities such as San Diego and Los Angeles. Formerly a desert, Imperial Valley is now 200,000 hectares (about 500,000 acres) of irrigated agricultural land. Farmers can use the money earned from water sales to update their water systems to use water more efficiently.

A 2008 study by researchers at the Scripps Institution of Oceanography and the University of California–San Diego predicts that climate change will lead to 10% to 30% less flow in the Colorado River within 50 years. If this happens, Lake Mead's water levels could soon be too low to allow the Hoover Dam to produce electricity. The Bureau of Reclamation projects that electricity production there will be down almost 50% by May 2016. Dam engineers are installing turbines that function more efficiently at low water levels.

Drought in Texas   At the most severe point of Texas's historic 2011 drought, 97% of the state, including at least part of each county, was experiencing extreme or exceptional drought conditions. Although rains have resumed and Texas has emerged from the worst of the drought, it has not fully recovered from the drastic event. In early 2014, half the state was still under some level of drought conditions, and many lakes—especially in central Texas—were less than half full. These conditions are exacerbated by the expanding water needs of the state's rapidly growing population. Scientists believe the lingering drought conditions serve as an example of what's to come as climate change proceeds.

of Hawaii. Florida and many coastal regions in the Northeast and Mid-Atlantic states also have saltwater intrusion.

The Ogallala Aquifer   According to the U.S. Geological Survey, the High Plains cover 6% of U.S. land but produce more than 15% of its wheat, corn, sorghum, and cotton and almost 40% of its livestock. To achieve this productivity, this area requires approximately 30% of the irrigation water used in the United States. Farmers on the High Plains rely on water from the **Ogallala Aquifer**, the largest groundwater deposit in the world (**Figure 13.16**).

In some areas, farmers are drawing water from the Ogallala Aquifer as much as 40 times faster than nature replaces it; at current usage rates the aquifer could be 70% depleted by 2060. Withdrawals have lowered the water table more than 30 m (100 ft) in places, and higher pumping costs associated with these depletions have made it too expensive to irrigate. When farmers revert to dryland farming in these semiarid regions, they risk economic and ecological ruin during droughts (see the discussion of the Dust Bowl in Chapter 14). Areas where the Ogallala is most shallow have experienced recent population declines as farms fail during dry spells. Those areas where the Ogallala is deepest may not experience water shortages during the 21st century. Hydrologists (scientists who deal with water supplies) predict that groundwater will eventually drop in all areas of the Ogallala to a level uneconomical to pump. Their goal is to postpone that day through water conservation, including the use of water-saving irrigation systems.

GROUNDWATER

Roughly half the population of the United States uses groundwater for drinking. Many large cities, including Tucson, Miami, San Antonio, and Memphis, have municipal well fields and depend entirely or almost entirely on groundwater for their drinking water. In addition, many rural homes have private wells for their water supply. Groundwater is also used for industry and agriculture. Approximately 40% of the water used for irrigation in the United States comes from groundwater. Increased groundwater consumption since the 1950s has diminished groundwater levels in many areas of heavy use across the United States. **Figure 13.15** shows areas where aquifer depletion is particularly critical. Groundwater has been overdrawn for irrigation in arid and semiarid areas.

In certain coastal areas of Louisiana and Texas, the removal of too much groundwater has resulted in the intrusion of salt water from the Gulf of Mexico. Saltwater intrusion from the Pacific Ocean has occurred along parts of the California coast, along coastal areas of Puget Sound in Washington State, and in certain areas

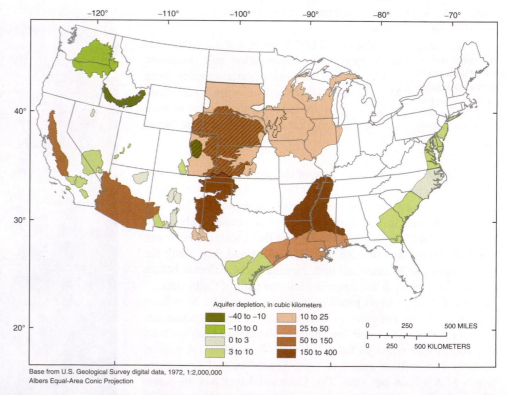

Base from U.S. Geological Survey digital data, 1972, 1:2,000,000
Albers Equal-Area Conic Projection

Figure 13.15   **Aquifer depletion, 1900–2008.** Aquifer depletion is a widespread problem in the United States, particularly in the High Plains, California, southern Arizona, and parts of the South. Hatched area indicates overlap of two aquifers with different depletion levels. (Adapted from USGS *Groundwater Depletion in the United States* (1900–2008))

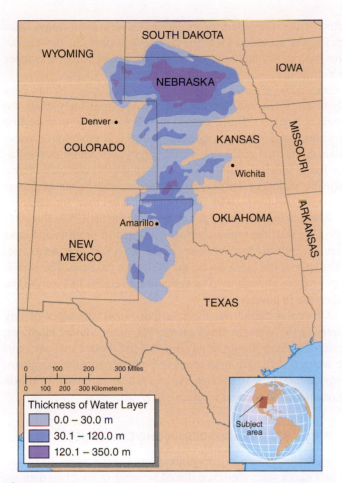

Figure 13.16 **Ogallala Aquifer.** This massive deposit of groundwater lies under eight midwestern states, with extensive portions in Texas, Kansas, and Nebraska. Water in the Ogallala Aquifer takes hundreds or even thousands of years to renew after it is withdrawn to grow crops and raise cattle. (USGS)

REVIEW

1. Which U.S. regions have the most severe water-scarcity problems?

Global Water Problems

LEARNING OBJECTIVES

- Explain the relationship between climate change and available water.
- Briefly describe each of the following international water problems: drinking-water problems, population growth and water problems, the Rhine River basin, the Aral Sea, and potential international conflicts over water rights.

Data on global water availability and use indicate that the amount of fresh water on the planet is adequate to meet human needs, even taking population growth into account. These data, though, do not consider the *distribution* of water resources in relation to human populations. For example, citizens of Bahrain, a tiny island nation in the Persian Gulf, have no freshwater supply and must rely on desalination of ocean water.

Per capita water use varies greatly from country to country and from continent to continent, depending on the size of the human population and the available water supply. South America and Asia receive more than one-half of the world's renewable fresh water (by precipitation). Although South America has more available water per person than Asia does, it does not have the potential to support as many people as its water supply would suggest. Most of South America's precipitation falls in the Amazon River basin, which has poor soil and is unsuitable for large-scale agriculture. In contrast, because most of the precipitation in Asia falls on land suitable for agriculture, the water supply supports more people.

Humans need an adequate supply of water year-round. In some places, **stable runoff**, the portion of runoff from precipitation available throughout the year, is low even though total runoff is quite high. India's wet season—June to September—produces 90% of its annual precipitation. Most of the water that falls during India's monsoon quickly drains away into rivers and is unavailable during the rest of the year. Thus, India's stable runoff is low.

> **stable runoff** The share of runoff from precipitation that can be depended on every month.

Variation in annual water supply is an important factor in certain areas of the world. The African Sahel region (see Figure 4.22) has wet years and dry years, and the lack of water during the dry years limits human endeavors during the wet years. Since the late 1960s, the Sahel has experienced an ongoing drought that has had a devastating impact on the people and wildlife living in the region.

WATER AND CLIMATE CHANGE

Climate change will play an important role in future freshwater availability, as precipitation is expected to increase in some areas while it drops in others. Changes in rainfall may lead to abrupt changes in available surface water, since runoff is influenced by geologic factors such as soil permeability and biological factors such as amount of vegetation. One recent study suggests that a 10% decrease in rainfall in one part of Africa will lead to a 17% reduction in runoff, while the same precipitation decrease in another area will lead to a 50% reduction in runoff. This means that climate change will impact water supply more severely in some areas than in others. The study concludes that predicted variations in rainfall due to climate change will affect available surface water for one-fourth of the African continent by 2100.

Climate change will affect available fresh water in other ways as well. The type of precipitation is important: Earlier, we noted how reduced snowfall in the Rocky Mountains and Sierra Nevada will impact water availability in the American West and Southwest. Sea-level rise—caused by thermal expansion and surface ice melt—has already caused saltwater intrusion into drinking-water sources for certain low-lying island nations.

DRINKING-WATER PROBLEMS

Many inhabitants of developing countries have insufficient water to meet the most basic drinking and household needs. The water exists—only about 1% of the Earth's water would suffice for the

ENVIRONEWS

Climate Change and Crops in Southern Africa

Researchers have carried out extensive modeling and analyses to predict future effects of climate change on agriculture in southern Africa, with findings published by the International Food Policy Research Institute. The rising temperatures and erratic precipitation associated with climate change are considered a real threat to the region's crop yields. Hot, dry conditions reduce soil moisture content and curb the growing season, effects that are expected to cause declines in average yields of major crops such as maize and sorghum. The increased variability in precipitation is proving particularly threatening to crop production. Most farms in the region are small operations, and farmers depend on rainfall to water their crops. Adaptations to changing conditions may include planting different, more drought-resistant crops.

entire human population. However, this water is not available to many people, who have to spend large amounts of money or travel great distances to secure the water they need. Individual governments, the United Nations, the World Bank,[1] nongovernmental organizations (NGOs), and civic organizations all sponsor water projects in developing countries.

The World Health Organization (WHO) estimates that 748 million people lack access to safe drinking water and about 2.5 billion are without access to a satisfactory means of domestic wastewater and fecal waste disposal. These people risk disease because sewage or industrial wastes contaminate the water they consume. WHO estimates that 80% of human illness results from insufficient water supplies and poor water quality caused by lack of sanitation. Although many affected countries have installed or are installing public water systems, population increases tend to overwhelm efforts to improve the water supply.

POPULATION GROWTH AND WATER PROBLEMS

As the world's population continues to increase, global water problems will become more serious. Asia has the world's largest available water resources—36% of the Earth's total. However, it is also home to 60% of the world's people, and the people and the water are not always in the same places. In India, 20% of the world's population has access to only 4% of the world's fresh water, and approximately 8000 Indian villages have no local water. The water supply to some Indian cities—Madras, for example—is so severely depleted that water is rationed from a public tap. Extracting groundwater faster than its rate of recharge has caused water tables to fall 1 to 3 m (3 to 10 ft) per year in parts of Punjab and Haryana.

Water supplies are precarious in much of China, owing to population pressures. The water table across much of the North China Plain, with a population more than twice that of the United States, is falling 2 to 3 m (6 to 10 ft) per year. More than one-third of the wells in Beijing have gone dry. Much of the water in the

Yellow River is diverted for irrigation, leaving downstream areas with little or no water. Over the past several decades, the Yellow River has run dry hundreds of kilometers inland before it reaches the Yellow Sea.

Iraq faces a particular geographical challenge: Headwaters of both the Tigris and the Euphrates Rivers originate outside the country's borders. While current conflicts in Iraq overshadow the water issue, shortages in both quality and quantity of water remain an internal challenge. Water supply will continue to influence Iraq's relations with neighboring countries, especially those upstream, including Turkey, Syria, Jordan, and Saudi Arabia as well as Iran, with which Iraq has had armed conflicts for decades.

Pakistan faced an extended drought during the 1990s and early 2000s. In 2001, the water shortages resulted in protests and riots stemming from rivalry between certain provinces over water use, particularly of the Indus River, the lifeline in Pakistan. The inability to produce enough food in agricultural regions has resulted in an increase in poverty.

Mexico is facing the most serious water shortages of any country in the Western Hemisphere. The main aquifer supplying Mexico City is dropping as much as 3.5 m (about 11 ft) per year. The water table is falling as much as 3 m per year in Guanajuato, an agricultural state in Mexico. As of 2012, an estimated 2 million Mexicans lacked access to water.

SHARING WATER RESOURCES AMONG COUNTRIES

Surface water is often an international resource. Around 260 of the world's major watersheds are shared between at least two nations. International cooperation is required to manage rivers that cross international borders.

The Rhine River Basin   The drainage basin for the Rhine River in Europe is primarily in five highly developed and densely populated countries—Switzerland, Germany, France, Luxembourg, and the Netherlands (**Figure 13.17**). Traditionally, Switzerland, Germany, and France used water from the Rhine for industrial purposes and then discharged polluted water back into the river. The Netherlands—which in 2010 got 82% of its water from beyond its borders—then had to clean up the water so its citizens could drink it. Today, these countries recognize that international cooperation is essential to conserve and protect the supply and quality of the Rhine River water.

In 1950, the five countries formed the International Commission for Protection of the Rhine (ICPR) to deal with water issues relating to the Rhine River, but little was accomplished for several decades. River quality began to improve in the mid-1970s, largely in response to international reports on the river's poor condition.

In 1986, a severe chemical spill in Switzerland dumped 30 tons of dyes, herbicides, fungicides, insecticides, and mercury into the river. The spill galvanized the ICPR, which initiated a 15-year Rhine Action Plan. It eliminated some major pollution sources, and the water in the Rhine River today is almost as pure as drinking water. Long-absent fishes have returned to the river, including Atlantic salmon, which returned in 1990 after a 30-year absence. The ICPR is currently working on bank restoration, flood control, and cleaning up remaining pollutants.

---

[1] The World Bank makes loans to developing countries for projects it thinks will lessen poverty and encourage development.

Figure 13.17 **Rhine River basin.** The Rhine River drains large areas in five European countries—Switzerland, Germany, France, Luxembourg, and the Netherlands. (The green area represents the drainage basin.) Water management of such a river requires international cooperation.

The Aral Sea The Aral Sea, which straddles Kazakhstan and Uzbekistan (both parts of the former Soviet Union), suffers from the same problem as Mono Lake in California (**Figure 13.18**). In the 1950s, the Soviet Union began diverting water from the Amu Darya and the Syr Darya, the two rivers that feed into the Aral Sea, to irrigate surrounding desert areas. By the early 1980s, irrigation for growing cotton had diverted more than 95% of the Aral Sea's inflow.

From 1960 to 2000 the Aral Sea, once the world's fourth-largest freshwater lake, declined in area by more than 50%. At one point, its total volume was down as much as 80%. Much of its biological diversity disappeared—all 24 fish species originally found there are gone.

About 35 million people live in the Aral Sea's watershed. Millions have developed health problems ranging from tuberculosis to severe anemia, and their death rate from respiratory illnesses is among the world's highest. Kidney disease and various cancers are on the rise. International health experts have begun to assess which medical problems are due to the Aral Sea's environmental problems. Toxic salt storms caused by winds that whip the salt on the receding shoreline into the air may be responsible for many of these chronic conditions. Since the 1950s, such storms have increased 60-fold. The wind carries salt hundreds of kilometers from the Aral Sea; where it is deposited, it reduces the productivity of the land.

Immediately following the breakup of the Soviet Union in 1991, plans to save the Aral Sea faltered as responsibility for its rescue shifted from Moscow to the five central Asian countries that share the Aral Basin: Uzbekistan, Kazakhstan, Kyrgyzstan,

Worldsat International, Inc.

(a) 1976

Courtesy NASA Earth Observatory

(b) 2013

Figure 13.18 **Aral Sea.** The satellite images show changes in the Aral Sea across nearly four decades. As water was diverted for irrigation, the sea level subsided.

Turkmenistan, and Tajikistan. In 1994, the five nations established a fund to prevent the complete disappearance of the Aral Sea. The World Bank and the UN Environment Programme approved a grant to the five countries to help address the environmental problems of the area.

At this time, recovery of the Aral Sea is mixed. The World Bank sponsored the **Syr Darya Control and Northern Aral Sea Project**, with goals of ecological restoration and commercial fishery recovery. Due in part to construction of the Kok-Aral Dam, the Northern Aral Sea experienced more than a 30% increase in surface area between 2003 and 2010, salinity levels were cut in half from 1991 values, and some fish species have returned. The Southern Aral Sea, however, continues to shrink and become saltier, and no restoration efforts are in place there.

Potentially Volatile International Water Situations   Conflict over water is nothing new. In 2009, **Peter Gleick** of the Pacific Institute for Studies in Development, Environment, and Security developed a chronology of water conflicts, beginning in 3500 B.C.E., which is now updated regularly. (See examples in **Table 13.2**.) The conflicts range from attacks on dams to intentional water contamination to stabbings. Because so many watersheds and aquifers are shared between two or more countries, the 21st century may well see countries facing one another in armed conflict over water rights.

Humans remove so much water from the Amu Darya, Ganges, Indus, Nile, Yellow, and Colorado Rivers that their channels run dry during at least some parts of the year. Tensions are high along the Mekong River basin, shared by Laos, Thailand, and Vietnam, and the Indus River basin, shared between Pakistan and India. India and Bangladesh quarrel over the Ganges River. Slovakia and Hungary both depend on the Danube River. Developing cooperative international agreements on shared water resources is an urgent global issue.

The Jordan River, which supplies water to Israel, Jordan, the West Bank, and the Gaza Strip, experienced a 90% reduction in flow between 1960 and 2010. Water use continues to increase because of population growth, agriculture, and industry. A collaborative study by the U.S. National Academy of Sciences, the Israel Academy of Sciences and Humanities, the Palestine Academy for Sciences and Technology, and the Royal Scientific Society of Jordan concluded that the future outlook in the region is one of significant water stress. Participants urged their respective governments to cooperate in developing conservation measures such as reusing wastewater. Nonetheless, differential access between Israeli settlers on the West Bank, who use four to five times as much water as do neighboring Palestinians, may be a significant source of conflict.

Northeastern Africa has a serious water-use situation: the Nile River. Egypt uses most of the Nile's water (and has for millennia), even though 10 nations share the Nile River basin. Ethiopia and Sudan are expanding their use of the Nile River's flow to meet the demands of their rapidly growing populations; these actions could imperil Egypt's freshwater supply at a time when its population is increasing. The United Nations engineered an international water-use agreement among the Nile River countries to help diffuse this potentially dangerous water situation, and the 10 nations in the region have formed the Nile Basin Initiative to review past agreements and develop future ones.

REVIEW

1. What is the relationship between climate change and global water problems?

2. Which international water problems have been most amenable to management? Least amenable?

Water Management

LEARNING OBJECTIVES

- Define *sustainable water use*.
- Contrast the benefits and drawbacks of dams and reservoirs, using the Columbia River to provide specific examples.
- Briefly describe two methods of desalination.

People have always considered water different from other resources. Coal and gold are owned privately and sold as free-market goods, but people variously view water as a public resource or

---

Table 13.2   Some Historical and Recent Water Conflicts

Date	Conflict	Description
2500 B.C.E.	Military tool	King Urlama of Lagash diverts flow of water from rival kingdom Umma (Lagash and Umma were both kingdoms in what is now Iraq).
1187 C.E.	Military tool	Saladin defeats European Crusaders, in part by depriving them of access to water. This includes filling wells with debris and eliminating villages that could have supported the Crusaders.
1672	Military tool	The Dutch breach their protective dikes to prevent Spanish armies from invading by land. Dikes around Amsterdam and other Dutch cities are designed both to keep out seawater in peacetime and deter potential invaders in times of war.
1850s	Development dispute/terrorism	When a dam is built to provide water for factories in New Hampshire, locals who object to the effect on their water supplies attack the dam.
1907–1913	Terrorism/development dispute	The aqueduct from Owens Valley to Los Angeles is bombed multiple times by people objecting to the large-scale shift in allocation of this water resource.
1969	Military target	Israel attacks the East Ghor Canal in Jordan to prevent diversion of water from the Yarmouk River.
1991	Military target	During the First Gulf War, Iraq destroys desalinization facilities in Kuwait.
2003–2007	Military tool, target/terrorism	Wells in Sudan and Darfur are destroyed and poisoned as part of civil war–related violence.
2009	Development dispute	Ethiopian village attacked in disagreement over a water pipe.
2010	Development dispute	Water dispute between Pakistani tribes leads to more than 100 deaths.
2012	Military target	Water pipeline to city of Aleppo severely damaged during Syrian Civil War; 3 million residents suffer serious water shortages.

Source: Peter H. Gleick (2011). *Water Conflict Chronology.* Pacific Institute for Studies in Development, Environment, and Security; additional updates from the Pacific Institute online chronology.

a private resource. Historically, both in much of the United States and in many other countries, water rights were bound up with land ownership. As more and more users compete for the same water, state or provincial governments increasingly make allocation decisions. Some countries have separated land and water ownership so that water rights are sold separately.

Because rivers usually flow through more than one governmental jurisdiction, all affected parties must develop agreements about the management of a river or other shared water resource. Such interstate or transboundary cooperation permits comprehensive rather than piecemeal management. In addition, these arrangements divide the water fairly among the jurisdictions, which then apportion their respective shares to individual users according to an established set of priorities.

Groundwater management is more complicated, in part because the extent of local groundwater supplies is often not known. Groundwater management includes issuing permits to drill wells, limiting the number of wells in a given area, and restricting the amount of water pumped from each well.

The price of water varies, depending on how it is used. Historically, domestic use is most expensive and agricultural use is least expensive. Consumers rarely pay directly for the entire cost of water, which includes its transportation, storage, and treatment. State and federal governments heavily subsidize water costs, so that we pay for some of the cost of water indirectly, through taxes. Increasingly, state and local governments are adjusting the price of water as a mechanism to help ensure an adequate supply of water. Raising the price of water to reflect the actual cost generally promotes a more efficient use of water.

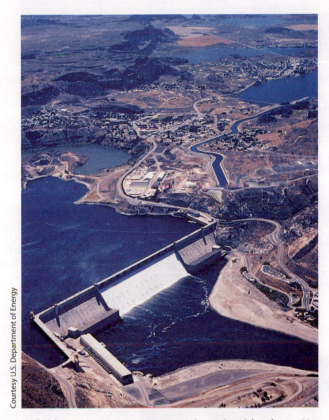

Courtesy U.S. Department of Energy

Figure 13.19 **Grand Coulee Dam on the Columbia River.** Shown are the dam and part of its reservoir, the Franklin D. Roosevelt Lake. Dams help to regulate water supply, storing water produced in times when precipitation is plentiful for use during dry periods. The many beneficial uses of dams include electricity generation and flood control, but they destroy the natural river habitat and are expensive to build.

PROVIDING A SUSTAINABLE WATER SUPPLY

sustainable water use The use of water resources in a fashion that does not harm the essential functions of the hydrologic cycle or the ecosystems on which present and future humans depend.

The main goal of water management is to provide a sustainable supply of high-quality water. **Sustainable water use** means humans use water resources carefully so that water is available for future generations and for existing nonhuman needs.

Dams and Reservoirs Dams ensure a year-round supply of water in areas with seasonal precipitation or snowmelt. Dams confine water in reservoirs, from which the flow is regulated (**Figure 13.19**). Dams have other benefits, particularly the generation of electricity (recall the discussion of dams and hydroelectric power in Chapter 12). Many people, however, feel that the drawbacks of dams, including the environmental costs, far outweigh any benefits they provide.

In recent years scientists have come to understand how dams alter river ecosystems, both above and below the dam. Heavy deposition of sediment occurs in the reservoir behind the dam, and the water that passes over the dam does not have its normal sediment load. As a result, the river floor downstream of the dam is scoured, producing a deep-cut channel that is a poor habitat for aquatic organisms.

The Glen Canyon Dam, built in 1963, has profoundly affected the Colorado River in the Grand Canyon National Park. Prior to the dam's construction, powerful spring floods carried sediment

that formed beaches and sandbars, providing nesting sites for birds and shallow waters for breeding fishes. The regulated flow of water since the Glen Canyon Dam was constructed changed the ecosystem, to the detriment of some of the Grand Canyon's wildlife. To rectify some of the changes to the river, the Bureau of Reclamation has flooded the Grand Canyon several times, beginning in 1996. Although these floods are small in comparison to some of the natural floods of the past, the sediment-laden floodwater rebuilds beaches and sandbars that are continually eroding (**Figure 13.20**).

The Columbia River The Columbia River, the fourth-largest river in North America, illustrates the impact of dams on natural fish communities. Its watershed covers an area the size of France, spanning seven states and two Canadian provinces. Of the more than 100 dams within the Columbia River system, 19 are major generators of hydroelectric power. The Columbia River system supplies municipal and industrial water to several major urban areas, including Boise, Portland, Seattle, and Spokane. More than 3 million hectares (7.8 million acres) of agricultural land are irrigated with the Columbia's waters (the river and its tributaries), and commercial ships navigate 800 km (500 mi) of the river.

Several interests have competing uses for Columbia River water. Conservationists think that water should be seasonally released from the dams to simulate spring snowmelt. Farmers want

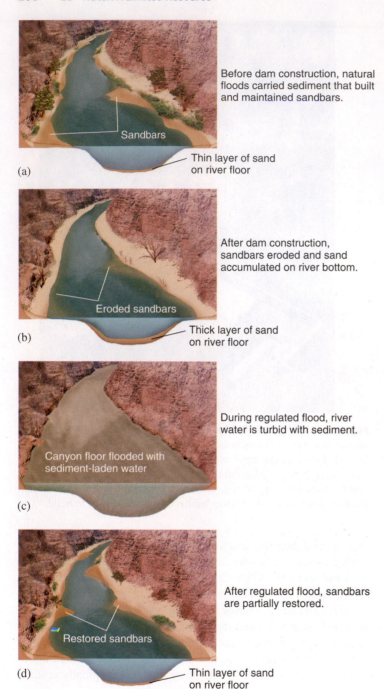

(a) Before dam construction, natural floods carried sediment that built and maintained sandbars.

Sandbars

Thin layer of sand on river floor

(b) After dam construction, sandbars eroded and sand accumulated on river bottom.

Eroded sandbars

Thick layer of sand on river floor

(c) During regulated flood, river water is turbid with sediment.

Canyon floor flooded with sediment-laden water

(d) After regulated flood, sandbars are partially restored.

Restored sandbars

Thin layer of sand on river floor

Figure 13.20   How periodic flooding of the Grand Canyon helps restore the riverbanks. (USGS)

to save the plentiful snowmelt water for irrigation during summer months. The hydropower industry wants to save the water to generate electricity during the winter months, its time of peak demand.

As is often the case in natural resource management, a particular use of the Columbia River system may have a negative impact on other uses. The dams that generate electricity and control floods have adversely affected fish populations, particularly the migratory salmon that spawn in the upper reaches of rivers and streams.

The young offspring (smolts) migrate to the ocean, returning years later to their place of birth to reproduce and die.

The salmon population in the Columbia River system is only a fraction of what it was before the watershed was developed. Dams, along with sedimentation and loss of shade due to logging, have adversely impacted how salmon eggs develop. Projects to rebuild salmon populations have not been particularly effective.

To protect some of the remaining natural salmon habitats, several streams in the Columbia River system are off-limits for dam development. Underwater screens and passages are being installed at dams to steer smolts away from turbine blades. Trucks and barges transport some of the young fish around dams, while others swim safely over the dam because the electrical generators are periodically turned off to allow passage. In addition, many dams have fish ladders to allow some of the adult salmon to bypass the dams and continue their upstream migration (**Figure 13.21**).

In 1999, the National Marine Fisheries Service (NMFS) extended the protection of the Endangered Species Act (ESA) to all species of salmon and steelhead trout found in northwest rivers from the Canadian border to northern California, and eastward to Montana. This action, the largest implementation of the ESA ever, affects public and private lands and includes both rural areas and major cities such as Portland, Oregon.

The Missouri River   The Missouri River flows from Montana to St. Louis, Missouri, where it joins the Mississippi River and flows on to the Gulf of Mexico. The Army Corps of Engineers has constructed six dams on the Missouri River; these dams provide both benefits and problems for people living along the river. Since 1987, the Corps has increased water flow over the northern dams to protect downstream navigation, including the shipping of approximately 4 million tons of cargo each year. People who live downstream count on the river water for irrigation, electrical power, and individual water consumption.

The claims on and usage preferences for Missouri River water are many, complex, and often contested. The area along the northern Missouri River depends on the river for its multimillion-dollar fishing and tourism industry. Farmers want additional dikes and levees to protect their crops on the floodplains, whereas environmentalists want the river restored to its natural state as much as possible. Native Americans with claims to water rights along portions of the river want to use the water in a variety of ways, from generating hydroelectric power to irrigating cropland.

The Missouri River Association of States and Tribes, a coalition of river-basin states and Native American tribes, has the unenviable job of working with the Corps to meet the demands of the various competing interest groups as they decide the river's future. The association recognizes that the river does not belong to any single group and that it must prioritize uses of the river in a way that will at least partially address needs and concerns of environmental groups, farmers, hydroelectric producers, Native Americans, and fishing and tourism interests.

Water Diversion Projects   One way to increase the natural supply of water to a particular area is to divert water from areas

Figure 13.21 Fish ladder. This ladder is located at the Bonneville dam on the Columbia River in Oregon. Fish ladders help migratory fishes to bypass dams in their migration upstream.

where it is in plentiful supply by pumping water through a system of aqueducts. Much of southern California receives its water supply via aqueducts from northern California (**Figure 13.22**). Water from the Colorado River is also diverted into southern California by aqueducts.

Large-scale water diversion projects are controversial and expensive. The Central Arizona Project, which pumps water 540 km (336 mi) from the Colorado River to Phoenix and Tucson, was completed at a cost of almost $4 billion. As you learned earlier, a river or other body of water is damaged when a major portion of its water is diverted. Pollutants that would have been diluted in the normal river flow reach higher concentrations when much of the flow is removed. Fishes and other organisms may decline in number and diversity. Although no one denies that people must have water, opponents of water diversion projects contend that serious water conservation efforts would eliminate the need for additional large-scale water diversion.

Desalination Seawater and salty groundwater are made fit to drink through **desalination** (or **desalinization**). There are two major approaches to desalination: distillation systems and membrane/filtration systems. In **distillation**, salt water is heated until the water evaporates, leaving behind a crust of salt. The water vapor is then condensed to produce fresh water. **Reverse osmosis**, the most common type of membrane/filtration system, involves forcing salt water through a membrane permeable to water but not to salt. Reverse osmosis removes about 97% of the salt from water.

desalination The removal of salt from ocean or brackish (somewhat salty) water.

Although costs have declined considerably, desalination remains expensive because it requires a large energy input. Recent advances in reverse osmosis technology have increased its efficiency so that it requires much less energy than distillation. Other expenses involved in desalination projects include the cost of transporting the desalted water from the site of production to where it is used. The cost of desalination, excluding transport costs, can vary from approximately $0.25 to $3.00 per m³ (tap water typically

Figure 13.22 Water diversion in southern California. Largely desert, southern California relies on water diversion for the water needs of its millions of inhabitants. The California Water Project includes 1042 km (648 mi) of aqueducts to transfer large quantities of water to southern California. This map also shows some of the main reservoirs of the California Water Project.

costs $0.10 to $0.15 per m³), depending on what technology is used and how salty the water is to begin with. Removing salt from seawater costs three to five times more than removing salt from brackish water.

The disposal of salt produced by desalination is also a concern, since dumping it back into the ocean near productive coastal areas could harm marine organisms. In addition, ocean water desalination requires the intake of huge amounts of water. This water can contain large numbers of microorganisms and fish, leading to severe localized disruption.

In 2011, there were almost 16,000 desalination plants worldwide in 150 countries. Total global capacity in 2011 was 66.5 million m³ per day—more than twice the capacity in 2000. Because other freshwater sources are scarce, desalination is a huge industry in North Africa and the Middle East. In Saudi Arabia, which has the world's largest plant, desalination accounts for 60% of drinking water. The largest plant in the United States is in Tampa Bay, Florida.

REVIEW

1. What is sustainable water use?

2. What are some major problems and solutions associated with dams on the Columbia River?

3. What is desalinization? Reverse osmosis?

Water Conservation

• Give examples of water conservation by agriculture, industry, and individual homes and buildings.

Population and economic growth place an increased demand on Earth's water supply. Today there is more competition than ever among water users with different priorities, and water conservation measures are necessary to guarantee sufficient water supplies. Most water users use more water than they really need, whether it is for agricultural, industrial, or direct personal consumption. With incentives, these users will lower their rates of water consumption. Many studies have shown that programs combining increased prices for water, improved technology, and effective educational tools motivate consumers to conserve water.

REDUCING AGRICULTURAL WATER WASTE

Irrigation generally makes inefficient use of water. Traditional irrigation methods practiced for more than 5,000 years involve flooding the land or diverting water to fields through open channels. Plants absorb about 40% of the water applied to the soil by flood irrigation; the rest usually evaporates into the atmosphere or seeps into the ground.

One of the most important innovations in agricultural water conservation is **microirrigation**, also called **drip** or **trickle irrigation**, in which pipes with tiny holes bored in them convey water directly to individual plants (**Figure 13.23**). Microirrigation substantially reduces the water needed to irrigate crops, usually by 40% to 60% compared to center-pivot irrigation or flood irrigation, and it also reduces the amount of salt left in the soil by irrigation water.

microirrigation
A type of irrigation that conserves water by piping it to crops through sealed systems.

Another important water-saving measure in irrigation is the use of lasers to level fields, allowing a more even water distribution. As a laser beam sweeps across a field, a field grader receives the beam and scrapes the soil, leveling it. Because farmers must use extra water to ensure that plants growing at higher elevations of a field receive enough, laser leveling of a field reduces the water required for irrigation.

A technology that has advanced significantly in the past few years is a combination of **low-energy precision application (LEPA)** irrigation and **geographic information systems (GIS)**. LEPA involves dragging hoses across fields in a computer-controlled pattern. Water is released only when and where needed. Less water pumped and sprayed means less energy used, less evaporation, and less runoff of unused water. GIS uses signals from satellites to identify locations for irrigation within 20 cm (8 in.) or less of the target plants. LEPA and GIS irrigation in parts of Texas has allowed an eightfold water use reduction.

The use of sound water management principles in agriculture reduces water consumption. Traditionally, western farmers were

Figure 13.23   Microirrigation. Cutaway view of soil shows a small tube at the root line. Tiny holes in the tube deliver a precise amount of water directly to roots, eliminating much of the waste associated with traditional methods of irrigation. Photographed in Fresno, California.

Courtesy U.S. Department of Agriculture

allotted specific amounts of water at specific times, with a "use it or lose it" philosophy. This approach encourages waste. Instead, a field's water needs should be carefully monitored (by measuring rainfall and soil moisture) to determine when to irrigate and how much water to apply. These water management strategies effectively reduce overall water consumption.

Although advances in irrigation technology are improving the efficiency of water use, many challenges remain. For one thing, sophisticated irrigation techniques are expensive. Few farmers in highly developed countries, let alone subsistence farmers in developing nations, can afford to install them. Another challenge is that irrigation needs to make much greater use of recycled wastewater instead of fresh water that could be used for direct human consumption.

REDUCING INDUSTRIAL WATER WASTE

Electric power generators and many industries require water (recall from Chapters 10 to 12 that power plants heat water to form steam, which turns the turbines). In the United States, five major industries—chemical products, paper and pulp, petroleum and coal, primary metals, and food processing—consume almost 90% of industrial water. Water use by these industries does not include water used for cooling purposes.

Stricter pollution-control laws provide some incentive for industries to conserve water. Industries usually recapture, purify, and reuse water to reduce their water use and their water treatment costs. For example, in 2010, Jackson Family Wines in California implemented a water recycling system estimated to eventually save the winery up to 6 million gallons of water annually and to greatly reduce energy usage.

It is likely that water scarcity, in addition to more stringent pollution-control requirements, will encourage further industrial recycling. The potential for industries to conserve water by recycling is enormous.

REDUCING MUNICIPAL WATER WASTE

Like industries, regions and cities recycle or reuse water to reduce consumption. Homes and buildings can be modified to match water quality to water use, for example, by collecting and storing **gray water**.[2] Gray water is water that was already used in sinks, showers, washing machines, and dishwashers. Gray water is recycled to flush toilets, wash cars, or sprinkle lawns (**Figure 13.24**).

gray water Water that has already been used for a relatively nonpolluting purpose, such as showers, dishwashing, and laundry; gray water is not potable, but it can be reused for toilets, plants, or car washing.

In contrast to water recycling, wastewater reuse occurs when water is collected and treated before being redistributed for use. Israel probably has the world's most highly developed system of treating and reusing municipal wastewater. Israel does this out of necessity because all of its possible freshwater sources are already tapped. Reclaimed water is used for irrigation, leaving higher-quality fresh water for cities. Used water contains pollutants, but most of these are nutrients from treated sewage and are beneficial to crops.

In addition to recycling and reuse, cities decrease water consumption through other conservation measures. These include consumer education of both children and adults, use of water-saving household fixtures, and development of economic incentives to save water (see You Can Make a Difference: Conserving Water at Home). These measures successfully pull cities through dry spells; they are effective because individuals are willing to conserve for the common good during water crisis periods.

Increasingly, cities are examining ways to encourage individual water conservation methods. The installation of water meters in residences in Boulder and New York City reduced water consumption by about one-third. Before the installation, homeowners were charged a flat fee, regardless of their water use. For many apartment dwellers, water use is included in the rent; charging each apartment for its water use provides the incentive to use water more efficiently. A city might also encourage water conservation by offering a rebate to any homeowner who installs a conserving device such as a water-saving toilet, faucet, or showerhead. Building codes that specify installation of such water conservation fixtures help reduce municipal water consumption.

Some cities are investing in systems to collect and store rainwater. A collection system from roofs of buildings, for example, can provide a substantial amount of water that would normally drain into a city's sewage system. Sun Valley Park, California, has developed a project to collect rainwater that floods certain streets during heavy downpours, clean it, and inject it into the Los Angeles aquifer for use at a later time.

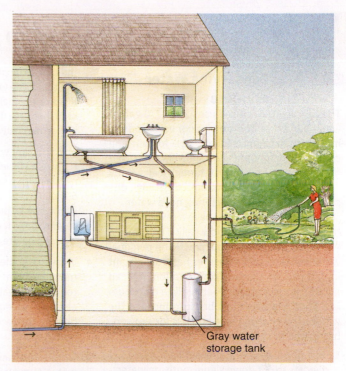

Figure 13.24 **Recycling water.** Individual homes and buildings can be modified to collect and store gray water—water already used for showering or washing clothes or dishes. This gray water is used when clean water is not required—for example, in flushing toilets, washing cars, and watering lawns.

Increasing the price of water to reflect its true cost promotes water conservation. As water prices rise, people quickly learn to conserve water. For example, charging more for water during dry periods encourages individuals to conserve water. Although the average cost of water to consumers rose during the 1990s and early 2000s, many U.S. cities still did not charge consumers what the water actually cost them. Costs, however, are now on the rise, in large part to cover needed repairs in municipal water systems.

Water supply systems (pipes and water mains) in many urban areas are old and leaky. In fact, the average U.S. city loses about one-fourth of its piped water to leaks. Repairing water mains and pipes improves *water accountability*, the efficiency of water use. As part of its aggressive water efficiency program, which began in the late 1980s, the Massachusetts Water Resources Authority actively detects and repairs leaks for the Greater Boston area. The resulting water conservation has saved taxpayers millions of dollars.

REVIEW

1. What are the most promising water conservation measures for agriculture, industry, and individual homes and businesses?

---

[2] Permits to install gray water systems vary from state to state. Arizona and other states with severe water shortages are more flexible about allowing gray water systems.

YOU CAN MAKE A DIFFERENCE

Conserving Water at Home

The average U.S. citizen uses 265 L (70 gal) of water per day at home on indoor uses: approximately 27% to flush toilets, 22% for washing clothes, 19% for baths and showers, 16% used from faucets, and 14% lost through leaks. Many appliances, such as dishwashers, garbage disposals, and washing machines, need water.

As a water user, you have a responsibility to use water carefully and wisely. The cumulative effect of many people practicing personal water conservation measures has a significant impact on overall water consumption. You can adopt these measures yourself. The bathroom is a good place to start because most of the water used in an average home is for showers, baths, and flushing toilets (see cartoon):

1. Install water-saving showerheads and faucets to cut down significantly on water flow. Low-flow showerheads, for example, reduce water flow from 5 to 9 gal per minute to 2.5 gal per minute. Replacing one old showerhead brings a home $30 to $50 each year in water and energy savings. You can also save water by replacing washers on leaky faucets.

2. Install a low-flush toilet or use a water displacement device in the tank of a conventional toilet. Low-flush toilets require only 2 gal or less per flush, compared with 5 to 9 gal for conventional toilets. To save water with a conventional toilet, fill an empty plastic laundry bottle with water and place it in the tank to displace some of the water. Don't put the bottle where it will interfere with the flushing mechanism; don't add bricks to the tank, because they dissolve over time and can cause costly plumbing repairs.

3. An important way to conserve water at home is to fix leaky fixtures. For example, a toilet with a silent leak could waste 30 to 50 gal of water each day. You can test for a silent leak by putting food coloring in the reserve tank. If the color shows up in the toilet bowl before you flush, you have a leak.

4. If you are in the market for a washing machine, high-efficiency washing machines require less water than traditional models and also require less energy and less detergent. Always adjust the water level to match the size of the load.

5. Modify your personal habits to conserve water. Avoid leaving the faucet running. Allowing the faucet to run while shaving consumes an average of 20 gal of water; you will use only 1 gal if you simply fill the basin with water or run the water only to rinse your razor. You may save as much as 10 gal of water a day by wetting your toothbrush and then turning off the tap while you brush your teeth, as opposed to running the water during the entire process. Also, most of us take too long to shower. Time yourself the next time you take a shower, and if it is 10 minutes or longer, work on reducing your shower time.

6. Surprisingly, you will save water by using a dishwasher, which typically consumes about 12 gal per run, instead of washing dishes by hand with the tap running—but only if you run the dishwasher with a full load of dishes. That 12 gal of water is used regardless of whether the dishwasher is full or half-empty.

Remember that wasting water costs you money. Conserving water at home reduces your water bill and heating bill: If you are using less hot water, you are using less energy to heat that water.

"It's your father's idea. He wants you to take quicker showers."

An unconventional approach to water conservation. What more practical measures might you take to reduce your personal water consumption in the bathroom? In the kitchen?

ENERGY AND CLIMATE CHANGE

Impacts on Water Resources

A significant—and troubling—positive feedback loop may exist between climate change and hydropower. As climate changes due to greenhouse gas emissions, we are observing shifts in precipitation patterns and expect them to continue. In some places, reduced precipitation will mean less hydropower, as in the case of Lake Mead, in the Colorado River basin (see the figure). Hydropower does not produce greenhouse gases, but some of its potential replacements (coal, oil, and natural gas) do. If we need to move water farther as a result of droughts, some of the energy required for pumping will likely come from fossil fuels as well. In this case, climate change could indirectly lead to *increased* greenhouse gases.

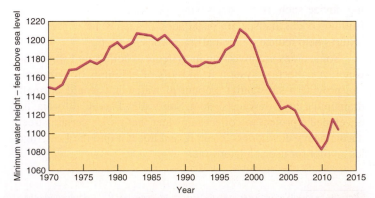

Minimum water level in Lake Mead. Over the past decade and a half, drought and water demand have diminished the water level in Lake Mead. How might this impact agricultural water? Water for generating electricity?
Source: U.S. Bureau of Reclamation

REVIEW OF LEARNING OBJECTIVES WITH SELECTED KEY TERMS

• **Describe the structure of a water molecule and explain how hydrogen bonds form between adjacent water molecules.**

Many of the properties of water, such as its high heat capacity and high dissolving ability, are the result of its **polarity**—one end of the molecule has a positive charge and the other end has a negative charge. The negative end of one molecule is attracted to the positive end of another, forming a **hydrogen bond** between the molecules.

• **Describe surface water and groundwater.**

Surface water comes from precipitation that remains on the surface of the land as **runoff** and does not seep down through the soil. **Groundwater** is the supply of fresh water under Earth's surface that is stored in underground **aquifers**; the **water table** is the upper surface of groundwater.

• **Describe the role of irrigation in world water consumption.**

The greatest user of water worldwide is agriculture, for irrigation. Irrigation accounts for 70% of the world's total water consumption.

• **Define *floodplain* and explain and evaluate flood control efforts, using the Mississippi River basin as an example**

A **floodplain** is the area bordering a river channel that has the potential to flood. Flood damage is exacerbated by the deforestation of hillsides and mountains and by development on floodplains. Major flooding occurs periodically on the Mississippi, including in 2011, when the river's elaborate flood control system was heavily tested. Many ecologists believe allowing a more natural regimen of flooding would better restore the floodplain and promote wetland development.

• **Relate some of the problems caused by overdrawing surface water and aquifer depletion.**

When surface water is overdrawn, the organisms in freshwater ecosystems suffer; also, natural wetlands dry up and estuaries become saltier. **Aquifer depletion** is the removal of groundwater faster than it can be recharged by precipitation or melting snow. Aquifer depletion from porous rocks can cause **subsidence**, or sinking land. In some areas, **sinkholes** occur when droughts or excessive pumping of water causes a lowering of the water table. Aquifer depletion can cause **saltwater intrusion**, the movement of seawater into a freshwater aquifer located near the coast.

• **Relate the background for the following U.S. water problems: Mono Lake, the Colorado River basin, drought in Texas, and the Ogallala Aquifer.**

Mono Lake had surface water diverted to Los Angeles, lowering its water level and increasing its salinity. In 1994 California decided that Mono Lake would be returned almost to its original level. The Colorado River basin supplies water to much of the southwestern United States as well as northwestern Mexico. Demand, especially due to urbanization, matches or exceeds supply, and climate change appears to be reducing the precipitation that fills the Colorado River. Historic drought conditions in Texas are exacerbated by the state's rapidly growing population. Aquifer depletion of the Ogallala Aquifer on the High Plains has lowered the water table in some places by more than 30 m, and experts predict that groundwater in the Ogallala will eventually drop to a level uneconomical to pump. These cases demonstrate how balancing limited water supply with demands from existing users, new users, and ecosystems can be a challenge.

• **Explain the relationship between climate change and available water.**

Climate change is expected to lead to changes in precipitation that, due to biological and geological factors, may have severe impacts in some areas. Other effects of climate change on fresh water include reduced snowpack and saltwater intrusion from sea-level rise.

• **Briefly describe each of the following international water problems: drinking-water problems, population growth and water problems, the Rhine River basin, the Aral Sea, and potential international conflicts over water rights.**

People in many developing countries lack access to safe drinking water and wastewater disposal. Population growth is outstripping water supplies in countries such as India, China, and Mexico. The Rhine River basin spans five European countries, which cooperate to conserve and protect the supply and quantity of Rhine River water. For decades, water flowing into the Aral Sea was overdiverted for irrigation of farmland, and airborne salt from the dry lakebed likely harmed the health of people living nearby. Parts of the Aral Sea have improved over the past decade, while other sections continue to decline. Water disputes have historically resulted in conflicts. International tensions over water rights for such rivers as the Mekong, Indus, Ganges, Tigris and Euphrates, Jordan, and Nile could contribute to or result in armed conflicts.

• **Define *sustainable water use*.**

Sustainable water use is the use of water resources in a fashion that does not harm the essential functions of the hydrologic cycle or the ecosystems on which present and future humans depend.

• **Contrast the benefits and drawbacks of dams and reservoirs, using the Columbia River to provide specific examples.**

Dams ensure a year-round supply of water in areas that have seasonal precipitation or snowmelt. The Columbia River, which has more than 100 dams, is used for shipping, hydroelectric power, and municipal and industrial water. Dams along the Columbia River have adversely affected salmon populations.

• **Briefly describe two methods of desalination.**

Desalination is the removal of salt from ocean or brackish (somewhat salty) water. One method of desalination is **distillation**, heating salt water until water evaporates, leaving behind salt; the water vapor is then condensed. Membrane/filtration methods are more energy-efficient than distillation. The main method is **reverse osmosis**, in which salt water is forced through a membrane permeable to water but not to salt.

• **Give examples of water conservation by agriculture, industry, and individual homes and buildings.**

Certain agricultural techniques can significantly cut agricultural water consumption. **Microirrigation** is a type of irrigation that conserves water by piping it to crops through sealed systems. Water conservation, including recycling and reuse, can reduce both industrial and municipal water consumption. **Gray water**, which is water that has already been used for light household applications, can substitute for fresh water where drinkability is not required.

Critical Thinking and Review Questions

1. Explain why the poor in many countries have to pay more for their water than do the wealthy.

2. What does this diagram represent? Explain how this process affects the properties of water.

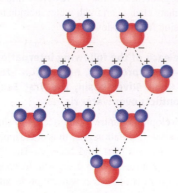

3. Describe several threats to surface water and groundwater supplies.

4. Discuss the dissolving ability of water as it relates to ocean salinity and to water pollution.

5. Should we allow housing on the floodplain of a river? Should taxpayers provide federal disaster assistance for those who choose to live on floodplains? Explain your answers.

6. Why might an ecologist consider a system of levees, spillways, and other flood control structures detrimental to a river system?

7. Explain the relationship between global climate change and fresh water.

8. Are water supply problems due to too many people or too much consumption per person in the United States? In less affluent countries? Explain the differences (if any).

9. Briefly describe the complexity of international water use, using the Rhine River, the Aral Sea, or Iraq as an example.

10. Explain how water resource problems might contribute to economic or political instability.

11. What are some of the advantages of dams? The disadvantages?

12. Describe several types of irrigation that reduce water use. With these techniques available, why would farmers still use traditional irrigation methods?

13. Which industries consume the most water? Discuss one way to use industrial water more sustainably.

14. Use the following graph to compare residential water use in the United States, Sweden, and the Netherlands, as reported in a past study. How do the countries differ in their water use habits? What particular changes might be made within each country to conserve water?

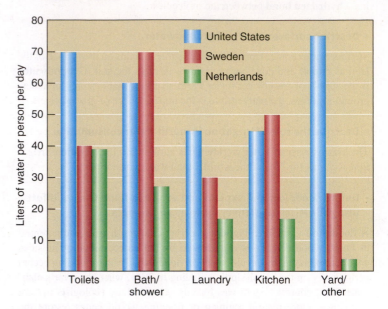

Residential water use in three countries with developed water systems. (Data from J. Kindler and C. S. Russell. *Modeling Water Demands*. Toronto: Academic Press [1984]. As reported in P. H. Gleick, "Basic Water Requirements for Human Activities: Meeting Basic Needs." *Water International*, Vol. 21 [1996].)

15. One strategy for inland cities like Denver or Phoenix to increase their water supply would be to fund desalination plants hundreds of miles away in California and trade that water for an increased share of Colorado River water. Explain why this might be a cost-effective solution, and discuss whether it seems like a fair or politically viable approach.

16. Develop a brief water conservation plan for your own personal daily use.

 ## Food for Thought

Problems with water availability affect food production around the world. Compare water issues as they relate to agriculture in Africa, North America, and the Middle East; how are they similar, and how do they differ? What questions might a new farmer consider if he or she hopes to limit or avoid water constraints? Discuss two ways the farmer might use agricultural water more sustainably.

14

Soil Resources

Clearing a mudslide on the Prithvihighway from Kathmandu to Pokhara during the monsoon.

Anders Blomqvist / Getty Images

In March of 2014, a mudslide covered the small town of Oso, Washington. Over 20 residents disappeared beneath the combination of moving soil, water, trees, and debris from the community. This particular mudslide was precipitated by heavy rainfall, soil erosion exacerbated by a history of logging, and a weak streambank at the base of the supporting hillside. Earthquakes, hurricanes, volcanic eruptions, heavy rainfall (**see photograph**), and other natural events can cause mudslides. In Colorado in 2013, wildfires created a situation ripe for mudslide, as mountains were left unprotected by burned vegetation. Human activities such as logging, mining, and construction also make soil vulnerable to erosion events.

Surface mining, such as mountaintop removal coal mining, increases the risks of mudslides in several ways. Mountaintop removal coal mining is the process of exposing coal by removing the mountain (trees, soil, rock layers) above the coal; trees, soil, and rock are then dumped off the mountainside into valleys and streams. First, removal of vegetation leaves soil exposed to heavy rains and eliminates roots that stabilize soil. Second, mountaintop removal buries existing streambeds, changing water channels in unpredictable ways. Altered slopes are

less stable than the original slopes. Unlike soil on an undisturbed mountaintop, soil and rock debris (called *overburden*) piled by mining equipment does not have a stable, settled structure. People living downstream from mountaintop removal sites rarely have voice, power, or vote in the mining process. When mudslides or coal waste spills result at these sites, the term *natural disaster* deflects blame from the human actions that made erosion a likely outcome.

Mudslides are a forceful example of soil erosion through the action of water. Wind can also cause soil erosion, as in the 1930s Dust Bowl (see Case in Point in this chapter), when drought and poor agricultural practices in the U.S. Midwest resulted in soil loss so great that blown soil reduced air quality in Europe. Soil is not usually considered a deadly weapon, but when soil moves, human lives are often at risk.

At the same time, many of the most agriculturally productive areas are places where soil has accumulated over time. River valleys and deltas are graced with deep, fertile soils, carried from eroded lands upstream. Many of the world's great cities are located in these areas where soil has landed after a long journey, where waterways facilitate transport of products and where rich soils facilitate food

growth. Soil structures are the foundation of human activity and civilization. Soil instability threatens the foundations of human communities, even when mudslides are unlikely.

In this chapter we discuss soil as a valuable natural resource on which humans and many other organisms depend. Many human activities cause or accentuate soil problems, such as erosion and nutrient mineral depletion. The goals of soil conservation are to minimize soil erosion and maintain soil fertility so that this resource can be used in a sustainable fashion. Conserving our soil resources is critical to human survival because more than 99% of our food comes from the land.

In Your Own Backyard...
Visit http://websoilsurvey.nrcs.usda.gov/app/HomePage.htm. Click on "archived soil surveys" to see a list of soil surveys by state. Click on your state, then on your county. Select the manuscript pdf and scroll down to "physiography." What is the geologic history for your area of the county?

The Soil System

LEARNING OBJECTIVES

- Define *soil* and identify the state factors involved in formation of the soil system.
- List the four components of the soil system and explain the ecological significance of each.
- Describe the various soil horizons.
- Name at least two ecosystem services performed by soil organisms and briefly discuss nutrient cycling.

soil The uppermost layer of Earth's crust, which supports terrestrial plants, animals, and microorganisms.

Soil is the relatively thin surface layer of Earth's crust consisting of mineral and organic matter modified by the natural actions of weather, wind, water, and organisms. It is easy to take soil for granted. We walk on it throughout our lives but rarely stop to think about how important it is to our survival.

Soil supports virtually all terrestrial food webs. Vast numbers and kinds of organisms, mainly microorganisms, inhabit soil and depend on it for shelter, food, and water. Plants anchor themselves in soil, and from it they receive essential nutrient minerals and water. Because we depend on plants for our food, humans could not exist without soil either (**Figure 14.1**).

Soil represents an evolving system in which a range of organisms and physical processes interact. The following sections describe portions of the soil system.

STATE FACTORS: SOIL-FORMING FACTORS

State factors determine the state of the soil system. The five main state factors are parent material, climate, topography, organisms, and time. Soil is formed from *parent material*, rock that is slowly broken down, or fragmented, into smaller and smaller particles by biological, chemical, and physical **weathering processes** in nature. It takes a long time, sometimes thousands of years, for rock to disintegrate into finer and finer mineral particles. To form 2.5 cm (1 in.) of topsoil may require between 200 and 1000 years. Time is

Figure 14.1   Agricultural soil. This soil on a hillside in Dongchuan, China, supports growth of many crops. Soil is an important natural resource on which humans and terrestrial organisms rely.

also required for organic material to accumulate in the soil. Soil formation is a continuous process that involves interactions between Earth's solid crust and the biosphere. The weathering of parent material beneath already-formed soil continues to add new soil. The thickness of soil varies from a thin film on young lands, near the North and South Poles as well as on the slopes near the tops of mountains, to more than 3 m (10 ft) on old lands, such as savannas.

Organisms and climate both play essential roles in weathering, sometimes working together. When plant roots and other soil organisms respire, they produce carbon dioxide, CO_2, which diffuses into the soil and reacts with soil water to form carbonic acid, H_2CO_3. Organisms such as lichens produce other kinds of acids. These acids etch tiny cracks in the rock; water then seeps into these cracks. If the parent material is located in a temperate climate, the alternate freezing and thawing of the water during the winter causes the cracks to enlarge, breaking off small pieces of rocks. Small plants send their roots into the larger cracks, fracturing the rock further.

Topography, a region's surface features, such as the presence or absence of mountains and valleys, is also involved in soil formation. Steep slopes often have little or no soil on them because soil and rock are continually transported down the slopes by gravity; runoff from precipitation tends to amplify erosion on steep slopes. Moderate slopes and valleys, on the other hand, may encourage the formation of deep soils.

Humans are often considered a state factor because of their global influence on environmental processes such as the carbon and nitrogen cycles (see Chapter 4). Human activities that affect the soil system include agriculture, urbanization, mining, pollution, deforestation, and forestation.

SOIL COMPOSITION

The soil system is composed of four distinct parts: mineral particles, which make up about 45% of a "typical" soil; organic matter (under 5%); water (about 25%); and air (about 25%). Soil occurs in layers, each of which has a certain composition and special properties. The plants, animals, fungi, and microorganisms that inhabit soil interact with it, and nutrient minerals cycle from the soil to organisms, which use them in their biological processes. When the organisms die, bacteria and other soil organisms decompose the dead material, returning the nutrient minerals to the soil.

The mineral portion, which comes from weathered rock, constitutes most of the soil. It provides anchorage and essential nutrient minerals for plants as well as pore space for water and air. Because different rocks are composed of different minerals, soils vary in mineral composition and chemical properties. Rocks rich in aluminum form acidic soils, whereas rocks with silicates of magnesium and iron form soils that may be deficient in calcium, nitrogen, and phosphorus. Even soils formed from the same kind of parent material may not develop in the same way because other factors such as weather, topography, and organisms differ.

The age of a soil affects its mineral composition. In general, older soils are more weathered and lower in certain essential nutrient minerals. Large portions of Australia, Africa, South America, and India have old, infertile soils. In contrast, in geologically recent time, glaciers passed across much of the Northern

Hemisphere, pulverizing bedrock and forming the parent material of fertile soils. At their greatest extent, these ice sheets covered about 10 million km² (4 million mi²) of North America, extending south as far as the Ohio and Missouri Rivers. Essential nutrient minerals are readily available in these geologically young soils and in young soils formed in areas of volcanic activity.

Litter (dead leaves and branches on the soil's surface), animal dung, and dead remains of plants, animals, and microorganisms in various stages of decomposition constitute *soil organic material*. Microorganisms, particularly bacteria, nematodes, and fungi, gradually decompose this material. During organic matter decomposition, essential nutrient mineral ions are released into the soil, where they may be bound to soil particles, absorbed by plant roots, or leached through the soil by water. Organic matter increases the soil's water-holding capacity by acting much like a sponge, and also increases aeration. For these reasons, gardeners often add organic matter to soils, because organic matter fosters optimum air and water conditions for root growth.

The black or dark-brown organic material that remains after much decomposition has occurred is **humus**. Humus, which is not a single chemical compound but a mix of many organic compounds, binds to nutrient mineral ions and holds water. On average, humus persists in agricultural soil for about 20 years. Certain components of humus, however, may persist in the soil for hundreds or even several thousands of years. Although humus is somewhat resistant to further decay, a succession of microorganisms gradually reduces it to the products of decomposition: carbon dioxide, water, and nutrient minerals. Detritus-feeding animals such as earthworms, termites, and ants also help break down humus.

Soil has numerous pore spaces around and among the soil particles. The pore spaces occupy roughly 50% of a soil's volume and are filled with varying proportions of water (called **soil water**) and air (called **soil air**) (**Figure 14.2**); both are necessary to produce a moist but aerated soil that sustains plants and other soil-dwelling organisms. Generally speaking, water is held in the smaller pores (less than 0.05 mm in diameter), whereas air is found in the larger pores. After a prolonged rain, almost all the pore spaces may be filled with water, but water drains rapidly from the larger pore spaces, drawing air from the atmosphere into those spaces.

Soil water originates as precipitation, which drains downward, or as groundwater, which rises upward from the water table (see Chapter 13). Soil water contains low concentrations of dissolved nutrient mineral salts that enter the roots of plants as they absorb the water. Water not bound to soil particles or absorbed by roots **leaches**, or percolates (moves down), through the various layers of the soil, carrying dissolved minerals with it. The deposition of leached material in the lower layers of soil is known as **illuviation**. Iron and aluminum compounds, humus, and clay are some illuvial materials that gather in the subsurface portion of the soil. Some substances completely leach out of the soil because they are so soluble they migrate down into the groundwater. Water can also move upward in the soil, transporting dissolved materials with it, such as when the water table rises.

Soil air contains the same gases as atmospheric air, although they are usually present in different proportions. As a result of cellular respiration by soil organisms, the soil generally contains more carbon dioxide and less oxygen than atmospheric air. (Recall

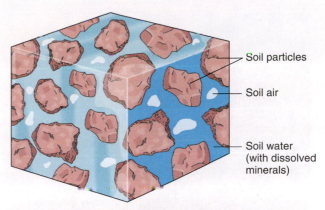

(a) In a wet soil, most of the pore space is filled with water.

- Soil particles
- Soil air
- Soil water (with dissolved minerals)

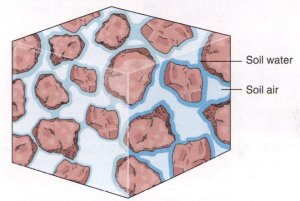

(b) In a dry soil, a thin film of water is tightly bound to soil particles, and soil air occupies most of the pore space.

- Soil water
- Soil air

Figure 14.2   Pore space.

from Chapter 3 that cellular respiration uses oxygen and produces carbon dioxide.) Among the important gases in soil air are oxygen, required by soil organisms for cellular respiration; nitrogen, used by nitrogen-fixing bacteria (see Chapter 4); and carbon dioxide, involved in soil weathering.

SOIL HORIZONS

A deep vertical cut through many soils reveals that they are organized into distinctive horizontal layers called **soil horizons**. A **soil profile** is a vertical section from surface to parent material, showing the soil horizons (**Figure 14.3**). While soil horizons often have a general pattern, the specific horizons that develop are a consequence of the interactions among state factors during soil development or human disturbance.

> **soil horizons** The horizontal layers into which many soils are organized, from the surface to the underlying parent material.

The uppermost layer of soil, the **O-horizon**, is rich in organic material. Plant litter accumulates in the O-horizon and gradually decays. In desert soils the O-horizon is often completely absent, but in certain organically rich soils it may be the dominant layer.

Just beneath the O-horizon is the topsoil, or **A-horizon**, which is dark and rich in accumulated organic matter and humus. Plant roots and soil organisms are plentiful in this horizon. In some soils, a heavily leached **E-horizon** develops between the A- and B-horizons.

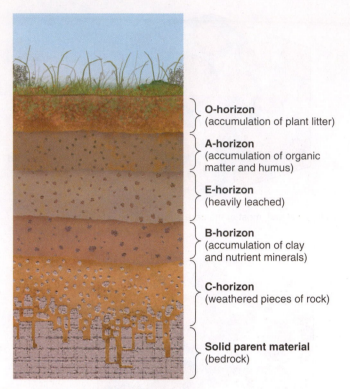

O-horizon
(accumulation of plant litter)

A-horizon
(accumulation of organic matter and humus)

E-horizon
(heavily leached)

B-horizon
(accumulation of clay and nutrient minerals)

C-horizon
(weathered pieces of rock)

Solid parent material
(bedrock)

Figure 14.3   Generalized soil profile. Each horizon has its own chemical and physical properties; each soil has its own set of horizons.

The **B-horizon**, the lighter-colored subsoil beneath the A-horizon, is often a zone of accumulation in which nutrient minerals that leached out of the topsoil and litter accumulate. It is typically rich in iron and aluminum compounds and clay.

Beneath the B-horizon is the C-horizon, which contains weathered pieces of rock and borders the unweathered solid parent material. The C-horizon is below the extent of most roots and is often saturated with groundwater.

SOIL ORGANISMS

Although soil organisms are usually hidden underground, their numbers are huge. Millions of microorganisms, including bacteria, fungi, algae, nematodes, and protozoa, may inhabit just one teaspoon of fertile agricultural soil. Many other organisms colonize the soil ecosystem, including plant roots, insects such as termites and ants, earthworms, moles, snakes, and groundhogs (**Figure 14.4**). Most numerous in soil are bacteria, which number in the hundreds of millions per gram of soil. Scientists have identified about 170,000 species of soil organisms, but thousands remain unidentified. Moreover, scientists know little about the roles of most soil organisms, in part because most do not grow well in laboratories where thorough study is possible.

ecosystem services Important environmental benefits that ecosystems provide to people, including clear air to breathe, clean water to drink, and fertile soil in which to grow crops.

Soil organisms provide several essential **ecosystem services**, such as maintaining soil fertility by decaying and cycling organic material, preventing soil erosion, breaking down toxic

pollutants, cleansing water, and affecting the composition of the atmosphere.

Worms are important organisms living in soil. Among the most familiar soil inhabitants, earthworms ingest soil and obtain energy and raw materials by digesting some of the compounds that make up humus. **Castings**, bits of soil that have passed through the gut of an earthworm, are deposited near the soil surface. In this way, nutrient minerals from deeper layers in the soil are brought to upper layers. Earthworm tunnels serve to aerate the soil, and the worms' waste products and corpses add organic material to the soil.

Ants live in the soil in enormous numbers, constructing tunnels and chambers that aerate it. Members of soil-dwelling ant colonies forage on the surface for bits of food, which they carry back to their nests. Not all this food is eaten, and its eventual decomposition helps increase the organic matter in the soil. Many ants are also indispensable in plant reproduction because they bury seeds in the soil. Some ants grow fungi underground, which help hold soil moisture and increase organic matter.

The properties of soil affect plant growth, although most plants tolerate a wide range of soil types. In turn, the types of plants that grow in soil affect it. Are the plants growing in a certain locality because of the soil found there, or do the plants determine the soil's type? Different soil systems support both scenarios.

One important symbiotic relationship in the soil occurs between fungi and the roots of vascular plants. These fungi, called **mycorrhizae**, help plants absorb adequate amounts of essential nutrient minerals from the soil. The threadlike body of the fungal partner, its **mycelium**, extends into the soil well beyond the roots. Nutrient minerals absorbed from the soil by the fungus are transferred to the plant, whereas food produced by photosynthesis in the plant is delivered to the fungus. Mycorrhizal fungi enhance the growth of plants (see Figure 5.17). Absence of mycorrhizal fungi from soil following a natural or human-induced disturbance retards the reestablishment of many tree species and other plants, such as orchids.

E N V I R O N E W S

Soil Fungi and Plant Communities

Scientists have yet to determine the specific factors that influence plant diversity, but two teams of researchers emphasize the importance of soil-dwelling fungi in shaping plant communities. Researchers in Switzerland and Ontario reported that plant diversity in a given location is strongly influenced by the variety of mycorrhizae present there. In controlled experiments simulating both European and North American landscapes, replicates of the same plant community were grown in the presence of from 1 to 14 mycorrhizal species. The scientists observed that plant communities grown in the presence of 8 or more mycorrhizal species maintained the greatest diversity. When plants grew with fewer kinds of fungi, the structure and composition of plant communities fluctuated considerably, depending on the number and composition of mycorrhizal species present. Researchers think that mycorrhizal diversity plays a significant role in determining plant diversity as well as ecosystem variability and productivity.

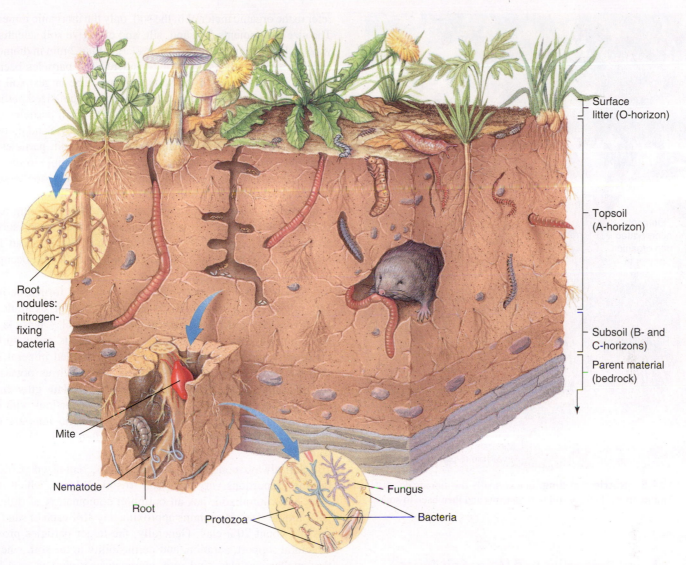

Surface
litter (O-horizon)

Topsoil
(A-horizon)

Root
nodules:
nitrogen-
fixing
bacteria

Subsoil (B- and
C-horizons)

Parent material
(bedrock)

Mite

Nematode

Root

Protozoa

Fungus

Bacteria

Figure 14.4   Soil organisms. The biodiversity in fertile soil includes plants, algae, fungi, earthworms, nematodes, insects, spiders and mites, bacteria, and burrowing animals such as moles, snakes, and groundhogs.

NUTRIENT CYCLING

In a normally functioning ecosystem, the relationships among soil and the organisms that live in and on it ensure soil fertility. Essential nutrient minerals such as nitrogen and phosphorus are cycled from the soil to organisms and back again to the soil (**Figure 14.5**; also see discussion of the nitrogen and phosphorus cycles in Chapter 4). Decomposition is part of **nutrient cycling**. Bacteria and fungi decompose plant and animal detritus and wastes, transforming large organic molecules into small inorganic molecules, including carbon dioxide, water, and nutrient minerals; the nutrient minerals are released into the soil to be used again.

Abiotic (nonliving) processes are also involved in nutrient cycling. Although leaching causes some nutrient minerals to be lost from the soil ecosystem to groundwater, weathering

nutrient cycling The pathway of various nutrient minerals or elements from the environment through organisms and back to the environment.

of the parent material replaces much or all of this loss. In addition, dusts carried in the atmosphere for hundreds or thousands of kilometers help replace nutrient minerals in certain soils. Hawaiian rainforest soils, for example, receive dust inputs from central Asia, a distance more than 6000 km (about 4000 mi) away.

REVIEW

1. How do weathering processes affect soil formation?
2. What are the four components of soil, and how is each important?
3. What are soil horizons? Distinguish between the O-horizon and A-horizon.
4. Name two ecosystem services that soil organisms perform.

refer to the organic material in the soil, only the inorganic minerals. The size assignments for sand, silt, and clay give soil scientists a way to classify soil texture. Particles larger than 2 mm in diameter, called gravel or stones, are not considered soil particles because they do not have any direct value to plants. The largest soil particles (0.05 to 2 mm in diameter) are **sand**, medium-sized particles (0.002 to 0.05 mm in diameter) are **silt**, and small particles (less than 0.002 mm in diameter) are **clay** (**Figure 14.6**). Sand particles are large enough to be seen easily with the eye; silt particles are about the size of flour particles and are barely visible to the eye; clay particles are only seen under an electron microscope. A soil's texture affects many of a soil's properties, which in turn influence plant growth. Clay is particularly important in determining many soil characteristics because clay particles have the greatest surface area of all soil particles. If the surface areas of 500 g (about 1 lb) of clay particles were laid out side by side, they would occupy 1 hectare (2.5 acres).

Soil minerals are often present in charged forms, or **ions**. Mineral ions may be positively charged (K^+, for example) or negatively charged (OH^-, for example). Each clay particle has predominantly negative electrical charges on its outer surface that attract and reversibly bind positively charged mineral ions (**Figure 14.7**). Many of these mineral ions, such as potassium (K^+) and magnesium (Mg^{2+}), are essential for plant growth and are "held" in the soil for plant use by their interactions with clay particles. In contrast, negatively charged mineral ions are usually not held as tightly in the soil and are often washed out of the root zone.

Soil always contains a mixture of different-sized particles, but the proportions vary from soil to soil. A **loam**, which is an ideal agricultural soil, has an optimum combination of different soil particle sizes. It contains approximately 40% each of sand and silt and about 20% clay. Generally, the larger particles provide structural support, aeration, and permeability to the soil, whereas the smaller particles bind into aggregates, or clumps, and hold nutrient minerals and water (**Table 14.1**). Soils with larger proportions of sand are not as desirable for most plants because they do not hold mineral ions or water as well. Plants grown in such soils are more susceptible to mineral deficiencies and drought. Soils with larger proportions of clay are not as desirable for most plants because they provide poor drainage and often do not contain enough oxygen. Clay soils used for agriculture tend to get compacted, which reduces the number of pore spaces that water and air can fill.

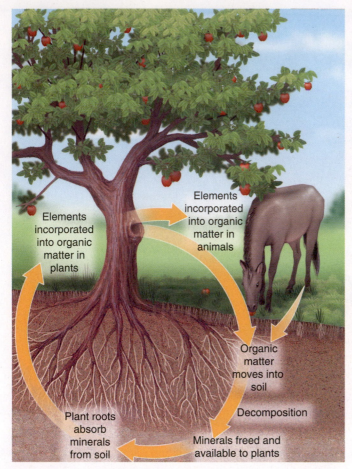

Figure 14.5 **Nutrient cycling.** In a normally functioning ecosystem, mineral nutrients cycle from soil to organisms and then back to soil.

Soil Properties and Major Soil Types

LEARNING OBJECTIVES

- Briefly describe soil texture and soil acidity.
- Distinguish among spodosols, alfisols, mollisols, aridisols, and oxisols.

Texture and acidity are two parameters that characterize soils. Soil **texture** refers to the relative proportions of different-sized inorganic mineral particles of sand, silt, and clay; soil texture does not

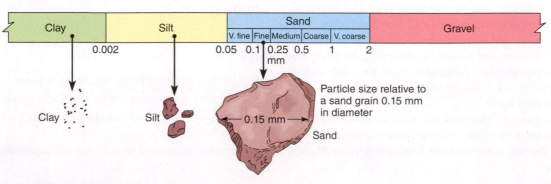

Figure 14.6 **Relative sizes of soil particles.**

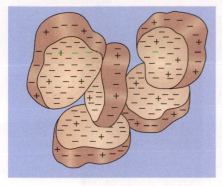

(a) The outer surfaces of clay particles are predominantly negatively charged.

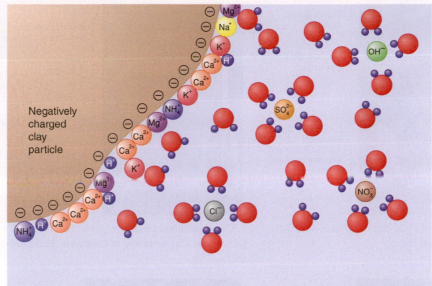

(b) In this close-up of a clay particle and the thin film of water (red oxygen spheres with two blue hydrogens attached) around it, note the large number of positively charge ions attracted to the surface of the clay particle. Also note how the positive (blue) ends of water molecules surround the negatively charged ions dissolved in solution.

Figure 14.7   **Availability of nutrient minerals.**

Table 14.1   Soil Properties Affected by Soil Texture

	Soil Texture Type		
Soil Property	Sandy Soil	Loam	Clay Soil
Aeration	Excellent	Good	Poor
Drainage	Excellent	Good	Poor
Nutrient mineral–holding capacity	Low	Medium	High
Water-holding capacity	Low	Medium	High
Workability (tillage)	Easy	Moderate	Difficult

SOIL ACIDITY

Soil acidity is measured using the pH scale, which extends from 0 (extremely acidic) through 7 (neutral) to 14 (extremely alkaline). (See Appendix I for a discussion of pH.) The pH of most soils ranges from 4 to 8, but some soils are outside this range. Soil pH affects plants partly because the solubility of certain nutrient minerals varies with differences in pH; that is, soil nutrients are available to plants or unavailable depending on the pH. Plants absorb soluble mineral elements but cannot absorb insoluble forms. At a low pH, the aluminum and manganese in soil water are more soluble, and the roots sometimes absorb them in toxic concentrations. Certain mineral salts essential for plant growth, such as calcium phosphate, become less soluble and less available to plants at a higher pH.

Soil pH greatly affects the leaching of nutrient minerals. The more acidic a soil the less capacity it has to bind positively charged ions. As a result, certain nutrient mineral ions essential for plant growth, such as potassium (K^+), leach more readily from an acidic soil. The optimum soil pH for most plant growth is 6.0 to 7.0 because most nutrient minerals needed by plants are available in that pH range.

Soil pH affects plants and, in turn, is influenced by plants and other soil organisms. Litter composed of the needles of conifers contains acids that leach into the soil, lowering its pH. The decomposition of humus and the cellular respiration by soil organisms also decrease the pH of soil.

Acid precipitation is a type of air pollution in which human-produced sulfuric and nitric acids fall to the ground as acid rain, acid sleet, acid snow, or acid fog. Acid precipitation alters soil chemistry, damages plants, and causes forest decline in some areas (see Chapter 19). Studies of acid rain at Hubbard Brook, New Hampshire, show that the effects of acid rain can be reversed by adding calcium to soil, which helps restore the pH balance.

MAJOR SOIL GROUPS

Variations in climate, local vegetation, parent material, underlying geology, topography, and soil age result in thousands of soil types throughout the world. These soils differ in color, depth, mineral content, acidity, pore space, and other properties. **Soil taxonomy** is the method of classification of these soils into 12 distinctive *orders*; each order is subdivided into many different *series*. In the United States alone, more than 20,000 soil series are known.

Here we focus on five common soil orders: spodosols, alfisols, mollisols, aridisols, and oxisols. Regions with colder climates, ample precipitation, and good drainage typically have soils called **spodosols**, with distinct layers (**Figure 14.8a**). A spodosol usually forms under a coniferous forest and has an O-horizon of acidic litter composed primarily of needles; an ash-gray, acidic, leached E-horizon; and a dark-brown, illuvial B-horizon. Spodosols do not make good farmland because they are too acidic and are nutrient-poor because of leaching.

Temperate deciduous forests grow on **alfisols**, soils with a brown to gray-brown A-horizon (**Figure 14.8b**). Precipitation is

(a) **Spodosol** in cold (subarctic) climate. Characteristic of evergreen forests.

Thin or absent A-horizon

Leached, acidic E-horizon

Dark-brown illuvial B-horizon; rich in organic matter and aluminum and iron oxides

Photo courtesy of USDA NRCS

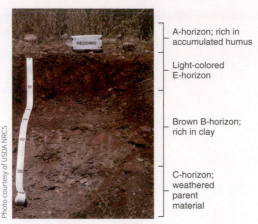

(b) **Alfisol** in humid climate with hot summers and cold winters. Characteristic of temperate, deciduous forest.

A-horizon; rich in accumulated humus

Light-colored E-horizon

Brown B-horizon; rich in clay

C-horizon; weathered parent material

Photo courtesy of USDA NRCS

(c) **Mollisol** in semiarid climate with hot summers and cold winters. Characteristic of temperate grassland.

Thick dark A-horizon; rich in humus

Thick B-horizon; rich in calcium carbonate in deeper parts

Photo courtesy of USDA NRCS

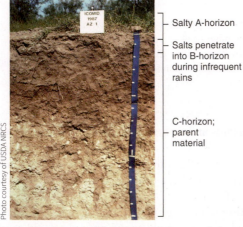

(d) **Aridisol** in arid (dry) climate. Characteristic of desert.

Salty A-horizon

Salts penetrate into B-horizon during infrequent rains

C-horizon; parent material

Photo courtesy of USDA NRCS

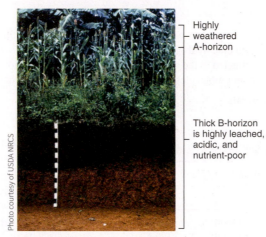

(e) **Oxisol** in warm, wet climate. Characteristic of tropical rain forest.

Highly weathered A-horizon

Thick B-horizon is highly leached, acidic, and nutrient-poor

Photo courtesy of USDA NRCS

Figure 14.8 **Five representative soil orders.** As climate change occurs, plant life on each of these soils will change. How might the changes in plant life affect the characteristics of soil orders?

great enough to wash much of the clay and soluble nutrient minerals out of the A- and E-horizons and into the B-horizon. When the deciduous forest is intact, soil fertility is maintained by a continual supply of decaying plant litter. When the soil is cleared for farmland, soil fertility drops unless it is supplemented with organic or inorganic fertilizer.

Mollisols, found primarily in temperate, semiarid grasslands, are fertile soils (**Figure 14.8c**). They possess a thick, dark-brown to black A-horizon rich in humus. Some soluble nutrient minerals remain in the upper layers because precipitation is not great enough to leach them into lower layers. Most of the world's grain crops are grown on mollisols. These soils provide an excellent example of the involvement of organisms in soil formation: without the deep-rooting grasses, mollisols would not form.

Aridisols are found in arid regions of all continents. The lack of precipitation in these deserts precludes much leaching, and the lack of lush vegetation precludes the accumulation of much organic matter. As a result, aridisols do not usually have distinct layers of leaching and illuviation. Some aridisols have a salic (salty) A-horizon (**Figure 14.8d**). Some aridisols provide rangeland for grazing animals, and crops can be grown on aridisols if water is conserved for irrigation.

Oxisols, which are low in nutrient minerals, exist in tropical and subtropical areas with ample precipitation (**Figure 14.8e**). Little organic material accumulates on the forest floor (O-horizon) because leaves and twigs are rapidly decomposed in the hot, humid climate. The A-horizon is enriched with humus derived from the rapidly decaying plant parts. The B-horizon, which is quite thick, is highly leached, acidic, and nutrient-poor. Oddly enough, lush tropical rain forests grow on oxisols. Most of the nutrient minerals in tropical rain forests are locked up in the vegetation rather than in the soil. As soon as plant and animal remains touch the forest floor, they promptly begin to decay, and plant roots quickly reabsorb the nutrient minerals. Even wood, which may take years to decompose in temperate soils, decomposes in a matter of months in tropical rain forests, largely by subterranean termites.

REVIEW

1. What is soil texture?
2. What is the optimum pH of most soils? What happens if the soil pH falls outside this range?
3. Which of the five soil orders (spodosols, alfisols, mollisols, aridisols, and oxisols) is associated with deserts? With tropical rain forests? With semiarid grasslands?

Environmental Problems Related to Soil

LEARNING OBJECTIVES

- Define *sustainable soil use*.
- Explain the impacts of soil erosion, mineral depletion, soil salinization, and desertification on plant growth.
- Describe the American Dust Bowl and explain how a combination of natural and human-induced factors caused this disaster.

Richard Hansen/Science Source

(a) Water erosion in a field in San Simeon, California. The branchies gullies will continue to deepen unless checked by some type of erosion control.

Humans disrupt soil systems that would be balanced (i.e., functioning normally) in nature. Human activities often cause or exacerbate soil problems, including erosion, mineral depletion of the soil, soil salinization, and desertification, all of which occur worldwide

sustainable soil use The wise use of soil resources, without a reduction in soil fertility, so that the soil remains productive for future generations.

(**Figure 14.9**). Understanding how soil systems work is essential to mitigating these disruptive effects and promoting **sustainable soil use.** Soil used in a sustainable way renews itself by natural processes, sometimes fostered by human management, year after year.

SOIL EROSION

Water, wind, ice, and other agents promote **soil erosion**. Water and wind are particularly effective in moving soil from one place to another (**Figure 14.10**). Rainfall loosens soil particles, which are

soil erosion The wearing away or removal of soil, especially topsoil, from the land.

then transported away by moving water. Wind loosens soil and blows it away, particularly if the soil is barren and dry. Soil erosion is a natural process accelerated by human activities.

Erosion reduces the amount of soil in an area and therefore limits the growth of plants. Erosion causes a loss of soil fertility because essential nutrient minerals and organic matter in the soil are removed. As a result of these losses, the productivity of eroded agricultural soil drops.

© Joel Day/Alamy

(b) Wind erosion on a farm in northern Maryland. When soil is dry, wind picks up topsoil and can move it long distances. Machinery or foot traffic can exacerbate wind erosion.

Figure 14.10   Soil erosion.

Humans often accelerate soil erosion with poor soil management practices. Although soil erosion is often caused by poor agricultural practices, agriculture is not the only culprit. Removal of natural plant communities, such as during the construction of roads and buildings, and unsound logging practices, such as clearcutting large forested areas, accelerate soil erosion.

Soil erosion has an impact on other natural resources as well. Sediment that gets into streams, rivers, and lakes affects water quality and fish habitats. If the sediment contains pesticide and fertilizer residues, they further pollute the water. When forests are removed within the watershed of a hydroelectric power facility, accelerated soil erosion causes the reservoir behind the dam to fill with sediment much faster than usual. This process results in a reduction of electricity production at that facility.

Sufficient plant cover limits the amount of soil erosion. Leaves and stems cushion the impact of rainfall, and roots help hold the soil in place. Although soil erosion is a natural process, abundant plant cover makes it negligible in many natural ecosystems.

Soil Erosion in the United States and the World Every five years the Natural Resources Conservation Service (NRCS),

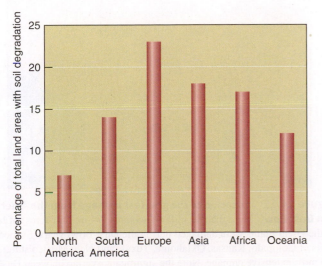

Figure 14.9   Soil degradation. This graph shows the degree of soil degradation (eroded, desertified, or salinized soil) by continent. (*World Atlas of Desertification*, UN Environment Programme, 2002.)

formerly called the Soil Conservation Service, measures the rate of soil erosion at thousands of sites across the United States. It also uses satellite data and models to estimate annual soil erosion. These measurements and estimates indicate that erosion remains a serious threat to cultivated soils in many regions throughout the United States, particularly in parts of southern Iowa, northern Missouri, western and southern Texas, and eastern Tennessee. The good news is that soil erosion on all U.S. croplands declined about 55% between 1982 and 2007, although the Environmental Working Group suggests that erosion has been underestimated both before and after 2007.

Water erosion is particularly severe in the Midwestern grain belt along the Mississippi and Missouri Rivers, as well as in the Central Valley of California and in the hilly Palouse River region of the Pacific Northwest. The NRCS estimates that about 25% of U.S. agricultural lands are losing topsoil faster than natural soil-forming processes regenerate it. This loss is often so gradual that even farmers fail to notice it. A severe rainstorm may wash away 1 mm (0.04 in.) of soil, which seems insignificant until the cumulative effects of many storms are taken into account. Twenty years of soil erosion amounts to the loss of about 2.5 cm (1 in.) of soil, an amount that could take hundreds of years for natural soil-forming processes to replace.

Soil erosion is a significant problem worldwide. Although estimates vary widely, depending on what assumptions are made, soil erosion results in an annual loss of as much as 75 billion metric tons (83 billion tons) of topsoil around the world. Soil erosion is greatest in certain parts of Asia, Africa, and Central and South America. In India and China, soil experts estimate that erosion causes an annual loss of as much as 4.5 billion metric tons and 5.5 billion metric tons of soil, respectively. These two countries have 13% of the world's total land area, from which they must feed 2.6 billion people—more than 35% of the world's human population.

CASE IN POINT

THE AMERICAN DUST BOWL

Semiarid lands, such as the Great Plains of North America, have low annual precipitation and are subject to periodic droughts. Prairie grasses, the plants that grow best in semiarid lands, are adapted to survive droughts. Although the above-ground portions of the plant may die, the root systems survive several years of drought. When the rains return, the root systems send up new leaves. Soil erosion is minimal because the dormant but living root systems hold the soil in place and resist the assault by wind and water.

The soils of semiarid lands are often of high quality, owing largely to the accumulation over many centuries of a thick, rich humus. These lands are excellent for grazing and for growing crops on a small scale. Problems arise when large areas of land are cleared for crops or when animals overgraze the land. The removal of the natural plant cover opens the way for climate conditions to "attack" the soil, and it gradually deteriorates from the onslaught of hot summer sun, occasional violent rainstorms, and wind. If a prolonged drought occurs under such conditions, disaster can strike.

The effects of wind on soil erosion were vividly experienced over a wide region of the central United States during the 1930s (**Figure 14.11**). Throughout the late 19th and early 20th centuries, much of the native grasses were removed to plant wheat. Then, between 1930 and 1937, the semiarid lands stretching from Oklahoma and Texas into Canada received 65% less annual precipitation than was normal. The rugged native prairie grasses could have survived these conditions, but not the wheat. The prolonged drought caused crop failures, which left fields barren and particularly vulnerable to wind erosion.

Winds from the west swept across the barren, exposed soil, causing dust storms of incredible magnitude. Topsoil from Colorado, Texas, Oklahoma, and other prairie states was blown eastward for hundreds of kilometers. Women hanging out clean laundry in Georgia went outside later to find it covered with dust. Bakers in New York City and Washington, D.C., had to keep freshly baked bread away from open windows so that it would not get dirty. The dust even discolored the Atlantic Ocean several hundred kilometers off the coast. On April 14, 1937, known as Black Sunday, the most severe dust storm in U.S. history darkened the sky and blotted out the sun.

The Dust Bowl years occurred during the Great Depression, and ranchers and farmers quickly went bankrupt. Many abandoned

Figure 14.11   **Location of the greatest damage from the American Dust Bowl.** More than 30 million hectares (74 million acres) of land in the Great Plains were damaged during the Dust Bowl years. Shaded parts of Colorado, Kansas, Oklahoma, Texas, and New Mexico suffered the most extensive damage. Note the significant overlap between Dust Bowl erosion and the extent of the Ogallala Aquifer (Figure 13.16).

their dust-choked land and dead livestock and migrated west to the promise of California (**Figure 14.12**). The plight of these dispossessed farmers is movingly portrayed in the novel *The Grapes of Wrath*, written in 1939 by **John Steinbeck**, for which he won the Pulitzer Prize in 1940.

When the rains finally arrived in the Great Plains, many areas were too eroded to support agriculture. In the years following the Dust Bowl, the U.S. Soil Conservation Service planted fence rows of shrubs and trees (to slow the winds) and seeded many badly eroded areas with native prairie grasses. Agriculture returned, largely because irrigation protected against crop loss from droughts. However, the Ogallala Aquifer, which provides irrigation water for this region, is being drained faster than its rate of replacement (see Chapter 13). When farmers are forced to abandon irrigation and revert to dryland farming, only careful management practices will prevent Dust Bowl conditions from returning during the next drought.

Although the United States no longer has a dust bowl, the Great Plains are still subject to droughts and soil erosion. For example, a severe five-year drought in the early 2000s ruined crops in parts of Montana, Wyoming, Colorado, Kansas, Texas, and New Mexico. Dust storms (brownouts) reoccurred on the High Plains, reminding people of the "Dirty Thirties." ■

NUTRIENT MINERAL DEPLETION

In a natural ecosystem, essential nutrient minerals cycle from the soil to organisms, particularly plants, which absorb them through their roots. When those organisms die and microorganisms decompose them, the essential nutrient minerals are released into the soil, where they become available for use by organisms again. An agricultural system disrupts this pattern of nutrient cycling when the crops are harvested. Much of the plant material that contains nutrient minerals is removed from the cycle, so it fails to decay and release its nutrient minerals back to the soil. Over time, agricultural soil loses its fertility (**Figure 14.13**).

Worldwide, more than 1 billion people depend on agricultural soils that are not productive enough to adequately support them.

Unsound farming methods, extensive soil erosion, and desertification contribute to nutrient mineral depletion. In addition, the needs of a rapidly expanding population exacerbate soil problems worldwide.

In tropical rain forests, the climate, the typical soil type (oxisols), and the removal by humans of the natural forest community result in a particularly severe type of mineral depletion. Soils found in tropical rain forests are somewhat nutrient-poor because the nutrient minerals are stored primarily in the vegetation. Any nutrient minerals that are released as dead organisms decay in the soil and are promptly reabsorbed by plant roots and their mycorrhizae. If this did not occur, the heavy rainfall would quickly leach the nutrient minerals away. Nutrient reabsorption by vegetation is so effective that tropical rainforest soils support luxuriant plant growth despite the relative infertility of the soil, as long as the forest remains intact.

When a forest is cleared to sell the wood or make way for crops or rangeland, the efficient nutrient cycling in the forest is disrupted. Removal of the vegetation that so effectively stores the forest's nutrient minerals allows them to leach out of the system. Crops can be grown on these soils for only a few years before the small mineral reserves in the soil are depleted. When cultivation is abandoned, the forest may eventually return to its original state, provided forest is nearby to serve as a source of seeds. The regrowth of forest is a slow process, however, and not all forest ecosystems return after being destroyed. (Chapter 17 discusses deforestation in detail.)

SOIL SALINIZATION

Soils found in arid and semiarid regions often contain high natural concentrations of inorganic compounds as mineral salts (see Figure 14.8d). In these areas, the amount of water that drains into deeper soil is minimal because the little precipitation that falls quickly evaporates, leaving behind the salt. In contrast, humid climates have enough precipitation to leach salts out of soils and into waterways and groundwater.

Irrigation of agricultural fields often results in their becoming increasingly saline (salty), an occurrence known as **salinization** (**Figure 14.14**). Irrigation water contains small amounts of dissolved salts (fresh water always contains some salts). The continued application of such water, season after season, year after year, leads to the gradual accumulation of salt in the soil. When the water evaporates, the salts are left behind, particularly in the upper layers of the soil, which are the layers that are most important for agriculture. Given enough time, the salt concentration can rise to such a high level that plants are poisoned or their roots dehydrated. Also, when irrigated soil becomes waterlogged, capillary movement may carry salts from groundwater to the soil surface, where they are deposited as a crust of salt.

salinization The gradual accumulation of salt in a soil, often as a result of improper irrigation methods.

DESERTIFICATION

Asia and Africa have large land areas with extensive soil damage, and in both places rapid population growth compounds the problem. Consider the Sahel, a broad band of semiarid land that

Figure 14.12 **Dust Bowl remnants.** A car and a house mark and date a farm abandoned in the Dust Bowl.

Sissee Brimberg & Cotton Coulson, Keenpress / Getty Images

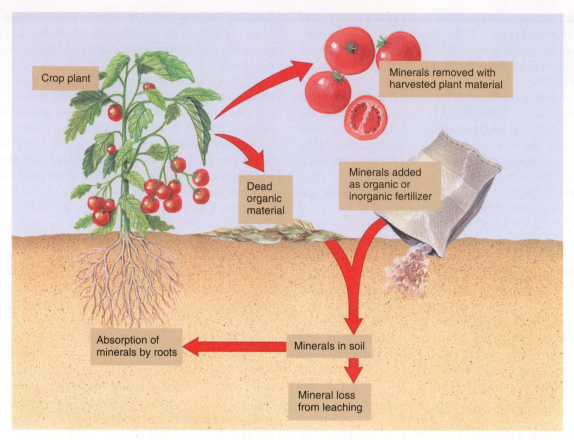

Figure 14.13   Mineral depletion in agricultural soil. In agriculture, much of the plant material is harvested. Because the nutrient minerals in the harvested portions are unavailable to the soil, the nutrient cycle is disrupted, and fertilizer must be periodically added to the soil.

Jim Richardson / Getty Images

Figure 14.14   Soil salinization. A Colorado farm soil with white salt residues from irrigation.

K. Tumanowicz / Science Source

Figure 14.15   Desertification. Livestock in Mali, Africa, have eaten all the ground cover. The dead trees were stripped of branches to feed livestock and provide firewood. Overexploitation by desperately poor people, in combination with extended drought, is increasing the desert area in the Sahel.

stretches across Africa just south of the Sahara Desert and includes all or parts of many countries (see Figure 4.22). The Sahel normally experiences periodic droughts, but for the past 40 years there has been a sustained rainfall deficit—that is, substantially less precipitation than normal. During droughts, the soil cannot support as many crops or grazing animals. Despite the drought, the Sahelians must use their land to grow crops and animals for food or they will starve. The soil is so overexploited under these

circumstances that it supports fewer and fewer people; the day is approaching when much of the Sahel could become unproductive desert (**Figure 14.15**).

The degradation of once-fertile rangeland or forest into nonproductive desert caused partly by soil erosion, forest removal,

desertification Degradation of once-fertile rangeland, agricultural land, or tropical dry forest into nonproductive desert.

overcultivation, and overgrazing is called **desertification**. To reclaim the land would require restricting its use for many years so it could recover. If these measures were taken, the Sahelians would have no means of obtaining food. Land degradation, including desertification, is discussed in greater detail in Chapters 17 and 18.

REVIEW

1. What is sustainable soil use?
2. How would you distinguish between soil erosion and nutrient mineral depletion?
3. What was the American Dust Bowl?

Soil Conservation and Regeneration

LEARNING OBJECTIVES

- Summarize how conservation tillage, crop rotation, contour plowing, strip cropping, terracing, cover crops, shelterbelts, and agroforestry minimize erosion and mineral depletion of the soil.
- Briefly describe the Conservation Reserve Program.

Although agriculture may cause or accelerate soil degradation, good soil conservation practices promote sustainable soil use. Conservation tillage, crop rotation, cover crops, contour plowing, strip cropping, terracing, shelterbelts, and agroforestry help to minimize erosion and mineral depletion of the soil. Land badly damaged by soil erosion and mineral depletion can be successfully restored, but it is a costly, time-consuming process.

CONSERVATION TILLAGE

Conventional methods of tillage, or working the land, include spring plowing, in which the soil is cut and turned in preparation for planting seeds, and harrowing, in which the plowed soil is leveled, seeds are covered, and weeds are removed. Conventional tillage prepares the land for crops, but in removing all plant cover, it greatly increases the likelihood of soil erosion. Conventionally tilled fields contain less organic material and generally hold less water than undisturbed soil.

An increasing number of farmers have adopted **conservation tillage**, which causes little disturbance of the soil (**Figure 14.16**).

conservation tillage A method of cultivation in which residues from previous crops are left in the soil, partially covering it and helping to hold it in place until the newly planted seeds are established.

During planting, special machines cut a narrow furrow in the soil for seeds. Several types of conservation tillage fit different areas of the country and different crops. The most extreme of these, **no-tillage**, does not involve any tilling (that is, no plowing or disking) of the soil. Conservation tillage is one of the fastest-growing trends in U.S. agriculture. Over 60% of U.S. soybeans are currently farmed using conservation tillage; the remainder is planted using conventional tillage. In comparison,

Figure 14.16   Conservation tillage. Decaying residues from the previous year's crop (rye) surround young soybean plants in a field in Iowa. Conservation tillage reduces soil erosion by as much as 70% because plant residues from the previous season's crop protect soil from wind and water erosion.

less than 7% of the world's farmland is planted using conservation tillage methods.

In addition to reducing soil erosion, conservation tillage increases the organic material in the soil, which, in turn, improves the soil's water-holding capacity. Decomposing organic matter releases nutrient minerals more gradually than when conventional tillage methods are employed. Farmers who adopt no-tillage save on fuel costs, machinery wear and tear, and labor time when they do not plow their land. However, use of conservation tillage requires new equipment, new techniques, and greater use of herbicides to control weeds. Research to develop alternative methods of weed control for use with conservation tillage is under way. (Chapter 18 discusses *sustainable agriculture*, which includes conservation tillage and the other soil conservation practices presented in this chapter.)

CROP ROTATION

Farmers who practice effective soil conservation measures often use a combination of conservation tillage and **crop rotation**. When the same crop is grown continuously, pests for that crop tend to accumulate to destructive levels, so crop rotation lessens insect damage and disease. Many studies have shown that continuously growing the same crop for many years

crop rotation The planting of a series of different crops in the same field over a period of years.

depletes the soil of certain essential nutrient minerals faster and makes the soil more prone to erosion. Crop rotation is therefore effective in maintaining soil fertility and in reducing soil erosion.

A typical crop rotation would be corn → soybeans → oats → alfalfa. Soybeans and alfalfa, both of which are members of the legume family, increase soil fertility through their association with bacteria that fix atmospheric nitrogen into the soil. Thus, soybeans and alfalfa provide nutrients for the grain crops they alternate with in crop rotation. Organic farmers often use crop rotations of seven or more crops in a cycle, further breaking pest cycles and reducing demand for the same arrays of crop nutrients.

CONTOUR PLOWING, STRIP CROPPING, AND TERRACING

Hilly terrain must be cultivated with care because it is more prone to soil erosion than flat land. Contour plowing, strip cropping, cover crops, and terracing help control erosion of farmland with variable topography.

contour plowing Plowing that matches the natural contour of the land.

In **contour plowing**, furrows run around hills rather than in straight rows. **Strip cropping**, a special type of contour plowing, produces alternating strips of different crops (**Figure 14.17a**). For example, alternating a row crop such as corn with a closely sown crop such as wheat reduces soil erosion. Even more effective control of soil erosion is achieved when strip cropping is done in conjunction with conservation tillage. **Cover crops** are grown between seasons of other crops (**Figure 14.17b**) and help protect soil during seasons when it would otherwise be bare of plants.

strip cropping Crops laid out in alternating, narrow strips along natural contours, as on a hillside.

cover crop A crop which is planted primarily to protect soil from erosion after the harvest of one crop and before the next harvestable crop is planted.

Farming is undesirable on steep slopes, but if it must be done, **terracing** produces level areas and thereby reduces soil erosion (**Figure 14.17c**). Nutrient minerals and soil are retained on the horizontal platforms instead of being washed away. Soils are preserved in a somewhat similar manner in low-lying areas that are diked to make rice paddies. The water forms a shallow pool, retaining sediments and nutrient minerals.

terracing A soil conservation measure that involves building dikes on hilly terrain to produce level areas for agriculture.

PRESERVING SOIL FERTILITY

The two main types of fertilizer are organic and commercial inorganic. *Organic fertilizers* include such natural materials as animal manure, crop residues, bone meal, and compost. Organic fertilizers are chemically complex, and their exact compositions vary. The nutrient minerals in organic fertilizers become available to plants only as the organic material decomposes. For that reason, organic fertilizers are slow-acting and long-lasting. (For a discussion of compost, see You Can Make a Difference: Composting and Mulching. Also see discussion of municipal solid waste composting in Chapter 23.)

Commercial inorganic fertilizers are manufactured from chemical compounds, and their exact compositions are known. Because they are soluble, they are immediately available to plants. However, commercial inorganic fertilizers are available in the soil for only a short period because they quickly leach away or may volatilize into the air. In sustainable soil management, use of manufactured fertilizers is avoided or limited. First, because of their high solubility, commercial inorganic fertilizers are mobile and often leach into groundwater or surface runoff, polluting the water. Second, manufactured fertilizers do not improve the water-holding capacity of the soil as organic fertilizers do. Another advantage of

(a) Strip cropping is evident on this farm. Crop rotation in such strips often include a legume, which contributes nitrogen to future crops. Photographed in Wisconsin.

(b) A vineyard in autumn, after the harvest. Napa Valley, California.

(c) Terracing hilly or mountainous areas, such as these rice fields in Guangxi, China, reduces the amount of soil erosion. However, some slopes are so steep that they are vulnerable to extensive erosion if not left covered by natural vegetation.

Figure 14.17   Strip cropping, terracing, and cover cropping. These techniques help control erosion on hilly terrain that is farmed.

YOU CAN MAKE A DIFFERENCE

Composting and Mulching

Gardeners often dispose of grass clippings, leaves, and other plant refuse by bagging it for garbage collection or burning it. But these materials are a valuable resource for making **compost**, a natural soil and humus mixture that improves both soil fertility and soil structure. Grass clippings, leaves, weeds, sawdust, coffee grounds, ashes from the fireplace or grill, shredded newspapers, potato peels, and eggshells are just some of the materials that can be composted, or transformed by microbial action into compost.

To make compost, spread a 15- to 30-cm (6- to 12-in.) layer of grass clippings, leaves, or other plant material in a shady area, sprinkle it with an organic garden fertilizer or a thin layer of farm animal manure, and cover it with several inches of soil. Continually add layers of more organic debris. Water the mixture thoroughly, and turn it over with a pitchfork each month to aerate it. Although it is possible to make compost on the open ground, an enclosure allows temperatures to build from the heat generated by microbial action. (Organic material decomposing efficiently in a compost heap has a really hot core.) An enclosed compost heap is also less likely to attract animals. The enclosure should allow some air flow and should allow water to drain out of the material.

When the compost is uniformly dark in color, is crumbly, and has a pleasant, "woodsy" odor, it is ready to use. The time it takes for decomposition will vary from one to six months, depending on the climate, the materials in the mixture, and how often it is turned and watered.

Whereas compost is mixed into soil to improve the soil's fertility, **mulch** is placed on the surface of soil, around the bases of plants (see the photograph). Mulch helps control weeds and increases the amount of water in the upper levels of the soil by reducing evaporation. It lowers the soil temperature in the summer and extends the growing season slightly by providing protection against cold in the fall. Mulch decreases soil erosion by lessening the amount of precipitation runoff.

Although mulches can consist of inorganic materials such as plastic sheets or gravel, organic mulches of compost, grass clippings, straw, chopped corncobs, or shredded bark have the added benefit of increasing the organic content of the soil. Grass clippings are an effective mulch in a lawn, where they slowly enhance lawn fertility. Some gardeners prefer mulches of more expensive materials, such as shredded bark, because they take longer to decompose and are more attractive.

Courtesy U. S. Department of Agriculture

Mulch. Mulch discourages the growth of weeds and helps keep the soil moist. Organic mulches such as this shredded bark have the added benefit of increasing soil fertility as they slowly decay.

---

organic fertilizers is that, in ways not completely understood, they change the types of organisms that live in the soil, sometimes suppressing microorganisms that cause certain plant diseases. Commercial inorganic fertilizers are a source of nitrogen-containing gases (nitrous and nitric oxides) that are air pollutants and greenhouse gases. Finally, the production of commercial inorganic fertilizers requires a great deal of energy, which is largely obtained from our declining reserves of fossil fuels.

SOIL RECLAMATION

Soil reclamation involves two steps: (1) stabilizing the land to prevent further erosion and (2) restoring the soil to its former fertility. The United States has largely reversed the effects of the 1930s Dust Bowl, and China has reclaimed badly eroded land in Inner Mongolia (northern China). To stabilize the land, the bare ground is seeded with plants that eventually grow to cover the soil, holding it in place. The plants start to improve the quality of the soil almost immediately, as dead portions are converted to humus. The humus holds nutrient minerals in place and releases them a little at a time; humus also improves the water-holding capacity of the soil.

One of the best ways to reduce the effects of wind on soil erosion is to plant **shelterbelts** to lessen the impact of wind (**Figure 14.18**). Restoration of soil fertility to its original level is a slow process. During soil recovery, use of the land must be restricted. Disaster is likely if the land is put back to agricultural use before the soil has completely recovered. But restriction of land use for a period of years is sometimes difficult to accomplish. Landowners often object to government dictates about how to manage their lands, and soil erosion in poorer regions of the world is often driven by farmers trying to produce enough food to satisfy basic needs.

shelterbelt A row of trees planted as a windbreak to reduce soil erosion of agricultural land.

AGROFORESTRY

Organizations such as the International Centre for Research in Agroforestry, established in Nairobi, Kenya, are developing techniques to lessen the environmental degradation of the Sahel and other tropical areas. One of their research aims is to use **agroforestry**—land use practices in

agroforestry Concurrent use of forestry and agricultural techniques on the same land area to improve degraded soil and offer economic benefits.

© David Wall/Alamy

Figure 14.18   Aerial view of shelterbelts surrounding row crops. Trees protect the soil from the wind, while also protecting the crops. Photographed in South Island, New Zealand.

which trees and crops are planted together to improve soil fertility in degraded soils.

For example, nitrogen-fixing acacias and other trees might be intercropped with traditional crops such as millet and sorghum. Other crops planted in agroforestry include shade coffee, cocoa, jatropha (a biofuel crop) and bananas. The trees grow for many years and provide several environmental benefits, such as reducing soil erosion, regulating the release of rainwater into groundwater and surface waters, and providing habitat for the natural enemies of crop pests. Acacia trees fix nitrogen, thereby improving soil fertility. When the leaves fall off the trees, they gradually decompose, returning mineral nutrients to the soil. The leaf layer also improves the soil's ability to hold moisture (less moisture evaporates from leaf-covered soil). Over time, the degraded land slowly improves. The result is higher crop yields. When the trees are so tall that they shade out the crops, the forest provides the farmers with food (such as fruits and nuts), fuelwood, lumber, and other forestry products.

SOIL CONSERVATION POLICIES IN THE UNITED STATES

During the late 1920s and early 1930s, **Hugh H. Bennett**, a soil scientist in the U.S. Department of Agriculture, spoke out about the dangers of soil erosion. A report he published in 1928 was largely ignored until the disastrous effects of the Dust Bowl years focused attention on soil as a valuable natural resource. The **Soil Conservation Act** of 1935 authorized the formation of the Soil Conservation Service (now called the Natural Resources Conservation Service, or NRCS); its mission is to work with U.S. citizens to conserve natural resources on private lands. To that end, the NRCS assesses soil damage and develops policies to improve and sustain soil resources.

The **Food Security Act (Farm Bill)** of 1985 contained provisions for two main soil conservation programs: a conservation compliance program and the Conservation Reserve Program. The conservation compliance program requires farmers with highly erodible land to develop and adopt a five-year conservation plan for their farms that includes erosion-control measures. If they don't comply, they lose federal agricultural subsidies such as price supports.

The **Conservation Reserve Program (CRP)** is a voluntary subsidy program that pays farmers to stop producing crops on highly erodible farmland. It requires planting native grasses or trees on such land and then "retiring" it from further use for 10 to 15 years. During that period the land may not be grazed, nor may the grass be harvested for hay. The CRP has benefited the environment. Since its inception in 1985, annual loss of soil on CRP lands planted with grasses or trees has been reduced from an average of 7.7 metric tons of soil per hectare (8.5 tons per acre) to 0.6 metric ton per hectare (0.7 ton per acre). Because the vegetation is not disturbed once it is established, it provides wildlife habitat. Small and large mammals, birds of prey, and ground-nesting birds such as ducks have increased in number and kind on CRP lands. The reduction in soil erosion has improved water quality and enhanced fish populations in surrounding rivers and streams.

The future of the Conservation Reserve Program is unclear. Historically, farmers are more likely to practice soil conservation during hard financial times and periods of agricultural surpluses, both of which translate into lower prices for agricultural products. When prices are high, with a good market for agricultural products, farmers have more incentive to put every parcel of land into production, including marginal, highly erodible lands. Because food prices have risen in recent years, and federal mandates support converting corn crops into ethanol fuel, some farmers are beginning to grow crops on former CRP lands.

REVIEW

1. What is crop rotation? Strip cropping?
2. How can degraded soils be reclaimed? What is the Conservation Reserve Program?

REVIEW OF LEARNING OBJECTIVES WITH SELECTED KEY TERMS

• **Define *soil* and identify the state factors involved in formation of the soil system.**

Soil is the uppermost layer of Earth's crust, which supports terrestrial plants, animals, and microorganisms. Soil formation involves interactions among five main **state factors**: parent material, climate, topography, organisms, and time. Biological, chemical, and physical **weathering processes** slowly break parent material into smaller and smaller particles.

• **List the four components of the soil system and explain the ecological significance of each.**

The soil system is composed of inorganic nutrient minerals, organic materials, soil air, and soil water. The inorganic portion, which comes from weathered parent material, provides anchorage and essential nutrient minerals for plants as well as pore space for water and air. Organic material decomposes, releasing essential nutrient mineral ions into the soil, where they may be absorbed by plant roots. **Soil air** and **soil water** produce a moist but aerated soil that sustains plants and other soil-dwelling organisms.

- **Describe the various soil horizons.**

Soil horizons are the horizontal layers into which many soils are organized, from the surface to the underlying parent material. The uppermost layer of soil, the **O-horizon**, contains plant litter that gradually decays. Just beneath the O-horizon is the topsoil, or **A-horizon**. In some soils, a heavily leached **E-horizon** develops between the A- and B-horizons. The **B-horizon** beneath the A-horizon is often a zone of **illuviation** in which nutrient minerals that leached out of the topsoil and litter accumulate. Beneath the B-horizon is the **C-horizon**, which contains weathered pieces of rock and borders the unweathered solid parent material.

- **Name at least two ecosystem services performed by soil organisms and briefly discuss nutrient cycling.**

Soil organisms are important in two **ecosystem services**: forming soil and cycling nutrient minerals. **Nutrient cycling** is the pathway of various nutrient minerals or elements from the environment through organisms and back to the environment. In nutrient cycling the nutrient minerals removed from the soil by plants are returned when plants and animals die and are decomposed by soil microorganisms.

- **Briefly describe soil texture and soil acidity.**

The **texture** of a soil refers to the relative proportions of **sand**, **silt**, and **clay**. The pH of most soils ranges from 4 (acidic) to 8 (slightly alkaline), but some soils are outside this range. The soil properties of texture and acidity affect a soil's water-holding capacity and nutrient availability, which, in turn, determine how well plants grow.

- **Distinguish among spodosols, alfisols, mollisols, aridisols, and oxisols.**

Regions with colder climates, ample precipitation, and good drainage typically have **spodosols** with distinct layers. Temperate deciduous forests grow on **alfisols**, soils with a brown to gray-brown A-horizon. **Mollisols**, found primarily in temperate, semiarid grasslands, are fertile. **Aridisols** are found in arid regions of all continents. **Oxisols** are low in nutrient minerals and exist in tropical and subtropical areas with ample precipitation.

- **Define** *sustainable soil use.*

Sustainable soil use is the wise use of soil resources, without a reduction in soil fertility, so that the soil remains productive for future generations.

- **Explain the impacts of soil erosion, mineral depletion, soil salinization, and desertification on plant growth.**

Soil erosion is the wearing away or removal of soil from the land. Mineral depletion occurs in all soils that are farmed. **Salinization** is the gradual accumulation of salt in a soil, often as a result of improper irrigation methods. **Desertification** is degradation of once-fertile rangeland, agricultural land, or tropical dry forest into nonproductive desert. Soil erosion, mineral depletion, salinization, and desertification limit the growth of plants.

- **Describe the American Dust Bowl and explain how a combination of natural and human-induced factors caused this disaster.**

The 1930s Dust Bowl in the western United States is an example of accelerated wind erosion caused by use of marginal land for agriculture. Much of the native grasses were removed to plant wheat. Then, for several years the semiarid lands received less annual precipitation than was normal. The prolonged drought caused crop failures. Winds from the west swept across the barren, exposed soil, causing dust storms.

- **Summarize how conservation tillage, crop rotation, contour plowing, strip cropping, cover crops, terracing, shelterbelts, and agroforestry minimize erosion and mineral depletion of the soil.**

Conservation tillage is a method of cultivation in which residues from previous crops are left in the soil, partially covering it and helping to hold it in place until the newly planted seeds are established. **Crop rotation** is the planting of a series of different crops in the same field over a period of years, thereby maintaining soil fertility. **Contour plowing** reduces soil erosion because it matches the natural contour of the land. In **strip cropping**, a type of contour plowing, different crops are laid out in alternating, narrow strips along natural contours, as on a hillside. **Cover crops** are grown between seasons of other crops and help protect soil during seasons when it would otherwise be bare of plants. **Terracing** is a soil conservation measure for hilly terrain, in which dikes are built to produce level areas for agriculture. A **shelterbelt** is a row of trees planted as a windbreak to reduce soil erosion of agricultural lands. **Agroforestry** is the use of both forestry and agricultural techniques to improve degraded areas and offer economic benefits.

- **Briefly describe the Conservation Reserve Program.**

The **Conservation Reserve Program**, which is voluntary, pays farmers to stop producing crops on highly erodible farmland.

Critical Thinking and Review Questions

1. Why do we consider soil to be a system?

2. Explain the roles of weathering, organisms, climate, and topography in soil formation.

3. What are the four distinct parts of the soil system?

4. Which soil horizons are most prone to erosion? How might your answer be significant to farmers?

5. How do organisms contribute to nutrient cycling in the soil system?

6. How does the presence of various-sized particles (sand, silt, and clay) affect soil characteristics?

7. Give an example of how plants affect soil pH. Give an example of how soil pH affects plants.

8. Which two of these soil orders are best suited for agriculture: spodosol, alfisol, mollisol, aridisol, and oxisol? Why?

9. The pie chart shows an overview of the world's total land area with respect to its suitability for agriculture. How could soil that is too dry for agriculture be converted to cropland? How could soil that is too wet for agriculture be converted to cropland? Explain why each of the remaining categories in the pie chart cannot be converted to agricultural land.

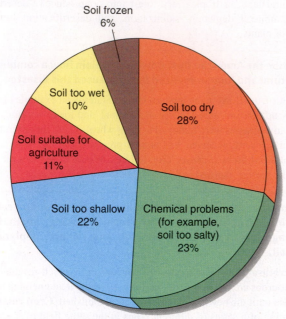

Soil frozen
6%

Soil too wet
10%

Soil suitable for
agriculture
11%

Soil too dry
28%

Soil too shallow
22%

Chemical problems
(for example,
soil too salty)
23%

One Planet, Many People: Atlas of Our Changing Environment. UN Environment Programme, 2005.

10. How does sustainable soil use protect the agricultural soil system?

11. Describe two ways in which nutrient minerals are lost from the soil.

12. Where does eroded soil go after it is transported by water, wind, or ice?

13. How is human overpopulation related to world soil problems?

14. The American Dust Bowl is sometimes portrayed as a "natural" disaster brought on by drought and high winds. Present a case for the view that this disaster was not caused by nature as much as by humans.

15. Distinguish among conservation tillage, crop rotation, contour plowing, strip cropping, terracing, shelterbelts, and agroforestry as methods of sustainable soil use.

16. President Franklin D. Roosevelt once sent a letter to the state governors in which he said, "A nation that destroys its soils, destroys itself." Would he have supported the Conservation Reserve Program? Explain your answer.

17. Does conversion of undisturbed land to farmland impact the global carbon cycle and climate change? Explain your answer.

Food for Thought

Philosophy and science both inform the process of biodynamic farming, a system invented by Rudolf Steiner. Orchards, vineyards, gardens, crops, and livestock can all be grown in a biodynamic system. Soil nutrition is a key part of biodynamic management.

Biodynamic farming aims to replicate natural ecosystems as closely as possible, using specific plants as organic fertilizers and fermenting the plants in combination with animal products. Where is the nearest biodynamic farm in your area? What kinds of products are grown there?

Mineral Resources

Aerial view of the Summitville, Colorado, mine.

The **General Mining Law of 1872** was intended to encourage settlement in the sparsely populated western states. It allows companies or individuals, whether U.S. citizens or foreigners, to stake mining claims on federal land. They then purchase the land for a fraction of its value, extract the valuable hardrock minerals such as gold, silver, copper, lead, or zinc, and keep all the profits. In return, a percentage of profits on lumber, coal, oil, and natural gas obtained from federal land is paid to the government.

The General Mining Law contains no provisions for environmental protection, such as the replacement of topsoil and vegetation or the reestablishment of biological habitat. As a result, hardrock mining has left a legacy of ravaged land, poisoned water, and lifeless ecosystems throughout the West. Acid draining from loose rocks produced during the mining process has made many streams and rivers totally lifeless.

More than 50 of the estimated 500,000 abandoned mines in the United States are designated Superfund sites. The U.S. government—that is, taxpayers—must finance the cleanup of these sites (see Chapter 23 for more on Superfund sites). For example, after a mining company extracted $105 million of gold from a mine in Summitville, Colorado (**see photograph**), it declared bankruptcy, leaving behind an environmental disaster. The EPA designated the Summitville mine as a Superfund site and has provided more than $210 million in cleanup efforts there, focusing on decontaminating polluted soil, preventing further contamination of creeks and streams, and protecting local farmland. Cleanup of all Superfund mining sites will cost billions of dollars.

In 1872, the same year that President Ulysses S. Grant signed the General Mining Law, Yellowstone was designated the first U.S. national park. During the 1990s, a proposed mine in Montana less than 5 km (3 mi) from the Yellowstone border threatened its pristine condition. A Canadian company (Noranda, Inc.) had the rights to this land, which was surrounded by federal wilderness lands, and the company planned to establish a gold, silver, and copper mine there. Opponents of the mine, including the superintendent of Yellowstone, worried that pollution from the mine would contaminate the Yellowstone River and the park. President Clinton stopped the mine by presidential order and negotiated with the Canadian company. In 1997 Noranda agreed to sell the property to the U.S. government.

During the early 2000s, mining reform was a contentious issue in Congress. A 2005 bill proposed that U.S. and foreign mining companies could purchase claims within national parks, national wilderness areas, and other federal lands. Public outcry led to this bill's failure. Legislative efforts in 2007 and 2009 to impose royalties on mining companies that would be used for site cleanup failed to pass through Congress.

In this chapter you will consider the distribution and abundance of minerals as well as the environmental damage from obtaining and processing them. You will also examine options for the future, when mineral deposits become depleted.

> ***In Your Own Backyard...***
> *Select a mineral that you use in your everyday life (for example, steel, aluminum, nickel, or gold) and research how and where it is mined.*

Introduction to Minerals

LEARNING OBJECTIVES

- Define *minerals* and explain the difference between high-grade and low-grade ores, as well as between metallic and nonmetallic minerals.
- Describe several natural processes that concentrate minerals in Earth's crust.
- Distinguish between surface mining and subsurface mining.
- Briefly describe how mineral deposits are discovered, extracted, and processed.

Minerals are such an integral part of our lives that we often take them for granted. Steel, an essential building material, is a blend of iron and other metals. Beverage cans, aircraft, automobiles, and buildings all contain aluminum. Electrical and communications wiring contain copper, which readily conducts electricity. The concrete used in buildings and roads is made from sand and gravel as well as cement, which contains crushed limestone. Sulfur, a component of sulfuric acid, is an indispensable industrial mineral with many applications in the chemical industry. It is used to make plastics and fertilizers and to refine oil. Tantalum, a rare, hard metal that is resistant to corrosion, has become important to the production of capacitors in a range of electronic devices. Other important minerals include platinum, mercury, manganese, and titanium.

> **mineral** An inorganic solid, occurring naturally in or on Earth's crust, with characteristic chemical and physical properties.

The human need and desire for minerals have influenced the course of history. Phoenicians and Romans explored Britain in a search for tin. One of the first metals that humans used, tin came into its own during the Bronze Age (3500 to 1000 B.C.E.), when tin and copper were combined to produce the tougher and more durable alloy, bronze. The desire for gold and silver was directly responsible for the Spanish conquest of the New World. In 1849, a gold rush in California led to the arrival of a large number of settlers from the eastern United States. More recently, the lure of gold in Amazonian and Indonesian rain forests has contributed to the destruction of indigenous homelands and ecosystems.

Earth's minerals are elements or (usually) compounds of elements that have precise chemical compositions. For example, **sulfides** are mineral compounds in which certain elements are combined chemically with sulfur, and **oxides** are mineral compounds in which elements are combined chemically with oxygen.

Rocks, which are aggregates, or mixtures, of one or more minerals, fall into three categories, based on how they formed: igneous, sedimentary, and metamorphic. **Igneous rocks** form when magma rises from the mantle and cools. **Sedimentary rocks** form when small fragments of weathered, eroded rocks (or marine organisms) are deposited, compacted, and cemented together. **Metamorphic rocks** form when intense heat and pressure alter igneous, sedimentary, or other metamorphic rocks.

> **rock cycle** The cycle of rock transformation that involves all parts of Earth's crust.

In the **rock cycle**, rock moves from one physical state or location to another (**Figure 15.1**). The rock cycle continually forms, modifies, transports, and destroys igneous, sedimentary, and metamorphic rocks. For example, igneous rocks form deep in Earth's crust and are uplifted to the surface. Weathering and erosion wear these rocks down to sediments, which eventually solidify into sedimentary rock. The rock cycle is similar to other cycles of matter, such as the carbon and hydrologic cycles (see Chapter 4). However, rocks are formed and move through the environment much more slowly than the elements of other cycles.

Rocks have varied chemical compositions. An **ore** is rock that contains a large enough concentration of a particular mineral to be profitably mined and extracted. **High-grade ores** contain relatively large amounts of particular minerals, whereas **low-grade ores** contain lesser amounts.

Minerals are metallic or nonmetallic (**Table 15.1**). **Metals** are minerals such as iron, aluminum, and copper, which are malleable, lustrous, and good conductors of heat and electricity. **Nonmetallic minerals**, such as sand, stone, salt, and phosphates, lack these characteristics.

MINERAL DISTRIBUTION AND FORMATION

Certain minerals, such as aluminum and iron, are relatively abundant in Earth's crust. Others, including copper, chromium, and molybdenum, are relatively scarce. Abundance does not necessarily mean that the mineral is easily accessible or profitable to extract. It is possible that you have gold and other expensive minerals in your own backyard. However, unless the concentrations are large enough to make them profitable to mine, they will remain there.

Like other natural resources, mineral deposits in Earth's crust are distributed unevenly. Some countries have extremely rich mineral deposits, whereas others have few or none. Although iron is widely distributed, Africa has less iron than the other continents. Many copper deposits are concentrated in North and South America, particularly in Chile and the United States, whereas most of Asia (other than Indonesia) has a relatively small amount of copper. Much of the world's tin is in China and Indonesia, and most chromium reserves are in South Africa. We discuss the international implications of the unequal distribution of important minerals later in the chapter.

Formation of Mineral Deposits Concentrations of minerals within Earth's crust are the result of several natural processes, including magmatic concentration, hydrothermal processes, sedimentation, and evaporation. As magma (molten rock) cools and solidifies deep in Earth's crust, it often separates into layers, with the heavier iron-containing rock settling on the bottom and the lighter silicates (rocks containing silicon) rising to the top. Varying concentrations of minerals are often found in the different rock layers. This layering, called **magmatic concentration**, is responsible for some deposits of iron, copper, nickel, chromium, and other metals.

Hydrothermal processes involve water that was heated deep in Earth's crust. This water seeps through cracks and fissures and dissolves certain minerals in the rocks. The minerals are then carried along in the hot water solution. The dissolving ability of the water is greater if chlorine or fluorine is present because these elements react with many metals (such as copper) to form water-soluble salts (e.g., copper chloride). When the hot solution encounters sulfur, a common element in Earth's crust, a chemical reaction between the metal salts and the sulfur produces metal sulfides. Because metal sulfides are not soluble in water, they

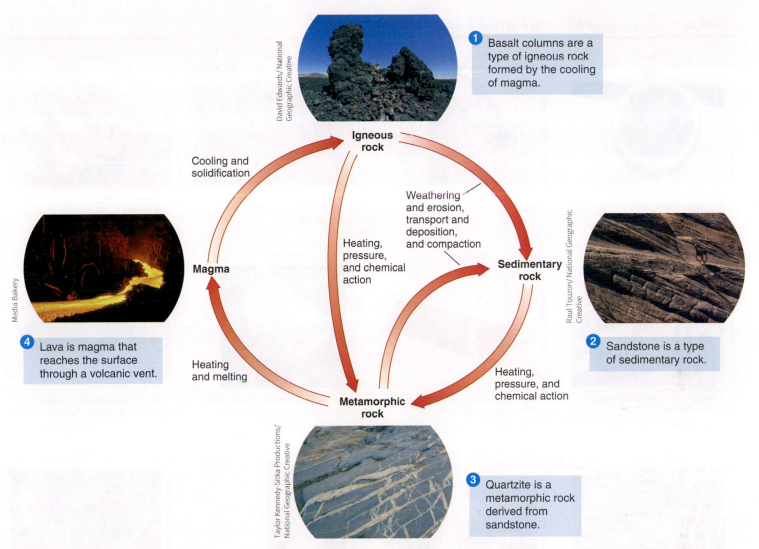

Figure 15.1   **The rock cycle.** Rocks do not remain in their original form forever. This highly simplified diagram shows how rocks cycle from one form to another.

form deposits by settling out of the solution. Hydrothermal processes are responsible for deposits of minerals such as gold, silver, copper, lead, and zinc.

The chemical and physical weathering processes that break rock into finer and finer particles are important not only in soil formation (see Chapter 14) but also in the production of mineral deposits. Weathered particles are transported by water and deposited as sediment on riverbanks, deltas, and the seafloor in a process called **sedimentation**. During their transport, certain minerals in the weathered particles dissolve in the water. They later settle out of solution. When the warm water of a river meets the cold water of the ocean, for example, settling occurs because less material dissolves in cold water than in warm water. Sedimentation has formed important deposits of iron, manganese, phosphorus, sulfur, copper, and other minerals.

Significant amounts of dissolved materials accumulate in inland lakes and in seas that have no outlet or only a small outlet to the ocean. If these bodies of water dry up by **evaporation**, a large amount of salt is left behind. Over time, it may be covered

with sediment and incorporated into rock layers. Evaporation has formed significant deposits of common table salt, borax, potassium salts, and gypsum.

HOW MINERALS ARE FOUND, EXTRACTED, AND PROCESSED

The process of making mineral deposits available for human consumption occurs in several steps. First, a particular mineral deposit is located. Second, mining extracts the mineral from the ground. Third, the mineral is processed, or refined, by concentrating it and removing impurities. During the fourth and final step, the purified mineral is used to make a product. As you will see, each of these steps has environmental implications.

Discovering Mineral Deposits   Geologists employ a variety of instruments and measurements to help locate valuable mineral deposits. Aerial or satellite photography sometimes discloses geologic formations associated with certain types of mineral deposits. Instruments that measure Earth's magnetic field and gravity reveal

Table 15.1   Some Important Minerals and Their Uses.*

Aluminum

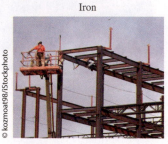

Aircraft, motor vehicles, packaging (cans, foil), water treatment

Chromium

Chrome plate, dyes and paints, steel alloys (cutlery)

Cobalt

Corrosion and wear-resistant alloys, pigments (cobalt blue)

Gold

Jewelry, money, restorative dentistry

Iron

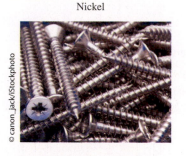

Steel (alloy of iron) buildings and machinery

Magnesium

Beverage cans, electronic devices, firecrackers, flares

Mercury

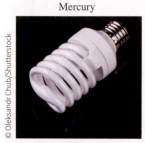

Industrial chemicals, electric and electronic applications, batteries

Molybdenum

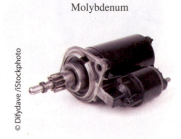

High-temperature alloys for aircraft, industrial motors

Nickel

Coins, metal plating, alloys with various uses

Potassium

Fertilizers, photography

Silver

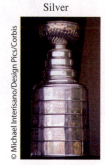

Jewelry, silverware, photography, electronics

Titanium

Alloy in steel and other industrial alloys; pigment in paints, plastics

Zinc

Galvanizing steel, alloys (brass), anode in alkaline batteries

Gypsum (CaSO₄-2H₂O)

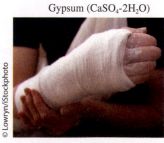

Drywall, plaster of Paris, Soil conditioner

Silicon

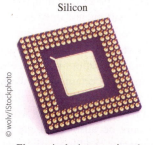

Electronic devices, semiconductors, natural stone, glass, concrete

Sulfur

Industrial chemicals, insecticides, gunpowder, vulcanized tires

*Gypsum, silicon, and sulfur are nonmetals. All other minerals shown are metals.

certain deposits. Seismographs, used to detect earthquakes, also provide clues about mineral deposits. Geologists analyze these data, along with their knowledge of Earth's crust and how minerals are formed, to estimate locations of possible mineral deposits. Once these sites are identified, mining companies drill or tunnel for mineral samples and analyze their composition.

Geologists use data from depth-measuring devices to produce detailed three-dimensional maps of the seafloor. From sophisticated computer analysis of these maps, geologists can estimate the types and amounts of deposits on the ocean floor.

Extracting Minerals   The depth of a particular deposit determines whether surface or subsurface mining will be used. In

surface mining The extraction of mineral and energy resources near Earth's surface by first removing the soil, subsoil, and overlying rock strata.

subsurface mining The extraction of mineral and energy resources from deep underground deposits.

overburden Soil and rock overlying a useful mineral deposit.

surface mining, minerals are extracted near the surface, whereas in subsurface mining, minerals too deep to be removed by surface mining are extracted. Surface mining is more common because it is less expensive than subsurface mining. Because even surface mineral deposits occur in rock layers beneath Earth's surface, the overlying soil and rock layers, called **overburden**, must first be removed, along with the vegetation growing in the soil. Then giant power shovels scoop the minerals out.

Mining companies have two options for surface mining—open-pit surface mining and strip mining. Iron, copper, stone, and gravel are usually extracted by **open-pit surface mining**, in which a giant hole is dug (**Figure 15.2a**). Large holes formed by open-pit surface mining are called quarries. In **strip mining**, a trench is dug to extract the minerals (**Figure 15.2b**). Then a new trench is dug parallel to the old one; the overburden from the new trench is put into the old trench, creating a hill of loose rock called a **spoil bank**.

Subsurface mining, which is underground, may be done with a shaft mine or a slope mine. A **shaft mine** is a direct vertical shaft

spoil bank A hill of loose rock created when the overburden from a new trench is put into the already excavated trench during strip mining.

to the vein of ore (**Figure 15.2c**). The ore is broken up underground and then hoisted through the shaft to the surface in buckets. A **slope mine** has a slanting passage that makes it possible to haul the broken ore out of the mine in carts rather than hoisting it up in buckets (**Figure 15.2d**). Sump pumps keep the subsurface mine dry, and a second shaft is usually installed for ventilation.

Subsurface mining disturbs the land less than surface mining, but it is more expensive and more hazardous. Miners face risk of death or injury from explosions or collapsing walls, and prolonged breathing of dust in subsurface mines can result in lung disease. (Chapter 11 discusses coal mining and its hazards.)

Processing Minerals   Processing metallic minerals often

smelting The process in which ore is melted at high temperatures to separate impurities from the molten metal.

involves **smelting**. Purified copper, tin, lead, iron, manganese, cobalt, or nickel smelting is done in a chimney-like *blast furnace*. **Figure 15.3** shows a blast furnace used to smelt iron. Iron ore, limestone rock, and coke (modified coal used as an industrial fuel) are added at the top of the furnace, while heated air or oxygen is added at the bottom. Chemical

James L. Amos/ National Geographic Creative

(a) This open-pit copper mine in Utah is the largest human-made excavation in the world.

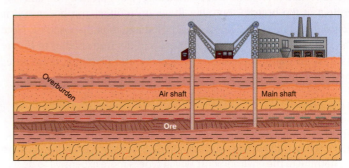

(b) Strip mining removes overburden along narrow strips to reach the ore beneath.

(c) In a shaft mine, a hole is dug straight through the overburden to the ore, which is removed up through the shaft in buckets.

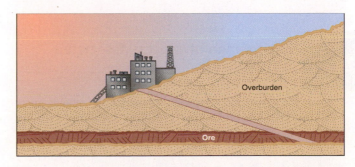

(d) In a slope mine, an entry to the ore is dug at an angle so that the ore can be hauled out in carts.

Figure 15.2   **Types of mining operations.**

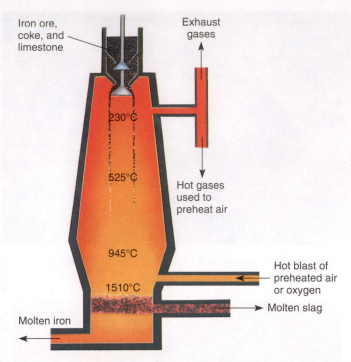

Figure 15.3 **Blast furnace.** Such towerlike furnaces are used to separate metal from impurities in the ore. The energy for smelting comes from a blast of heated air.

reactions take place throughout the furnace as the ore moves downward: The iron ore reacts with coke to form molten iron and carbon dioxide, whereas the limestone reacts with impurities in the ore to form a molten mixture called slag. Both molten iron and slag collect at the bottom, but slag floats on molten iron because it is less dense than iron. The slag is cooled and then disposed of. Note the vent for exhaust gases near the top of the iron smelter in Figure 15.3. If air pollution–control devices are not installed, toxic gases such as sulfur oxides are emitted during smelting.

REVIEW

1. What is the difference between high-grade and low-grade ores?
2. Which type of mining—surface mining or subsurface mining—is more harmful to the environment? Which type is more expensive to do?
3. How does magmatic concentration form mineral deposits?
4. How are mineral deposits discovered?

Environmental Impacts Associated with Minerals

LEARNING OBJECTIVES

- Explain the environmental impacts of mining and refining minerals, including a brief description of acid mine drainage.
- Explain how mining lands can be restored.

There is no question that the extraction, processing, and disposal of minerals harm the environment. Mining disturbs and damages the land, and processing and disposal of minerals pollute the air, soil, and water. As noted in the discussion of coal in Chapter 11,

pollution can be controlled and damaged lands can be fully or partially restored, but these remedies are costly. Historically, the environmental cost of extracting, processing, and disposing of minerals has not been incorporated into the actual price of mineral products to consumers.

Most highly developed countries have regulatory mechanisms to minimize environmental damage from mineral consumption, and many developing countries are in the process of putting regulations in place. Such mechanisms include policies to prevent or reduce pollution, restore mining sites, and exclude certain recreational and wilderness sites from mineral development.

MINING AND THE ENVIRONMENT

Mining, particularly surface mining, disturbs large areas of land. In the United States, current and abandoned metal and coal mines occupy an estimated 9 million hectares (22 million acres). Because mining destroys existing vegetation, this land is particularly prone to erosion, with wind erosion causing air pollution and water erosion polluting nearby waterways and damaging aquatic habitats.

Open-pit mining of gold and other minerals uses huge quantities of water. As miners dig deeper into the ground to obtain the ore, they eventually hit the water table and have to pump out the water to keep the pit dry. In northern Nevada, scientists from the U.S. Geological Survey surveyed several wells and measured a drop in the water table of as much as 305 m (1000 ft). This drop, which took place during the 1990s, was linked to gold mining in the region. (Nevada has numerous gold mines that provide the United States with half of its gold.) At the same time, the Western Shoshone tribe living in the area noticed that springs important to their culture began to dry up. The region's farmers and ranchers are concerned that gold mining is depleting the groundwater they need for irrigation. The lowering of the water table in the desert ecosystem can also affect fishes and other organisms that rely on the rare pools of water formed where this groundwater reaches the surface. Ranchers, farmers, environmentalists, and others want the mining companies to reinject the water into the ground after they have pumped it out.

Mining affects water quality. According to the Worldwatch Institute, mining has contributed to the contamination of at least 19,000 km (11,800 mi) of streams and rivers in the United States. Rocks rich in minerals often contain high concentrations of heavy metals such as arsenic and lead. When rainwater seeps through sulfide minerals exposed in mine wastes, sulfuric acid is produced that in turn dissolves other toxic substances, such as lead, arsenic, and cadmium, in the spoil banks of coal and metal ore mines. These acids and highly toxic substances, called **acid mine drainage**, are washed into soil and water, including groundwater, by precipitation runoff (**Figure 15.4**). When such acids and toxic compounds make their way into nearby lakes and streams, they adversely affect the numbers and kinds of aquatic life. Rapid drainage during thunderstorms or a spring snowmelt produces "toxic pulses" of poisonous water that are particularly harmful to waterfowl, fish, and other wildlife in the watershed. Two examples of the many acid mine drainage sites in North America are the Berkeley Pit Superfund site near Butte, Montana, and Britannia Beach in British Columbia, Canada.

acid mine drainage Pollution caused when sulfuric acid and toxic dissolved materials, such as lead, arsenic, and cadmium, wash from mines into nearby lakes and streams.

Figure 15.4   Acid mine drainage contaminating a stream.
Shown is the characteristic orange acid runoff that contains sulfuric acid contaminated with lead, arsenic, cadmium, silver, and zinc. Photographed in Preston County, West Virginia

© Thomas R. Fletcher/Alamy

Table 15.2   Ore and Waste Production for Selected Minerals

Mineral	Amount of Mined Ore (Million Tons)	Percentage of Ore That Becomes Waste During Refining*
Iron ore	2958	60
Copper	1663	99
Gold	745	99.99
Lead	267	97.5
Aluminum	128	81

*Data do not include the overburden of rock and soil that originally covered the ore deposits.

Source: Adapted from Table 6.4 on page 117 in G. Gardner, et al. *State of the World*, 2003. New York: W.W. Norton & Company (2003) and based on data from U.S. Geological Survey and *Worldwatch*.

Cost–Benefit Analysis of Mine Development   Environmental economists suggest that before a decision is made to develop a mine, a cost–benefit analysis should be performed that includes the benefits of the mine in dollar terms versus the benefits in dollar terms of preserving the land intact for wildlife habitat, ranchers, farmers, indigenous people, watershed protection, and recreation. This economic analysis should include the cost of damage to the environment caused by extracting and processing mineral resources. The evaluation should take into account the fact that, over time, the benefits of the mine will decline as the mineral ore is exhausted, whereas the benefits of the natural environment will likely increase, in part because natural areas are becoming rarer as the number of developed areas increases. When a cost–benefit analysis of this type is performed, it may indicate that present and future benefits of preserving the land are greater than present and future benefits of developing a mine.

ENVIRONMENTAL IMPACTS OF REFINING MINERALS

On average, approximately 80% or more of mined ore consists of impurities that become wastes after processing (**Table 15.2**). These wastes, called **tailings**, are usually left in giant piles on the ground or in ponds near the processing plants (**Figure 15.5a**). The tailings contain toxic materials such as cyanide, mercury, uranium, and sulfuric acid. Left exposed in this way, these toxic substances contaminate the air, soil, and water (**Figure 15.5b**). Heavy metals

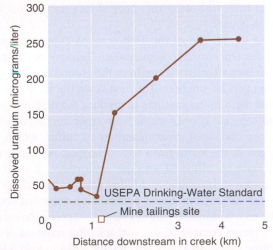

©Ron Chapple Stock/Alamy

(a) This aerial view shows tailing ponds for mineral wastes, located in Utah.

(b) Dissolved uranium from tailings contaminates a creek downstream of a former uranium and copper mine in southeastern Utah.

Figure 15.5   Environmental impact of disposed tailings.

Otton, J.K., Zielinski, R.A., and Horton, R.J. 2010. Geology, Geochemistry, and Geophysics of the Fry Canyon Uranium/Copper Projects Site, Southeastern Utah —Indications of Contaminant Migration. U.S. Geological Survey Scientific Investigations Report 2010-5075.

in mine tailings at the Bunker Hill Superfund site in northern Idaho, for example, have leached (washed) into the south fork of the Coeur d' Alene River, killing fishes and waterfowl.

Smelting plants have the potential to emit large quantities of air pollutants during mineral processing. One impurity in many mineral ores is sulfur. Unless expensive pollution-control devices are added to smelters, the sulfur escapes into the atmosphere, where it forms sulfuric acid. The environmental implications of the resulting acid precipitation are discussed in Chapter 19. Pollution-control devices for smelters are the same as those used when sulfur-containing coal is burned—scrubbers and electrostatic precipitators (see Figure 19.9).

Other contaminants found in many ores include the heavy metals lead, cadmium, arsenic, and zinc. These elements have the potential to pollute the atmosphere during the smelting process. Cadmium, for example, is found in zinc ores, and emissions from zinc smelters are a major source of environmental cadmium contamination. In humans, cadmium is linked to high blood pressure; diseases of the liver, kidneys, and heart; and certain types of cancer. In addition to airborne pollutants, smelters emit hazardous liquid and solid wastes that can cause soil and water pollution. Pollution-control devices prevent such hazardous emissions, although the toxic materials captured must be safely disposed of or they will still cause environmental pollution.

CASE IN POINT

COPPER BASIN, TENNESSEE

Copper Basin, Tennessee, in the southeast corner of Tennessee, near its borders with Georgia and North Carolina, is a historic example of environmental degradation caused by smelting. Until relatively recently, the Copper Basin area progressed from lush forests to a panorama of red, barren hills baking in the sun (**Figure 15.6a**). Few plant or animal species could be found—just 130 km² (50 mi²) of hills with deep ruts gouged into them. How did this degradation occur?

During the middle of the 19th century, copper ore was discovered near Ducktown in southeastern Tennessee. Copper mining companies extracted the ore from the ground and dug vast pits to serve as open-air smelters. They cut down the surrounding trees and burned them in the smelters to produce the high temperatures needed to separate copper metal from other contaminants in the ore. The ore contained great quantities of sulfur, which reacted with oxygen in the air to form sulfur dioxide. As sulfur dioxide from the smelters billowed into the atmosphere, it reacted with water, forming sulfuric acid that fell as acid precipitation.

The deforestation and acid precipitation triggered ecological ruin of the area within a few short years. Acid precipitation quickly killed any plants attempting a comeback after removal of the forests. Because plants no longer covered the soil and held it in place, soil erosion cut massive gullies into the gently rolling hills. Of course, the forest animals disappeared with the plants, which had provided their shelter and food. The damage did not stop there. Soil eroding from the Copper Basin, along with acid precipitation, ended up in the Ocoee River, killing its entire aquatic community.

Beginning in the 1920s and 1930s, several government agencies, including the Tennessee Valley Authority and the U.S. Soil

Conservation Service, tried to replant a portion of the area. They planted millions of loblolly pine and black locust trees as well as shorter ground-cover grasses and legume plants that tolerate acid conditions, but most of the plants died. The success of such efforts was marginal until the 1970s, when land reclamation specialists began using new techniques such as application of seed and time-released fertilizer by helicopter. These plants had a greater survival rate, and as they became established, their roots held the soil in place. Leaves dropping to the ground contributed organic material to the soil. The plants provided shade and food for animals such as birds and field mice, which slowly began to return.

Today, reclamation of Copper Basin continues under a 2001 agreement between the state of Tennessee and the U.S. Environmental Protection Agency, with the goal of having the entire area under plant cover (**Figure 15.6b**). (Of course, the forest ecosystem's full recovery will take at least a century or two.) Water treatment efforts are also restoring the Ocoee River, where fish and invertebrates are returning. ■

(a) Air pollution from a copper smelter in Tennessee killed the vegetation, and then water erosion carved gullies into the hillsides. This unrestored section of Copper Basin was photographed in 1973.

(b) A section of Copper Basin that has been reforested.

Figure 15.6 **Environmental devastation near Ducktown, Tennessee.**

RESTORATION OF MINING LANDS

When a mine is no longer profitable, the land can be reclaimed, or restored to a seminatural condition, as has been done to most of the Copper Basin in Tennessee. The goals of reclamation include preventing further degradation and erosion of the land, eliminating or neutralizing local sources of toxic pollutants, and making the land productive for purposes other than mining. Restoration also makes such areas visually attractive (**Figure 15.7**).

A great deal of research is available on techniques of restoring lands degraded by mining, called **derelict lands**. Restoration involves filling in and grading the derelict land to its natural contours, then planting vegetation to hold the soil in place. The establishment of plant cover is not as simple as throwing a few seeds on the ground. Often the topsoil is completely gone or contains toxic levels of metals, so special types of plants that tolerate such a challenging environment must be used. According to experts, the main limitation on the restoration of derelict lands is not a lack of knowledge but a lack of funding.

The **Surface Mining Control and Reclamation Act** of 1977 requires reclamation of areas that were surface-mined for coal (see Chapter 11). However, no federal law is in place to require restoration of derelict lands produced by other kinds of mines. Recall from the chapter introduction that the General Mining Law makes no provision for reclamation.

Creative Approaches to Cleaning Up Mining Areas   Although wetlands are widely known to provide beneficial wildlife habitats, few people realize the potential of wetlands to help clean up former mining lands. Wetlands tend to trap sediments and pollutants that enter them from upstream areas, so the quality of water resources located downstream from wetlands is improved. Although a single wetland provides these benefits, a series of wetlands constructed in the affected drainage basin is much more effective.

Consider the area around Butte, Montana, where copper was mined for 100 years. This area is the largest Superfund site in the United States. Its soil and water are contaminated with copper, zinc, nickel, cadmium, and arsenic. Many cleanup technologies are being developed and tested in Butte, including the design and construction of artificial wetlands. As contaminated water seeps into the wetland, bacteria consume the sulfur draining from the mines, making the water less acidic. As the water becomes less acidic, zinc and copper precipitate (settle out of solution) and enter the sediments. Constructed wetlands typically take 50 to 100 years to neutralize the acid enough for aquatic life to return to rivers and streams downstream from acid mine drainage. This time estimate is based on observations of more than 800 wetland systems constructed at coal mining sites in Appalachia, the region in the eastern United States that encompasses the central and southern Appalachian Mountains.

Creating and maintaining wetlands is expensive, although it is cost effective when compared to using lime to reduce the water's acidity. Recently, scientists at Butte have tried an inexpensive approach that involves cow manure, a "resource" readily available in Montana. Dumping cow manure onto mining lands causes the pH of the mine drainage water to increase as bacteria consume the manure. Toxic materials precipitate out of the basic water, just as they do in the artificial wetlands, improving the water quality.

Scientists are also using plants to remove heavy metals from former mining lands. **Phytoremediation** is the use of specific plants to absorb and accumulate toxic materials such as nickel from the soil. Although most plants do not tolerate soils rich in nickel, some plants, such as twist flower (*Streptanthus polygaloides*), thrive on it. This species is a hyperaccumulator, a plant that absorbs high quantities of a metal and stores it in its cells. The plants can be grown on nickel-contaminated land, harvested, and hauled to a hazardous waste site for disposal. Alternatively, the plants are burned, and nickel is obtained from the ashes. Phytoremediation has great potential to decontaminate mining and other hazardous waste sites and to extract valuable metals from soil in an environmentally benign way. (See Chapter 23 for further discussion of phytoremediation.)

AP Photo/Roger Alford

Figure 15.7   Reclaimed coal-mined land. Prior to reclamation, this 68-acre site in Kentucky contained refuse and barren land associated with coal mining there.

REVIEW

1. What are three harmful environmental effects of mining and processing minerals?

2. Are mining lands usually restored when the mine is no longer profitable to operate? Why or why not?

Minerals: An International Perspective

LEARNING OBJECTIVES

- Describe the relationship between a country's level of industrialization and its mineral consumption.
- Distinguish between mineral reserves and mineral resources.

The economies of industrialized countries require the extraction and processing of large amounts of minerals to make products. Most of these highly developed countries rely on the mineral deposits in developing countries, having long since exhausted their own supplies. As developing countries become more industrialized, their own mineral requirements increase correspondingly, adding further pressure to a nonrenewable resource. In fact, humans have consumed more minerals since World War II than were consumed in the previous 5000 years, from the beginning of the Bronze Age to the middle of the 20th century (**Figure 15.8**).

Mining in the United States has clearly caused many serious environmental problems. The problems in developing countries that rely on mining for a significant part of their economies are as great as or greater than those faced by highly developed countries. The governments in developing countries lack the financial resources and infrastructure to deal with acid mine drainage and other serious environmental problems caused by hardrock mining. To complicate the issue, foreign companies often have significant mining interests in developing countries. For example, France, Germany, Great Britain, Japan, Russia, Spain, and the United States have been involved at various times during the past two centuries in mining (some would say exploiting) ores containing tin, zinc, copper, and lead in Bolivia. The mining district of Bolivia currently faces a catastrophic environmental nightmare from decades of mining abuse. Yet the Bolivian government does not address the issue because mining is the predominant industry in Bolivia.

WORLDWIDE MINERAL PRODUCTION AND CONSUMPTION

At one time, most of the highly developed countries had rich resource bases, including abundant mineral deposits that enabled them to industrialize. In the process of industrialization, they largely depleted their domestic reserves of minerals and as a result must increasingly turn to developing countries. This is particularly true for European countries, Japan, and, to a lesser extent, the United States.

The level of mineral consumption varies widely between highly developed and developing countries. The United States and Canada, which have not quite 5% of the world's population, consume about 25% of many of the world's metals. It is too simplistic, however, to divide the world into two groups, the mineral consumers (highly developed countries) and the mineral producers (developing countries). Although many developing countries do lack any significant mineral deposits, the world's 10 most resource-rich nations—in terms of value of metal and ore reserves—are not all the wealthiest: South Africa, Russia, Australia, Canada, Brazil, China, Chile, the United States, Ukraine, and Peru.

Mineral production and consumption in China are increasing dramatically as China industrializes. For example, in 2013 China produced 44% of the world's primary aluminum (aluminum obtained from ores, not from recycling). China also consumed almost all of this aluminum, making it both the world's largest producer and biggest consumer of primary aluminum.

Because industrialization increases the demand for minerals, developing countries that at one time met their mineral needs with domestic supplies become increasingly reliant on foreign supplies as economic development occurs. South Korea is one such nation. During the 1950s it exported iron, copper, and other minerals. South Korea has experienced dramatic economic growth from the 1960s to the present and, as a result, must now import iron and copper to meet its needs.

MINERAL DISTRIBUTION VERSUS CONSUMPTION

The metallic element chromium provides a useful example of global versus national distribution and consumption. Chromium is used to make vivid red, orange, yellow, and green pigments for paints; chrome plate; and, combined with other metals, certain types of hard steel. Chromium has no known substitute for many of its important applications, including jet engine parts. Industrialized nations that lack significant chromium deposits, such as the United States, must import essentially all of their chromium. South Africa is one of only a few countries with significant deposits of chromium. Zimbabwe and Turkey also export chromium. Although world reserves of chromium are adequate for the immediate future, the United States and several other industrialized countries are utterly dependent on a few countries for their chromium supplies.

Many industrialized nations have stockpiled strategically important minerals to reduce their dependence on potentially unstable suppliers. The United States and others have stockpiles of strategic minerals such as titanium, tin, manganese, chromium, platinum, and cobalt, mainly because these metals are critically important to industry and defense; in the United States, stockpiles have been decreasing since 1994.

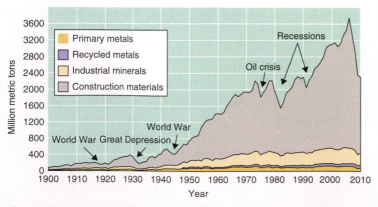

Figure 15.8 **Annual U.S. use of raw nonfuel minerals, 1900–2010.**

Source: USGS National Minerals Information Center 2012

Recently, China began stockpiling *rare earth metals*, which are 17 elements such as dysprosium and terbium that are important in high-technology applications like hybrid-car batteries, wind turbines, and laser-guided missiles. China, which controls about 80% of the global supply of rare earth metals, has also reduced its exports to other countries. China's chokehold on supply triggered steep price increases in the minerals in 2010–2011, but a more recent lack of global demand and the development of new sources led to price drops. Despite their name, some rare earth metals are not rare, but the ones needed in high-technology industries are relatively rare.

EVALUATING OUR MINERAL SUPPLIES

Will we run out of supplies of important minerals? To address this question, we must first examine how large the global supply of

mineral reserves Mineral deposits that have been identified and are currently profitable to extract.

mineral resources Any undiscovered mineral deposits or known deposits of low-grade ore that are currently unprofitable to extract.

various mineral reserves and mineral resources is. **Mineral reserves** are currently profitable to extract, whereas **mineral resources** are potential resources that may be profitable to extract in the future. The combination of a mineral's reserves and resources is its **total resources** or **world reserve base**.

Estimates of mineral reserves and resources fluctuate with economic, technological, and political changes. If the price of a mineral on the world market falls, certain borderline mineral

reserves may slip into the mineral resource category; increasing prices may restore them to the mineral reserve category. When new technological methods decrease the cost of extracting ores, deposits ranked in the mineral resource category are reclassified as mineral reserves. If the political situation in a country becomes so unstable that mineral reserves cannot be mined, they are reclassified as mineral resources; a later shift toward political stability may cause the minerals to be placed on the mineral reserve list again.

It is extremely difficult to forecast future mineral supplies. In the 1970s, projections of escalating demand and impending shortages of many important minerals were commonplace. There are three reasons that none of these shortages actually materialized. First, new discoveries of major deposits occurred in recent decades—iron and aluminum deposits in Brazil and Australia, for example. Second, plastics, synthetic polymers, ceramics, and other materials replaced metals in many products. Third, a global economic slump resulted in a lower consumption of minerals. However, it is always possible that changes in the world economic situation—such as the rapid economic development of China—will contribute to future mineral shortages.

Economic factors aside, the prediction of future mineral needs is difficult because it is impossible to know when or if there will be new discoveries of mineral reserves or replacements for minerals (such as plastics). It is impossible to know when or if new technological developments will make it economically feasible to extract minerals from low-grade ores.

With these reservations in mind, most experts currently think mineral supplies, both metallic and nonmetallic, will be adequate during the 21st century. However, several important minerals—mercury, tungsten, and tin, for example—may become increasingly scarce during that period. Another reasonable projection is that the prices of even relatively plentiful minerals, such as iron and aluminum, will increase during your lifetime. The eventual depletion of large, rich, and easily accessible deposits of these metals means that we will have to mine and refine low-grade ores, which will be more expensive.

REVIEW

1. How does a country's mineral consumption tend to change as it becomes more industrialized?
2. How do mining experts differentiate between mineral reserves and mineral resources?

Increasing the Supply of Minerals

LEARNING OBJECTIVE
- Briefly discuss efforts to discover new mineral supplies.

As a resource becomes scarce, efforts intensify to discover new supplies, to conserve existing supplies, and to develop new substitutes for it. Although many reserves have been discovered and exploited, unknown deposits may be found. In addition, the development of advanced mining technologies may make it possible to exploit known resources that are too expensive to develop using existing techniques.

LOCATING AND MINING NEW DEPOSITS

Many known mineral reserves have not yet been exploited. Although Indonesia is known to have many rich mineral deposits, its thick forests and mosquitoes that carry the malaria parasite have made accessibility to these deposits difficult. Large deposits of iron, copper, cobalt, gold, and lithium exist in Afghanistan—enough minerals to transform the country's economy. Years of war, however, make investment in Afghanistan unlikely, at least in the short term. Both northern and southern polar regions have had little mineral development. This is due in part to a lack of technology for mining in frigid environments. Normal offshore drilling rigs cannot be used in Antarctic waters because the shifting ice formed during its harsh winter would tear the rigs apart. As new technologies become available, increasing pressure will be exerted to mine in northern Canada, Siberia, and Antarctica.

Exploitation of the rich mineral deposits in Siberia is planned, although new technologies must be developed to make this feasible. Some of the ore deposits in Siberia have unusual combinations of minerals (e.g., potassium combined with aluminum) that cannot be separated using existing technology.

Is there a possibility that currently unknown mineral deposits will be discovered at some future time? The U.S. Geological Survey thinks that undiscovered mineral deposits may exist, particularly in developing countries where detailed geologic surveys are not yet available. A detailed survey of the western portion of South America along the Andes Mountains could reveal significant mineral deposits. Geologists presume that minerals will be found in the Amazon Basin, although in many ways the rain forest and thick overlying alluvial layers of the river basin make these deposits as inaccessible as those in Antarctica. Logistic problems hamper examination of certain areas deep in the rain forest to assess the likelihood of deposits being present. As in other regions, mining in the Amazon Basin would pose a grave environmental threat.

Geologists consider it likely that deep deposits buried 10 km (6.2 mi) or more in Earth's crust will someday be discovered and exploited. The special technology required to mine deposits that deep is not yet available.

MINERALS IN ANTARCTICA

To date, no substantial mineral deposits have been found in Antarctica, although smaller amounts of valuable minerals have been discovered. Geologists think it likely that major deposits of valuable metals and oil are present and that they will be discovered in the future. Nobody owns Antarctica, and many nations are involved in negotiations on the future of this continent and its possible mineral wealth.

The **Antarctic Treaty**, an international agreement in effect since 1961, limits activity in Antarctica to peaceful uses such as scientific studies. Twenty-six nations are voting members to the Antarctic Treaty. During the 1980s, nearly a decade of delicate negotiations resulted in a pact that would have permitted exploitation of Antarctica's minerals. The pact required unanimous agreement for ratification. In 1989, several countries refused to support the pact because of concerns that any mineral exploitation would damage Antarctica's environment. As a result of these concerns, an international agreement, the **Environmental Protection Protocol to the Antarctic Treaty**, or the **Madrid Protocol**, was established. The protocol, which went into effect in 1990, includes a moratorium on mineral exploration and development for a minimum of 50 years. It designates Antarctica and its marine ecosystem as a "natural reserve dedicated to peace and society."

Why be concerned about Antarctica's environment? For one thing, polar regions are extremely vulnerable to human activities. Even scientific investigations and tourists, with their trash, pollution, and noise, have negatively affected the wildlife, such as emperor penguins, leopard seals, and blue whales, along Antarctica's coastline. No one doubts that large-scale mining operations would wreak havoc on such a fragile environment.

Maintaining Antarctica in a pristine state is important because this continent plays a pivotal role in regulating many aspects of the global environment, such as global changes in sea level. Studying the natural environment of Antarctica helps scientists gain valuable insights into such important environmental issues as global climate change and stratospheric ozone depletion (**Figure 15.9**).

MINERALS FROM THE OCEAN

Seawater, which covers approximately three-fourths of our planet, contains many dissolved minerals. The total amount of minerals available in seawater is staggeringly high, but their concentrations are low. Currently, sodium chloride (common table salt), bromine, and magnesium are profitably extracted from seawater. It may be possible in the future to profitably extract other minerals from seawater and concentrate them, but current mineral prices and technology make this impossible now.

Large deposits of minerals lie on the ocean floor. **Manganese nodules**—small rocks the size of potatoes that contain manganese and other minerals, such as copper, cobalt, and

Figure 15.9 **Research in Antarctica.** Scientists set up a GPS receiver in west Antarctica as part of the west Antarctic GPS Network (WAGN) project. The WAGN project is measuring crustal movements surrounding the west Antarctic ice sheet. These movements in the lithosphere could affect the future of the ice sheet and its corresponding effect on sea level. How is this research related to global climate change?

nickel—are widespread on the ocean floor, particularly in the Pacific (**Figure 15.10**). Estimates of manganese nodules are quite large. Nodules in the Pacific Ocean may contain as much as 6 billion metric tons (6.6 billion tons) of manganese, more than the total amount found in land reserves. Dredging manganese nodules from the ocean floor would adversely affect sea life, and the current market value for these minerals would not cover the expense of obtaining them using existing technology. Despite these concerns, many experts think that deep-sea mining will be feasible in a few decades, and several industrialized nations such as the United States have staked out claims in a region of the Pacific known for its large number of nodules. To date, none have been mined.

Such potential exploitations of the ocean floor are controversial. Many people think it is inevitable that minerals will be mined from the floor of the deep sea, but others think the seabed should be declared off limits because of the potential ecological havoc that mining could cause on the diverse life-forms inhabiting the ocean floor. Sea urchins, sea cucumbers, sea stars, acorn worms, sea squirts, sea lilies, and lamp shells are but a few of the animals known to inhabit the seabed environment.

These problems have been considered since the 1960s, when industrialized countries first expressed an interest in removing manganese nodules from the ocean floor. Their interest triggered the formation of an international treaty, the **U.N. Convention on the Law of the Sea (UNCLOS)**. UNCLOS, which became effective in 1994, is generally considered a "constitution for the ocean" that protects its resources. As of 2013, 166 countries had ratified this treaty. (The United States has not yet ratified UNCLOS but voluntarily observes its provisions.)

The provisions of UNCLOS are not binding for territorial waters, only for international waters, so seabed mining is not prohibited in territorial waters. For example, hydrothermal vent systems in deep territorial waters off Papua New Guinea contain gold, zinc, copper, and silver in higher concentrations than are found in land mines. It seems likely that ocean mining will become technologically feasible and profitable sometime in the next few years. Nautilus Minerals, Inc. is planning to mine one of Papua New Guinea's hydrothermal vents soon.

ADVANCED MINING AND PROCESSING TECHNOLOGIES

We have already mentioned that special technologies will be needed to mine minerals in inaccessible areas such as polar regions

(a) These polymetallic nodules are concretions of about 20% manganese, as well as iron, copper, cobalt, nickel, and other metals. These were taken from the Pacific Ocean at a depth of about 4 km (2.4 mi).

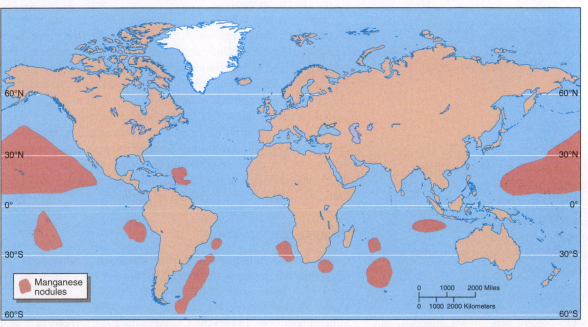

(b) Location of known manganese nodules in the ocean. Although these potato-sized nodules have enticed mining companies, it is not yet commercially feasible to obtain them. (Adapted from P.A. Rona, "Resources of the Sea Floor." *Science*, Vol. 299. January 31, 2003)

Figure 15.10 **Manganese nodules.**

and deposits deep in the ground. Making use of large, low-grade mineral deposits throughout the world will also require the development of special techniques. As minerals grow scarcer, economic and political pressure to exploit low-grade ores will increase. Obtaining high-grade metals from low-grade ores is an expensive proposition, in part because a great deal of energy must be expended to obtain enough ore. Future technology may make such exploitation more energy-efficient, thereby reducing costs.

Even if advanced technology makes obtaining minerals from low-grade ores feasible, other factors (e.g., poor access to water needed to extract and process the minerals) might limit exploitation of this potential source. The environmental costs may be too high, because obtaining minerals from low-grade ores causes greater land disruption and produces far more pollution than does the development of high-grade ores.

Biomining   Microorganisms are used to extract minerals from some low-grade ores. Microorganisms have proved efficient for copper mining, allowing the U.S. copper industry to become more competitive internationally. When mixed with sulfuric acid, a bacterium (*Thiobacillus ferrooxidans*) promotes a chemical reaction that leaches copper into an acidic solution, releasing larger quantities of the metal more efficiently than traditional methods.

Other important applications of **biomining** are emerging. Although it is still in the development stage, treating low-grade gold ores with bacteria such as *Thiobacillus* allows a 90% recovery of gold, compared to 75% recovery for the more expensive and energy-intensive conventional methods. Phosphates, used primarily for fertilizers and additives in some manufactured goods, are traditionally extracted by inefficient burning at high temperatures or by wasteful acid treatment processes. New biological processes extract phosphates at room temperature. Some researchers are developing biomining techniques to remove and recycle phosphates from wastewater.

REVIEW

1. What is a problem that limits mining known mineral deposits in Indonesia? In Siberia?

Using Substitution and Conservation to Expand Mineral Supplies

LEARNING OBJECTIVES

- Summarize the conservation of minerals by reuse, recycling, and changing our mineral requirements.
- Explain how sustainable manufacturing and dematerialization contribute to mineral conservation.

Much of our civilization's technology depends on minerals, and certain minerals may be unavailable or quite limited in the future. Therefore, we should extend existing mineral supplies as far as possible through substitution and conservation (**Figure 15.11**).

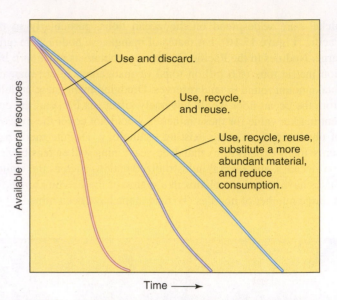

Figure 15.11   Depletion times of a nonrenewable resource such as copper. Three assumptions are made; in each case, we start with the same quantity of mineral reserves available for production. The remaining reserves level off as the resource becomes scarcer and more costly. This graph does not take into consideration the discovery of new reserves or improved mining technologies.

FINDING MINERAL SUBSTITUTES

The substitution of more abundant materials for scarce minerals is an important goal of manufacturing. Economics partly drives the search for substitutes; one effective way to cut production costs is to substitute an inexpensive or abundant material for an expensive or scarce one. In recent years, plastics, ceramic composites, and high-strength glass fibers have been substituted for scarcer materials in many industries.

Earlier in the 20th century, tin was a critical metal for can making and packaging; since then, other materials have been substituted for tin, including plastic, glass, and aluminum. The amounts of lead and steel used in telecommunications cables have decreased dramatically during the past 35 years, and the amount of plastics has had a corresponding increase. In addition, glass fibers have replaced copper wiring in telephone cables.

Although substitution extends mineral supplies, it is not a cure-all for dwindling resources. Certain minerals have no known substitutes. For example, platinum catalyzes many chemical reactions important in industry. So far, no other substance has been found with the catalyzing abilities of platinum.

MINERAL CONSERVATION

Conservation, including both reuse and recycling, extends mineral supplies. The **reuse** of items such as beverage bottles, which are collected, washed, and refilled, is one way to extend mineral resources. In **recycling**, used items such as beverage cans and scrap iron are collected, remelted, and reprocessed into new products. In addition to promoting specific conservation

reuse Conservation of the resources in used items by using them over and over again.

recycling Conservation of the resources in used items by converting them into new products.

techniques such as reuse and recycling, public awareness and attitudes about resource conservation can be modified to encourage low waste.

Reuse   When the same product is used over and over again, both mineral consumption and pollution are reduced. The benefits of reuse are greater than those of recycling (see Chapter 23). Recycling a glass bottle requires crushing it, melting the glass, and forming a new bottle. Reusing a glass bottle simply requires washing it, which typically expends less energy than recycling. Reuse is a national policy in Denmark, where nonreusable beverage containers are prohibited.

Several countries and states have adopted beverage container deposit laws, which require consumers to pay a deposit, usually a nickel or dime, for each beverage bottle or can that they purchase. The deposit is refunded when the container is returned to the retailer or to special redemption centers. Unredeemed deposits are generally used to provide revenue for environmental programs such as hazardous waste cleanups. In addition to encouraging reuse and recycling, thereby reducing mineral resource consumption, beverage container deposit laws save tax money by reducing litter and solid waste. Countries that have adopted beverage container deposit laws include the Netherlands, Germany, Norway, Sweden, and Switzerland, as well as parts of Canada and the United States.

Recycling   A large percentage of the products made from minerals—such as cans, bottles, chemical products, electronic devices, and batteries—is typically discarded after use. The minerals in some of these products—batteries and electronic devices, for instance—are difficult to recycle. Minerals in other products, such as paints containing lead, zinc, or chromium, are lost through normal use. However, the technology exists to recycle many other mineral products. Recycling of certain minerals is already a common practice throughout the industrialized world, including the United States, but there is much room for improvement (**Table 15.3**).

Recycling has several advantages beyond extending mineral resources. It saves unspoiled land from the disruption of mining, reduces the amount of solid waste that must be disposed of, and decreases energy consumption and pollution. Recycling an aluminum beverage can saves the energy equivalent of about 180 mL

Table 15.3   Recycling Rates for Metals in the United States, 2011

Mineral	Percent Recycled
Aluminum	60
Copper	34
Iron and Steel	63
Lead	73
Magnesium	53
Nickel	42
Zinc	29

Source: USGS *2011 Minerals Yearbook*.

(6 oz) of gasoline. Recycling aluminum reduces the emission of aluminum fluoride, a toxic air pollutant produced during the processing of aluminum ore.

About 67% of the aluminum cans in the United States were recycled in 2012, a large increase reported by professional and recycling organizations. The aluminum industry, local governments, and private groups have established thousands of recycling centers across the country. It takes approximately six weeks for a used can to be melted, re-formed, filled, and put back on a supermarket shelf. Clearly, even more recycling is possible. It may be that today's sanitary landfills will become tomorrow's mines, as valuable minerals and other materials are extracted from them. (In many countries, sanitary landfills are already treated as mines.)

Changing Our Mineral Requirements   We can reduce mineral consumption by becoming a low-waste society. U.S. citizens have developed a "throwaway" mentality wherein damaged or unneeded articles are discarded. Industries looking for short-term economic profits encourage this attitude, even though the long-term economic and environmental costs of it are high. We consume fewer resources if products are durable and repairable. Laws such as those requiring a deposit on beverage containers reduce consumption by encouraging reuse and recycling.

The throwaway mentality is also evident in manufacturing industries. Traditionally, industries consume raw materials and produce goods *and* a large amount of waste that is simply discarded (**Figure 15.12a**). Increasingly, manufacturers are finding that the waste products from one manufacturing process can become the raw materials for another industry. By selling these "wastes," industries gain additional profits and lessen the amounts of materials that must be discarded.

The chemical and petrochemical industries were among the first businesses to minimize wastes by converting them into useful products. For example, some chemical companies buy aluminum wastes from other companies and convert the aluminum in the wastes to aluminum sulfate, a chemical used to treat municipal water supplies. Such waste minimization is known as **sustainable manufacturing** (**Figure 15.12b**; also see Meeting the Challenge: Industrial Ecosystems). Sustainable manufacturing requires that companies provide information about their waste products to other industries. However, many companies are reluctant to reveal the kinds of wastes they produce because their competitors may deduce trade secrets from the nature of their wastes. This difficulty will have to be overcome before sustainable manufacturing is fully implemented.

sustainable manufacturing A manufacturing system based on minimizing industrial waste.

Dematerialization   As products evolve, they tend to become lighter in weight and often smaller. Washing machines manufactured in the 1960s were much heavier than comparable machines manufactured today. The same is true of other household appliances, automobiles, and electronic items. This decrease in the weight of products over time is known as **dematerialization**. Ideally, dematerialization is beneficial to the environment because

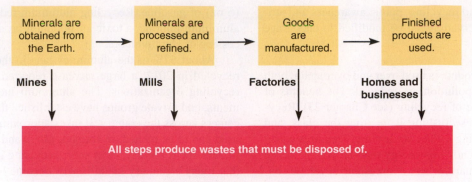

(a) Massive amounts of solid waste are produced at all steps in the traditional flow of minerals, from mining the mineral to discarding the used-up product.

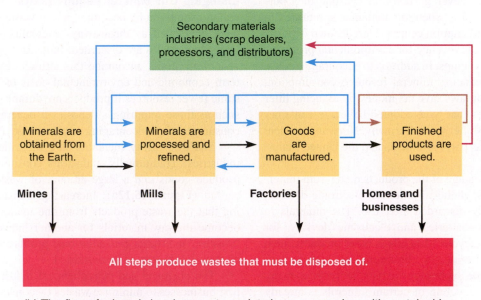

(b) The flow of minerals in a low-waste society is more complex, with sustainable manufacturing, consumer reuse, and consumer recycling practiced at intermediate steps.

Arrow key: —— Sustainable manufacturing
—— Consumer reuse
—— Consumer recycling

Figure 15.12 **Mineral flow in an industrial society.**

it reduces the quantity of waste during both production and consumption.

Although dematerialization gives the appearance of reducing consumption of minerals and other materials, it sometimes has the opposite effect. Smaller and lighter products may be of lower quality. Because repairing broken lightweight items is difficult and may cost more than the original products, retailers and manufacturers encourage consumers to replace rather than repair the items.

Although the weight of materials being used to make each item has decreased, the number of such items being used in a given period may actually have increased.

REVIEW

1. Explain the difference between reuse and recycling.

2. What is sustainable manufacturing?

MEETING THE CHALLENGE

Industrial Ecosystems

Traditional industries operate in a one-way, linear fashion: natural resources from the environment → products → wastes dumped back into environment. However, natural resources such as minerals and fossil fuels are present in finite amounts, and the environment has a limited capacity to absorb waste. The field of **industrial ecology** has emerged to address these issues. An extension of the concept of sustainable manufacturing, industrial ecology seeks to use resources efficiently and regards "wastes" as potential products. Industrial ecology tries to create **industrial ecosystems** that compare in many ways to natural ecosystems.

Consider the industrial ecosystem in Kalundborg, Denmark, which consists of an electric power plant, an oil refinery, a pharmaceutical plant, a wallboard factory, a sulfuric acid producer, a cement manufacturer, fish farming, horticulture (greenhouses), and area homes and farms. At first glance, these entities seem to have little in common, but they are linked to one another in ways that resemble a food web in a natural ecosystem (see the figure). In this industrial ecosystem, the wastes produced by one company are sold to another company as raw materials for their processes, in a manner analogous to nutrient cycling in nature.

The coal-fired electric power plant originally cooled its waste steam and released it into the local fjord. The steam is now supplied to the oil refinery and the pharmaceutical plant, and additional surplus heat produced by the power plant warms greenhouses, the fish farm, and area homes. The need for 3500 oil-burning home-heating systems was eliminated as a result.

Surplus natural gas from the oil refinery is sold to the power plant and the wallboard factory. The power plant now saves tons of coal each year by burning the less expensive natural gas. Before selling the natural gas, the oil refinery removes excess sulfur from it, as required by air pollution–control laws. This sulfur is sold to a company that uses it to manufacture sulfuric acid.

To meet environmental regulations, the power plant installed pollution-control equipment to remove sulfur from its coal smoke. This sulfur, in the form of calcium sulfate, is sold to the wallboard plant and used as a substitute for gypsum, which is naturally occurring calcium sulfate. The fly ash produced by the power plant goes to the cement manufacturer for use in road building.

Local farmers use the sludge from the fish farm as a fertilizer for their fields. The fermentation vats at the pharmaceutical plant also generate a high-nutrient sludge used by local farmers. Most pharmaceutical companies discard this sludge because it contains living microorganisms, but the Kalundborg plant heats the sludge to kill the microorganisms, converting a waste material into a commodity.

These interactions did not spring into existence at the same time; each represents a separately negotiated deal. It took years to develop the entire industrial ecosystem. Although these examples of industrial cooperation were initiated for economic reasons, each has distinct environmental benefits, from energy conservation to a reduction of pollution.

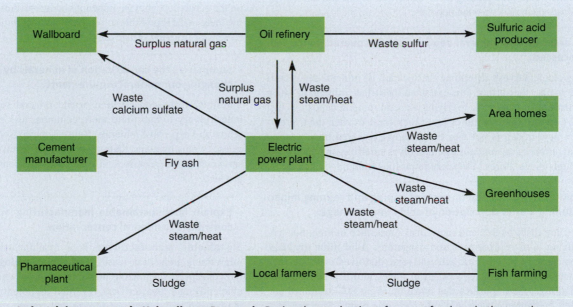

Industrial ecosystem in Kalundborg, Denmark. During the production of energy, food, and other products, resource recovery is maximized and waste production is minimized.

REVIEW OF LEARNING OBJECTIVES WITH SELECTED KEY TERMS

- **Define *minerals* and explain the difference between high-grade and low-grade ores, as well as between metallic and nonmetallic minerals.**

A **mineral** is an inorganic solid, occurring naturally in or on Earth's crust, with characteristic chemical and physical properties. **High-grade ores** contain relatively large amounts of particular minerals; **low-grade ores** contain lesser amounts. Minerals may be **metals**, such as iron, aluminum, and copper, or **nonmetallic minerals**, such as phosphates, salt, and sand.

- **Describe several natural processes that concentrate minerals in Earth's crust.**

Magmatic concentration is the formation of mineral deposits as liquid magma separates into layers, cools, and solidifies. **Hydrothermal processes** form mineral deposits as hot water dissolves minerals from rock, and the solution seeps through cracks until the minerals encounter sulfur, react to produce metal sulfides, cool, and settle out of solution. In **sedimentation**, certain minerals dissolve in water and later settle out of solution on riverbanks, deltas, and the seafloor. In **evaporation**, dissolved materials in lakes with no outlet form mineral deposits when the water evaporates and the materials are left behind.

- **Distinguish between surface mining and subsurface mining.**

Surface mining is the extraction of mineral and energy resources near Earth's surface by first removing the soil, subsoil, and overlying rock strata. In **strip mining**, a type of surface mining, a trench is dug to extract the minerals. **Subsurface mining** is the extraction of mineral and energy resources from deep underground deposits.

- **Briefly describe how mineral deposits are discovered, extracted, and processed.**

Detailed geologic surveys determine the location of mineral deposits. If the deposit is near the surface, it is extracted by surface mining, in which the **overburden**, soil and rock overlying a useful mineral deposit, is first removed. A **spoil bank** is a hill of loose rock created when the overburden from a new trench is put into the already excavated trench during strip mining. Processing minerals often involves **smelting**, in which ore is melted at high temperatures to separate impurities from the molten metal.

- **Explain the environmental impacts of mining and refining minerals, including a brief description of acid mine drainage.**

Surface mining disturbs the land more than subsurface mining, but subsurface mining is more expensive and dangerous. **Acid mine drainage** is pollution caused when sulfuric acid and toxic dissolved materials such as lead, arsenic, and cadmium wash from mines into nearby lakes and streams. **Tailings**, the impurities that make up about 80% of mined ore, are often left in giant piles near mineral-processing plants. Mercury, cyanide, uranium, and sulfuric acid leach into the soil and water from tailings. Unless pollution-control devices are used, smelting plants may emit large amounts of air pollutants during mineral processing.

- **Explain how mining lands can be restored.**

Derelict lands are extensively damaged due to mining but can be restored to prevent further degradation and to make the land productive for other purposes. Land reclamation is expensive, and no federal law exists to require restoration of derelict lands other than those produced by coal mines.

- **Describe the relationship between a country's level of industrialization and its mineral consumption.**

Highly developed countries consume a disproportionate share of the world's minerals, but as developing countries industrialize, their need for minerals increases. China's mineral production and consumption are increasing rapidly.

- **Distinguish between mineral reserves and mineral resources.**

Mineral reserves are mineral deposits that have been identified and are currently profitable to extract. **Mineral resources** are any undiscovered mineral deposits or known deposits of low-grade ore that are currently unprofitable to extract.

- **Briefly discuss efforts to discover new mineral supplies.**

Mineral deposits will probably be discovered in some developing countries when they are surveyed geologically. Minerals that possibly exist in Antarctica may be mined in the future, but current international law protects Antarctica from such activities. Mineral deposits in ocean water and on the ocean floor may be mined in the future. Advanced mining technology may make it possible to profitably extract minerals from inaccessible regions or from low-grade ores.

- **Summarize the conservation of minerals by reuse, recycling, and changing our mineral requirements.**

Substitution and conservation extend mineral supplies. Manufacturing industries try to substitute more common, less expensive minerals for scarce and expensive minerals. **Reuse** is the conservation of the resources in used items by using them over and over again. **Recycling** is the conservation of the resources in used items by converting them into new products.

- **Explain how sustainable manufacturing and dematerialization contribute to mineral conservation.**

Sustainable manufacturing is a manufacturing system based on minimizing industrial waste. **Industrial ecology** is an extension of the concept of sustainable manufacturing, in which resources are used efficiently and "wastes" are regarded as potential products. **Dematerialization** is the decrease in size and weight of a product as a result of technological improvements over time. Dematerialization reduces consumption only if products are durable and easily and inexpensively repaired.

Critical Thinking and Review Questions

1. What is the difference between rocks and minerals? Between metals and nonmetallic minerals?

2. Distinguish between surface and subsurface mining, between open-pit and strip mines, and between shaft and slope mines.

3. How does the cartoon below relate to the General Mining Law of 1872? How does it reflect the environmental effects of strip mining versus subsurface mining?

4. What is overburden? A spoil bank? Smelting? Tailings?

5. Explain why it is more environmentally damaging to obtain minerals from low-grade ores than to extract them from high-grade ores.

6. Historically, the cost of environmental damage arising from mining and processing minerals was not included in the price of consumer products. Do you think it should be? Why or why not?

7. How did Copper Basin, Tennessee, become an environmental disaster? What efforts have aided in its recovery?

8. Compare mineral consumption in the United States, China, and a poorer country such as Nigeria. Which has experienced the most rapid increase in mineral consumption in recent years? Why?

9. Why do we have to distinguish between mineral resources and mineral reserves when making projections about how long a specific mineral will be available? Why it is difficult to obtain an accurate appraisal of total resources for a particular mineral?

10. What are manganese nodules? Why have these known deposits not been mined?

11. Have mineral deposits been mined in Antarctica? Why or why not?

12. Sketch mineral flow in a traditional industrialized society. Make sure you include mines, mills, factories, homes, and businesses. Now sketch mineral flow in a low-waste industrialized society. Which diagram is more complex, and why?

13. Some people in industry argue that the planned obsolescence of products, which means they must be replaced often, creates jobs. Others think that the production of smaller quantities of durable, repairable products would generate jobs and stimulate the economy. Explain each viewpoint.

14. How does the industrial ecosystem at Kalundborg, Denmark, resemble a natural ecosystem? Would this approach produce more or fewer greenhouse gases than if the same energy and products were produced using traditional industrial practices? Explain.

15. Examine the graph, which shows the increase in world use of electricity for mining and refining virgin aluminum since the 1950s. Given that the production of aluminum has become more energy-efficient, how do you account for the current global increase in electricity used in aluminum production?

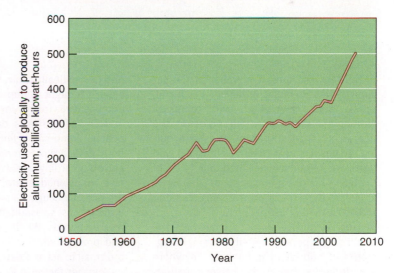

(Adapted from: *Vital Signs 2007–2008*. New York: W.W. Norton & Company, 2007)

Food for Thought

Mining and farming can often be linked. What are some examples of mined minerals that are important to the production of food and livestock? What environmental effects might poor mining practices have on agriculture?

16

Biological Resources

Tiger cub in Bandhavgarh National Park, India.

The tiger, the largest member of the cat family, lives in India, Indonesia, Thailand, Russia, and a few other Asian countries. Although a large, powerful predator, the tiger is in danger of becoming extinct. During the past century, the areas in which wild tigers are found have declined to only 7% of the tiger's historical range. About 3200 tigers are left in the wild, down from an estimated 100,000 a century ago.

More wild tigers—more than half the world's tiger population—live in small fragments of habitat in India than anywhere else. India has hundreds of tiger sanctuaries, including the Bandhavgarh National Park (see photograph). However, in 2005 park rangers discovered that all 20 tigers in India's Sariska Tiger Reserve had disappeared. (A few tigers have since been reintroduced to the reserve.) Tiger populations in Bandhavgarh and other Indian reserves are also declining precipitously.

Although tigers are protected by law, they are illegally hunted to meet a growing demand for tiger skins, bones, organs, and meat. In traditional Asian medicine, tiger bone wine, made by steeping tiger bones in alcohol for an extended period, is thought to be a potent health tonic. Similarly, tiger penis soup is thought to improve sexual capacity, and tiger eyes to prevent convulsions.

Many of the laws passed against tiger poaching and smuggling of tiger skins and body parts are not enforced. The number of patrols guarding against poachers in tiger reserves is inadequate, and few poachers are actually caught. Instead, poachers have in many cases assaulted or even murdered reserve guards. When poachers are caught and prosecuted, the fines they receive are too small to act as a deterrent.

Another factor in tiger decline is loss of habitat. As the human population continues to grow, more wild land is altered for human needs, leaving tiger habitat increasingly fragmented. Humans also hunt the same animals that tigers do, and existing tigers have fewer potential prey in their fragmented habitats.

Clearly, a concerted commitment is needed at local, national, and international levels if wild tiger populations are to be saved. Like many environmental problems, the challenge of tiger conservation is complex: Science, economics, social standards, cultural beliefs and traditions, political factors, human population growth, and national and international policies can all influence the fate of wild tiger populations.

In this chapter we first examine the importance of all forms of life, and then we consider extinction, which is an increasing threat to tigers and many other organisms. Finally, we explore how to preserve biological resources and save at least some endangered species from disappearing forever.

In Your Own Backyard...
Visit www.fws.gov/endangered/ and identify your state to search for endangered species located near you. Scroll down to see all of them. How many endangered animal species are there? How many plants?

Biological Diversity

LEARNING OBJECTIVES

- Define *biological diversity* and distinguish among genetic diversity, species richness, and ecosystem diversity.
- Describe several important ecosystem services provided by biological diversity.

A **species** is a group of more or less distinct organisms that are capable of interbreeding with one another in nature to produce fertile offspring but do not interbreed with other organisms. We do not know exactly how many species exist. In fact, biologists now realize how little we know about Earth's diverse organisms. Scientists estimate there may be as few as 5 million or as many as 100 million species. According to the International Union for Conservation of Nature (IUCN), about 1.7 million species have been scientifically named and described to date, including nearly 310,000 plant species, 65,000 vertebrate animal species, and 1 million insect species. About 10,000 new species are identified each year (**Figure 16.1**).

The variation among organisms is referred to as **biological diversity** or **biodiversity,** but the concept includes much more than simply the number of species, which is known as **species richness**.

biological diversity The number, variety, and variability of Earth's organisms; consists of three components: genetic diversity, species richness, and ecosystem diversity.

Biological diversity occurs at all levels of biological organization, from populations to ecosystems. It takes into account **genetic diversity**, the genetic variety *within* all populations of that species. Biological diversity also includes **ecosystem diversity**, the variety of ecosystems found on Earth: the forests, prairies, deserts, coral reefs, lakes, coastal estuaries, and other ecosystems of our planet. Ecosystem diversity also encompasses the variety of interactions among organisms in natural communities. For example, a forest community with its trees, shrubs, vines, herbs, insects, worms, vertebrate animals, fungi, bacteria, and other microorganisms has greater ecosystem diversity than a wheat field.

Figure 16.1   A newly discovered species. This wallaby (*Dorcopsulus* sp. nov.) is the smallest kangaroo known in the world. Discovered in Indonesia, it is 60 cm (about 2 ft) long. Conservation International announced the discovery in 2010.

WHY WE NEED ORGANISMS

Humans depend on the contributions of thousands of species for their survival. In traditional (nonmodern) societies, these contributions are direct: Plants, animals, and other organisms are the sources of food, clothing, and shelter. In industrialized societies, most people do not hunt for their morning breakfasts or cut down trees for their shelter and firewood. Nevertheless, we still depend on organisms.

Although all societies make use of many kinds of plants, animals, fungi, and microorganisms, most species have not been evaluated for their potential usefulness. At least 250,000 of the 310,000 known plant species have not been assessed for their industrial, medicinal, or agricultural potential. The same is true of most of the millions of microorganisms, fungi, and animals.

Most people do not think of insects as an important biological resource, but insects are instrumental in several important ecological and agricultural processes, including pollination of crops, weed control, and insect pest control. In addition, many insects produce unique chemicals that may have important applications for human society. Bacteria and fungi provide us with foods, antibiotics, and other medicines, as well as important biological processes such as nitrogen fixation (see Chapter 4). Biological diversity represents a rich, untapped resource for future uses and benefits, and many as-yet-unknown species may someday provide us with products. A reduction in biological diversity decreases this treasure prematurely and permanently.

Ecosystem Services and Species Richness   The living world is a complex system. Each ecosystem is composed of many separate parts, the functions of which are organized and integrated to maintain the ecosystem's overall performance. The activities of all organisms are interrelated; we are linked and dependent on one another and on the physical environment, often in subtle ways. When one species declines, other species linked to it may decline or increase in number.

Consider, for example, the role of alligators in the environment (**Figure 16.2**). The American alligator helps maintain populations of smaller fishes by eating the gar, a fish that preys on them. Alligators dig underwater holes that other aquatic organisms use during droughts, when the water level is low. The nest mounds they build are enlarged each year and eventually form small islands colonized by trees and other plants. In turn, the trees on these islands support heron and egret populations. The alligator habitat is maintained in part by underwater "gator trails," which help clear out aquatic vegetation that might eventually form a marsh.

Plants, animals, fungi, and microorganisms are instrumental in many environmental processes essential to human existence. Forests are not just a potential source of lumber; forests provide watersheds from which we obtain fresh water, control the number and severity of local floods, and reduce soil erosion. (See the Chapter 14 opener for a vivid example of the consequences of erosion, a landslide in Washington State.) Many flowering plant species depend on insects to transfer pollen for reproduction. Animals, fungi, and microorganisms help keep the populations of various species in check so that the numbers of one species do not

© W. Perry Conway / CORBIS

Figure 16.2 **Role of alligators in the environment.** The American alligator (*Alligator mississippiensis*) plays an integral, but often subtle, role in its natural ecosystem. Photographed in the Everglades.

increase enough to damage the stability of the entire ecosystem. Soil dwellers, from earthworms to bacteria, develop and maintain soil fertility for plants. Bacteria and fungi perform the crucial task of decomposition, which allows nutrients to cycle in the ecosystem. All these processes are **ecosystem services**, important environmental benefits that ecosystems provide to people, such as clean air to breathe, clean water to drink, and fertile soil in which to grow crops. Ecosystem services maintain the living world, including human societies, and we are completely dependent on them. (Table 5.1 summarizes some important ecosystem services.)

The loss of only a few species from an ecosystem endangers the other organisms in unpredictable ways. Just as a car runs less smoothly if some parts are missing, the removal of species from a community makes an ecosystem run less smoothly. If enough species are removed, the entire ecosystem will change. Species richness within an ecosystem provides the ecosystem with resilience, that is, the ability to recover from environmental changes or disasters (see the discussion of species richness and community stability in Chapter 5).

Genetic Reserves   The maintenance of a broad genetic base is critical for each species' long-term health and survival. Consider economically important crop plants. During the 20th century, plant scientists developed genetically uniform, high-yielding varieties of important food crops such as wheat. It quickly became apparent, however, that genetic uniformity resulted in increased susceptibility to pests and disease.

By crossing the "superstrains" with more genetically diverse relatives, disease and pest resistance can be reintroduced into such plants. A corn blight fungus that ruined the corn crop in the United States in 1970 was brought under control by crossing the cultivated, highly uniform U.S. corn varieties with genetically diverse ancestral varieties from Mexico. When some of the genes from Mexican corn were incorporated into the U.S. varieties, the U.S. strains became resistant to the corn blight fungus. (The global decline in domesticated plant and animal varieties is discussed in Chapter 18.)

Scientific Importance of Genetic Diversity   *Genetic engineering*, the incorporation of genes from one organism into a different species (see Figure 18.11), makes it possible to use the genetic resources of organisms on a wide scale. The gene for human insulin, for example, was engineered into bacteria. These bacteria subsequently become tiny chemical factories, manufacturing at a relatively low cost the insulin required in large amounts by diabetics. Though not without controversy, genetic engineering, available since the mid-1970s, has provided new vaccines, more productive farm animals, and agricultural plants with greater disease resistance.

Although we have the skills to transfer genes from one organism to another, we do not have the ability to *make* genes that encode for specific traits. Genetic engineering depends on the broad base of genetic diversity from which it obtains genes. It has taken hundreds of millions of years for **evolution** to produce the genetic diversity found in organisms living on our planet today. This diversity may hold solutions to today's problems and to future problems we have not begun to imagine. We would be unwise to allow such an important part of our heritage to disappear.

Medicinal, Agricultural, and Industrial Importance of Organisms   The genetic resources of organisms are vitally important to the pharmaceutical industry, which incorporates into its medicines many hundreds of chemicals derived from organisms. From extracts of cherry and horehound for cough medicines to certain ingredients of periwinkle and mayapple for cancer therapy, derivatives of plants play important roles in the treatment of illness and disease (**Figure 16.3**). Many of the natural products taken directly from marine organisms, such as tunicates, red algae, mollusks, corals, and sponges, are promising anticancer or antiviral drugs. The AIDS (acquired immune

© Florapix/Alamy

Figure 16.3 **Medicinal value of the rosy periwinkle.** The rosy periwinkle (*Catharanthus roseus*) produces chemicals effective against certain cancers. Drugs (e.g., Vincristine) from the rosy periwinkle have increased the chance of surviving childhood leukemia from about 5% to more than 95%.

deficiency syndrome) drug AZT (azidothymidine), for example, is a synthetic derivative of a compound from a sponge. The 20 best-selling prescription drugs in the United States are either natural products, natural products that are slightly modified chemically, or synthetic drugs whose chemical structures were originally obtained from organisms.

The agricultural importance of plants and animals is indisputable, because we must eat to survive. However, the kinds of foods we eat are limited compared with the total number of edible species. Many species that are nutritionally superior to our common foods probably exist. Winged beans are a tropical legume from Southeast Asia and Papua New Guinea. Because the seeds of the winged bean contain large quantities of protein and oil, they may be the tropical equivalent of soybeans. Almost all parts of the plant are edible, from the young, green fruits to the starchy storage roots.

Modern industrial technology depends on a broad range of products from organisms. Plants supply oils and lubricants, perfumes and fragrances, dyes, paper, lumber, waxes, rubber and other elastic latexes, resins, poisons, cork, and fibers. Animals provide wool, silk, fur, leather, lubricants, waxes, and transportation, and they are important in medical research.

Insects secrete a large assortment of chemicals that represent a wealth of potential products. Certain beetles produce steroids with birth-control potential, fireflies produce a compound that may be useful in treating viral infections, and some fly species show promise as a source of new antibiotics. Centipedes secrete a fungicide that could help protect crops. Because biologists estimate that perhaps 90% of all insects have not yet been identified, insects represent an important potential biological resource.

Aesthetic, Ethical, and Spiritual Value of Organisms   Organisms not only contribute to human survival and physical comfort; they also provide mental health benefits, recreation, inspiration, and spiritual solace. Our natural world is a thing of beauty largely because of the diversity of living forms found in it. Artists have attempted to capture this beauty in drawings, paintings, sculpture, and photography; poets, writers, architects, and musicians have created works reflecting and celebrating the natural world. Several studies of people living in urban environments have indicated that parks and green spaces promote psychological well-being, including the ability to concentrate better and to lower levels of anxiety.

The strongest ethical consideration involving the value of organisms is how humans perceive themselves in relation to other species. Traditionally, many human cultures have viewed themselves as superior beings, subduing and exploiting other forms of life for their benefit. An alternative view is that organisms have intrinsic value in and of themselves and that, as stewards of the life-forms on Earth, humans should watch over and protect their existence (see Chapter 2 for a discussion of *environmental ethics*).

REVIEW

1. What is biological diversity?

2. What are three examples of ecosystem services provided by biological diversity?

Extinction and Species Endangerment

LEARNING OBJECTIVES

- Define *extinction* and distinguish between background extinction and mass extinction.
- Contrast threatened and endangered species, and list four characteristics common to many endangered species.
- Define *biodiversity hotspots* and explain where most of the world's biodiversity hotspots are located.
- Describe four human causes of species endangerment and extinction, and explain which cause is the most important.
- Explain how invasive species endanger native species.

Extinction, the death of a life-form, occurs when the last individual member of a species dies. Extinction is an irreversible loss: Once a species is extinct, it will never reappear. Biological extinction appears to be the eventual fate of all species, much as death is the eventual fate of all individuals. Biologists estimate that for every 2000 species that have ever lived, 1999 of them are extinct today.

> **extinction** The death of the last individual of a species.

During the time span in which organisms have occupied Earth, a continuous, low-level extinction of species, or **background extinction**, has occurred. At certain periods in Earth's history, maybe five or six times, there has been a second kind of extinction, **mass extinction**, in which numerous species disappeared during a relatively short period of geologic time. The course of a mass extinction episode may have taken millions of years, but that is a short time compared with the time that life has existed on this planet, which is estimated at about 3.5 billion years.

The causes of past mass extinctions are not well understood, but biological and environmental factors were probably involved. A major climate change could have triggered the mass extinction of species. Marine organisms are particularly vulnerable to temperature changes; if Earth's temperature changed just a few degrees, it is likely that many marine species would become extinct. It is possible that mass extinctions of the past were triggered by catastrophes, such as the collision of Earth with a large asteroid or comet. The impact could have forced massive quantities of dust into the atmosphere, blocking the sun's rays and cooling the planet.

Although extinction is a natural biological process, it is greatly accelerated by human activities. The burgeoning human population has spread into almost all areas of Earth. Whenever humans invade an area, the habitats of many organisms are disrupted or destroyed, which contributes to their extinction.

Currently, Earth's biological diversity is disappearing at an unprecedented rate (**Figure 16.4**). Conservation biologists estimate that species are presently becoming extinct at a rate of 100 to 1000 times the natural rate of background extinctions. For example, IUCN listed nearly 20,000 species as threatened with extinction in 2012, including 13% of birds, 25% of mammals, and 41% of amphibians. About 9400 species of plants are currently threatened with extinction (see You Can Make a Difference: Addressing Declining Biological Diversity).

Figure 16.4 **Representative endangered or extinct species.**
Officials at the U.S. Fish and Wildlife Service estimate that more than 500 U.S. species have gone extinct during the past 200 years. Of these, roughly 250 have become extinct since 1980.

ENDANGERED AND THREATENED SPECIES

The legal definition of an **endangered species**, as stipulated in the Endangered Species Act, is a species in imminent danger of extinction throughout all or a significant portion of its range. (The area in which a particular species is found is its **range**.) A species is endangered when its numbers are so severely reduced that it is in danger of becoming extinct without human intervention.

endangered species A species that faces threats that may cause it to become extinct within a short period.

A species is defined as **threatened** when extinction is less imminent but its population is quite low. The legal definition of a threatened species is a species likely to become endangered in the foreseeable future, throughout all or a significant portion of its range.

threatened species A species whose population has declined to the point that it may be at risk of extinction.

Endangered and threatened species represent a decline in biological diversity because as their numbers decrease, their genetic variability is severely diminished. Long-term survival and evolution depend on genetic diversity, so a decline in genetic diversity adds to the risk of extinction for endangered and threatened species, as compared to species that have greater genetic variability.

YOU CAN MAKE A DIFFERENCE

Addressing Declining Biological Diversity

Your children and grandchildren are faced with inheriting a biologically impoverished world, but people dedicated to preserving our biological heritage can reverse the trend toward extinction. You do not have to be a biologist to make a contribution; some of the most important contributions come from outside the scientific arena. Following is a partial list of actions that would help maintain the biological diversity that is our heritage:

1. A political commitment to protect organisms is necessary because no immediate or short-term economic benefit is obtained from conserving species. This commitment must take place at all levels, from local to international. Lawmaking will not ensure the protection of organisms without strong public support. Increasing public awareness of the importance of biological diversity is critical. As individuals, we can educate ourselves on biodiversity issues of particular importance in our regions.

2. Providing publicity on species conservation issues costs money. Private funds raised by organizations such as the Sierra Club, the Nature Conservancy, and the World Wildlife Fund support such endeavors, but more money is needed. As individuals, we

can join and actively support conservation organizations.

3. Before an endangered species can be saved, its numbers, range, ecology, biological nature, and vulnerability to environmental changes must be determined. Basic research provides this information. We cannot preserve a given species effectively until we know how large to make a protected habitat and what features are essential in its design. As individuals, we can inform state and national politicians of our desire to have conservation research funded with tax dollars.

4. A worldwide system of protected parks and reserves that includes every major ecosystem must be established. The protected land would provide other benefits in addition to the preservation of biological diversity. It would safeguard the watersheds that supply water, serve as a renewable source of important biological products in areas with multiple uses, and provide people with unspoiled lands for aesthetic and recreational enjoyment. As individuals, we can help establish parks by writing to national lawmakers.

5. Establishing protected areas will not prevent biological impoverishment if we continue to pollute the planet, because it is impossible to protect parks and refuges

from threats such as global climate change. Strong steps must be taken to curb the practice of dumping CO_2 and other pollutants into the air, soil, and water—for human health and well-being, and for the well-being of species so important to ecosystem stability. Specific recommendations on how we as individuals can reduce pollution are discussed in Chapters 19 through 23.

6. Developing nations in the tropics, the repositories of most of Earth's genetic diversity, do not have much money to spend on conservation. Their governments are consumed with human problems such as overpopulation, disease, and foreign debts. Participating in ecotourism is one way to help such countries appreciate the importance of the biological resources they possess. Ecotourists pay to visit natural environments and view native species. Well-executed ecotourism conserves natural areas and improves the well-being of local people. Additionally, developed nations can provide economic incentives to developing countries by forgiving or reducing their debts in exchange for their support of local conservation efforts, including the protection of endangered species.

The choices we make and the policies we support at all levels of government can impact ecosystems and species around the world.

Characteristics of Endangered Species   Many endangered species share certain characteristics that seem to have made them more vulnerable to extinction. These characteristics include having an extremely small (localized) range; requiring a large territory; living on islands; having low reproductive success, often the result of a small population size or low reproductive rates; needing specialized breeding areas; and having specialized feeding habits.

Many endangered species have a limited natural range, which makes them particularly prone to extinction if their habitat is altered. The Tiburon mariposa lily consists of a single population growing on a hilltop near San Francisco. Development of that area would almost certainly cause the extinction of this species.

Species that require extremely large territories to survive—often because they are tertiary consumers at the top of the food web—may be threatened with extinction when all or part of their territory is modified by human activity. The California condor, a scavenger bird that lives off carrion and requires a large, undisturbed territory of hundreds of square kilometers in which to find adequate food, is slowly recovering from the brink of extinction. In 1983, the California condor population reached a low of 22 birds, and from 1987 to 1992, it was no longer found in nature (**Figure 16.5**). A program to reintroduce zoo-bred California

Frans Lanting / MINT Images / Science Source

Figure 16.5   California condor. California condors (*Gymnogyps californianus*), which have up to a 3-meter (10-foot) wingspan, are critically endangered largely because development has reduced the size of their wilderness habitat. Condors also face other environmental threats. Photographed in Big Sur, California.

condors into the wild began in 1992. By the end of 2013, there were 412 condors, more than half of them living in the wild in California (including Pinnacles National Park; see Chapter 17) and adjacent areas in Mexico and Arizona. The condor story is not an unqualified success, however, because they have yet to become a self-sustaining population that replaces its numbers without augmentation by captive breeding. Currently, the greatest threat to the condor's recovery is lead toxicity; condors ingest lead pellets when they eat prey shot by hunters.

Many species **endemic** to certain islands (i.e., they are not found anywhere else in the world) are endangered. These organisms often have small populations that cannot be replaced by immigration if their numbers are destroyed. Because they evolved in isolation from competitors, predators, and disease organisms, island species have few defenses when such organisms are introduced, usually by humans. It is not surprising that of the 171 bird species that have become extinct in the past few centuries, 155 of them lived on islands.

In ecological terms, *island* refers not only to any landmass surrounded by water but also to any isolated habitat surrounded by an expanse of unsuitable territory. Accordingly, a small patch of forest surrounded by agricultural and suburban lands is considered an island. **Habitat fragmentation**, the breakup of large areas of habitat into small, isolated patches (i.e., islands), is a major threat to the long-term survival of many species. (National parks are discussed as islands in Chapter 17.) For a species to survive, its members must be present within their range in large enough numbers for males and females to mate. The minimum population density and size that ensure reproductive success vary from one type of organism to another. For all organisms, if the population density and size fall below a critical minimum level, the population declines, becoming susceptible to extinction.

Endangered species often share other characteristics. Some have low reproductive rates. The female blue whale produces a single calf every other year, and no more than 6% of swamp pinks, an endangered species of small flowering plant, produce flowers in a given year (**Figure 16.6**). Some endangered species breed only in specialized areas; the Kemp's ridley sea turtle, for example, lays its eggs on a single beach in Mexico.

Highly specialized feeding habits may endanger a species. In nature, the giant panda eats only bamboo. Periodically, all the bamboo plants in a given area flower and die; when this occurs, panda populations face starvation. Like many other endangered species, giant pandas are also endangered because their habitat has been fragmented into small islands. China's approximately 1600 wild giant pandas live in isolated habitats that occupy a small fraction of their historical range. China, in partnership with the World Wildlife Fund (WWF), has established roughly 40 panda reserves.

WHERE IS DECLINING BIOLOGICAL DIVERSITY THE GREATEST PROBLEM?

Declining biological diversity is a concern throughout the United States, but it is most serious in the states of Hawaii (where 63% of species are at risk) and California (where about 29% of species are at risk). Hawaii has lost hundreds of species and has more species listed as endangered than any other state. At least two-thirds of Hawaii's native forests are gone.

Jeffrey Lepore/Science Source

Figure 16.6   The swamp pink. This endangered species lives in the boggy areas of the eastern United States. Photographed in Killens Pond State Park, Delaware. Why might this species be particularly vulnerable to climate change? (*Hint:* How might its habitat change as the climate changes?)

As serious as declining biological diversity is in the United States, it is even more serious abroad, particularly in tropical rain forests. Although tropical rain forests—found in South and Central America, central Africa, and Southeast Asia—cover only 7% of Earth's surface, as many as 50% of all species inhabit them. Using remote-sensing surveys, scientists have determined that approximately 1% of tropical rain forests are being cleared or severely degraded each year. The forests are making way for human settlements, banana and oil palm plantations, oil and mineral explorations, and other human activities. (Tropical rain forests are discussed in Chapters 6 and 17.)

Earth's Biodiversity Hotspots   In 1988, ecologist **Norman Myers** of Oxford University coined the term **biodiversity hotspots**. In 2000, using plants as their criteria, Myers and ecologists at Conservation International identified 25 biodiversity hotspots around the world. A substantive reanalysis has now delineated a total of 34 hotspots (**Figure 16.7**). More than 50% of all species of vascular plants live within the hotspots. An estimated 42% of the world's terrestrial vertebrate species and 29% of its freshwater fish are endemic to these regions. Many humans—nearly 20% of the world's population—live in the hotspots. Many hotspots are tropical forests, 10 are mostly or solely islands, and others are isolated by deserts or mountain ranges.

biodiversity hotspots Relatively small areas of land that contain an exceptional number of endemic species and are at high risk from human activities.

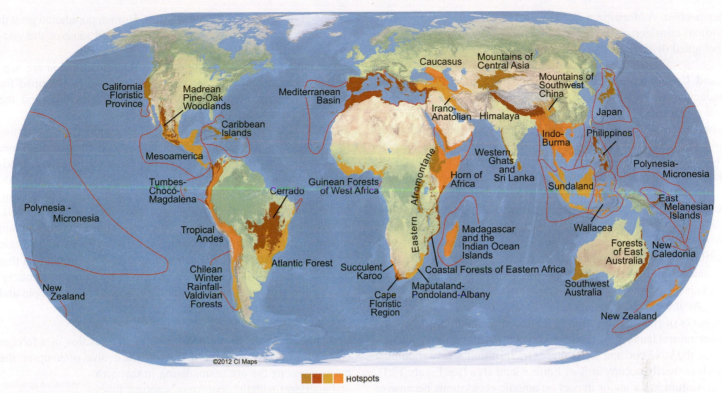

Figure 16.7   Biodiversity hotspots. These 34 hotspots, which are rich in endemic species, are at great risk from human activities. (Map by Conservation International)

Many biologists recommend that conservation planners focus on preserving land in these hotspots to reduce the mass extinction of species currently under way. Not all biologists agree. Some critics think that concentrating most of our efforts on the 34 biodiversity hotspots causes us to neglect species living in other habitats, such as deserts, grasslands, tundra, and temperate forests, all of which are also at risk.

HUMAN CAUSES OF SPECIES ENDANGERMENT

In 2001, the United Nations requested a **Millennium Ecosystem Assessment** to gather scientific information about ecosystem changes and the effects of these changes on human well-being. Several reports were published in 2005 based on the work of more than 1300 scientists from 95 countries. One of these reports, the *Biodiversity Synthesis Report*, examined the important links between ecosystem health and biological diversity. The report stated that biological diversity is declining rapidly due to several factors.

Scientists generally agree that the single greatest threat to biological diversity is land use change, which causes loss of habitat. The spread of invasive species, overexploitation, and pollution (including climate change from CO_2 pollution) are also important. Underlying these direct causes of declining biological diversity are human population increase; increasing economic activity; increased use of technology; and social, political, and cultural factors (**Figure 16.8**). All of these direct and indirect factors interact in complex ways, and so it is most effective to deal with the problem of declining biological diversity using a systems

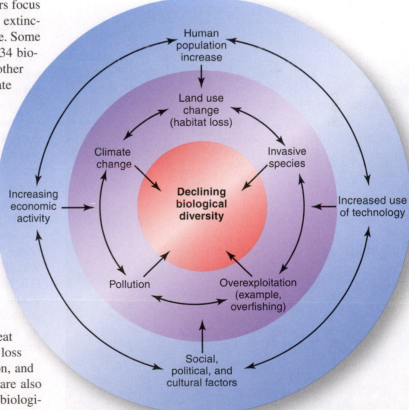

Figure 16.8   Causes of declining biological diversity. In this highly simplified diagram, indirect causes (blue) interact with and amplify the effects of one another and of direct causes (purple).

perspective. Addressing a single factor, such as overexploitation, without considering the other factors that may amplify declining biological diversity is probably doomed to failure.

Land Use Change   Most species facing extinction today are endangered because of the destruction, fragmentation, or degradation of habitats by human activities. We demolish or alter habitats when we build roads, parking lots, bridges, and buildings; log forests for timber; and clear forests to graze domestic animals or grow crops (**Figure 16.9**; also see Figure 17.11). We drain marshes to build on aquatic habitats, thus converting them to terrestrial ones, and we flood terrestrial habitats when we build dams with their reservoirs. Exploration and mining of minerals, including fossil fuels, disrupt the land and destroy habitats. Habitats are altered by outdoor recreation, including off-road vehicles, hiking off-trail, golfing, skiing, and camping. Because most organisms are utterly dependent on a particular type of environment, habitat destruction reduces their biological range and ability to survive.

As the human population has grown, the need for increased amounts of food has resulted in a huge conversion of forests and other natural lands into croplands and permanent pastures. According to the UN Food and Agriculture Organization, total agricultural lands currently occupy 36% of Earth's land area (see Figure 17.1). Agriculture has a major impact on aquatic ecosystems because of the diversion of water for irrigation.

Little habitat remains for many endangered species. The grizzly bear, for example, occupies about 2% of its original habitat in the lower 48 states of the United States. Human population growth and the extraction of resources have destroyed most of the grizzly's wilderness habitat.

Habitat destruction, fragmentation, and degradation are happening around the world. As entire habitats are transformed for human purposes, many species are becoming extinct, and the genetic diversity within many surviving species is declining.

Africa provides a vivid example of a systems issue—the conflict between humans and species such as elephants over land use. African elephants are nomads that require a lot of natural landscape in which to forage for the hundreds of kilograms of food that each consumes daily. In Africa, people are increasingly pushing into the elephants' territory to grow crops and graze farm animals. A study of 25 African regions with both wild areas and human settlements found that when human density increases to a certain level, the elephants migrate out of the area. The problem is that the wild areas to which elephants can move are steadily shrinking. One of the great challenges is finding a way to allow people and elephants to coexist in an increasingly crowded world.

Invasive Species   **Biotic pollution**, the introduction of a foreign species into an ecosystem in which it did not evolve, often upsets the balance among the organisms living in that area and interferes with the ecosystem's normal functioning. Unlike other forms of pollution, which may be cleaned up, biotic pollution is usually permanent. The foreign species may compete with native species for food or habitat or may prey on them. Generally, an introduced competitor or predator has a greater negative effect on local organisms than do native competitors or predators. Foreign species whose introduction causes economic or environmental harm are called **invasive species** (**Figure 16.10**).

> **invasive species**
> Foreign species that spread rapidly in a new area where they are free of predators, parasites, or resource limitations that may have controlled their populations in their native habitat.

Although invasive species may be introduced into new areas by natural means, humans are usually responsible for such introductions, either knowingly or unknowingly. The water hyacinth was deliberately brought from South America to the United States because it has lovely flowers. Today it has become a nuisance in Florida waterways, clogging them to such an extent that boats cannot easily move, and crowding out native species.

Islands are particularly susceptible to the introduction of invasive species. The brown tree snake was accidentally introduced in Guam, an island in the West Pacific, shortly after the end of World War II. Thought to have arrived from the Solomon Islands on a U.S. Navy ship, the brown tree snake thrived and is now estimated to number about 2 million. It consumed rain-forest birds in large numbers; as a result, 9 of Guam's 12 native species of forest birds are extinct in nature. The snakes have also decimated Guam's small reptiles and mammals. In 2013, the U.S. Department of Agriculture began experimental efforts to control the brown snake population by parachute-dropping dead mice laced with acetaminophen, which is toxic to the snake but not to humans.

Overexploitation   Sometimes species become endangered or extinct as a result of deliberate efforts to eradicate or control their numbers. Many of these species prey on game animals or livestock. Ranchers, hunters, and government agents have reduced

Figure 16.9   Land use change. These soybean fields surround a small sliver of tropical rain forest. Photographed in Mato Grosso, Brazil.

© John Lee/Aurora Photos

Asian long-horned beetle	Asian tiger mosquito	Brazilian pepper tree	Brown tree snake	Caulerpa
European wild boar	European gypsy moth	Fire ant	Formosan termite	Japanese mud snail
Kudzu	Northern snakehead	Nutria	Puerto Rican frog	Purple loosestrife
Rosy wolf snail	Sea lamprey	Tamarisk	Water hyacinth	Zebra mussel

Figure 16.10   Invasive species. Selected examples of the more than 6500 established foreign species that have been accidentally or deliberately introduced into the United States.

populations of large predators such as the wolf and grizzly bear. Predators of game animals and livestock are not the only animals vulnerable to human control efforts. Some animals are killed because their lifestyles cause problems for humans. The Carolina parakeet, a beautiful green, red, and yellow bird endemic to the southern United States, was extinct by 1920, exterminated by farmers because it ate fruit and grain crops.

Unregulated hunting, or overhunting, was a factor contributing to the extinction of certain species in the past but is now strictly controlled in most countries. The passenger pigeon was one of the most common birds in North America in the early 1800s, but a century of overhunting, coupled with habitat loss, resulted in its extinction in the early 1900s. Unregulated hunting was one of several factors that caused the near-extinction of the American bison. Bison were decimated by the U.S. Army, which killed them to disrupt the food supply of the Plains Indians, and by commercial hunters, who killed them for their hides and tongues (considered a choice food) as well as meat for the work crews of railroad companies.

Illegal commercial hunting, or poaching, endangers many larger animals, such as the tiger, cheetah, and snow leopard, whose beautiful furs are quite valuable (**Figure 16.11**). Rhinoceroses are

© Terry Whittaker / Alamy

Figure 16.11   Illegal trade in products made from endangered species. Photographed in Myanmar.

slaughtered primarily for their horns, used for ceremonial dagger handles in the Middle East and for purported medicinal purposes in Asian medicine. Bears are killed for their gallbladders, used in Asian medicine to treat ailments ranging from indigestion to heart disease. Endangered American turtles are captured and exported illegally to China, where they are killed for food. Caimans (reptiles similar to crocodiles) are killed for their skins and made into shoes and handbags. Although all these animals are legally protected, the demand for their products on the black market has caused them to be hunted illegally.

In West Africa, poaching has contributed to the decline in lowland gorilla and chimpanzee populations. The meat (called *bushmeat*) of these rare primates and other protected species, such as anteaters, elephants, and mandrill baboons, provides an important source of protein for indigenous people (see Figure 7.3). Bushmeat is also sold to urban restaurants.

Commercial harvest is the collection and sale of live organisms from nature. Commercially harvested organisms end up in zoos, aquaria, biomedical research laboratories, circuses, and pet stores. Several million birds are commercially harvested each year for the pet trade, but unfortunately, many of them die in transit, and many more die from improper treatment after they are in their owners' homes. Although it is illegal to capture endangered animals from nature, there is a thriving black market, mainly because collectors in the United States, Canada, Europe, and Japan are willing to pay large amounts to obtain a variety of species, particularly rare tropical birds (**Figure 16.12**). The U.S. **Wild Bird Conservation Act** of 1992 imposed a moratorium on importing rare bird species. Poaching data collected before and after 1992 indicated a drop in poaching rates after the law went into effect.

Animals are not the only organisms threatened by excessive commercial harvest. Many unique and rare plants have been collected from nature to the point that they are endangered. These include carnivorous plants, wildflower bulbs, certain cacti, and orchids.

Pollution Human-produced acid rain, stratospheric ozone depletion, and climate change degrade even wilderness habitats that are otherwise natural and undisturbed. Acid rain is thought to have contributed to the decline of large stands of forest trees and the biological death of many freshwater lakes. Because ozone in the upper atmosphere shields the ground from a large proportion of the sun's harmful ultraviolet (UV) radiation, ozone depletion in the upper atmosphere represents a threat to all terrestrial life.

Climate warming, caused in part by an increase in atmospheric CO_2 released when fossil fuels are burned, is another threat. Overwhelming evidence indicates that recent climate changes have already affected biological diversity. Further climate change is expected to increase the rate of extinction, particularly in certain regions such as polar areas and alpine habitats. Such habitat modifications particularly reduce the biological diversity of species with extremely narrow and rigid environmental requirements (discussed further in Chapter 20).

Excessive fertilizer use has contributed to high levels of nutrients in soil and aquatic ecosystems. Other types of pollutants that affect organisms include industrial chemicals, agricultural pesticides, organic pollutants from sewage, antibiotics and hormones from agriculture and human prescriptions, acid mine drainage seeping from mines, thermal pollution from the heated wastewater of industrial plants, and plastics (**Figure 16.13**). The effects of various forms of pollution on biological diversity are discussed throughout the text.

CASE IN POINT

DISAPPEARING FROGS

Of all this chapter's examples of species in trouble, frogs and other amphibians deserve special notice. Many scientists think amphibians are **indicator species** that provide an early warning of environmental damage with the potential to affect other species. Amphibians also merit attention because precipitous declines in amphibian populations are occurring around the world. These

Figure 16.12   Illegal animal trade. These parrots, captured illegally in the South American jungle, were confiscated at an animal black market in Peru.

Figure 16.13   Plastic pollution harms wildlife. The plastic from a broken fishing net has ensnared a stellar sea lion (*Eumetopias jubatus*), and part of the net is cutting into the seal's neck. Photographed in Alaska.

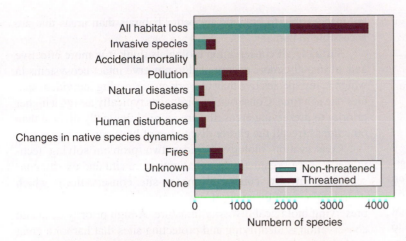

(a) The IUCN has identified a range of threats to global amphibian species.

(b) Pollution and parasites are implicated in frog deformities, developmental abnormalities such as the extra leg on this plains leopard frog (*Rana blairi*). Photographed in Leavenworth County, Kansas.

Figure 16.14   **Frogs in trouble.**

animals face many documented threats, some of which are particularly acute to threatened species (**Figure 16.14a**).

Amphibians, represented by about 7000 species of frogs, toads, and salamanders, have existed as a group for more than 350 million years. Despite their evolutionary resilience, amphibians are remarkably sensitive environmental indicators in both aquatic and terrestrial ecosystems. Amphibians typically spend part of their life in water and part on land. Most frogs lay gelatinous and unprotected eggs in ponds and other pools of standing water, and tadpoles (immature frogs) undergo metamorphosis in water, maturing into adult frogs that live on land. As adults, frogs breathe primarily through their permeable skin. This moist, absorptive skin makes frogs susceptible to environmental contaminants.

Since the 1970s, many of the world's frog populations have dwindled or disappeared. At least 42% of known amphibian species are declining, and approximately 165 species are believed to be extinct. Habitat loss is the greatest threat to amphibians, but researchers have observed that the declines are not limited to areas with obvious habitat destruction. Some remote, pristine locations also show dramatic declines in amphibians. Biologists are not certain what is causing these declines, and it appears no single factor is responsible. Factors with scientific support include pollutants, infectious diseases, and climate change.

Agricultural chemicals are implicated in amphibian declines in California's Sierra Nevada. Frog populations on the eastern slopes are relatively healthy, but about eight species are declining on the western slopes, where prevailing winds carry residues of pesticides from the Central Valley, a huge agricultural region. Agricultural chemicals may also be contributing to amphibian declines in Maryland and in Ontario, Canada.

A *chytrid* (fungus) is responsible for massive die-offs of more than 100 amphibian species around the world. Climate change may be exacerbating chytrid-induced amphibian deaths. At certain altitudes and temperatures, the chytrid has infected and killed 85% of the amphibians.

Amphibian Deformities The discovery in some areas of frogs and other amphibians with deformities adds another layer of complexity to the amphibian crisis. Deformed frogs—with extra legs, extra toes, eyes located on the shoulder or back, deformed jaws, bent spines, missing legs, missing toes, and missing eyes—usually die before they reproduce (**Figure 16.14b**). Predators easily catch frogs with extra or missing legs. Since the discovery of deformed frogs in Minnesota in 1995, amphibian deformities have been found in most U.S. states and on four continents.

Several factors may produce amphibian deformities. Some pesticides affect normal development in frog embryos. Also, infecting tadpoles with a *trematode* (a parasitic flatworm) causes the developing adults to exhibit limb deformities. Multiple environmental stressors, such as habitat loss, disease, and air and water pollution, may interact synergistically with one another to cause deformities. For example, an amphibian stressed by pesticide residues or drought may be more susceptible to a parasite. In 2008, scientists confirmed a positive link between the level of atrazine (a herbicide) in water samples and the number of trematodes infecting northern leopard frogs living in the water. ◾

REVIEW

1. How do you distinguish between background extinction and mass extinction events?

2. What are endangered species? Threatened species?

3. What are biodiversity hotspots? Where are most biodiversity hotspots located?

4. What is the most significant cause of species endangerment and extinction?

5. How can an invasive species endanger native species?

Conservation Biology

LEARNING OBJECTIVES

- Define *conservation biology* and compare in situ and ex situ conservation.
- Describe restoration ecology.

conservation biology The scientific study of how humans impact organisms and of the development of ways to protect biological diversity.

What strategies should we develop to cope with declining biological diversity? The broad field of **conservation biology** addresses these concerns. Conservation biology ranges from studying the processes that influence a decline in biological diversity, to protecting and restoring populations of endangered species, to preserving entire ecosystems and landscapes. Conservation biologists develop models, design experiments, and perform fieldwork to address this range of questions.

Conservation biologists have demonstrated that a single large area of habitat, which has the potential to support large populations, is generally more effective at safeguarding an endangered species than several habitat fragments, each with the potential to support small populations. A large area of habitat also has the potential to support greater species richness than several habitat fragments.

It is better if areas of habitat for a given species are located close together rather than far apart. If an area of habitat is isolated from other areas, individuals of a species may not effectively disperse from one habitat to another. Because the presence of humans adversely affects many species, areas that lack roads or are inaccessible to humans are better habitats than areas that are accessible to humans.

Virtually all conservation biologists think it is more effective and, ultimately, more economical to preserve intact ecosystems in which many species live than to work on preserving individual species one at a time. Conservation biologists typically assign a higher priority to preserving areas that are more biologically diverse than other areas (recall the earlier discussion of biodiversity hotspots).

Conservation biology includes two problem-solving techniques to save organisms from extinction: in situ and ex situ conservation. **In situ conservation** (on-site conservation), which includes the establishment of parks and reserves, concentrates on preserving biological diversity in nature. A high priority of in situ conservation is identifying and protecting sites that harbor a great deal of diversity. With increasing demands on land, in situ conservation cannot guarantee the preservation of all types of biological diversity. Sometimes only ex situ conservation can save a species.

Ex situ conservation (off-site conservation) involves conserving biological diversity in human-controlled settings. The breeding of captive species in zoos (such as the condors discussed earlier in the chapter) and the seed storage of genetically diverse plant crops are examples of ex situ conservation.

IN SITU CONSERVATION: PROTECTING HABITATS

Protecting animal and plant habitats—that is, conserving and managing the ecosystem as a whole—is the single best way to preserve biological diversity. Because human activities have adversely affected the sustainability of many of Earth's ecosystems, direct conservation management of protected areas is often required (**Figure 16.15**).

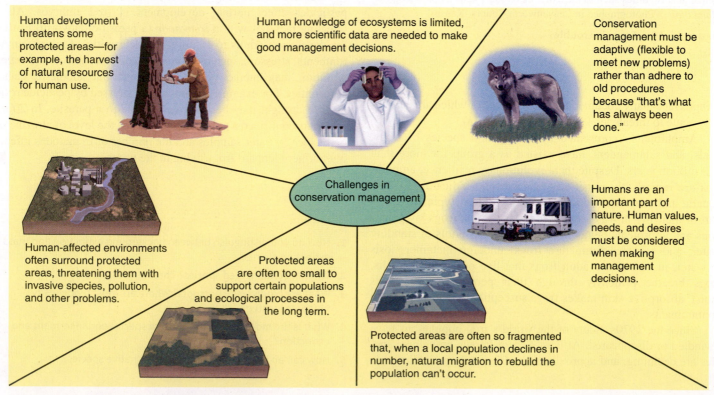

Human development threatens some protected areas—for example, the harvest of natural resources for human use.

Human knowledge of ecosystems is limited, and more scientific data are needed to make good management decisions.

Conservation management must be adaptive (flexible to meet new problems) rather than adhere to old procedures because "that's what has always been done."

Challenges in conservation management

Humans are an important part of nature. Human values, needs, and desires must be considered when making management decisions.

Human-affected environments often surround protected areas, threatening them with invasive species, pollution, and other problems.

Protected areas are often too small to support certain populations and ecological processes in the long term.

Protected areas are often so fragmented that, when a local population declines in number, natural migration to rebuild the population can't occur.

Figure 16.15   Some challenges in conservation management.

Many nations appreciate the need to protect their biological heritage and have set aside areas for biological habitats. Ecuador, Venezuela, Denmark, and the Dominican Republic have established protected areas totaling more than 30% of their land. Austria, Germany, New Zealand, Slovakia, Bhutan, and Belize have more than 20% of their land areas protected.

Currently, more than 100,000 national parks, marine sanctuaries, wildlife refuges, forests, and other protected areas exist worldwide. These encompass a total area almost as large as Canada. Some of these areas were set aside to protect specific endangered species. The world's first such refuge was established in 1903 at Pelican Island, Florida, to protect the brown pelican. Today the U.S. National Wildlife Refuge System has land set aside in more than 560 refuges. Although the bulk of the protected land is in Alaska, refuges exist in all 50 states.

Many protected areas have multiple uses that sometimes conflict with the goal of preserving species. National parks provide for recreation, and national forests may be open for logging, grazing, and mineral extraction. The mineral rights in many wildlife refuges are privately owned, and some wildlife refuges have played host to military exercises.

Protected areas are not always effective in preserving biological diversity, particularly in poorer countries, often where biological diversity is greatest, because there is little money or expertise to manage them. Such sites, where governments are unable or unwilling to enforce conservation laws, are vulnerable to logging, farming, mining, and poaching.

Another shortcoming of the world's protected areas is that many are in lightly populated mountain areas, tundra, and the driest deserts, places that often have spectacular scenery but relatively few kinds of species. Such remote areas are often designated reserves because they are unsuitable for commercial development. In contrast, ecosystems in which biological diversity is greatest often receive little protection. Protected areas are urgently needed in tropical rain forests, the tropical grasslands and savannas of Brazil and Australia, and dry forests widely scattered around the world. Desert organisms are underprotected in northern Africa and Argentina, and many islands and temperate river basins need protection.

Worldwide, about 11.5% of Earth's land area—nearly 19 million km² (7.3 million mi²)—has been set aside to protect biological diversity. However, many existing protected areas are too small or too isolated from other protected areas to effectively conserve species. Also, the habitats for more than 700 highly endangered mammals, birds, reptiles, and amphibians are not included in the global network of protected areas.

Connecting Fragmented Habitats Conservation biologists have proposed setting aside **habitat corridors**, which are strips of habitat that connect isolated habitat fragments. Habitat corridors, also called *wildlife corridors*, allow wild organisms to move from one fragment to another so they can feed, mate, and recolonize habitats should local extinctions occur. Some habitat corridors are very small, such as an underpass or overpass that allows wildlife to cross a road safely (**Figure 16.16a**). Other habitat corridors are large (e.g., several kilometers of forest) and connect separate wildlife reserves (see Meeting the Challenge: Dealing Effectively with Fragmented Habitats; also see **Figure 16.16b** for an example of proposed habitat corridors).

(a) Close-up of wildlife overpass. Without habitat corridors under or over highways, grizzly bears and other wildlife must cross dangerous roads or avoid moving to new areas. Photographed in Banff National Park, Alberta, Canada.

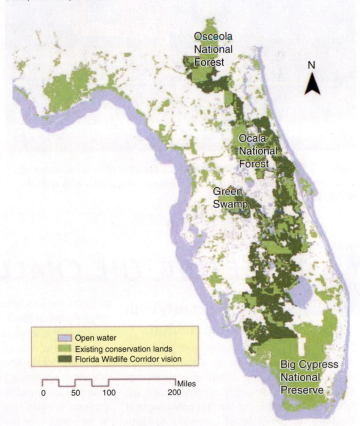

(b) Proposed habitat corridors to link Florida wilderness areas, as designed by the ongoing Florida Wildlife Corridor project.

Figure 16.16   Habitat corridors.

RESTORING DAMAGED OR DESTROYED HABITATS

Although preserving habitats is an important part of conservation biology, the realities of our world, including the fact that the land-hungry human population continues to increase, dictate a variety of other conservation measures. Scientists can reclaim disturbed lands and convert them into areas with high biological diversity.

Restoration ecology, in which the principles of ecology are used to help return a degraded environment to a more functional and sustainable one, is an important part of in situ conservation. (**Figure 16.17**)

restoration ecology The study of the historical condition of a human-damaged ecosystem, with the goal of returning it as close as possible to its former state.

Since 1934, the University of Wisconsin–Madison Arboretum has carried out one of the world's most famous examples of restoration ecology. During that time, several distinct natural communities of Wisconsin were carefully developed on damaged agricultural land. These communities include a tallgrass prairie, a dry prairie, and several types of pine and maple forests. Today hundreds of species of native plants, birds, mammals, and insects live in the restoration site.

Restoration of disturbed lands creates biological habitats and has additional benefits such as the regeneration of soil damaged by agriculture or mining. The disadvantages of restoration include the expense and the amount of time it requires to restore an area. Even so, restoration is an important aspect of conservation biology, because it is thought that restoration will reduce extinctions.

Figure 16.17   Prairie restoration. Principles of restoration ecology were applied to restoring a midwestern farm, overgrown with trees, to prairie habitat.

© NatPar Collection / Alamy

ZOOS, AQUARIA, BOTANICAL GARDENS, AND SEED BANKS

Zoos, aquaria, and botanical gardens often play a critical role in saving individual species on the brink of extinction. Eggs may be collected from nature, or the remaining few wild animals may be captured and bred in zoos, aquaria, and other research environments.

Special techniques, such as artificial insemination and embryo transfer, are used to increase the number of wild animal offspring. In **artificial insemination**, sperm collected from a suitable male of a rare species is used to artificially impregnate a female, perhaps located in another zoo in a different city or even in another country. In **embryo transfer**, a female of a rare species is treated with fertility drugs, which cause her to produce multiple eggs. Some of these eggs are collected, fertilized with sperm, and surgically implanted into a female of a related but less rare species, who later gives birth to offspring of the rare species. Another technique involves hormone patches to stimulate reproduction in endangered birds; the patch is attached under the female bird's wing.

MEETING THE CHALLENGE

Dealing Effectively with Fragmented Habitats

Collaboration among diverse sets of stakeholders is crucial to meeting the challenge of designing and implementing a system of habitat corridors among fragmented habitats. Two recent examples of corridor planning that illustrate the principles of successful collaboration are located on two different continents. The *Mesoamerican Biological Corridor* connects fragments of tropical forests from Mexico through Panama, and the *Bossou-Nimba Green Corridor Project* is located in Africa in the country of Guinea.

These corridor projects involve a diverse array of intergovernmental partnerships that incorporate conservation biology, citizen science, and ecological economics. Both corridor projects evolved around the habitat needs of large mammal species. For the Mesoamerican Corridor, the focus was on creating a habitat network large enough to facilitate jaguar movement, and for the Green Corridor, the emphasis was on maintaining gene flow among isolated chimpanzee populations. Thus, both corridor projects are tied to a **flagship species**, which is an organism that captures the public's imagination.

Because connecting fragments of tropical forest habitat requires planting trees, both projects also involved *ecological restoration*. Public acceptance of corridor construction was enhanced in several ways. In Costa Rica, the Mesoamerican Corridor is widely used to promote ecotourism, but individual landowners are also compensated for replanting trees through tax credits that are linked to a reduction in global carbon dioxide levels and to improved water quality. In Guinea, forest restoration stimulates local economies because the planted trees produce fruits (oranges, mangoes, and mandarin oranges) and nuts.

The challenge of monitoring how jaguars, chimpanzees, and other wildlife species use these corridors is being met through commitments to environmental education. Each corridor project has developed a system of promoting **citizen science**, whereby people living adjacent to the corridor are trained and become responsible for annual biodiversity surveys within the protected corridor. Citizen scientists also report the occurrence of destructive uses of corridor land. Thus, neighbors—rather than police, military, or international entities—enforce laws against poaching and timber harvesting.

Restoring the Drew University Forest Preserve

Environmental science students at New Jersey's Drew University learn for themselves how to restore ecosystems. The campus's forest ecosystem had become degraded by invasive plant species, an unchecked deer population, and habitat fragmentation. A carefully planned and implemented restoration program activated teams of students and other volunteers to build a special deer exclosure fence, remove invasive plants, and replant with native plant species.

The collaborative effort included partnerships with the U.S. Fish and Wildlife Service, the New Jersey Audubon Society, private industries, and local naturalists. Biodiversity in the preserve has increased substantially as the native plants have established. The site serves as a prime educational tool for both the university and the community, and is the focus of much ecological research.

Drew University students extend their restoration efforts beyond the forest preserve. An annual "Fern Fest" event during Earth Week replaces a portion of the campus lawn with native ferns and wildflowers, with the goal of increasing biodiversity and further restoring the former forest ecosystem. A rain garden and native meadow have also been constructed, and a campus-wide policy directs the use of native species for all other plantings.

Scientists have learned to successfully breed endangered whooping cranes in captivity. The captive population now totals 157 cranes. Four wild whooping crane populations currently exist—in Florida (nonmigratory), Louisiana (nonmigratory), Texas-Canada (migratory), and Wisconsin-Florida (migratory). In late 2013, the total number of wild cranes was 454. Scientists are trying to enhance existing flocks and establish additional flocks in the wild by soft-releasing cranes (for a nonmigratory flock) or flying them behind an ultralight aircraft (for a migratory flock). (In *soft release*, the birds are initially placed in a specially constructed pen to help them adjust before being released into the wild.) The Louisiana population was established by soft release in 2011 and produced eggs in 2014, the first in the state in 75 years.

Attempting to save a species on the brink of extinction is expensive, and only a small proportion of endangered species can be saved. Because zoos, aquaria, and botanical gardens do not have the space or money to save all endangered species, conservation biologists must prioritize which species they will attempt to save. Clearly, it is more cost-effective to maintain existing natural habitats so that species will never become endangered in the first place.

Reintroducing Endangered Species to Nature The ultimate goal of captive-breeding programs is to produce offspring in captivity and then release them into nature so that wild populations can be restored. However, only 1 of every 10 reintroductions using animals raised in captivity is successful. What guarantees that a reintroduced population will survive?

Whether such reintroductions actually succeed has been evaluated only in recent years. The Hawaiian goose, or nene (pronounced *nay-nay*), was down to 30 individuals in the 1960s. It was reintroduced to the islands of Hawaii (the Big Island) and Maui beginning in the 1970s, but although hundreds of birds have been released, self-sustaining nene populations have been slow to develop. Apparently, some of the same factors that originally caused the nene's extinction in nature, including habitat destruction and foreign predators such as the Indian mongoose, are responsible for the failure of the reintroduced birds. On Kauai, a group of captive nenes escaped during a hurricane and are successfully reestablishing themselves, with a 2011 estimated population of 1400–1600 birds. Kauai is less developed than Maui or the Big Island and does not have the Indian mongoose. The statewide nene population was estimated at 2500 in 2011, including a fairly stable population on Maui, but it remains dependent on the regular release of captive-bred birds.

Before attempting a reintroduction, conservation biologists now undertake a feasibility study. This study includes determining what factors originally caused the species to become extinct in nature, whether these factors still exist, and whether any suitable habitat still remains.

If the animal to be reintroduced is a social animal, a small herd is usually released together. This is accomplished by first placing the herd in a large, semiwild enclosure that is somewhat protected from predators but that requires the herd to obtain its own food. When the herd's behavior begins to resemble the behavior of wild herds, it is released.

Sometimes it is impossible to teach critical survival skills to animals raised in captivity. The effort to reintroduce captive-raised thick-billed parrots to the Chiricahua Mountains of Arizona was canceled in 1993 because all 88 birds released between 1986 and 1993 died or disappeared. Wild, thick-billed parrots are loud, sociable birds whose flocking instinct contributes to their survival because individual birds loudly announce the presence of danger, such as hawks or other birds of prey, to the group. Those parrots raised in captivity lacked such social behavior, and despite efforts to teach them to stay together, they separated from the flock after they were released.

Once animals are released, they must be monitored. If any animals die, their cause of death is determined so scientists can search for ways to prevent unnecessary deaths in future reintroductions.

Seed Banks More than 100 seed collections called **seed banks**, or *gene banks*, exist around the world and collectively hold more than 3 million samples at low temperatures. The Svalbard Global Seed Vault, sometimes called the Doomsday Vault, stores known varieties of 150 widely grown crop plants (**Figure 16.18**). The Millennium Seed Bank Partnership of the Royal Botanic Garden in Kew stores seeds from 10% of the world's known wild plant species. Seed banks offer the advantage of storing a large amount of plant genetic material in a small space. Seeds stored in seed banks are safe from habitat destruction, climate change, and general neglect. There have even been some instances of using seeds from seed banks to reintroduce to nature a plant species that had become extinct.

Seeds do not remain alive indefinitely and must be germinated periodically so that new seeds can be collected. Growing, harvesting, and returning seeds to storage are the most expensive steps in storing plant material in seed banks. (Cryopreservation

Figure 16.18 **Svalbard Global Seed Vault.** Located in Svalbard, Norway, this seed bank stores seeds of some 1.5 million varieties of the 150 most important species of crop plants. How might climate change impact the success of using seeds from this seed bank to replace a locally extinct variety?

in liquid nitrogen at –160°C, or –256°F, is being developed for certain kinds of seeds. Seeds stored at this temperature survive for longer periods than seeds stored at warmer temperatures.) Because accidents such as fires or power failures can result in the permanent loss of the genetic diversity in the seeds, biologists typically subdivide seed samples and store them in several seed banks.

Perhaps the most important disadvantage of seed banks is that plants stored in this manner remain stagnant in an evolutionary sense. They do not evolve in response to changes in their natural environments, such as climate warming. As a result, they may be less fit for survival when they are reintroduced into nature. Despite their shortcomings, seed banks are increasingly viewed as an important method of safeguarding seeds for future generations.

Many individuals, small businesses, and organizations (e.g., the Sustainable Mountain Agriculture Center in Berea, Kentucky), act on their own to preserve seeds and cultivate traditional vegetable varieties no longer grown widely on farms. Often referred to as *heirloom vegetables*, these open-pollinated species typically have more flavor than their hybrid counterparts.

CONSERVATION ORGANIZATIONS

Conservation organizations are an essential part of the effort to maintain biological diversity. These groups help educate policymakers and the public about the importance of biological diversity. In certain instances, they serve as catalysts and galvanize public support for important biodiversity preservation efforts. They provide financial support for conservation projects, from basic research to the purchase of land that is a critical habitat for a particular species or group of species.

The IUCN assists countries with hundreds of conservation biology projects. The IUCN and other conservation organizations are currently assessing how effective established wildlife refuges are in maintaining biological diversity. For example, researchers are investigating the minimum size that a protected area requires to avoid species encroaching from surrounding areas. The WWF

and Brazil's National Institute for Amazon Research Biological Dynamics are conducting a long-term study on the effects of habitat fragmentation on the Amazonian rain forest by using 1-hectare, 10-hectare, and 100-hectare forest fragments, along with controls of identically sized sections in intact forest. Preliminary data indicate that the smaller forest fragments do not maintain their ecological integrity. For example, many bird species that live in the central Amazon are not found in the smaller habitat fragments.

In addition, the IUCN and the WWF have identified major conservation priorities by determining which biomes and ecosystems do not have protected areas. The IUCN maintains a data bank on the status of the world's species; its material is published in the *IUCN Red Data Books* about organisms and their habitats.

REVIEW

1. What is conservation biology?
2. Is restoration ecology an important aspect of in situ or ex situ conservation?

Conservation Policies and Laws

LEARNING OBJECTIVE
• Briefly describe the U.S. Endangered Species Act.

In 1973 the **Endangered Species Act (ESA)** was passed in the United States, authorizing the U.S. Fish and Wildlife Service (FWS) to protect endangered and threatened species in the United States and abroad. Many other countries now have similar legislation. The ESA requires a detailed study of a species to determine if it should be listed as endangered or threatened. Currently, more than 1500 species in the United States are listed as endangered or threatened. The ESA provides legal protection to listed species to help reduce their danger of extinction. This act makes it illegal to sell or buy any product made from an endangered or threatened species.

The ESA requires the FWS to select critical habitats and design a detailed recovery plan for each species listed. The recovery plan includes an estimate of the current population size, an analysis of what factors contributed to its endangerment, and a list of activities to help the population recover. Currently, 1145 U.S. endangered or threatened species have approved recovery plans.

The ESA was updated in 1982, 1985, and 1988. It is considered one of the strongest pieces of U.S. environmental legislation, in part because species are designated as endangered or threatened entirely on biological grounds. Currently, economic considerations cannot influence the designation of endangered or threatened species. Biologists generally agree that as a result of passage of the ESA in 1973, fewer species became extinct than would have if the law had not been passed.

The ESA is one of the most controversial pieces of environmental legislation. It does not provide compensation for private property owners who suffer financial losses because they cannot develop their land if a threatened or endangered species lives there. The ESA has also interfered with some federally funded development projects.

The ESA was scheduled for congressional reauthorization in 1992 but has been entangled since then in political wrangling between conservation advocates and those who support private property rights. Conservation advocates think the ESA does not do enough to save endangered species, whereas those who own land on which rare species live think the law goes too far and infringes on property rights. Another contentious issue is the financial cost of the law. Critics say that federal and state governments spend too much, given the little bit of environmental gain that the ESA accomplishes. Some critics—notably business interests and private property owners—view the ESA as an impediment to economic progress.

Those who defend the ESA point out that of 34,000 past cases of endangered species versus economic development, only 21 cases were not resolved by some sort of compromise. Compromise is crucial to the success of saving endangered species because, according to the U.S. General Accounting Office, more than 90% of endangered species live on at least some privately owned lands. Some critics of the ESA think the law should be changed so that private landowners are given economic incentives to help save endangered species living on their lands. For example, tax cuts for property owners who are good land stewards could make the presence of endangered species on their properties an asset instead of a liability.

Defenders of the ESA agree that the law is not perfect. Relatively few endangered species have recovered enough to be delisted—that is, removed from protection of the ESA. However, the FWS reports that hundreds of listed species are stable or improving; it expects as many as several dozen additional species to be delisted in the next decade or so (**Figure 16.19**). However, many species are considered *conservation reliant*—that is, they may never recover sufficiently to be delisted.

Figure 16.19 **Delisting of the brown pelican.** The brown pelican population, endangered due to hunting, habitat destruction, and the pesticide DDT, had recovered sufficiently by 2009 to remove it from the threatened and endangered species list.

The ESA is geared more to saving a few popular or unique endangered species rather than the much larger number of less glamorous species, such as fungi and insects, that perform valuable ecosystem services. Yet it is the less glamorous organisms that play central roles in ecosystems and contribute the most to their functioning.

Conservationists would like the ESA to be strengthened in such a way as to manage whole ecosystems and maintain complete biological diversity rather than attempt to save endangered species as isolated entities. This approach would offer collective protection to many declining species rather than to single species.

HABITAT CONSERVATION PLANS

The 1982 amendment of the ESA provided a way to resolve conflicts between protection of endangered species and development interests on private property: **habitat conservation plans (HCPs)**. HCPs vary greatly, from small projects to regional conservation and development plans.

Habitat conservation plans allow a landowner to "take" (injure, kill, or modify the habitat of) a rare species if the "taking" doesn't threaten the survival or recovery of the threatened or endangered species on that property. If a landowner sets aside land as habitat for the rare species, he or she then has the right to develop part of the property without threat of legal action by the FWS. Conservationists point out that HCPs do not provide any promise of recovery of rare species. In some cases, conservationists are concerned that HCPs may actually contribute to a species' extinction.

INTERNATIONAL CONSERVATION POLICIES AND LAWS

The **World Conservation Strategy**, a plan designed to conserve biological diversity worldwide, was formulated in 1980 by the IUCN, the WWF, and the UN Environment Programme. In addition to providing guidelines for conserving biological diversity, the World Conservation Strategy seeks to preserve the vital ecosystem services on which all life depends for survival and to develop sustainable uses of organisms and the ecosystems that they compose.

The **Convention on Biological Diversity** was produced by the 1992 Earth Summit to decrease the rate of extinction of the world's endangered species. This treaty requires that each signatory nation inventory its own biodiversity and develop a **national conservation strategy**, a detailed plan for managing and preserving the biological diversity of that specific country. Despite an increase in conservation efforts since the Earth Summit, however, the loss of biological diversity is not declining.

The exploitation of endangered species is somewhat controlled at the international level by the **Convention on International Trade in Endangered Species of Wild Flora and Fauna (CITES)**, which went into effect in 1975. Originally drawn up to protect endangered animals and plants considered valuable in the highly lucrative international wildlife trade, CITES bans the hunting, capturing, and selling of endangered or threatened species and regulates the trade of organisms listed as potentially threatened. Unfortunately, enforcement of this treaty varies from country to country. Even where enforcement exists, the penalties aren't severe. As a result, illegal trade in rare, commercially valuable species continues.

The African elephant, discussed earlier in the chapter, bears out these problems. Listed as an endangered species since 1989 to halt the slaughter of elephants driven by the ivory trade, the species seemed to have recovered in southern Africa. However, poaching rebounded in the 2000s, and an estimated 65% of forest elephants were killed between 2002 and 2013. According to wildlife specialists, by 2013 about 9% of the African elephant population was being killed for ivory each year. This level of slaughter is greater than the level in 1989, when the ban went into effect. Much of the illegally obtained ivory is sold online.

The goals of CITES often stir up controversy over such issues as who actually owns the world's wildlife and whether global conservation concerns take precedence over competing local interests. These conflicts often highlight socioeconomic differences between wealthy consumers of CITES products and poor people who trade in endangered organisms.

REVIEW

1. What is the Endangered Species Act and what has it accomplished?

Wildlife Management

LEARNING OBJECTIVE

• Distinguish between conservation biology and wildlife management.

Wildlife management is an applied field of conservation biology that focuses on the continued productivity of

wildlife management
The application of conservation principles to manage wild species and their habitats for human benefit or for the welfare of other species.

plants and animals. Wildlife management programs often have different priorities than those of conservation biology. In contrast to conservation biology, which often focuses on threatened or endangered species, most attention in wildlife management is focused on common organisms. Wildlife management includes the regulation of hunting and fishing and the management of food, water, and habitat.

The natural predators of many game animals have been largely eliminated in the United States. As a result of the near-disappearance of predators such as wolves, the populations of animals such as squirrels, ducks, and deer sometimes exceed the carrying capacity of their environment (see Chapter 5). When this occurs, the habitat deteriorates, and many animals starve to death. Sport hunting effectively controls the overpopulation of game animals, provided restrictions are observed to prevent overhunting. Laws in the United States determine the time of year and length of hunting seasons for various species, as well as the number, sex, and physical size of each species that may be harvested.

Wildlife managers manipulate the plant cover, food, and water supplies of a specific animal's habitat. Because different animals predominate in different stages of ecological succession, controlling the stage of ecological succession of an area's vegetation encourages the presence of certain animals and discourages the presence of others (see Chapter 5). Quail and ring-necked pheasant are found in grassy, open areas characteristic of early-succession stages. Moose, deer, and elk predominate in partially open forest, such as a meadow adjacent to a forest; the meadow provides food, and the forest provides protective cover. Other animals, such as grizzly bears and bighorn sheep, require undisturbed vegetation. Wildlife managers control the stage of succession with techniques such as planting certain types of vegetation, burning the undergrowth with controlled fires, and building artificial ponds.

MANAGEMENT OF MIGRATORY ANIMALS

International agreements are established to protect migratory animals. Ducks, geese, and shorebirds spend their summers in Canada and their winters in the United States and Central America. During the course of their annual migrations, which usually follow established routes, or **flyways**, they must have areas in which to rest and feed. Wetlands, the habitat of these animals, must be protected in both their winter and summer homes.

Arctic Snow Geese The Arctic snow goose has become a major challenge for wildlife managers because its population expanded rapidly beginning in the 1980s. This goose breeds in large colonies along coastal salt marshes of the Arctic during the short Arctic summer. The population migrates south during autumn and traditionally wintered in salt marshes along the Texas and Louisiana coasts. The snow goose has successfully expanded its winter range into Arkansas, Mississippi, Oklahoma, New Mexico, and northern Mexico, largely because the geese obtain seeds and other food from agricultural lands. Because snow geese have expanded their winter range so successfully, more of the adults survive to return to the Arctic. Their adaptability to human-induced changes in the environment has enabled them to avoid *density-dependent factors* (i.e., lack of food during winter months) that would normally keep the population in check. The huge population of snow geese has damaged much of the Arctic's fragile coastal ecosystem because the geese forage there for a variety of plants and insects (**Figure 16.20**).

Figure 16.20 **Damage caused by snow geese.** The only part of this once-lush salt marsh that survived the onslaught of snow geese was this small patch enclosed in protective fencing to prevent the geese from foraging. The rest of the marsh was reduced to a wasteland. Photographed in northern Manitoba, Canada, along the Hudson Bay.

HUDSON BAY PROJECT/KRT/Newscom

Wildlife managers want to avoid a massive die-off of geese in the Arctic, a catastrophe that is unavoidable if the goose population is not brought under control. To reduce population numbers, U.S. and Canadian wildlife managers have increased the "taking" of snow geese by sport hunters. However, hunting has not had an appreciable effect on population size, which continues to grow. Wildlife managers are looking at other options to reduce the number of snow geese to sustainable levels. For example, the geese could be commercially harvested for human consumption.

MANAGEMENT OF AQUATIC ORGANISMS

Fishes with commercial or sport value must be managed to ensure they are not overexploited to the point of extinction. Freshwater fishes such as trout and salmon are managed in several ways. Fishing laws regulate the time of year, size of fish, and maximum allowable catch. Natural habitats are maintained to maximize population size. Ponds, lakes, and streams may be restocked with young hatchlings from hatcheries.

Traditionally, the ocean's resources have been considered common property, available to the first people to exploit them. As a result, commercial fishing is severely reducing the number of marine fishes. (Chapter 18 discusses this dwindling resource.)

During the 19th and 20th centuries, many whale species were harvested to the point of **commercial extinction**, meaning that so few remained that it was unprofitable to hunt them. Although commercially extinct species still have living representatives, their numbers are so reduced that they are endangered. In 1946, the International Whaling Commission set an annual limit on killed whales for each whale species in an attempt to secure sustainable whale populations. Unfortunately, these limits were set too high, resulting in further population declines during the next 20 years (review the concept of sustainable harvest in Chapter 5). Conservationists began to call for a global ban on commercial whaling; such a moratorium went into effect in 1986.

Scientists have since monitored whale populations and concluded that overall the ban is working. The populations of most whales, such as humpbacks and bowheads, seem to be growing. One species, the Pacific gray whale, has recovered sufficiently to

be removed from the endangered and threatened species lists. The North Atlantic right whale and southern blue whales, however, are still endangered. In 1994, the International Whaling Commission established the Southern Ocean Whale Sanctuary in Antarctic waters, where many of the world's great whales feed and reproduce. This vast sanctuary, which bars commercial hunting, would continue to exist should the current ban on whaling ever be lifted.

Despite international pressure, Japan, Norway, and Iceland do not honor either the global ban on commercial whaling or the Southern Ocean Whale Sanctuary. Japan has justified its continuing whale harvests by saying that the whales are killed for "scientific purposes," although the whale meat from these harvests is sold in Japanese markets and restaurants (**Figure 16.21**).

REVIEW

1. What is wildlife management? How do the goals of wildlife management and conservation biology differ?

Figure 16.21   **Dead minke whales.** These whales were captured in the Southern Ocean and slaughtered ostensibly for research purposes. They are aboard the Japanese research factory ship *Nisshin Maru*.

ENERGY AND CLIMATE CHANGE

Assisted Colonization

Biologists have documented that climate warming has caused many plant and animal species to shift their ranges toward the poles or to higher elevations. The ranges of other species have contracted in response to climate change, and some species have gone locally extinct. Global climate is expected to continue changing rapidly in the current century, and this fact combined with extensive habitat fragmentation of the landscape means that species and even whole ecosystems will be at risk. It is not known how long various vulnerable species can survive in the wild.

Some scientists are considering a highly controversial intervention called **assisted colonization**, in which species at risk of extinction

are moved to areas where they have not been found before. This technique must be done with care, because it is possible that the introduced species could become a pest problem or carry diseases to organisms already living at the new site. In some cases, assisted colonization might not be feasible because no suitable habitat for the at-risk species remains. The cost of assisted colonization could also be prohibitively expensive. Despite the problems, assisted colonization may be the only viable option to reduce the loss of biological diversity from rapid climate change.

REVIEW OF LEARNING OBJECTIVES WITH SELECTED KEY TERMS

- **Define** *biological diversity* **and distinguish among genetic diversity, species richness, and ecosystem diversity.**

Biological diversity is the number and variety of Earth's organisms; it consists of three components: **genetic diversity** (the variety within a species), **species richness** (the number of species), and **ecosystem diversity** (variety within and among ecosystems).

- **Describe several important ecosystem services provided by biological diversity.**

Ecosystem services are important environmental benefits that ecosystems provide to people. Bacteria and fungi perform the important ecosystem service of decomposition. Forests provide watersheds from which we obtain fresh water. Insects transfer plant pollen for reproduction. Soil organisms maintain soil fertility. Plant roots anchor in the soil and reduce soil erosion.

- **Define** *extinction* **and distinguish between background extinction and mass extinction.**

Extinction is the elimination of a species from Earth. **Background extinction** is the continuous, low-level extinction of species. **Mass extinctions** are episodes in Earth's history in which numerous species became extinct in a relatively short period.

- **Contrast threatened and endangered species, and list four characteristics common to many endangered species.**

A **threatened species** is a species whose population has declined to the point that it may be at risk of extinction. An **endangered species** is a species that faces threats that may cause it to become extinct within a short period. Endangered and threatened species often have limited natural **ranges**, low population densities, low reproductive rates, or specialized food or reproduction requirements.

- **Define** *biodiversity hotspots* **and explain where most of the world's biodiversity hotspots are located.**

Biodiversity hotspots are relatively small areas of land that contain an exceptional number of **endemic** species and are at high risk from human activities. **Norman Myers** and Conservation International have identified 34 biodiversity hotspots around the world; many are tropical, and 10 are mostly or entirely islands.

- **Describe four human causes of species endangerment and extinction, and explain which cause is the most important.**

Habitat destruction is the most significant cause of declining biological diversity because it reduces a species' biological range. **Habitat**

fragmentation is a type of habitat destruction in which large areas of habitat are broken into small, isolated patches. Other causes of declining biological diversity are the introduction of foreign species, overexploitation, and pollution.

- **Explain how invasive species endanger native species.**

Invasive species are foreign species that spread rapidly in a new area where they are free of predators, parasites, or resource limitations that may have controlled their population in their native habitat. Invasive species often upset the balance among the organisms living in that area—for example, by competing with native species for food or habitat—and interfere with an ecosystem's normal functioning.

- **Define** *conservation biology* **and compare in situ and ex situ-conservation.**

Conservation biology is the scientific study of how humans affect organisms and of the development of ways to protect biological diversity. Efforts to preserve biological diversity in nature, called **in situ conservation**, include establishing parks, wildlife sanctuaries and refuges, and other protected areas. **Ex situ conservation**, which includes captive breeding and storing genetic material, occurs in human-controlled settings.

- **Describe restoration ecology.**

Restoration ecology is the study of the historical condition of a human-damaged ecosystem, with the goal of returning it as close as possible to its former state. Restoration ecology is an important part of in situ conservation.

- **Briefly describe the U.S. Endangered Species Act.**

The **Endangered Species Act (ESA)** authorizes the U.S. Fish and Wildlife Service (FWS) to protect from extinction endangered and threatened species, both in the United States and abroad. The FWS selects critical habitats and designs detailed recovery plans for each species listed. **Habitat conservation plans (HCPs)** help resolve ESA conflicts between conservation and development interests on private lands.

- **Distinguish between conservation biology and wildlife management.**

Wildlife management is the application of conservation principles to manage wild species and their habitats for human benefit or for the welfare of other species. In contrast to conservation biology, which often focuses on threatened or endangered species, most attention in wildlife management is focused on common organisms.

Critical Thinking and Review Questions

1. Is biological diversity a renewable or nonrenewable resource? Why could it be seen both ways?

2. Aldo Leopold once wrote, "To keep every cog and wheel is the first precaution of intelligent tinkering." How does his statement relate to biological systems? (Aldo Leopold, *A Sand County Almanac.* Oxford University Press, Inc. [1991])

3. Describe five important ecosystem services provided by living organisms.

4. How is the current period of mass extinction different from all previous periods of mass extinction?

5. List four characteristics common to many endangered species.

6. Review the following table and calculate the percent change in number of U.S. threatened or endangered species listed between 2005 and 2014, both total and for each group. What is the total percent change? Which groups of organisms have experienced the greatest declines? Suggest reasons for these trends.

U.S. Organisms Listed as Endangered or Threatened, 2005–2014

Type of organism	2005	2014
Mammals	78	90
Birds	90	97
Reptiles	36	36
Amphibians	21	31
Fishes	114	154
Insects	44	70
Clams	70	88
Flowering plants	715	839
Other plants and animals	96	119
Total	1264	1524

Data from U.S. Fish and Wildlife Service

7. What are biodiversity hotspots? Why are so many of them isolated?

8. According to the cartoon, what is one of the main causes of habitat loss in terrestrial ecosystems?

9. What invasive species are problems in your area? How do they affect local ecosystems, including native species? How are the invasive species being controlled?

10. Give examples of in situ and ex situ conservation.

11. The following graph shows the effects of predation over a four-week period on young bay scallops in three different marine seagrass environments. How does habitat fragmentation (patchiness) affect the percentage of young bay scallops lost to predation? Suggest a possible reason for this effect.

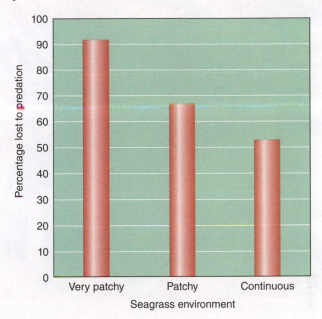

(After data from E.A. Irlandi, W.G. Ambrose, and B.A. Orando, "Landscape Ecology and the Marine Environment: How Spatial Configuration of Seagrass Habitat Influences Growth and Survival of the Bay Scallop," *Oikos* 72 [1995].)

12. If you had the assets and authority to take any measure to protect and preserve biological diversity but could take only one, what would it be?

13. What are two advantages of using the principles of restoration ecology to return a degraded environment to a more functional one? How can these be applied to small, local sites?

14. What U.S. law provides legal protection to species to reduce their danger of extinction? How are scientific considerations and economic value weighed in these listings? Explain your answer.

15. Why is the Arctic snow goose such a challenge for U.S. wildlife managers? For Canadian wildlife managers?

16. When people discuss the impact of global climate change, they usually mention polar bears or other charismatic animals. However, according to scientists at Botanic Gardens Conservation International, half of all plant species could be threatened by climate change. Most of these plants live on islands or mountains. Offer an explanation for why this is the case.

Food for Thought

The quest for food in the United States has resulted in many dramatic changes to our natural landscapes. In what ways has agriculture contributed to species endangerment and to habitat damage and destruction? How might the wildlife in your town or region have been affected by agriculture? Consider some approaches that food producers might take to avoid these historical effects. How might a backyard gardener protect species and habitats?

17

Land Resources

California condor (*Gymnogyps californianus*), a species which is resident in Pinnacles National Park.

In 2013, a new U.S. national park was designated: Pinnacles National Park, in the California chaparral habitat. This park brings the total number of national parks in the United States to 59.

Pinnacles National Park is one of few habitats where the native California condor can be reliably seen. This species, once at a population of nine remaining individuals, was removed from the wild, eggs were hatched and reared in captivity, and birds were released to the wild. Many of these birds now live in the volcanic rock formations at Pinnacles. Like some other new national parks (Congaree National Park, in South Carolina, for example), Pinnacles had previously been a national monument. (National monuments are similar to national parks except that they are given less funding and therefore less protection.) Over 400 bee species also make use of the habitat at Pinnacles, many of them solitary. Salamanders, butterflies, toads, and many other birds also make their home at Pinnacles. The park offers visitors such recreational activities as hiking, camping, rock climbing, and caving.

Designating a national park does not happen overnight. Theodore Roosevelt established Pinnacles as a national monument in 1908, because of its caves and unique rock formations. The Civilian Conservation Corps established trails in the area and guided visitors to the caves in the 1930s. The area was designated a national park by President Barack Obama in January 2013.

Congaree National Park, established in 2003, was first designated as a national monument in 1976. Congaree is also an International Biosphere Reserve, one of about 500 international areas of significant ecological importance; this designation by the United Nations Educational, Scientific, and Cultural Organization suggests that the host country for this reserve should make strong efforts to preserve it.

Great Sand Dunes National Park and Preserve in Colorado was formerly a national monument that was redesignated a national park in 2004. The tallest sand dunes in North America are located here, on the eastern side of the San Luis Valley. Some dunes are as tall as 230 m (750 ft) above the valley floor. They formed as rain and wind eroded the surrounding mountains and prevailing winds pushed the eroded sand into dunes. As the winds blow, the dunes continually change shape, but they remain more or less in the same position. For recreation, the park offers nature walks, short tours, hiking, and camping, including limited camping on the dunes themselves.

National parks preserve the land as much as possible so that future generations can appreciate and enjoy the beauty of fast-disappearing natural areas. Though many national parks can be crowded in peak seasons, even a short hike leads visitors to great beauty and solitude in these vast and impressive landscapes.

In Your Own Backyard
What national park is closest to where you live? Check online to find out what activities, species, and landscapes are available there.

Land Use

Most of Earth's land area has a low density of humans. These sparsely populated **nonurban** or **rural** lands include forests, grasslands, deserts, and wetlands. Most people living in rural areas have jobs directly connected with natural resources, such as farming or logging. The many **ecosystem services** performed by rural lands enable the majority of humans to live in concentrated urban environments. Maintaining parcels of undisturbed land adjacent to agricultural and urban areas provides vital ecosystem services such as wildlife habitat, flood and erosion control, and groundwater recharge. Undisturbed land breaks down pollutants and recycles wastes. Natural environments provide homes for organisms. One of the best ways to maintain biological diversity and to protect endangered and threatened species is by preserving or restoring the natural areas to which these organisms are adapted. (Table 5.1 summarizes some important ecosystem services.)

Scientists use undisturbed rural lands as a benchmark, or point of reference, to determine the impacts of human activity. Geologists, zoologists, botanists, ecologists, and soil scientists are some of the scientists who use undisturbed rural lands for scientific inquiry. These areas provide ideal settings for educational experiences in science as well as history because they demonstrate the way the land was when humans originally settled here.

Unspoiled natural areas are important for their recreational value, providing places for hiking, camping, swimming, boating, rafting, bird watching, sport hunting, and fishing. Wild areas are important to the human spirit. Forest-covered mountains, rolling prairies, barren deserts, and other undeveloped areas are aesthetically pleasing and also help us recover from the stresses of urban and suburban living. We can escape the tensions of the civilized world by retreating, even temporarily, to the solitude of natural areas.

WORLD LAND USE

The extent of human land use is increasingly having global impacts. Currently, humans use an estimated 4% of the world's total land area for cities and 36% for agriculture—that is, for raising crops and livestock (**Figure 17.1**). Another 30% of the land surface consists of rock, ice, tundra, and desert—areas considered unsuitable for long-term human use. This leaves 30% of the land surface as natural ecosystems, such as forests and oceans, that could potentially be developed for human purposes. As we have discussed, these natural ecosystems provide many valuable ecosystem services important to human survival. Yet urban areas, croplands, and pastures have expanded worldwide in recent decades in response to the need to support human populations.

LAND USE IN THE UNITED STATES

About 55% of the land in the United States is privately owned by citizens, corporations, and nonprofit organizations, and about

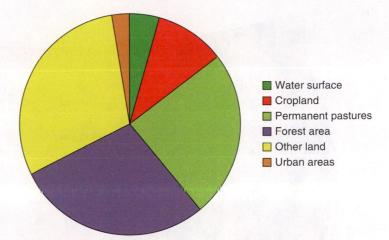

Figure 17.1 **World land use.** Note that 3.5% of the world's total land is used for cities, and about 36% for agriculture (cropland and pasture land). (WorldStat, Columbia University Socioeconomic Data and Applications Center)

Legend:
- Water surface
- Cropland
- Permanent pastures
- Forest area
- Other land
- Urban areas

3% by Native American tribes. The rest is owned by the federal government (about 35% of U.S. land) and by state and local governments (about 7% of U.S. land). Government-owned land encompasses all types of ecosystems, from tundra to desert, and includes land that contains important resources such as minerals and fossil fuels, land that possesses historical or cultural significance, and land that provides critical biological habitat. Most federally owned land is in Alaska and 11 western states (**Figure 17.2**). It is managed primarily by four agencies, three in the U.S. Department of the Interior—the Bureau of Land Management (BLM), the Fish and Wildlife Service (FWS), and the National Park Service (NPS)—and one in the U.S. Department of Agriculture—the U.S. Forest Service (USFS) (**Table 17.1**).

Managing Private Land The American landscape is rapidly changing as it is converted to urban, suburban, and rural residential development. In 2007 the National Resources Inventory of the U.S. Department of Agriculture and the National Resources Conservation Service projected that an area larger than the size of New England could be converted from undeveloped land to residential development by the 2030s. Significant population growth, increasing numbers of retirees, and the consumption of more land per capita are all factors in this projected increase.

Many environmental concerns converge in the matter of land use. Pollution, population issues, preservation of our biological resources, mineral and energy requirements, and production of food are all tied to land use. Economic factors often dominate land use decisions, including the way privately owned land is taxed. Sometimes forest or agricultural land located near urban and suburban areas is taxed as potential urban land. Because of the higher taxes on this land, its owners fall under greater pressure to sell it, which ultimately hastens its development. However, if such land is taxed as forest or farmland, the lower taxes are an incentive for owners to hold onto the land and maintain it in its undeveloped condition. Thus, economic factors largely control land use.

Public Planning of Land Use People in many cities may live surrounded by high-rises and factories or by tree-lined streets

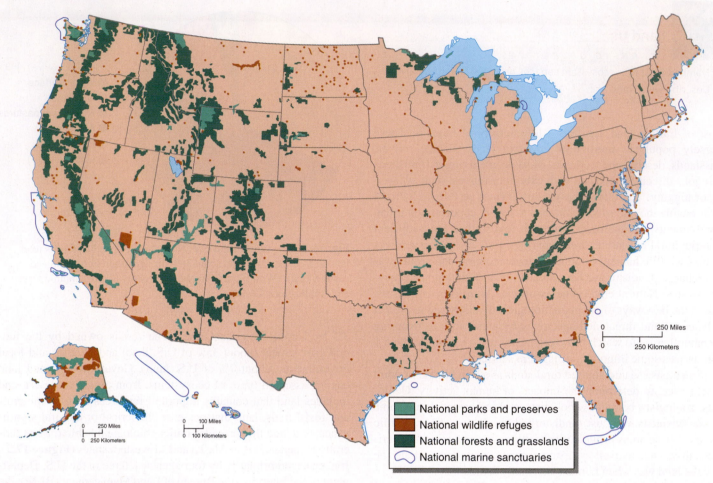

Figure 17.2 **Selected federal lands.** Shown are national parks and preserves, national wildlife refuges, national forests and grasslands, and national marine sanctuaries in the United States. Note the preponderance of federal lands in western states and Alaska. Other federal lands, such as military installations and research facilities, are not included, though many of these function effectively as wildlife sanctuaries. (National Geographic Maps)

Table 17.1   Administration of Federal Lands

Agency	Land Held	Primary Uses	Area in Millions of Hectares (Acres)
Bureau of Land Management (Dept. of Interior)	National resource lands	Mining, livestock grazing, oil and natural gas extraction	102 (253)
U.S. Forest Service (Dept. of Agriculture)	National forests	Logging, recreation, conservation of watersheds, wildlife habitat, mining, livestock grazing, oil and natural gas extraction	78 (193)
U.S. Fish and Wildlife Service (Dept. of Interior)	National wildlife refuges	Wildlife habitat; also logging, hunting, fishing, mining, livestock grazing, oil and natural gas extraction	38 (93)
National Park Service (Dept. of Interior)	National Park System	Recreation, wildlife habitat	34 (84)
Other—includes Department of Defense, Corps of Engineers (Dept. of the Army), and Bureau of Reclamation (Dept. of Interior)	Remaining federal lands	Military uses, wildlife habitat	23 (57)

Source: U.S. Department of Interior, U.S. Department of Agriculture, and U.S. Department of Defense

interspersed with open parkland. Regardless of the municipal land uses, city officials probably have a land use plan that includes zoning. However, land use plans seldom take into account all aspects of land as a resource both before and after development. The philosophy of most land use plans is that development is good because it increases the tax base, even though the revenue from these taxes is usually consumed providing services to the developed area.

Land use decisions are complex because they have multiple effects. If a tract of land is developed for housing, then roads, sewage lines, hospitals, and schools must be built nearby to accommodate the influx of people. Services are also required, such as new restaurants and shopping areas, which take up more land.

Ideally, public planning of land use should take into account all repercussions of the proposed land use, not just its immediate effects. It helps to begin with an inventory of the land, including its soil type, topography, types of organisms, endangered or threatened species, and historic or archaeological sites. By doing such an inventory, a public planning commission attempts to understand the value of the land as it currently exists, as well as its potential value after any proposed change.

In addition to providing people with open space for recreation and mental health, undeveloped land provides ecosystem services that must be recognized. These benefits should be compared with the possible economic benefits of development. In the long term, the best use of land may not be the use that provides immediate economic gain.

If the land will ultimately be developed, a well-designed development plan will be comprehensive. It will indicate which areas will remain open space, which will remain agricultural, and which will be zoned for high-, medium-, and low-density housing. (Chapter 9 contains additional discussion of land use patterns and zoning.)

REVIEW

1. What are ecosystem services?
2. What percentage of world land do urban areas occupy? How much is used for agriculture?

range in size from tiny islands to portions of national parks (42% of wilderness areas are in national parks), national forests (33% of wilderness areas), and national wildlife refuges (22% of wilderness areas). The Big Gum Swamp wilderness in Florida is only 5528 hectares (13,660 acres), whereas the Selway-Bitterroot Wilderness in Idaho is more than 530,000 hectares (1.3 million acres). Areas designated as wilderness are given the highest protection of any public lands (**Figure 17.3**). Wilderness areas are located within other types of federal lands, and the agencies that manage those lands also manage the wilderness areas. Thus, four government agencies—the NPS, USFS, FWS, and BLM—oversee the 758 wilderness areas that encompass 44.1 million hectares (109 million acres) of land.

Although mountains are the most common land safeguarded by this system, representative examples of other ecosystems have also been set aside, including tundra, desert, and wetlands. More than one-half of the lands in the National Wilderness Preservation System lie in Alaska, and western states contain much of the remaining lands. Because few sites untouched by humans exist in the eastern states, requirements were modified in 1975 so that the wilderness designation could be applied to certain federally owned lands where forests are recovering from logging.

Millions of people visit U.S. wilderness areas each year, and some areas are overwhelmed by this traffic: eroded trails, soil and water pollution, air pollution from cars, litter and trash, and human congestion predominate over quiet, unspoiled land. Government agencies now restrict the number of people allowed into each wilderness area at one time so that human use does not seriously affect the wilderness. Some of the most popular wilderness areas may require more intensive future management, such as building trails, outhouses, cabins, and campsites. These amenities are not encountered in true wilderness, posing a dilemma between wilderness preservation and human use and enjoyment of wild lands.

Limiting the number of human guests in a wilderness area does not control all factors that threaten wilderness. **Invasive species** that

Wilderness, Parks, and Wildlife Refuges

LEARNING OBJECTIVES

- Describe the following federal lands and current issues of concern for each: wilderness, national parks, and national wildlife refuges.

wilderness A protected area of land in which no human development is permitted.

Wilderness encompasses regions where the land and its community of organisms are not greatly disturbed by human activities and where humans visit but do not permanently inhabit. The **Wilderness Act** of 1964 authorized the U.S. government to set aside federally owned land that retains its primeval character and lacks permanent improvements or human habitation as part of the **National Wilderness Preservation System**. These federal lands

Theodore Clutter / Science Source

Figure 17.3 **The Selway-Bitterroot Wilderness area in Idaho.**

become established in wilderness have the potential to upset the balance among native species. The white pine blister rust, a foreign fungus that kills white pine trees, has invaded the wilderness in the northern Rocky Mountains. Wilderness managers are concerned that declining white pine populations could harm the population of grizzly bears in the region. (Pine seeds are a major part of the grizzlies' diet.) The Wilderness Act specifies both the preservation of natural conditions and the avoidance of intentional ecological management. In this example, the only way to preserve as much as possible of the original wilderness may be to intentionally manipulate the white pine population by breeding and planting fungus-resistant trees.

Large tracts of wilderness, many in Alaska, have been added to the National Wilderness Preservation System since passage of the Wilderness Act in 1964. People who view wilderness as a non-renewable resource support the designation of wilderness areas. Increasing the amount of federal land in the National Wilderness Preservation System is opposed by groups who operate businesses on public lands (such as timber, mining, ranching, and energy companies) and by their political representatives.

NATIONAL PARKS

In 1872 Congress established the world's first national park, Yellowstone National Park, in federal lands in the territories of Montana and Wyoming. The purpose of the park was to protect this land of great scenic beauty and biological diversity in an unimpaired condition for present and future generations. The **National Park System (NPS)** was originally composed of such large, scenic areas in the West as the Grand Canyon, Yosemite Valley, and Yellowstone (**Figure 17.4**). Today the National Park System has more cultural and historical sites—battlefields and historically important buildings and towns—than places of scenic wilderness. Additions to the NPS are made through acts of Congress, although the president has the authority to establish national monuments on federally owned lands.

Figure 17.4 Morning Glory Pool at Yellowstone. Boardwalks at Yellowstone protect hot springs from visitors, and protect visitors from hot springs. Yellowstone National Park has abundant geologic activity, in addition to its wolves and other wildlife.

The NPS was created in 1916 as a new federal bureau in the Department of the Interior and given the responsibility of administering the national parks and monuments. The NPS currently administers 397 sites, 59 of which are national parks. The total acreage administered by the NPS is 34.2 million hectares (84.4 million acres).

One of the primary roles of NPS is to teach people about the natural environment, management of natural resources, and history of a site by providing nature walks and guided tours of its parks. Exhibits along roads and trails, evening campfire programs, museum displays, and lectures are other educational tools. The popularity and success of U.S. national parks have led many other nations to follow the example of the United States in establishing national parks. Today, the UN Environment Programme identifies more than 3500 national parks, as defined by the International Union for the Conservation of Nation (IUCN), in nearly 100 countries. As in the United States, these parks usually have multiple roles, from providing biological habitat to facilitating human recreation.

Threats to U.S Parks All the problems plaguing urban areas are found in popular national parks during peak seasonal use, including crime; vandalism; litter; traffic jams; and pollution of the soil, water, and air. In addition, thousands of resource violations, from cutting live trees and collecting plants, minerals, and fossils to defacing historical structures with graffiti and setting fires, are investigated in national parks each year. Park managers have had to reduce visitor access to environmentally fragile park areas degraded from overuse.

Facilities at some of the largest, most popular parks, such as Yosemite, the Grand Canyon, and Yellowstone, were last upgraded some 30 years ago. Entrance fees account for about $132 million of the $2.2 billion a year the NPS spends. Although steps were taken to make parks more self-sufficient, they still depend on general tax revenues to pay for their operations.

Some national parks have imbalances in wildlife populations that involve declining populations of many species of mammals, including bears, white-tailed jackrabbits, and red foxes. Grizzly bears in national parks in the western United States are threatened. Grizzly bears require large areas of wilderness, and the human presence in national parks may adversely affect them. Many parks are too small to support grizzlies. Fortunately, so far grizzly bears have survived in sustainable numbers in Alaska and Canada.

Other mammal populations, notably elk, have proliferated. Elk in Yellowstone National Park's northern range increased from a population of 3100 in 1968 to a record high of 19,000 in 1994; populations have since dropped to roughly 4000 in 2012. Ecologists have documented that the elk reduce the abundance of native vegetation, such as willow and aspen, and seriously eroded stream banks. The reintroduction of wolves to Yellowstone, which began in 1995, is helping to reduce and redistribute the elk population (see the Chapter 5 introduction), and particularly keeping elk from overgrazing stream banks.

Pollution does not respect park boundaries, and parks are increasingly becoming *islands* of natural habitat surrounded by human development. Development on the borders of national parks limits the areas in which wild animals may range, forcing them into isolated populations. Ecologists have found that when environmental stressors occur, several small "island" populations are more likely to become threatened than a single large population occupying a sizable range (see Chapter 16).

Natural Regulation A park management policy called **natural regulation** was introduced in Yellowstone National Park and many other U.S. national parks in 1968. Under natural regulation, the population of the elk herd in Yellowstone is allowed to fluctuate naturally because of varying weather conditions and predator populations such as grizzlies and wolves. Park managers do not try to maintain the herd at a consistent level by culling or artificially propagating elk. Because fires are an integral part of the Yellowstone ecosystem, wildfires in the park are not suppressed unless they threaten people or buildings. Park managers may intervene in other situations, such as to control invasive species or to restore native species (such as wolves).

natural regulation
A park management policy that involves letting nature take its course most of the time, with corrective actions undertaken as needed to adjust for changes caused by pervasive human activities.

WILDLIFE REFUGES

The **National Wildlife Refuge System**, established in 1903 by President Theodore Roosevelt, is the most extensive network of lands and waters committed to wildlife habitat in the world. The National Wildlife Refuge System contains 560 refuges, with at least one in each of the 50 states, and encompasses 60.7 million hectares (150 million acres) of land. The various refuges represent all major ecosystems found in the United States, from tundra to temperate rain forest to desert, and are home to some of North America's most endangered species, such as the whooping crane. The mission of the National Wildlife Refuge System, which the FWS administers, is to preserve lands and waters for the conservation of fishes, wildlife, and plants of the United States. Wildlife-dependent activities, such as hunting, fishing, wildlife observation, photography, and environmental education, are permitted on parts of some wildlife refuges as long as they are compatible with scientific principles of fish and wildlife management.

REVIEW

1. What is the U.S. National Wilderness Preservation System? Which state has the most wilderness land?

2. What are some current concerns of the National Park System?

3. What is the purpose of the National Wildlife Refuge System?

Forests

LEARNING OBJECTIVES

- Define *sustainable forestry* and explain how monocultures are related to forestry.
- Describe national forests and current issues of concern.
- Define *deforestation*, including clear-cutting, and describe the main causes of tropical deforestation.
- Explain how conservation easements help private landowners protect forests from development.

Forests, important ecosystems that provide many goods and services to support human society, occupy about one-fourth of Earth's total land area. Timber harvested from forests is used for fuel, construction materials, and paper products. Forests supply nuts, mushrooms, fruits, and medicines and provide employment for millions of people worldwide. They also offer recreation and spiritual sustenance in an increasingly crowded world.

Forests provide a variety of beneficial ecosystem services. Forests influence local and regional climate conditions. If you walk into a forest on a hot summer day, you will notice that the air is cooler and moister than it is outside the forest. This is the result of a biological cooling process called *transpiration*, in which water from the soil is absorbed by roots, transported through plants, and then evaporated from their leaves and stems. Transpiration provides moisture for clouds, eventually resulting in precipitation (**Figure 17.5**). Thus, forests help maintain local and regional precipitation.

Forests play an essential role in regulating global biogeochemical cycles, such as those for carbon and nitrogen. Photosynthesis

Up to 75% water recycled by transpiration and evaporation in a forest

In a forest 25% or more of water seeps into ground or runs off to rivers, streams, and lakes

Figure 17.5 Role of forests in the hydrologic cycle. Forests return most (up to 75%) of the water that falls as precipitation to the atmosphere by transpiration. When an area is deforested, almost 100% of precipitation is lost as runoff.

by Earth's approximately 1 trillion canopy trees removes large quantities of heat-trapping carbon dioxide from the atmosphere and fixes it into carbon compounds. According to the Millennium Ecosystem Assessment, approximately as much carbon as is in the atmosphere is stored in trees. Forests thus act as carbon "sinks" that help mitigate global climate change. At the same time, oxygen, which almost all organisms require for cellular respiration, is released into the atmosphere.

Tree roots hold vast tracts of soil in place, reducing erosion and mudslides (the 2014 mudslide in Oso, Washington, mentioned in Chapter 14 was likely caused by deforestation). Forests protect watersheds because they pull deep water into upper soil layers; they absorb, hold, and slowly release water. This moderation of water flow provides a more regulated flow of water downstream, even during dry periods, and helps control floods and droughts. Forest soils remove impurities from water, improving its quality. In addition, forests provide a variety of essential habitats for many organisms, such as mammals, reptiles, amphibians, fishes, insects, lichens and fungi, mosses, ferns, conifers, and numerous kinds of flowering plants.

FOREST MANAGEMENT

When forests are managed for timber production, their species composition and other characteristics are altered from their natural condition. Specific varieties of commercially important trees are planted, and those trees not as commercially desirable are thinned out or removed. *Traditional forest management* often results in low-diversity forests. In the southeastern United States, many tree plantations of young pine that are grown for timber and paper production are **monocultures** (**Figure 17.6**; see also Chapter 6). Tree plantations, however, have the potential to benefit remaining natural forests, provided that remaining forests are conserved and protected and that the plantations themselves do not replace natural forests.

monoculture
Ecological simplification in which only one type of plant is cultivated over a large area.

In recognition of the many ecosystem services performed by natural forests, many foresters now practice **sustainable forestry**, or **ecologically sustainable forest management**. Sustainable forestry maintains a mix of forest trees, by age and species, rather than a monoculture. This broader approach seeks to conserve forests for the long-term commercial harvest of timber and nontimber forest products. Sustainable forestry also attempts to sustain biological diversity by providing improved habitats for a variety of species, to prevent soil erosion and improve soil conditions, and to preserve watersheds that produce clean water. Effective sustainable forest management involves cooperation among environmentalists; loggers; farmers; indigenous people; and local, state, and federal governments.

sustainable forestry
The use and management of forest ecosystems in a way that meets the needs of the current generation without compromising the ability of future generations to use the forests.

When logging adheres to sustainable forestry principles, unlogged areas are set aside as sanctuaries for organisms, along with **habitat corridors**—protected zones that connect isolated unlogged or undeveloped areas (see Figure 16.16). The purpose of habitat corridors is to provide escape routes should they be needed

Figure 17.6   Tree plantation. This intensively managed pine plantation is a monoculture, with trees of uniform size and age. Such plantations have little biodiversity and provide limited habitat, but they supplement the harvesting of trees in wild forests to provide the United States with the timber it requires. Photographed in Georgia.

and to allow animals to migrate so that they can maintain healthy, diverse genetic populations.

The actual methods of ecologically sustainable forest management that distinguish it from traditional forest management are gradually being developed. These vary from one forest ecosystem to another, in response to different ecological, cultural, and economic conditions. In Mexico, many sustainable forestry projects involve communities that are economically dependent on forests. Because trees have such long life spans, scientists and forest managers of the future will judge the results of today's efforts.

Harvesting Trees   According to the UN Food and Agriculture Organization (FAO), about 3.4 million cubic meters of wood (for fuelwood, timber, and other products) are harvested annually. The five countries with the greatest tree harvests are the United States, Canada, Russia, Brazil, and China; these countries currently produce more than half the world's timber. About 50% of harvested wood is burned directly as fuelwood or used to make charcoal. (Partially burning wood in a large kiln from which air is excluded converts the wood into charcoal.) Most fuelwood and charcoal are used in developing countries (discussed shortly). Highly developed countries consume more than three-fourths of the remaining 50% of harvested wood for paper and wood products.

Loggers harvest trees in several ways—by selective cutting, shelterwood cutting, seed tree cutting, and clear-cutting (**Figure 17.7**). **Selective cutting**, in which mature trees are cut individually or in small clusters while the rest of the forest remains intact, allows the forest to regenerate naturally. The trees left by selective cutting produce seeds that germinate to fill the void.

(a) In selective cutting, the older, mature trees are selectively harvested from titme to time, and the forest regenerates itsels naturally.

(b) In shelterwood cutting, less desirable and dead trees are harvested. As younger trees mature, they produce seedlings, which continue to grow as the now-mature trees are harvested.

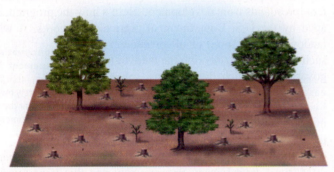

(c) Seed tree cutting nvolves the removal of all but a few trees, which are allowed to remain, providing seeds for natural regeneration.

Calvin Larsen/Science Source

(d) Aerial view of a large patch of clear-cut forest in Washington state. The lines are roads built to haul away the logs.

Figure 17.7   Harvesting trees.

Selective cutting has fewer negative effects on the forest environment than other methods of tree harvest, but it is not as profitable in the short term because timber removal from the forest requires greater amounts of labor.

The removal of all mature trees in an area over an extended period is **shelterwood cutting**. In the first year of harvest, undesirable tree species and dead or diseased trees are removed. The forest is then left alone for perhaps a decade, during which the remaining trees continue to grow and new seedlings become established. During the second harvest, many mature trees are removed, but some of the largest trees are left to shelter the young trees. The forest is then allowed to regenerate on its own for perhaps another decade. A third harvest removes the remaining mature trees, but by this time a healthy stand of younger trees is replacing the mature ones. Little soil erosion occurs with this method of tree removal, even though more trees are removed than in selective cutting.

clear-cutting A logging practice in which all the trees in a stand of forest are cut, leaving just the stumps.

In **seed tree cutting**, almost all trees are harvested from an area; a scattering of desirable trees is left behind to provide seeds for the regeneration of the forest. **Clear-cutting** is harvesting timber by removing all trees from an area. After the trees are removed by clear-cutting, the area is either allowed to reseed and regenerate itself naturally or is planted with one or more specific varieties of trees. Timber companies prefer clear-cutting because it is the most cost-effective way to harvest trees. Clear-cutting destroys biological habitats and increases soil erosion, particularly on sloping land. Sometimes the land is so degraded from clear-cutting that reforestation does not take place; whereas lower elevations are usually regenerated successfully, higher elevations are often difficult to regenerate.

DEFORESTATION

The most serious problem facing the world's forests is **deforestation**. According to *The Global Forest Resources 2010*, compiled by the FAO, forests shrank more than 13 million hectares (32 million acres) *annually* between 2000 and 2010. This amounts to a net 10-year loss equivalent to an area the size of Costa Rica. This estimate of forest loss does not take into account remaining forests that have been thinned or degraded by overharvesting, declining biological diversity, and reduced soil fertility.

deforestation The temporary or permanent clearance of large expanses of forest for agriculture or other uses.

Causes of deforestation include fires caused by drought, mining, land-clearing practices, expansion of agriculture, construction of roads in forests, tree harvests, and insects and diseases. When forests are converted to other land uses, they no longer make valuable contributions to the environment or to the people who depend on them. Forest destruction threatens people whose cultural and physical survival depends on the forests.

Deforestation results in decreased soil fertility through rapid leaching of the essential mineral nutrients found in most forest soils. Uncontrolled soil erosion, particularly on steep deforested slopes, affects the production of hydroelectric power as silt builds up behind dams. Increased sedimentation of waterways caused by soil erosion harms downstream fisheries. In drier areas, deforestation contributes to the formation of deserts (discussed shortly). When a forest is removed, the total amount of surface water that flows into rivers and streams actually increases, sometimes causing floods. Because the forest no longer regulates this water flow, the affected region experiences alternating periods of flood and drought.

Deforestation contributes to the extinction of many species. Many tropical species, in particular, have limited ranges within a forest, so they are especially vulnerable to habitat modification and destruction. Migratory species, including birds and butterflies, suffer from deforestation.

Deforestation contributes to regional climate changes. Trees release substantial amounts of moisture into the air; about 97% of the water that roots absorb from the soil is evaporated directly into the atmosphere. This moisture falls back to Earth in the hydrologic cycle (see Chapter 4). When a large portion of forest is removed, rainfall may decline and droughts may become more common in that region. Studies suggest that the local climate has become warmer and drier in parts of Brazil where huge tracts of the rain forest were burned. As the forest continues to shrink, it is possible that the changing regional climate will no longer be able to support the forest. Thus, large parts of what had once been tropical rain forest could become savanna.

Deforestation also contributes to an increase in global temperature, ocean acidification, and other global changes by releasing carbon originally stored in the trees into the atmosphere as carbon dioxide. Carbon dioxide enables the air to retain heat. The carbon in forests is released immediately if the trees are burned or more slowly when unburned parts decay. If trees are harvested and logs are removed, roughly one-half of the forest carbon remains as dead materials (branches, twigs, roots, and leaves) that decompose, releasing carbon dioxide. When an old-growth forest is harvested, researchers estimate that it takes about 200 years for the replacement forest to accumulate the amount of carbon that was stored in the original forest.

FOREST TRENDS IN THE UNITED STATES

In recent years, most temperate forests in the Rocky Mountains, Great Lakes region, and New England and other eastern states

have been holding steady or even expanding. Expanding forests are the result of *secondary succession* on abandoned farms (see Chapter 5), commercial planting (tree plantations) on both private and public lands, and government protection. Although the returning forests generally do not have the biological diversity of original forests, many forest organisms have successfully become reestablished in regenerated forests.

More than half of U.S. forests are privately owned by individuals and corporations (**Figure 17.8**). Many private owners are under economic pressure, such as high property taxes, to subdivide the land and develop tracts of it for housing or shopping malls (see Meeting the Challenge: Preserving Forests in the Eastern States). Projected conversion of forests to agricultural, urban, and suburban lands over the next 40 years will probably have the greatest impact in the South, where more than 85% of forest is privately owned and logging is largely unregulated.

The **Forest Legacy Program** is a provision of the 1990 **Farm Bill** that helps private landowners protect environmentally important forestlands from development. Here's how the program, managed by the U.S. Department of Agriculture, works. A willing landowner sells some or all ownership rights, such as the right to develop the land, to the U.S. government, which then holds a **conservation easement**. The landowner usually continues to live and/or work on the property and can sell those rights not already sold to the government to other private individuals. All future property owners must honor the provisions of the conservation easement. In 2014, the Forest Legacy Program reported that it now protects more than 2.3 million acres of private forests threatened by development.

conservation easement A legal agreement that protects privately owned forest or other property from development for a specified number of years.

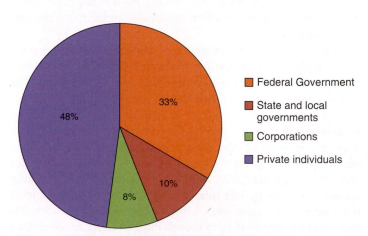

Figure 17.8 **Forest ownership in the United States.** Most forests (56%) are privately owned, either by individuals or corporations. (Data obtained from USDA/US Forest Service, 2006)

Legend:
- Federal Government
- State and local governments
- Corporations
- Private individuals

(33% Federal Government, 10% State and local governments, 8% Corporations, 48% Private individuals)

MEETING THE CHALLENGE

Preserving Forests in the Eastern States

Timber companies own most of the large tracts of privately owned forests in the eastern United States. Many of these temperate deciduous forests are becoming available for purchase because developers are willing to pay huge amounts of money for them. The timber companies find that selling the land is worth far more than they can earn from logging it. After purchasing the forest property, the developers cut down the trees and build residential communities, golf courses, and hunting clubs. Once development has occurred, there is no chance for the forest to ever regenerate.

Conservation organizations are buying some of these properties to preserve the forest. The mission of the Nature Conservancy, which was founded in 1951, is to "preserve the plants, animals, and natural communities that represent the diversity of life on Earth by protecting the lands and waters they need to survive." The Nature Conservancy has about 1 million members who contribute funds to purchase land and aquatic ecosystems. This organization typically collaborates with individual landowners, corporations (such as timber companies), state and federal governments, and indigenous people in a collaborative effort to preserve natural areas. Great Sand Dunes National Park, discussed in the chapter opener, was conserved in large part through the work of the Nature Conservancy.

The Nature Conservancy and other conservation organizations have only enough money to purchase a small fraction (less than 2%) of the eastern forests that will probably be sold in the next few years. However, Nature Conservancy scientists prioritize their "wish list" to select the ecosystems that are most uncommon and that may provide habitat for endangered or threatened species. The lands encompassing those ecosystems are the ones that the Nature Conservancy works to obtain. Frequently, a deal is brokered with the timber companies so they can continue logging the ecologically valuable forests for five years or so. But at the end of the logging cycle, the land is allowed to undergo secondary succession and revert to forest. More important, the land is no longer available for economic development.

The **Nature Conservancy Appalachian Initiative** will protect a large portion of eastern forest that is considered one of the healthiest, most diverse temperate deciduous forests in the world. Saving landscape in the central Appalachians from mountaintop removal coal mining and from development not only preserves biological diversity in the forest but also protects water quality and aquatic organisms in the Chesapeake Bay and its tributaries. (The forest is part of the bay's watershed.) In addition, this initiative will provide safe drinking water for cities such as Baltimore, Harrisburg, Richmond, and Washington, D.C.

U.S. National Forests According to the USFS, the United States has 155 national forests encompassing 78 million hectares (193 million acres) of land, mostly in Alaska and western states. The USFS manages most national forests, with the remainder overseen by the BLM. National forests were established to provide U.S. citizens with the maximum benefits of natural resources such as fish, wildlife, and timber. Multiple uses include timber harvest; livestock forage; water resources and watershed protection; mining; hunting, fishing, and other forms of outdoor recreation; and habitat for fishes and wildlife. Recreation, which increased dramatically in national forests during the 1990s and early 2000s, ranges from camping at designated campsites to backpacking in the wilderness. Visitors swim, boat, picnic, and observe nature in national forests. With so many possible uses of national forests, conflicts inevitably arise, particularly between timber interests and those who wish to preserve the trees for other purposes.

Road building is a contentious issue in national forests, in part because the USFS builds taxpayer-funded roads that allow private logging companies access to the forest to remove timber. The money the USFS charges for timber concessions does not cover the cost for taxpayer-funded roads. Taxpayers also subsidize logging operations on lands the BLM manages. About 697,000 km (433,000 mi) of logging roads have been built throughout U.S. national forests. Road building in national forests is environmentally destructive if improper construction accelerates soil erosion and mudslides (particularly on steep terrain) and causes water pollution in streams. Biologists are concerned that so many roads fragment wildlife habitat and provide entries for disease organisms and invasive species.

Another issue in national forests is clear-cutting. We now examine the clear-cutting debate in the Tongass National Forest, which is located along Alaska's southeastern coast.

CASE IN POINT

TONGASS NATIONAL FOREST

Despite its northern location, the Tongass National Forest is one of the world's few temperate rain forests (**Figure 17.9**; also see Chapter 6 for a description of temperate rain forests). It is one of the wettest places in the United States. This moisture supports old-growth forest of giant Sitka spruce, yellow cedar, and western hemlock, some of which are 700 years old. Stream banks are lined with bent willows, ferns, mosses, and other vegetation. This forest, the largest in the National Forest System, provides habitat for a wealth of wildlife, such as grizzly bears and bald eagles.

About 70% of the old-growth forest in the Tongass has been logged. The Tongass is a prime logging area because a single large Sitka spruce may yield as much as 10,000 board feet of high-quality timber. The logging industry forms the basis of much of the local economy. Regeneration of mature forest after it was clear-cut

Richard Cummins/Getty Images

Figure 17.9   Tongass National Forest. This temperate rain forest is in southeastern Alaska along the Pacific Ocean.

is slow, on the order of several centuries. On Vancouver Island to the south of the Tongass, the trees in temperate rain forest that was clear-cut in 1911 have regrown to only 20% of their original size.

As in most national forests, it is expensive to log in the Tongass. To cover the high costs of operating, timber interests such as pulp mills rely on obtaining the timber from the federal government at below-market prices. This right was granted in 1954 by a contract that expired in the 1990s. In 1990, some members of Congress tried to pass the Tongass Timber Reform Act to force timber interests to pay market prices, but the legislation was bitterly opposed by other members of Congress. The compromise agreement, reached in 1997, provided timber to the mills at market prices. As a result of this legislation, clear-cut logging continued in the Tongass but at lower rates than in the past.

In 1999, the Tongass Land Management Plan of 1997 was modified after several dozen appeals were filed against it. The 1999 decision, called the *Modified 1997 Forest Plan*, protects an additional 100,000 acres of old-growth forest from logging. This brings the total protected area in the Tongass to 234,000 acres. (The total area of the Tongass National Forest is 17 million acres.) The modified plan increases timber harvest rotations from 100 years to 200 years in specially designated wildlife areas. This change reduces the impact of forest fragmentation and protects the Sitka black-tailed deer population, used for food by native tribes. The 1999 decision specifies that road density will be reduced to protect the habitat of wolves, bears, and other wildlife. It reduces the maximum quantity of timber harvested on a sustainable basis from 267 million board feet (the 1997 value) to 187 million board feet. This quantity is considered sufficient to meet the requirements of timber operators in the region.

The USFS officially adopted the controversial *Roadless Area Conservation Rule* in 2000, during the Clinton administration.

A series of lawsuits challenging the rule ensued, initiated mainly by logging, mining, and gas and oil interests. In 2001, a U.S. district judge blocked the roadless rule, and plans proceeded to log several areas of Tongass National Forest that lack roads. In 2005, the George W. Bush administration issued new rules making it easier to build roads in pristine parts of national forests for logging, mining, and other development; these rules put the decision to log in the hands of individual states.

In 2008, a new amendment to the Forest Plan for the Tongass Forest was implemented. Data were collected for five years, and as of 2013, public input regarding the next five-year plan was being gathered. Whether Tongass Forest will be increasingly logged or conserved depends on the input and influence of various interest groups. ◼

TRENDS IN TROPICAL FORESTS

There are two main types of tropical forests: tropical rain forests and tropical dry forests. In places where the climate is warm and moist throughout the year—with about 200 or more cm (at least 79 in.) of precipitation annually—**tropical rain forests** prevail. Tropical rain forests are found in Central and South America, Africa, and Southeast Asia, but almost half of them are in just three countries: Brazil, the Democratic Republic of the Congo, and Indonesia (**Figure 17.10**).

In tropical areas where annual precipitation is less than in tropical rain forests but still enough to support trees, including regions subjected to a wet season and a prolonged dry season, **tropical dry forests** occur. During the dry season, tropical trees shed their leaves and remain dormant, much as temperate trees do during the winter. India, Kenya, Zimbabwe, Egypt, and Brazil are examples of countries that have tropical dry forests.

Most of the remaining undisturbed tropical forests, which lie in the Amazon and Congo River basins of South America and Africa, are being cleared and burned at a rate unprecedented in human history. Tropical forests are also being destroyed at an extremely rapid rate in southern Asia, Indonesia, Central America, and the Philippines.

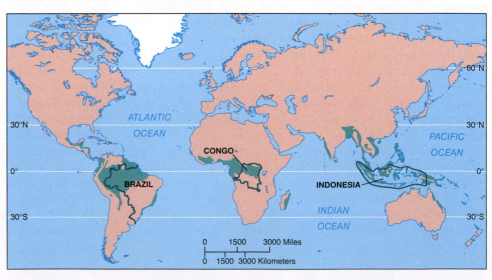

Figure 17.10   Distribution of tropical rain forests. Rain forests (green areas) are located in Central and South America, Africa, and Southeast Asia. Many of the remaining tropical rain forests are highly fragmented.

ENVIRONEWS

Shopping in the Rain Forest

The mostly contiguous Brazilian Amazon rain forest is interrupted by five Amazonian cities (Belém, Manaus, Porto Velho, Macapa, and Rio Branco) whose populations exceed 300,000 people. These city dwellers clamor for the amenities associated with economic development, and the arrival of shopping malls has answered their needs. Like malls in other parts of the world, Amazonian malls have modern architecture, air conditioning, and a variety of local and international stores in which to spend money for consumer goods. So far, the malls have been built on land that was deforested decades ago and so are not contributing to deforestation. However, an ever-expanding network of roads connects the malls to smaller cities and towns throughout the Amazon region, and these roads do accelerate the pace of deforestation. Most environmentalists recognize that shopping malls are part of a global economic trend. Instead of opposing urban shopping malls, many environmentalists are focusing on ways to bring sustainable development to the rural people living in the rain forest.

National Remote Sensing Centre Ltd/Science Source

Figure 17.11 **Satellite photograph of human settlements along a portion of road in Brazil's tropical rain forest.** Numerous smaller roads extend perpendicularly from the main roads. As the farmers settle along the roads, they clear out more forest (dark green) for their croplands and pastures (tans and pinks). How does road building and its accompanying deforestation contribute to climate change?

Why Are Tropical Rain Forests Disappearing?

Several studies show a strong correlation between population growth and deforestation. More people need more food, so the forests are cleared for agricultural expansion. Tropical deforestation is a problem that cannot be attributed simply to population pressures, however. The main causes of deforestation vary from place to place, and a variety of economic, social, and government factors interact to cause deforestation (**Figure 17.11**).

Keeping in mind that tropical deforestation is complex, three agents—subsistence agriculture, commercial logging, and cattle ranching—are considered the most immediate causes of deforestation. Other reasons for the destruction of tropical rain forests include mining and dam construction for hydroelectric power, which inundates large areas of forest.

Subsistence Agriculture. **Subsistence agriculture**, in which a family produces just enough food to feed itself, accounts for more than half of tropical deforestation. In many developing countries where tropical rain forests occur, the majority of people do not own the land they live and work on. In Brazil, 5% of the farmers own 70% of the land. Most subsistence farmers have no place to go except into the forest, which they clear to grow food.

Subsistence farmers often follow loggers' access roads until they find a suitable spot. They first cut down the trees and allow them to dry, then they burn the area and plant crops immediately after burning; this is known as **slash-and-burn agriculture** (discussed further in Chapter 18). The yield from the first crop is often quite high because the nutrients that were in the burned trees are now available in the soil. However, soil productivity declines at a rapid rate, and subsequent crops are poor. In a short time, the people farming the land must move to a new part of the forest and repeat the process. Cattle ranchers often claim the abandoned land for grazing because land not rich enough to support crops can still support livestock.

Slash-and-burn agriculture done on a small scale, with plenty of forest to shift around in so that periods of 20 to 100 years occur between cycles, may be sustainable. The forest regrows rapidly after a few years of farming. But when human population density is high, the land is not allowed to lie uncultivated long enough to recover. Globally, at least 200 million subsistence farmers obtain a living from slash-and-burn agriculture, and the number is growing. Moreover, only half as much forest is available today as was available 50 years ago.

Commercial Logging. Vast tracts of tropical rain forests, particularly in Southeast Asia, are harvested for export abroad. Most tropical countries allow commercial logging to proceed at a much faster rate than is sustainable. Unmanaged tropical deforestation, rather than contributing to economic development, depletes a valuable natural resource faster than it can regenerate for sustainable use.

Cattle Ranching and Agriculture for Export. Some tropical deforestation is carried out to provide open rangeland for cattle (**Figure 17.12**). Cattle ranching, which employs relatively few local people (ranching does not require much labor), is particularly important in Latin America. Some of the beef raised on these ranches, often owned by foreign companies, is exported to highly developed countries, although much is consumed locally. After the forests are cleared, cattle graze on the land for perhaps 20 years, after which time the soil fertility is depleted. When this occurs, shrubby plants, known as **scrub savanna**, may take over the range.

Figure 17.12 **Deforestation in Brazil.** This pasture was a rain forest cleared to graze cattle. One rainforest tree was left in the field—a Brazil nut tree.

Figure 17.13 **Logging in Canada's boreal forest.** About 80% of Canada's forest products are exported to the United States. Photographed in Alberta, Canada.

A considerable portion of forestland is cleared for plantation-style agriculture, which produces export crops such as citrus fruits, palm fruit (for margarine), bananas, and soybeans (see Figure 16.9).

Why Are Tropical Dry Forests Disappearing? Tropical dry forests are being destroyed at a rapid rate, primarily for fuelwood. About half of the wood consumed worldwide is used as heating and cooking fuel by much of the developing world. The unsustainable use of wood has led to a fuelwood crisis in many developing countries. The 2 billion or so people that the FAO says cannot get enough fuelwood to meet basic needs, such as boiling water and cooking, are at risk of waterborne infectious diseases (see Table 21.1 for a list of some human diseases transmitted by polluted water). Often the wood cut for fuel is converted to charcoal, which is then used to power steel, brick, and cement factories. Charcoal production is extremely wasteful: 4.0 metric tons (4.5 tons) of wood produce enough charcoal to fuel an average-sized iron smelter for only six minutes.

BOREAL FORESTS AND DEFORESTATION

Although tropical forests have been depleted extensively, they are not the only forests at risk. Extensive logging of certain boreal forests began in the late 1980s and early 1990s and continues today. Boreal forests occur in Alaska, Canada, Scandinavia, and northern Russia. Coniferous evergreen trees such as spruce, fir, cedar, and hemlock dominate these northern forests. Boreal forests comprise the world's largest biome, covering about 11% of Earth's land area.

Boreal forests, harvested primarily by clear-cut logging, are currently the primary source of the world's industrial wood and wood fiber. The annual loss of boreal forests is estimated to encompass an area twice as large as the Amazonian rain forests of Brazil.

Canada is the world's biggest timber exporter (**Figure 17.13**). About 1 million hectares (2.5 million acres) of Canadian forests are logged annually, and most of Canada's forests are under logging tenures. (*Tenures* are agreements between provinces and

companies that give companies the right to cut timber.) Based on harvest quotas, logging in Canada was widely considered unsustainable until recently. In 2010, logging companies and environmental groups brokered an agreement in which logging was suspended in more than 300,000 km^2 (about 116,000 mi^2) of boreal forest, an area the size of Great Britain. An additional 385,000 km^2 (about 149,000 mi^2) were designated for restricted logging using sustainable guidelines.

Extensive tracts of Siberian forests in Russia are also harvested, although exact estimates are unavailable. Alaska's boreal forests are at risk because the U.S. government may increase logging on public lands (recall the earlier discussion of logging in Alaska's temperate rain forest).

REVIEW

1. What is sustainable forestry?
2. Why is the fact that U.S. national forests were created for multiple uses often a contentious issue?
3. What is deforestation?
4. How do conservation easements protect forests from development?

Rangelands and Agricultural Lands

LEARNING OBJECTIVES

- Describe public rangelands and current issues of concern.
- Define *desertification* and explain its relationship to overgrazing.
- Discuss trends in U.S. agricultural land, such as encroachment of suburban sprawl.

Rangelands are grasslands, in both temperate and tropical climates, that serve as important areas of food production for humans by providing fodder for livestock such as cattle, sheep, and goats (**Figure 17.14**). Rangelands may be mined for minerals and

rangeland Land that is not intensively managed and is used for grazing livestock.

Nancy Gift

Figure 17.14 **Rangeland.** When the carrying capacity of rangeland is not exceeded, it is a renewable resource, useful both for ranchers and wildlife, like this pronghorn antelope. Photographed in Wyoming.

energy resources, used for recreation, and preserved for biological habitat and for soil and water resources. The predominant vegetation of rangelands includes grasses, forbs (small herbaceous plants other than grasses), and shrubs. Rangelands differ from pastures in that pasture plants have been cultivated specifically for grazing by livestock, whereas rangeland plant communities are native, not cultivated, and are naturally suitable for grazing animals.

RANGELAND DEGRADATION AND DESERTIFICATION

Grasses, the predominant vegetation of rangelands, have a *fibrous root system*, in which many roots form a diffuse network in the soil to anchor the plant. Plants with fibrous roots hold the soil in place quite well, thereby reducing soil erosion. Grazing animals eat the leafy shoots of the grass, and the fibrous roots continue to develop, allowing the plants to recover and regrow to their original size.

Carefully managed grazing is beneficial for grasslands. Because the vegetation is adapted to grazing, the removal of mature vegetation by grazing animals stimulates rapid regrowth. At the same time, the hooves of grazing animals moving in groups disturb the soil surface enough to allow rainfall to more effectively reach the root systems of grazing plants and the seeds of new plants. Several studies around the world have reported that moderate levels of grazing encourage greater plant diversity. Rangeland health can be maintained by management of the movement and number of grazing animals.

When the **carrying capacity** (see Chapter 5) of a rangeland is exceeded and recovery times are not sufficient for healthy regrowth after grazing, grasses and other plants are said to be **overgrazed**. When plants die, the exposed soil is susceptible to erosion. Sometimes other plants that can tolerate the depleted soil invade an overgrazed area. In parts of the Texas Hill Country that were overgrazed, junipers (which are not good forage food) replaced the lush grasses. Unpalatable or poisonous plants can successfully invade overgrazed pasture, because these plants remain uneaten by foraging animals.

overgrazing Destruction of vegetation caused by so many animals grazing too close to the soil or repeatedly grazing the plants before complete root and shoot regrowth.

Most of the world's rangelands lie in semiarid areas that have natural extended droughts. During dry periods, the carrying capacity of the rangeland is considerably lower because the lack of precipitation reduces plant productivity. Native grasses in these drylands can survive a severe drought: The aerial portion of the plant dies back, but the extensive root system remains alive and holds the soil in place. When the rains return, the roots help support new shoot growth.

When extended drought results in a need for longer recovery times or smaller numbers of grazers, once-fertile rangeland can convert to desert if stocking patterns are not changed. The reduced grass cover caused by overgrazing allows winds to erode the soil, reducing organic matter and soil seedbanks needed for ecosystem recovery. Even when the rains return, **land degradation** is so extensive that the rangeland may not recover. Water erosion removes the little bit of remaining topsoil, and the sand left behind forms dunes. This progressive degradation, which induces unproductive desertlike conditions on formerly productive rangeland (or tropical dry forest), is called **desertification**. It reduces the agricultural productivity of economically valuable land, forces many organisms out, and threatens endangered species.

land degradation The natural or human-induced process that decreases the future ability of the land to support crops or livestock.

desertification Degradation of once-fertile rangeland, agricultural land, or tropical dry forest into nonproductive desert.

Worldwide, desertification seems to be on the increase. Each year since the mid-1990s, the United Nations estimates that 3560 km² (1374 mi²) of land—an area about the size of Rhode Island—has turned into desert. However, many people are working to reverse this trend. The Green Belt Movement, founded by Wangari Maathai of Kenya, helps people to plant and nurture trees in areas at risk of desertification. The Chipko Movement in India, like the Green Belt Movement, particularly encouraged and empowered women to help prevent deforestation.

We do not know the extent to which desertification results from natural fluctuations in climate versus population pressures and human activities. Currently, many ecologists who study desertification do not think human activities are the main cause of desert encroachment. For example, several years of satellite data support the hypothesis that natural climate variation is the primary factor responsible for the shifting boundary between the southern edge of the Sahara Desert and the African Sahel. Ecologists acknowledge that human activities and overpopulation contribute to degradation in the Sahel but say that fluctuations in climate are causing the desert to advance southward into the Sahel (during drier years) and then retreat (during wetter years). Of course, climate change is also a likely factor, and in this way, desertification of rangelands relates directly to deforestation, since deforestation contributes to climate change.

About 30% of the total human population lives in the Sahel or other rangelands that border deserts. One of the consequences of desertification is a decline in agricultural productivity, and studies have shown that when an area cannot feed its people, they emigrate. For example, more Senegalese live in certain regions of France than in their native villages in Africa. On the other hand, forcing nomadic peoples, such as the Tuareg of the Sahara, to adopt more settled lifestyles may exacerbate desertification by forcing them to reintroduce animals back to a grazed area before plants are again suitable for grazing. According to the UN Environment Programme, about 135 million people worldwide are in danger of displacement as a result of desertification. The World Bank, which estimates that desertification costs $42.3 billion per year in economic losses, is beginning to emphasize programs with sustainable agricultural systems for dryland areas.

RANGELAND TRENDS IN THE UNITED STATES

Rangelands make up approximately 30% of the total land area in the United States and occur mostly in the western states. Of this, approximately two-thirds is privately owned. Much of the private rangeland is under increasing pressure from developers, who want to subdivide the land into lots for houses and condominiums. To preserve the open land, conservation groups often pay ranchers for conservation easements that prevent future owners from developing the land. An estimated 400,000 hectares (1 million acres) of private rangeland are now protected by conservation easements.

Excluding Alaska, there are at least 89 million hectares (220 million acres) of public rangelands. The BLM, guided by the **Taylor Grazing Act** of 1934, the **Federal Land Policy and Management Act** of 1976, and the **Public Rangelands Improvement Act** of 1978, manages most of the approximately 69 million hectares (170 million acres) of public rangelands. The USFS manages an additional 20 million hectares (50 million acres).

Overall, the condition of public rangelands in the United States has slowly improved since the low point of the Dust Bowl in the 1930s (see Chapter 14). Much of this improvement is attributed to fewer livestock being permitted to graze the rangelands after passage of the Taylor Grazing Act in 1934. Better livestock management practices, such as controlling the distribution of animals on the range through fencing or herding, as well as scientific monitoring, have contributed to rangeland recovery. But restoration is slow and costly, and more is needed. Rangeland management includes seeding in places where plant cover is sparse or absent, conducting controlled burns to suppress shrubby plants, constructing fences to allow rotational grazing, controlling invasive weeds, and protecting habitats of endangered species. Most livestock operators today use public rangelands in a way that results in their overall improvement.

Grazing Fees on Public Rangelands The federal government distributes permits to private livestock operators that allow them to use public rangelands for grazing in exchange for a fee. (In 2012 the monthly grazing fee was $1.35 per animal on lands managed by the BLM and USFS. The comparable monthly cost on private land is around $15.) The permits are held for many years and are not open to free-market bidding by the general public—that is, only ranchers who live in the local area are allowed to obtain grazing permits.

Some environmental groups are concerned about the ecological damage caused by overgrazing of public rangelands and want to reduce the number of livestock animals allowed to graze. They want public rangelands managed for other uses, such as biological habitat, recreation, and scenic value, rather than exclusively for livestock grazing. To accomplish this goal, they would like to purchase grazing permits and then set aside the land. At the very least, some groups, like the Center for Biological Diversity, would like to see grazing fees increased to a level closer to the open market rate.

Some economists have joined environmentalists in criticizing the management of federal rangelands. According to policy analysts at Taxpayers for Common Sense, in 2010 taxpayers contributed at least $115 million more than the grazing fees collected on public rangelands. This money is used to manage and maintain the rangelands, including installing water tanks and fences, and to fix the damage from overgrazing. Taxpayers for Common Sense and other free-market groups want grazing fees to cover all costs of maintaining herds on publicly owned rangelands.

AGRICULTURAL LANDS

The United States has more than 120 million hectares (300 million acres) of **prime farmland**, land that has the soil type, growing conditions, and available water to produce food, forage, fiber, and oilseed crops. Certain areas of the country have large amounts of prime farmland. For example, 90% of the *Corn Belt*, a corn-growing region in the Midwest encompassing parts of six states, is considered prime farmland. Not all prime farmland is used to grow crops; approximately one-third contains roads, pastures, rangelands, forests, feedlots, and farm buildings.

Traditionally, farming was a family business. However, larger agribusiness conglomerates that operate less labor intensively are replacing the family farm. As of 2010, there were 2.2 million farms in the United States, as compared to 6.4 million in 1920. In the same period, the average farm size increased from 60 hectares (148 acres) to 169 hectares (418 acres).

Much of prime U.S. agricultural land is falling victim to urbanization and suburban sprawl by being converted into parking lots, housing developments, and shopping malls—the same kinds of development threatening private rangelands. Both natural ecosystems and agricultural lands adjacent to urban areas are being developed (**Figure 17.15**). In certain areas of the United States, loss of rural land is a significant problem. According to the American Farmland Trust, the top five U.S. farm areas threatened

AP Photo/York Daily Record, Bil Bowden

Figure 17.15 **Suburban spread onto agricultural land.** Homes and businesses occupy land that was once cornfields in York County, Pennsylvania.

by population growth and urban/suburban spread are California's Central Valley, south Florida, California's coastal region, the Mid-Atlantic Chesapeake region (Maryland to New Jersey), and the North Carolina Piedmont. More than 160,000 hectares (400,000 acres) of prime U.S. farmland are lost each year.

The 1996 Farm Bill included funding for the establishment of a national *Farmland Protection Program.* (Many states and local jurisdictions also have farmland protection programs.) This voluntary program helps farmers keep their land in agriculture. The farmers sell conservation easements that prevent them or future landowners from converting their land to nonagricultural uses. Such easements can protect farmland in areas where local zoning laws might prioritize development over agricultural uses. The conservation easements are in effect from a minimum of 30 years to forever. As with other conservation easements, the farmers retain full rights to use their property—in this case, for agricultural purposes. Funding of the Farmland Protection Program is subject to annual appropriations from Congress.

REVIEW

1. How do rangelands differ from pasture? What is the carrying capacity of a rangeland?

2. How is desertification related to rangeland management?

3. What human activity is threatening many acres of prime farmland?

Wetlands and Coastal Areas

LEARNING OBJECTIVE

• Describe current threats to freshwater and coastal wetlands.

wetlands Lands that are usually covered by shallow water for at least part of the year and that have characteristic soils and water-tolerant vegetation.

Wetlands are lands transitional between aquatic and terrestrial ecosystems (see Figure 6.16). People used to think the only benefit of wetlands was to provide habitat for migratory waterfowl and other wildlife. In recent years, the many ecosystem services that wetlands perform have been recognized, and estimates of their economic value have increased accordingly. Wetlands recharge groundwater and reduce damage from flooding because they hold excess water when rivers flood their banks. Wetlands improve water quality by trapping and holding nitrates and phosphates from fertilizers, and they help cleanse water that contains sewage, pesticides, and other pollutants. Wetlands provide habitat for many species listed as endangered or threatened; these include up to half of all fish species, one-third of all bird species, and one-sixth of all mammal species listed on the U.S. endangered or threatened species lists. Freshwater wetlands produce many commercially important products, including wild rice, blackberries, cranberries, blueberries, and peat moss. They are sites for fishing, hunting, boating, bird watching, photography, and nature study.

Many human activities threaten wetlands, including drainage for agriculture or mosquito control and dredging for navigation. Other threats include channelization and construction of dams, dikes, or seawalls for flood control; filling in for solid waste

disposal, road building, and residential and industrial development; conversion for aquaculture (fish farming); mining for gravel, phosphate, and fossil fuels; and logging. In the United States, wetlands have been shrinking an estimated 24,300 hectares (60,000 acres) per year since 1985. This represents a slower rate of loss than in previous years. In the contiguous 48 states, more than half of the more than 89.4 million hectares (221 million acres) of wetlands that originally existed during colonial times are gone. Most of the loss since the 1950s has been caused by farmers converting wetlands to cropland. The greatest percentage declines in wetlands have occurred in the agricultural states of Ohio, Indiana, Iowa, and California. The greatest total acreage declines in wetlands have occurred in Florida, Louisiana, and Texas.

The loss of wetlands is legislatively controlled by a section of the 1972 **Clean Water Act**. This legislation, up for renewal since 1997, does a reasonably good job of protecting coastal wetlands but a poor job of protecting inland wetlands—and most wetlands occur inland. The **Emergency Wetlands Resources Act** of 1986 authorized the FWS to inventory and map U.S. wetlands. These maps have many uses, such as planning protection for drinking-water supplies, siting development projects, and making recovery plans for endangered species.

Since 1989, the United States has attempted to prevent any new net loss of wetlands by conserving existing wetlands and restoring some that were lost. Development of wetlands is allowed only if a corresponding amount of previously converted wetlands is restored. However, not all wetland restorations are successful, and there is no routine tracking of compliance. For example, the Sweetwater Marsh was constructed along San Diego Bay in 1984 by the California Department of Transportation, which was legally required to do so when it destroyed a similar marsh during road construction. One of the main purposes of the reconstructed marsh was to provide habitat for the light-footed clapper rail, an endangered species (**Figure 17.16**). The clapper rail never became established in the reconstructed marsh. Ecologists determined that the marsh grass in Sweetwater Marsh was too short for the bird to use for nesting. That species of marsh grass grows significantly taller in natural marshes, where the sediments retain enough nitrogen

Figure 17.16 **Light-footed clapper rail.** This endangered species is found in certain California salt and brackish marshes.

to fertilize it. The reconstructed marsh's sediments, which were dredged from San Diego Bay's shipping channel and obtained from an old urban dump, are too sandy to retain nitrogen.

Thus, the policy of preventing wetlands loss is only partially successful, and wetlands loss continues, though at a slower rate. Two factors complicate the wetlands policy: (1) confusion and dissent about the definition of wetlands, which was not spelled out in the original Clean Water Act, and (2) the question of who owns wetlands. In 1989, a team of government scientists developed a comprehensive, scientifically correct definition of wetlands. (The definition is technical and beyond the scope of this text.) It provoked an outcry from farmers and real estate developers, who perceived it as an economic threat to their property values. Largely in response to their criticisms, politicians attempted to narrow the definition of wetlands several times during the 1990s, removing marginal wetlands that were not as wet as swamps or marshes. This narrower definition ignored decades of wetlands research and excluded about one-half of existing U.S. wetlands from protection.

In 1992, Congress asked the National Research Council to help settle the issue because it had become so controversial. The council published its wetlands study in 1995, urging Congress to put the wetlands debate back on a more scientific footing. The council recommended that because shallow wetlands and intermittent wetlands perform the same ecosystem services as swamps and marshes, they should be regulated by the same principles. The council study generated an even greater debate in Congress, illustrating that more complete scientific information does not necessarily solve problems when stakes are high and values diverge. During the late 1990s and early 2000s, a series of federal court cases involving wetland protection resulted in conflicting decisions. To further complicate the matter, in 2006 the Supreme Court asked the U.S. Army Corp of Engineers to determine which wetlands deserve federal protection under the Clean Water Act. In 2014, the EPA ruled that upstream waters must be protected to support wetland health.

The federal government owns less than 25% of wetlands in the lower 48 states; the remaining 75% is privately owned. This means that private citizens control whether wetlands are protected and preserved or developed and destroyed. Because of the traditional rights of private land ownership in the United States, landowners resent the federal government's telling them what they may or may not do with their properties. Property-rights advocates in Congress side with landowners in thinking the government has overprotected wetlands, though many of these property rights advocates have economic ties to development and mining groups.

Congress authorized the establishment of the *Wetlands Reserve Program* (WRP) under the **Food Security Act** of 1985 and its amendments. The WRP is a voluntary program that seeks to restore and protect privately owned freshwater wetlands that were previously drained, such as those drained for conversion to cropland. Participants are offered financial incentives to restore the wetlands, and they can establish conservation easements to protect them. If a property owner establishes a permanent easement, the WRP pays all restoration costs and an amount equal to the agricultural value of the land. As with other conservation easements, the landowner continues to control access to the land. The Natural Resources Conservation Service administers the WRP, which is funded annually by Congress and is subject to budget cuts.

Buildings within flood zones are subject to higher flood insurance rates. Development and behavior of private landowners may therefore be influenced through flood zone maps, which were redrawn by the *Federal Emergency Management Agency* (FEMA) in 2013. Development may be hindered in flood zone areas, which may help protect many wetlands, both inland and in coastal areas.

COASTLINES

Coastal wetlands, also called saltwater wetlands, provide food and protective habitats for many aquatic animals. They could be considered the ocean's nurseries because so many marine fishes and shellfish spend the first parts of their lives there. In addition to being highly productive, coastal wetlands protect coastlines from erosion, enhance water quality, and reduce damage from hurricanes and other natural events. Intact coastal wetlands act as buffer zones that moderate the effects of tides and storms. Sandy beaches along intact coastal wetlands are alternately washed away and replenished by wave action.

Historically, tidal marshes and other coastal wetlands were regarded primarily as mosquito breeding grounds. Coastal wetlands throughout the world have been drained, filled in, or dredged out to turn them into "productive" endeavors such as industrial parks, housing developments, and marinas. Agriculture contributes to the demise of coastal wetlands, which are drained for their rich soil. Other human endeavors that cause the destruction of coastal wetlands include timber harvesting and fish farming (see Figure 18.19). In the United States, people have belatedly recognized the importance of coastal wetlands in protecting against shoreline erosion, and Congress has passed some legislation to slow their destruction. Many people became more willing to protect coastal wetlands from development (or redevelopment) following Hurricane Sandy in 2012, which devastated many densely populated areas of the eastern United States.

Seawalls and Sand Dunes During the past century, shoreline erosion has increased along many coastlines. Many property owners protect their shoreline by adding retaining walls or rebuilding sand dunes. When seawalls are built parallel to the beach, however, the beach rapidly erodes away, including beach on properties adjacent to the wall. Most legislation protects beaches at the shoreline from seawalls by *rolling easements*, which prioritize public access to the shore over property owners' rights to build walls. Few laws protect shoreline along bays and sounds, and walls are rapidly replacing these tidal shorelines in many areas of the United States. However, sand dunes also show great potential for decreasing coastal flooding (**Figure 17.17**); the force of tides and waves can be absorbed by sand, thereby decreasing the impact inland. Because sea levels have begun to rise due to global climate change and are expected to rise even more rapidly as the 21st century progresses, protection and construction of sand dunes can have an important role in protecting coastal areas.

Coastal Demographics Many coastal areas are overdeveloped, highly polluted, and overfished. Although more than 50 countries have coastal management strategies, their focus is narrow and usually deals only with the economic development of a thin strip of land that directly borders the ocean. Coastal management

Jacques LOIC/Getty Images

Figure 17.17   **Storm damage on a coastal dune.** This boardwalk, on a beach on the Atlantic Coast of France, demonstrates the force of a coastal storm. Without sand dunes, the force of storms is directed inland, often in developed areas.

plans generally do not integrate the management of both land and offshore waters, nor do they take into account the main reason for coastal degradation—human numbers. Perhaps 3.8 billion people—more than half of the world's population—live within 150 km (93 mi) of a coastline. Demographers project that three-fourths of all humans—perhaps as many as 6.0 billion—will live in that area by 2025. Many of the world's largest cities are situated in coastal areas, and these cities are currently growing more rapidly than noncoastal cities.

The United States is not immune to environmental destruction along its coasts. In fact, 14 of the 20 largest U.S. cities and 19 of the country's 20 most densely populated counties lie along coasts. These population statistics do not take into account the additional effects of seasonal visitors to coastal resorts. By 1997, 14% of U.S. coasts were developed, and the percentage is expected to rise to 25% by 2025.

E N V I R O N E W S
O N   C A M P U S
Restoring Damaged Habitats

Regardless of your course of study in college, imagine getting the opportunity to minor in *restoration ecology* and work during your senior year with a student-faculty team to restore a nearby degraded site. Such an opportunity is available to students at the University of Washington (UW), which formed the interdisciplinary UW Restoration Ecology Network in 1999. Each year, faculty in the program meet with potential clients—such as local governments, schools, utilities, or community groups—to select an appropriate site for a year-long capstone project. Students participating in the project must have previously taken courses in restoration ecology and be familiar with how it links to their major field of study. Each student team is assembled from a variety of majors to provide interdisciplinary perspectives. Past projects have ranged from restoring riparian areas (stream banks) to controlling soil erosion to developing bird habitats. The capstone projects allow students to gain some real-world experience with ties to their academic majors.

Coastal management strategies must be developed that take into account projections of climate change, human population growth, and population distribution. Such comprehensive management plans will be difficult to formulate and execute because they must regulate coastal development and prevent resource degradation, both on land and in offshore waters. The key to successful planning is local community involvement. If the public understands the importance of natural coastal areas, people may become committed to the sustainable inhabitation of coastlines.

REVIEW

1. List at least five causes of wetlands destruction.

Conservation of Land Resources

LEARNING OBJECTIVE

- Name at least three of the most endangered ecosystems in the United States.

When wilderness was abundant, people did not observe or appreciate the many ecosystem services provided by forests, prairies, rangelands, and wetlands. Our ancestors considered natural areas as an unlimited resource to exploit. They appreciated prairies as valuable agricultural land and forests as immediate sources of lumber and eventual farmland. This outlook was practical as long as there was more land than people needed. But as the population increased and the amount of available land decreased, it became necessary to consider land as a limited resource. Increasingly, as we have come to understand the great value of ecosystem services, the emphasis has shifted from exploitation to preservation and restoration of the remaining natural areas.

Although all types of ecosystems must be conserved, several are in particular need of protection. The USGS Biological Resources Discipline and the Defenders of Wildlife commissioned studies that developed a numerical ranking of the most endangered ecosystems in the United States. They used four criteria:

1. The area lost or degraded since Europeans colonized North America

2. The number of current examples of a particular ecosystem, or the total area

3. An estimate of the likelihood that a given ecosystem will lose a significant area or be degraded during the next 10 years

4. The number of threatened and endangered species living in that ecosystem

Table 17.2 lists the 10 most endangered U.S. ecosystems based on these criteria. Examples include the south Florida landscape, southern Appalachian spruce-fir forests, and longleaf pine forests and savannas. As these ecosystems are lost and degraded, the organisms that compose them decline in number and in genetic diversity. Researchers have also found that some rare types of soils are threatened. Soils can take hundreds of years to develop but can be eroded very quickly when disrupted, threatening individual species that rely on these specialized soils. Conservation strategies

Table 17.2   The 10 Most Endangered Ecosystems in the United States

Ecosystems (in order of priority)
South Florida landscape
Southern Appalachian spruce-fir forests
Longleaf pine forests and savannas
Eastern grasslands, savannas, and barrens
Northwestern grasslands and savannas
California native grasslands
Coastal communities in lower 48 states and Hawaii
Southwestern riparian communities
Southern California coastal sage scrub
Hawaiian dry forest

Source: R. F. Noss, M. A. O'Connell, and D. D. Murphy. *The Science of Conservation Planning: Habitat Conservation Under the Endangered Species Act*. Island Press: World Wildlife Fund (1997). Reprinted with permission.

that set aside ecosystems are the best way to preserve an area's biodiversity (as well as its soil).

As discussed in this chapter, government agencies, private conservation groups, and private citizens have begun to set aside natural areas for permanent preservation. Such activities ensure that our children and grandchildren will inherit a world with wild places and other natural ecosystems.

REVIEW

1. What are three U.S. ecosystems that need protection?

ENERGY AND CLIMATE CHANGE

Native foods of the Amazonian Rain Forest

During the 21st century, both deforestation and global climate change will interact to affect the tropical rain forests of the Amazon River basin (Amazonia). As economic development proceeds in Brazil and other Amazonian countries, more forest will be converted to rangeland and cropland. This deforestation releases much of the carbon that had been stored in the trees into the atmosphere. Many plant and animal species that have traditionally been used as food sources will no longer grow wild.

REVIEW OF LEARNING OBJECTIVES WITH SELECTED KEY TERMS

- **Name at least four ecosystem services provided by natural areas.**

Natural areas provide many **ecosystem services**, including watershed management, soil erosion protection, climate regulation, and wildlife habitat.

- **Describe world land use.**

Globally, about 40% of the world's land areas are for human use (urban areas and agriculture); 30% are rock, ice, tundra, and desert; and 29% are natural ecosystems such as forests.

- **Describe the following federal lands and current issues of concern for each: wilderness, national parks, and national wildlife refuges.**

Wilderness is a protected area of land in which no human development is permitted. Problems in the **National Wilderness Preservation System** include overuse of some areas and the introduction of **invasive species**. National parks have multiple roles, including recreation and ecosystem preservation; problems include overuse, high operating costs, imbalances in wildlife populations, and development of lands adjoining park boundaries. The **National Wildlife Refuge System** contains habitat for the conservation of fishes, wildlife, and plants of the United States.

- **Define** *sustainable forestry* **and explain how monocultures are related to forestry.**

Sustainable forestry is the use and management of forest ecosystems in an environmentally balanced and enduring way. Sustainable forestry seeks to conserve forests for timber harvest, sustain biological diversity, prevent soil erosion, and preserve watersheds. A **monoculture** represents ecological simplification in which only one type of plant is cultivated over a large area; tree plantations that are monocultures have little species richness and are not a desirable practice for sustainable forestry.

- **Describe national forests and current issues of concern.**

National forests have multiple uses, including timber harvest, livestock forage, water resources and watershed protection, mining, recreation, and habitat for fishes and wildlife. Issues in national forests include confrontations over multiple uses, building logging roads with general tax revenues, and cutting old-growth forest.

- **Define** *deforestation*, **including clear-cutting, and describe the main causes of tropical deforestation.**

Deforestation is the temporary or permanent clearance of large expanses of forest for agriculture or other uses. Deforestation causes decreased soil fertility and increased soil erosion, impairs watershed functioning, contributes to the extinction of species, and contributes to regional and global climate changes. **Clear-cutting** is a logging practice in which all the trees in a stand of forest are cut, leaving just the stumps. **Tropical rain forests** and **tropical dry forests** are cut to provide temporary agricultural land, obtain timber, obtain open rangeland for cattle, and supply firewood and charcoal.

- **Explain how conservation easements help private landowners protect forests from development.**

A **conservation easement** is a legal agreement that protects privately owned forest or other property from development for a specified number of years. In the **Forest Legacy Program**, the landowner grants a conservation easement to the U.S. government.

- **Describe public rangelands and current issues of concern.**

Rangeland is land that is not intensively managed and is used for grazing livestock. Provided the management of grazing animals is balanced with the rangeland's **carrying capacity**, the rangeland remains a renewable resource. About two-thirds of U.S. rangelands are privately owned; the federal government owns much of the remainder. The federal government allows private livestock operators to use public rangelands for grazing; environmental groups want public rangelands managed for other uses, such as wildlife habitat, recreation, and scenic value.

- **Define *desertification* and explain its relationship to overgrazing.**

Desertification is degradation of once-fertile rangeland or tropical dry forest into nonproductive desert. **Overgrazing** is the destruction of vegetation caused by so many grazing animals consuming the plants in a particular area that the land cannot recover. Overgrazing results in barren, exposed soil susceptible to erosion; if this **land degradation** continues, it contributes to desertification.

- **Discuss trends in U.S. agricultural land, such as encroachment of suburban sprawl.**

Agricultural lands are former forests or grasslands that were plowed for cultivation. Increasingly, agribusiness conglomerates farm large tracts of land and operate more cost-effectively than do family farmers on small pieces of land. In certain areas, agricultural lands are threatened by expanding urban and suburban areas.

- **Describe current threats to freshwater and coastal wetlands.**

Wetlands are lands that are usually covered by shallow water for at least part of the year and that have characteristic soils and water-tolerant vegetation. Freshwater wetlands provide habitat for many organisms, purify natural bodies of water, and recharge groundwater. Despite their many ecosystem services, wetlands are often drained or dredged for other purposes, such as conversion to agricultural land. Coastlines include tidal marshes and other coastal wetlands. Ecosystem services provided by coastal wetlands include food and habitat for many wildlife species and protection from coastal erosion and storm damage. Failing to recognize the importance of coastal wetlands has led to their destruction and their development for residential and industrial purposes.

- **Name at least three of the most endangered ecosystems in the United States.**

According to the Biological Resources Discipline and the Defenders of Wildlife, the three most endangered ecosystems in the United States are the south Florida landscape, southern Appalachian spruce-fir forests, and longleaf pine forests and savannas.

Critical Thinking and Review Questions

1. Name at least four ecosystem services provided by nonurban lands. Why is it difficult to assign economic values to many of these benefits?

2. What are the main types of federally owned land in the United States? What uses are permitted on each type of land? What are current issues of concern for each type?

3. Do you think additional federal lands should be added to the wilderness system? Why or why not?

4. Suppose a valley contains a small city surrounded by agricultural land. The valley is encircled by mountain wilderness. Explain why the preservation of the mountain ecosystem would support both urban and agricultural land in the valley.

5. How would you expect a park manager of a national park that adheres to natural regulation to respond to a lightning-induced forest fire? To the establishment of a noxious invasive weed species?

6. How does a healthy forest affect a region's hydrologic cycle? The carbon cycle?

7. Why is deforestation a serious global environmental problem?

8. What are four important causes of tropical deforestation?

9. What are the environmental effects of clear-cutting on steep mountain slopes? On tropical rain forest land?

10. Describe the Forest Legacy Program.

11. Distinguish between rangeland degradation and desertification.

12. Explain how certain tax and zoning laws increase the conversion of prime farmland to urban and suburban development.

13. What is a wetland? Why is the scientific definition of wetlands controversial?

14. What are three ecosystem services that wetlands provide?

15. Describe current wetlands protection policies in the United States and explain their strengths and weaknesses.

16. Should private landowners have control over what they wish to do to their land? How would you as a landowner handle land-use decisions that might affect the public? Present arguments for both sides of this issue.

17. Explain the irony in the following cartoon.

© Harley Schwadron

"WE BELIEVE WE WILL BE ABLE TO TEAR DOWN THE ENTIRE ECOSYSTEM IN AN ENVIRONMENTALLY FRIENDLY MANNER."

Food for Thought

Food plants from the Amazon rain forest include several plants that are familiar to us already: *açaí*, avocado, banana, coconuts, fig, grapes, lemon, mango, oranges, and pineapple. Others, like cupuaçu, a pulpy, white fruit with a fuzzy brown skin, are still unknown outside the region. How might our consumption of Amazon food plants support rainforest peoples? What kinds of problems might result from increased markets for rainforest food products?

Lew Robertson / Getty Images

Harvesting *açaí*.

18

Nancy Gift

Food
Resources

A rooftop garden. Chicago youth learning to grow vegetables on a rooftop garden.

Although human population is growing, the amount of land and water available for food production is not increasing. In fact, land area for food production has decreased, as many urban areas, like Chicago, are built on land ideally suited for agriculture. Urban agriculture can address two challenges simultaneously: the loss of agricultural land, and the decrease in food availability faced by urban residents.

Rooftop gardens, which are a type of *green roof*, have additional benefits for urban areas. Many urban areas experience flooding, since water cannot percolate into soil covered by buildings and pavement. The soil on a green roof holds water temporarily, enabling it to be absorbed by plant roots, and reducing the capacity needed for stormwater storage and management. When food

is grown on a green roof, urban residents can garden in safety away from crime and traffic.

In the past, a far larger fraction of the US population participated in farming. As populations have increased in urban areas, fewer people practice agriculture. Urban agriculture offers opportunities for intensive food production and for teaching about food production, which increases **food access** for urban residents. Food access is the ability to grow, earn, or purchase one's own food. At the Gary Comer Youth Center in Chicago (see photograph), food access is increased for young people by teaching them to grow vegetables on a rooftop garden, and by teaching them to prepare meals in a kitchen inside the building below.

Urban gardens can provide work experience and job opportunities for some of the

teens. Other students, working with nearby universities, learn about soil science. Younger children exposed to gardening are typically more willing to try new foods, particularly fruits and vegetables they grow themselves. At some urban gardens, students learn about marketing by selling excess produce at farm stands. These farm stands are often located in neighborhoods with little other access to farmer's markets or fresh produce.

The production and distribution of food, both locally and globally, is one of our greatest challenges. Agricultural skills need to be practiced in new places and by new growers if we are to meet this challenge. In this chapter, we discuss food security, nutritional quality, agricultural practices and technologies, and sustainable agricultural practices.

In Your Own Backyard...
How much access do you have to nutritious foods? If you eat mainly on campus, do your options include locally raised foods, such as fresh produce and plant-based proteins?

World Food Security

- Explain how famines differ from chronic hunger.
- Define *world grain stocks* and explain how they are a measure of world food security.
- Identify the leading cause of hunger.

The UN Food and Agriculture Organization (FAO) reported in 2013 that 842 million people lack access to the food needed for healthy, productive lives. Most of these malnourished people live in rural areas of the poorest developing countries. However, even in the United States *food insecurity*, the state of living in fear of going hungry, is a widespread problem, with 49.0 million people in food-insecure households.

Worldwide, an estimated 182 million children under the age of five are seriously underweight for their age. People can get enough calories in their diets but still be malnourished because they are not receiving enough specific essential nutrients, such as protein or vitamin A. Adults suffering from malnutrition are more susceptible to disease than those who are well fed. In addition to suffering from poor physical development and increased disease susceptibility, children who are malnourished do not grow normally. Because malnutrition affects cognitive development, malnourished children typically do not perform well in school.

Overeating arises when people consume food energy or nutrients in excess of that required. Overeating does not always deliver an excess of nutrients; often, it results from a diet high in saturated (animal) fats, sugar, and salt, but low in fruits and vegetables. Overeating can result in obesity, high blood pressure, and an increased likelihood of such disorders as diabetes and heart disease. Many human studies show a correlation between diets high in animal fat and red meat and certain kinds of cancer (for example, colon and prostate). While overeating is most common among people in more developed nations, it is also emerging in developing countries, particularly in urban areas where people have adopted Western diets. The emergence of overeating in countries that also have widespread hunger is known as the **nutrition transition**.

Producing enough food to feed the world's people is the largest challenge in agriculture today, and the challenge grows more difficult each year because the human population is continually expanding. As shown in **Figure 18.1**, annual grain production increased from 1961 to 2013. However, the world population increased by just over 4 billion during that period, so the amount of grain per person has not increased appreciably. Also, the amount of available grain per person varies greatly from one country to another. In the United States, it is about 1.2 metric tons (1.3 tons) per person, most of which is fed to livestock, whereas in Zimbabwe, it is about 90 kg (200 lb) per person.

Global food production can be increased in the short term, although whether this increase is sustainable is questionable. Part of the long-term solution to the food supply problem is stabilization of the human population. Most promising in this regard are observations of the human *demographic transition*—the fact that reductions in birth rates seem to follow reductions in death rates by

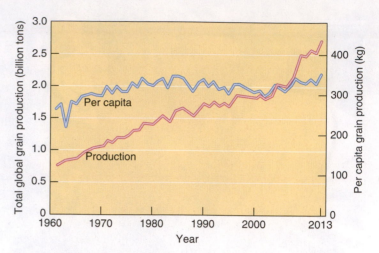

Figure 18.1 **Total world grain production and grain production per person, 1961–2013.** Total world grain production increased from about 0.75 billion tons in 1961 to just over 2.7 billion tons in 2013. The amount of grain produced per person has not changed significantly in the past 50 years. (FAO and other UN data)

a generation or two (see Figure 8.5). The most cost-effective way to reduce birth rates is to educate women and girls; birth rates drop when women have high literacy rates. Therefore, helping people in developing countries get enough to eat, attend school, live in safe environments, and have access to good medical care may be the most promising way to stabilize the human population.

FAMINES

Drought, war, flood, or some other catastrophic event may result in **famine**, a temporary but severe food shortage. Throughout human history, famine has struck one or more regions of the world every few years. Usually, a natural event, such as drought or flood, is accompanied by political instability—an approaching army may seize cropland, for example. The developing nations of Africa and Asia are most at risk. The worst African famine in history, caused in part by widespread drought, occurred from 1983 to 1985. Hardest hit were Ethiopia and Sudan, where 1.5 million people died of starvation. The people living in this region lacked sufficient money to purchase food and did not have stored food reserves to protect them against several years of crop failures.

In the early 1990s, drought and civil unrest in Somalia resulted in famine for an estimated 2 million Somalis. More than 2 million North Koreans died from starvation or hunger-related illnesses during a famine in the late 1990s. The North Korean famine was caused by several years of floods and drought, coupled with a dictatorship that refused to acknowledge the famine or accept aid from other countries. In 2005, drought and an invasion of locusts threatened 3.6 million people in Niger with famine. Between 2010 and 2012, famine in Somalia killed over 200,000 people, including 10% of the country's children under five years of age.

Famines get a great deal of media attention because of the huge and obvious amount of human suffering they cause. However, many more people die each year from endemic malnutrition than from the starvation associated with famine.

MAINTAINING GRAIN STOCKS

food security The goal of all people having access to sufficient, safe, nutritious food at all times.

world grain stocks The amounts of rice, wheat, corn, and other grains stored by governments from previous harvests as a cushion against poor harvests and rising prices.

Food security is the goal of all people having access at all times to adequate amounts and kinds of food needed for healthy, active lives. Although food security is difficult to assess, **world grain stocks,** also called *world grain carryover stocks*, provide us with one measure of food security. World grain stocks have decreased since their all-time highs in the mid-1980s and late-1990s (**Figure 18.2**). The world grain stock in 2010 was estimated at 72 days' supply.

When food is scarce and prices increase, the risk of political instability is a real concern in poor nations, where many people spend most of their income on foods. World grain carryover stocks dropped to 64 days in 2007, and food prices increased in 2008 to historic levels, causing food riots and street demonstrations in more than a dozen countries in Africa, Asia, Central and South America, and the Middle East. Likewise, food prices surged in early 2011, though they declined somewhat in 2012–2014. Some of these price spikes were caused by financial speculation; agricultural economists are researching regulatory policies that would limit the impact of price fluctuations on those least able to afford food.

World grain stocks have dropped for several reasons. Environmental conditions such as rising temperatures, falling water tables, and droughts have caused poor harvests. Many severe weather events have occurred—record heat waves, severe droughts, and numerous wildfires—as Earth's climate changes. Heat waves, in particular, reduce regional grain *yields* (the amount of a food crop produced per unit of land), contributing to price volatility in both domestic and international food markets (as, for example, when policymakers decide to restrict food exports to other countries to provide for local consumption when harvests are bad). As the United States and other countries search for gasoline substitutes to reduce dependency on foreign oil, corn yields will be increasingly diverted to ethanol production (to blend with gasoline) instead of to food and animal feed. World

grain stocks have also dropped because consumption of beef, pork, poultry, and eggs has increased in China and other developing countries, where some people are becoming more affluent and can afford to increase animal protein in their diets. According to Norman Myers, a scientist at Oxford University, if each of China's people were to eat just one extra chicken per year, the grain required for animal feed to raise those chickens would equal all the grain exports of Canada. (Canada is the world's second-largest exporter of grain.)

Over the last few decades, grain has been used increasingly to feed animals rather than people. While grain can be a financially efficient way to feed livestock, grain-fed cattle (as well as sheep and goats) require more energy to produce, most of it from fossil fuels. Also, the meat from grain-fed cattle is higher in saturated fat, which is detrimental to human health. Pasture-fed livestock yield healthier meat and require less fossil fuel energy to produce.

ECONOMICS, POLITICS, AND FOOD SECURITY

The leading cause of hunger and famine in the world is neither lack of total food nor inefficient distribution. The observation that led to Indian economist **Amartya Sen**'s Nobel Prize in economics in 1998 is that the leading cause of famine is the type of government: Democratic governments are more likely to get their people fed in difficult times than are totalitarian regimes. In addition, government inefficiency and bureaucratic red tape add to food problems, sometimes making it difficult to distribute the food to the hungriest people and to ensure that those who need it get it instead of those who do not need it. (Sometimes dishonest government officials, military personnel, or civilians sell food intended for hungry people for personal profit.) Thus, getting food to the people who need it is largely a political problem.

One solution to food problems that is often suggested is for developing countries to shift to more local food production and consumption—that is, their food should be produced in the area where it is consumed. If food production matches market demands in developing countries, people are fed and economic growth occurs as agriculture generates incomes and employment. At the same time, developing countries can participate in **globalization** of the food supply by growing high-value crops, such as spices or coffee, for export to the United States and other more wealthy countries. The right balance of local food production with participation in the global economy may be difficult to achieve, especially if governments pressure citizens to produce exports rather than use agricultural efforts to feed themselves.

globalization The process by which people around the world are increasingly linked through economics, communication, transportation, governance, and culture.

POVERTY AND FOOD

Despite the fact that enough food is currently produced to feed the world's people with an adequate but not generous diet, the harsh reality is that food is not shared equally by all people. The root cause of **food insecurity** is poverty, not

food insecurity a state of fear of going hungry or that one will not be able to acquire sufficient food for oneself and one's family.

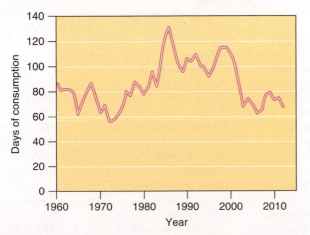

Figure 18.2 World grain stocks, 1960–2012. World grain stocks should be at least 70 days' worth to allow a cushion against poor harvest and to help provide stability in world grain prices. (USDA data from Earth Policy Institute)

lack of available food. Many of the poorest people—those living in developing countries in Asia and Africa—do not own land on which to grow food and do not have sufficient money to purchase food. More than 1.3 billion people in developing countries have incomes so low that they cannot afford to eat enough food or enough of the right kinds of food.

Women in particular are valuable in the global effort to grow and distribute food. In some regions, women's agricultural efforts produce 80% of basic nutrition. When women's wages increase, the money typically goes directly toward buying food for their family. Therefore, increasing pay equity and economic opportunities for women have a strong role in reducing food insecurity and hunger. Women's contributions to overall food production are often undercounted, since food raised by women often feeds their own households, rather than being sold (and counted) in the marketplace.

Worldwide, chronic hunger is more common in rural areas than in urban areas. The lack of rural infrastructure, such as roads, contributes to hunger. Poor rural people living in areas without roads have few employment opportunities to improve their incomes. Studies in China and India have shown that when the government builds roads into rural areas, poverty and chronic hunger are reduced because people use the roads to find jobs with higher wages.

Infants, children, and the elderly are particularly susceptible to poverty and chronic hunger. The greatest *proportion* of chronically hungry people is in sub-Saharan Africa, whereas the largest *number* of hungry people is in Asia (**Figure 18.3**). Poverty and hunger are not restricted to developing countries, however. Poor, hungry people are also found in the United States, Canada, Europe, and Australia.

World food problems are many, as are their solutions. Any effective solution to world food problems should include increasing food production, promoting economic development, and improving food distribution. The ultimate solution to hunger, however, may be ensuring education and financial opportunities for women and small-scale farmers.

REVIEW

1. Which is a more pervasive global issue—famine or chronic hunger? Explain your answer.

2. What are world grain stocks?

3. How do education and political structures affect hunger in a region?

Food Production

LEARNING OBJECTIVES

- Name the three most important food crops and explain why having just three plant species to provide almost half of the calories people consume is a potential problem.
- Contrast industrialized agriculture with subsistence agriculture and describe three kinds of subsistence agriculture.

SOURCES OF FOOD

Biologists have identified more than 330,000 species of plants. Of these, slightly more than 100 provide about 90% of the food that humans consume, either directly or indirectly. (Humans consume foods indirectly when cereal grains are used to feed livestock that humans eat as meat.) Just 15 species of plants provide the bulk of food for humans (**Table 18.1**).

Of these plants, three cereal grains—rice, wheat, and corn—provide about half of the calories that people consume. Our

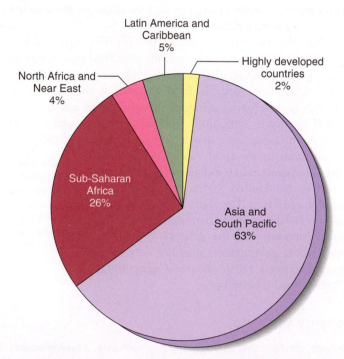

Figure 18.3 **Locations of the world's undernourished people.** Most of the world's undernourished people live in Asia, but sub-Saharan Africa has the highest proportion of chronically hungry people. (Adapted from "The Growing Problem," in the food section of *Nature*, Vol. 466 [July 29, 2010].)

Table 18.1 The 15 Most Important Food Crops in Terms of Production

Plant Crop	Type of Crop	2012 World Production* (1000 tons)
Sugar cane	Sugar plant (stem)	2,020,031
Corn (maize)	Cereal grain	961,289
Rice, paddy	Cereal grain	793,376
Wheat	Cereal grain	739,513
Potatoes	Ground crop (tuber)	402,133
Sugar beet	Sugar plant (root)	297,476
Cassava	Ground crop (root)	289,451
Soybeans	Legume	266,585
Tomatoes	Fruit (annual herb)	178,347
Barley	Cereal grain	146,482
Watermelons	Fruit (vine)	116,153
Sweet potatoes	Ground crop (root)	113,698
Bananas	Fruit (tree)	112,428
Onions, dry	Ground crop (root)	91,328
Apples	Fruit (tree)	84,193

*Based on the 20 highest-producing countries for a specific agricultural commodity.
Source: World production data from FAO.

dependence on so few species of plants for the bulk of our food puts us in a vulnerable position. Should disease or some other factor wipe out one of the important food crops, severe food shortages could occur. Tens of thousands of kinds of plants have been used as sources of food at one time or another, many of which could be developed into important food sources. Animals provide us with foods that are particularly rich in protein, including fishes, shellfish, meat, eggs, milk, and cheese. Cows, sheep, pigs, chickens, turkeys, geese, ducks, goats, and water buffalo are the most important types of about 80 species of livestock. Insects are not considered to be food in the United States, but they serve as an important source of animal protein elsewhere, with 2 billion people worldwide eating insects on a regular basis.

Animal products account for 40% of the calories people consume in highly developed countries but only 5% of the calories consumed in developing countries. **Figure 18.4** compares meat consumption in India, China, Africa, Western Europe, Central America, and the United States. Meat consumption has risen between 1961 and 2012, especially in China, Central America, and Western Europe; rises in meat consumption closely track agricultural greenhouse gas emissions. While in some of the 20 least developed countries, annual meat consumption remains below 10 kg per person, residents of some more developed countries consume 80 kg per person annually.

Though nutritious, livestock is an expensive source of food because animals are inefficient converters of plant food. Of every 100 calories of plant material a cow consumes, it burns off approximately 86 in its normal metabolic functioning. That means humans consume only 14 calories out of 100 (14%) stored in the cow. Meat consumption is high in affluent societies, so large portions of the crops grown in highly developed countries are used to produce livestock animals for human consumption. Almost half of the cereal grains grown in highly developed countries are used to feed livestock (see You Can Make a Difference: Vegetarian Diets).

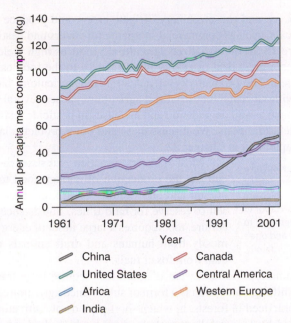

Figure 18.4   Annual meat consumption in selected regions, 1961–2002. North American meat consumption levels are the highest, and have risen somewhat over the 40-year period. In Europe and China, meat consumption levels have risen steeply, and less sharply in Central America. Meat consumption levels in Africa and India have remained relatively constant. (FAO, World Resources Institute)

THE MAIN TYPES OF AGRICULTURE

Agriculture can be roughly divided into two types: industrialized agriculture and subsistence agriculture. Most farmers in highly developed countries and some in developing countries practice **industrialized agriculture**, or *high-input agriculture*. Industrialized agriculture relies on large inputs of capital and energy, in the form of fossil fuels,

> **industrialized agriculture** Modern agricultural methods, which require a large capital input and less land and labor than traditional methods.

YOU CAN MAKE A DIFFERENCE

Vegetarian Diets

Vegetarians do not eat the flesh of any animal, including that of fish and poultry. People embrace vegetarian diets for many reasons. Balanced vegetarian diets provide good nutrition without high levels of saturated fats or cholesterol, both of which cause health problems such as heart disease and obesity. Some studies in the United States indicate that people who are vegetarians live longer, healthier lives than nonvegetarians.

Some people become vegetarians because they are morally or philosophically opposed to killing animals, even for food. Certain religious

groups, notably Hindus and Seventh Day Adventists, exclude animal products from their diets. Other people become vegetarians out of their sense of responsibility for land use and its wide repercussions. In general, fewer plants are required to support vegetarians than to support meat eaters.[1]

---

[1]For people living in marginal lands, such as semiarid grasslands, the soil will not support extensive crop production but will support livestock that consume native plants for forage. In these areas, it is more efficient to grow livestock animals for food than to grow crop plants.

The amount of usable energy in the food chain is decreased approximately 90% by adding an additional level—that is, the animals we eat—to the chain (see the discussion of pyramids of energy in Chapter 3). The actual percentage varies because not all animals are alike in their efficiency at turning food into meat. Simply stated, if everyone ate less meat, more food would be available for human consumption. Some people are reluctant to switch to a vegetarian diet because they fear they will not get enough protein. (Plant foods generally have a lower percentage of protein than do animal foods.) However, consumption of high protein plants can help address this concern.

to produce and run machinery, irrigate crops, and produce agro-chemicals, such as commercial inorganic fertilizers and pesticides (**Figure 18.5**). Industrialized agriculture produces high yields but requires a great deal of energy input relative to the food calories produced. The productivity of industrialized agriculture has not been without costs. Industrialized agriculture causes several problems, such as soil degradation and increases in pesticide resistance in agricultural pests. We discuss these and other problems later in the chapter and in Chapter 22.

subsistence agriculture
Traditional agricultural methods, which are dependent on labor to produce enough food to feed oneself and one's family.

Most farmers practice **subsistence agriculture**, the production of enough food to feed oneself and one's family, with little left over to sell or reserve for hard times. Subsistence agriculture, too, requires a large input of energy, but mostly from humans and draft animals rather than from fossil fuels.

Some types of subsistence agriculture require large tracts of land. Shifting cultivation is a form of subsistence agriculture, typically practiced in forests, in which short periods of cultivation are followed by longer fallow periods, during which the land is left uncultivated and reverts to forest. Slash-and-burn agriculture, one of several distinct types of shifting cultivation, involves clearing small patches of tropical forest to plant crops (see Chapter 17). Slash-and-burn agriculture is land-intensive; because tropical soils lose their productivity quickly when they are cultivated, farmers using slash-and-burn agriculture must move from one area of forest to another every three years or so. Nomadic herding, in which livestock is supported by land too arid for successful crop growth, is another type of land-intensive subsistence agriculture. Nomadic herders must continually move their livestock to find adequate food for them. Shifting cultivation and nomadic herding can be sustainable as long as available land area is sufficient to allow full regrowth between periods of cultivation or grazing.

Intercropping is a form of intensive subsistence agriculture that involves growing a variety of plants simultaneously on the same field. When certain crops are grown together, they produce higher yields than when they are grown as **monocultures**. One reason for higher yields is that different pests are found on each crop, and intercropping discourages the buildup of any single pest species to economically destructive levels. Native Americans practiced intercropping in a system called The Three Sisters when they planted corn, bean, and squash seeds in the same mound of

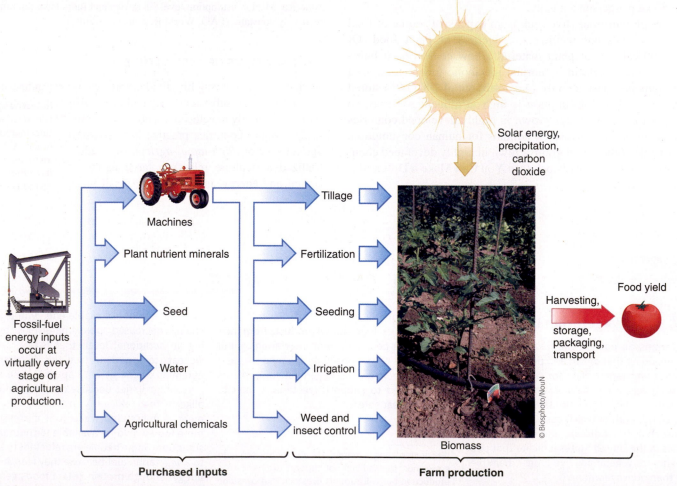

Figure 18.5 **Energy inputs in industrialized agriculture.** How do the huge amounts of fossil fuels used in industrialized agriculture impact the climate? (Adapted from G.H. Heichel, "Agricultural Production and Energy Resources." *American Scientist*, Vol. 64 [January/February 1976]; photo from Biosphoto/NouN/Peter Arnold/Photolibrary.)

soil (**Figure 18.6**). Because the root systems of these plants grow to different depths, they do not compete with one another for water and essential minerals. In addition, the protein-rich bean crop fixes nitrogen that fertilizes the corn and squash plants naturally. **Polyculture** is a type of intercropping in which several kinds of plants that mature at different times are planted together. In polyculture practiced in the tropics, fast- and slow-maturing crops are often planted together so that crops are harvested throughout the year. For example, vegetable crops and cereal grains, which mature first, might be planted with papayas and bananas, which mature later.

In developed countries, many people practice small-scale forms of subsistence agriculture, growing polyculture vegetable gardens and sometimes keeping a few chickens or other small livestock. Generally, these small subsistence agriculture practices do not fully supply a family's food needs. Many small-scale agriculture operations are managed by people who also work at wage-earning jobs (see Case in Point, Grow Appalachia).

REVIEW

1. What are the three most important food crops?
2. What are three differences between industrialized agriculture and subsistence agriculture?

Figure 18.6   The Three Sisters. In this Native American cropping system, each crop has a nutritional and agronomic role. Corn nutritionally provides high energy and a living trellis for beans; beans add protein and provide soil nitrogen; squash provides vitamins and covers the ground, preventing weed growth.

Labels on figure: Beans, Corn, Squash

Challenges of Producing More Crops and Livestock

LEARNING OBJECTIVES

- Contrast the goals in selecting modern agricultural seed with the selection goals for traditional, heirloom varieties.
- Explain the roles of hormones and antibiotics in industrialized livestock production.
- Identify potential benefits and problems of genetic engineering.

Around the world, we find many kinds of agriculture and types of food. Despite this diversity, industrialized agriculture has caused an overall trend toward greater uniformity in the plants and animals we eat.

THE EFFECT OF DOMESTICATION ON GENETIC DIVERSITY

Wild plant and animal populations usually have a lot of **genetic diversity** — that is, variation in their genes, units of hereditary information that specify certain traits. Genetic diversity contributes to a species' long-term survival by providing the variation that enables each population to adapt to changing environmental conditions. During the **domestication** of plants and animals, much of this genetic diversity is lost because the farmer selects for propagation only those plants and animals with the most desirable agricultural characteristics, including taste, yield, and ability to survive transport and storage. Many of the high-yielding crops produced by modern agriculture are genetically uniform, and only a few varieties of vegetable crops are grown in the United States. Similarly, dairy cattle and poultry in the United States have low genetic diversity. The familiar black-and-white Holsteins, known for their ability to produce high volumes of milk, comprise 91% of U.S. dairy cattle, and white leghorn chickens produce almost all of the white eggs consumed in the United States.

genetic diversity Variation in a population's genes—units of hereditary information that specify certain traits.

domestication The process of taming wild animals or adapting wild plants to serve humans; domestication markedly alters the characteristics of the domesticated organisms.

THE GLOBAL DECLINE IN DOMESTICATED PLANT AND ANIMAL VARIETIES

Although domestication contributes in general to less genetic diversity than is found in wild relatives, farmer-breeders around the world have selected for specific traits, developing many local varieties of each domesticated plant and animal. Southern Appalachian gardeners have selected hundreds of varieties of beans and kept these varieties distinct for generations; throughout the Americas, native farmers have selected hundreds of varieties of corn for diverse climates (**Figure 18.7**). A traditional variety, sometimes called heirloom (plants) or heritage (animals), is adapted to the climate where it was bred and contains a unique combination of traits conferred by its unique combination of genes.

Until the 1940s, agricultural yields in various countries, both highly developed and developing, were generally equal. Advances

Figure 18.7 **Genetic diversity in corn.** The variation in corn kernels is evidence of the genetic diversity in the species *Zea mays*. This variation is not evident in the genetically uniform corn grown by modern agricultural methods.

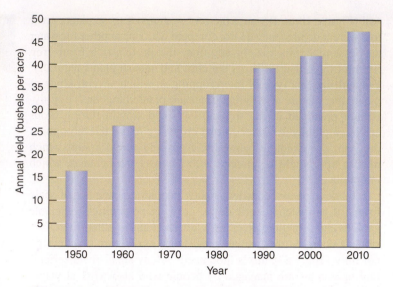

Figure 18.8 **Average U.S. wheat yields, 1950–2010.** Each year shown is actually an average of three years to minimize the effects that poor weather conditions might have in a given year. Similar increases in yield occurred in other grain crops. (USDA)

by research scientists since then have dramatically increased food production in highly developed countries (**Figure 18.8**). Greater knowledge of plant nutrition has resulted in fertilizers that promote high yields. The use of pesticides to control insects, weeds, and disease-causing organisms has improved crop yields. Selective breeding programs have resulted in agricultural plants more suited to mechanical harvesting, such as wheat plants with larger, heavier grain heads (for higher yield). Because of the weight of the heads, other traits were gradually incorporated into wheat, such as shorter, thicker stalks, which prevent the plants from falling over during storms.

These breeding programs were so successful in increasing crop yields during the 1960s that the change was called the **Green Revolution**. During the Green Revolution, agronomists encouraged traditional farmers throughout food insecure regions to replace their old varieties of grain with high-yielding varieties. In traditional agricultural systems, however, newer plant varieties sometimes fail to meet other community needs, such as providing straw for livestock bedding. The Green Revolution appeared to be highly successful at first, but some of its success has been reevaluated as we begin to understand the potential value of some traditional crop varieties.

Currently, many people are working to save heirloom seeds, both by storing and labeling existing seed, and by planting them both for food and to produce more seed. The modern varieties, which are bred for uniformity and maximum production, are generally more susceptible to insect pests and disease and less able to adapt to environmental changes, including climate change. When farmers and gardeners abandon traditional varieties in favor of more modern ones, the former varieties frequently face extinction. This represents a great loss in genetic diversity, because each variety's characteristic combination of genes gives it distinctive nutritional value, size, color, flavor, resistance to disease, and adaptability to different climates and soil types.

To preserve older, more diverse varieties of plants and livestock, many countries collect **germplasm**, which includes seeds, plants, and plant tissues of traditional crop varieties and the sperm and eggs of traditional livestock breeds. The International Plant Genetics Resources Institute in Rome, Italy, is the organization that oversees plant germplasm collections worldwide. National germplasm collections range in size from the U.S. National Plant Germplasm System in Colorado, which holds almost half a million varieties, to the national gene bank in the African country Malawi, which holds about 8,000 varieties of native crops and fruits. (See the section on seed banks in Chapter 16 for more information about plant germplasm collections.)

germplasm Any plant or animal material that may be used in breeding.

Efforts are under way to increase food security in low-income, food-deficient countries. During the 1990s and the first decade of the 2000s, the FAO initiated programs for farmers in 19 nations, most in Africa. Participating farmers are given genetically improved seeds, commercial inorganic fertilizers, and pesticides, and are trained in improved agricultural techniques. These farmers then provide demonstrations to neighboring farmers on several goals: how to increase and diversify food production, reduce water use, control pests, and protect the soil and other natural resources. Food production and environmental conservation goals also drive very different efforts from organizations such as Heifer International and Navdanya (see EnviroNews, Capacity Building). These organizations help farmers manage their land and traditional livestock and seedstocks to their best ability, and give opportunity to pass their skills on to other local farmers.

E N V I R O N E W S

Capacity Building

Part of the solution to problems of world hunger involves capacity building, creating a framework within which people can work toward improvements in health and well-being. Among the most significant hurdles to breaking the cycle of human suffering is empowering local communities to become self-reliant and self-sustaining. For example, *Heifer International* works to provide impoverished communities with livestock for eggs, protein, field labor, and wool. Each family that receives livestock agrees to "pay it forward" by donating newborn livestock to another impoverished family in the community.

Navdanya, which means "nine seeds," is an organization founded by Vandana Shiva in India. Navdanya organizers have helped set up indigenous seed banks, trained farmers in sustainable agriculture, and established fair trade marketing networks. By rejuvenating indigenous agricultural knowledge, Navdanya helps farmers protect both heritage seed stocks and maintain **food sovereignty**, the ability of a group of people to control and manage their own food sources for community benefit.

Many communities engage in capacity building by supporting farmers' markets, which offer opportunities for consumers to purchase a variety of agricultural products and for farmers to hear and see directly from consumers about products. For a community to support a farmers' market, designated public spaces, public transportation, and marketing efforts can help farmers and consumers meet their mutual needs.

Capacity building also includes short-term commitments to alleviating world hunger. One of the UN World Food Programme's missions is to respond to sudden food emergencies. For example, it raised more than $1 million to support food aid for Japan in the wake of the 2011 earthquake and tsunami. The money was raised within 36 hours of the disaster, and coordinated efforts to deliver food aid to Japanese in heavily impacted areas were under way within two days.

With contributions by Keith S. Summerville, Drake University

CASE IN POINT

GROW APPALACHIA

The Appalachian region is a mostly forested, mountainous region rich in a variety of natural resources (coal, forests, water, wildlife) and cultural traditions. Like many regions with fossil fuel-dependent economies, the gap between wealth and poverty is wide. According to the Appalachian Regional Commission, unemployment lingers above 10%; per capita income is nearly 15% lower than the national average; poverty rates exceed 20% of the rural population. The land can be rich and productive in small patches, but due to steep slopes and variable soil drainage, large-scale agriculture is not a suitable practice.

Historically, people in Appalachia relied on home gardens, hunting, and small-scale livestock production, as well as harvesting of forest resources such as chestnuts, hickory nuts, mushrooms, and ginseng. Cash income came from small tobacco fields as well as surplus of these subsistence harvests. However, land losses in the Great Depression coincided with the arrival of chestnut blight, which destroyed a source of food, feed, and income in the region. As farms throughout the United States have grown

larger, small-scale farmers have been pushed to take off-farm jobs. Gradually, gardening, foraging, and other subsistence skills have been lost, affecting household economies and personal health. Reliance on cheap, processed foods has left many Appalachian people with illnesses such as diabetes and heart disease. Appalachia, home to hundreds of heritage seed varieties and a rich diversity of wild, native food plants, contains many food deserts, where fresh, healthy food is difficult to attain at any price.

Grow Appalachia was created in 2009 through funding from **John Paul Dejoria**, co-founder and owner of John Paul Mitchell Systems (JPMS) and Patron Tequila, to address the problem of food security in Appalachia. Tommy Callahan, a colleague of Dejoria's, told Dejoria about his experience growing up in Harlan County, Kentucky, where food insecurity is still pervasive and healthy food is frequently unavailable. Dejoria began cultivating a unique vision for tackling Appalachian food insecurity: supporting gardeners with information, seed, tools, and skill-building (**Figure 18.9**). Gardening provides both healthy food and exercise, as well as an opportunity for supplemental income.

Through Grow Appalachia, Dejoria began collaborating with **Berea College** in Berea, Kentucky, to develop a program that would both meet needs and leverage existing community strengths. Director **David Cooke**—a West Virginia native and a lifelong gardener—answered the call and has been responsible for developing the Grow Appalachia program and its partnerships ever since. In 2010, Grow Appalachia participants at four partner sites in Kentucky grew food for over 2500 people. By 2013, Grow Appalachia had expanded to numerous counties across southern Appalachia. Participants included 19,500 people growing over 1,151,000 pounds of food. The program facilitated jobs in central Appalachia through gardening projects with Grow Appalachia, and participants sold more than $54,000 in produce. In 2013, small livestock (laying hens and honeybees) were offered to program participants. Fresh eggs from just a few hens can add a substantial amount of high-quality protein to a

Courtesy Grow Appalachia

Figure 18.9 Grow Appalachia farmer. Large-scale gardening is a solution to problems of food insecurity and food access. This grower is using a tiller provided by Grow Appalachia to grow food for his family, with an opportunity to sell or share any excess produce.

family's diet, and excess eggs can be sold. The benefits of bees are numerous, and locally produced honey sells at a premium. Through providing the initial infrastructure investment, Grow Appalachia jump-started food production in the region and made rural gardening both possible and profitable. ■

SUPPLEMENTS FOR LIVESTOCK

The use of hormones and antibiotics increases animal growth rates. **Hormones**, usually administered by ear implants, regulate livestock bodily functions and promote faster growth. Although U.S. and Canadian farmers use hormones, the European Union (EU) currently bans all imports of hormone-treated beef because of health concerns for human consumers. For example, girls in developed countries start puberty over a year earlier than girls did in the 1960s. EU regulators cite a few studies suggesting that these hormones or their breakdown products, both found in trace amounts in meat and meat products, could cause cancer or affect the growth of young children. In 1999, an international scientific committee, organized by the FAO and World Health Organization (WHO), examined the hormone issue in detail and concluded that the traces of hormones found in beef are safe because they are quite low compared to normal hormone concentrations in the human body.

Modern agriculture has embraced the routine addition of low doses of **antibiotics** to the feed for pigs, chickens, and cattle. These animals typically gain 4% to 5% more weight than animals that do not receive antibiotics, presumably because they must expend less energy to fight infections. According to the *New England Journal of Medicine*, 40% of the 25,000 tons of antibiotics produced annually in the United States is used in livestock operations, particularly those in which large numbers of animals are confined in small areas. Most of these antibiotics are administered continuously to healthy animals.

Several studies link the widespread use of antibiotics in humans and livestock to the increasing resistance of bacteria to antibiotics. As a result, many antibiotics used in livestock are being phased out of use by the U.S. Food and Drug Administration (FDA). The development of bacterial resistance to antibiotics is an example of evolution. Bacteria are continually evolving, even inside the bodies of human and animal hosts. When an antibiotic is used to treat a bacterial infection, a few bacteria may survive because they are genetically resistant to the antibiotic, and they pass these genes to future generations. As a result, the bacterial population contains a larger percentage of antibiotic-resistant bacteria than before (**Figure 18.10**). The Worldwatch Institute reports that antibiotic resistance has been documented in more than 20 kinds of potentially harmful bacteria (the tuberculosis bacterium is one example), and some bacterial strains are resistant to every antibiotic known, a total of more than 100 drugs.

Because increasing evidence indicates that the use of antibiotics in agriculture reduces their medical effectiveness for humans, in 2003 WHO recommended that routine use of antibiotics in livestock be eliminated. Many European countries have stopped administering low doses of antibiotics for growth promotion in livestock. In the United States, the FDA ruled in 2013 that medically important antibiotics will, in three years, no longer be allowed for use in animal feed, though cooperation with the guidelines by the livestock industry will be voluntary.

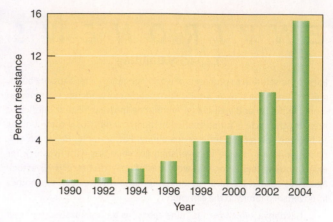

Figure 18.10 **Evolution of antibiotic resistance.** Shown is the increasing resistance of *E. coli* in blood and cerebrospinal infections to the antibiotic ciprofloxacin. (Data from D. Livermore, Health Protection Agency's Resistance Monitoring and Reference Laboratory, United Kingdom.)

GENETIC MODIFICATION

Genetic engineering is a controversial technology that has begun to revolutionize medicine and agriculture. The agricultural goals of genetic engineering are not new. Using traditional breeding methods, farmers and scientists have developed desirable characteristics in crop plants and agricultural animals for centuries. However, these techniques necessitate 15 years or more to incorporate genes for disease resistance into a particular crop plant. Genetic engineering has the potential to accomplish the same goal in a fraction of that time.

> **genetic engineering**
> The manipulation of genes, for example, by taking a specific gene from a cell of one species and placing it into a cell of an unrelated species, where it is expressed.

Genetic engineering differs from traditional breeding methods in that desirable genes from any organism can be used, not just those from the species of the plant or animal being improved. If a gene for disease resistance found in soybeans would be beneficial in tomatoes, the genetic engineer can splice the soybean gene into the tomato plant (**Figure 18.11**). Traditional breeding methods could not do this because soybeans and tomatoes belong to separate groups of plants and do not interbreed.

Genetic engineering may produce **genetically modified (GM)** food plants that will be more nutritious because they will contain all the essential amino acids. Genetic engineering may produce crops rich in beta-carotene, which the body uses to make vitamin A. According to WHO, 250 million of the world's children are at risk of vitamin A deficiency. Vitamin A deficiency causes poor vision, protein deficiency (vitamin A helps the body absorb and use amino acids), and an impaired immune system.

In 2000, an international team of scientists reported that they had successfully engineered rice grains to produce beta-carotene. This so-called golden rice has the potential to improve world health because about half the world's people eat rice as their staple food, and unmodified rice is a poor source of many vitamins, including vitamin A. On the other hand, the best long-term answer to vitamin A deficiency is a more fully balanced diet, with foods naturally rich in beta-carotene.

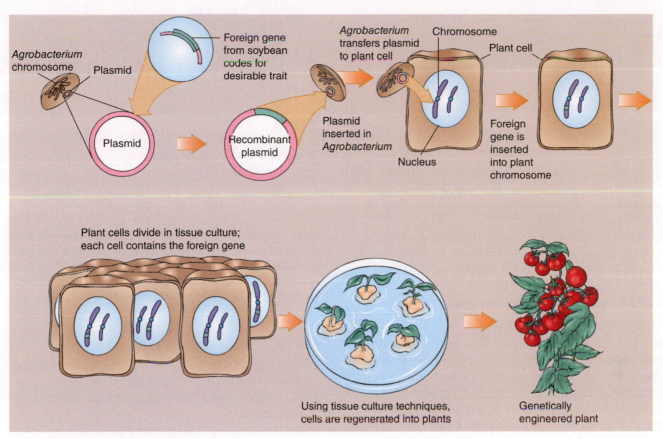

Figure 18.11   Genetic engineering. This example of genetic engineering uses a plasmid, a small circular molecule of DNA (genetic material) found in many bacteria. The plasmid of the bacterium *Agrobacterium* introduces desirable genes from another organism into a plant. After the foreign DNA is spliced into the plasmid, the plasmid is inserted into *Agrobacterium*, which then infects plant cells in culture. The foreign gene is inserted into the plant's chromosome, and genetically modified plants are then produced from the cultured plant cells.

Scientists are developing crop plants resistant to insect pests, viral diseases, heat, cold, herbicides, salty or acidic soils, and drought. For example, they have inserted a gene from a virus into yellow squash and zucchini, thereby making them resistant to viral diseases that yellow their leaves and reduce crop yields. As another example, researchers at the U.S. Department of Agriculture (USDA) identified a gene in rye that codes for a protein that prevents plant roots from absorbing aluminum, a metal more soluble in acidic soils. (Aluminum is a natural component of the inorganic minerals of soils, but it is normally insoluble and is not absorbed by plant roots.) Crops grown in acidic soils often contain toxic levels of aluminum. Acidic soils are widespread in the tropics; in Latin America, 51% of all soils are acidified. Incorporation of an anti-aluminum gene into crop plants such as wheat would allow them to be grown in such soils. Some crops, called pharm crops, produce medicines or vaccines. The range of GM crops is wide and still growing.

Several hundred private genetic engineering firms, as well as thousands of scientists in colleges, universities, and government research labs around the world, are involved in agricultural genetic engineering. Although a great deal of research must be done before we fully understand either the risks or benefits of genetic engineering, GM crops have already transformed agriculture (**Figure 18.12**). The first GM crops were approved for commercial planting in the United States in the early 1990s, and today

the United States is the world's top producer of GM crops. The FDA regulates most of these crops.

A complete analysis of the costs and benefits of *long-term* planting of GM crops remains to be done, in part because more research on the environmental impacts of GM crops is needed. To that end, strict guidelines exist in areas of genetic engineering research that could possibly affect the environment. Much research is currently being conducted to assess the effects of introducing GM crops whose foreign genes might spread to non-GM plants and be incorporated into their genetic makeup. Each GM organism has unique characteristics that might cause an environmental hazard under certain conditions.

The **Cartagena Protocol on Biosafety**, an outgrowth of the 1992 UN Convention on Biological Diversity, lessens the threat of gene transfer from GM organisms to their wild relatives by providing appropriate procedures for the handling and use of GM organisms. The protocol went into effect in 2003. (Although the United States has neither signed nor ratified the protocol, it has participated as an active observer. Other major GM crop exporters, such as Argentina and Canada, also have not ratified the protocol.)

The Backlash Against GM Foods   During the late 1990s and early 2000s, opposition to GM crops increased in many countries in Europe and Africa. The EU has mostly rejected the use of GM crops, with only one GM crop approved as of early 2014. Some opposition

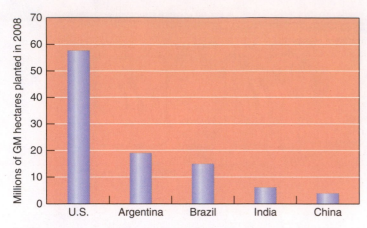

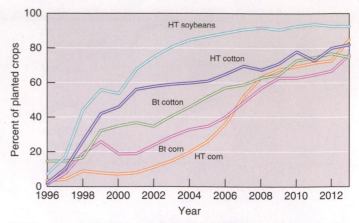

(a) The world's top producers of GM crops. Currently, about half the world's GM crops are planted in the United States. Most GM crops are designed to be herbicide-resistant (63%); another large proportion is insect resistant (18%). (Data from WorldWatch *Vital Signs* 2010.)

Figure 18.12 **Global production of GM crops.**

(b) The proportion of GM crops in the United States has risen sharply since the first GM crops were released in 1996; the vast majority of corn, cotton, and soybean planted in the United States are now GM. HT crops are herbicide tolerant; Bt crops produce insecticide, originally derived from *Bacillus thuringiensis*. (USDA, Economic Research Service)

may have been the result of economic considerations, such as protecting the market for homegrown foods by banning imports, and some opposition was based on scientific concerns. One concern is that the inserted genes could spread in an uncontrolled manner from GM crops to weeds or wild relatives of crop plants (**Figure 18.13**). The spread of genes into natural systems could cause considerable disruption, and the decision to use GM crops should address this risk. Human health risks have not been proven, though animal studies suggest a potential for digestive system impacts from a GM foods diet. Critics also worry that some consumers might develop food allergies to GM foods, although scientists routinely screen new GM crops for allergenicity. Finally, farmers who do not choose to grow GM crops still seek protections for the genetic integrity of their crops against the possibility of pollen transfer.

1. What types of plant traits are likely to be found in heirloom seeds? What traits are likely in modern agricultural seed?

2. Describe two risks of routine hormone and antibiotic use in livestock production.

3. Give one advantage and one risk of GM crops.

The Environmental Impacts of Agriculture

LEARNING OBJECTIVES

• Describe the environmental impacts of industrialized agriculture, including land degradation and habitat fragmentation.

Figure 18.13 **Potential gene crossover between a crop and a weed.**
Sorghum is a drought-tolerant grain crop that freely hybridizes with johnsongrass (*Sorghum halapense*), opening a potential for genetic material from GM sorghum to spread to wild plant populations.

The practices of industrialized agriculture have resulted in several environmental problems that impair the ability of nonagricultural terrestrial and aquatic ecosystems to provide essential **ecosystem services** (see Table 5.1). The environmental effects of industrialized agriculture raise questions about the long-term sustainability of intensive agriculture (**Figure 18.14**).

Industrialized agriculture has an increasing *carbon footprint* that contributes to global climate change. Agriculture produces large amounts of three greenhouse gases—carbon dioxide (from fossil fuels), methane (from animal digestion), and nitrous oxide (from fertilizer use). Food production is the single largest contributor of greenhouse gases (83%) in the food pipeline. Other, much smaller contributors are transportation and packaging. The agricultural use of fossil fuels and pesticides also produces other types of air pollution.

Untreated animal wastes and agricultural chemicals such as fertilizers and pesticides cause water pollution that reduces biological diversity, harm fisheries, and lead to outbreaks of nuisance species. According to the

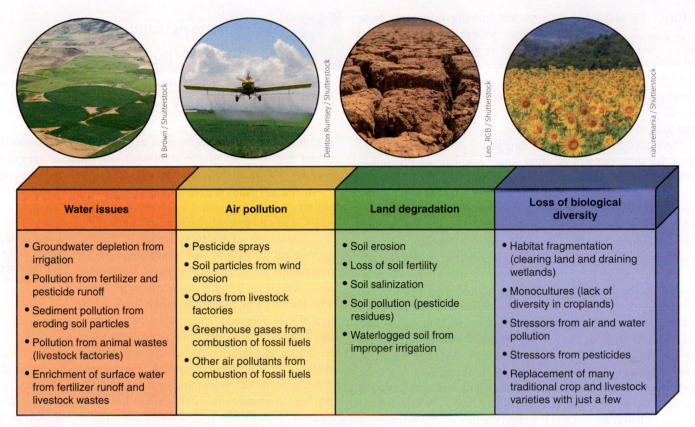

Water issues	Air pollution	Land degradation	Loss of biological diversity
• Groundwater depletion from irrigation • Pollution from fertilizer and pesticide runoff • Sediment pollution from eroding soil particles • Pollution from animal wastes (livestock factories) • Enrichment of surface water from fertilizer runoff and livestock wastes	• Pesticide sprays • Soil particles from wind erosion • Odors from livestock factories • Greenhouse gases from combustion of fossil fuels • Other air pollutants from combustion of fossil fuels	• Soil erosion • Loss of soil fertility • Soil salinization • Soil pollution (pesticide residues) • Waterlogged soil from improper irrigation	• Habitat fragmentation (clearing land and draining wetlands) • Monocultures (lack of diversity in croplands) • Stressors from air and water pollution • Stressors from pesticides • Replacement of many traditional crop and livestock varieties with just a few

Figure 18.14   Selected environmental effects of industrialized agriculture.

Environmental Protection Agency, agricultural practices are the single largest cause of surface-water pollution in the United States. Water pollution from agriculture is particularly significant in Midwest states such as Iowa, Wisconsin, and Illinois. Some of these contaminants flow into the Mississippi River and, from there, into the Gulf of Mexico, which has suffered greatly reduced marine populations as a result. Some agricultural chemicals have been detected in water deep underground, as well as in surface waters. Nitrates from animal wastes and commercial inorganic fertilizers are probably the most widespread groundwater contaminant in agricultural areas.

Industrialized agriculture has favored the replacement of traditional family farms by large agribusiness conglomerates. In the United States, most cattle, hogs, and poultry are now grown in feedlots and livestock factories. In livestock factories, thousands of animals are confined to small pens in buildings the size of football fields. Such large concentrations of animals create many environmental problems, including air and water pollution. At hog factories, manure is often stored in deep lagoons that have the potential to pollute the soil, surface water, and groundwater. This happened during Hurricane Fran in 1996, when manure from 22 large animal waste lagoons in North Carolina spilled onto the floodplain and into streams, causing major fish kills. People living near livestock factories dislike the odor, which often exceeds federal and state guidelines for emissions and causes their property values to decline.

Many insects, weeds, and disease-causing organisms have developed or are developing resistance to pesticides, which forces farmers to apply progressively larger quantities. Residues of pesticides contaminate our food supply and reduce the number and diversity of beneficial microorganisms in the soil. Fishes and other aquatic organisms are sometimes killed by pesticide runoff into lakes, rivers, and estuaries. Farm chemicals also harm amphibian populations (see the Case in Point in Chapter 16).

Land degradation is a reduction in the potential productivity of land (see section on carrying capacity in Chapter 5). Soil erosion, which is exacerbated by large-scale mechanized operations, causes a decline in soil fertility, and the sediments lost by erosion damage water quality. The USDA estimates that about one-fifth of U.S. cropland is vulnerable to soil erosion damage, and soil erosion is an even greater problem in some developing nations. Examples of other types of degradation are compaction of soil by heavy farm machinery and waterlogging and salinization of soil from improper irrigation methods.

land degradation The natural or human-induced process that decreases the future ability of the land to support crops or livestock.

Crop production requires enormous amounts of water. According to the Worldwatch Institute, 1000 tons of water are needed to produce 1 ton of grain. Worldwide, irrigation consumes almost 70% of the total fresh water that humans withdraw from aquifers and surface waters. Some agricultural regions remove water from aquifers faster than it is recharged by precipitation, lowering water tables. The huge Ogallala Aquifer under Nebraska, Kansas, Texas, and other states is one of the best-known examples of overdrawing an aquifer for agriculture (see Figure 13.16). Most of the water in the Ogallala is ancient, left by melting glaciers at the end of the last ice age. Thus, the Ogallala Aquifer is largely a nonrenewable resource. Lack of water increasingly affects agricultural yields in parts of Africa, Australia, and Asia. As a result of mismanagement of water resources, vast areas of irrigated land have become too waterlogged or too salty to grow crops.

habitat fragmentation
The breakup of large areas of habitat into small, isolated patches.

Clearing grasslands and forests and draining wetlands to grow crops have resulted in **habitat fragmentation** that reduces biological diversity (see Chapter 16). Many species have become endangered or threatened as a result of habitat loss caused by agriculture. The most dramatic example of habitat loss in North America is tallgrass prairie, more than 90% of which was converted to agriculture.

The United States had agricultural surpluses during the 1980s, in part because farmers brought large amounts of previously unused land into production. Unfortunately, much of this land was marginal as agricultural land because it was prone to soil erosion caused by intermittent floods or frequent droughts (and therefore wind erosion when the ground cover was removed). Agricultural expansion onto floodplains can also lead to greater flooding problems. Harvesting crops from highly erodible land is ecologically unsound and cannot be done indefinitely. Some of the marginal farmlands in the United States are now retired from use (see the discussion of the Conservation Reserve Program in Chapter 14).

Other countries have paid a high price for cultivating land highly prone to erosion. The former Soviet Union began to cultivate large areas of marginal land during the 1950s. Although the initial production of cereal crops was high, much of this land had to be abandoned by the 1980s. The annual per capita food production in Haiti, a poor Caribbean nation, is half what it was in 1950 as a result of both population growth and lower production on eroded soils.

From the 1960s to the 21st century, the amount of land being irrigated for agriculture was greatly increased. About 70% of the world's total irrigated land is found in Asia, and the amount of irrigated land there continues to expand each year. Since 1995, however, the amount of irrigated land in the rest of the world has remained steady. In Europe and Oceania, the amount of irrigated land has actually decreased. This change is due to the increasing cost of irrigation, the depletion of aquifers, the abandonment of salty soil, and the diversion of irrigation water to residential and industrial uses.

REVIEW

1. What are the major environmental problems associated with industrialized agriculture?

Solutions to Agricultural Problems

LEARNING OBJECTIVES

• Define *sustainable agriculture* and contrast it with industrialized agriculture.

Farming practices and techniques exist that ensure a sustainable output at yields comparable to those of industrialized agriculture. Farmers who practice industrialized agriculture can adopt these alternative agricultural methods, which require less fossil fuel and are less damaging to the environment. Advances are also being made in sustainable subsistence agriculture.

sustainable agriculture
Agricultural methods that maintain soil productivity and a healthy ecological balance while having minimal long-term impacts.

Sustainable agriculture integrates modern agricultural techniques with traditional farming methods from agriculture's past. Sustainable agriculture is modeled after natural ecosystems, with their high biological diversity, biodegradation of materials, and maintenance of soil fertility. To this end, sustainable agriculture relies on beneficial biological processes and environmentally friendly chemicals that disintegrate quickly and do not persist as residues in the environment. A sustainable farm consists of field crops, trees that bear fruits and nuts, small herds of livestock, and even tracts of forest (**Figure 18.15**). Such diversification protects farmers against unexpected changes in the marketplace. Breeding disease-resistant crop plants and maintaining animal health without reliance on antibiotics are important parts of sustainable agriculture. Water and energy conservation are practiced in sustainable agriculture.

Instead of using large quantities of chemical pesticides, sustainable agriculture controls pests by enhancing natural predator-prey relationships. For example, apple growers in Maryland monitor and encourage the presence of ladybird beetles in their orchards because these insects feed voraciously on European red mites, a major pest of apples. As a general rule, sustainable agriculture tries to maintain biological diversity on farms as a way to minimize pest problems. Providing hedgerows (rows of shrubs) between fields provides a habitat for birds and other insect predators.

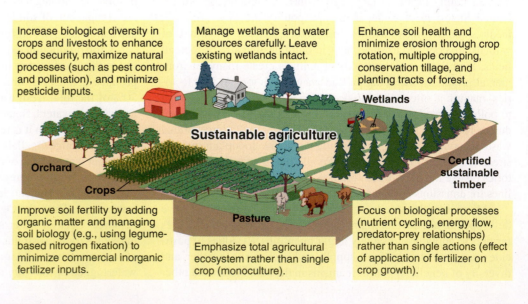

Figure 18.15 **Some goals of sustainable agriculture.** Natural ecosystems provide a model for sustainable agriculture.

Increase biological diversity in crops and livestock to enhance food security, maximize natural processes (such as pest control and pollination), and minimize pesticide inputs.

Manage wetlands and water resources carefully. Leave existing wetlands intact.

Enhance soil health and minimize erosion through crop rotation, multiple cropping, conservation tillage, and planting tracts of forest.

Improve soil fertility by adding organic matter and managing soil biology (e.g., using legume-based nitrogen fixation) to minimize commercial inorganic fertilizer inputs.

Emphasize total agricultural ecosystem rather than single crop (monoculture).

Focus on biological processes (nutrient cycling, energy flow, predator-prey relationships) rather than single actions (effect of application of fertilizer on crop growth).

Orchard • Crops • Wetlands • Certified sustainable timber • Pasture • Sustainable agriculture

Crop selection helps control pests without heavy pesticide use. In parts of Oregon, apples are grown without major pest problems, but insects often infest peaches, whereas in western Colorado, apples have major pest problems but peaches do well. Therefore, apples would be the preferred crop for sustainable agriculture in Oregon, as would peaches in Colorado.

An important goal of sustainable agriculture is to preserve the quality of agricultural soil. Crop rotation, conservation tillage, and contour plowing help control erosion and maintain soil fertility (see Figures 14.16 and 14.17a). Sloping hills converted to mixed-grass pastures erode less than do hills planted with field crops, thereby conserving the soil and supporting livestock.

A combination of manure and crop rotations with legumes is environmentally superior to intensive agricultural methods that use commercial inorganic fertilizers to supply nitrogen. Animal manure added to soil decreases the need for high levels of commercial inorganic fertilizers and cuts costs. Using biological nitrogen fixation (by planting legumes) lessens the need for nitrogen fertilizers.

Sustainable agriculture is not a single program but a series of programs adapted for specific soils, climates, and farming requirements. Some sustainable farmers—those who use a system of **integrated pest management (IPM)**—reduce unnecessary pesticide use through such practices as crop rotation, continual monitoring for potential pest problems, use of disease-resistant varieties, and biological pest controls (see Chapter 22).

Other sustainable farmers practice **organic agriculture** and use no commercial inorganic fertilizers or pesticides. According to guidelines established by the **Organic Food Production Act** in 1990, *organic foods* are crops grown in soil free of commercial inorganic fertilizers and pesticides for at least three years. If the land that the crops are grown on has been inspected, private or state agencies label it *certified organic*, which specifies the highest standards. Cattle and other livestock labeled *USDA organic* are not treated with antibiotics or hormones and are fed organic feed grown without commercial inorganic fertilizers or pesticides. Animals certified *humanely raised and handled* get fresh air and exercise and are not raised in cramped pens. Federal standards for certification of organically grown food went into effect in 2002, replacing standards that varied from state to state. Some farmers use organic practices but do not seek certification due to the costs of the certification process.

In growing recognition of the environmental problems associated with industrialized agriculture, more and more mainstream farmers are trying some methods of sustainable agriculture (see Meeting the Challenge: Certified Environmental Management Systems for Agriculture). These methods cause fewer environmental problems to the agricultural ecosystem, or **agroecosystem**, than industrialized agriculture. Britain's Royal Society describes the trend from intensive techniques that produce high yields to methods that focus on long-term sustainability of the soil as the **second green revolution**. Such a revolution will require changing priorities in agricultural research, for example, to develop crops that can tolerate drought, heat, and pests.

Some researchers are studying ways to make former rainforest land retain its productivity for longer periods than is usual in shifting cultivation. Consider Papua New Guinea, a small island nation in which approximately 80% of the people are subsistence farmers. Research scientists in this country have developed methods to deal with some of the most troublesome problems associated with shifting cultivation: soil erosion, declining fertility, and attacks by insects and diseases. Their research has helped forest plots remain productive for longer periods. Heavy mulching with organic material, such as weed and grass clippings, has lessened soil infertility and erosion. The composted mulch is piled into rows that follow the contours of the land, further reducing erosion. Several crops are planted together, reducing insect damage. One of the crops is always a legume (such as beans), which helps restore nitrogen fertility to the soil. An extension program demonstrates these methods to rural farmers.

REVIEW

1. What are some features of a sustainable farm?

MEETING THE CHALLENGE

Certified Environmental Management Systems for Agriculture

Crop producers, agriculture extension agencies, commodity traders, and environmentalists all affect the management of agriculture. The farming landscape is thus the product of two dynamics: producer-ecological interactions and producer-consumer interactions. Producer-ecological interactions determine variables such as soil health, crop yield, pest damage, and water quality. Producer-consumer interactions determine how much revenue farm products generate and what crops are produced from year to year. Because each region in the United States has different ecological and economic constraints on crop growth, a one-size-fits-all prescriptive approach to farm policy may prove unwieldy as a tool for sustainable agriculture.

An alternative mechanism that is meeting the challenge of sustainable agriculture is the **Certified Environmental Management Systems for Agriculture** (CEMSA) program. Pioneered by the Iowa Soybean Association, a CEMSA is simply a specific application of environmental impact analysis. Businesses use environmental impact analysis to determine how practices are managed with respect to negative and positive environmental impacts. For farmers, a CEMSA is a process inventory of everything a producer consumes or discharges during the production and harvest of crops. This includes energy use, chemical applications for pest and nutrient management, soil and water quality impacts, and wildlife mortality. Farmers that enroll in the CEMSA program use their process inventories to identify inefficiencies and waste, to target best practices to mitigate erosion and sources of pollution, and to adopt alternative land use practices to conserve biological diversity.

The ultimate goal of CEMSA participants is to meet two parallel outcomes—ensure that the farmer produces a profitable yield while simultaneously reducing negative environmental impacts. The CEMSA program is open to all farmers, with levels of commitment evidenced by increasingly stringent management plans that emphasize greater commitment to preserving Iowa's soil, water quality, and nonfarm biodiversity. In essence, the CEMSA program *treats each farm like an ecosystem* and, in doing so, attempts to close nutrient and energy cycles so that the farm becomes self-sustaining.

Contributed by Keith S. Summerville, Drake University

Fisheries of the World

LEARNING OBJECTIVES

- Provide two reasons why fish stocks are overexploited.
- Contrast fishing and aquaculture, and describe the environmental challenges of each activity.

The ocean contains valuable food resources. About 90% of the world's total marine catch is fishes, with clams, oysters, squid, octopus, and other mollusks representing an additional 6% of the total catch. Crustaceans, including lobsters, shrimp, and crabs, make up about 3%, and marine algae constitute the remaining 1%.

Fishes and other seafood are highly nutritious because they contain easily digestible, high-quality protein (protein with a good balance of essential amino acids). Humans obtain approximately 5% of the total protein in their diets from fish and other seafood. In certain areas, particularly in developing countries that border the ocean, seafood makes a much larger contribution to the total protein in the human diet.

Fleets of fishing vessels obtain most of the world's marine catch. In addition, numerous types of fish are captured in shallow coastal waters and inland waters. According to the FAO, the world annual fish harvest increased substantially from 1950 (19.2 million tons) to 2011, when the world fish catch was 99.6 million tons.

PROBLEMS AND CHALLENGES FOR THE FISHING INDUSTRY

No nation lays legal claim to the open ocean. Consequently, resources in the ocean are more susceptible to overuse and degradation than are resources on the land, which individual nations own and for which they feel responsible (see the section on the tragedy of the commons in Chapter 1). Water pollution, another challenge to fishing, is discussed in Chapters 6 and 21.

The most serious problem for marine fisheries is that many marine species have been *overexploited*, to the point that their numbers are severely depleted (**Figure 18.16**). Large predatory fish such as tuna, marlin, and swordfish have declined by 90% since the 1950s, according to Canadian researchers who analyzed data from ocean and coastal regions around the world. Each fish species has a maximum sustainable harvest level (see Chapter 5); if a particular species is overharvested, its numbers drop and harvest is no longer economically feasible.

According to the FAO, 29.9% world's fish stocks are overexploited, with catch levels below their biological and ecological potential. The three areas with the largest number of depleted fish stocks are the eastern and northwestern Atlantic Ocean and the Mediterranean Sea. Fisheries have experienced such pressure for two reasons: (1) the growing human population requires protein in its diet, leading to a greater demand; (2) technological advances in fishing gear have made it possible to fish so efficiently that every single fish is often removed from an area.

Sophisticated fishing equipment includes sonar, radar, computers, airplanes, and even satellites to locate fish schools

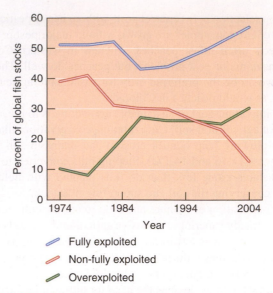

Figure 18.16 **Trends in exploitation of fish stocks, 1974–2009.** In the 1970s, only around 10 percent of global fish stocks were overexploited; but by 2009, nearly a third of fish stocks were overexploited, and less than 15 percent were non-fully exploited. (FAO)

(**Figure 18.17**). Some boats set out **longlines**, fishing lines with thousands of baited hooks; each longline is up to 130 km (80 mi) long. **Purse seine nets** are huge nets, as long as 2 km (more than 1 mi), set out by small powerboats to encircle large schools of tuna and other fishes; after the fishes are completely surrounded, the bottom of the net is closed to trap them. A **trawl bag** is a weighted, funnel-shaped net pulled along the bottom of the ocean to catch bottom-feeding fishes and shrimp; as much as 27 metric tons (30 tons) of fishes, shrimp, and other seafood are caught in a single net. Trawl bags, some of which are large enough to hold 12 Boeing 747s, destroy the ocean floor habitat. **Drift nets** are plastic nets up to 64 km (40 mi) long that entangle thousands of fish and other marine organisms. Although most countries have banned drift nets, they are still used illegally.

Fishermen tend to concentrate on a few fish species with high commercial value, such as menhaden, salmon, tuna, and flounder, while other species, collectively called **bycatch**, are unintentionally caught and then discarded. Bottom trawling and bottom seining in particular were responsible for high rates of bycatch and habitat destruction on the seafloor. However, increased bycatch regulation and improved net technologies have significantly reduced bycatch rates over the last decade.

bycatch Fishes, marine mammals, sea turtles, seabirds, and other animals caught unintentionally in a commercial fishing catch.

In response to overharvesting, many nations have extended their limits of jurisdiction to 320 km (200 mi) offshore. This action removed most fisheries from international use, because more than 90% of the world's fisheries are harvested in relatively shallow waters close to land. This policy was supposed to prevent overharvesting by allowing nations to regulate the amounts and types of fish and other seafood harvested from their waters. However, many countries have a policy of **open management**, in which all

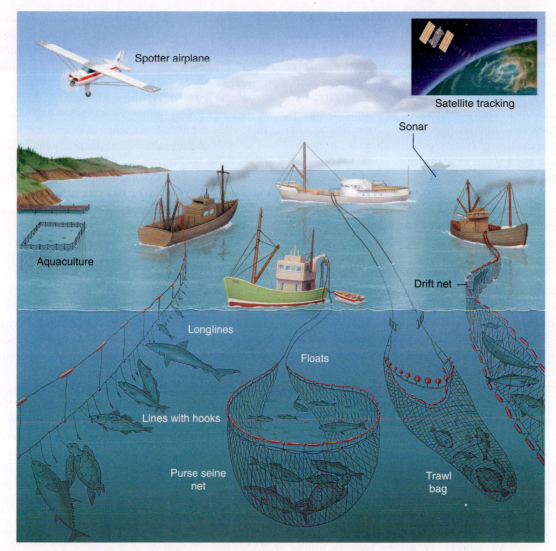

Figure 18.17 **Modern commercial fishing methods.** Modern methods of harvesting fish are so effective that many fish species have become rare. The depth of longlines is adjusted to catch open-water fishes, such as sharks and tuna, or bottom fishes, such as cod and halibut. Purse seines catch anchovies, herring, mackerel, tuna and other fishes that swim near the water's surface. Trawl bags catch cod, flounder, red snapper, scallops, shrimp, and other fishes that live on or near the ocean floor; these have been modified in recent years to do less damage to ocean floor habitat. Drift nets catch salmon, tuna, and other fishes that swim in open waters.

fishing boats of that country are given unrestricted access to fish in national waters.

The **Magnuson Fishery Conservation Act**, which went into effect in 1977, regulates marine fisheries in the United States. This law established eight regional fishery councils, each of which developed a management plan for its region. Until 1996, the act was not particularly successful because managers were often pressured to set quotas too high, and the National Marine Fisheries Service estimated that more than one-third of U.S. fish stocks were being fished at higher levels than could be sustained. In 1996, the act was updated to protect "essential fish habitat" for more than 600 fish species, reduce overfishing, rebuild the populations of overfished species, and minimize bycatch. Fishing quotas, restrictions of certain types of fishing gear, limits on the number of fishing boats, and closure of fisheries during spawning periods are some of the management

tools used to reduce overfishing. The 2007 updates to the law strengthened controls on illegal and unreported fishing in U.S. waters. This act has led to improved sustainable fishing levels on 67% of U.S. fish stocks.

AQUACULTURE: FISH FARMING

Aquaculture is more closely related to agriculture on land than it is to the fishing industry just described. Aquaculture is carried out in both fresh water and marine water; the cultivation of marine organisms is sometimes called **mariculture**. To optimize the quality and productivity of their "crops," aquaculture farmers control the diets, breeding cycles, and environmental conditions of their ponds or enclosures.

aquaculture The growing of aquatic organisms (fish, shellfish, and seaweeds) for human consumption.

Aquaculturists try to reduce pollutants that might harm the organisms they are growing, and they keep them safe from potential predators.

Although aquaculture is an ancient practice that probably originated in China several thousand years ago, its enormous potential to provide food has been appreciated only recently. Aquaculture contributes variety to the diets of people in highly developed countries. Inhabitants of developing countries benefit even more from aquaculture: It may provide them with much-needed protein and even serve as a source of foreign exchange when they export such delicacies as aquaculture-grown shrimp.

According to the FAO, world aquaculture production reached 59.9 million metric tons in 2010 (**Figure 18.18**). Important aquaculture crops include fish, shrimp, seaweeds, oysters, mussels, clams, lobsters, and crabs. Currently, aquaculture is the fastest-growing type of food production in the world, and three out of every five fish destined for human consumption comes from a fish farm. The nation with the largest aquaculture harvest is China, which accounts for about 60% of total world production.

Recently a $900-million-a-year industry, aquaculture in the United States has begun to decline due to price competition from other global aquaculture producers. All striped bass and rainbow trout available at U.S. retail markets are produced by aquaculture, as are more than half the fresh salmon served in the United States. Catfish, tilapia, salmon, shrimp, and oysters are the most important types of seafood grown by aquaculture in the United States.

Aquaculture differs from fishing in several respects. For one thing, although highly developed countries harvest more fishes from the ocean, developing countries produce much more seafood by aquaculture. One reason for this is that developing nations have an abundant supply of cheap labor, a requirement of aquaculture because it is labor-intensive. Another difference between fishing and aquaculture is that the limit on the size of a catch in fishing is the size of the natural population, whereas the limit on aquacultural production is largely the size of the area in which they are grown.

In addition to being carried out inland, aquaculture is practiced in estuaries and in the ocean, both near the shore and offshore. Other uses of coastlines compete with aquaculture for available space. Developing countries that grow shrimp by aquaculture cut down the coastal mangroves that provide so many important environmental benefits (**Figure 18.19**; also see Chapter 6). Many marine fish breed in the tangled roots of mangroves, and an expansion of shrimp farming could contribute to a decline in marine fish populations.

Offshore aquaculture facilities are becoming increasingly common. These "ocean ranches," which increasingly use cutting-edge technologies such as submersible cages with robotic surveillance, may avoid damaging coastlines but often lack pollution-restricting legal oversight. In addition, disease has become a significant factor in offshore aquaculture facilities.

Because so many fish are concentrated in a relatively small area, aquaculture produces wastes that pollute the adjacent water and harm other organisms. Parasites from salmon fish farms, for example, can cause infestations in wild salmon populations. Aquaculture causes a net loss of wild fish because many of the farmed fish are carnivorous. Sea bass, for example, may eat up to 5 kg of wild fish to gain 1 kg of weight. Antibiotics used in aquaculture persist in water, and can lead to increased populations of antibiotic-resistant bacteria.

One of the most important limits on aquaculture's potential is the receptivity of animals to the domestication process itself. Land animals such as cows, pigs, and sheep were domesticated over thousands of years. During this time, there were undoubtedly failed attempts to domesticate other animals, which for one reason or another could not be domesticated. The same is true of aquaculture. Aquatic organisms that are social and do not exhibit territoriality or aggressive behavior are possible candidates for domestication.

REVIEW

1. How does overexploitation of fish stocks lead to reduced food availability?

2. What are some of the harmful environmental effects associated with aquaculture? With fishing?

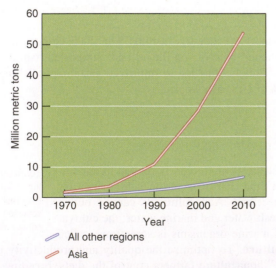

Figure 18.18 **Global aquaculture production 1970–2010.** Aquaculture production has risen globally, with production in Asia far outpacing aquaculture in all other continents combined. (FAO)

Figure 18.19 **Aquaculture in a coastal mangrove forest.** The dikes enclose shrimp ponds. Aquaculture of shrimp is the single largest factor responsible for mangrove habitat losses worldwide. Photographed in Thailand.

REVIEW OF LEARNING OBJECTIVES WITH SELECTED KEY TERMS

• *Explain how famines differ from chronic hunger.*

People who are chronically hungry lack access to the food they need to have healthy, productive lives. Crop failures caused by drought, flood, or some other catastrophe, combined with war or political instability, may result in temporary but severe food shortages called **famines**.

• *Define* world grain stocks *and explain how they are a measure of world food security.*

World grain stocks are the amounts of rice, wheat, corn, and other grains stored by governments from previous harvests as a cushion against poor harvests and rising prices. According to the United Nations, world grain stocks supply a measure of **food security** if they do not fall below the minimum amount of 70 days' supply.

• **Identify the leading cause of hunger.**

The leading cause of hunger is poverty, not food shortage.

• **Name the three most important food crops and explain why having just three plant species to provide almost half of the calories people consume is a potential problem.**

Three cereal grains—rice, wheat, and corn—provide about half of the calories that people consume. Should disease or some other factor wipe out one of these important food crops, severe food shortages could occur.

• **Contrast industrialized agriculture with subsistence agriculture and describe three kinds of subsistence agriculture.**

Industrialized agriculture requires a large capital input (for fossil fuels, equipment, and agricultural chemicals) and less land and human labor than traditional methods. **Subsistence agriculture** depends on labor (human and animal) and a large amount of land to produce enough food to feed oneself and one's family. **Shifting cultivation**, in which short periods of cultivation are followed by long periods of the land lying fallow, and **nomadic herding**, in which herders wander freely over rangelands, are examples of subsistence agriculture. **Polyculture** is a type of intensive subsistence agriculture in which several crops that mature at different times are grown together.

• **Contrast the goals in selecting modern agricultural seed with the goals for selecting traditional, heirloom varieties.**

Traditional seed selection was primarily for taste and resistance to pests, with storability as a lesser concern. Modern agricultural seeds are selected with uniformity and transportability as top priority traits; a variety of other market-driven traits are also considered.

• **Explain the roles of hormones and antibiotics in industrialized livestock production.**

Hormones regulate livestock bodily functions and promote faster growth. The routine addition of low doses of **antibiotics** to the livestock feed causes these animals to gain more weight than animals that do not receive antibiotics, but antibiotic use in livestock also offers opportunities for resistant bacteria to infect people.

• **Identify potential benefits and problems of genetic engineering.**

Genetic engineering may produce food plants that are resistant to insect pests and viral diseases; or tolerant of drought, heat, cold, herbicides, or salty soil. **Genetically modified (GM)** organisms can also cross with wild species and increase insect or weed resistance to pesticides.

• **Describe the environmental impacts of industrialized agriculture, including land degradation and habitat fragmentation.**

Land degradation is the natural or human-induced process that decreases the future ability of the land to support crops or livestock. Pesticides and commercial inorganic fertilizers cause air, water, and soil pollution. Soil erosion causes a decline in soil fertility as well as downstream sediment pollution. Many insects, weeds, and disease-causing organisms have developed resistance to pesticides, forcing farmers to apply larger quantities. Irrigation consumes huge quantities of fresh water. Expanding the amount of agricultural land has resulted in **habitat fragmentation**, which reduces biological diversity.

• **Define** *sustainable agriculture* **and contrast it with industrialized agriculture.**

Sustainable agriculture consists of agricultural methods that maintain soil productivity and a healthy ecological balance while having minimal long-term impacts. Unlike the **monocultures** of industrialized agriculture, the sustainable farm consists of field crops, trees that bear fruits and nuts, small herds of livestock, and tracts of forest. Sustainable agriculture avoids the continual use of antibiotics, large quantities of chemical pesticides, and high levels of commercial inorganic fertilizers.

• **Provide two reasons why fish stocks are overexploited.**

Overexploitation of fish stocks results from two factors: (1) The growing human population requires protein in its diet, leading to a greater demand; (2) technological advances in fishing gear have made it possible to fish so efficiently that every single fish is often removed from an area. Each fish species has a maximum sustainable harvest level; if a particular species is overexploited, its numbers drop, and harvest is no longer economically feasible.

• **Contrast fishing and aquaculture, and describe the environmental challenges of each activity.**

Because of the growing human population and technological advances in fishing gear, fishing has overexploited or depleted many of the world's fish stocks. **Aquaculture** is the growing of aquatic organisms for human consumption. Although aquaculture has an enormous potential to provide food, it causes environmental problems, such as loss of coastlines and water pollution.

Critical Thinking and Review Questions

1. What groups of people (age, gender) are usually most affected by undernutrition and famine? Why?

2. Why does food insecurity occur? What are the links among poverty, food insecurity, and population issues?

3. Distinguish between organic agriculture and sustainable agriculture.

4. What are two environmental problems associated with industrialized agriculture?

5. Describe some traits which consumers and farmers might desire in crops (see photo).

6. Describe the environmental problems associated with farming each of these areas: tropical rain forests, hillsides, semiarid, and arid regions.

7. Give at least three examples of ways that industrialized agriculture could be made more sustainable.

8. What is the problem with open management of ocean fisheries?

9. Explain why aquaculture is more like agriculture than it is like traditional fishing.

10. How does a sustainable agricultural system resemble a natural ecosystem?

11. Climate scientists have observed a feedback loop involving soil moisture: Increasing temperatures dry the soil, and the dried soil exacerbates the heat. Is this an example of a positive feedback loop or a negative feedback loop? Explain your answer.

Courtesy Grow Appalachia

Food for Thought

Investigate the range of food offerings in your campus dining hall. How well represented are fruits, vegetables, carbohydrates, whole grains, and proteins? Notice also the type of food actually selected by students relative to what is offered. Do students tend to select foods based on proportional availability (they take whatever is offered in greatest abundance), nutritional value (they take only the most nutritious foods regardless of abundance), information about the source of the food (local farm? college farm?) or some other criteria? What kinds of food are most often wasted, and left on plates? Does your dining hall compost food waste? What can you do to help make the dining hall food system "sustainable"?

ROMEO GACAD/AFP/Getty Images

Air Pollution

Forest burning in Sumatra, 2013. This fire was one of thousands of fires, many of them intentionally set, in Sumatran rain forests in 2013.

For the past decade, Indonesia, Malaysia, Thailand, and other areas in Southeast Asia have suffered from periods of severe air pollution. Although air pollution in many places is associated with industry and transportation, agriculture and **silviculture** (commercial production of forest products) are often the culprits in Southeast Asia. Farmers set fires to clear forested areas for use as plantations. These fires can burn uncontrolled for days and weeks, creating a smoky haze that covers entire countries and causing extremely hazardous conditions.

Throughout Southeast Asia, serious smoke problems have been an annual problem since 1997. In 2006, two cargo ships collided in a Singapore harbor due to poor visibility. The following year, Malaysia, Singapore, and Indonesia agreed to reduce fires by 75% by 2012 and 90% by 2025, but despite increased law enforcement, the 2012 goal was not met. In 2014, more than 3,000 individual fires were thought to have been intentionally set on the island of Sumatra (**see photo**). These caused respiratory damage to as many as 50,000 people in Indonesia, as well as property and other damage. It is unclear whether the resulting agricultural benefits exceeded the costs associated with the smoke.

Our changing climate—both increased temperature and shifting precipitation—has led to increased smoke from wildfires. In California, wildfires endanger homes and spread smoke over suburban and urban populations. Likewise, extreme temperatures likely attributable to climate change destroyed forests across Russia, where plumes of smoke traveled hundreds of miles and enveloped Moscow in a haze.

Smoke, whether from fireplaces or forests, is hazardous. Small particles settle deep in the lungs, causing infection as well as reduced ability to breathe. Incomplete combustion of carbon-containing materials releases volatile organic chemicals and carbon monoxide, which interfere with oxygen uptake. Smoke can contain toxic amounts of phosphorus, potassium, nitrogen, sulfur, and metals. In addition, fires release substantial amounts of carbon dioxide—the most significant greenhouse gas.

Smoke is even a problem around San Francisco, California, which has some of the strictest air quality regulations in the world. On still winter nights, thousands of fireplaces provide comfort to people in their homes, raising smoke and carbon monoxide levels outside. And in the late summer, legal agriculture-related fires are set throughout California's Central Valley.

One of the earliest recorded environmental laws can be traced to 13th-century London: a ban on burning soft coal that resulted in dense, acidic smoke. Yet London's air quality continued to be notoriously bad for the ensuing seven centuries. Smoke is a major pollutant, from indoor wood smoke in Kenya to fluorine-laden coal smoke in central China to pesticide-laden smoke from California's wildfires.

In this chapter, we explore a range of sources, types, and effects of air pollutants, including smoke. We will see that, while technological improvements have reduced air pollution substantially in some areas, increased fuel use and industry continue to cause environmental damage, especially in less-developed countries.

In Your Own Backyard...
Does agricultural burning take place where you live? If so, for which crops?

The Atmosphere as a Resource

LEARNING OBJECTIVE

- Describe the composition of the atmosphere.

The atmosphere is a gaseous envelope surrounding Earth (see Figure 4.11). Excluding water vapor, four gases comprise most of the atmosphere: nitrogen (N_2, 78.08%), oxygen (O_2, 20.95%), argon (Ar, 0.93%), and carbon dioxide (CO_2, 0.04%). Other gases and particles, including those we call **pollutants**, occur in much smaller concentrations. The two atmospheric gases most important to humans and other organisms are carbon dioxide and oxygen. During *photosynthesis*, plants, algae, and certain bacteria use carbon dioxide to manufacture sugars and other organic molecules; this process produces oxygen. During *cellular respiration*, most organisms use oxygen to break down food molecules and supply themselves with chemical energy; this process produces carbon dioxide. Nitrogen gas is an important component of the nitrogen cycle. The atmosphere performs additional **ecosystem services**, namely, blocking Earth's surface from much of the ultraviolet (UV) radiation coming from the sun, moderating the climate, and redistributing water in the hydrologic cycle.

Living at the interface of Earth's surface and atmosphere, humans think of the atmosphere as an unlimited resource, but we should reconsider. **Ulf Merbold**, a German space shuttle astronaut, felt differently about the atmosphere after viewing it in space (**Figure 19.1**). "For the first time in my life, I saw the horizon as a curved line. It was accentuated by the thin seam of dark blue light—our atmosphere. Obviously, this wasn't the 'ocean' of air I had been told it was so many times in my life. I was terrified by its fragile appearance."

REVIEW

1. What are the primary constituents of the atmosphere?

Figure 19.1   The atmosphere. The "ocean of air" is an extremely thin layer compared to the size of Earth. In this photo the atmosphere is a thin blue layer that separates the planet from the blackness of space.

SPL/Science Source

Types and Sources of Air Pollution

LEARNING OBJECTIVES

- List the seven major classes of air pollutants, including ozone and hazardous air pollutants, and describe their effects.
- Explain how ozone-related air quality has changed in southern California over the past half century.

Air pollution consists of gases, liquids, or solids present in the atmosphere in high enough levels to harm humans, other organisms, or materials. Although air pollutants can come from natural sources, such as forest fires and volcanoes, human activities release many substances into the atmosphere and make a major contribution to air pollution. Some of these substances are harmful when they precipitate (form a solid) and settle on land and surface waters, whereas other substances are harmful because they alter the chemistry of the atmosphere. From the standpoint of human health, probably more significant than the overall human contribution to air pollution is the fact that much of the air pollution released by humans is concentrated in densely populated urban areas.

air pollution Various chemicals added to the atmosphere by natural events or human activities in high enough concentrations to be harmful.

Although many different air pollutants exist, we will focus on the seven most important types from a regulatory perspective: particulate matter, nitrogen oxides, sulfur oxides, carbon oxides, hydrocarbons, ozone, and air toxics (**Table 19.1**). Air pollutants are often divided into two categories, primary and secondary (**Figure 19.2**). **Primary air pollutants** are harmful chemicals that enter directly into the atmosphere. The major ones are carbon oxides, nitrogen oxides, sulfur dioxide, particulate matter, and hydrocarbons. **Secondary air pollutants** are harmful chemicals that form from other substances released into the atmosphere. Ozone and sulfur trioxide are secondary air pollutants because both are formed by chemical reactions that take place in the atmosphere.

primary air pollutant A harmful substance, such as soot or carbon monoxide, that is emitted directly into the atmosphere.

secondary air pollutant A harmful substance formed in the atmosphere when a primary air pollutant reacts with substances normally found in the atmosphere or with other air pollutants.

MAJOR CLASSES OF AIR POLLUTANTS

Particulate matter consists of thousands of different solid and liquid particles suspended in the atmosphere, and includes solid particulate matter (dust) and liquid suspensions (mists). Particulate matter includes soil particles, soot, lead, asbestos, sea salt, and sulfuric acid droplets. It reduces visibility by scattering and absorbing sunlight. Urban areas receive less sunlight than rural areas, partly as a result of greater quantities of particulate matter in the air. Particulate matter corrodes metals, erodes buildings and sculptures when the air is humid, and soils clothing and draperies.

Particulate matter can be dangerous to health in two ways. First, it may contain materials—such as heavy metals, asbestos, or organic chemicals—that have toxic or carcinogenic effects. These toxins, upon contacting or being absorbed into the body, have a range of effects. Second, extremely small particles, even if not toxic, can damage lung tissue and arteries. Microscopic particles

Table 19.1   Major Air Pollutants

Pollutant	Composition	Primary or Secondary	Characteristics
Particulate Matter			
Dust	Variable	Primary	Solid particles
Lead	Pb	Primary	Solid particles
Sulfuric acid	H_2SO_4	Secondary	Liquid droplets
Nitrogen Oxides			
Nitrogen dioxide	NO_2	Primary	Reddish-brown gas
Sulfur Oxides			
Sulfur dioxide	SO_2	Primary	Colorless gas with strong odor
Carbon Oxides			
Carbon monoxide	CO	Primary	Colorless, odorless gas
Carbon dioxide*	CO_2	Primary	Colorless, odorless gas
Hydrocarbons			
Methane	CH_4	Primary	Colorless, odorless gas
Benzene	C_6H_6	Primary	Liquid with sweet smell
Ozone	O_3	Secondary	Pale-blue gas with acrid odor
Air Toxics			
Chlorine	Cl_2	Primary	Yellow-green gas

*Discussed in Chapter 20.
Source: Environmental Protection Agency. Compiled by authors.

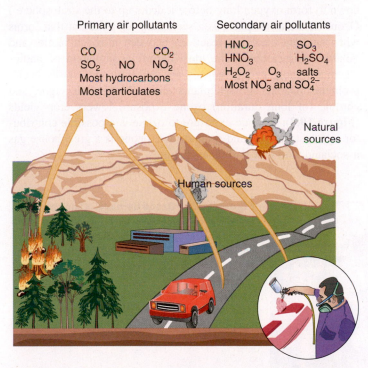

Figure 19.2   Primary and secondary air pollutants. Primary air pollutants are emitted, unchanged, from a source directly into the atmosphere, whereas secondary air pollutants are produced in the atmosphere from chemical reactions involving primary air pollutants or other substances normally found in the atmosphere.

may be classified as PM-10 (particular matter less than 10 μm [micrometers] in diameter) or PM-2.5 (particular matter less than 2.5 μm in diameter). The Environmental Protection Agency (EPA) samples microscopic particulate matter at about 1000 locations around the United States so that it can better understand its composition, which varies with location and season.

Lead, a soft metal that is used in industrial and chemical processes, has a variety of health impacts. Acute lead poisoning rarely results from outdoor exposure, but chronic effects can include permanently reduced cognitive ability, behavioral problems, slowed growth, hearing problems, and headaches. Airborne lead can be a problem both when it is inhaled and when it settles—in water and on surfaces, including foods.

Nitrogen oxides are gases produced by the chemical interactions between atmospheric nitrogen and oxygen when a source of energy, such as combustion of fuels, produces high temperatures. Collectively known as NO_x, nitrogen oxides consist mainly of nitric oxide (NO), nitrogen dioxide (NO_2), and nitrous oxide (N_2O). Nitrogen oxides inhibit plant growth and, when breathed, aggravate health problems such as asthma, a disease in which breathing is wheezy and labored because of airway constriction. They are involved in the production of photochemical smog and acid deposition (when nitrogen dioxide reacts with water to form nitric acid and nitrous acid). Nitrous oxide is associated with global warming (nitrous oxide traps heat in the atmosphere and is therefore a **greenhouse gas**) and depletes ozone in the stratosphere. Nitrogen oxides cause metals to corrode and textiles to fade and deteriorate.

Sulfur oxides are gases produced by the chemical interactions between sulfur and oxygen. Sulfur dioxide (SO_2), a colorless, nonflammable gas with a strong, irritating odor, is a major sulfur oxide emitted as a primary air pollutant. Another major sulfur oxide is sulfur trioxide (SO_3), a secondary air pollutant that forms when sulfur dioxide reacts with oxygen in the air. Sulfur trioxide, in turn, reacts with water to form another secondary air pollutant, sulfuric acid. Sulfur oxides cause acid deposition, and they corrode metals and damage stone and other materials. Sulfuric acid and sulfate salts produced in the atmosphere from sulfur oxides damage plants and irritate the respiratory tracts of humans and other animals.

Carbon oxides are the gases carbon monoxide (CO) and carbon dioxide (CO_2). Carbon monoxide, a colorless, odorless, and tasteless gas produced in larger quantities than any atmospheric pollutant except carbon dioxide, is poisonous and interferes with the blood's ability to transport oxygen. Carbon dioxide, also colorless, odorless, and tasteless, is a greenhouse gas; its buildup in the atmosphere is associated with global climate change.

Hydrocarbons are a diverse group of organic compounds that contain only hydrogen and carbon; the simplest hydrocarbon is methane (CH_4). Small hydrocarbon molecules are gaseous at room temperature. Methane is a colorless, odorless gas that is the principal component of natural gas. (The odor of natural gas comes from sulfur compounds deliberately added so that humans can indirectly detect the presence of the explosive methane gas by smelling the sulfur-containing compounds.) Medium-sized hydrocarbons such as benzene (C_6H_6) are liquids at room temperature, although many are volatile and evaporate readily. The largest hydrocarbons, such as the waxy substance paraffin, are solids at room temperature. The many different hydrocarbons have a variety of effects on human and animal health. Some cause no adverse effects, others injure the respiratory tract, and still others cause cancer. All except methane are important in the production of photochemical smog. Methane is a potent greenhouse gas linked to global climate change.

Ozone (O_3) is a form of oxygen considered an essential component in one part of the atmosphere but a pollutant in another. In the **stratosphere**, which extends from 12 to 50 km (7.5 to 30 mi) above Earth's surface, oxygen reacts with UV radiation coming from the sun to form ozone. Stratospheric ozone prevents much of the solar UV radiation from penetrating to Earth's surface. Unfortunately, certain human-made pollutants (chlorofluorocarbons, or CFCs) react with stratospheric ozone, breaking it down into molecular oxygen, O_2.

> **ozone** A pale blue gas that is an essential component that screens out UV radiation in the upper atmosphere (stratosphere) but is a pollutant in the lower atmosphere (troposphere).

Unlike stratospheric ozone, ozone in the **troposphere**—the layer of atmosphere closest to Earth's surface—is a human-made air pollutant. (Ground-level ozone does not replenish the ozone depleted from the stratosphere because ground-level ozone breaks down to form oxygen long before it drifts up to the stratosphere.) Ozone in the troposphere is a secondary air pollutant that forms when sunlight catalyzes reactions between nitrogen oxides and volatile hydrocarbons. The most harmful component of photochemical smog, ozone reduces air visibility and causes health problems. Ozone stresses plants and reduces their vigor, and chronic (of long duration) ozone exposure lowers crop yields (**Figure 19.3**). Chronic exposure to ozone is a possible contributor to forest decline, and ground-level ozone is a greenhouse gas associated with global climate change.

(a) (b)

TED SPIEGEL/National Geographic Creative

TED SPIEGEL/National Geographic Creative

Figure 19.3 **Ozone damage.** Compare the grape leaf grown in clean air (a) with the one damaged by ozone (b). Note the yellow/brown color and black spots on the damaged leaf. In contrast, the healthy leaf is a dark and even green, indicating that it has more chlorophyll. Plants exposed to ozone pollution exhibit other symptoms, including reduced root growth and a lowered productivity. Grapes exposed to ozone pollution exhibit a substantial reduction in yield.

Most of the hundreds of other air pollutants—such as chlorine, hydrochloric acid, formaldehyde, radioactive substances, and fluorides—are present in low concentrations, although it is possible to have high local concentrations of specific pollutants. Some of these air pollutants, known as **hazardous air pollutants** (**HAPs**), or **air toxics**, are potentially harmful. HAPs come from many industrial, commercial, and residential activities, including bakeries, distilleries, dry cleaners, furniture makers, gasoline service stations, hospitals, auto paint shops, print shops, and barbeques.

> **hazardous air pollutants** Air pollutants that are potentially harmful and may pose long-term health risks to people who live and work around chemical factories, incinerators, or other facilities that produce or use them.

SOURCES OF OUTDOOR AIR POLLUTION

The main human sources of primary air pollutants are transportation (**mobile sources**) and industries (**stationary sources**) (**Figure 19.4**), although intentional fires can also contribute significantly. Automobiles and trucks, known as mobile sources, generate significant quantities of nitrogen oxides, carbon oxides, particulate matter, and hydrocarbons as a result of the combustion of gasoline. Although diesel engines in trucks, buses, trains, and ships consume less fuel than other types of combustion engines, they produce more air pollution. One heavy-duty truck emits as much particulate matter as 150 automobiles.

Electric power plants and other industrial facilities, known as stationary sources, emit most of the particulate matter and sulfur oxides released in the United States; they emit sizable amounts of nitrogen oxides, hydrocarbons, and carbon oxides. The combustion of fossil fuels, especially coal, is responsible for most of these emissions. The top three industrial sources of toxic air pollutants—that is, chemicals released into the air that are fatal to humans—are the chemical industry, the metals industry, and the paper industry.

URBAN AIR POLLUTION

Air pollution localized in urban areas, where it reduces visibility, is often called **smog**. The word *smog* was coined at the beginning of the 20th century for the smoky fog that was so prevalent in London because of coal combustion. Smoke pollution is sometimes called **industrial smog**. The principal pollutants in industrial smog are sulfur oxides and particulate matter. The worst episodes of industrial smog typically occur during winter months, when combustion of household fuel such as heating oil or coal is high.

In December 1952, 4000 Londoners died in the world's worst industrial smog incident. An additional 8000 people died within the next two months, possibly due to the lingering effects of the smog, although the exact causes of death for these people have never been explained.

Because of air quality laws and pollution-control devices, industrial smog is generally not a significant problem in highly developed countries today, but it is often serious in many communities and industrial regions of developing countries.

Another important type of smog is **photochemical smog**. This brownish-orange smog is called photochemical because sunlight initiates several chemical reactions that collectively form its ingredients. First noted in Los Angeles in the 1940s, photochemical smog is generally worst during the summer months. Both nitrogen oxides and hydrocarbons are involved in its formation. One of the photochemical reactions occurs among nitrogen oxides (largely from automobile exhaust), volatile hydrocarbons, and oxygen in the atmosphere to produce ground-level ozone; this reaction requires solar energy (**Figure 19.5**). The ozone formed in this way then reacts with other air pollutants, including hydrocarbons, to form more than 100 different secondary air pollutants that injure plant tissues, irritate eyes, and aggravate respiratory illnesses in humans.

> **photochemical smog** A brownish-orange haze formed by chemical reactions involving sunlight, nitrogen oxide, and hydrocarbons.

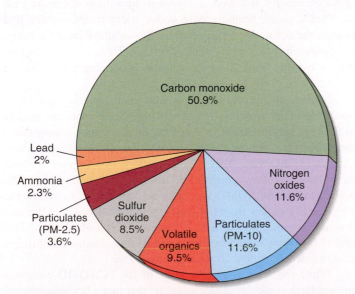

(a) Air pollutants. The concentrations of nitrogen oxides and volatile organics are an indirect measure of ozone, which is a secondary air pollutant formed in their presence.

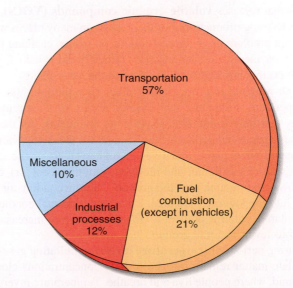

(b) Sources of air pollutants. Transportation and industrial fuel combustion (such as electric power plants) are major contributors of pollutants.

Figure 19.4   Air pollutants in the United States. (Environmental Protection Agency)

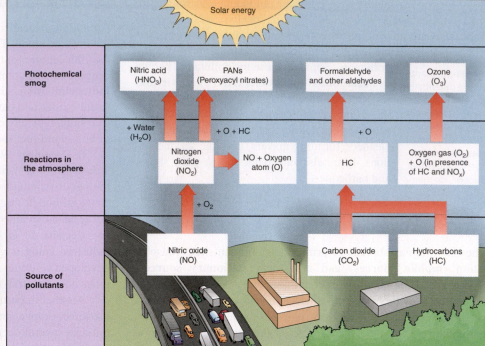

(a) Smog in Los Angeles is sometimes so bad it blocks out the sun.

(b) Several chemical reactions result in the complex mixture known as smog: ozone, peroxyacyl nitrates (PANs), nitric acid, and various organic compounds such as formaldehyde.

Figure 19.5   Photochemical smog.

CASE IN POINT

EFFORTS TO REDUCE OZONE IN SOUTHERN CALIFORNIA

Many human sources contribute to the ingredients of photochemical smog. Automobiles, as well as emissions from gasoline stations and oil refineries, are major significant contributors. But any process that releases **volatile organic compounds (VOCs)** contributes to smog production. For this reason, in many places where ozone is a problem there are restrictions on paints, cleaning products, dry cleaners, and even bakeries (when bread is baked, yeast byproducts that are VOCs are released to the atmosphere).

Both weather and topography affect air pollution. Variation in temperature during the day usually results in air circulation patterns that help dilute and disperse air pollutants. As the sun increases surface temperatures, the air near the ground is warmed. This heated air expands and rises to higher levels in the atmosphere (warm air is lighter and more buoyant than cool air), causing a low-pressure area near the ground. The surrounding air then moves into the low-pressure area. Thus, under normal conditions, air circulation patterns prevent toxic pollutants from increasing to dangerous levels near the ground.

During periods of **temperature inversion** polluting gases and particulate matter remain trapped in high concentrations close to the ground, where people live and breathe. Temperature inversions usually persist for only a few hours before being broken up by solar heating that warms the air near the ground. Sometimes, however, atmospheric stagnation caused by a stalled high-pressure air mass allows a temperature inversion to persist for several days.

Certain types of topography (surface features) increase the likelihood of temperature inversions. Cities located in valleys, near the coast, or on the leeward side of mountains (the side toward which the wind blows) are prime candidates for temperature inversions. The Los Angeles Basin is a plain that lies between the Pacific Ocean and mountains to the north and east. During the summer, the sunny climate produces a layer of warm, dry air at upper elevations. However, a region of upwelling occurs just off the Pacific coast, bringing cold ocean water to the surface and cooling the ocean air. As this cool air blows inland over the basin, the mountains block its movement further. Thus, a layer of warm, dry air overlies cool air at the surface, producing a temperature inversion.

Shortly after World War II, a number of governmental organizations began working to manage historically poor air quality in the area surrounding Los Angeles. In 1977, they consolidated their efforts by creating the South Coast Air Quality Management District (SCAQMD). This special agency was needed because air pollutants follow geographic, not political, boundaries. The borders of the SCAQMD represent an "air basin" in which the air generally moves from the west (the Pacific Ocean) to the east (toward Arizona and Nevada). It covers an area of nearly 18,000 km^2 (7,000 mi^2), including part or all of four southern California counties, where almost 15 million people live.

While the rules and regulations of the SCAQMD comprise several large books, evidence suggests that just a few major regulations have been responsible for substantial improvements in air quality over the past five decades (**Figure 19.6**). In 1976, after a decade of improvement, ozone levels in the air basin still exceeded the federal "safe" standard of 0.12 ppm on 194 days and the state

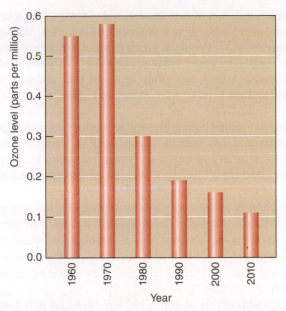

Figure 19.6 **Peak ozone concentrations in southern California, 1960–2010.** Peak ozone is the highest level of ozone recorded on any single day during the year. Average daily ozone, number of days above federal and state standards, and other measures show similar patterns. Air quality has improved steadily over the past half century but still presents a health threat. (South Coast Air Quality Management District)

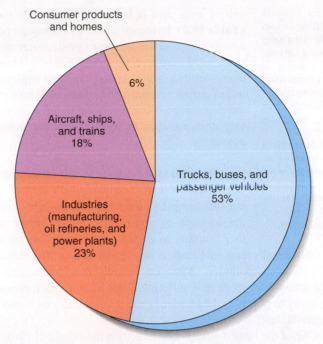

Figure 19.7 **Sources of smog in Los Angeles.** Trucks, buses, and passenger vehicles account for more than half of the emissions that produce smog.

standard of 0.09 ppm on 237 days. This means that the ozone levels were considered unsafe nearly two-thirds of the year! Ozone levels appeared to have reached a plateau from 1998 to 2003, but they dropped from 2003 to 2010.

Since ozone is created by a combination of high temperature, sunlight, hydrocarbons, water vapor, and oxides of nitrogen, some of those have to be controlled to reduce ozone. Only two of these components, hydrocarbons and oxides of nitrogen, can be controlled. Regulations to limit ozone formation have required restrictions on everything from combustion engines and oil refineries to charcoal lighter fluid and spray paints (**Figure 19.7**). ■

REVIEW

1. What are some effects of each of the seven major classes of air pollutants?

2. How and why have tropospheric ozone concentrations changed in southern California since World War II?

Effects of Air Pollution

LEARNING OBJECTIVES

• Describe the adverse health effects of specific air pollutants.

• Explain why children are particularly susceptible to air pollution.

Air pollution injures organisms, reduces visibility, and attacks and corrodes materials such as metals, plastics, rubber, and fabrics. The respiratory tracts of animals, including humans, are particularly harmed by air pollutants, which worsen existing medical conditions such as chronic lung disease, pneumonia, and cardiovascular

problems (**Figure 19.8**). Most forms of air pollution reduce the overall productivity of crop plants, and when combined with other environmental stressors, such as extreme temperatures or prolonged droughts, air pollution causes plants to decline and die. Air pollution is involved in acid deposition, global climate changes, and stratospheric ozone depletion.

AIR POLLUTION AND HUMAN HEALTH

Generally speaking, exposure to even low levels of pollutants such as ozone, sulfur oxides, nitrogen oxides, and particulate matter can

Figure 19.8 **Caution: Air pollution is hazardous to your health.** How does this cartoon depict a system out of balance?

emphysema A disease in which the air sacs (alveoli) in the lungs become irreversibly distended, decreasing the efficiency of respiration and causing breathlessness and wheezy breathing.

chronic bronchitis A disease in which the air passages (bronchi) of the lungs become permanently inflamed, causing breathlessness and chronic coughing.

irritate eyes and inflame the respiratory tract (**Table 19.2**). Evidence shows that many air pollutants suppress the immune system, increasing susceptibility to infection. In addition, evidence continues to accumulate that exposure to air pollution during respiratory illnesses may result in the development later in life of chronic respiratory diseases, such as **emphysema** and **chronic bronchitis**.

Health Effects of Specific Air Pollutants

Both sulfur dioxide and particulate matter irritate the respiratory tract and, because they cause the airways to constrict, actually impair the lungs' ability to exchange gases. People suffering from emphysema and asthma are sensitive to sulfur dioxide and particulate pollution. Nitrogen dioxide also causes airway constriction and, in people suffering from asthma, an increased sensitivity to pollen and dust mites (microscopic animals found in household dust).

One of the largest studies conducted on the effects of particulate pollution on human health, conducted by the American Lung Association, tracked the causes of death for more than 319,000 people in all 50 states. The study compared mortality data with air pollution levels—specifically, PM-2.5—in their communities. (Coal-fired power plants and diesel engines are the main emitters of tiny airborne soot particles.) Because so many people were included in the study, scientists could discount the effects of tobacco, obesity, diet, and other disease factors related to death rates. The study found that people who live and work in the country's most polluted areas are more likely to die prematurely from specific kinds of heart disease than those living in U.S. cities with the cleanest air. The soot is inhaled and lodges in the lung tissue, causing an inflammation reaction that triggers a number of processes that clog arteries leading to the heart.

Carbon monoxide binds irreversibly with iron in the blood's hemoglobin, eliminating its ability to transport oxygen. At medium concentrations, carbon monoxide causes headaches and fatigue.

As the concentration of carbon monoxide increases, reflexes slow down and drowsiness occurs: at a certain high level, carbon monoxide causes death. People at greatest risk from carbon monoxide include pregnant women, infants, and those with heart or respiratory diseases. A four-year study in seven U.S. cities—Chicago, Detroit, Houston, Los Angeles, Milwaukee, New York, and Philadelphia—linked carbon monoxide concentrations in the air to increases in hospital admissions for congestive heart failure.

Ozone and the volatile compounds in smog are irritants that cause a variety of health problems, including burning eyes, coughing, and chest discomfort. Ozone brings on asthma attacks, suppresses the immune system, and can even lead to premature death, although usually only among individuals who are already ill. A 2010 study by the Congressional Research Service suggested that changing the maximum allowable ozone standard from 75 ppb to 60 ppb could save over 1000 lives each year and reduce school days missed by children made ill by ozone by as many as a million.

The health effects of about 150 hazardous air pollutants produced by motor vehicles, businesses, and industries have not been widely studied, although long-term exposure to certain air toxics has been linked to cancer. The Environmental Defense Fund estimates that 360 people out of every million Americans develop cancer as a result of air toxics, although the cancer rate varies widely from one place to another.

Children and Air Pollution As is true of essentially all environmental stressors, air pollution is a greater health threat to children than it is to adults. Lungs develop throughout childhood, and air pollution can restrict lung development, making children more vulnerable to health problems later in life. In addition, a child has a higher metabolic rate than an adult and needs more oxygen. To obtain this oxygen, a child breathes more air—about two times as much air per pound of body weight as an adult. This means that a child breathes more air pollutants into the lungs. A 1990 study in which autopsies were performed on 100 Los Angeles children who died for unrelated reasons found that more than 80% had early-stage lung disease.

Table 19.2 Health Effects of Several Major Air Pollutants

Pollutant	Source	Effects
Particulate matter	Industries, electric power plants, motor vehicles, construction, agriculture	Aggravates respiratory illnesses; long-term exposure may cause increased incidence of chronic conditions such as bronchitis; linked to heart disease; suppresses immune system; some particles, such as heavy metals and organic chemicals, may cause cancer or other tissue damage
Nitrogen oxides	Motor vehicles, industries, heavily fertilized farmland	Irritate respiratory tract; aggravate respiratory conditions such as asthma and chronic bronchitis
Sulfur oxides	Electric power plants and other industries	Irritate respiratory tract; same effects as particulates
Carbon monoxide	Motor vehicles, industries, fireplaces	Reduces blood's ability to transport oxygen; headache and fatigue at lower levels; mental impairment or death at high levels
Ozone	Formed in atmosphere (secondary air pollutant)	Irritates eyes; irritates respiratory tract; produces chest discomfort; aggravates respiratory conditions such as asthma and chronic bronchitis

A 10-year study of about 5000 children in 12 communities in southern California examined the effects of chronic exposure to air pollution on children's developing lungs. Results from this study, which ended in 2001, indicate that children who live in high-ozone areas and participate in sports are more likely to develop asthma than children who live there but do not participate in sports. In addition, results indicate that children who breathe the most polluted air (higher concentrations of nitrogen dioxide, particulate matter, and acid vapor) have less lung growth than children who breathe cleaner air. If the children moved to areas with less particulate air pollution, their lung development increased, but if they moved to areas with worse particulate air pollution, their lung development decreased.

Exposure to lead is also a particular concern for young children. It can slow development and lead to permanent reductions in mental ability. While lead levels have been reduced substantially over the past several decades, in 2008 the EPA reduced its acceptable level from 1.5 µg per m^3 to 0.15 µg per m^3, based on evidence of the effects lead has on the very young.

REVIEW

1. What is the general effect of air pollution on the body's immune system?

2. Why are children particularly susceptible to the effects of air pollution?

Controlling Air Pollution in the United States

LEARNING OBJECTIVES

- Provide several examples of air pollution–control technologies.
- Summarize the effects of the Clean Air Act on U.S. air pollution.

There is bad and good news about air pollution in the United States. The bad news is that many locations throughout the country still have unacceptably high levels of one or more air pollutants. Moreover, most health experts estimate that air pollution causes the premature deaths of thousands of people in the United States each year. The good news is that, overall, air quality has improved since 1970. This improvement in air quality is largely due to the U.S. Clean Air Act and state-level air quality initiatives.

CONTROLLING AIR POLLUTANTS

Historically, "command-and-control" technologies have been used to reduce emissions. Usually, this means equipment that limits the emissions after they have been generated. However, this technological approach can be more expensive than alternatives such as changing industrial processes to reduce emissions.

Smokestacks fitted with *electrostatic precipitators*, fabric filters, *scrubbers*, or other technologies remove particulate matter (**Figure 19.9**). In addition, particulate matter is controlled by careful land-excavating activities, such as sprinkling water on dry soil being moved during road construction.

Several methods remove sulfur oxides from flue (chimney) gases, but it is often less expensive simply to switch to a low-sulfur fuel such as natural gas or even to a non–fossil-fuel energy source such as solar energy. Sulfur can be removed from fuels before they are burned, as in coal gasification (see Figure 11.19).

Gasoline is extremely volatile, and gasoline vapors can be a major source of VOCs. To reduce these emissions, gasoline sellers in most urban (and many rural) parts of the world require some form of **vapor recovery**. *Phase I vapor recovery* involves underground storage tanks at gas stations (**Figure 19.10**). As one hose from a delivery truck fills the underground tank, another returns the vapors in the tank—which otherwise would be vented to the atmosphere—to the truck. The truck then returns to the gasoline depot, where the vapors are either combusted or condensed into gasoline. *Phase II vapor recovery* involves removing vapor from gas tanks in cars as the gas is pumped in. These vapors are usually returned to the underground tank for removal in the phase I process.

> **vapor recovery** The removal of unburned gasoline vapors from gasoline containers, including underground tanks at gas stations and automobile gas tanks. Recovered vapors are either burned or condensed.

Lower combustion temperature in automobile engines reduces the formation of nitrogen oxides. Mass transit reduces automobile use, thereby decreasing nitrogen oxide emissions. Nitrogen oxides produced during high-temperature combustion processes in industry can be removed from smokestack exhausts. The release of nitrogen oxides from cultivated fields to which nitrogen fertilizers were applied is reduced significantly when no-tillage is practiced.

Advanced furnaces and engines burn more cleanly, reducing production of both carbon monoxide and hydrocarbons. **Catalytic converters**, used immediately following combustion, oxidize most unburned gases, and the use of properly functioning catalytic converters to treat auto exhaust reduces carbon monoxide and volatile hydrocarbon emissions by about 85%. Careful handling of petroleum and hydrocarbons, such as gasoline, paint thinner, and lighter fluid, reduces air pollution from spills and evaporation.

Reducing the sulfur content in gasoline from its current average of 330 parts per million (ppm) could significantly reduce air pollution. (Parts per million is the number of molecules of a particular pollutant found in a million molecules of air, water, or some other material.) Sulfur clogs catalytic converters so that they cannot effectively remove emissions from automobile exhaust. California, Canada, and the European Union have restricted the allowable amount of sulfur in gasoline.

Minivans, sport-utility vehicles (SUVs), and light pickup trucks, which account for almost 50% of new passenger vehicles, currently do not have the same federal standards as automobiles. These larger vehicles produce more than twice the pollution of a car. However, beginning in model year 2009, the **corporate average fuel economy**, or **CAFE**, standard required that the average fuel economy of all covered vehicles sold by a single company be 27.5 miles per gallon (mpg); this average has increased each subsequent year. One strategy to improve mileage is to shift to low carbon fuels (see Meeting the Challenge: States Adopt Low Carbon Fuel Standard).

Many states, such as California, New York, New Jersey, and Connecticut, now require diesel trucks and buses to undergo emissions tests similar to those that these states have required for automobiles for many years. Diesel exhaust, particularly particulate matter, is viewed as a toxic pollutant. Public health advocates maintain that breathing diesel exhaust may contribute to asthma, an increased incidence of lung cancer, and other lung diseases.

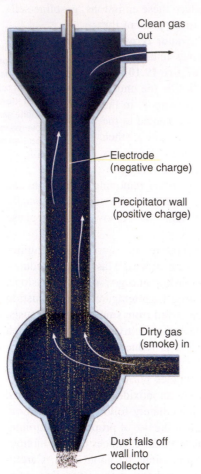

Clean gas out

Electrode (negative charge)

Precipitator wall (positive charge)

Dirty gas (smoke) in

Dust falls off wall into collector

(a) In an electrostatic precipitator, the electrode imparts a negative charge to particles in the dirty gas. These particles are attracted to the positively charged precipitator wall and then fall off into the collector.

© Russell Gordon/DanitaDelimont.com

(b) When emissions are not controlled, facilities like this aluminum smelter in China release large amounts of particulate matter.

Harrison Shull/Aurora Photos

(c) With a scrubber system in place, most particulate matter is removed, leaving only steam. Photographed in South Ashville, NC.

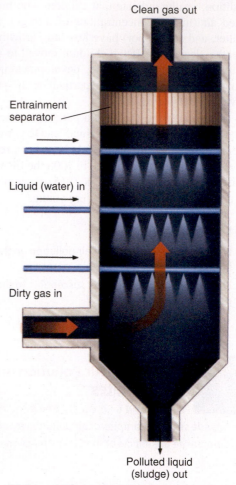

Clean gas out

Entrainment separator

Liquid (water) in

Dirty gas in

Polluted liquid (sludge) out

(d) In a scrubber, mists of water droplets trap particulates in the dirty gas. The toxic dust produced by electrostatic precipitators and the polluted sludge produced by scrubbers must be safely disposed of or they will cause soil and water pollution. (a and d adapted from M. D. Joesten and J. L. Wood, *World of Chemistry*, 2nd ed. Philadelphia: Saunders College Publishing [1996])

Figure 19.9   **Electrostatic precipitator and scrubber.**

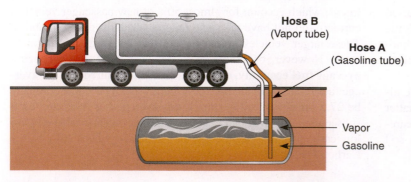

Hose B (Vapor tube)

Hose A (Gasoline tube)

Vapor

Gasoline

Figure 19.10   **Phase I vapor recovery.** Phase I vapor recovery removes vapors from the empty space above the gasoline in underground gasoline tanks.

MEETING THE CHALLENGE

States Adopt Low Carbon Fuel Standard

Historically, many of the world's environmental regulations have begun with a few states or cities deciding to deal with their own environmental problems and generating solutions that were subsequently adopted by other states, the United States as a whole, and other countries. This innovation continues with California's decision to adopt a **low carbon fuel standard (LCFS)** for fuel sold in the state beginning in 2020. Several states, including Florida and Oregon, as well as the Canadian province of British Columbia, have adopted similar approaches.

The California standard, established by an executive order of the governor, is based on two concerns: global climate change and reliance on foreign oil. Currently, oil is the energy source for 96% of the state's transportation and contributes 40% of its carbon emissions. And while California produces some oil, it imports the majority of what it uses, making it one of the largest oil importers in the world.

The goal of the LCFS is to reduce the **carbon intensity**, or amount of carbon produced per unit of energy, of fuel used in passenger vehicles by 10% or more by 2020. This means that alternative fuels—including biomass, hydrogen, and electricity—would represent a larger share of the fuel supply. Natural gas, a fossil fuel, could also play a role, since it has a lower carbon intensity than does gasoline.

Reducing the carbon content of fuels simultaneously reduces other air pollutants with local or region effects. Ground-level ozone, oxides of nitrogen, and particulates are the most impacted, but some reductions in oxides of sulfur can be expected as well. The expected effect on ozone is twofold. First, less carbon also means fewer of the precursors to ozone—recall that ozone is a secondary pollutant, produced by a combination of oxides of nitrogen and carbon-containing volatile organic compounds. Second, reducing carbon should mean less atmospheric warming. Since ozone formation requires not only chemicals but also high temperatures, less warming will mean fewer days with high enough temperatures to allow ozone formation.

THE CLEAN AIR ACT

The first air quality legislation in the United States was the Air Pollution Control Act of 1955. However, the **Clean Air Act** of 1970 (with updates in 1977 and 1990) set the standard for modern air quality regulation. This law authorizes the EPA to set limits on the amount of specific air pollutants permitted everywhere in the United States. Individual states are responsible for meeting deadlines for air pollution standards. States may pass stronger pollution controls than the EPA authorizes, but they cannot mandate weaker limits than those stipulated in the Clean Air Act.

Based on the Clean Air Act, the EPA has established maximum acceptable concentrations for lead, particulate matter, sulfur dioxide, carbon monoxide, nitrogen oxides, and ozone. The most dramatic improvement is in the amount of lead in the atmosphere, which showed a 98% decrease between 1970 and 2006, primarily because of the switch from leaded to unleaded gasoline. Atmospheric levels of the other pollutants are also reduced (**Figure 19.11**). For example, between 1970 and 2012, sulfur dioxide emissions declined 80%. During this same period, energy consumption increased by about 50%.

Although air quality is gradually improving, over 80 million metric tons of pollutants are emitted into U.S. air each year. The atmosphere in many urban areas still contains higher levels of pollutants than are recommended on the basis of health standards, although air quality regulations over the past several decades have led to significant improvements in many metropolitan areas (**Table 19.3**). In EPA language, these cities are classified as nonattainment areas for one or more criteria air pollutants. Nonattainment areas are classified on the basis of how badly polluted they are. This classification ranges from marginal (somewhat polluted and relatively easy to clean up) to extreme (so polluted that it will take many years to clean up the air). The EPA estimates that over 200 million Americans currently live in nonattainment areas.

The Clean Air Act and its amendments required progressively stricter controls of motor vehicle emissions. The provisions of the

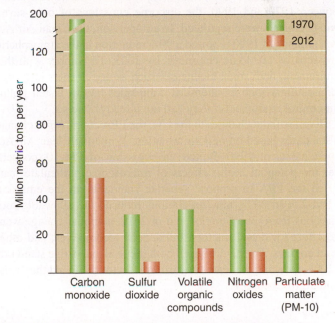

Figure 19.11   Emissions in the United States, 1970 and 2012. Carbon monoxide, sulfur dioxide, volatile organic compounds (many of which are hydrocarbons), particulate matter, and nitrogen oxides showed decreases. PM-10 applies to particles less than or equal to 10 μm (10 micrometers) in size. Since 1990, the EPA has also monitored PM-2.5, which are small particles less than or equal to 2.5 μm. (Air Quality Planning and Standards, Office of Air and Radiation, EPA.)

Clean Air Act Amendments of 1990 include the development of "superclean" cars, which emit lower amounts of nitrogen oxides and hydrocarbons, and the use of cleaner-burning gasoline in the most polluted cities in the United States. These changes were phased in gradually by the year 2000. More recent automobile models do not produce as many pollutants as older models. Yet despite the increasing percentage of newer automobile models on the road, air

Table 19.3 U.S. Urban Areas with the Worst Air Quality in 1999 (Ozone Nonattainment Areas), and Conditions in the Same Locations in 2009.

	1999	2013
Los Angeles South Coast Air Basin, California	Extreme	Extreme
Chicago, Gary, and Lake County, Illinois–Indiana	Very severe	Marginal
Houston, Galveston, and Brazoria, Texas	Very severe	Marginal
Milwaukee and Racine, Wisconsin	Very severe	No longer listed
New York City, northern New Jersey, and Long Island, New York –New Jersey–Connecticut	Very severe	Marginal
Baltimore, Maryland	Severe	Moderate
Philadelphia, Wilmington, Trenton, Pennsylvania–New Jersey –Delaware–Maryland	Severe	Marginal
Sacramento, California	Severe	Severe
San Joaquin Valley, California	Severe	No longer listed
Ventura County (between Santa Barbara and Los Angeles), California	Severe	Serious

quality has not improved in some areas of the United States because of the large increase in the number of cars being driven.

The Clean Air Act Amendments of 1990 focus on industrial airborne toxic chemicals in addition to motor vehicle emissions. Between 1970 and 1990, the airborne emissions of only seven toxic chemicals were regulated. In comparison, the Clean Air Act Amendments of 1990 required a 90% reduction in the atmospheric emissions of 189 toxic chemicals by 2003. To comply with this requirement, both small businesses such as dry cleaners and large manufacturers such as chemical companies had to install pollution-control equipment if they had not already done so.

Even without changes to the law, the EPA continues to change its standards based on new information. While the Clean Air Act Amendments of 1990 limited the emission of PM-10, concern over the potential health effects of microscopic particulate matter led the EPA to propose separate standards on the emission of PM-2.5. In 2007, new regulations reduced the allowable soot emissions from diesel trucks, and in 2012, a similar change went into effect for diesel tractors, bulldozers, locomotives, and other diesel engines. The EPA revised both the lead and ozone standards in 2008, reducing the lead standard by a factor of 10 and the ozone standard by about 5%.

REVIEW

1. What are some technologies that reduce air pollution? Some policies?

2. What is the U.S. Clean Air Act, and how has it reduced outdoor air pollution?

Ozone Depletion in the Stratosphere

LEARNING OBJECTIVES

- Define *stratospheric ozone thinning* and explain how chlorofluorocarbons and other chemicals reduce stratospheric ozone.
- Describe some of the harmful effects of ozone depletion.
- Explain the policy response to ozone depleting chemicals.

Ozone (O_3) is a form of oxygen that is a human-made pollutant in the troposphere but a naturally produced, essential component in the stratosphere, which encircles our planet some 10 to 45 km (6 to 28 mi) above the surface. The relatively high concentrations of ozone in the stratosphere form a layer that shields the surface from much of the **ultraviolet (UV) radiation** coming from the sun (**Figure 19.12**).

ultraviolet (uv) radiation That part of the electromagnetic spectrum with wavelengths just shorter than visible light.

Scientists divide UV radiation into three bands: UV-A (with wavelengths of 320 to 400 nm), UV-B (280 to 320 nm), and UV-C (200 to 280 nm). (A nanometer, abbreviated nm, is one-billionth of a meter.) The shorter the wavelength, the more energetic and more dangerous UV radiation is. Fortunately, oxygen and ozone in the atmosphere absorb all incoming UV-C (the most lethal wavelengths). The ozone layer absorbs most incoming UV-B radiation. UV-A is not affected by ozone, and most of it reaches the surface.

STRATOSPHERIC OZONE THINNING

The ozone layer over Antarctica thins naturally for a few months each year. In 1985, however, **stratospheric ozone thinning** was first observed to be greater than could be explained by natural causes. This increased thinning, which occurs each September, is commonly referred to as the *ozone hole* (**Figure 19.13**). During the 1990s, the ozone-thin area continued to grow. By 2000 it had reached the record size of 29.2 million km² (11.4 million

stratospheric ozone thinning The accelerated destruction of ozone in the stratosphere by human-produced chlorine- and bromine-containing chemicals.

mi²). A smaller thinning has also been detected in the stratospheric ozone layer over the Arctic. In addition, world levels of stratospheric ozone have decreased for several decades. According to the National Center for Atmospheric Research, ozone levels over Europe and North America dropped almost 10% between the 1970s and the early 2000s. Since about 2005, data suggest a slow but consistent increase in stratospheric ozone.

Both chlorine- and bromine-containing substances catalyze ozone destruction. The primary chemicals responsible for release of chlorine in the stratosphere, thus causing ozone depletion, are

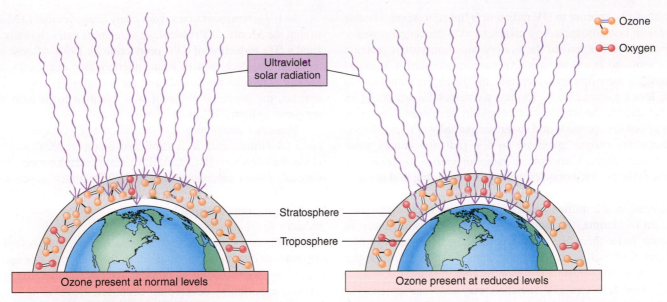

Figure 19.12   Stratospheric ozone layer.

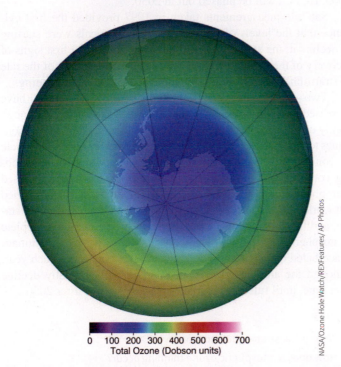

NASA/Ozone Hole Watch/REXFeatures/AP Photos

Figure 19.13   Ozone thinning. A computer-generated image of part of the Southern Hemisphere, taken in October 2013, reveals ozone thinning (bluish-purple area over Antarctica). The ozone-thin area is not stationary but moves about as a result of air currents.

chlorofluorocarbons (CFCs) Human-made organic compounds of carbon, chlorine, and fluorine that had many industrial and commercial applications but were banned because they attack the stratospheric ozone layer.

chlorofluorocarbons (CFCs). Chlorofluorocarbons were used as propellants for aerosol cans, as coolants in air conditioners and refrigerators (e.g., Freon), as foam-blowing agents for insulation and packaging, and as solvents.

Halons, methyl bromide, methyl chloroform, and carbon tetrachloride also release chlorine or bromine and thus lead to ozone depletion. Halons are used as fire retardants and methyl bromide is used as a pesticide. Methyl chloroform and carbon tetrachloride are industrial solvents.

The evidence linking CFCs and other human-made compounds to stratospheric ozone destruction includes laboratory measurements, atmospheric observations, and calculations by computer models. In 1995, the Nobel Prize in chemistry was awarded to **Sherwood Rowland**, **Mario Molina**, and **Paul Crutzen**, the scientists who first explained the connection between the thinning ozone layer and chemicals such as CFCs.

CFCs and other chlorine-containing compounds released at ground level slowly drift up to the stratosphere, where UV radiation breaks them down, releasing chlorine. Similarly, the breakdown of halons and methyl bromide releases bromine. The hole in the ozone layer that was discovered over Antarctica occurs annually between September and November (spring in the Southern Hemisphere). At this time, two important conditions are present: Sunlight returns to the polar region, and the **circumpolar vortex** develops—a mass of cold air that circulates around the southern polar region and isolates it from the warmer air on the rest of the planet.

The cold air causes polar stratospheric clouds to form; these clouds contain ice crystals to which chlorine and bromine adhere, making them available to destroy ozone. The sunlight catalyzes the chemical reaction in which chlorine or bromine breaks ozone molecules apart, converting them into oxygen molecules. The chemical reaction that destroys ozone does not alter the chlorine or bromine, and one chlorine or bromine atom can break down many thousands of ozone molecules. When the circumpolar vortex breaks up, the ozone-depleted air spreads northward, diluting ozone levels in the stratosphere over South America, New Zealand, and Australia.

THE EFFECTS OF OZONE DEPLETION

With depletion of the ozone layer, higher levels of UV radiation reach Earth's surface. A study conducted in Toronto from 1989 to 1993 showed that wintertime levels of UV-B increased more than 5% each year as a result of lower ozone levels. A study in New Zealand showed that summertime levels of peak UV-B radiation were 12% higher during the 1998–1999 summer than during similar periods a decade earlier.

Excessive exposure to UV radiation is linked to several health problems in humans, including cataracts, skin cancer, and weakened immunity. The lens of the eye contains transparent proteins that are replaced at a slow rate. Exposure to excessive UV radiation damages these proteins; over time, the damage accumulates so that the lens becomes cloudy, forming a cataract. Cataracts can be cured by surgery, but millions of people in developing countries cannot afford the operation and so remain partially or totally blind.

Excessive, chronic exposure to UV radiation causes most cases of skin cancer. Ultraviolet B radiation causes mutations, or changes, in the deoxyribonucleic acid (DNA) residing in skin cells. Such changes gradually accumulate and may lead to skin cancer. Globally, about 2.2 million cases of skin cancer occur each year. Malignant melanoma, the most dangerous type of skin cancer, is increasing faster than any other type of cancer (**Figure 19.14**). Some forms of malignant melanoma spread rapidly through the body and may cause death a few months after diagnosis.

Increased levels of UV radiation may also disrupt ecosystems. For example, the productivity of Antarctic phytoplankton, the microscopic drifting algae that are the base of the Antarctic food web, has declined from increased exposure to UV radiation. In the Antarctic, increased DNA mutations in ice-fish eggs and larvae (young fish) were matched to increased levels of UV radiation. Because organisms live in interdependent ecosystems, a negative effect on one species has ramifications throughout the system. High levels of UV radiation may also damage crops and forests. For example, exposure to higher levels of UV radiation makes cucumber crops more susceptible to disease.

FACILITATING THE RECOVERY OF THE OZONE LAYER

In 1978 the United States, the world's largest user of CFCs, banned the use of CFC propellants in products such as antiperspirants and hair sprays. Although this ban was a step in the right direction, it did not solve the problem. Most nations did not follow suit, and besides, propellants represented only the tip of the iceberg in terms of CFC use.

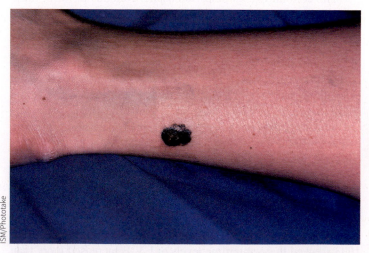

Figure 19.14 **Malignant melanoma on the back of the leg.** These tumors of pigmented cells sometimes, but not always, arise from pre-existing moles. Early diagnosis is important because this form of skin cancer spreads to other parts of the body and can be fatal.

In 1987, representatives from many countries met in Montreal to sign the **Montreal Protocol**, an agreement that originally stipulated a 50% reduction of CFC production by 1998. After scientists reported that decreases in stratospheric ozone occurred over the heavily populated midlatitudes of the Northern Hemisphere in all seasons, the Montreal Protocol was modified to include stricter measures to limit CFC production.

Industrial companies that manufacture CFCs quickly developed substitutes, such as hydrofluorocarbons (HFCs) and hydrochlorofluorocarbons (HCFCs). HFCs do not attack ozone, although they are potent greenhouse gases. HCFCs attack ozone but are not as destructive as the chemicals they have replaced. HFCs and HCFCs are transitional substances that will be used only until industry develops substitutes.

Production of CFCs, carbon tetrachloride, and methyl chloroform was completely phased out in the United States and other highly developed countries in 1996, except for a relatively small amount exported to developing countries. Developing countries phased out CFC use in 2005. Methyl bromide was phased out in 2005. HCFCs will be phased out in 2030.

Satellite measurements taken in 1997 provided the first evidence that the levels of ozone-depleting chemicals were starting to decline in the stratosphere. In the early 2000s, the first signs of recovery of the ozone layer were evident: Measurement of the rate of stratospheric ozone depletion indicated that it was declining.

Two chemicals—CFC-12 and halon-1211—may have increased and still represent a threat to ozone recovery. Although highly developed countries no longer manufacture CFC-12, it continues to leak into the atmosphere from old refrigerators and vehicle air conditioners discarded in those countries. Moreover, developing countries such as China, India, and Mexico have increased their production of CFC-12. In contrast, halon use has been phased out worldwide since 2006. Unfortunately, CFCs are extremely stable, and those being used today probably will continue to deplete stratospheric ozone for at least 50 years. Scientists expect human-exacerbated ozone thinning to reappear over Antarctica each year, although the area and degree of thinning will gradually decline over time, until full recovery takes place sometime after 2050.

REVIEW

1. What is the stratospheric ozone layer?

2. How does stratospheric ozone thinning occur?

3. How did the Montreal Protocol reverse ozone depletion?

Acid Deposition

LEARNING OBJECTIVES

- Explain how acid deposition develops and describe some of its effects.
- Define *forest decline* and explain its possible causes.
- Describe the challenges associated with reducing acid deposition.

What do fishless lakes in the Adirondack Mountains, recently damaged Mayan ruins in southern Mexico, and dead trees in the Czech Republic have in common? The answer is that these damages are

acid deposition
Sulfur dioxide and nitric oxide emissions react with water vapor in the atmosphere to form acids that return to the surface as either dry or wet deposition.

the result of acid precipitation or, more properly, **acid deposition**. It includes sulfuric and nitric acids in precipitation (**wet deposition**) as well as dry particles of sulfuric acid and nitric acid that settle out of the air (**dry deposition**).

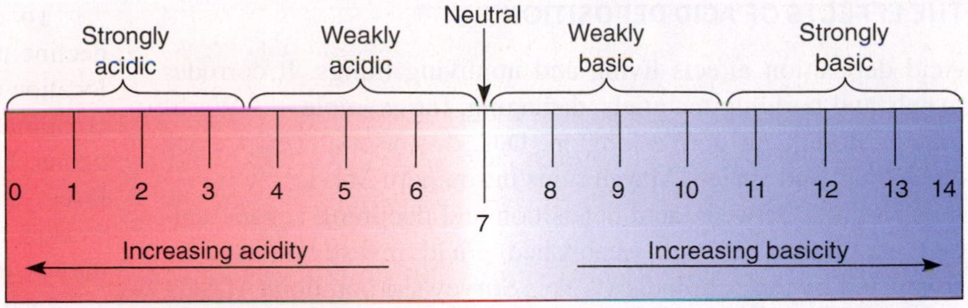

Strongly acidic | Weakly acidic | Neutral | Weakly basic | Strongly basic

0 1 2 3 4 5 6 7 8 9 10 11 12 13 14

Increasing acidity → ← Increasing basicity

Figure 19.15 **The pH scale.** This scale, which includes values between 0 and 14, expresses the concentrations of acidic and basic solutions. Pure water, which is neutral, has a pH of 7. Each decrease of one pH unit represents a tenfold increase in acidity.

HOW ACID DEPOSITION DEVELOPS

Robert Angus Smith, a British chemist, coined the term *acid rain* in 1872 after he noticed that buildings in areas with heavy industrial activity were being worn away by rain. Until recently, industrialized countries in the Northern Hemisphere had been hurt the most by acid deposition, especially the Scandinavian countries, Central Europe, Russia, and North America. More recently, largely due to high-sulfur coal, acid rain has become a major concern in China, where SO_2 releases doubled between 1995 and 2005. While SO_2 emissions in China have diminished since 2005, they remain high.

The **pH scale**, which runs from 0 to 14, expresses the relative degree of acidity or basicity of a substance (**Figure 19.15**). A pH of 7 is neither acidic nor basic, whereas a pH less than 7 indicates an acidic solution. The pH scale is logarithmic, so a solution with a pH of 6 is 10 times more acidic than a solution with a pH of 7. Similarly, a solution with a pH of 5 is 10 times more acidic than a solution with a pH of 6 and 100 times more acidic than a solution with a pH of 7. A solution with a pH greater than 7 is basic, or alkaline.

For purposes of comparison, distilled water has a pH of 7, tomato juice has a pH of 4, vinegar has a pH of 3, and lemon juice has a pH of 2. Normally, rainfall is slightly acidic (with a pH from 5 to 6) because CO_2 and other naturally occurring compounds in the air dissolve in rainwater, forming dilute acids. However, the pH of precipitation in the northeastern United States averages 4 and is often 3 or even lower.

Acid deposition occurs when sulfur dioxide and nitrogen oxides are released into the atmosphere, combine with moisture to form acids, and then are deposited on land through rain, snow, or condensate (dew) (**Figure 19.16**). Motor vehicles are a major source of nitrogen oxides. Coal-burning power plants, large smelters, and industrial boilers are the main sources of sulfur dioxide emissions and produce substantial amounts of nitrogen oxides as well. Sulfur dioxide and nitrogen oxides, released into the air from tall smokestacks, are carried long distances by winds. Tall smokestacks were an early attempt to control local air pollution—under the premise that "the solution to pollution is dilution." Tall smokestacks allow England to "export" its acid deposition problem to the Scandinavian countries and the midwestern United States to "export" its acid emissions to New England and Canada.

During their stay in the atmosphere, sulfur dioxide and nitrogen oxides react with water to produce dilute solutions of sulfuric acid (H_2SO_4), nitric acid (HNO_3), and nitrous acid (HNO_2). Acid deposition returns these acids to the ground, causing the pH of surface waters and soil to decrease.

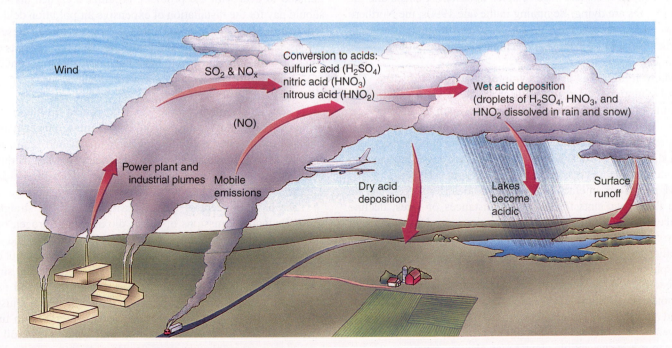

Figure 19.16 **Acid deposition.** Sulfur dioxide and nitrogen oxide emissions react with water vapor in the atmosphere to form acids that return to the surface as either dry or wet deposition.

THE EFFECTS OF ACID DEPOSITION

Acid deposition affects living and nonliving things. It corrodes metals and building materials, damaging, for example, the Washington Monument in Washington, D.C., historic sites in Venice and Rome, and ancient Mayan ruins in southern Mexico.

The link between acid deposition and declining aquatic animal populations is well established. Field investigations were conducted by the Adirondack Lakes Survey Corporation (ALSC), a nonprofit group that works with various universities, the EPA, and various state and local organizations. Of the 1469 Adirondack lakes and ponds examined, 352 had pH values of 5.0 or less, and 346 of those acidified lakes and ponds had no fish populations. Toxic metals such as aluminum dissolve in acidic lakes and streams and enter food webs. This increased concentration of toxic metals may explain how acidic water adversely affects fishes.

Although impacts on fish have received the lion's share of attention regarding acid deposition, other animals are also harmed. Several studies have found that birds living in areas with pronounced acid deposition were much more likely to lay eggs with thin, fragile shells that break or dry out before the chicks hatch. The inability to produce strong eggshells was attributed to reduced calcium in the birds' diets. Calcium is unavailable because in acidic soils it becomes soluble and is washed away, with little left for plant roots to absorb. A smaller amount of calcium in plant tissues means a smaller amount of calcium in the insects and snails that eat the plants. Less calcium is therefore available to the birds that eat these insects and snails.

Forest surveys in the Black Forest of southwestern Germany indicate that up to 50% of trees in the areas surveyed are dead or severely damaged. (Monitoring the loss of leaves and needles from trees determines the damage.) The same is true for trees in many other European forests. More than half of the red spruce trees in the mountains of the northeastern United States have died since the mid-1970s, and sugar maples in eastern Canada and the United States are dying. Beginning in the late 1990s, the Northern Hardwood Damage Survey has observed and mapped tree death at higher elevations of the Appalachian Mountains from Georgia to Maine.

Many living trees exhibit symptoms of **forest decline**. The general symptoms of forest decline are reduced vigor and growth, but some plants exhibit specific symptoms, such as yellowing of needles in conifers. Forest decline is more pronounced at higher elevations, possibly because most trees growing at high elevations are at the limits of their normal range and are less vigorous and more susceptible to stressors of any type (**Figure 19.17a**).

Many factors interact to decrease the health of trees, and no single factor accounts for the recent instances of forest decline (**Figure 19.17b**). Although acid deposition correlates well with areas experiencing tree damage, it is only partly responsible. Several other human-induced air pollutants are implicated, including tropospheric (surface-level) ozone and toxic heavy metals such as lead, cadmium, and copper. Power plants, ore smelters, refineries, and motor vehicles produce these pollutants in addition to the sulfur and nitrogen oxides that interact to form acid deposition. Insects and weather factors such as drought and severe winters (cold and wind can injure susceptible plants) may also be important.

To complicate matters further, the actual causes of forest decline may vary from one tree species to another and from one location to another. Thus, forest decline appears to result from the combination of multiple stressors—acid deposition, tropospheric ozone, UV radiation (which is more intense at higher altitudes), insect attack, drought, and so on. When one or more stressors weaken a tree, then an additional stressor, such as air pollution, may be decisive in causing its death.

One way in which acid deposition harms plants is well established: Acid deposition alters the chemistry of soils, which affects the development of plant roots as well as their uptake of dissolved minerals and water from soil. Essential plant minerals such as calcium and potassium readily wash out of acidic soil, whereas others, such as nitrogen, become available in larger amounts. Heavy metals such as manganese and aluminum dissolve in acidic soil water, becoming available for absorption in toxic amounts. A study completed in 1989 in Central Europe, which has experienced greater forest damage than North America, found a strong correlation between forest damage and soil chemistry altered by acid deposition.

MANAGING ACID DEPOSITION

One reason acid deposition is so hard to combat is that it does not occur only in the locations where the gases that cause it are emitted. Acid deposition does not recognize borders between states or countries; it is entirely possible for sulfur and nitrogen oxides released in one spot to return to the ground hundreds of kilometers from their source.

The United States has wrestled with this issue. Coal burning in several Midwest and Eastern states—Illinois, Indiana, Missouri, Ohio, Pennsylvania, Tennessee, and West Virginia—produces between 50% and 75% of the acid deposition that contaminates New England and southeastern Canada. When legislation was formulated to deal with the problem, arguments ensued about who should pay for the installation of expensive air pollution devices to reduce emissions of sulfur and nitrogen oxides. Despite these difficulties, and thanks to a cap-and-trade approach that restricts emissions of sulfur and nitrogen to the atmosphere, the total amount of SO_2 generated in the United States has declined since the 1980s (**Figure 19.18**).

Pollution-abatement issues are magnified in international disputes. For example, gases from coal-burning power plants in England move eastward with prevailing winds and return to the surface as acid deposition in Sweden and Norway. Similarly, emissions from mainland China produce acid deposition in Japan, Taiwan, and North and South Korea.

The basic concept of acid deposition control is straightforward: Reducing emissions of sulfur dioxide and nitrogen oxides curbs acid deposition. Simply stated, if sulfur dioxide and nitrogen oxides are not released into the atmosphere, they cannot come down as acid deposition. Installation of scrubbers in the smokestacks of coal-fired power plants (see Figure 19.9) and use of clean-coal technologies to burn coal without excessive emissions effectively diminish acid deposition (see Chapter 11). In turn, a decrease in acid deposition prevents surface waters and soil from becoming more acidic than they already are.

(a) These trees, photographed in the Black Forest, Germany, show signs of forest decline due to acid rain.

Despite the fact that the United States, Canada, and many European countries have reduced emissions of sulfur, acid precipitation remains a serious problem. Acidified forests and bodies of water have not recovered as quickly as hoped. Many northeastern streams and lakes, such as those in New York's Adirondack Mountains, remain acidic. A primary reason for the slow recovery is probably that the past 30 or more years of acid rain have profoundly altered soil chemistry in many areas. Essential plant minerals such as calcium and magnesium have leached from forest and lake soils. Because soils take hundreds or even thousands of years to develop, it may be decades or centuries before they recover from the effects of acid rain.

Many scientists are convinced that ecosystems will not recover from acid rain damage until substantial reductions in nitrogen oxide emissions occur, a trend that seems to be under way. Nitrogen oxide emissions are harder to control than sulfur dioxide emissions because motor vehicles produce a substantial portion of nitrogen oxides. Engine improvements may reduce nitrogen oxide emissions, but as the population continues to grow,

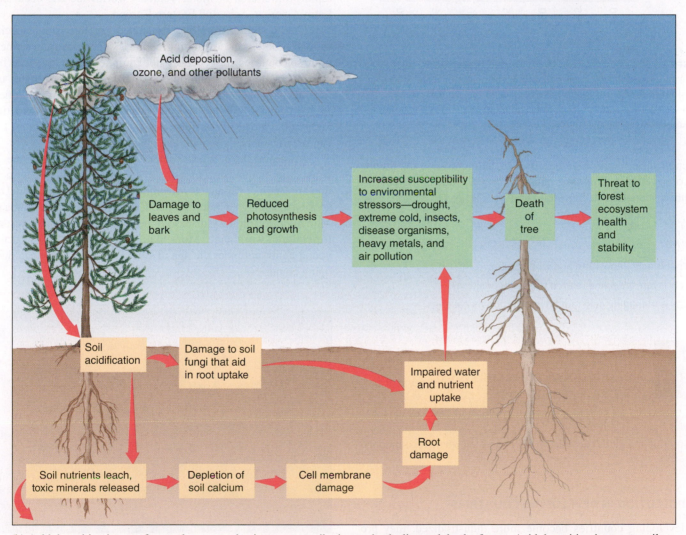

(b) Acid deposition is one of several stressors that interact, contributing to the decline and death of trees. Acid deposition increases soil acidity, causing certain essential mineral ions, such as calcium, to leach out of the soil.

Figure 19.17   Forest decline.

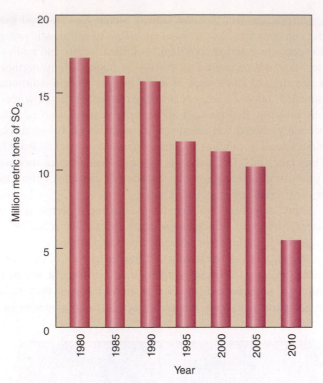

Figure 19.18 **SO₂ emitted in the United States, 1980–2010.** Sulfur dioxide (SO₂) is a major contributor to acid rain. Total sulfur emissions in the United States have been declining, initially as the result of strict SO₂ emission controls and more recently as a result of a cap-and-trade regulatory approach. (U.S. DOE, 2014)

the engineering gains may be offset by an increase in the number of motor vehicles. Dramatic cuts in nitrogen oxide emissions will require a reduction in high-temperature energy generation, especially in gasoline and diesel engines.

REVIEW

1. What is acid deposition and what causes it?
2. How is forest decline related to acid deposition?
3. Why is managing acid rain an international challenge?

Air Pollution around the World

LEARNING OBJECTIVES

- Explain why air pollution is generally worse in developing countries than in highly developed countries.
- Describe the global distillation effect and tell where it commonly occurs.

AIR POLLUTION IN DEVELOPING COUNTRIES

As developing countries become more industrialized, they produce more air pollution. The leaders of most developing countries believe they must become industrialized rapidly to compete economically with highly developed countries. Environmental quality is usually a low priority in the race to develop. Less expensive, outdated technologies are often adopted, and air pollution laws,

where they exist, are not enforced. Thus, air quality is deteriorating rapidly in many developing countries.

Many cities in China have so many smokestacks releasing coal smoke (coal is burned to heat many homes) that residents see the sun only a few weeks of the year (**Figure 19.19**). The rest of the time residents are choked in a haze of orange-colored coal dust. In other developing countries, such as India and Nepal, biomass (wood or animal dung) is burned indoors, often in stoves with little or no outside ventilation, thereby exposing residents to serious indoor air pollution. Scientists have determined that one of the principal causes of acute respiratory infections, a serious worldwide health threat, is exposure to the pollutants in biomass fuels when they are burned indoors.

The growing number of automobiles in developing countries contributes to air pollution, particularly in urban areas. Many vehicles in these countries are 10 or more years old and have no pollution-control devices. Motor vehicles produce about 60% to 70% of the air pollutants in urban areas of Central America and 50% to 60% in urban areas of India. Since the mid-1990s, the most rapid proliferation of motor vehicles worldwide has occurred in Latin America, Asia, and Eastern Europe.

Lead pollution from heavily leaded gasoline is an especially serious problem in developing countries. The gasoline refineries in these countries are generally not equipped to remove lead from gasoline. (The same situation occurred in the United States until federal law mandated that the refineries upgrade their equipment.) In Cairo, children's blood lead levels are more than two times higher than the level considered at-risk in the United States. Lead can retard children's growth and cause brain damage.

According to a World Health Organization study, the five worst cities in the world in terms of exposing children to air pollution are Beijing, China; New Delhi, India (Beijing and New Delhi are tied); Santiago, Chile; Mexico City, Mexico; and Ulaanbaatar, Mongolia. Respiratory disease is now the leading cause of death for children worldwide. More than 80% of these deaths occur in young children (under the age of five) who live in cities in developing countries.

Figure 19.19 **Air pollution in Liaoning Province, China.** Coal smoke pollutes the air above workers' houses in Liaoning Province, China. All forms of pollution are increasing threats as China becomes industrialized. China has many of the worlds' most polluted cities, including Beijing.

CASE IN POINT

AIR POLLUTION IN BEIJING, NEW DELHI, AND MEXICO CITY

Los Angeles, California, was once known for the worst smog in the world. However, while L.A. smog is still some of the worst in the United States, six decades of regulation have significantly reduced the problem. Now Beijing and New Delhi are tied for having the worst air quality in the world. Mexico City, which just a few years ago had the worst air in the world, continues to have severe air quality problems. Many of the world's worst cities for air pollution can be found in China and India.

In Beijing, automobiles, dust from construction sites, and electricity generated with coal outside of town are the biggest culprits. In preparation for hosting the 2008 summer Olympics, Beijing took steps to curb its air pollution. However, the improvements were temporary, and even as Beijing passes some regulations designed to improve air quality, other rules undermine those efforts. Beijing was once known as a town where bicycles ruled; now bicycles are banned from certain parts of the city to improve the flow of automobile traffic.

Not even on the list of the 10 worst cities for air pollution a few years ago, New Delhi is now tied for worst (**Figure 19.20**). A 2014 World Health Organization report found that in addition to other pollutants, residents of New Delhi are exposed to an annual average of 153 micrograms of small particles per cubic meter of air. A major source of this pollution is the rapid growth of motor vehicles with inefficient engines and no emissions control devices. Indian officials argue that the annual average is a misleading number, since India's monsoon weather patterns mean that air pollution has high seasonal variability. Even so, air pollution is thought to lead to tens of thousands of premature deaths in New Delhi each year.

Mexico City has moved from worst to fourth worst on the world listing over the past decade. This is not because its air pollution problem has improved but because pollution elsewhere has gotten so much worse. Average visibility has dropped from 11 km (7 mi) in the 1940s, when surrounding snow-capped volcanoes were commonly seen, to 1.6 km (1 mi). Mexico City's air pollution is due in part to its large population growth in the past several decades (the city grew from 5.4 million in 1960 to 21.2 million in 2014) and in part to its location. Mexico City is in a bowl-shaped valley ringed on three sides by mountains; winds coming in from the open northern end are trapped in the valley. Air quality is at its worst from October to January, largely as a result of temperature inversions caused by seasonal variations in atmospheric conditions.

In 2014, the city had more than 6 million passenger vehicles (up from half that number in 2000), fueled at over 400 gasoline stations, as well as about 36,000 businesses, which the Mexican government says release over 4 million metric tons (4.5 million tons) of pollutants into the air each year. Mexican gasoline contains a lot of contaminants, and the average automobile is 10 years old and produces more pollutants than do newer cars. The air contains particles of dried fecal matter from the millions of gallons of sewage dumped onto land near the city. In addition, liquefied petroleum gas—a major source of energy for cooking and heating—escapes unburned into the atmosphere from thousands of leaks, increasing the level of hydrocarbons in the city's air. Simply breathing the air in Mexico City is equivalent to smoking two packs of cigarettes a day.

During the 1990s, Mexico embarked on an ambitious plan to improve Mexico City's air quality. It spent more than $5 billion to replace old buses, taxis, delivery trucks, and cars with cleaner vehicles, such as those with catalytic converters. Mexico switched to unleaded gasoline and reforested some of the nearby hillsides to reduce particulate matter produced by wind erosion. Driving restrictions apply when the air quality is particularly poor, and exhaust emissions are periodically checked on autos.

In addition, Pemex, Mexico's national oil company, has upgraded its refineries and increased gas imports from the United States, which produces a cleaner fuel. An old, polluting oil refinery within the city limits was closed, and several large industries installed pollution-control devices. ■

LONG-DISTANCE TRANSPORT OF AIR POLLUTION

Certain hazardous air pollutants are distributed globally by atmospheric transport in a process known as the **global distillation effect**. The air toxics involved in the global distillation effect are persistent compounds, such as polychlorinated biphenyls (PCBs, industrial compounds) and dichlorodiphenyltrichloroethane (DDT, a pesticide), that do not readily break down and so accumulate in the environment. Many of these persistent compounds are restricted or even banned by many countries. Yet because they are volatile, they move through the atmosphere. The pathway of movement is generally from warmer developing countries, where they are still used, to colder, highly developed countries, where they condense and are deposited on land and surface water (**Figure 19.21**). A recent study suggests that a decrease in sea ice associated with climate change will increase the amount of air pollution in far northern regions, a phenomenon referred to as *arctic haze*.

global distillation effect The process whereby volatile chemicals evaporate from land as far away as the tropics and are carried by air currents to higher latitudes, where they condense and fall to the ground.

Figure 19.20   Smog in New Delhi, India. A busy intersection in New Delhi shows the variety of vehicles on the roads, and the limited visibility caused by particulate matter.

Adam Jones/Science Source

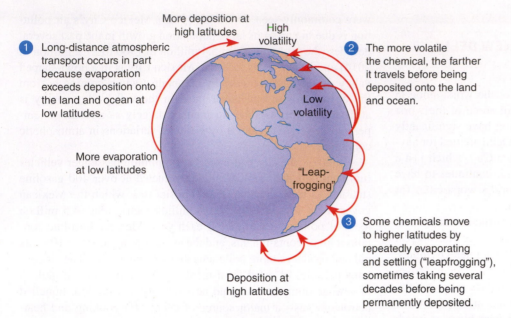

Long-distance atmospheric transport occurs in part because evaporation exceeds deposition onto the land and ocean at low latitudes.

More deposition at high latitudes

High volatility

Low volatility

More evaporation at low latitudes

"Leap-frogging"

Deposition at high latitudes

The more volatile the chemical, the farther it travels before being deposited onto the land and ocean.

Some chemicals move to higher latitudes by repeatedly evaporating and settling ("leapfrogging"), sometimes taking several decades before being permanently deposited.

Figure 19.21 **Global distillation effect.** (Adapted from F. Wania, and D. Mackay. "Tracking the Distribution of Persistent Organic Pollutants." *Environmental Science and Technology,* Vol.30 [1996])

Many industrialized countries remain highly contaminated by persistent compounds despite their restricted use. The effect is more pronounced where it is colder—that is, at higher latitudes and higher elevations. Dangerous levels of certain persistent toxic compounds have been measured in the Yukon (northwestern Canada) and in other pristine arctic regions. These chemicals enter food webs and become concentrated in the body fat of animals at the top of the food chain (see the discussion of biological magnification in Chapter 7). Fishes, seals, polar bears, and arctic people such as the Inuit are particularly vulnerable. An Inuit who consumes a single bite of raw whale skin ingests more PCBs than scientists think should be consumed in a week. The level of PCBs in the breast milk of an Inuit woman is five times higher than that in the milk of women who live in southern Canada.

The Indian Ocean has generally been considered one of the world's cleanest areas because the countries that surround it are not heavily industrialized. During a six-week study in the winter of 1999, scientists from the National Science Foundation and the Scripps Institution of Oceanography reported that a large portion of the Indian Ocean was covered by hazy, polluted air. The pollution extended over 9.5 million km² (3.8 million mi²), an area the size of the United States. It is thought that prevailing winds during the winter monsoon blow the pollution, which includes particulate matter and sulfur droplets, from the Indian subcontinent, China, and Southeast Asia. The hazy area is expected to increase as industry develops in these areas.

Pollution also travels from one continent to another. Certain atmospheric conditions (i.e., a low-pressure system over the Aleutians and a high-pressure system near Hawaii) cause a strong wind toward North America that allows air pollution from Asia to cross the Pacific Ocean. In 1997, scientists from the University of Washington detected carbon monoxide, particulate matter, and PANs in the atmosphere over the western United States. Computer models suggested that these pollutants were produced in Asia six days earlier. In 1998, more definitive evidence of pollution from Asia affecting air quality in North America occurred when a major dust storm in China produced a visible cloud of particulate matter

that was tracked by satellite across the Pacific Ocean. The polluted air was analyzed when it reached the United States a few days later and found to contain arsenic, copper, lead, and zinc from ore smelters in Manchuria.

REVIEW

1. Is air pollution worse in highly developed or in developing countries? Why?

2. What is the global distillation effect? What kinds of air pollutants are involved in the global distillation effect?

Indoor Air Pollution

LEARNING OBJECTIVE

- Describe the major sources of indoor air pollution.
- Explain why tobacco smoke and radon are considered major indoor air pollutants.

The air in enclosed places such as automobiles, homes, schools, and offices may have significantly higher levels of air pollutants than the air outdoors. In congested traffic, levels of harmful pollutants such as carbon monoxide, benzene, and airborne lead may be several times higher inside an automobile than in the air immediately outside. The concentrations of certain indoor air pollutants may be two to five times greater—and sometimes more than 100 times—than outdoors. Indoor pollution is of particular concern to urban residents because they may spend as much as 90% to 95% of their time indoors. The EPA considers indoor air pollution as one of the top five environmental health risks in the United States.

SOURCES AND EFFECTS OF INDOOR AIR POLLUTION

Because illnesses caused by indoor air pollution usually resemble common ailments such as colds, influenza, or upset stomachs, they are often not recognized. The most common contaminants of indoor air are radon (discussed shortly), cigarette smoke, carbon monoxide, nitrogen dioxide (from gas stoves), formaldehyde (from carpeting, fabrics, and furniture), household pesticides, lead, cleaning solvents, ozone (from photocopiers), and asbestos (**Figure 19.22**). In addition, the reaction of indoor ozone, generally present at lower levels than outdoors, with volatile chemicals in air fresheners, aromatherapy candles, and cleaning agents forms secondary air pollutants such as formaldehyde.

Viruses, bacteria, fungi (yeasts, molds, and mildews), dust mites, pollen, and other organisms or their toxic parts are important forms of indoor air pollution often found in heating, air-conditioning, and ventilation ducts. Excessive indoor dampness exacerbates indoor microbial growth (particularly fungal growth), dust mite populations, and cockroach and rodent infestations. A report by the U.S. Institute of Medicine showed that the presence of mold in a damp indoor environment is linked to upper respiratory tract (nose and throat) symptoms, including wheezing and coughing, and to asthma symptoms in people who already have asthma. There was also suggestive evidence of an association between molds, a damp indoor environment, and illness in the lower respiratory tracts in otherwise healthy children.

Asthma was considered a rare disease until the middle of the 20th century, and it is far more common in industrialized nations than in developing countries. Since 1970, the number of people in the United States who suffer from asthma has more than doubled, to more than 20 million; 11 million asthma sufferers are children. Health officials are worried by this trend, which is due in part to indoor air pollution. Indoor exposure to different air pollutants contributes to the development and exacerbation of asthma. Exactly what pollutant(s) is/are causing the increase in asthma is unknown, although some evidence suggests that exposure to allergens (substances that stimulate an allergic reaction) such as dust mites and cockroach feces is a major cause.

A variety of approaches can monitor or reduce indoor air pollution. Carbon monoxide monitors are increasingly common, especially in homes with natural gas appliances. An alarm lets residents know about unsafe levels. One of the best strategies to reduce indoor air pollution is to keep surfaces (especially floors, walls, and ducts) clean and dry. Filters, both free-standing and those that are part of heating and air conditioning, can capture particulates but should be cleaned or replaced regularly.

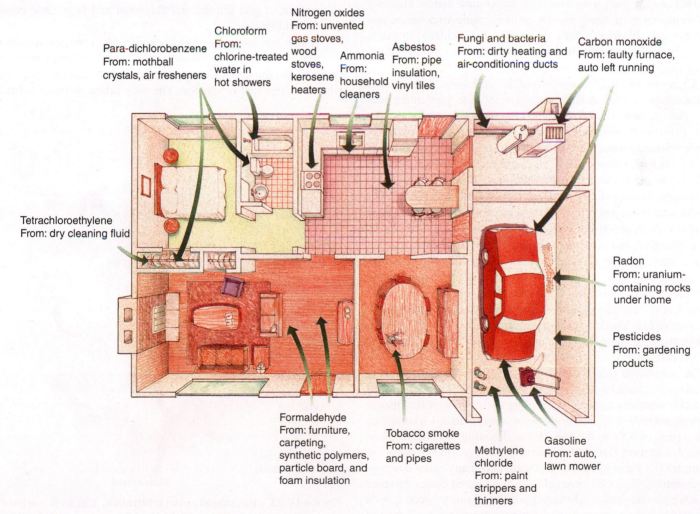

Figure 19.22   Sources of household air pollution. Homes may contain higher levels of toxic pollutants than outside air, even near polluted industrial sites.

TOBACCO SMOKE AND RADON

Smoking, which causes serious diseases such as lung cancer, emphysema, and heart disease, is responsible for the premature deaths of nearly half a million people in the United States each year. Cigarette smoking annually causes about 120,000 of the 140,000 deaths from lung cancer in the United States. Smoking also contributes to heart attacks, strokes, male impotence, and cancers of the bladder, mouth, throat, pancreas, kidney, stomach, voice box, and esophagus. It also causes substantial property damage through fires, burns, and smoke odor and discoloration.

Cigarette smoke is a mixture of air pollutants that includes hydrocarbons, carbon dioxide, carbon monoxide, particulate matter, cyanide, and a small amount of radioactive materials that come from the fertilizer used to grow the tobacco plants. Smokers exhale tobacco smoke into the air we all breathe. Passive smoking, which is the nonsmokers' chronic breathing of smoke from cigarette smokers, also increases the risk of cancer, especially in business settings (bars, casinos, restaurants) and homes. For this reason, it has been banned as a workplace hazard in many locations. Passive smokers suffer more cancer, respiratory infections, allergies, and other chronic respiratory diseases than other nonsmokers. Passive smoking is particularly harmful to infants and young children, pregnant women, the elderly, and people with chronic lung disease. When parents of infants smoke, the infant has double the chance of pneumonia or bronchitis in its first year of life. Smoking during pregnancy adversely affects fetal development, leading to lower birth weights and smaller head circumferences.

A worldwide trend is that more total people in developing countries are smoking, even while fewer people in highly developed countries are smoking. According to the Centers for Disease Control, about 18% of U.S. adults were smokers in 2013, compared with a peak of 41% in the mid-1970s. Smoking has also declined in Japan and most European countries. However, more people are taking up the habit in Brazil, Pakistan, and many other developing countries. Tobacco companies in the United States promote smoking abroad, and a substantial portion of our tobacco crop is exported. Cigarette sales in developing countries have increased by 80% since 1990.

The World Health Organization (WHO) estimates that, worldwide, over 5 million people die each year of smoking-related causes, and it wants a global ban on tobacco advertising. To meet this goal, WHO developed the Framework Convention on Tobacco Control, which calls for a ban on cigarette advertising, higher taxes on tobacco products, and restrictions on smoking in public places. This treaty went into effect for signatory nations in 2005. (The United States had not ratified this treaty as we went to press.)

In the United States, Canada, and other highly developed countries, bans on smoking in many public places—including government buildings, restaurants, college campuses, and airplanes—have substantially reduced both smoking and exposure to smoke. Although fewer U.S. citizens are smoking, certain groups in our society still have high numbers of tobacco addicts, including certain minority groups and those with the least education. A need exists to continue educating these groups, as well as all young people (more than 1 million U.S. children and teenagers take up smoking each year), about the dangers of smoking before they become addicted.

Radon is another serious indoor air pollutant in many places in highly developed countries. Radon seeps through the ground and enters buildings, where it sometimes accumulates to dangerous levels (**Figure 19.23**). Although radon is also emitted into the atmosphere, it gets diluted and dispersed and is of little consequence outdoors.

Radon and its decay products emit alpha particles, a form of ionizing radiation that is damaging to tissue but cannot penetrate very far into the body. Consequently, radon harms the body only when it is ingested or inhaled. The radioactive particles lodge in

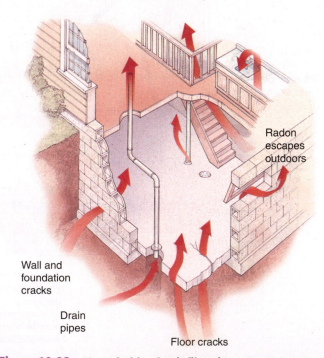

Radon escapes outdoors

Wall and foundation cracks

Drain pipes

Floor cracks

Figure 19.23 **Household radon infiltration.** Cracks in basement walls or floors, openings around pipes, and pores in concrete blocks provide some of the entries for radon.

the tiny passages of the lungs and damage surrounding tissue. There is compelling evidence, based primarily on several studies of uranium miners, that inhaling large amounts of radon increases the risk of lung cancer. Other studies suggest that people who are exposed to relatively low levels of radon over an extended time are at risk for lung cancer.

In 1998, the National Research Council of the National Academy of Sciences released an extensive evaluation of radon's effects on human health. It estimated that residential exposure to radon causes 12% of all lung cancers—between 15,000 and 22,000 cases of lung cancer annually. Cigarette smoking exacerbates the risk from radon exposure; about 90% of radon-related cancers occur among current or former smokers.

According to the EPA, about 6% of U.S. homes have high enough levels of radon to warrant corrective action—that is, a radon level above 4 picocuries per liter of air. (As a standard of reference, outdoor radon concentrations range from 0.1 to 0.15 picocurie per liter of air.) The highest radon levels in the United States occur in homes on a geologic formation, the Reading Prong, which runs across southeastern Pennsylvania into northern New Jersey and New York. Iowa has the most pervasive radon problem, where 71% of the homes tested in 1989 had radon levels high enough to warrant corrective action.

Ironically, efforts to make our homes more energy-efficient have increased the hazard of indoor air pollutants, including radon. Drafty homes waste energy but allow radon to escape outdoors so that it does not build up inside. Testing for radon can be inexpensive, and corrective actions are generally reasonably priced. Radon concentrations in homes are minimized by sealing basement concrete floors and by ventilating crawl spaces and basements.

REVIEW

1. What are the major sources of indoor air pollution?

2. Why are tobacco smoke and radon considered to be particularly hazardous indoor air pollutants?

REVIEW OF LEARNING OBJECTIVES WITH SELECTED KEY TERMS

- **Describe the composition of the atmosphere.**

Excluding water vapor and trace gases including air **pollutants**, the atmosphere comprises four gases: nitrogen, oxygen, argon, and carbon dioxide. The two atmospheric gases most important to living organisms are carbon dioxide and oxygen. Nitrogen gas is an important component of the nitrogen cycle.

- **List the seven major classes of air pollutants, including ozone and hazardous air pollutants, and describe their effects.**

The main classes of air pollutants produced by human activities are particulate matter, nitrogen oxides, sulfur oxides, carbon oxides, hydrocarbons, ozone, and hazardous air pollutants. **Particulate matter** corrodes metals, erodes buildings, and soils fabrics. **Nitrogen oxides** are associated with photochemical smog and acid deposition; nitrous oxide is associated with global climate change as well as stratospheric ozone depletion; nitrogen oxides corrode metals and fade textiles. **Sulfur oxides** are associated with acid deposition and corrode metals and damage stone and other materials. **Carbon oxides** include carbon monoxide, which is poisonous, and carbon dioxide, a **greenhouse gas**. **Hydrocarbons** include methane, which is a greenhouse gas; some hydrocarbons are dangerous to human health. **Ozone** is a pollutant in the lower atmosphere (the **troposphere**) and an essential component that screens out UV radiation in the upper atmosphere (the **stratosphere**). In the troposphere ozone reduces air visibility, causes health problems, stresses plants, and is a greenhouse gas. **Hazardous air pollutants** are potentially harmful and can pose long-term health risks to people who live and work around chemical factories, incinerators, or other facilities that produce or use them.

- **Explain how ozone-related air quality has changed in southern California over the past century.**

Since the middle of the last half century, California has passed regulations reducing emissions (including **volatile organic compounds**) from many sources. These restrictions had reduced ozone concentrations in the area more than threefold by 2012.

- **Describe the adverse health effects of specific air pollutants.**

Air pollutants irritate the eyes, inflame the respiratory tract, and suppress the immune system. Sulfur dioxide, particulate matter, and nitrogen dioxide constrict airways, impairing the lungs' ability to exchange gases. Carbon monoxide combines with hemoglobin and reduces its ability to transport oxygen; carbon monoxide poisoning can cause death. Adults at greatest risk from air pollution include those with heart and respiratory diseases.

- **Explain why children are particularly susceptible to air pollution.**

Air pollution is a greater health threat to children than it is to adults, in part because air pollution impedes lung development. Children with weaker lungs are more likely to develop respiratory problems, including chronic respiratory diseases.

- **Provide several examples of air pollution–control technologies.**

Electrostatic precipitators use electrodes to impart a negative charge to particulates in the dirty gas. These particles are attracted to the positively charged precipitator wall and then fall off into the collector. In a scrubber, mists of water droplets trap particulates in the dirty gas. The toxic dust produced by electrostatic precipitators and the polluted sludge produced by scrubbers must be safely disposed of or they will cause soil and water pollution. Phase I and II **vapor recovery** capture gasoline vapors that would otherwise be released into the atmosphere. **Catalytic converters** reduce the amount of hydrocarbons and carbon monoxide released in automobile exhaust. Reducing sulfur content of gasoline, increasing fuel mileage, and requiring regular emissions inspections can all reduce pollution associated with cars, trucks, and buses.

- **Summarize the effects of the Clean Air Act on U.S. air pollution.**

Air quality in the United States has slowly improved since passage of the **Clean Air Act** in 1970. This law authorizes the EPA to set limits on how much of specific air pollutants are permitted in the United States. The most dramatic improvement has been the decline in lead in the atmosphere, although levels of sulfur oxides, nitrogen oxides, ozone, carbon monoxide, volatile organic compounds (many of which are hydrocarbons), and particulate matter have also been reduced.

• **Define** *stratospheric ozone thinning* **and explain how chlorofluorocarbons and other chemicals reduce stratospheric ozone.**

Stratospheric ozone thinning is the accelerated destruction of ozone, a naturally occurring gas, in the stratosphere by human-produced chlorine- and bromine-containing chemicals. With depletion of the ozone layer, higher levels of **ultraviolet (UV) radiation** reach Earth's surface. **Chlorofluorocarbons (CFCs)** are human-made organic compounds of carbon, chlorine, and fluorine that had many industrial and commercial applications. Catalyzed by sunlight, chlorofluorocarbons and other compounds, including halons, methyl bromide, methyl chloroform, carbon tetrachloride, and nitrous oxide, can break protective ozone molecules in the stratosphere apart, converting them into oxygen molecules.

• **Describe some of the harmful effects of ozone depletion.**

In humans, excessive exposure to UV radiation causes cataracts, weakened immunity, and skin cancer. Increased levels of UV radiation may disrupt ecosystems, such as the Antarctic food web, because the negative effect of UV radiation on one species has ramifications throughout the ecosystem.

• **Explain the policy response to ozone depleting chemicals.**

The United States banned some ozone depleting chemicals in the 1980s. However, it was not until the Montreal Protocol created an international strategy to phase out different chemicals that progress was made. The most damaging chemicals have now been banned worldwide, and the ozone layer is recovering.

• **Explain how acid deposition develops and describe some of its effects.**

Acid deposition is a type of air pollution that includes acid that falls from the atmosphere as precipitation or as dry acidic particles. Acid deposition occurs when sulfur dioxide and nitrogen oxides are released into the atmosphere. These pollutants react with water to produce sulfuric acid, nitric acid, and nitrous acid. Acid deposition kills aquatic organisms and may harm forests. Acid deposition attacks materials such as metals and stone.

• **Define** *forest decline* **and explain its possible causes.**

Forest decline is a gradual deterioration and often death of many trees in a forest; air pollution and acid deposition contribute to forest decline in many areas. No single factor accounts for forest decline, which appears to result from the combination of multiple stressors—acid deposition, tropospheric ozone, UV radiation (which is more intense at higher altitudes), insect attack, drought, climate change, and so on.

• **Describe the challenges associated with reducing acid deposition.**

Reducing acid deposition is a challenge in part because so many different stationary and mobile sources contribute to it. In addition, the deposition often occurs far away from where the pollutants are release.

• **Explain why air pollution is generally worse in developing countries than in highly developed countries.**

Air quality is deteriorating in developing countries. Rapid industrialization, a growing number of automobiles in those countries, and lack of emissions standards are contributing to air pollution, particularly in urban areas.

• **Describe the global distillation effect and tell where it commonly occurs.**

The **global distillation effect** is the process in which volatile chemicals evaporate from land as far away as the tropics and are transported by winds to higher latitudes, where they condense and fall to the ground. Volatile chemicals contaminate some remote polar regions as a result of the global distillation effect.

• **Describe the major sources of indoor air pollution**

Air pollution comes from a variety of sources, including household chemicals (including cleaners and pesticides), human and pet hair, appliances, and furniture, as well as tobacco products and radon-containing rock formations.

• **Explain why tobacco smoke and radon are considered major indoor air pollutants.**

Tobacco smoke contains many hazardous chemicals and causes many diseases in smokers and passive smokers. Indoor tobacco smoke can be a significant workplace hazard to nonsmokers who spend large amounts of time in smoky areas. **Radon**, which enters houses from the surrounding bedrock, is a carcinogen. Radon concentrations can vary greatly from region to region.

Critical Thinking and Review Questions

1. The atmosphere of Earth has been compared to the peel covering an apple. Explain the comparison.

2. List the seven main kinds of air pollutants and briefly describe their sources and effects.

3. Distinguish between primary and secondary air pollutants.

4. What are industrial smog and photochemical smog, and how do they differ?

5. Why might global warming lead to more photochemical smog, even if emissions of nitrogen oxides and volatile organic chemicals remain constant?

6. Which is a more stable atmospheric condition, cool air layered over warm air or warm air layered over cool air? Explain. Which condition is a temperature inversion?

7. What urban areas have the worst air pollution in the world? Is this likely to change in the near future? Why or why not?

8. Distinguish between the benefits of the ozone layer in the stratosphere and the harmful effects of ozone at ground level.

9. The figure below depicts the thickness of the stratospheric ozone layer (measured in Dobson units) above New Zealand. In what year did the average yearly column of ozone above New Zealand first drop below 300 Dobson units?

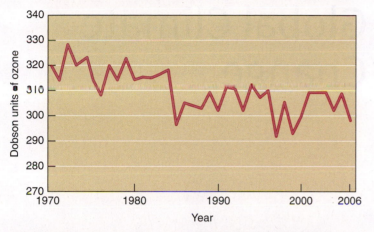

Average yearly ozone column over New Zealand, 1970–2006.
(New Zealand National Institute of Water and Atmospheric Research)

10. The amount of ozone-depleting chemicals released to the atmosphere have been reduced substantially since the late 1980s. Why has the concentration of stratospheric ozone not increased? Do we expect it to?

11. Discuss the harmful effects of acid deposition on forests, materials, aquatic organisms, and soils.

12. Why is the global distillation effect likely to become an increasingly challenging problem in the future?

13. One of the most effective ways to reduce the threat of radon-induced lung cancer is to quit smoking. Explain.

14. Conserving energy by reducing the rate at which indoor air is replaced with outdoor air can improve energy efficiency (less heating and cooling) but can contribute to indoor air pollution. Explain how a systems approach to building design might help solve this problem.

15. Explain why switching away from gasoline-powered vehicles can reduce local, regional, and global environmental problems, including climate change.

Food for Thought

The opening essay in this chapter described the problem of air pollution in Southeast Asia from burning associated with agriculture. Agricultural burning takes place in many parts of the world, including to clear land, destroy diseased trees, and eliminate wastes like rice plant stems and sugar cane. Research agricultural burning near where you live. What are the rules that limit the practice? Who is exposed, when, and how much? What are alternatives to burning?

20

Global Climate Change

Reforestation project in Ghana. Since deforested soil does not hold water as well as that of established forests, these reforestation workers in Ghana must water plantings by hand until they become established.

Anthropogenic (human-caused) climate change is an established phenomenon. Within the scientific community, the question is no longer whether climate change will occur, but at what rate, with what effects, and what, if anything, we can do about it. The biggest driver of climate change is an increase in atmospheric carbon dioxide (CO_2), which is generated primarily through burning fossil fuels.

Robert Socolow and **Stephen Pacala** of the Princeton University Carbon Mitigation Initiative suggest that somewhere near but below a doubling of atmospheric carbon from preindustrial levels lies the "boundary separating the truly dangerous consequences of emissions from the merely unwise." They propose that to keep future atmospheric carbon below a doubling of the preindustrial level (in the early 1800s, the atmosphere contained 600 billion tons of carbon) would require 7 billion fewer tons generated each year by 2056 than are currently expected.

Many people, seeing the enormity of this challenge, despair of finding a solution. Socolow and Pacala, however, propose that reductions can be thought of in terms of "wedges," each of which would result in a 1-billion-ton-per-year reduction by 2056. A combination of any seven wedges would put us on a path to avoid the critical doubling of CO_2. They identify 15 technologies in five categories, any one of which could serve as one of the seven wedges. Four of these wedges are:

Increase the fuel economy of 2 billion cars from 30 to 60 mpg. Two billion cars are expected to be on the world's roads by 2056, each traveling an average of 10,000 miles per year. If the typical car operates at 60 mpg, there will be 1 billion fewer tons of carbon generated each year than if they operate at 30 mpg.

Install carbon capture and storage at 800 large coal-fired power plants. Currently, the carbon dioxide produced from burning coal is released to the atmosphere. Instead, this carbon could be captured and stored. If 90% of the carbon released each year is captured and stored, 1 billion fewer tons will be released to the atmosphere.

Stop all deforestation. Deforestation worldwide currently releases 2 billion tons of carbon to the atmosphere each year. However, it is expected to slow to 1 billion tons per year without intervention. Thus, to achieve a 1-billion-ton wedge below the expected amount would require either a complete cessation of deforestation or a balance between deforestation and reforestation (see photo).

Add twice today's nuclear energy output to displace coal. Nuclear energy currently accounts for about 17% of the world's electricity. Increasing the amount of nuclear power alone will not affect the amount of carbon dioxide generated. However, replacing existing coal-fired power plants with nuclear power facilities would reduce greenhouse gases. Similarly, we could replace coal with solar or wind power.

In this chapter we examine climate change, which is a truly global challenge. Climate change is perhaps the best example of the systemic nature of environmental problems. The interaction of economics, politics, energy, agriculture, and human values with the natural world has led to climate change. Making changes in economics, politics, energy, agriculture, and human behavior such as those associated with these wedges will be necessary to curb climate change.

In Your Own Backyard

Many communities are preparing to reduce their reliance on gasoline-powered vehicles by preparing infrastructure to charge electric vehicles and/or fuel hydrogen vehicles. Is either of these being developed near where you live?

Introduction to Climate Change

Several factors cause Earth to have a climate amenable to life as we understand it. These include the amount of energy Earth receives from the sun, the distribution of water around the planet, the location and topography of land masses, the tilt of Earth on its axis, the reflectivity of the surface, and the content of Earth's atmosphere. Of these, most vary over thousands or millions of years. Only two—minor fluctuations in solar intensity and the content of the atmosphere—vary over decades. And of these two, only the content of the atmosphere can account for the changes in temperature scientists have observed over the past few centuries.

Earth's average temperature is based on daily measurements taken at several thousand land-based meteorological stations around the world, as well as data from weather balloons, orbiting satellites, transoceanic ships, and hundreds of sea-surface buoys with temperature sensors. The year 2011 was tied with 2005 as the hottest year since the mid-1880s, and data show that the 20 warmest years in recorded history have occurred since 1990 (**Figure 20.1**). According to the National Oceanic and Atmospheric Administration (NOAA), global temperatures in those years may have been the highest in the last millennium. (Although widespread thermometer records have been assembled only since the mid-19th century, scientists reconstruct earlier temperatures using indirect climate evidence in tree rings, lake and ocean sediments, stalagmites, small air bubbles in ancient ice, and coral reefs.)

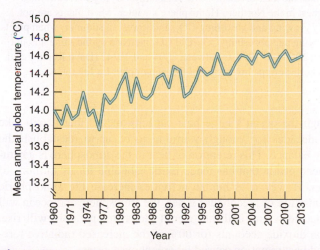

Figure 20.1 Mean annual global temperature, 1960–2013. Data are presented as surface temperatures (°C) for 1960 through 2013. The measurements, which naturally fluctuate, clearly show the warming trend of the last several decades. (The dip in global temperatures in the early 1990s, caused by the eruption of Mount Pinatubo in 1991, is discussed later in the chapter.) (Global Land-Ocean Temperature Index, Goddard Institute of Space Studies, NASA)

Other evidence confirms the increase in global temperature. Several studies have documented that phenological spring in the Northern Hemisphere now comes about six days earlier than it did in 1959, and autumn is delayed five days. (*Phenological spring* is determined by when buds of specific plants open, and autumn, by when leaves of specific trees turn color and fall.) Since 1949, the United States has experienced an increased frequency of extreme heat-stress events, which are very hot, humid days and nights during summer months; medical records indicate that heat-related deaths among the elderly and other vulnerable people increase during these events. In the past few decades, the rate of sea-level rise has increased. For most of the 20th century, the rate of increase was about 1.5 to 2 cm/decade—it is now at about 3 cm/decade. Glaciers worldwide have retreated, and extreme weather events such as severe rainstorms have occurred with increasing frequency in certain regions.

Scientists around the world began researching climate change over a century ago. As the evidence has accumulated, those most qualified to address the issue have concluded that temperatures have increased over the past century, that it is extremely unlikely that natural causes can explain the warming, and that human-produced greenhouse gases are the most plausible explanation for the warming that has occurred. Further, the remainder of this century will experience significant additional climate change, and human activities will be largely responsible for this change.

In response to the growing scientific agreement about both climate change and its human cause, governments around the world, through the United Nations, organized the Intergovernmental Panel on Climate Change (IPCC) in 1988. With input from and review by hundreds of climate experts, the IPCC provides definitive scientific statements about global climate change. The IPCC reviews all the published literature, especially that published over the previous five years, and summarizes the current state of knowledge and uncertainty as they relate to global climate change. The most recent IPCC report (its fifth), issued in 2014, confirmed earlier findings that human-produced air pollutants have caused most of the climate warming observed over the last 50 years. Depending on the assumed **emissions scenario** (a prediction about the amounts, rates, and mix of future greenhouse gases) and on the intensity of the climatic response, the IPCC report projects a 0.2°C (0.4°F) increase in global average temperature in each of the next two decades. By the year 2100, depending on whether and how much we can control greenhouse gas emissions, temperatures are expected to have risen anywhere from 1.8 to 4.0°C (3.2 to 7.2°F). Based on reconstructions of Earth's past climates, such warming would make the Earth warmer during the 21st century than in several tens of millions of years.

The fifth IPCC report also projected that it is very likely that higher maximum temperatures and more hot days will be experienced over nearly all land areas. A similar level of confidence exists for projections of higher minimum temperatures, fewer frost days, fewer cold days, an increase in the heat index, and more intense precipitation events over many areas.

THE CAUSES OF GLOBAL CLIMATE CHANGE

Carbon dioxide (CO_2) and certain other trace gases, including methane (CH_4), nitrous oxide (N_2O), and chlorofluorocarbons (CFCs), and tropospheric ozone (O_3), are accumulating in the atmosphere as a result of human activities (**Table 20.1**). Tropospheric ozone (O_3) has also increased and while estimates vary, it has probably increased by about 50% since the middle of the 18th century). All of these are **greenhouse gases**,

greenhouse gas A gas that absorbs infrared radiation; carbon dioxide, methane, nitrous oxide, chlorofluorocarbons, and tropospheric ozone are all greenhouse gases.

or gases that absorb radiated heat from the sun, thereby increasing the temperature of the atmosphere. Additional, but minor, greenhouse gases include carbon tetrachloride, methyl chloroform, chlorodifluoromethane (HCFC-22), sulfur hexafluoride, trifluoromethyl sulfur pentafluoride, fluoroform (HFC-23), and perfluoroethane.

The concentration of atmospheric carbon dioxide grew from about 288 **parts per million (ppm)** approximately 200 years ago (before the Industrial Revolution began) to 400 ppm in 2014 (**Figure 20.2**). Burning carbon-containing fossil fuels—coal, oil, and natural gas—accounts for most of the human contribution to total carbon dioxide. Land conversion, such as when tracts of tropical forests are logged or burned, also releases CO_2 and causes an increase in the atmospheric CO_2 concentration.

Not only does the burning of vegetation release CO_2 into the atmosphere, but it also reduces the capacity of the biosphere to

infrared radiation Radiation that has wavelengths longer than that of visible light but shorter than that of radio waves; most of the energy absorbed by Earth is radiated as infrared radiation, which greenhouse gases can absorb.

remove and store carbon in roots and tree trunks by photosynthesis. Scientists estimate that without aggressive efforts to reduce carbon emissions, during the second half of the 21st century the concentration of atmospheric CO_2 will reach or exceed double what it was in the 1700s.

Because these gases absorb **infrared radiation**—that is, radiated heat from the sun—higher greenhouse gas concentrations lead to warming and global climate change. This occurs because the absorption of heat slows its eventual reradiation into space, thereby warming the lower atmosphere.

radiative forcing The capacity of a gas to affect the balance of energy that enters and leaves Earth's atmosphere; measured in units of power per unit area, usually watts per square meter (W/m²).

The capacity of various gases to affect the balance of energy entering and leaving the atmosphere is referred to as **radiative forcing**. A major fraction of the trapped heat is transferred to the ocean and raises its temperature as well,

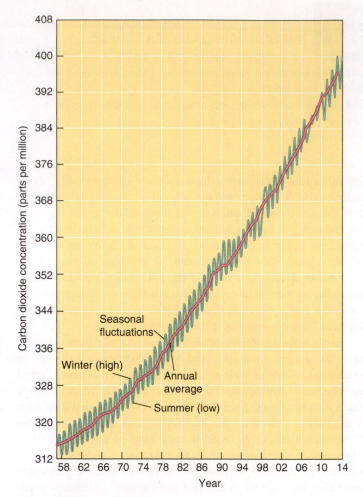

Figure 20.2 **Carbon dioxide (CO_2) in the atmosphere, 1958–2014.** The annual mean concentration of atmospheric CO_2 has increased steadily since 1958, when measurements began at the Mauna Loa Observatory in Hawaii. This location was selected because it is far from urban areas where factories, power plants, and motor vehicles that emit CO_2 might bias the measurements. The seasonally varying concentrations are highest in the Northern Hemisphere's winter, when plants are not actively growing and absorbing CO_2, and lowest in summer, when plants are growing and absorbing CO_2. Did CO_2 increase by the same amount between 1960 and 1970 as between 2004 and 2014? (Dave Keeling and Tim Whorf, Scripps Institution of Oceanography, La Jolla, CA)

Table 20.1   **Increases in Selected Atmospheric Greenhouse Gases, Preindustrial Times to the Present**

Gas	Estimated Preindustrial Concentration*	2014 Concentration
Carbon dioxide	288 ppm**	401 ppm
Methane	848 ppb***	1874 ppb
Nitrous oxide	285 ppb	324 ppb
Chlorofluorocarbon-12	0 ppt****	531 ppt
Chlorofluorocarbon-11	0 ppt	238 ppt

*The preindustrial value is for the 17th and 18th centuries. There have been significant variations, as, for example, over the course of the ice ages.
**ppm = parts per million.
***ppb = parts per billion.
****ppt = parts per trillion.
Source: Carbon Dioxide Information Analysis Center, Environmental Sciences Division, Oak Ridge National Laboratory (historical estimates), NOAA Annual Greenhouse Gas Index (2014 data).

although the ocean's large heat capacity means that it will take many decades for enough warming to occur to reestablish the energy balance. This retention of heat in the atmosphere is a natural phenomenon that has made Earth habitable for its millions of species. However, as human activities increase the atmospheric concentration of greenhouse gases, the atmosphere and ocean will continue to warm, and the overall global temperature will rise. Carbon dioxide accounts for 60% of the increased radiative forcing and heat retention caused by greenhouse gases.

Because CO_2 and other gases slow the loss of heat generated by the incoming solar radiation in a manner parallel to how the glass enclosure reduces energy loss in a greenhouse, the natural trapping of heat in the atmosphere is often referred to as the **greenhouse effect**, and the gases that absorb infrared radiation are called greenhouse gases. Greenhouse gases accumulating in the atmosphere as a result of human activities are thus causing an **enhanced greenhouse effect (Figure 20.3)**.

The levels of the other trace gases associated with global climate change are also rising. Every time you drive your car, the combustion of gasoline in the car's engine releases CO_2 along with other pollution-creating gases. Decomposition of carbon-containing organic material by anaerobic bacteria in moist places as varied as rice paddies, sanitary landfills, and the intestinal tracts of cattle and other large animals (humans included) is a major source of methane (CH_4). Various industrial processes, land use conversion, and the use of fertilizers produce nitrous oxide. CFCs are refrigerants released into the atmosphere from old, leaking refrigerators and air conditioners. CFC emissions have decreased, but the very long lifetime of emissions in the past—from a variety of sources including aerosol spray cans and foam insulation—means that they will continue to contribute to future climate change. Over the past decade, CFC concentrations in the atmosphere have begun to decrease. Water vapor, which is also a greenhouse gas, exerts a **positive feedback** on the climate that amplifies warming. Warmer temperatures cause greater evaporation from the ocean and a higher concentration of atmospheric water vapor, which, in turn, causes warmer air and ocean temperatures, causing further evaporation.

Although current rates of fossil-fuel combustion and deforestation are high, causing the CO_2 level in the atmosphere to increase markedly, scientists think the warming trend will be slower than the increasing level of CO_2 might indicate. The reason is that it requires more heat to raise the temperature of the ocean than of the air (recall that water has a high heat capacity).

In addition, the atmosphere is well mixed, while the ocean is stratified, so the ocean takes longer than the atmosphere to absorb heat. For this reason, climate scientists expect that ocean warming will be more pronounced in the 21st century than it was in the 20th century, and recent ocean surface temperature data confirm this prediction.

POLLUTANTS THAT COOL THE ATMOSPHERE

One of the complications that makes the rate and extent of global climate change difficult to predict is that other air pollutants, known as atmospheric aerosols, tend to cool the atmosphere in what is called the **aerosol effect**. Aerosols, which come from both natural and human sources, are tiny particles so small they remain suspended in the troposphere for days, weeks, or months. Because sulfate particles are efficient at scattering radiation, a sulfate-laden haze tends to cool the planet by reflecting some of the incoming sunlight back into space, away from Earth.

aerosol effect Atmospheric cooling that occurs where and when aerosol pollution is the greatest.

Temperature observations indicate that sulfur-laden haze has significantly moderated warming in some of the industrialized parts of the world. By contrast, sooty aerosols generally absorb radiation, so they tend to warm the planet. In the atmosphere, complex mixtures of aerosols make the actual aerosol effect on the climate relatively uncertain, although it will likely exert an overall cooling influence.

The sulfur dioxide emissions that produce sulfur-laden haze come mainly from the stacks of the same power plants that are responsible for much of the CO_2 emissions. In addition, volcanic eruptions inject sulfur-containing particles into the atmosphere. The explosion of Mount Pinatubo in the Philippines in June 1991 was the largest volcanic eruption in the 20th century. The force of this eruption injected massive amounts of sulfur into the stratosphere (the layer of the atmosphere above the troposphere), where

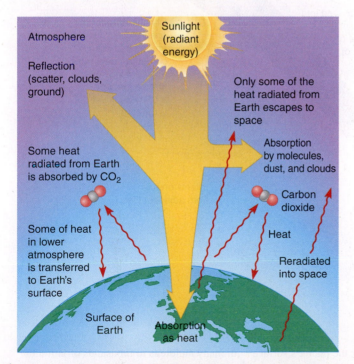

Figure 20.3   Enhanced greenhouse effect. The buildup of carbon dioxide (CO_2) and other greenhouse gases absorbs some of the outgoing infrared (heat) radiation, warming the atmosphere. Some of the heat in the warmed atmosphere is transferred back to Earth's surface, warming the land and ocean. The percent of incoming radiation absorbed is increasing, while the percent reflected is decreasing.

the particles stay aloft longer (up to a few years) than do the aerosols emitted into the troposphere. Because a sulfur-laden layer in the stratosphere reduces the amount of sunlight that reaches Earth's surface (although not the total amount that reaches the planet), this eruption caused a temporary period of global cooling. Compared to the rest of the 1990s, 1992 and 1993 global temperatures were relatively cool (see Figure 20.1).

Overall, the effect of the greenhouse gas increase is considerably more potent than that of the sulfur-laden haze. The increased concentrations of some greenhouse gases will persist in the atmosphere for years to hundreds of years, whereas human-produced sulfur emissions remain for only days, weeks, or months. And carbon dioxide and other greenhouse gases warm the planet 24 hours a day, whereas sulfur haze cools the planet only during the daytime. In addition, because sulfur emissions are a respiratory irritant and cause acid deposition (see Chapter 19), most nations are trying to reduce their sulfur emissions, not maintain or increase them.

MODELING THE FUTURE CLIMATE

Many interacting factors, such as winds, clouds, ocean currents, and albedo (a measure of reflectivity; ice has a higher **albedo** than does asphalt; see Chapter 4), affect the complex climate system, each exerting its influence on the climate. Because interactions among the atmosphere, ocean, and land are too complex and too large to construct in a testing laboratory, climate scientists develop simulation models using powerful computers to test how the Earth system works. Such models use well-established physical laws along with approximations for treating small-scale features of the climate to represent the effects of competing processes and thereby to describe Earth's climate system in numerical terms (**Figure 20.4**). Models can be used to explore and analyze past climate events and, in the most advanced models, to project future warming and suggest the consequences (what-if scenarios) of warming on the biosphere and its life-support systems.

A climate model is only as good as its representation of the physical laws and processes. Global climate change models have been refined in recent years and are now capable of representing many features of the current climate and the climate of the last few centuries. However, limitations do remain, particularly in the representation of clouds and changes that are likely to occur as the climate changes. If global climate change leads to more low-lying clouds, they would reflect some of the incoming sunlight and decrease the amount of warming, acting as **negative feedback**.

On the other hand, if global climate change leads to a greater number of high, thin cirrus clouds, they would only reflect a little more solar radiation but would trap a lot more infrared radiation, intensifying the warming (i.e., serving as a positive feedback

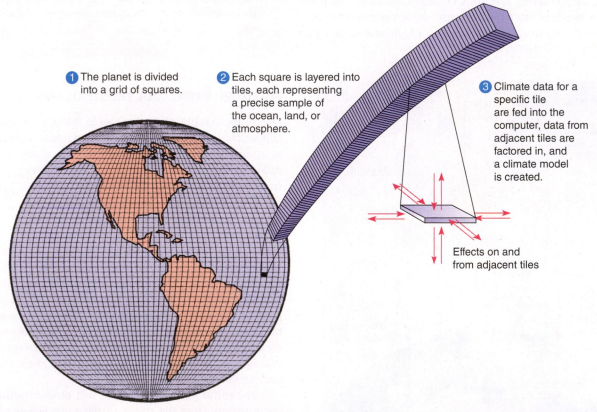

① The planet is divided into a grid of squares.

② Each square is layered into tiles, each representing a precise sample of the ocean, land, or atmosphere.

③ Climate data for a specific tile are fed into the computer, data from adjacent tiles are factored in, and a climate model is created.

Effects on and from adjacent tiles

Figure 20.4　Building a climate model. Climate models typically cover the planet's surface with many hundreds to thousands of latitude-longitude tiles, much like the tiles of a mosaic. At each location, the tiles are piled atop each other, creating stacks of a few dozen tiles extending up into the upper atmosphere and down into the ocean depths. The model's program considers how hour-by-hour (or even minute-by-minute) changes in sunlight, temperature, air pressure, currents or winds, and water vapor in one tile affect conditions in that tile and in each of the adjacent tiles. The computer continually performs calculations, taking into account the fundamental laws involving conservation of mass, momentum, and energy. These calculations can stretch out to centuries, specifying, as appropriate, any changes in the concentrations of greenhouse gases, solar radiation, or some other parameter.

mechanism). As additional case studies are carried out and as new understanding about these and other uncertainties becomes available, the predictions of the models are viewed with greater confidence.

Most climate models are used to project what the climate will be like a few decades or a century from now. One model, developed at Princeton University, has been used to examine the implications of climate warming five centuries from now. The model simulation assumed that limitations on emissions will be implemented in time to stabilize the CO_2 concentration in 2050 at double the **preindustrial CO_2 level**. It presents a dramatic scenario of a warmer climate that future generations will have to live with. Among other changes, the model projects that the rising sea level would inundate the southern end of Florida from Key Largo to Fort Lauderdale. Average summer temperatures in the southeastern and mideastern states (north to Pennsylvania) would increase from the current 27°C (80°F) to 31°C (87°F), but because this warm air would hold more moisture, the average temperature would feel more like 36°C (97°F).

Climate models present humans with a potential ethical dilemma. How do we balance scientific uncertainty about the rate and extent of climate change with the similar uncertainty about the economic impacts of reducing greenhouse gas emissions? The IPCC has established a "business as usual" scenario, which estimates the amount of carbon dioxide that will be released over the next century if economies develop as expected with no intentional attempts to reduce emissions on a large scale. This scenario projects a doubling of atmospheric carbon by around 2050. Knowing this baseline allows us to think about what combination of strategies will most effectively and efficiently avoid this doubling.

Models also take into account variability in the sun's power output. The sun is a dynamic system, and the energy reaching Earth varies over time. Climate modelers calculate possible effects of this variability but conclude that it cannot explain much, if any, of the climate change observed over the past several hundred years.

UNPREDICTABLE AND EXTREME CLIMATE CHANGE

Our current knowledge of global climate is so incomplete that unanticipated effects from a globally warmed world will undoubtedly occur. Some of these effects are simply unpredictable—that is, there will be complete surprises. Others are conceptually predictable, but there are thresholds (as in the case of dying coral reefs, discussed later in the chapter) and tipping points, and we don't know when they might occur.

As an example of a tipping point, there could be a disruption of the ocean conveyor belt, which transports heat around the globe (see Figure 4.16). The ocean conveyor belt delivers heat from the tropics into the northern part of the North Atlantic Ocean. Some of this heat is transferred to the atmosphere, thereby helping to warm Europe and adjacent lands as much as 10°C (18°F). As the warm North Atlantic water transfers heat to the atmosphere, it cools, sinks, and flows southward. The cooler, sinking water carries some of the CO_2 from the atmosphere deep into the ocean, where, through incompletely understood mechanisms, much of the carbon is sequestered (stored).

Models based on the behavior of the ocean conveyor belt during past episodes of climate warming—immediately following the ice ages, for example—suggest that an abrupt climate change could occur. Climate warming, with its associated freshwater melting off the Greenland ice sheet, could weaken or even shut down the ocean conveyor belt in as short a period as a decade. Changes in the ocean conveyor belt could cause major cooling in Europe even as greater global warming occurs elsewhere. In addition, a weakened ocean conveyor belt would not sequester as much carbon in the ocean, leading to a positive feedback loop: Less CO_2 stored in the ocean would mean more CO_2 in the atmosphere, which would cause additional atmospheric warming, which, in turn, would cause the ocean conveyor belt to weaken even further.

In 2014, researchers at NASA concluded that a large portion of the West Antarctic ice sheet is undergoing a breakup that appears to be irreversible, at least over the next several decades. Water warming the edge of the ice sheet doesn't just melt that ice, but in doing so destabilizes the entire ice sheet. Because the ice sheet sits on land, any ice that melts will cause sea levels to rise worldwide.

Parts of North America experienced an unusually cold winter in 2013–2014, which many climatologists interpret as an effect of our changing climate. During the winter, cold air from the poles can move towards the equator in irregular patterns, a phenomenon called a **polar vortex**. The difference in temperature between the poles and the equator determines how far cold air can travel. Over the past few decades, the poles have warmed more quickly than has the equator, reducing the temperature differential. This allows the polar vortex to significantly reduce temperatures at middle latitudes, such as what much of the northern United States experienced in January and February 2014.

In this case overall warming of Earth's climate led to unusually cold weather in the short term across a large region. These same areas experienced higher than normal summertime temperatures both before and after the extreme polar vortex cold spells. Shortly after the polar vortex in the northern United States, the west coast of the United States experienced record springtime high temperatures.

Climate models project expected or most likely outcomes and ranges of possible outcomes. The outcomes usually reported represent a range that modelers feel is reasonably likely to include the actual outcome. Sometimes these ranges include best cases that might be somewhat troubling and worst cases that might cause serious disruptions. For example, a 0.5°C (0.9°F) increase in average summer temperature at some location might not make much difference, whereas a 4°C (7.2°F) increase certainly would.

In addition, climate model outputs often include possible extreme cases, such as a 6°C (11°F) increase in global annual average temperatures or a 6-m (19-ft) sea-level rise. As you read the next section on the effects of climate change, keep in mind that there may be effects beyond what is described here and that there are likely to be surprises we cannot currently predict.

REVIEW

1. What is the enhanced greenhouse effect, and what are the five main greenhouse gases that contribute to it?

2. How do climate models project future climate conditions?

3. Why are unpredictable and extreme climate changes important?

The Effects of Global Climate Change

LEARNING OBJECTIVES

- Differentiate between sea-level rise due to melting ice and that due to the thermal expansion of water.
- Describe how climate change impacts the physical environment.
- Give examples of effects of climate change on organisms, including humans.

Global climate change directly or indirectly affects many physical and biological systems. Many effects have already been observed—temperature increases, shifts in plant and animal habitats, and sea-level rise. Climate researchers expect that these changes will continue in the future, and they anticipate some new changes. In addition, researchers expect surprises—that is, they know changes will occur that we cannot predict.

In this section, we consider some observed and potential effects of global climate change, including changes in sea level, precipitation patterns (including the frequency and intensity of storms), ecosystems, human health, agriculture, and wildfires (**Figure 20.5**). A more complete, but still not exhaustive, list of impacts of climate change would also include changes in forests (and thus the timber industry), tourism and recreation, and coastal infrastructure.

MELTING ICE AND RISING SEA LEVELS

The IPCC projects a sea-level rise of 18 to 59 cm (0.6 to 1.9 ft) by 2100, while noting that it could be much more. Two factors cause sea-level rise. Water, like other substances, expands as it warms. The IPCC reports that during the 20th century, sea level rose about 0.2 m (8 in.), much of it due to **thermal expansion**. Thermal expansion contributes more than half of sea-level rise. The current

rate of thermal expansion leads to about 3 mm of sea-level rise per year, and that rate is increasing. In addition, sea level rises due to the retreat of glaciers and thawing of ice at the South Pole. Water absorbs more heat than does ice, which is highly reflective. As a consequence, melting ice has a positive feedback effect on heating: Water absorbs more heat, which causes more ice to melt.

The area of ice-covered ocean in the Arctic has decreased significantly over the past several decades. The average latitude of the southern edge of arctic ice (between 71°N and 72°N latitude during the 1970s) has retreated northward, to 75°N. Sonar measurements from naval submarines operating under the ice indicate that the remaining arctic ice pack has thinned rapidly, losing 40% of its volume in less than three decades.

Mountain glaciers around the world are also melting at accelerating rates, contributing to sea-level rise. Qori Kalis Glacier, the largest glacier in the Peruvian Andes, has almost completely melted, having retreated about 60 m (200 ft) a year for the past 25 years. The Gangotri Glacier in India is retreating at a similar rate. According to the National Park Service, in 1850 Glacier National Park had 150 glaciers; today, only 25 are big enough to be considered functional (larger than 25 acres in area; **Figure 20.6**). Glacier regression models predict that they will probably be gone by 2030.

The Greenland ice sheet (the world's second-largest expanse of landbound ice) is losing over 250 km^3 (57 mi^3) of ice each year, up from estimates of 44 km^3 (11 mi^3) in 2002 and 8.3 km^3 (2 mi^3) in the period 1993–1998. In 2014, a report in *Nature Climate Change* indicated that this melt is contributing at least 0.5 mm (0.02 in.), and possibly as much as 3.2 mm (0.13 in.), to sea-level rise each year. If half of the Greenland ice sheet were to melt, sea level would rise several meters.

CASE IN POINT

IMPACTS IN FRAGILE AREAS

The Eskimo Inuit, the indigenous people of Alaska's and Canada's far north, pursue a way of life dictated by the frigid climate. Effects of global climate change are altering the traditional Inuit way of life. Many populations of wildlife, which the Inuit harvest for food, are smaller or displaced. Other changes that threaten subsistence livelihoods include reduced snow cover, shorter river ice seasons, and thawing of permafrost. Warmer temperatures increase the risk of water supplies being contaminated, as bacteria move more freely through thawed soil. Larger thawed areas could lead to the collapse of bridges, buildings, roads, and oil pipelines.

Glimpses of these observed and potential changes are gained from the region's natives, who report changes linked to warmer temperatures: drying tundra; thinner and retreating sea ice; warmer winters; and changes in the numbers, distribution, and migration of some wildlife species. Climate change data support the Inuits' observations. Scientists analyzing lake-sediment cores to explore climate changes for the past 400 years note that the greatest warming trend occurred from 1840 to the late 20th century. With temperatures projected to rise even faster in the 21st century, scientists caution that the relatively undisturbed Arctic may be particularly susceptible to the effects of climate change.

Figure 20.5 **Wildfire in California, September 2014.** While California has always experienced wildfires, major fires, like the one shown here, have been rare. Beginning in the mid-1980s, wildfires in the western United States have become more frequent, have lasted longer, and have occurred over a longer fire season. Changes in land use may have some effect, but the main driver appears to be climate warming, due to both temperature increases and snowmelt earlier in the spring.

ALVIN JORNADA/EPA / Newscom

(a)   (b)

(c)   (d)

Figure 20.6   **Grinnell Glacier, a monitor of climate change.** As with many other glaciers worldwide, Grinnell Glacier, in Glacier National Park, is decreasing rapidly due to increasing temperature. Here, the glacier is shown photographed in 1938 (a), 1981 (b), 1998 (c), and 2005 (d).

Melting ice that drains into the ocean raises the sea level, but what about landbound melting ice? Evidence indicates that **permafrost**, the permanently frozen subsoil characteristic of the tundra and boreal forests of Alaska, Canada, Russia, China, and Mongolia, is thawing. Permafrost provides the foundation on which tundra plants and forest trees are anchored and on which houses and roads are built. As the permafrost thaws, this foundation collapses. Near Fairbanks, Alaska, hundreds of homes and telephone poles are sinking at odd angles into the ground (**Figure 20.7**). Thawing permafrost also releases methane and other greenhouse gases—another positive feedback.

While melting ice and thawing permafrost are important in the extreme north, sea-level rise has begun to impact small island nations. In 1999, two uninhabited islands (Tebua Tarawa and Abanuca) in the South Pacific were submerged under rising water.

Figure 20.7   **Thawing permafrost.** These buildings are sinking as the permafrost on which they were built thaws. Thawing permafrost causes ground subsidence, erosion, and landslides.

In 2001, the 11,000 residents of nearby Tuvalu announced that they had to evacuate because the rise in sea level had caused lowland flooding, harming their water supply and food production. New Zealand has offered to allow some Tuvaluans to emigrate there each year. Small island nations such as the Maldives, a chain of 1200 islands in the Indian Ocean, are considered highly vulnerable to a rise in sea level. About 80% of the Maldives is less than 1 m (39 in.) above sea level, and the country's highest point is only 2 m above sea level. As sea levels rise, storm surges could easily sweep over entire islands.

CHANGES IN PRECIPITATION PATTERNS

Computer models indicate that, as global climate change occurs, precipitation patterns will change, causing some areas to have more frequent droughts (**Figure 20.8**). At the same time, heavier snow and rainstorms are projected to cause more frequent flooding in other areas. Some recent flooding, including heavy floods in England, has most likely been caused by changing climate.

Changes in precipitation patterns are likely to affect the availability and quality of fresh water in many locations. Arid or semi-arid areas, such as the Sahel region just south of the Sahara Desert (see Figure 4.22), may have the most troublesome water shortages as the climate changes. Closer to home, experts predict water shortages in the American West because warmer winter temperatures will cause more precipitation to fall as rain rather than snow; melting snow currently provides 70% of stream flows in the West during summer months. The UN Security Council began meetings in 2007 to consider the security implications of climate change–related drought.

The frequency and intensity of storms over warm surface waters appear likely to increase. In 1998, NOAA developed a computer model examining how global climate change might affect hurricanes, and since Hurricane Katrina devastated New Orleans in 2005, there has been increased interest in the relationship between climate change and the intensity and frequency of hurricanes. Severe flooding in Australia in late 2010, caused by Cyclone Tasha, reminds us that this is a global phenomenon.

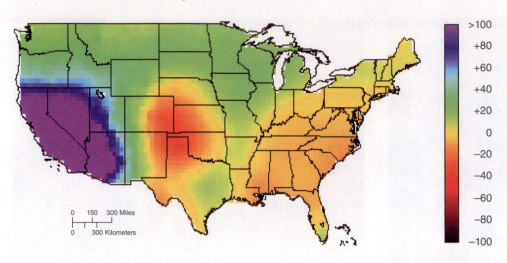

Figure 20.8 **Effect of climate warming on precipitation.** This model shows one scenario (the Canadian Model Scenario) of how warmer global temperatures might alter precipitation in the United States over the next 100 years. Colors indicate the percent change in annual precipitation per century. Locate the state you live in. Will it be wetter, drier, or about the same? (U. S. Climate Change Research Program)

Recent research suggests that a sea-surface temperature 2.2°C (4.0°F) warmer than today's will result in hurricanes with higher maximum wind speeds and increased total precipitation. Changes in storm intensity are expected because as Earth warms, more water evaporates, which in turn releases more energy into the atmosphere (recall the discussion of water's heat of vaporization in Chapter 13). This energy can generate more powerful storms. While the number of hurricanes each year does not appear to have changed over the past several decades in response to climate change, the mean and peak intensities have increased in correlation to increased surface-water temperature.

As discussed in Chapter 4, the El Niño–Southern Oscillation (ENSO), the periodic warming (El Niño) and cooling (La Niña) of the tropical Pacific Ocean, affects precipitation and other aspects of the entire global climate system. Until recently, climate scientists could not predict whether human-induced global climate change would affect ENSO. The IPCC's most recent analysis predicts greater extremes of drying and heavy rainfall during El Niño events. Scientists are uncertain whether El Niño events will occur more frequently with global climate change.

EFFECTS ON ORGANISMS

An increasing number of studies report measurable changes in the biology of plant and animal species as a result of climate warming. Such effects range from earlier flowering times for plant species to changes in migration patterns of aquatic species. Changes are also evident in many populations, communities, and ecosystems. Other human-induced factors, such as pollution and changes in land use, exacerbate threats posed by climate change. Here we report on the results of several of the thousands of studies conducted to date.

Researchers determined that populations of zooplankton in parts of the California Current, which flows from Oregon southward along the California coast, had declined 80% since 1951, apparently because the current has warmed slightly. The decline in zooplankton in the California Current has affected the entire food web there, and populations of plankton-eating fishes and seabirds have declined.

As temperatures have risen in the waters around Antarctica during the past two decades, a similar decline in shrimp-like krill has contributed to a reduction in Adélie penguin populations (see

Case in Point: How Humans Have Affected the Antarctic Food Web, in Chapter 3). Because there are fewer krill, the birds do not get enough food. Warmer temperatures in Antarctica—during the past 50 years the average annual temperature on the Antarctic Peninsula has increased 2.6°C (4.5°F)—have contributed to reproductive failure in Adélie penguins. The birds normally lay their eggs in snow-free rocky outcrops. However, open water is now closer to the nesting ground, resulting in increased air moisture and increased snowfall. When the penguins incubate the eggs, this snow melts into cold pools of slush that kill the developing chick embryos.

Some species have shifted their geographic ranges in response to global climate change. One type of western butterfly (the Edith's checkerspot butterfly) has disappeared in the southern parts of its range, while its northern range has expanded about 160 km (about 100 mi) northward, establishing new colonies there. Similar studies conducted in Europe have found that 22 butterfly species out of the 35 species examined have shifted their ranges northward, anywhere from 32 to 240 km (20 to 150 mi). Scientists studying birds in Great Britain also report that the ranges of several dozen species have moved northward an average of 19 km (12 mi).

The ranges of some species are also shifting up in altitude. For over 100 years, researchers at the University of California at Berkeley have been collecting small mammal specimens from Yosemite National Park. This collection clearly shows that small mammals previously found only at relatively low elevations are now found at higher elevations; a steady upward shift has been documented during this time.

Biologists in the Netherlands have determined that in the past (as recently as 1980), trees leafed out, then winter moth caterpillars hatched, and then great tit eggs hatched, and the parents fed and successfully raised their chicks. Species react to warming differently: Some changes are triggered by temperature, others by seasonal changes in the length of the day. The interdependent "leaves/caterpillars/birds" system has become uncoupled because the peak number of caterpillars now occurs when the trees leaf out, which is before the chicks are hatched (egg-laying time for this species has not changed), to the detriment of the bird populations.

As warming accelerates in the 21st century, many species will undoubtedly become extinct, particularly those with narrow

Figure 20.9 **Polar bears are encountering difficulties surviving in a world with less contiguous (unbroken) polar ice.** What does this cartoon imply about human obligations to other animals?

© Media Bakery

(a) The corals in this image are normal and healthy.

Peter Scoones / Science Source

(b) Bleached corals. Coral researchers expect that many of Earth's corals will die from the combined effects of ocean warming and acidification.

Figure 20.10 **Effects of climate change on coral reefs.**

temperature requirements; those confined to small, specialized habitats; and those living in fragile ecosystems. Other species may survive in greatly reduced numbers and ranges. Ecosystems considered at greatest risk of species loss in the short term are coral reefs, mountain ecosystems, coastal wetlands, tundra, and polar areas (**Figure 20.9**).

Coral reefs are systems that include the corals, symbiotic organisms living within the reefs, and other organisms, including fish, that live, eat, and reproduce around the reefs (**Figure 20.10a**). Corals are impacted by climate change in two ways: acidification and temperature increases. The ocean absorbs about half of the CO_2 released into the atmosphere by human activities. This moderates the CO_2 in the atmosphere; if the ocean were not acting as a sink for CO_2, warming would be much more severe now than it is. However, CO_2 reacts with water to form carbonic acid, H_2CO_3. As more and more CO_2 is absorbed, it has made the ocean more acidic. Many organisms, such as corals, shellfish, and plankton, incorporate $CaCO_3$ into their protective shells, a chemical process that is sensitive to acidity.

Temperature-related coral bleaching occurs when water temperature exceeds a threshold, affecting the coral symbiotes and making them and the corals more susceptible to disease-causing organisms that healthy corals are normally resistant to (**Figure 20.10b**). Increased acidity exacerbates this effect—two parts of the delicately balanced ocean system change at once. The IPCC predicts that with a 2°C (3.6°F) increase in global mean annual temperature, most corals worldwide will experience bleaching, with widespread coral mortality occurring above an increase of 3°C (5.4°F). In 1998, scientists documented the most geographically extensive and severe epidemic of coral bleaching ever observed. About 10% of the world's corals died that year, in many cases from viral, bacterial, or fungal infections. (Coral bleaching is also discussed in Chapter 6.)

Biologists generally agree that global climate change will have an especially severe impact on plants because they cannot move about when environmental conditions change. Although wind and animals disperse seeds, sometimes over long distances, limits in seed dispersal rates can limit the speed of migration. During past climate warmings, such as during the glacial retreat that took place some 12,000 years ago, analysis of tree pollens indicates that species were able to migrate at rates of only 4 to 200 km (2.5 to 124 mi) per century.

If Earth warms the projected 1.8 to 4.0°C (3.2 to 7.2°F) during the 21st century, the ideal ranges for some temperate tree species (i.e., the environment where they grow the best) may shift northward as much as about 500 km (about 300 mi). The U.S. Department of Agriculture reports hardiness zones, based on the average low temperature in an area, to determine which plants will do well around the country. By 2006, these zones had shifted enough that they had to be updated—the first time that an update had been necessary (**Figure 20.11**). Moreover, soil characteristics, water availability, competition with other plant species, and habitat fragmentation will affect the rate at which plants can move into a new area.

Some species will come out of global climate change as winners, with greatly expanded numbers and range. Those organisms considered most likely to prosper include certain weeds, insect pests, and disease-carrying organisms common in a wide range of environments. A 3°C (5.4°F) increase in average surface temperature, for example, could allow the Mediterranean fruit fly, an economically important insect pest, to expand its range into northern Europe.

1990 Map

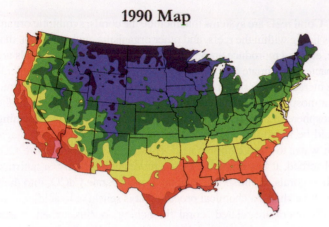

After USDA Plant Hardiness Zone Map, USDA Miscellaneous
Publication No. 1475, Issued January 1990.

2006 Map

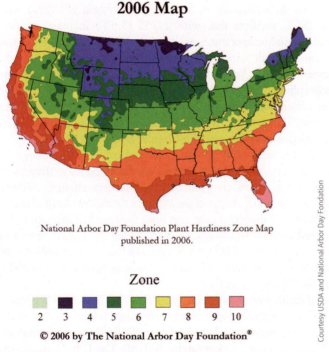

National Arbor Day Foundation Plant Hardiness Zone Map
published in 2006.

Zone

2   3   4   5   6   7   8   9   10

© 2006 by The National Arbor Day Foundation®

Courtesy USDA and National Arbor Day Fondation

**Figure 20.11   U.S. Department of Agriculture/National Arbor
Day Foundation hardiness zones, 1990 and 2006.** In 1990, the U.S.
Department of Agriculture published a "Plant Hardiness Zone Map" for
farmers and gardeners to identify which plants were best suited to their
locations. Climate change necessitated an updated map in 2006. A few
places have become colder, but shifts to more heat-tolerant plants have
been necessary in many others. Has the hardiness zone changed where
you live? (Courtesy USDA and National Arbor Day Foundation).

EFFECTS ON HUMAN HEALTH

Although the precise extent to which climate change contributes
to adverse health effects is uncertain, it is clear that climate change
can significantly impact human health and will have an increasing
effect in the future. The relationships between human health and
climate work at the systemic level and are both complex and inseparable. Some immediate health impacts are clear—for example,
the 2010 heat wave in Russia resulted in around 15,000 deaths and

cost the country as much as $15 billion in lost GDP. Most of the
health effects associated with climate change are indirect and have
multiple, interrelated causes; people in Russia are less prepared
for multiple days of extreme heat than are residents of Nevada or
Saudi Arabia.

Climate warming has also affected human (and animal) health
indirectly. The ranges of mosquitoes and other disease carriers
have expanded to the newly warm areas and could, in the absence
of other limiting factors, spread malaria, dengue fever, schistosomiasis, and yellow fever, as well as livestock diseases such
as bovine brucellosis. Changes in temperature and precipitation
could increase certain food and waterborne diseases (see Chapter 7). Warmer winters could actually reduce the rates of some
respiratory diseases (like the flu), but shifting of human populations into cities could counteract this effect. Also, local changes
in soil ecology, humidity, and other physical factors could affect
the range and prevalence of fungi, molds, and mildew. **Table
20.2** summarizes some observed and possible impacts of climate
change on diseases in North America.

EFFECTS ON AGRICULTURE

Agriculture, as demonstrated in Chapter 18, is a carefully manipulated ecological system. Consequently, the impacts of global climate change on agriculture are difficult to anticipate. Agricultural
productivity could increase because higher levels of CO_2 in the
atmosphere could allow a higher rate of photosynthesis. However,
many interacting factors are at work. How will growing seasons
change? Will certain pests become more or less of a problem?
How many of the types of crops traditionally grown in a given
area have to change?

The rise in sea level may cause water to inundate river deltas,
which are some of the world's best agricultural lands. Certain agricultural pests and disease-causing organisms will probably proliferate and reduce yields. Scientists think global climate change
will increase the frequency and duration of droughts, a problem
that will be particularly serious for countries with limited water
resources. It is likely that warmer temperatures will result in
decreases in soil moisture in many agricultural soils (warmer temperatures cause increased evaporation).

Another effect of climate warming on agriculture involves
nighttime temperatures, which have generally increased more than
daytime temperatures since 1950, when measurements began.
Changing nighttime temperatures will have positive effects on
some crops, but others—like tomatoes, which set their fruit only
if nighttime temperatures go below a certain level—will become
more difficult to grow. Other crops that require cool summers and/
or winter freezes include blueberries, sugar maples, apples, and
broccoli.

In 1999, a study at the National Science Foundation's Long-
Term Ecological Research Site in Colorado linked warmer nighttime temperatures to changes in the types and distribution of
grasses on the prairie. Most notably, weeds and nonnative grasses
have largely replaced buffalo grass, an important food for cattle
and other livestock. Range scientists point out that buffalo grass
can withstand continuous grazing, but the invading plants replacing buffalo grass may be more sensitive to grazing pressures.

Table 20.2   Some Known and Expected Effects of Climate Change on Human Diseases

Type of Disease	Known Effects	Predicted Effects
Vector-borne* disease (e.g., Lyme disease, malaria)	Warming-related expansion or shift of geographic range of disease vector. Precipitation-related expansion or shift of geographic range. Travel-related risk of disease. Changes in temperature, precipitation could lead to major ecosystem shifts, impacting disease burden and incidence.	Increased transmission of some diseases transmitted from animals to humans. Establishment of novel imported infectious diseases. Regional variability in changes; some areas will be impacted more than others.
Water- and food-borne disease (e.g., *E. coli*, *Salmonella*)	Temperature-related increase in survival of disease organisms. Precipitation-related shifts in geographic range of disease organisms. Increased temperature expected to shift northernmost tolerance of many diseases.	Increased intensity and frequency of water- and food-borne diseases, particularly in the northernmost parts of the United States and Canada.
Respiratory disease (e.g., influenza, *Streptococcus bacillus*)	Shorter/warmer winter could reduce disease incidence and burden. Migration of nonimmune populations to areas with endemic disease increases incidence. Air pollution changes could cause/exacerbate disease.	Reduced number of respiratory diseases with cold weather.** Increased respiratory diseases associated with dense populations.** Increased respiratory diseases associated with diminished air quality.**
Fungal diseases (e.g., *Blastomyces dermaditis*, *Cryptococcus gatti*)	Soil ecology changes could shift range of fungus populations. Warm, dry summers followed by warm, wet winters could increase fungus counts. Changes in temperature, precipitation, and humidity could expand geographic range of some fungi	Growing conditions that favor infectious fungal spores. Regional variability in changes of numbers and types of fungi.

*Vectors are the animals that carry a disease organism or a stage of a disease organism. Mosquitoes are a common vector, but many diseases require a mammalian vector (e.g., fox, deer, rat).
**It is not clear which of these effects will dominate.
Source: Greer, Ny, and Fisman, "Climate Change and Infectious Diseases in North America: The road ahead, *Canadian Medical Association Journal* (2008).

Ecologists suggest that such changes in the structure and dynamics of rangeland ecosystems worldwide could have a profound effect on livestock production.

On a regional scale, current models forecast that with modest warming, agricultural productivity will increase in some areas and decline in others. Canada and Russia may be able to increase their agricultural productivity in a warmer climate, whereas tropical and subtropical regions, where many of the world's poorest people live, will be hardest hit by declining agricultural productivity. Central America and Southeast Asia may experience some of the greatest declines in agricultural productivity.

In addition to these expected impacts of climate warming on agriculture, modern, energy-intensive agricultural methods may have to be altered so that they are less reliant on CO_2–producing fossil fuels (see Chapter 18). The manufacture of fertilizers, pesticides, and other agricultural chemicals requires a huge input of energy from fossil fuels, as does the production and use of modern farm equipment.

Climate-induced changes to agriculture will require cultural, economic, and infrastructure adaptation. Some farmers will do much better, while others will no longer be able to compete. The agriculture-related fraction of GDP will go up in some countries and down in others. The types of crops that do well—and therefore the diets of many people—will change. The extent to which such adaptation can keep pace with climate change is uncertain.

INTERNATIONAL IMPLICATIONS OF GLOBAL CLIMATE CHANGE

Dealing with global climate change is complicated by social, economic, and political factors that vary from one country to another. How will the global community deal with the environmental refugees of global climate change, such as those who might be affected by extreme weather events that lead to agricultural failures? Where will they go? Who will help them resettle? It will be difficult for all countries to develop a consensus on dealing with global climate change, partly because it will clearly have greater impacts on some nations than on others. All major nations must cooperate if we are to effectively address global climate change and its impacts.

Although highly developed countries are the primary producers of greenhouse gases, the rate of production by certain developing countries is rapidly increasing (even though their per capita emissions remain well below those for highly developed countries). In 2007, China surpassed the United States as the largest single contributor of CO_2, although the per capita rate in the United States remains about three times that of China. Furthermore, although highly developed countries have huge amounts of coastal infrastructure at risk, many developing countries may experience the greatest impacts of global climate change: Because developing countries have less technical expertise and fewer economic resources, they are least likely to be able to respond to the challenges of global climate change.

As the economies of developing countries progress, they are likely to follow the path of industrialized countries and consume more fossil fuels—and emit more greenhouse gases. Most of the increase in emissions would be a direct result of providing basic human needs for increasing populations. Because of their rapid population growth, developing countries are predicted to release higher levels of greenhouse gases than industrialized nations by 2020. This scenario may not unfold exactly as projected because some developing countries—such as China, India, Mexico, Saudi Arabia, South Africa, and Brazil—are making progress toward controlling greenhouse gas emissions.

Efforts to reduce carbon emissions include increasing energy efficiency, limiting fossil-fuel use, and developing alternative energy programs. Increased control of greenhouse gas emissions in developing countries may not be the result of specific climate change policies; instead, this may be a secondary benefit of these nations' efforts to meet their own social, economic, and health needs. Air pollution arising from the use of fossil fuels creates serious public health problems in developing countries, so curbing fossil-fuel consumption creates healthier living conditions locally while improving global air quality.

Highly developed and developing countries have different interests, needs, and perspectives. Most developing countries see increased use of fossil fuels as their route to industrial development and resist pressure from highly developed countries to decrease fossil-fuel consumption. Developing countries often ask why they should have to take actions to curb CO_2 emissions when the rich industrialized nations historically have been the main cause of the problem. Since 1950, the 20% of the world's population living in highly developed countries have produced 74% of the CO_2 emissions. Currently, highly developed countries produce about 10 times more CO_2 emissions per person than developing countries (**Figure 20.12**). Highly developed countries counter that the booming economic growth and much greater number of people living in developing countries will dominate global carbon production in the near future.

LINKS AMONG GLOBAL CLIMATE CHANGE, OZONE DEPLETION, AND ACID DEPOSITION

Environmental studies often examine a single issue, such as global climate change, acid deposition, or ozone depletion (see Chapter 19). Canadian researchers at the Experimental Lakes Area in Ontario, Canada, decided to take a different approach and explore the interactions of all three environmental problems simultaneously. They reported that organisms in North American lakes may be more susceptible to damage from ultraviolet (UV) radiation than the thinning of the ozone layer would indicate. The combined effects of acid deposition and climate warming may increase how far UV radiation penetrates lake water. Some of the possible effects of increased UV penetration are to disrupt photosynthesis in algae and aquatic plants and to cause sunburn damage (skin lesions) in fishes.

How far UV radiation penetrates lake water is related to the presence of dissolved organic compounds. This organic material, which is present even in the clearest, least polluted lakes, comes from the decomposition of dead organisms. Dissolved organic material acts like a sunscreen, absorbing UV radiation so that it penetrates only a few inches into the water.

A warmer climate increases evaporation, which reduces the amount of water flowing into a lake from the surrounding watershed. Because most of a lake's dissolved organic compounds wash into the lake in the stream flow, even a slightly drier climate reduces how much organic material is present in the lake. Therefore, UV radiation penetrates deeper into the lake as a result of climate warming.

Acid deposition also affects the amount of dissolved organic compounds in a lake. The presence of acid causes organic matter to clump and settle on the lake floor. The removal of organic material from the water allows UV radiation to penetrate farther.

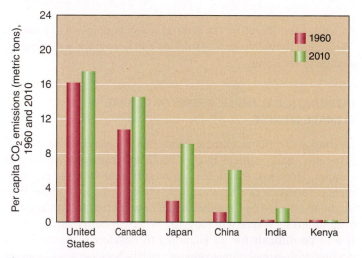

Figure 20.12 **Per capita carbon dioxide (CO_2) emission estimates for selected countries, 1960 and 2010.** For each country, the column on the left is the 1960 value and the column on the right is the 2010 value. In all cases except Kenya, per capita CO_2 emissions have grown. Currently, industrialized nations produce a disproportionate share of CO_2 emissions. As developing countries such as China and India industrialize, however, their per capita CO_2 emissions increase. China's per capita use has more than quadrupled in the past five decades. (World Bank)

1. How do melting ice and thermal expansion of water contribute to sea-level rise?

2. How has climate change impacted precipitation? What other future precipitation changes do we expect?

3. What are some effects of climate change already experienced by organisms, including humans?

Dealing with Global Climate Change

LEARNING OBJECTIVES

- Define *mitigation* and *adaptation* and provide some examples of each.
- Explain why reducing greenhouse gas emissions requires international efforts.

Our understanding of the changing global climate and its potential impact on human society and other species gives us many legitimate reasons to develop strategies to deal with this problem. Even if we immediately stopped greenhouse gas emissions (which we cannot do, given the importance to society of the energy associated with them), global temperature would continue increasing for several decades. Sea levels would continue to rise for centuries as the Earth's climate system adjusts to the influences of the greenhouse gases that have accumulated during the past two centuries. For the same reasons that the global climate system is slow to respond to increases in greenhouse gas concentrations, it will be slow to respond even if emissions are stopped quickly.

Although stopping climate change will require that we deal with all greenhouse gases, our focus will be on CO_2 because it is produced in the greatest quantities and has the largest influence (about 60%) of all the greenhouse gases. The increase in the CO_2 level caused by human activities will persist for centuries, so the emissions we are producing today will still be around in the 22nd century and beyond. The amount and severity of global climate change that will be left to future generations are thus directly related to the amount of additional greenhouse gases we add to the atmosphere during our lifetimes.

To avoid the most dangerous consequences of climate change, many studies indicate that the atmospheric CO_2 concentration needs to be stabilized at 550 ppm. This is roughly twice the concentration of atmospheric CO_2 that scientists estimate existed in the preindustrial world and only about 40% higher than the current CO_2 concentration.

There are two basic ways to manage global climate change: mitigation and adaptation (**Figure 20.13**). **Mitigation** focuses on limiting greenhouse gas emissions to moderate or postpone global climate change, thus buying time to pursue solutions that stop or reverse the change. **Adaptation** focuses on learning to live with the environmental changes and societal consequences brought about by global climate change.

Although some people have objected to the development of strategies to adapt to climate change because they feel this implies an acceptance that global climate change is certain, the widespread observations that the climate is changing make it clear that adaptation will be unavoidable. What is unclear is which changes we can and will prepare for, and which we will need to react to. How we deal with climate change is a decision with multigenerational consequences.

mitigation An action or actions that diminish the causes of climate change, thereby reducing the rate at which climate changes.

adaptation Preparatory actions that help humans tolerate the effects of a changing climate.

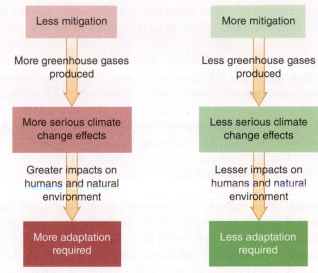

(a) If we make fewer efforts to mitigate climate change, we will have to adapt to more serious problems that impact food and water resources, biological diversity, and human health.

(b) If we take aggressive mitigation measures, the long-range changes in climate will be less and have less serious impacts on humans and the natural environment.

Figure 20.13   Relationship between mitigation and adaptation.

MITIGATION OF GLOBAL CLIMATE CHANGE

Because global warming is essentially a consequence of the choices that are made in the generation and use of energy, alternatives to fossil fuels have the potential to eventually halt the warming caused by CO_2 emissions. Alternatives to petroleum and natural gas (but not coal) are also likely to become necessary over coming decades because these fuels are present in limited amounts. Energy efficiency and conservation, discussed in Chapter 10, could help address a significant part of the climate change problem. Many alternatives to fossil fuels, including solar, nuclear, and wind energy, were discussed in Chapter 12.

Many studies indicate that energy use and greenhouse gas emissions can be significantly reduced with little cost to society by adopting the best currently available technologies and implementing certain policies to encourage their use. For example, increasing the efficiency of automobiles and appliances would reduce the use of fossil fuels and the output of CO_2. In California, regulatory programs that increase minimum energy-efficiency standards for appliances and buildings have cut per capita electricity use to half the U.S. average. The higher price of gasoline has led consumers to want cars with increased gas mileage, which carmakers have proven they can make profitably, as shown by the rising sales of hybrid vehicles.

Energy-pricing strategies, such as carbon taxes and the elimination of energy subsidies, are other policies that could help lead to reductions in emissions. A carbon tax could be levied on fossil fuels, based on the proportion of CO_2 emissions produced per unit of heat released when the fuel is burned (see Chapter 2). Because coal is carbon-intensive, it would have the highest carbon taxes of all fossil fuels. Alternatively, a *cap and trade* system, which limits the total amount of carbon worldwide but lets people buy and sell carbon emission rights, would be an economics-based solution.

Carbon Capture and Storage

In addition to taking steps to curb greenhouse gas emissions, many countries are investigating **carbon management**, ways to capture and store CO_2. Government incentives, such as providing research grants for the development of such technologies, most likely will be necessary to inspire innovations. As discussed in Chapter 11, carbon can be removed from the flue gases of coal-fired power plants, although this is a new and expensive technology.

Carbon capture and storage (CCS), also called **carbon sequestration**, would require a dramatic shift in where fossil fuels are used. Trying to remove the CO_2 from the engine in a gasoline-powered car or diesel train would be a huge logistical challenge: Capturing CO_2 is difficult, and transporting CO_2 as a gas or as calcium carbonate ($CaCO_3$) to some separate location would require additional energy. Most proposals for CCS involve generating electricity or hydrogen at a fixed location where the CO_2 can most easily be captured.

Sequestering Carbon in Trees

One way to mitigate global climate change involves removing atmospheric carbon dioxide from the air by planting and maintaining forests (recall the chapter introduction). Like other green plants, trees incorporate the carbon into organic matter in leaves, stems, and roots through the process of photosynthesis. Because trees typically live for 100 or more years, the carbon in their roots and stems remains sequestered away from the atmosphere for a relatively long time. Although estimates vary widely, planting trees could probably remove 10% to 15% of the excess CO_2 in the atmosphere; to do so would require enormous plantings. Scientists who have studied sequestering carbon in trees think this proposal might provide short-term benefits for the climate, although it is no substitute for cutting emissions of greenhouse gases.

Geoengineering

Considerable controversy surrounds the idea of using **geoengineering** to mitigate climate change. Geoengineering refers to projects that take place on a global scale. Some possibilities include seeding the ocean with iron to allow more carbon to be taken up by algae, building devices that extract massive amounts of carbon from the atmosphere, and ejecting sulfur particles into the atmosphere to reflect sunlight. One of the biggest concerns is that it is difficult to predict the efficacy of such large-scale projects, and they might either not succeed or cause unintended consequences. Proponents of geoengineering argue that the rapid rise in greenhouse gases was caused by humans and is thus an example of geoengineering and that a human intervention of an equivalent scale is appropriate.

ADAPTATION TO GLOBAL CLIMATE CHANGE

Because the overwhelming majority of climate experts think that significant human-induced global climate change is inevitable (the only question being how significant), government planners and social scientists are developing strategies to help various regions and sectors of society adapt to climate warming. One of the most pressing issues is rising sea level. People living in coastal areas could be moved inland, away from the dangers of storm surges. These people would become **climate change refugees**, people forced by climatic change to abandon their homes. This solution would have high societal and economic costs, especially given the increasing fraction of the population living near the ocean. Another alternative is the construction of dikes and levees to protect coastal land—an expensive option, but perhaps less so than relocating the people protected by such constructions. Rivers and canals that spill into the ocean would have to be channeled to prevent saltwater intrusion into fresh water and agricultural land.

We will also need to adapt to shifting agricultural zones. Many countries with temperate climates are evaluating semitropical crops to determine the best substitutes for traditional crops as the climate warms. Large lumber companies are currently developing drought-resistant strains of trees. Trees planted today will be harvested in the last half of the 21st century, when global climate change may be well advanced.

Adaptation to global climate change is currently under study at locations around the United States. Study groups typically consist of scientists and various representatives of municipal governments, state and federal agencies, local businesses, and community organizations. One of the potential problems identified in the New York City study group involves its sewer system. The waterways for storm runoff normally close during high tides to prevent salt water from the Atlantic Ocean from backing into the system. As the sea level rises in response to global climate change, the waterways will have to be shut during many low tides, which will greatly increase the risk of flooding during storms. (When the waterways are closed, excess water does not drain away.) City planners will have to rebuild the storm runoff system, which is an expensive proposition, or find some other way to prevent flooding. Evaluating such problems and finding and implementing solutions now will ease the future stresses of climate change.

INTERNATIONAL EFFORTS TO REDUCE GREENHOUSE GAS EMISSIONS

Despite all the posturing, the international community recognizes that it must stabilize CO_2 emissions. At least 192 nations, including the United States, have now signed the UN Framework Convention on Climate Change (UNFCCC) developed at the 1992 Earth Summit. Its ultimate goal was to stabilize greenhouse gas concentrations in the atmosphere at levels low enough to prevent dangerous human influences on the climate. The details of how that goal was to be accomplished were left to future conferences.

At the 1996 meeting of the parties to the UNFCCC held in Geneva, Switzerland, highly developed countries agreed to establish legally binding timetables to cut emissions of greenhouse gases. These timetables were decided at a meeting of representatives from 160 countries held in Kyoto, Japan, in 1997. By 2014, 192 countries had ratified the resulting **Kyoto Protocol**. This international treaty, which is legally binding, provides operational rules on reducing greenhouse gas emissions.

The U.S. government is involved in ongoing national and international climate policy activities. It supports a continuation of UNFCCC negotiations. Within the United States, policy is being driven by the administration's commitment to reducing greenhouse gases. In 2007, the Supreme Court ruled in *Massachusetts v. EPA* that greenhouse gases are covered under the Clean Air Act as an air pollutant, and in June 2014, the Obama administration established regulations for emissions from all coal-fired power plants.

REVIEW

1. What are some examples of mitigation and adaptation?
2. How do perspectives on climate change of highly developed countries differ from those of less developed countries?

REVIEW OF LEARNING OBJECTIVES WITH SELECTED KEY TERMS

- **Explain radiative forcing, greenhouse gases, and the enhanced greenhouse effect.**

Radiative forcing is the capacity of a gas to affect the balance of energy that enters and leaves the atmosphere. **Greenhouse gases** are the gases in Earth's atmosphere that absorb **infrared radiation.** The **enhanced greenhouse effect** is the additional warming produced by increased levels of gases that absorb infrared radiation (heat). The Intergovernmental Panel on Climate Change concluded in 2007 that human-produced greenhouse gases are the most likely cause of recent climate warming and that the world will almost certainly warm substantially during the 21st century.

- **Explain how climate models project future climate conditions.**

Climate models are computer models that describe the global climate as a system. The models divide the atmosphere and oceans into small, three-dimensional parts and evaluate the effects of changes in one part on adjacent parts. These models incorporate feedbacks that influence such factors as temperature, wind patterns, cloud moisture, and ice cover. Running these models based on different predicted levels of CO_2 leads to projections of possible future climate conditions. Models include **positive feedback**, in which a change in some condition triggers a response that intensifies the changed condition, and **negative feedback**, in which a change in some condition triggers a response that moderates the changed condition.

- **Describe the importance of extreme and unpredictable climate change.**

While we often look at expected results of climate models, the extreme or worst-case scenarios are often of much greater concern. Unpredictable changes are perhaps more problematic. It is almost certain that there will be surprises—things we cannot predict. But not knowing what the surprises will be makes it impossible to prepare for or mitigate them. The only way to avoid such surprises is to eliminate climate change.

- **Differentiate between sea-level rise due to melting ice and that due to the thermal expansion of water.**

More than half of recent and projected sea-level rise is associated with the fact that water expands as it heats up; this is known as **thermal expansion**. How quickly the ocean absorbs increased atmospheric heat will significantly influence the rate of sea-level rise. Melting of glaciers and land-based ice sheets is the other major contributor to sea-level rise, as water previously "stored" on land shifts to the ocean.

- **Describe how climate change impacts the physical environment.**

Globally, average ocean and land surface temperatures are increasing, although effects at individual locations can vary. Precipitation and other weather patterns have begun to shift, with increased flooding in some locations and increased intensity of tropical storms. Ice sheets and glaciers are melting. All of these phenomena are expected to intensify in the future.

- **Give examples of effects of climate change on organisms, including humans.**

Globally, organisms are affected by changes in temperature, precipitation, ocean acidification, sea-level rise, and other climate change–related effects. They are also affected by changes in other populations—for example, some organisms react more to the early arrival of spring than do others, which means that food webs can be disrupted. Humans are affected by heat waves, flooding, and sea-level rise, as well as changes to agriculture.

- **Define *mitigation* and *adaptation* and provide some examples of each.**

Mitigation is an action or actions that diminish the causes of climate change. Examples include burning less oil, natural gas, and coal; planting trees; and sequestering carbon dioxide. **Adaptation** consists of preparatory actions intended to diminish the effects of a changing climate. Examples include moving people away from shorelines and changing agricultural practices.

- **Explain why reducing greenhouse gas emissions requires international efforts.**

Climate change has global impacts. Reducing carbon emissions and adapting to a changed climate cannot succeed when only a few countries participate. One concern is that, in the short term, a country with strict CO_2 restrictions will be at a competitive disadvantage compared to a country without such restrictions. Thus, having international commitments may be necessary for success in any location.

Critical Thinking and Review Questions

1. One of Socolow and Pacala's wedges (see the introduction) involves reducing the number of miles driven by each car by 50% (they predict 2 billion cars worldwide by 2056). Considering what you know about how people drive now, how could people reduce their driving by 50%? How would this affect their lives?

2. How do greenhouse gases affect global climate conditions?

3. What are climate models? What inputs are used when building them? What can they tell us about future climate?

4. What two factors contribute to sea-level rise?

5. What effect has increased evaporation due to climate change had on regional weather patterns?

6. Austrian biologists who study plants growing high in the Alps found that plants adapted to cold mountain conditions migrated up the peaks as fast as 3.7 m (12 ft) every decade during the 20th century, apparently in response to climate warming. Assuming that warming continues during the 21st century, what will happen to the plants if they reach the tops of the mountains?

7. Suggest some ways in which seasonal changes in plant growth in Canada might affect songbird populations in Mexico.

8. What aspects of adaptation to global climate change would be easier for highly developed countries? For developing countries? Explain.

9. Based on what you've read in this chapter and the information on environmental economics in Chapter 2, explain how a global market for CO_2 permits might function.

10. Insurance companies that provide policies for hurricanes and other natural disasters may shift hundreds of millions of dollars of their investments from fossil fuels to solar energy. On the basis of what you have learned in this chapter, explain why insurance companies would consider such an investment in their best interest.

11. Some environmentalists contend that the wisest way to "use" fossil fuels is to leave them in the ground. How would this affect air pollution? Global climate change? Energy supplies?

12. What do the relationships among acid deposition, stratospheric ozone, and climate change tell us about the importance of thinking about the environment as a system?

13. **John Holdren**, science advisor to President Obama, argues that there are three possible human responses to climate change: mitigation, adaptation, and suffering. Discuss the implications of this for people today and people 50 years from now.

14. Sea level is expected to rise about 0.4 m (1.3 ft) by 2100 (the blue line in the figure below). However, the prediction is uncertain; different models suggest that we can expect anywhere from 0.2 m (green line) to 0.6 m (red line). Which, if any, of these increases should we prepare for? Explain your choice.

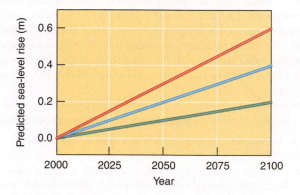

Food for Thought

The U.S. Department of Agriculture Hardiness Zones maps are likely to be updated again soon, and further shifts are expected. Consider foods that are currently produced by small farms and home gardeners in your area. How will their choices of what to grow change? What might disappear from your local farmers market? What will replace it? Will this impact your food choices?

21

Water Pollution

AP Photo/Gerry Broome

Coal ash spill in the Dan River. Coal ash swirls in the Dan River in Danville, Virginia, downstream from the North Carolina source of the major release, Duke Energy's coal ash basin, February 5, 2014.

Fossil-fuel consumption and use contaminate water in many places and in many ways. Coal, oil, and natural gas extraction all require water, often in large amounts. Oil and coal (and to a lesser extent liquefied natural gas) are transported around the world on rivers and the ocean. Power plants are invariably built near access to water, which can be used for cooling, processing, and waste management. All of these activities lead to both routine and accidental releases that affect human and ecosystem health.

Many types of water pollution are associated with oil. Oil spilled in the ocean can contaminate large areas of water and shoreline, with long-lasting impacts. Oil spilled on roads and in driveways can lead to large amounts of pollution running off into lakes, streams, and rivers. Leaky underground gasoline tanks can contaminate groundwater, and drilling for and refining oil invariably pollute groundwater and surface water.

Coal, too, can cause massive amounts of water pollution. Water passing through mined areas can pick up contaminants ranging from loose soil to highly acidic mine tailings. **Coal ash**, or fly ash, the solid material left over after coal is burned, can also cause considerable ecosystem damage. It typically contains toxic materials, including such heavy metals as nickel, arsenic, and mercury

as well as radioactive uranium, radium, and thorium. While some of the ash generated around the world is used in manufacturing (e.g., to make Portland cement), most is either stored in piles or ponds where it is generated—including hundreds of U.S. sites—or is hauled to a landfill. Two coal-related accidents in early 2014 illustrate the range of water pollution risks associated with coal use. In January 2014, 10,000 gallons of MCHM, a coal-processing chemical, leaked from a ruptured storage tank directly into West Virginia's Elk River just upstream of a major water distribution facility. Water supplies were interrupted and compromised for more than 250,000 people; long-term effects are unknown.

More than 30,000 gallons of coal ash poured into North Carolina's Dan River in early February 2014, originating from a faulty pipe under an abandoned Duke Energy power plant's coal ash basin. The third worst coal ash spill in U.S. history, the accident dumped a 70-mile plume of sludge and toxins downstream (see photograph).

The nation's most catastrophic coal ash spill occurred in late 2008, when 7 million m^3 (9.2 million yd^3) of coal ash—equivalent to more than 1 billion gallons—spilled from a holding pond at a Tennessee Valley Authority coal power plant in eastern Tennessee.

Triggered when heavy rains overwhelmed the pond, the spill spread liquid ash over 300 acres of land and into the Emory River. Water tests after the spill showed elevated lead and thallium levels, both of which are known to cause illness to humans. Several houses were damaged, and many plants and animals were killed. We will never be certain how big an effect the spill had on ecosystems or humans downstream.

So long as we use fossil fuels for energy, both routine activities like driving and accidents like oil and ash spills will continue to consume and pollute water. Improved technologies can help, but contamination will invariably occur. "Clean-coal" technologies may reduce carbon dioxide emissions but will have little impact on the adverse effects from coal mining and ash. Reducing reliance on fossil fuels is the best way to reduce the range of associated environmental impacts—including oil and fly ash spills, climate change, and related water pollution.

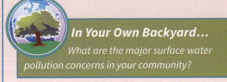

In Your Own Backyard...
What are the major surface water pollution concerns in your community?

Types of Water Pollution

Water pollution is a global problem that varies in magnitude and type of pollutant from one region to another. In many locations,

> **water pollution** Any physical or chemical change in water that adversely affects the health of humans and other organisms.

particularly in less developed countries, the main water pollution issue is lack of disease-free drinking water. Water pollutants are divided into eight categories: sewage, disease-causing agents, sediment pollution, inorganic plant and algal nutrients, organic compounds, inorganic

chemicals, radioactive substances, and thermal pollution. These eight types are not exclusive; for example, sewage can contain disease-causing agents, inorganic plant and algal nutrients, and organic compounds.

SEWAGE

The release of **sewage** into water causes several pollution problems. First, because it carries disease-causing agents, water polluted with sewage poses a threat to public health

> **sewage** Wastewater from drains or sewers (from toilets, washing machines, and showers); includes human wastes, soaps, and detergents.

(see the next section on disease-causing agents). Sewage also generates two serious environmental problems in water—enrichment and oxygen demand. **Enrichment**, the fertilization of a body of water, is caused by high levels of plant and

algal nutrients such as nitrogen and phosphorus. Microorganisms decompose sewage and other organic materials into carbon dioxide (CO_2), water, and similar inoffensive materials. This degradation process, known as **cellular respiration**, requires the presence of oxygen. Fishes and other organisms in healthy aquatic ecosystems also use oxygen. Since only limited amounts of oxygen can dissolve in water, in an aquatic ecosystem containing high levels of sewage or other organic material, the decomposing microorganisms use up most of the dissolved oxygen. This leaves little for fishes or other aquatic animals. At extremely low oxygen levels, fishes and other animals leave or die.

Sewage and other organic wastes are measured in terms of

> **biochemical oxygen demand (BOD)** The amount of oxygen microorganisms need to decompose biological wastes into carbon dioxide, water, and minerals.

their **biochemical oxygen demand (BOD)**, or **biological oxygen demand**. BOD is usually expressed as milligrams of dissolved oxygen per liter of water for a specific number of days at a given temperature. A large amount of sewage in water generates a high BOD, which robs the water of dissolved oxygen (**Figure 21.1**). When

dissolved oxygen levels are low, anaerobic (without oxygen) microorganisms produce compounds with unpleasant odors, further deteriorating water quality.

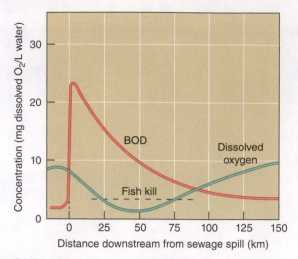

Figure 21.1 **Effect of sewage on dissolved oxygen and biochemical oxygen demand (BOD).** Note the initial oxygen depletion (green line) and increasing BOD (rust line) close to the sewage spill (at distance 0). The stream gradually recovers as the sewage is diluted and degraded. As indicated on the graph, fishes cannot live in water that contains less than 4 mg of dissolved oxygen per liter of water.

Eutrophication: An Enrichment Problem Lakes, estuaries, and slow-flowing streams that have minimal levels of nutrients are unenriched, or oligotrophic. An **oligotrophic** lake has clear water and supports small populations of aquatic organisms (**Figure 21.2a**). **Eutrophication** is the enrichment of a lake, estuary, or slow-flowing stream by inorganic plant and algal nutrients such as phosphorus; an enriched body of water is said to be **eutrophic**. The enrichment of water, which occurs gradually, results in an increased photosynthetic productivity. The water in a eutrophic lake is cloudy and usually resembles pea soup because of the presence of vast numbers of algae and cyanobacteria (**Figure 21.2b**).

Although eutrophic lakes contain large populations of aquatic animals, these organisms are different from those predominant in oligotrophic lakes. For example, an unenriched lake in the northeastern United States may contain pike, sturgeon, and whitefish (**Figure 21.2c**). All three are found in the deeper, colder part of the lake, where there is a higher concentration of dissolved oxygen. In eutrophic lakes, on the other hand, the deeper, colder levels of water are depleted of dissolved oxygen because when the excessive numbers of algae die, they settle to the lake's bottom and stimulate an increased amount of decay. Microorganisms that decompose the dead algae use up much of the lake's dissolved oxygen in the process. The lake floor has a high BOD from decomposition, and fishes such as pike, sturgeon, and whitefish die out and are replaced by warm-water fishes, such as catfish and carp, that tolerate smaller amounts of dissolved oxygen (**Figure 21.2d**).

Over vast periods, oligotrophic lakes, estuaries, and slow-moving streams naturally become eutrophic. As natural eutrophication occurs, these bodies of water are slowly enriched and grow shallower from the immense number of dead organisms that have settled in the sediments over a long period. Gradually, plants

(a) Crater lake, an oligotrophic lake in Oregon.

(b) A small eutrophic lake in the Catskill Mountains, New York.

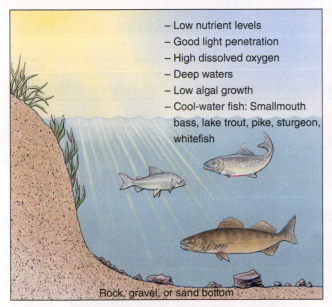

- Low nutrient levels
- Good light penetration
- High dissolved oxygen
- Deep waters
- Low algal growth
- Cool-water fish: Smallmouth bass, lake trout, pike, sturgeon, whitefish

Rock, gravel, or sand bottom

(c) An oligotrophic lake has a low level of inorganic plant and algal nutrients.

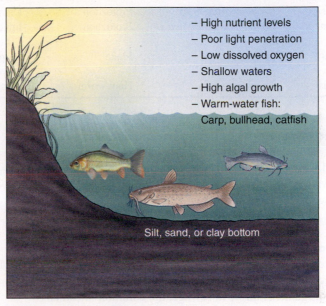

- High nutrient levels
- Poor light penetration
- Low dissolved oxygen
- Shallow waters
- High algal growth
- Warm-water fish: Carp, bullhead, catfish

Silt, sand, or clay bottom

(d) A eutrophic lake has a high level of these nutrients.

Figure 21.2   Oligotrophic and eutrophic lakes.

artificial eutrophication Overnourishment of an aquatic ecosystem by nutrients such as nitrates and phosphates; due to human activities such as agriculture and discharge from sewage treatment plants.

such as water lilies and cattails take root in the nutrient-rich sediments and begin to fill in the shallow waters, forming a marsh. Some human activities, however, greatly accelerate eutrophication. This fast, human-induced process is usually called **artificial eutrophication**, or *cultural eutrophication*, to distinguish it from natural eutrophication. Artificial eutrophication results from the enrichment of aquatic ecosystems by nutrients found predominantly in fertilizer runoff and sewage.

DISEASE-CAUSING AGENTS

Disease-causing agents are infectious organisms that cause diseases; they come from the wastes of infected individuals, entering water in sewage. Municipal wastewater usually contains many bacteria, viruses, protozoa, parasitic worms, and other infectious agents that cause human or animal diseases (**Table 21.1**). Typhoid, cholera, bacterial dysentery, polio, and infectious hepatitis are some of the more common bacterial or viral diseases transmitted through contaminated food and water. Most of these are rare in highly developed countries but major causes of death in less developed countries. However, many human diseases, such as acquired immune deficiency syndrome (AIDS), are not transmissible through water.

The vulnerability of public water supplies to waterborne disease–causing agents was dramatically demonstrated in 1993 when a microorganism (*Cryptosporidium*) contaminated the water supply in the greater Milwaukee area. About 370,000 people developed diarrhea, making it the largest outbreak of a waterborne disease ever recorded in the United States, and several people

Table 21.1 **Some Human Diseases Transmitted by Polluted Water**

Disease	Infectious Agent	Type of Organism	Symptoms
Cholera	*Vibrio cholerae*	Bacterium	Severe diarrhea, vomiting; fluid loss of as much as 20 quarts per day causes cramps and collapse
Dysentery	*Shigella dysenteriae*	Bacterium	Infection of the colon causes painful diarrhea with mucus and blood in the stools; abdominal pain
Enteritis	*Clostridium perfringens,* other bacteria	Bacterium	Inflammation of the small intestine causes general discomfort, loss of appetite, abdominal cramps, and diarrhea
Typhoid	*Salmonella typhi*	Bacterium	Early symptoms include headache, loss of energy, fever; later, a pink rash appears along with (sometimes) hemorrhaging in the intestines
Infectious hepatitis	Hepatitis virus A	Virus	Inflammation of liver causes jaundice, fever, headache, nausea, vomiting, severe loss of appetite, muscle aches, and general discomfort
Poliomyelitis	Poliovirus	Virus	Early symptoms include sore throat, fever, diarrhea, and aching in limbs and back; when infection spreads to spinal cord, paralysis and atrophy of muscles occur
Cryptosporidiosis	*Cryptosporidium* sp.	Protozoon	Diarrhea and cramps last up to 22 days
Amoebic dysentery	*Entamoeba histolytica*	Protozoon	Infection of the colon causes painful diarrhea with mucus and blood in the stools; abdominal pain
Schistosomiasis	*Schistosoma* sp.	Fluke	Tropical disorder of the liver and bladder causes blood in urine, diarrhea, weakness, lack of energy, repeated attacks of abdominal pain
Ancylostomiasis	*Ancylostoma* sp.	Hookworm	Symptoms are severe anemia and sometimes symptoms of bronchitis

with weakened immune systems died. Smaller outbreaks of contamination by salmonella and other bacteria have occurred in several cities and towns since the Milwaukee episode. Nonetheless, deaths from waterborne diseases remain rare in highly developed countries: Fewer than 10 deaths occur over a typical year in the United States.

Monitoring for Disease from Sewage Because sewage-contaminated water is a threat to public health, periodic tests are made for the presence of sewage in our water supplies. Although many different microorganisms thrive in sewage, the common intestinal bacterium *Escherichia coli* (*E. coli*) is typically used as an indication of the amount of sewage present in water and as an indirect measure of the presence of disease-causing agents. *E. coli* is perfect for monitoring sewage because it is not present in the environment except from human and animal feces, where it is found in large numbers. To test for the presence of *E. coli* in water, a **fecal coliform test** is performed (**Figure 21.3**). A

fecal coliform test A water quality test for the presence of fecal bacteria, which indicates a chance that pathogenic organisms may be present as well.

small sample of water is passed through a filter to trap all bacteria. The filter is then transferred to a petri dish that contains nutrients. After an incubation period, the number of greenish colonies present indicates the number of *E. coli*. Safe drinking water should contain no more than 1 coliform bacterium per 100 mL of water (about half a cup), safe swimming water should have no more than 200 per 100 mL of water, and general recreational water (for boating) should have no more than 2000 per 100 mL. In contrast, raw sewage may contain several million coliform bacteria per 100 mL of water. Although most strains of coliform bacteria do not cause disease, the fecal coliform test is a reliable way to indicate the likely presence of pathogens, or disease-causing agents, in water.

When dangerous levels of fecal coliform bacteria are discovered in a stream or other body of water, it is important to determine

(a) A water sample is first passed through a filtering apparatus. The filter disk is then placed on a medium that supports coliform bacteria for a period of 24 hours.

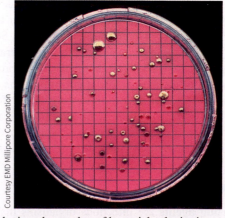

(b) After incubation, the number of bacterial colonies is counted. Each colony of *Escherichia coli* arose from a single coliform bacterium in the original water sample.

Figure 21.3 **Fecal coliform test.**

the source of contamination. Finding the source is not always easy because coliform bacteria reside in the intestinal tracts of many animals. The contamination could be coming from human wastes, such as from septic systems that are not operating effectively (discussed later in the chapter); from animal feedlots; or even from the droppings of raccoons, birds, and other wildlife. **Bacterial source tracking (BST)** allows investigators to identify subtle differences in strains of *E. coli* on the basis of their animal hosts.

SEDIMENT POLLUTION

Clay, silt, sand, and gravel can be suspended and carried in water. When a river flows into a lake or ocean, its flow velocity decreases, and the sediments often settle out. Over time, as sediments accumulate, new land is formed. A river delta is a flat, low-lying plain created from deposited sediments. River deltas, with their abundant wildlife and waterways for trade routes, have always been important settlement sites for people. Today, river deltas are among the most densely populated areas in the world. Sediments are also deposited on land when a river overflows its banks during a flood.

Sediments are not necessarily pollutants; they are important, for example, in regenerating soils in agricultural areas and providing essential nutrients to wetland areas. **Sediment pollution** occurs when excessive amounts of suspended soil particles eventually settle out and accumulate on the bottom of a body of water. Sediment pollution comes from erosion of agricultural lands, forest soils exposed by logging, degraded stream banks, overgrazed rangelands, strip mines, and construction. Control of soil erosion reduces sediment pollution in waterways.

Sediment pollution reduces light penetration, covers aquatic organisms, brings insoluble toxic pollutants into the water, and fills in waterways. When sediment particles are suspended in the water, they make the water turbid (cloudy), which in turn decreases the distance that light penetrates (**Figure 21.4a**). Because the base of the food web in an aquatic ecosystem consists of photosynthetic algae and plants that require light for photosynthesis, turbid water lessens the ability of producers to photosynthesize. Extreme turbidity reduces the number of photosynthesizing organisms, which in turn causes a decrease in the number of aquatic organisms that feed on the primary producers (**Figure 21.4b, c**). Sediment that settles out of the water and forms a layer over coral reefs or shellfish beds can clog the gills and feeding structures of many aquatic animals.

Sediments adversely affect water quality by carrying toxic chemicals, both inorganic and organic, into the water. The sediment particles provide surface area to which some insoluble, toxic compounds adhere, so that when sediments get into water, the toxic chemicals get in as well. Disease-causing agents are also transported into water via sediments.

On the basis of the growing recognition that most of the toxic pollutants in water are stored in

© Frans Lanting Studio/Alamy

(a) A turbid stream in Costa Rica carries a heavy sediment load.

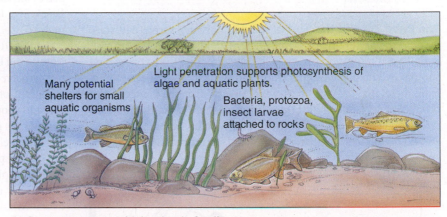

Many potential shelters for small aquatic organisms

Light penetration supports photosynthesis of algae and aquatic plants.

Bacteria, protozoa, insect larvae attached to rocks

(b) Stream ecosystem with low level of sediment.

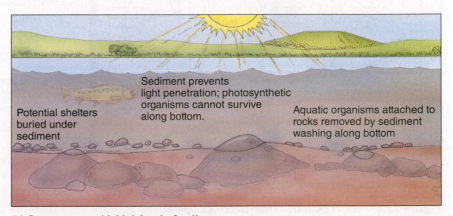

Potential shelters buried under sediment

Sediment prevents light penetration; photosynthetic organisms cannot survive along bottom.

Aquatic organisms attached to rocks removed by sediment washing along bottom

(c) Same stream with high level of sediment.

Figure 21.4 **Sediment pollution.**

and released from sediments, the U.S. Environmental Protection Agency (EPA) evaluated data from more than 21,000 sampling stations in 1363 U.S. watersheds. The resulting survey determined that sediments in 7% of watersheds are so seriously contaminated with toxic pollutants that eating fishes from those waterways would potentially threaten human health.

When sediments settle out of solution, they fill in waterways. This problem is particularly serious in lakes and channels through which ships must pass. Thus, sediment pollution may adversely affect the shipping industry.

INORGANIC PLANT AND ALGAL NUTRIENTS

Chemicals such as nitrogen and phosphorus that stimulate the growth of plants and algae, known as **inorganic plant and algal nutrients**, are essential for the normal functioning of healthy ecosystems but are harmful in larger concentrations. Nitrates and phosphates come from such sources as human and animal wastes, plant residues, atmospheric deposition, and fertilizer runoff from agricultural and residential land. Inorganic plant and algal nutrients encourage excessive growth of algae and aquatic plants. Although algae and aquatic plants are the base of the food web in aquatic ecosystems, their excessive growth disrupts the natural balance between producers and consumers and causes other problems, including enrichment and bad odors. In addition, high BOD occurs when the excessive numbers of algae die and are decomposed by bacteria.

The Dead Zone in the Gulf of Mexico   Every spring and summer, fertilizer runoff from midwestern fields and manure runoff from livestock operations in such states as Iowa, Wisconsin, and Illinois eventually find their way into the Mississippi River and, from there, into the Gulf of Mexico. Livestock in the United States produce approximately 20 times the feces and urine that humans do, yet many of these wastes are not covered by water quality laws and do not go to sewage treatment plants (**Figure 21.5**).

Courtesy University of Wisconsin-Extension and the Wisconsin Department of Natural Resources

Figure 21.5   Water pollution from livestock operations. How is it that a dairy farm in Wisconsin can contribute to fish deaths off the coast of Louisiana? Why is the effect seasonal?

Nitrogen and phosphorus from the Mississippi River are largely responsible for a huge **dead zone** in the Gulf of Mexico (**Figure 21.6**). The dead zone extends from the seafloor up into the water column, sometimes to within a few meters of the surface. Floods, droughts, and temperature change its size and shape. It generally persists from March or April, as snowmelt and spring rains flow from the Mississippi River into the Gulf, to September. It is most severe in June, July, and August. Although the size of the dead zone varies with weather conditions, overall it seems to be growing. In 2013, the Gulf of Mexico dead zone covered about 15,100 km² (5,800 mi²), an area the size of Connecticut.

> **dead zone** A section of the ocean or a sea in which oxygen has been depleted to the point that most animals and bacteria cannot survive; often caused by runoff of chemical fertilizers or plant and animal wastes.

Other than bacteria that thrive in oxygen-free environments, no life exists in the dead zone. The water does not contain enough dissolved oxygen to support fishes or other aquatic organisms. Fishes, shrimp, and other active swimmers avoid the area, while bottom dwellers such as sea stars, brittle stars, worms, and clams suffocate and die.

This low-oxygen condition, known as **hypoxia**, occurs when algae grow rapidly because of the presence of nutrients in the water. Dead algae sink to the bottom and are decomposed by bacteria, which deplete the water of dissolved oxygen, leaving too little for other sea life. In 2010, hypoxia was reported in more than 400 coastal areas around the world. The dead zone in the Gulf of Mexico is one of the largest in the ocean but is smaller than the dead zone in the Black Sea. Ocean warming associated with global climate change may be exacerbating these dead zones.

In 2001, the EPA recommended reductions of nitrogen in the Mississippi River by 30% by 2015. Such reductions, however, require that farms reduce their use of fertilizer. (The Mississippi River drains all or part of 31 states and 2 Canadian provinces, and its watershed contains more than half of all U.S. farms.) In 2004, the EPA added phosphorus pollution as a cause of the dead zone and recommended a focus on cleaning up phosphorus as well as nitrogen.

Other potential inorganic nutrient sources, such as sewage treatment plants and airborne nitrogen oxides from automobile emissions, also must be addressed. Restoring former wetlands in the Mississippi River watershed would reduce the nitrate and phosphate load from fertilizers entering the Gulf of Mexico. The EPA recognizes that the dead zone problem is immense in scope and will take billions of dollars and decades of effort to fix.

ORGANIC COMPOUNDS

Organic compounds are chemicals that contain carbon atoms; a few examples of natural organic compounds are sugars, amino acids, and oils. Most of the thousands of organic compounds found in water are human-produced chemicals; these synthetic chemicals include pharmaceuticals, pesticides, solvents, industrial chemicals, and plastics. (Several examples of synthetic organic compounds sometimes found in polluted water are given in **Table 21.2**). Some organic compounds seep from landfills into surface water and groundwater. Others, such as pesticides, leach downward through the soil into groundwater or get into surface water via runoff from farms and residences. Some industries dump organic compounds directly into waterways.

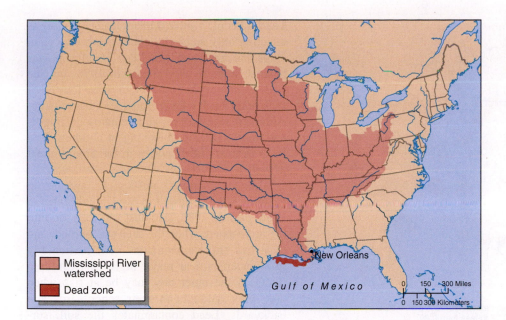

Figure 21.6 The dead zone in the Gulf of Mexico. Pollution from the Mississippi River's water basin is responsible for the dead zone (red). Climate modelers expect more rain in the central United States over the next several decades. Do you think this will impact the dead zone? How?

Legend:
- Mississippi River watershed
- Dead zone

New Orleans

Gulf of Mexico

0   150   300 Miles
0 150 300 Kilometers

In 2006, the U.S. Geological Survey (USGS) conducted the most recent study of **volatile organic compounds (VOCs)** in the nation's groundwater. Researchers found one or more VOCs in 20% of the over 3000 samples taken from domestic and public wells for the study. Fifteen types of VOCs, including trihalomethanes (including chloroform and bromoform), refrigerants, MTBE (a gasoline additive), and toluene (a component of gasoline), were present in more than 1% of the wells. **Figure 21.7** shows the distribution of VOCs in aquifers around the United States.

A similar USGS study in 2002 tested 139 U.S. streams for 95 different synthetic organic compounds, including antibiotics,

ibuprofen, acetaminophen, insect repellents, antimicrobial substances, fragrances, caffeine, and steroids such as hormones from birth-control pills and hormone therapy. Low concentrations—in the range of **parts per billion**—of 82 of these organic compounds were detected in 80% of the streams studied.

The effects on human health of ingesting drinking water containing traces of these VOCs and synthetic organic chemicals are generally unknown. However, the hormone contaminants cause problems in many aquatic organisms (see Chapter 7). Of the organic chemicals found in the 2002 study, 33 are suspected **endocrine disrupters** that cause hormonal effects (see Chapter 7).

Table 21.2   Some Synthetic Organic Compounds Found in Polluted Water

Compound	Some Reported Health Effects
Aldicarb (pesticide)	Attacks nervous system
Benzene (solvent)	Associated with blood disorders (bone marrow suppression); leukemia
Carbon tetrachloride (solvent)	Possibly causes cancer; liver damage; may also attack kidneys and vision
Chloroform (solvent)	Possibly causes cancer
Dioxins (TCDD) (chemical contaminants)	Some cause cancer; may harm reproductive, immune, and nervous systems
Ethylene dibromide (EDB) (fumigant)	Probably causes cancer; attacks liver and kidneys
Polychlorinated biphenyls (PCBs) (industrial chemicals)	Attack liver and kidneys; possibly cause cancer
Trichloroethylene (TCE) (solvent)	Probably causes cancer; induces liver cancer in mice
Vinyl chloride (plastics industry)	Causes cancer

Source: Adapted from the International Agency for Research on Cancer, an agency of the World Health Organization.

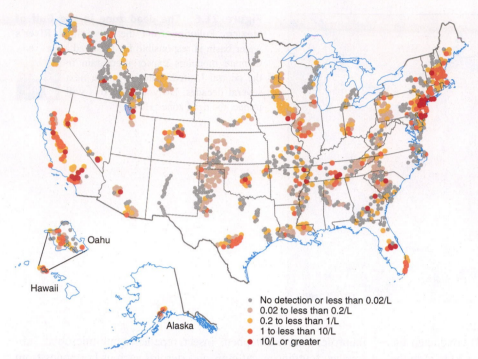

- ● No detection or less than 0.02/L
- ● 0.02 to less than 0.2/L
- ● 0.2 to less than 1/L
- ● 1 to less than 10/L
- ● 10/L or greater

Figure 21.7 **Volatile organic compounds in U.S. groundwater.** Volatile organic compounds (VOCs) are found in groundwater around the United States. While often found in low concentrations (<0.02 μg/L), concentrations exceed 10 μg/L in some public and domestic wells. ("Volatile Organic Compounds in the Nation's Ground Water and Drinking-Water Supply Wells," U.S. Geological Survey Circular 1292, 2006 [USGS])

There are several ways to control the presence of organic compounds in our water. Everyone, from individual homeowners to large factories, should take care to prevent organic compounds from ever finding their way into water. Alternative organic compounds, which are less toxic and degrade more readily so that they are not as persistent in the environment, can be developed and used. Tertiary water treatment, considered later in this chapter, eliminates many synthetic organic compounds in water.

INORGANIC CHEMICALS

Inorganic chemicals are contaminants that contain elements other than carbon; examples include acids, salts, and heavy metals. Inorganic chemicals do not easily degrade, or break down. When they are introduced into a body of water, they remain there for a long time. Many inorganic chemicals find their way into both surface water and groundwater from sources such as industries, mines, irrigation runoff, oil drilling, and urban runoff from storm sewers. Some of these inorganic pollutants are toxic to aquatic organisms. Their presence may make water unsuitable for drinking or other purposes.

Here we consider the heavy metals lead and mercury, two inorganic chemicals that sometimes contaminate water and accumulate in the tissues of humans and other organisms (see the discussion of bioaccumulation and biological magnification in Chapter 7). Arsenic, another heavy metal, is discussed later in this chapter.

Lead People used to think of lead poisoning as affecting only inner-city children who ate paint chips that contained lead. Lead-based paint was banned in the United States in 1978, but the EPA estimates that more than three-fourths of U.S. homes still contain some lead-based paint. Although lead-based paint remains an important source of lead poisoning in children, lead lurks in many other places in the environment as well.

Lead-containing anti-knock agents in gasoline were outlawed in the United States in 1986; prior to 1986, lead dust was released into the atmosphere when the fuel was burned, and that lead still contaminates soil, particularly in inner cities near major highways. Children living in the inner city may be at risk when they play outdoors in their schoolyards and backyards. Lead contaminates the soil, surface water, and groundwater when incinerator ash from solid waste facilities is dumped into ordinary sanitary landfills. Factories that lack pollution-control devices release lead into the air, from which it can settle on soil or water. We can ingest lead from pesticide and fertilizer residues on produce, from food cans soldered with lead, and even from certain types of dinnerware on which our food is served. Low amounts of lead also originate from natural sources such as volcanoes and windblown dust.

Millions of U.S. residents, many of them children, have damaging levels of lead in their bodies. At least 2% of U.S. children age one to five have blood lead levels that exceed 10 micrograms (μg) per deciliter (dL) of blood. A blood lead level above 10 μg per dL is considered dangerous, and recent data suggest that even lower concentrations of lead are harmful to the brain.

The three groups of people at greatest risk from lead poisoning are middle-aged men, pregnant women, and young children. Middle-aged men with high levels of lead are more likely to develop **hypertension**, or high blood pressure. High lead levels in pregnant women increase the risk of miscarriages, premature deliveries, and stillbirths. Children with even low levels of lead in their blood may suffer from a variety of mental and physical impairments, including partial hearing loss, hyperactivity, attention deficit, lowered IQ, and learning disabilities. Male juvenile delinquency and high bone lead concentrations have been linked, and in 2001 the American Medical Association reported a link between murder rates and lead levels in the air.

According to the EPA, in the mid-1990s more than 10% of all large and medium-sized municipal water supplies contained lead levels that exceeded the maximum permitted by the Safe Drinking Water Act (discussed later in this chapter). In addition, tap water often contains higher levels of lead than exist in municipal water supplies; the extra lead comes from the corrosion of old lead water pipes or of lead solder in newer pipes. Some treatments to clean water can have the unfortunate side effect of causing more lead to leach from pipes and solder.

Mercury   Mercury is a metal that vaporizes at room temperature; this characteristic poses special environmental challenges. Small amounts of mercury occur naturally in the environment, but most mercury pollution comes from human activities.

According to the EPA, coal-fired power plants release the largest amount (40%) of mercury into the environment. Coal contains traces of mercury that vaporize and are released into the atmosphere with the flue gases when the coal is burned. This mercury then moves from the atmosphere to the water via precipitation. Technology exists to control mercury emissions from coal-burning power plants, but it is expensive, and the trapped mercury would have to be properly disposed of in a hazardous waste landfill or it could recontaminate the environment. After many years of legal wrangling, in 2011 the EPA issued the Mercury and Air Toxics Standards, the first national standards to regulate mercury emissions from power plants, as well as the release of other toxins.

Municipal waste and medical waste incinerators also release mercury (when incinerators burn materials containing mercury). Fluorescent lights and thermostats are examples of municipal wastes that contain mercury, whereas thermometers and blood-pressure cuffs are examples of medical waste. The EPA now regulates mercury emissions from municipal and medical incinerators. Many hospitals, recognizing the risks posed by mercury, have switched away from mercury-containing equipment.

Significant amounts of mercury are released into the environment during the smelting of metals such as lead, copper, and zinc. Mercury is used in a variety of industrial processes, such as chemical plants that manufacture chlorine and caustic soda. Some of this mercury vaporizes, thereby entering the atmosphere. In addition, when industries release their wastewater, some metallic mercury may enter natural bodies of water along with the wastewater. Mercury sometimes enters water by precipitation after household trash containing batteries, paints, and plastics is burned in incinerators.

Once mercury enters a body of water, it settles into the sediments, where bacteria convert it to methyl mercury compounds, a more toxic form that readily enters the food web. Mercury bioaccumulates in the muscles of albacore tuna, swordfish, sharks, king mackerel, and marine mammals—the top predators of the open ocean. Human exposure to mercury is primarily from eating fishes and marine mammals containing high levels of mercury. At least 48 states in the United States have released health advisories on the human consumption of mercury-tainted seafood from lakes and reservoirs (**Figure 21.8**).

Methyl mercury compounds remain in the environment for a long time and are highly toxic to organisms, including humans. Prolonged exposure to methyl mercury compounds causes kidney disorders and severely damages the nervous and cardiovascular systems. The exposure of developing human fetuses to mercury is linked to a variety of conditions, such as diminished cognitive function, cerebral palsy, and developmental delays. Methyl mercury compounds are unusual in that they can cross the body's blood-brain barrier (many materials do not pass from the blood to the cerebrospinal fluid and brain). Low levels of mercury in the brain cause neurological problems such as headache, depression, and quarrelsome behavior.

Figure 21.8   Mercury contamination warning. Mercury is a common fish contaminant. Warnings like this one photographed in Florida are commonly found in popular fishing areas in the United States.

RADIOACTIVE SUBSTANCES

Radioactive substances contain atoms of unstable isotopes that spontaneously emit radiation. They get into water from several sources, some natural and others anthropogenic (human-caused). Examples of the latter include the mining and processing of radioactive minerals such as uranium and thorium. Many industries use radioactive substances; although nuclear power plants and the nuclear weapons industry use the largest amounts, medical and scientific research facilities also employ them. It is possible for radiation to inadvertently escape from any of these facilities, polluting the air, water, and soil. Radiation from natural sources, particularly radon, can contaminate groundwater.

Since the mid-1980s, low levels of radioactive substances have been measured in the wastewater of several sewage treatment plants in the United States. The EPA reports that radioactive materials may concentrate in sludge (a slimy solid mixture formed during the treatment of sewage). Guidelines from the EPA help municipal sewage treatment plants identify radioactive materials in sewage sludge and, when present, reduce or eliminate the contamination.

THERMAL POLLUTION

Thermal pollution occurs when heated water produced during certain industrial processes is released into waterways. Many industries, such as steam-generated electric power plants and nuclear power plants (Chapter 12), use water to remove excess heat from their operations. Afterward, the heated water is allowed to cool a little before it is returned to waterways, but its temperature is still warmer than it was originally. The result is that the waterway is warmed slightly.

Increasing the temperature of a lake, stream, or river leads to several chemical, physical, and biological effects. Chemical reactions, including decomposition of wastes, occur faster, depleting the water of oxygen. Moreover, less oxygen dissolves in warm water than in cool water, and the amount of oxygen dissolved in water has important effects on aquatic life (**Figure 21.9**). When the level of dissolved oxygen is lowered due to thermal pollution, a fish ventilates its gills more frequently to obtain enough oxygen. Gill ventilation, however, requires an increased consumption of oxygen. This situation puts a great deal of stress on the fish as it tries to obtain a greater supply of oxygen from a smaller supply dissolved in the water.

Other subtle changes may take place in the activities and behavior of aquatic organisms in thermally polluted water because temperature affects reproductive cycles, digestion rates, and respiration rates. At warmer temperatures, fishes require more food to maintain body weight. They typically have shorter life spans and smaller populations. In cases of extreme thermal pollution, fishes and other aquatic organisms die.

REVIEW

1. What are the eight main groups of water pollutants? Give an example of each main type.
2. What is biochemical oxygen demand (BOD)? How is BOD related to sewage?
3. What causes the dead zone in the Gulf of Mexico?

Water Quality Today

LEARNING OBJECTIVES

- Contrast point source pollution and nonpoint source pollution.
- Provide examples of agricultural, municipal, and industrial water pollution.

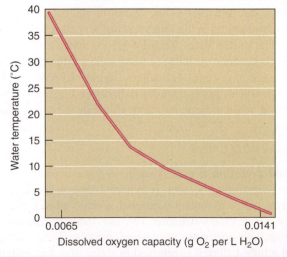

Figure 21.9 Dissolved oxygen in water at various temperatures.
As water temperature rises, its capacity to contain dissolved oxygen goes down. These data are for water in contact with air at 760 mm mercury pressure.

Water pollutants come from both natural sources and human activities. For example, some of the mercury that contaminates the biosphere is from natural sources in Earth's crust; the remainder comes from human activities. Nitrate pollution has both natural and human sources—the nitrate that occurs in soil and the inorganic fertilizers added to it, respectively. Although natural sources of pollution are sometimes of local concern, human-generated pollution is generally more widespread.

The sources of water pollution are classified into two types: point source pollution and nonpoint source pollution. **Point source pollution** is discharged into the environment through pipes, sewers, or ditches from specific sites such as factories or sewage treatment plants. Point source pollution is relatively easy to control legislatively, but accidents still occur (recall the chapter introduction).

> **point source pollution** Water pollution that can be traced to a specific origin.

The enormous damage sustained by a nuclear reactor in Fukushima, Japan, following a March 2011 earthquake and tsunami, produced point source pollution in the form of tainted reactor water. Long after the disaster, technicians struggled to prevent the release of radioactive water from the leaking reactor compound.

Nonpoint source pollution, also called *polluted runoff*, is caused by pollutants that enter bodies of water over large areas rather than at a single point. Nonpoint source pollution of water occurs when precipitation moves over and through the soil, picking up and carrying away pollutants that eventually are deposited in lakes, rivers, wetlands, groundwater, estuaries, and the ocean. Although nonpoint sources are diffuse, their cumulative effect is often huge.

> **nonpoint source pollution** Pollutants that enter bodies of water over large areas rather than being concentrated at a single point of entry.

Nonpoint source pollution includes agricultural runoff (such as fertilizers, pesticides, livestock wastes, and salt from irrigation), mining wastes (such as acid mine drainage), municipal wastes (such as inorganic plant and algal nutrients), and construction sediments. Soil erosion from fields, logging operations, eroding stream banks, and construction sites is a major cause of nonpoint source pollution.

Three major sources of human-induced water pollution that we now examine in greater detail are agriculture, municipalities (i.e., domestic activities), and industries. We then consider groundwater pollution and water pollution in other countries.

WATER POLLUTION FROM AGRICULTURE

According to the EPA, agriculture is the leading source of water quality impairment of surface waters nationwide: 72% of the water pollution in rivers is attributed to agriculture. As detailed in the previous section, agricultural practices produce several types of pollutants that contribute to nonpoint source pollution. Fertilizer runoff causes water enrichment. Animal wastes and plant residues in waterways produce high BODs and high levels of suspended solids as well as water enrichment. Amid growing concern about polluted runoff from animal wastes, the U.S. Department of Agriculture has developed guidelines to help the 450,000 livestock operations in the United States develop Comprehensive Nutrient Management Plans, focusing on six areas: feed management, manure and wastewater handling/storage, nutrient fertilizer management, land treatment practices, alternative manure and wastewater management, and record keeping.

Chemical pesticides used in agriculture may leach into the soil and from there into water. These chemicals are highly toxic and adversely affect human health as well as the health of aquatic organisms. The National Water Quality Assessment Program, an ongoing study of pesticides and their degradation products, indicates that pesticides are widespread in U.S. rivers, streams, and groundwater. More than 95% of the river and stream samples and almost 50% of groundwater samples examined by the program contained traces of at least one pesticide. More than 50% of streams sampled contained 5 or more pesticides, and about 10% contained 10 or more pesticides.

Soil erosion from fields and rangelands causes sediment pollution in waterways. In addition, some agricultural chemicals that are not very soluble in water, such as certain pesticides, find their way into waterways by adhering to sediment particles. Thus, soil conservation methods both conserve soil and reduce water pollution.

MUNICIPAL WATER POLLUTION

Although sewage is the main pollutant produced by cities and towns, municipal water pollution also has a nonpoint source: urban runoff from storm sewers (**Figure 21.10**; see You Can Make a Difference: Preventing Water Pollution on page 427). The water quality of urban runoff from city streets is often worse than that of sewage. Urban runoff carries salt from roadways, untreated garbage, animal wastes (especially from dogs), construction sediments, and traffic emissions (via rain that washes pollutants out of the air). It may often contain such contaminants as asbestos, chlorides, copper, cyanides, grease, hydrocarbons, lead, motor oil, organic wastes, phosphates, sulfuric acid, and zinc.

Nearly 800 U.S. cities, including New York, San Francisco, Pittsburgh, and Boston, have **combined sewer systems** in which human and industrial wastes are mixed with urban runoff from storm sewers before flowing into the sewage treatment plant. A problem arises following a heavy rainfall or a large snowmelt because even the largest sewage treatment plant can process only a given amount of wastewater each day. When too much water enters the system, the excess, known as **combined sewer overflow**, flows into nearby waterways without being treated. Combined sewer overflow, which contains raw sewage, has been illegal since passage of the Clean Water Act of 1972 (discussed shortly), but cities have only recently begun to address the problem. According to the EPA, 1.2 trillion gallons of combined sewer overflow are discharged into U.S. waterways every year.

Some cities with combined sewer systems, such as St. Paul, Minnesota, have installed two separate sewers, one for sewage

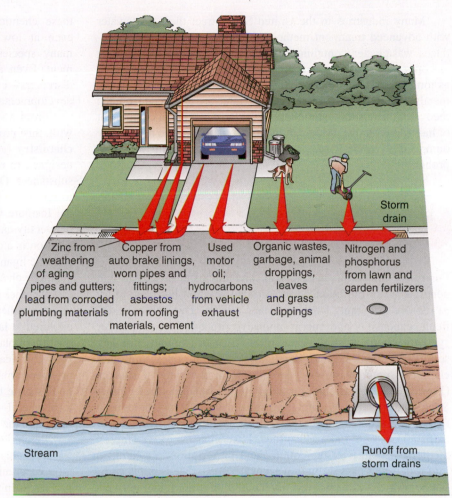

Figure 21.10   Urban runoff. The largest single pollutant in urban runoff is organic waste, which removes dissolved oxygen from water as it decays. Fertilizers cause excessive algal growth, which further depletes the water of oxygen. Other everyday pollutants include used motor oil, which is often illegally poured into storm drains, and heavy metals. These pollutants may be carried from storm drains on streets to streams and rivers.

and industrial wastes and one for urban runoff. However, such an installation is expensive and requires that every street be dug up. Other cities, such as Birmingham, Michigan, have kept their combined sewer systems but installed huge retention basins to hold the overflow until it is treated. The basin in Birmingham, which was installed in 1998, holds 21 million L (5.5 million gal). Such tanks are less expensive to install than separate sewer systems, but there are concerns that after several days of heavy rain or snow, the basin itself could overflow.

INDUSTRIAL WASTES IN WATER

Different industries generate different types of pollutants. Food-processing industries produce organic wastes that are readily decomposed but have a high BOD. Pulp and paper mills produce toxic compounds and sludge in addition to a high BOD. The paper industry, however, has begun to adopt new manufacturing methods, such as the production of paper without the use of chlorine as a bleaching agent, that produce significantly less toxic effluents.

Many industries in the United States treat their wastewater with advanced treatment methods. The electronics industry produces wastewater containing high levels of heavy metals such as copper, lead, and manganese but uses special techniques such as ion exchange and electrolytic recovery to reclaim those heavy metals. Metal plates with commercial value are produced from the recovered metals that would otherwise have become a component of hazardous sludge. Although most U.S. industries do not usually dump highly toxic wastes into water, accidental spills remain a problem.

CASE IN POINT

GREEN CHEMISTRY

For many industrial, agricultural, and domestic processes that generate water pollution, the traditional solution has been to try to remove pollutants from the waste stream. This can be a difficult, expensive, and energy-intensive process, particularly for synthetic chemicals that occur, often in small amounts, in many settings. Pesticides, pharmaceuticals, cosmetics, dyes, and other chemicals enter wastewater streams from a variety of sources (**Figure 21.11**). Caffeine, for example, is found in fish in Puget Sound (near Seattle, Washington) and musks—the molecules used to provide scent in deodorants, aftershaves, and perfumes—in catfish taken from Lake Mead (near Las Vegas, Nevada). Unused medicines, both over-the-counter and prescription, are dumped into toilets and pass through wastewater treatment systems without being altered or removed. Eventually, they are found in rivers and lakes—and in the plants and animals that live there.

These chemicals—there are thousands of them—occur in relatively small amounts compared to other pollutants we often worry about, such as oil, organic matter, and nitrogen. So why should we be concerned? First, they are difficult to remove. Most organic matter eventually breaks down into carbon dioxide; synthetic molecules, on the other hand, are not easily broken down. Indeed, as in the case of dyes, longevity is something that manufacturers *want* in their products. Second,

these chemicals can have substantial environmental impacts, even at low concentrations. Synthetic chemicals can disrupt many species' growth, especially their reproductive development. Even at low concentrations, some chemicals—endocrine disrupters—can affect the ratio of male to female fish and cause developmental damage, including deformities and cancer.

From a systems perspective, a variety of ways exist to deal with this problem. Several fall under the general term **green chemistry** (also known as sustainable chemistry), or chemistry designed to reduce or halt the use and production of hazardous substances. One approach of green chemistry is to find new ways to remove the chemicals from the waste stream. Researchers at the Institute for Green Oxidation Chemistry at Carnegie Mellon University have worked on this problem for years. One promising solution is a class of chemicals called TAMLs (tetra-amido macrocyclic ligands) that serve as catalysts to speed the breakdown of synthetic chemicals.

Another approach is to change chemical processes and use fewer synthetic chemicals. This branch of green chemistry is explored at laboratories around the world. For example, the production of plastic containers can involve the release of many complex chemicals. Alternatively, biodegradable containers can be produced from sugars, with few or no chemical byproducts. Such bottles have the added advantage that they can last long enough to be useful—but not for the decades that a plastic bottle would remain in the environment.

Perchloroethylene (or "perc") is a hydrocarbon responsible for the highly recognizable odor of dry cleaners. It is also a common water and air contaminant associated with a range of health problems. In parts of southern California, it is no longer legal to use perc in new or upgraded dry-cleaning facilities. Several companies now provide alternatives, including less damaging hydrocarbons, silicon-based cleaners, and even a new "wet-cleaning" approach that uses water and alternative soaps to clean delicate fabrics without harming them.

Green chemistry has the potential to change or replace many highly toxic industrial or household chemicals. The challenge is to find alternatives that work as well as and cost no more than traditional synthetic chemicals. ■

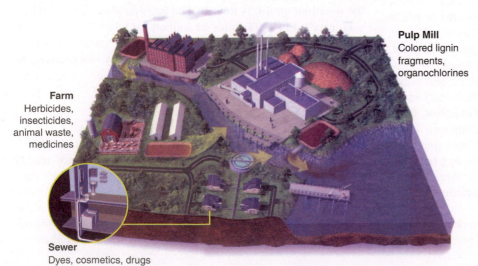

Farm
Herbicides, insecticides, animal waste, medicines

Pulp Mill
Colored lignin fragments, organochlorines

Sewer
Dyes, cosmetics, drugs

Figure 21.11   Sources of synthetic pollutants in water. Water pollution containing small amounts of synthetic chemicals can come from a variety of sources: industrial (e.g., pulp mills), agricultural, and municipal (e.g., sewers).

YOU CAN MAKE A DIFFERENCE

Preventing Water Pollution

Although individuals produce little water pollution, the collective effect of municipal water pollution, even in a small neighborhood, can be quite large. You can do many things to protect surface waters and groundwater from water pollution.

1. Many household chemicals, such as oven cleaners, mothballs, drain cleaners, and paint thinners, are quite toxic. Use such products sparingly, and try to substitute less hazardous chemicals wherever possible. When disposing of these chemicals, contact the solid waste management office in your county for information about local hazardous waste collection centers. Never put them down a drain or toilet, because they may disrupt your septic system or contaminate sewage sludge produced at municipal sewage treatment facilities. Never pour these chemicals on the ground because they may contaminate runoff when it rains. Here are some safer alternatives (note that these chemicals are not nontoxic, just less toxic than many commercial products):

   • Ammonia, to clean appliances and windows.

   • Bleach, to disinfect. Never mix ammonia and bleach because the mixture releases toxic chlorine gas.

   • Borax, to remove stains and mildew.

   • Baking soda, to remove stains, deodorize, and clean household utensils.

   • Vinegar, to clean surfaces, polish metals, and remove stains and mildew.

2. Never throw unwanted medicines down the toilet. Several studies have shown that traces of many drugs are present in tap water. In some places, pharmacies can collect unwanted medicines and return them to drug manufacturers.

3. Never pour used motor oil or antifreeze down storm drains or on the ground. Recycle these chemicals by dropping them off at a service station or local hazardous waste collection center.

4. Pick up pet waste and dispose of it in the garbage or toilet. If left on the ground, it eventually washes into waterways, where it can increase BOD and fecal coliform levels.

5. Drive less. Air pollution emissions from automobiles eventually enter ground water and surface water. Toxic metals and oil by-products deposited on the road by automobiles—are washed into surface waters by precipitation.

6. If you are a homeowner, replace some of your grass lawn with trees, shrubs, and ground covers, which absorb up to 14 times more precipitation and require little or no fertilizer. To reduce erosion, use mulch to cover bare ground (see Chapter 14).

7. Use fertilizer sparingly because excess fertilizer leaches into groundwater or waterways. If you hire a professional lawn care service to apply fertilizer and pesticides, use one that monitors soil fertility and applies chemicals only when they are needed.

8. Never apply fertilizer near a body of water. Always allow a buffer zone of at least 20 to 40 feet.

9. Make sure that gutters and downspouts drain onto water-absorbing grass or graveled areas instead of paved surfaces.

10. Clean up spilled oil, brake fluid, and antifreeze, and sweep sidewalks and driveways instead of hosing them off. Dispose of the dirt properly not into gutters or storm drains.

11. Likewise, do not let grass clippings or leaves wash into gutters or storm drains.

12. Use pesticides sparingly, both indoors and outdoors, and incorporate integrated pest management techniques (see Chapter 18). Dispose of unwanted pesticides at hazardous waste collection centers.

13. Replace paved driveways and sidewalks with porous surfaces, such as interlocking bricks or stones; build wood decks instead of concrete patios. These features allow precipitation to seep into the ground, decreasing runoff.

GROUNDWATER POLLUTION

Roughly half of the people in the United States obtain their drinking water from groundwater, which is also withdrawn for irrigation and industry. The most common groundwater pollutants, such as pesticides, fertilizers, and organic compounds, seep into groundwater from municipal sanitary landfills, underground storage tanks, backyards, golf courses, and intensively cultivated agricultural lands (**Figure 21.12**). More than 250,000 underground petroleum storage tanks may be leaking at service stations in the United States. Cleanup of leaking gas tanks is expensive—$500,000 or more per tank—but state gasoline taxes help pay for the cleanup in some states (such as California).

Nitrates sometimes contaminate shallow groundwater—30 m (100 ft) or less from the surface—with fertilizer being the most common source. High nitrate levels are a concern in some rural areas, where 80% to 90% of residents use shallow groundwater for drinking water. When nitrates get into the human body, they are converted to nitrites, which reduce the blood's ability to transport oxygen. This condition is one of the causes of cyanosis (the *blue baby* syndrome), a serious disorder in young children. The level of nitrate in drinking water is monitored for municipal systems, so it is generally not of concern. Users of well water should have water nitrate levels checked periodically.

Contamination of groundwater is a relatively recent environmental concern. People used to think that the underlying soil and rock through which surface water must seep to become groundwater filtered out any contaminants, thereby ensuring the purity of groundwater. This assumption proved false when groups began to monitor the quality of groundwater and discovered contaminants at certain sites. It appears that the natural capacity of soil and rock to remove pollutants from groundwater varies widely from one area to another.

Concern over groundwater safety has grown over the recent boom in hydraulic fracturing, known as fracking, a water-intensive process used to release natural gas and oil from underground rock formations (see Chapter 11). Many local conflicts have arisen over the potential contamination of drinking water by fracking chemicals.

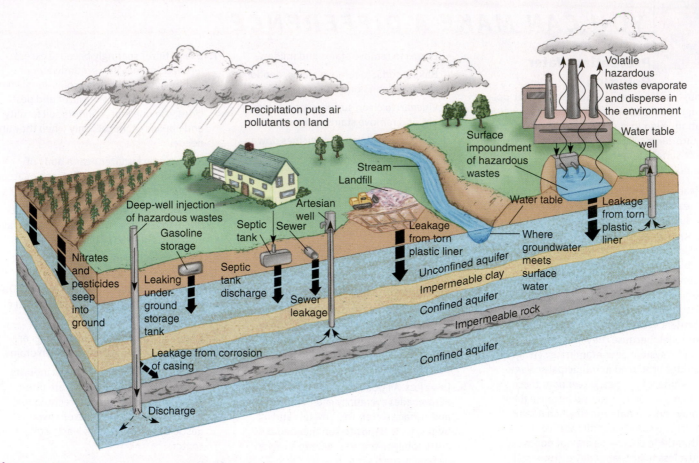

Figure 21.12 **Sources of groundwater contamination.** Agricultural practices, sewage (both treated and untreated), landfills, industrial activities, and septic systems are some of the sources of groundwater pollution. Once groundwater is contaminated, it does not readily cleanse itself by natural processes. We do not know the degree of groundwater contamination because access is difficult and because contaminants do not disperse quickly (groundwater moves very slowly). (Figure is not drawn to scale.)

Currently, most U.S. groundwater supplies are of good quality and do not violate standards established to protect human health. There are some local problems, however, that have led to well closures and raised public health concerns. For example, in 1996 Santa Monica, California, closed 7 of its 11 municipal wells when methyl tertiary butyl ether (MTBE), a gasoline additive that reduces tailpipe emissions, was detected in the groundwater. The groundwater became polluted primarily from leaking underground gas-storage tanks. MTBE, which may cause cancer, persists in groundwater for years. A 2006 study by the USGS found that MTBE was detected in groundwater samples around the United States. Some states now ban the addition of MTBE to gasoline.

Cleanup of polluted groundwater is costly, takes years, and in some cases is not technically feasible. Compounding the cleanup problem is the challenge of safely disposing of the toxic materials removed from groundwater, which, if not handled properly, could contaminate the groundwater once again.

WATER POLLUTION IN OTHER COUNTRIES

According to the World Health Organization, an estimated 768 million people lack access to safe drinking water, and about 2.5 billion people do not have access to adequate sanitation systems; most of these people live in rural areas of developing countries. Worldwide, more than 3.4 million people die of water-related illnesses each year, an estimated 800,000 of them children younger than five with diarrheal illnesses.

Municipal water pollution from sewage is a greater problem in developing countries, many of which lack water treatment facilities, than in highly developed ones. Sewage from many densely populated cities in Asia, Latin America, and Africa is dumped directly into rivers or coastal harbors.

Almost every nation in the world faces problems of water pollution. For an international perspective on water pollution, let's examine some specific issues in South America, Europe, Asia, and Africa.

Lake Maracaibo, Venezuela Lake Maracaibo in Venezuela is the largest lake in South America (**Figure 21.13**). It receives fresh water from several rivers, and water flows from it into the Caribbean Sea. Larger than the state of Connecticut, Lake Maracaibo suffers from the effects of oil pollution and human wastes as well as contamination from farms and factories. About 10,000 oil wells tap the oil and natural gas reserves under the shallow lake. An underwater network of old oil pipes about 15,400 km (9600 mi) long leaks oil into the lake.

is highly polluted. Little of the sewage and industrial waste produced by the 400 million people who live in the Ganges River basin is treated. Another major source of contamination is the 35,000 human bodies that are cremated annually in the open air in Varanasi, the holy city of the Hindus. (The Hindus cremate the body to free the soul; dumping the ashes into the Ganges increases the chances of the soul getting into heaven.) Incompletely burned bodies are dumped into the Ganges, where their decomposition adds BOD to the river. In addition, people who cannot afford cremation costs for their dead dump human remains into the river.

In 1985, the Indian government initiated the Ganga Action Plan, an ambitious cleanup project that was to include water treatment plants in 29 large cities and 32 electric crematoriums along the banks of the Ganges. Although the government has spent about $100 million constructing sewage treatment plants in major cities along the Ganges, most are not completed or are not working effectively. Most cities in the river basin still discharge raw sewage into the river. Costs have escalated, and many delays in the Ganga Action Plan have occurred. According to critics, the government plan has not worked in part because it has not tried to get people involved at the community level.

Figure 21.13 **Lake Maracaibo, Venezuela.** Drilling platforms are located throughout the lake, and pipelines join the oil wells to refineries on land. Shown are storage and export facilities in the foreground and the town of Cabimas in the background. An oil tanker is docked at the leftmost pier.

Fertilizers and other agricultural chemicals drain into the lake from nearby farms, providing the nutrients for an overgrowth of algae. Until recently, raw sewage from the city of Maracaibo's 2.1 million people and many smaller communities was discharged directly into the water, contributing to the nutrient overload. Modern sewage treatment facilities were installed during the 1990s to take care of human wastes, but Lake Maracaibo's other pollution problems must still be addressed.

Po River, Italy The Po River, which flows across northern Italy, empties into the Adriatic Sea. The Po, Italy's equivalent of the Mississippi River, is heavily polluted. Many cities, including Milan, with more than 4 million residents in its metropolitan area, historically have dumped treated and untreated sewage into the Po. Industry is responsible for half the pollutants that enter the river. Italian agriculture, including large poplar plantations, relies heavily on chemicals and is responsible for massive amounts of nonpoint source pollution. Soil erosion has resulted in so much sediment deposition at the mouth of the river that the Po River delta is advancing about 81 hectares (200 acres) into the Adriatic Sea each year.

About 17 million people—almost one-third of all Italians—live in the Po River basin. The health of many Italians is potentially threatened because the Po is the source of their drinking water. In addition, pollution from the Po has jeopardized tourism and fishing in the Adriatic Sea. Pollution has temporarily closed swimming at some beaches. Beginning in 2009, the Po Basin Water Board began to manage the river under an Integrated Management Plan. The cleanup of the Po will take place over a period of several decades.

Ganges River, India The Ganges River is a holy river that symbolizes the spirituality and culture of the Indian people. The river, widely used for bathing and washing clothes (**Figure 21.14**),

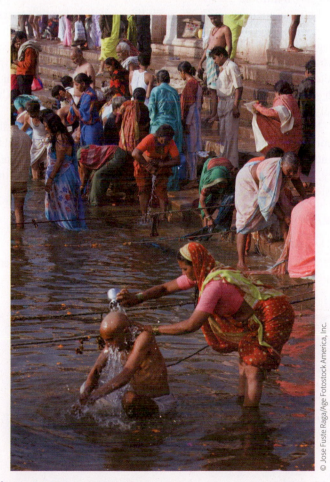

Figure 21.14 **Ganges River.** Bathing and washing clothes in the Ganges River are common practices in India. The river is contaminated by raw sewage discharged directly into the river at many locations.

Zimbabwe   A cholera outbreak throughout Zimbabwe in 2008–2010 led to over 4000 deaths from almost 100,000 cases of the disease. This epidemic was exacerbated by extreme poverty and shortages of food. In addition, many Zimbabweans are unaware of the cause of cholera—a bacterium found in water contaminated with human or animal wastes. Cholera infects the lower intestine, causing severe diarrhea and dehydration. Death can occur in as little as 18 hours after initial infection. Robert Mugabe, president of Zimbabwe, accused Western countries of intentionally causing a 2008 outbreak, and doctors and nurses protesting the government's lack of response were stopped by riot police in Harare, the Zimbabwean capitol.

Arsenic Poisoning in Bangladesh

During the 1980s, world health organizations, concerned about illnesses caused by drinking contaminated surface water in Bangladesh, funded the installation there of more than 2.5 million wells with hand pumps. Tragically, the groundwater in many of these wells is contaminated with high levels of naturally occurring arsenic. An estimated 57,000 people now suffer from chronic arsenic poisoning, which initially causes skin lesions, particularly on the hands and feet. Eventually, when the cumulative dose is large enough, arsenic poisoning leads to death from cancer. Health authorities do not know how many people are at risk. Estimates range from hundreds of thousands to as many as 70 million.

International agencies are currently providing funding to test all the wells. Meanwhile, the Bangladeshi people continue to drink the groundwater because they have no other option. Scientists have developed an inexpensive bucket filtration system to remove arsenic from well water using iron shavings. However, the arsenic-rich sediments produced by the filtration system would have to be safely disposed of if this procedure were to be widely adopted.

Bangladesh is not the only place in the world with naturally occurring arsenic contamination of groundwater. Arsenic is also found in shallow aquifers in parts of Argentina, Chile, China, Hungary, India, Mexico, Mongolia, Romania, Taiwan, Thailand, the western United States, and Vietnam.

REVIEW

1. How do point source and nonpoint source pollution differ?
2. What are some major agricultural, industrial, and municipal water pollutants?

Improving Water Quality

LEARNING OBJECTIVES

- Describe how most drinking water is purified in the United States, and discuss the chlorine dilemma.
- Distinguish among primary, secondary, and tertiary treatments for wastewater.

Water quality is improved by removing contaminants from the water supply before and after it is used. Technology assists in both processes.

PURIFICATION OF DRINKING WATER

The United States has nearly 60,000 municipal water facilities that serve over 300 million people. Surface-water sources of municipal water supplies include streams, rivers, and lakes. Often a dam is built across a river or stream to form an artificial lake, or reservoir. Reservoirs accumulate water when there is an adequate supply and store it for use during periods of drought.

In the United States, most municipal water supplies are treated before being used so that the water is safe to drink (**Figure 21.15**). Turbid water is treated with a chemical coagulant (aluminum sulfate) that causes the suspended particles to clump together and settle out. The water is then filtered through sand to remove remaining suspended materials as well as many microorganisms. A few cities, such as Cincinnati, pump the water through activated carbon granules to remove many of the dissolved organic compounds.

In the final purification step before distribution in the water system, the water is disinfected to kill any remaining disease-causing agents. The most common way to disinfect water is to add chlorine. A small amount of chlorine is left in the water to provide protection during its distribution through many kilometers of pipes. Other disinfection systems use ozone or ultraviolet (UV) radiation in place of chlorine.

The Chlorine Dilemma

During the 19th century, waterborne disease-causing organisms often contaminated drinking water supplies in the United States. The discovery that chlorine kills these organisms allowed 20th-century Americans to drink water with little fear of contracting typhoid, cholera, or dysentery. The addition of chlorine to our drinking water supply has undoubtedly saved millions of lives.

At the same time, chlorine by-products, formed when chlorine reacts with and oxidizes organic matter in treated wastewater, are tentatively linked to several kinds of cancer (rectal, pancreatic, and bladder), an increased risk of miscarriages, and the possibility of some rare birth defects. As a result, use of chlorine to disinfect drinking water has triggered a debate over the costs and benefits of chlorinating water. The concern is whether low levels of chlorine in drinking water present a long-term hazard.

Because there are few viable alternatives to chlorination, the EPA was initially reluctant to reduce the level of chlorine permissible in drinking water, despite the evidence of potential risks. The EPA did not want what happened in Peru to occur in the United States. In 1991, a terrible cholera epidemic swept much of Peru, infecting more than 300,000 people and killing at least 3500. This outbreak occurred after Peruvian officials decided to stop chlorinating much of the country's drinking water in response to the slightly increased cancer risk due to chlorination. Peru has since resumed chlorinating its drinking water.

After a detailed review of current evidence linking chlorine to cancer, the EPA proposed in 1994 that water treatment facilities reduce the maximum permissible level of chlorine in drinking water. One alternative to chlorination is to use *chloramine*, a disinfectant produced by combining chlorine with ammonia. Chloramine does not form potentially harmful by-products, although preliminary studies suggest that its use may cause an increase in lead levels in drinking-water systems. (Chloramine may make lead

(a) The water supply for a town may be stored in a reservoir, as shown, or obtained from groundwater.

(b) The water is treated before use so it is safe to drink.

(d) The quality of the wastewater is fully or partially restored by sewage treatment before the treated effluent is dispersed into a nearby body of water.

(c) After use, municipal sewer lines collect the wastewater.

Figure 21.15   Water treatment for municipal use.

atoms adhering to water pipes more soluble in the water.) Another alternative is to filter water through activated carbon granules, as is done in Cincinnati; one-third less chlorine is then needed in the final step. Much of Europe has adopted another alternative to chlorination, UV disinfection, which has also gained ground in U.S. cities as an additional level of treatment to kill microorganisms not eliminated by chlorine. In 2013, New York City launched the world's largest UV treatment facility.

Fluoridation   Small amounts of fluoride have been added to municipal drinking water since the mid-1940s to reduce tooth decay. (Fluoride is added to many toothpastes for the same reason.) This practice is somewhat controversial, with opponents questioning the safety and effectiveness of fluoride and supporters saying it is completely safe and effective in preventing decay. More than 40 years of research have failed to link fluoridation at the levels in U.S. drinking water to cancer, kidney disease, birth defects, or any other serious medical condition. Most dental health officials think fluoride is the main reason for the 50% to 60% decrease in tooth decay observed in children during the past several decades, based on comparisons of cavity rates in schoolchildren between cities with fluoridation and without.

As of 2012, almost 75% of the U.S. population received fluoridated water through public water supplies. Currently, fluoridation is more common in the eastern half of the country than in the western half, although it was mandated in California in 1995 and in Nevada in 2002.

MUNICIPAL SEWAGE TREATMENT

Wastewater, including sewage, usually undergoes several treatments at a sewage treatment plant to prevent environmental and public health problems. The treated wastewater is then discharged into rivers, lakes, or the ocean.

Primary treatment removes suspended and floating particles, such as sand and silt, by mechanical processes such as screening and gravitational settling (**Figure 21.16**, left side). The solid material that settles out at this stage is **primary sludge**. Primary treatment does little to eliminate the inorganic and organic compounds that remain suspended in the wastewater. The wastewater treatment facilities for about 11% of the U.S. population have primary treatment only.

Secondary treatment uses microorganisms (aerobic bacteria) to decompose the suspended organic material in wastewater (Figure 21.16, right side). One of the several types of secondary treatment is *trickling filters*, in which wastewater trickles through aerated rock beds that contain bacteria and other microorganisms, which degrade the organic material in the water. In another type of secondary treatment, the *activated sludge process*, wastewater is aerated and circulated through bacteria-rich particles; the bacteria degrade suspended organic material. After several hours, the particles and microorganisms are allowed to settle out, forming **secondary sludge**, a slimy mixture of bacteria-laden solids. Water that has undergone primary and secondary treatment is clear and free of organic wastes such as sewage. The wastewater treatment facilities for about 62% of the U.S. population have both primary and secondary treatments.

Even after primary and secondary treatments, wastewater still contains pollutants, such as dissolved minerals, heavy metals, viruses, and organic compounds (including pharmaceuticals, fragrances, and pesticides) (**Figure 21.17**). Advanced wastewater treatment methods, or **tertiary treatment**,

primary treatment Treating wastewater by removing suspended and floating particles by mechanical processes.

primary and secondary sludge The solids remaining after sewage treatment has been completed.

secondary treatment Treating wastewater biologically to decompose suspended organic material; secondary treatment reduces the water's biochemical oxygen demand.

tertiary treatment Advanced wastewater treatment methods that are sometimes employed after primary and secondary treatments.

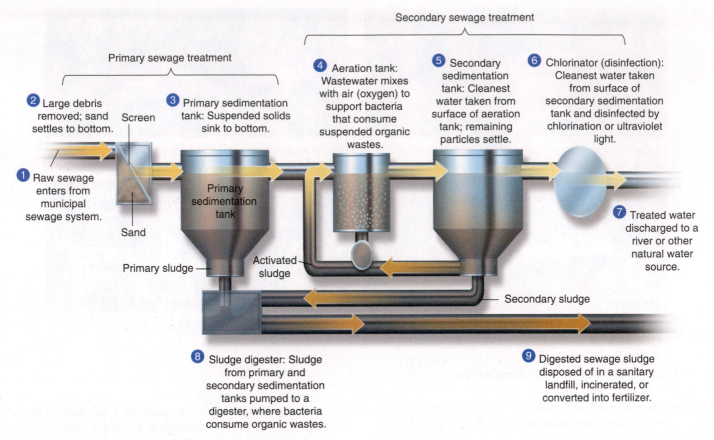

Figure 21.16   Primary and secondary sewage treatment. This sort of sewage treatment system is standard for municipalities in highly developed countries but often nonexistent in less developed ones.

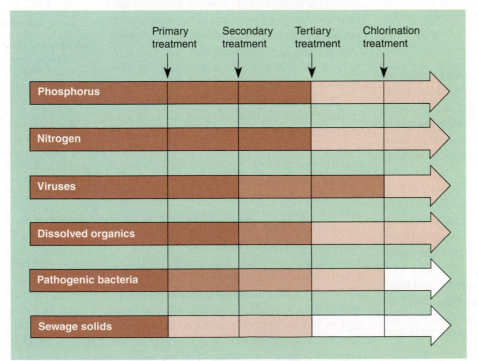

Figure 21.17   Effectiveness of primary, secondary, and tertiary sewage treatment. This figure shows the relative concentrations of various water pollutants after each treatment. The most intense color represents the greatest concentration of a given pollutant, whereas white means all of that specific pollutant is removed. Note how ineffective secondary treatment is in removing certain contaminants, such as phosphorus and nitrogen. Also note that even after tertiary treatment, some pollutants are still detectable, though at greatly reduced levels.

include a variety of biological, chemical, and physical processes. Tertiary treatment reduces phosphorus and nitrogen, the nutrients most commonly associated with enrichment. Tertiary treatment purifies wastewater for reuse in communities where water is scarce. The wastewater treatment facilities for about 27% of the U.S. population include tertiary treatments.

Disposal of Sludge   A major problem associated with wastewater treatment is disposal of the primary and secondary sludge formed during primary and secondary treatments. Five possible ways to handle sludge are anaerobic digestion, application to soil as a fertilizer, incineration, ocean dumping, and disposal in a sanitary landfill. In anaerobic digestion, the sludge is placed in large, circular digesters and kept warm (about 35°C, or 95°F), which allows anaerobic bacteria to break down the organic material into gases such as methane and CO_2. The methane can be trapped and burned to heat the digesters.

After a few weeks of digestion, the sludge resembles humus and can be used as a nutrient-rich fertilizer. Not all sludge can be used this way, in particular when sewer systems mix industrial waste, which may contain toxic substances, with household waste. Farmers have long used sludge to

the ocean. In 1988, Congress passed the **Ocean Dumping Ban Act**, which barred ocean dumping of sludge and industrial waste, beginning in 1991. Alternatively, sludge is disposed of in sanitary landfills (see Chapter 23). As landfill space becomes more costly, many cities are looking for other ways to handle sludge.

INDIVIDUAL SEPTIC SYSTEMS

Many private residences, particularly in rural areas, use individual septic systems for sewage disposal instead of using municipal sewage treatment. Household sewage is piped to the septic tank, where particles settle to the bottom (**Figure 21.18a**). Grease and oils form a scummy layer at the top, where bacteria decompose much of it. Wastewater containing suspended organic and inorganic material then flows into the drain field through a network of small, perforated pipes set in trenches of gravel or crushed stone (**Figure 21.18b**). The drain field is located just below the soil's surface, and bacteria decompose the remaining organic material in the well-aerated soil. The purified wastewater then percolates into the groundwater or evaporates from the soil.

Septic tank systems require care to operate properly. Household chemicals such as bleach and drain cleaners must be used sparingly because they could kill the bacteria that break down the organic wastes. A kitchen garbage disposal could overload the system. Every two to five years, depending on use, the sludge that collects at the bottom of the septic tank is removed and taken to a municipal sewage treatment plant for proper disposal. If a septic system is not maintained properly, it can malfunction or overflow, releasing bacteria and nutrients into groundwater or waterways.

fertilize hay and feed-grain crops. However, many farmers are reluctant to use it on crops for direct human consumption because consumers might not purchase the food grown in sludge out of concern that it may pose a threat to human health. Nonetheless, over half of dried municipal sludge generated in the United States is applied to land.

Although sludge can be used to condition soil, it is generally treated as a solid waste. Dried sludge is often incinerated, which may contribute to air pollution, although sometimes the heat is used constructively—to generate electricity, for example. In the past, coastal cities such as New York dumped their sludge into

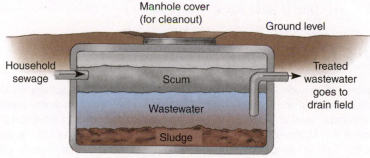

(a) The septic tank works much like primary treatment in municipal sewage treatment. Sewage from the house is piped to the septic tank, where particles settle to the bottom.

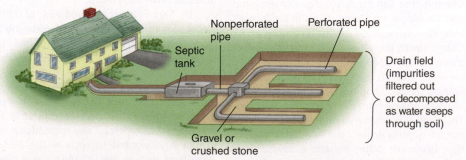

(b) Wastewater containing suspended organic and inorganic material flows into the drain field and gradually seeps into the soil.

Figure 21.18   Septic tank systems. Many private homes in rural areas use a septic tank and drainage field. Both are located underground.

ALTERNATIVE WASTEWATER TREATMENT SYSTEMS

A number of innovative wastewater treatment systems are designed to mimic natural processes to manage some wastewater, particularly sewage. *Constructed wetlands* channel contaminated water through a series of sand beds and plants. The sand beds filter the water, while the plants absorb many of the nutrients. Water leaving constructed wetlands can be as clean as water subjected to many tertiary treatment systems, and the plants can be harvested for biofuels or mulch (but usually not for human consumption). Elsewhere, *composting toilets* provide conditions under which human wastes break down slowly into a useful fertilizer, while at the same time eliminating disease and unpleasant odors.

REVIEW

1. How is most drinking water purified in the United States?
2. Why is chlorine added to drinking water? What is the potential problem with adding chlorine to drinking water?
3. Distinguish among primary sewage treatment, secondary sewage treatment, tertiary sewage treatment, and septic tanks.

Laws Controlling Water Pollution

LEARNING OBJECTIVES

- Compare the goals of the Safe Drinking Water Act and the Clean Water Act.
- Define *maximum contaminant level* and *national emission limitation*, and specify the water legislation each relates to.

Many governments have passed legislation to control water pollution. Point source pollutants lend themselves to effective control more readily than nonpoint source pollutants. Governments generally control point source pollution in one of two ways—by imposing penalties on polluters (a common approach in the United States) or by taxing polluters to pay for the cleanup (common in Japan).

Although most countries have passed laws to control water pollution, monitoring and enforcement are difficult, even in highly developed countries (see Meeting the Challenge: Using Citizen Watchdogs to Monitor Water Pollution). The United States has attempted to control water pollution through legislation since the passage of the **Refuse Act** of 1899, which was intended to reduce the release of pollutants into navigable rivers. The two federal laws that have the most impact on water quality today are the Safe Drinking Water Act and the Clean Water Act.

SAFE DRINKING WATER ACT

Prior to 1974, individual states set their own standards for drinking water, which, of course, varied a great deal from state to state. In 1974, the **Safe Drinking Water Act** (SDWA) was passed, which set uniform federal standards for drinking water to guarantee safe public water supplies throughout the United States. This law required the EPA to determine the **maximum contaminant level**, which is the maximum permissible amount of any water pollutant that might adversely affect human health. The EPA oversees the states to ensure that they adhere to the maximum contaminant levels for specific water pollutants. Private wells, which supply about 10% of U.S. residents, are not covered by the SDWA, although such wells are regulated in some states.

maximum contaminant level The upper limit for the concentration of a particular water pollutant in water intended for human consumption.

MEETING THE CHALLENGE

Using Citizen Watchdogs to Monitor Water Pollution

Insufficient staff and lack of funds prevent well-intentioned government agencies from effectively monitoring and enforcing environmental laws such as the Safe Drinking Water Act. For example, the San Francisco Bay Conservation and Development Commission is California's federally designated management agency charged with patrolling 1600 km (1000 mi) of shoreline and 1500 km² (600 mi²) of water to ascertain that no one is illegally polluting San Francisco Bay. In addition, it handles hundreds of cases arising from its monitoring activities. The agency is severely understaffed and cannot adequately protect the bay. San Francisco Bay, like most other aquatic ecosystems, is endangered by the combined impact of many different pollution sources. Thus, continual monitoring is necessary to ensure that many small polluters do not collectively do irreparable harm to the bay.

A growing number of private citizens have become actively involved in monitoring and enforcing environmental laws to protect waterways in their communities. Provisions in the Clean Water Act, the Safe Drinking Water Act, and other key environmental laws allow citizens to file suit when the government does not enforce the laws. Citizen action groups also pressure firms to clean up.

In San Francisco, hundreds of citizen watchdogs called Baykeepers monitor the bay from boats, airplanes, and helicopters. Law students at local universities advise the keepers on issues of litigation.

The San Francisco Baykeeper program is modeled after New York's Riverkeeper program, which was first organized in 1966 as a coalition of commercial and recreational fisherman who wanted to reclaim the Hudson River from its polluters. In 1983, using money obtained from successful lawsuits against polluters, the Riverkeeper program hired its first full-time riverkeeper. A network of community fishermen and environmentalists inform the riverkeeper of any suspicious activity on the river. The riverkeeper monitors water quality by boat, attends board meetings, educates the public, and employs litigation as a last resort. The riverkeeper position is described as part investigator, scientist, lawyer, lobbyist, and public relations agent.

More than 100 river, sound, bay, inlet, channel, and coast keeper groups have organized across the United States, from the Cook Inletkeeper in Alaska to the Emerald Coastkeeper on Florida's Gulf Coast. The umbrella organization for keeper groups is the Waterkeeper Alliance, which launched in 1992. Its philosophy is based on the idea that daily vigilance by citizens is required to protect a community's natural resources. The Waterkeeper Alliance helps new keeper programs organize, both in the United States and other countries. More than 200 keeper programs have been established on six continents, including in Bolivia, Canada, Mexico, Senegal, the Czech Republic, the United Kingdom, Iraq, India, China, Russia, and Australia.

Most water suppliers take few or no steps to prevent the contamination of the watershed or groundwater from which they draw. The vast majority of water utilities do not use modern water treatment technologies such as activated carbon granules or UV disinfection to reduce chemical contamination by pesticides, arsenic, and chlorine disinfection byproducts. Also, the average water pipe in the United States is 100 or more years old before it is replaced. Many aging pipes are cracked, which permits contaminated water to seep into them and increases the risk of waterborne diseases.

The Safe Drinking Water Act was amended in 1986 and again in 1996. The 1996 version requires municipal water suppliers to tell consumers what contaminants are present in their city's water and whether these contaminants pose a health risk. The act also requires that the EPA regularly update its regulations, which it last did in 2002.

CLEAN WATER ACT

The **Clean Water Act** affects the quality of rivers, lakes, aquifers, estuaries, and coastal waters in the United States. Originally passed as the Water Pollution Control Act of 1972, it was amended and renamed the Clean Water Act of 1977; additional amendments were made in 1981 and 1987. The Clean Water Act has two basic goals: to eliminate the discharge of pollutants in U.S. waterways and to attain water quality levels that make these waterways safe for fishing and swimming. Under the provisions of this act, the EPA is required to set up and monitor **national emission limitations**.

> **national emission limitation** The maximum permissible amount of a water pollutant that can be discharged from a sewage treatment plant, factory, or other point source.

Overall, the Clean Water Act effectively improved the quality of water from point sources, despite the relatively low fines it imposes on polluters. It is not hard to identify point sources, which must obtain permits from the **National Pollutant Discharge Elimination System (NPDES)** to discharge untreated wastewater.

According to the EPA, nonpoint source pollution is a major cause of water pollution. However, nonpoint source pollution is much more difficult and expensive to control than point source pollution. The 1987 amendments to the Clean Water Act expanded the NPDES to include nonpoint sources, such as sediment erosion from construction sites.

To date, U.S. environmental policies have not effectively addressed many nonpoint sources of pollution, which requires regulating land use, mining, agricultural practices, and many other activities. Such regulation necessitates the interaction and cooperation of many government agencies, environmental organizations, and private citizens, which can be enormously challenging.

The United States has improved its water quality in the past several decades, thereby demonstrating that the environment recovers once pollutants are eliminated. Much remains to be done, however. The most recent data available from the EPA's National Water Quality Inventory (collected by state between 2006 and 2012) indicated that 29% of the nation's rivers, 43% of its lakes, and 38% of its estuaries are too polluted for swimming, fishing, or drinking.

LAWS THAT PROTECT GROUNDWATER

Several federal laws attempt to control groundwater pollution. The Safe Drinking Water Act contains provisions to protect underground aquifers that are important sources of drinking water. In addition, the Safe Drinking Water Act regulates underground injection of wastes in an effort to prevent groundwater contamination. The **Resource Conservation and Recovery Act** deals with the storage and disposal of hazardous wastes and helps prevent groundwater contamination (see Chapter 23). Several miscellaneous laws related to pesticides, strip mining, and cleanup of abandoned hazardous waste sites indirectly protect groundwater. The many laws that directly or indirectly affect groundwater quality were passed at different times and for different reasons. These laws provide a disjointed and, at times, inconsistent protection of groundwater. The EPA makes an effort to coordinate all these laws, but groundwater contamination still occurs.

REVIEW

1. What are the main goals of the Safe Drinking Water Act? The Clean Water Act?

2. How do maximum contaminant levels and national emission limitations differ?

ENERGY AND CLIMATE CHANGE

Dead Zones

Dead zones appear to be increasing in some areas due to climate change, as warmer waters generally hold less oxygen. Unfortunately, one potential solution to climate change—more use of the biofuel ethanol (see Chapter 12)—also appears to contribute to at least one dead zone. Ethanol use in the United States has increased substantially over the past three decades (see the figure). Wastes associated with the production of ethanol from corn grown in the Midwest flow down the Mississippi River; researchers have concluded that these wastes are at least partly responsible for the large dead zone in the Gulf of Mexico. In contrast, scientists think that Hurricane Dolly in 2008 may have reduced the same dead zone by churning the water; the dead zone that year was half the size of the 2007 and 2009 dead zones. Climate scientists expect hurricanes to get stronger, which could mean more churning and thus smaller dead zones. Biofuel wastes and more intense hurricanes remind us that human activities have complex impacts on large-scale environmental systems.

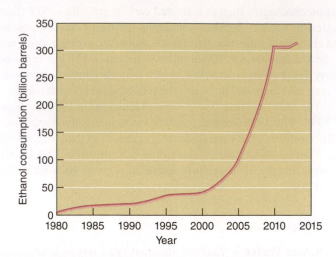

Ethanol consumption in the United States, 1980–2014. Over the past three and a half decades, ethanol consumption has increased substantially in the United States. Has ethanol use increased at a linear or exponential rate? (Monthly Energy Review, Energy Information Agency, April 2014)

REVIEW OF LEARNING OBJECTIVES WITH SELECTED KEY TERMS

- **List and briefly describe eight categories of water pollutants.**

Water pollution consists of any physical or chemical change in water that adversely affects the health of humans and other organisms. **Sewage** is the release of wastewater from drains or sewers (such as from toilets, washing machines, and showers); it includes human wastes, soaps, and detergents. **Disease-causing agents**, such as bacteria, viruses, protozoa, and parasitic worms, are transmitted in sewage. **Sediment pollution**, primarily from soil erosion, increases water turbidity, thereby reducing photosynthetic productivity in the water. **Inorganic plant and algal nutrients**, such as nitrogen and phosphorus, contribute to **enrichment**, the fertilization of a body of water. Many **organic compounds**, such as pesticides, pharmaceuticals, solvents, and industrial chemicals, are quite toxic to organisms. **Inorganic chemicals** include toxins such as lead and mercury. **Radioactive substances** include the wastes from mining, refining, and using radioactive metals. **Thermal pollution** occurs when heated water, produced during many industrial processes, is released into waterways.

- **Discuss how sewage is related to eutrophication, biochemical oxygen demand (BOD), and dissolved oxygen.**

Sewage supplies nutrients that contribute to eutrophication and a high biochemical oxygen demand. **Eutrophication**, the nutrient enrichment of lakes, estuaries, or slow-moving streams, results in high photosynthetic productivity, supporting an overpopulation of algae. Eutrophication kills fishes and causes a decline in water quality as these algae die and decompose. **Biochemical oxygen demand (BOD)** is the amount of oxygen needed by microorganisms to decompose biological wastes into carbon dioxide, water, and minerals. A large amount of sewage generates a high BOD, which lowers the level of dissolved oxygen in the water.

- **Define *dead zone*.**

A **dead zone** is a section of the ocean or a sea in which oxygen has been depleted to the point that most animals and bacteria cannot survive; it is often caused by runoff of chemical fertilizers or plant and animal wastes.

- **Contrast point source pollution and nonpoint source pollution.**

Point source pollution is water pollution that can be traced to a specific spot. **Nonpoint source pollution** consists of pollutants that enter bodies of water over large areas rather than being concentrated at a single point of entry.

- **Provide examples of agricultural, municipal, and industrial water pollution.**

Agriculture is the biggest contributor to surface-water pollution, including fertilizers, pesticides, plant and animal wastes, and eroded soil. Municipal waste can include surface runoff containing a variety of chemical and organic contaminants, as well as sewage. Industrial wastes vary, based on the type of industry, and can include a variety of chemicals, organic materials, and radioactive materials.

- **Describe how most drinking water is purified in the United States, and discuss the chlorine dilemma.**

Most municipal water supplies are treated before being used so that the water is safe to drink. Water is usually treated with aluminum sulfate to cause suspended particles to clump and settle out, filtered through sand, and disinfected by adding chlorine. There is concern that chlorine in drinking water creates health hazards, so alternatives, including UV radiation and ozonation, are being developed.

- **Distinguish among primary, secondary, and tertiary treatments for wastewater.**

Primary treatment is treating wastewater by removing suspended and floating particles by mechanical processes. **Secondary treatment** is treating wastewater biologically to decompose suspended organic material; secondary treatment reduces the water's biochemical oxygen demand. Primary and secondary treatment generate **primary** and **secondary sludge**, which consist of the solids remaining after sewage treatment has been completed. Some sludge can be safely used as fertilizer, while other sludge must be disposed of in landfills. **Tertiary treatment** is advanced wastewater treatment methods that are sometimes employed after primary and secondary treatments.

- **Compare the goals of the Safe Drinking Water Act and the Clean Water Act.**

The **Safe Drinking Water Act** sets uniform federal standards for drinking water to guarantee safe public water supplies throughout the United States. The **Clean Water Act** has two basic goals: to eliminate the discharge of pollutants into U.S. waterways and to attain water quality levels that make these waterways safe to fish and swim in.

- **Define *maximum contaminant level* and *national emission limitation*, and specify the water legislation each relates to.**

The **maximum contaminant level** is the upper limit for the concentration of a particular pollutant in water intended for human consumption. The EPA determines maximum contaminant levels for water pollutants that might affect human health as mandated by the Safe Drinking Water Act. The **national emission limitation** is the maximum permissible amount of a water pollutant that can be discharged from a sewage treatment plant, factory, or other point source. The Clean Water Act instructs the EPA to set up and monitor national emission limitations.

Critical Thinking and Review Questions

1. What is sustainable water use?

2. What is water pollution? Why is wastewater treatment an important part of sustainable water use?

3. Describe the relationship between biochemical oxygen demand and dissolved oxygen available for fish.

4. Distinguish between oligotrophic and eutrophic lakes.

5. How do midwestern farmers threaten the livelihood of fishermen in the Gulf of Mexico?

6. How does thermal pollution typically affect organisms? What causes this affect?

7. Contrast organic compounds and inorganic chemicals as types of water pollution.

8. Tell whether each of the following represents point source pollution or nonpoint source pollution: fertilizer runoff from farms, thermal pollution from a power plant, urban runoff, sewage from a ship, erosion sediments from deforestation.

9. What is the source of arsenic contamination of groundwater in Bangladesh?

10. Why is chlorine added to drinking water? Why does the Environmental Protection Agency recommend that public water treatment facilities find alternatives to chlorine?

11. What is removed during each of the three stages of wastewater treatment: primary, secondary, and tertiary? During which stage would you expect items to be recovered that were accidentally flushed, such as jewelry or coins?

12. What is sludge? How is it disposed of?

13. Taking a systems perspective, suggest how to prevent BOD from animal sewage and milk processing on a dairy farm from reaching an adjacent river.

14. In what ways might climate change make dead zones in the ocean worse? Better?

15. Which water pollution law regulates industrial waste discharge to a river?

16. The following graph reflects the monitoring of dissolved oxygen concentrations at six stations along a river. The stations are located 20 m apart, with A the farthest upstream and F the farthest downstream. Where along the river did a sewage spill occur? At which station would you most likely discover dead fish?

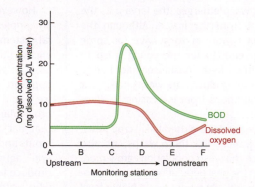

Food for Thought

Agriculture can be affected by water pollution but can also contribute to it. Which types of water pollution are most likely to affect food production, particularly in your region? Which types of water pollution are at least in some part generated by agriculture? How might sustainable food production reduce the risk of polluting water sources?

22

©Nigel Cattin/Alamy

Pest Management

Parasitized pest. Grain aphid (*Sitobion avenae*) parasitized and mummified by a parasitoid wasp (*Aphidius*). When pesticides are used to control pests, natural enemies of pests are also killed. Pest control methods can either foster or damage populations of beneficial insects.

Most people think of wasps as pest insects. Although wasps are indeed predators, many wasps are actually the enemies of our enemies: insect pests. Many wasps, like *Aphidius* (see photograph), kill pests by laying the eggs of their young inside herbivorous insects that consume crop plants. When the young wasp emerges, the larva eats the pest insect from the inside. Although the pest insect continues to survive for some time—long enough for the young wasp to reach adulthood—the pest insect dies before it is able to reproduce. The wasps keep pest insect populations from increasing to damaging levels.

When pesticides are used to kill crop pests, beneficial insects, like these wasps, are also killed. Since the wasp populations grow more slowly than the pest populations, the pest population can actually rebound to a higher level than before the pesticide was applied, sometimes within weeks after pesticide application. Other wildlife also consume crop pests. Many birds, bats, and amphibians are important predators of crop pests. Farmers and researchers have found that enhancing habitat for these beneficial organisms helps reduce crop damage by pests.

We are consumers—eating producers (plants) and other consumers (animals). We compete with other consumers, which we often call pests. We have a mutualistic relationship with bees and apple trees: We eat apples and honey, and we provide homes for the bees near the apples. We plant, water, and fertilize apples and protect them from other consumers (worms, birds) and competing plants (weeds).

However, while our relationship with bees and apples is similar to those found in natural systems, there are some profound differences in the details. We often dedicate large areas to growing a single crop—in this case, apple orchards. In a system without humans, we would find a combination of apples and other plants, as well as a range of animals that eat apples. We must rely on energy inputs and chemicals to maintain the "unnatural" agricultural system, and we send the produce long distances to sell to consumers who are too far away to pick the apples themselves. Parasitic wasps in apple orchards help control aphids, which damage apple trees and fruit. More commonly, chemical pesticides are used to kill weeds and insects.

Agriculture is the sector that uses the most pesticides worldwide—approximately 85% of the estimated 2.6 million metric tons (2.9 million tons) used each year. Highly developed countries use about three-fourths of all pesticides, but pesticide use is increasing most rapidly in developing countries.

We discuss here the types and uses of pesticides, their benefits, and their disadvantages. Pesticides have saved millions of lives by killing insects that carry disease and by increasing the amount of food we grow. Modern agriculture depends on pesticides to produce blemish-free fruits and vegetables while keeping labor costs low. Pesticides, however, cause environmental and health problems, and their harmful effects can outweigh their benefits. Pesticides rarely affect only the targeted pest species, and the balance of a natural system, such as predator-prey relationships, is upset. Certain pesticides concentrate at higher levels of the food chain. Humans who apply and work with pesticides may be at risk for pesticide poisoning (short term) and cancer (long term), and people who eat traces of pesticide on food are concerned about the long-term effects. In this chapter we discuss some alternatives to pesticides and examine the pesticide laws that protect our health and the environment.

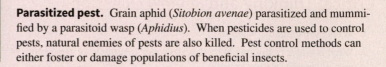

In Your Own Backyard…
Most homes deal with a range of pests, from dandelions in lawns to mice and ants. What pest management methods are used where you live?

What Is a Pesticide?

- Distinguish between narrow-spectrum and broad-spectrum pesticides, and describe the various types of pesticides, such as insecticides and herbicides.

Any organism that interferes in some way with human welfare or activities is a **pest**. Some plants, insects, rodents, bacteria, fungi, nematodes (microscopic worms), and other organisms compete with humans for food; other pests cause or spread disease. The definition of *pest* is subjective. A dandelion may be a pest to one lawn owner, but a food plant to someone else. People try to control pests, usually by reducing the size of the pest population. **Pesticides** are a common way of controlling pests, particularly in agriculture. Pesticides are grouped by their target organisms—that is, by the pests they are supposed to eliminate. **Insecticides** kill insects, **herbicides** kill plants, **fungicides** kill fungi, and **rodenticides** kill rodents, such as rats and mice. Although used as medicines, **antibiotics**, can also be considered a type of pesticide, because they kill bacteria.

The ideal pesticide would be a **narrow-spectrum pesticide** that would kill only the organism for which it was intended and not harm any other species. The perfect pesticide would be readily broken down, either by natural chemical decomposition or by biological organisms, into safe materials such as water, carbon dioxide, and oxygen. The ideal pesticide would stay exactly where it was put and would not move around in the environment.

Unfortunately, there is no such thing as an ideal pesticide. Most pesticides are **broad-spectrum pesticides**. Some pesticides do not degrade readily or else break down into compounds as dangerous as, if not more dangerous than, the original pesticide. Most pesticides move around in the environment, via air, soil, and especially water.

broad-spectrum pesticides Pesticides that kill a variety of organisms, including beneficial organisms, in addition to the target pest.

FIRST-GENERATION AND SECOND-GENERATION PESTICIDES

Before the 1940s, pesticides were of two main types, inorganic compounds (also called minerals) and organic compounds. Inorganic compounds that contain lead, mercury, copper, and arsenic are extremely toxic to pests but are not used much today, in part because of their chemical stability in the environment. Natural processes do not degrade inorganic compounds, which persist and accumulate in the soil and water. This accumulation poses a threat to humans and other organisms, which, like the target pests, are susceptible to poisoning by inorganic compounds.

Many plants, which have been fighting pests longer than humans, have evolved natural organic compounds that are poisonous to insects or inhibit growth of other plants. Such plant-derived pesticides are called **botanicals**. Examples of botanicals include nicotine from tobacco, pyrethrin from chrysanthemum flowers, and rotenone from roots of the derris plant, all of which are used to kill insects (**Figure 22.1**). Juglans, from black walnut trees, is a botanical herbicide. Botanicals, which are easily degraded by microorganisms, do not persist for long in the environment.

Figure 22.1   Pesticide derived from plants. Chrysanthemum flowers, shown here as they are harvested in Rwanda, are the source of the insecticide pyrethrin. Botanicals are plant chemicals used as pesticides.

However, they are highly toxic to aquatic organisms and to bees (beneficial insects that pollinate crops).

Synthetic botanicals are human-made insecticides produced by chemically modifying the structure of natural botanicals. An important group of synthetic botanicals is the pyrethroids, which are chemically similar to pyrethrin. Pyrethroids do not persist in the environment; they are slightly toxic to mammals and bees but very toxic to fishes. Allethrin is an example of a pyrethroid. However, neonicotinoids are a synthetic botanical insecticide family which is highly toxic to bees, despite the fact that they are chemically similar to a natural botanical (nicotine, from tobacco).

In the 1940s, chemical companies began to produce many synthetic organic pesticides. Earlier pesticides, both inorganic compounds and botanicals, are called **first-generation pesticides** to distinguish them from the vast array of synthetic poisons in use today, called **second-generation pesticides**. The insect-killing ability of dichlorodiphenyltrichloroethane (DDT), the first of the second-generation pesticides, was recognized in 1939 (**Figure 22.2**). Currently about 20,000 commercial pesticide products are registered; these consist of combinations of about 675 active chemical ingredients.

THE MAJOR GROUPS OF INSECTICIDES

Insecticides, the largest category of pesticides, are usually classified into groups based on chemical structure. Three of the most important groups of second-generation insecticides are the chlorinated hydrocarbons, organophosphates, and carbamates.

DDT is a **chlorinated hydrocarbon**, an organic compound containing chlorine. After DDT's insecticidal properties were recognized, many more chlorinated hydrocarbons were synthesized as

© Bettmann/CORBIS

Figure 22.2 **Applying DDT in 1945.** Pesticides such as DDT, shown as it was sprayed to control mosquitoes at New York's Jones Beach State Park in 1945, were used in ways that would be unacceptable now. The sign on the truck reads, in part, "D.D.T. Powerful Insecticide. Harmless to Humans." The harmful environmental effects of DDT were unknown to the public until many years later.

pesticides. Generally speaking, chlorinated hydrocarbons are broad-spectrum insecticides. Most are slow to degrade and may persist in the environment (even inside organisms) for many months or even years. They were widely used from the 1940s until the 1960s, but since then many have been banned or restricted, mainly because of problems associated with their persistence and impacts on humans and wildlife. Three chlorinated hydrocarbons still in use in the United States are endosulfan, lindane, and methoxychlor. Many people first became aware of the problems with pesticides in 1963, when Rachel Carson published *Silent Spring* (see Chapter 2).

Organophosphates, organic compounds that contain phosphorus, are more poisonous than other types of insecticides, and many are highly toxic to birds, bees, and aquatic organisms. The toxicity of many organophosphates in mammals, including humans, is comparable to that of some of our most dangerous poisons—arsenic, strychnine, and cyanide. Organophosphates do not persist in the environment as long as chlorinated hydrocarbons do. As a result, organophosphates have generally replaced the chlorinated hydrocarbons in large-scale uses such as agriculture, although many are not widely available to consumers because of their high level of toxicity. Methamidophos, dimethoate, and malathion are three examples of organophosphates. Some of these products have recently become restricted in their use due to their toxicity.

Carbamates, the third group of insecticides, are broad-spectrum insecticides derived from carbamic acid. Carbamates are generally not as toxic to mammals as the organophosphates, although they still show broad, nontarget toxicity. Two common carbamates are carbaryl and aldicarb.

THE MAJOR KINDS OF HERBICIDES

Herbicides are chemicals that kill or inhibit the growth of unwanted vegetation such as weeds in crops or lawns. Like insecticides, herbicides can be classified into groups on the basis of chemical structure, but this method is cumbersome because at least 12 different chemical

groups are used as herbicides. More simply, **selective herbicides** kill only certain types of plants, whereas **nonselective herbicides** kill all or most vegetation. Selective herbicides can be further classified according to the types of plants they affect. **Broad-leaf herbicides** kill plants with broad leaves but do not kill grasses; **grass herbicides** kill grasses but are safe for most other plants.

Two common herbicides with similar structures are *2,4-dichlorophenoxyacetic acid (2,4-D)* and *2,4,5-trichlorophenoxyacetic acid (2,4,5-T)*. Both were developed in the United States in the 1940s. These broad-leaf herbicides, disrupt the plants' natural growth processes; they kill plants such as dandelions but do not harm grasses. Many of the world's important crops, such as wheat, corn, and rice, are cereal grains (see Chapter 18), which are grasses. Both 2,4-D and 2,4,5-T kill weeds that compete with these crops, although 2,4,5-T is no longer used in the United States. The Environmental Protection Agency (EPA) banned most uses of 2,4,5-T in 1979 because of possible harmful side effects to humans that became apparent after its use in the Vietnam War. 2,4-D is no longer used by commercial lawn care companies due to toxicity to applicators, but it is still one of the most common ingredients in lawn weedkillers.

Glyphosate is currently a popular herbicide due to its nonselective action and its low toxicity to mammals. However, herbicides with glyphosate are highly toxic to amphibians, and extensive use of glyphosate on genetically engineered crops (like Roundup Ready® corn) has raised concerns about whether this herbicide is really as safe as herbicide company tests suggest. Glyphosate is safer than many herbicides, but "safer" is not the same as "safe."

REVIEW

1. How do narrow-spectrum pesticides differ from broad-spectrum pesticides?

Benefits and Problems with Pesticides

LEARNING OBJECTIVES

- Describe the benefits of pesticides in disease control and crop protection.
- Explain why monocultures are susceptible to pest problems.
- Summarize the problems associated with pesticide use, including genetic resistance; imbalances; persistence, bioaccumulation, and biological magnification; and mobility in the environment.
- Describe pesticide resistance and resistance management.

Each day a war is waged as farmers, struggling to produce bountiful crops, battle insects and weeds. Similarly, health officials fight their own war against the ravages of human diseases transmitted by insects.

Although pesticides have their benefits, they have several problems. First, many pest species evolve resistance after repeated exposure to pesticides. Second, pesticides affect numerous species, including beneficial ones, in addition to the target pests, generating imbalances in the ecosystem (including agricultural fields) and posing a threat to human health. Finally, the ability of some pesticides to resist degradation and to readily move around in the environment causes even more problems for humans and other organisms.

BENEFIT: DISEASE CONTROL

Insects transmit several devastating human diseases. Fleas and lice carry the microorganism that causes typhus in humans. Malaria, also caused by a microorganism, is transmitted to millions of humans each year by female *Anopheles* mosquitoes (**Figure 22.3**). According to the World Health Organization (WHO), in 2012 approximately 207 million people suffered from malaria, leading to over 620,000 deaths. Over 20% of childhood deaths in Africa are associated with malaria. Because only a few antimalarial drugs are available (mostly botanicals), the focus of controlling this disease is on reducing the mosquitoes that carry it. Important steps in mosquito control include limiting stagnant pools of water where mosquitoes reproduce and applying pesticides.

Pesticides, particularly DDT, have helped control mosquito populations, thereby reducing the incidence of malaria. Consider Sri Lanka. In the early 1950s, more than 2 million cases of malaria were reported in Sri Lanka each year. When spraying of DDT was initiated to control mosquitoes, malaria cases dropped to almost zero. When DDT spraying was discontinued in 1964, malaria reappeared almost immediately. By 1968, the country's annual incidence had increased to more than 1 million cases per year. Since then, DDT use has resumed in many places, but not in the huge amounts and broad applications that caused so much damage in the 1960s. Today, DDT-impregnated mosquito nets are used to keep the pests off people, and indoor walls are sprayed with DDT.

BENEFIT: CROP PROTECTION

Although it is difficult to make exact assessments, it is widely estimated that pests eat or destroy more than one-third of the world's crops. Interestingly, this general proportion has been true for centuries of human agriculture, despite changing practices of pest control. Given our expanding population and world hunger, many people hope that better control of agricultural pests is possible.

Pesticides reduce the amount of a crop lost through competition with weeds, consumption by insects, and diseases caused by plant **pathogens** (microorganisms, such as fungi and bacteria, that cause plant disease). Although many insect species are beneficial from a human viewpoint (two examples are honeybees, which pollinate crops, and ladybugs, which prey on crop-eating insects), many are considered pests. Of these, about 200 species have the potential to cause large economic losses in agriculture. For example, the Colorado potato beetle is one of many insects that voraciously consume the leaves of the potato plant, reducing the plant's ability to produce large tubers for harvest.

Pesticide use is usually justified economically, in that farmers save an estimated $3 to $5 in crops for every $1 that they invest in pesticides. In countries where pesticides are not used in appreciable amounts and crop health is compromised by poor agricultural practices, the losses due to agricultural pests can be considerable.

Why are agricultural pests found in such great numbers in our fields? Part of the reason is that agriculture is usually a **monoculture**, in which only one variety of one crop species is grown on large tracts of land (this term was also used, in reference to tree plantations, in Chapter 6). The cultivated field represents a simple ecosystem from which humans have removed much of the "system." In contrast, forests, wetlands, and other natural ecosystems are complex and contain many different species, including predators and parasites that control pest populations and plant species that pests do not use for food. A monoculture reduces the dangers and accidents that might befall a pest as it searches for food. A Colorado potato beetle in a forest would have a hard time finding anything to eat, but a 500-acre potato field is like a big banquet table set just for the pest. It eats, prospers, and reproduces. In the absence of many natural predators and in the presence of plenty of food, the population thrives and grows, and more of the crop becomes damaged.

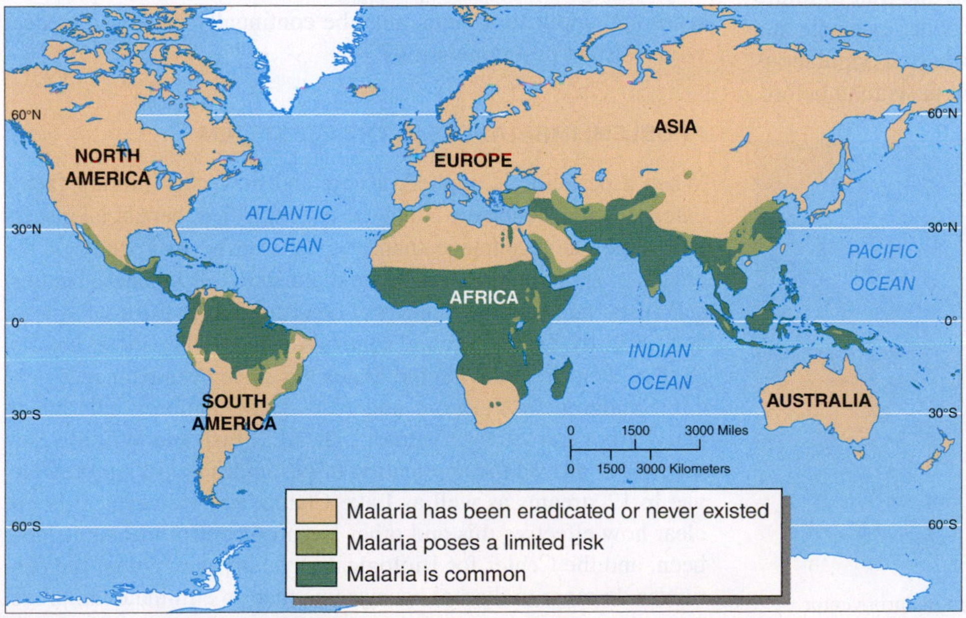

PROBLEM: EVOLUTION OF GENETIC RESISTANCE

The prolonged use of a particular pesticide can cause a pest population to develop **genetic resistance** to the pesticide. In the 60 years during which pesticides have been widely used, at least 520 species of insects and mites have evolved genetic resistance to certain pesticides. Many pests are now resistant to multiple pesticides, and at least 17 species, such as diamondback moths and palm thrips, are resistant to all major classes of insecticides that farmers are legally allowed to use on them. Insects are not the only pests to evolve genetic resistance; almost 200 plant

genetic resistance Any inherited characteristic that decreases the effect of a pesticide on a pest.

Figure 22.3 **Location of malaria.** Insecticides sprayed to control mosquitoes in these locations have saved millions of lives. Had insecticides not been used widely in agriculture as well, mosquitoes would also not have developed resistance to these insecticides so quickly.

Insects and other pests are constantly evolving. The short generation times (the period between the birth of one generation and that of another) and large populations characteristic of most pests favor rapid evolution, which allows the pest population to quickly adapt to the pesticides used against it (**Figure 22.4**). As a result, an insecticide that kills most of an insect population becomes less effective after prolonged use because the survivors and their offspring are genetically resistant. For example, after many years of DDT applications on walls to kill mosquitoes that land, mosquitoes have genetically evolved in their behavior: some populations of mosquitoes do not land on walls before seeking their next meal.

Resistance management is a useful approach to dealing with genetic resistance. It involves efforts to delay the evolution of genetic resistance in insect pests or weeds. Strategies of resistance management vary, depending on the pest species involved.

> **resistance management** Strategies for managing genetic resistance to maximize the period in which a pesticide is useful.

One strategy of resistance management for insect pests is to maintain a nearby "refuge" of untreated plants where the pest can avoid being exposed to the pesticide. Those insects or weeds that live and grow in the refuge remain susceptible to the pesticide. When susceptible insects migrate into the area being treated with insecticide, they mate with members of the genetically resistant population. Weeds in the refuge continue to contribute pollen to the larger weed population. This interbreeding delays the development of genetic resistance in the population as a whole.

After herbicides are applied, the field is scouted to see if any weed plants survived the herbicide application. These weeds are resistant to the herbicide, and they are removed from the field before they flower. Cultural methods that prolong the usefulness of herbicides include mechanically pulling weeds, planting seed that is certified free of weed seeds, avoiding importing manure from weedy fields onto cleaner ones, and rotating crops to keep weed populations from accumulating as easily. All of these methods can help managers avoid the pesticide treadmill, in which pesticide types and application rates must be continually increased to deal with resistant pest populations.

species are known to be resistant to herbicides. Some weeds, such as annual ryegrass and canary grass, may be resistant to all available herbicides.

How does genetic resistance to pesticides occur? Every time a pesticide is applied to control a pest, some pests survive. The survivors, because of certain genes they already possess, are genetically resistant to the pesticide, and they pass on this trait to future generations. Thus, *evolution*—any cumulative genetic change in a population of organisms—occurs, and subsequent pest populations contain larger percentages of pesticide-resistant pests than before.

PROBLEM: IMBALANCES IN THE ECOSYSTEM

When a pesticide is applied, a new and powerful limiting factor enters an often delicately balanced system. Often, beneficial insects experience more damage than do pest insects. In a study of the effects of spraying the insecticide dieldrin to kill Japanese beetles, scientists found a large number of dead animals in the treated area, such as birds, rabbits, ground squirrels, cats, and beneficial insects. (Use of dieldrin in the United States has since been banned.)

Similarly, in Vermont and New York, lamprey eels attack native trout and Atlantic salmon. The U.S. Fish and Wildlife Service has approved the application of pesticides to kill lamprey larvae in 12 streams as well as Lake Champlain. However, it is not clear how effective this and other lamprey-control measures have been, and the Center for Biological Diversity claims that the pesticides do more to disrupt ecosystems than to eliminate lampreys. Pesticides do not have to kill organisms to harm them. Quite often, the stress of pesticide exposure makes an organism more vulnerable to predators, diseases, or other stressors in its environment.

© Nigel Cattlin / Alamy

Figure 22.4   Mosquito feeding. South American malaria vector mosquito (*Anopheles albimanus*) feeding. Mosquitoes have a short life cycle, making it possible for insecticide resistant individuals to repopulate an area quickly after pesticide application.

Because the natural enemies of pests often starve or migrate in search of food after pesticide is sprayed in an area, pesticides are indirectly responsible for a large reduction in the populations of these natural enemies. Pesticides also kill natural enemies directly because predators consume a lot of the pesticide when consuming the pests. After a brief period, the pest population rebounds and gets larger than ever, partly because no natural predators are left to keep its numbers in check. Natural predator populations grow more slowly in numbers (see *K* selection in Chapter 5) than pest populations.

Despite a 33-fold increase in pesticide use in the United States since the 1940s, crop losses due to insects, diseases, and weeds have not changed appreciably (**Table 22.1**). Increasing genetic resistance to pesticides and the destruction of the natural enemies of pests provide a partial explanation. Changes in agricultural practices are also to blame; for example, crop rotation, a proven way of controlling certain pests, is not practiced as much today as it was several decades ago (see Chapter 14).

Creation of New Pests   In some instances, use of a pesticide has resulted in a pest problem that did not exist before. Creation of new pests—that is, turning minor pest organisms into major pests—is possible because the pesticide kills most of a certain pest's natural predators, parasites, and competitors, allowing the pest's population to rebound. The use of DDT to control certain insect pests on lemon trees was documented as causing an outbreak of a scale insect (a sucking insect that attacks plants) that was not a problem before spraying (**Figure 22.5**). In a similar manner, the European red mite became an important pest on apple trees in the northeastern United States, and beet armyworms became an important pest on cotton, both after the introduction of pesticides.

PROBLEMS: PERSISTENCE, BIOACCUMULATION, AND BIOLOGICAL MAGNIFICATION

As discussed in Chapter 7, some pesticides are extremely **persistent** in the environment and may take many years to break down into less toxic forms. When a persistent pesticide is ingested, it is stored, usually in fatty tissues. Over time, the organism may accumulate high concentrations of the pesticide, a phenomenon known as **bioaccumulation**.

Organisms at higher levels in food webs tend to have greater concentrations of bioaccumulated pesticide stored in their bodies than those lower in food webs. The increase in pesticide concentrations as the pesticide passes through successive levels of the food web is known as **biomagnification** (**Figure 22.6**; also see Figure 7.6).

bioaccumulation The buildup of a persistent pesticide or other toxic substance in an organism's body.

biological magnification The increased concentration of toxic chemicals such as certain pesticides in the tissues of organisms at higher trophic levels in food webs.

Table 22.1	Percentage of Crops Lost Annually to Pests in the United States		
Period	Insects	Diseases	Weeds
2006	13.0	12.5	12.0
1989–1999	13.0	12.0	12.0
1974	13.0	12.0	8.0
1951–1960	12.9	12.2	8.5
1942–1951	7.1	10.5	13.8

Source: USDA Agricultural Research Service

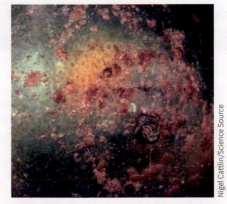

(a) Red scale on oranges.

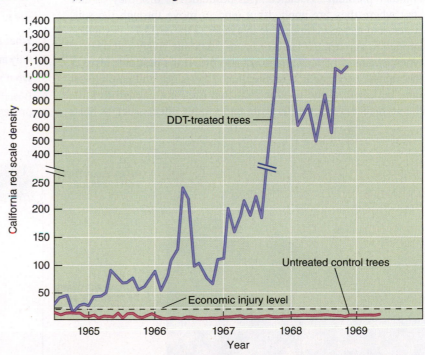

(b) A comparison of red scale populations on DDT-treated trees (blue line) and untreated trees under biological control (red line).

Figure 22.5   Pesticide use and new pest species. An infestation of red scale insects on lemons occurred after DDT was sprayed to control a different pest. Prior to DDT treatment, red scale did not cause significant economic injury to citrus crops.

Figure 22.6   **Peregrine falcon feeding chicks.** Biological magnification of DDT in the tissues of falcons and other birds of prey caused a decline in their reproductive success in the United States during the 1960s and early 1970s. DDT caused the birds to lay eggs with extremely thin, fragile shells, resulting in the chicks' deaths. (DDT use was banned in the United States in 1972.)

PROBLEM: MOBILITY IN THE ENVIRONMENT

Another problem associated with pesticides is that they do not stay where they are applied but tend to move through the soil, water, and air, sometimes long distances (**Figure 22.7**). Pesticides applied to agricultural lands or lawns wash into rivers and streams when it rains, which can harm aquatic life. If the pesticide level in the aquatic ecosystem is high enough, the fishes and amphibians may be killed. If the level is sublethal (i.e., not enough to kill), the animals may still suffer from undesirable effects such as bone

degeneration. These effects may decrease the animals' fitness and increase their chances of being preyed on.

Pesticide mobility is also a problem for humans. The most recent comprehensive study by the Environmental Protection Agency and the U.S. Geological Survey reported that pesticides are found throughout the year in most of the country's surface waters. The study also found at least one pesticide above the **water quality benchmark** (the level above which aquatic organisms are threatened) in 80% of urban streams. Another study by the Environmental Working Group (EWG), a private environmental organization, revealed that more than 14 million U.S. residents drink water containing traces of five widely used herbicides (alachlor, metachlor, atrazine, cyanazine, and simazine). The study concluded that 3.5 million people living in the Midwest, where those herbicides are used on corn and soybeans, face an elevated cancer risk. Since the study, the EPA has required a decrease in the use of these herbicides.

Another EWG study in California revealed how mobile pesticides can be in the atmosphere. Almost two-thirds of 100 air samples from California contained small amounts of pesticides that had drifted from farm fields. Air and water currents can carry pesticides around the world: A 2008 study sampled Antarctic krill, a keystone species in the Antarctic ecosystem (see Case in Point, Chapter 3). In each of 12 sites sampled, multiple organochlorine pesticides were found.

REVIEW

1. What are two important benefits of pesticide use?
2. Why are monocultures susceptible to pest problems?
3. What are persistence, bioaccumulation, and biomagnification?
4. Describe pesticide resistance and discuss some methods for preventing the problem.

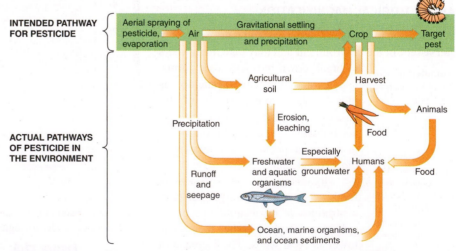

(a) A helicopter sprays pesticides on a crop—and everything else in its pathway—in California.

(b) The intended pathway of pesticides in the environment (shown at top of figure) is quite different from the actual pathways (shown at bottom).

Figure 22.7   **Intended versus actual pathways of pesticides.**

Risks of Pesticides to Human Health

LEARNING OBJECTIVES

- Discuss pesticide risks to human health, including short-term effects and long-term effects.
- Explain endocrine disruption.

Exposure to pesticides can damage human health. Pesticide poisoning caused by short-term exposure to high levels of pesticides can result in harm to organs and even death, whereas long-term exposure to lower levels of pesticides can cause certain types of cancer, or metabolic or reproductive problems. In addition, exposure to trace amounts of certain pesticides could disrupt the human endocrine (hormone) system. Chapter 7 discusses the elevated risk of pesticides to children (see Figure 7.10).

SHORT-TERM EFFECTS OF PESTICIDES

Pesticides poison tens of thousands of people in the United States each year. Most poisonings occur among farmworkers or others whose occupations involve daily contact with large quantities of pesticides. A person with a mild case of pesticide poisoning may exhibit symptoms such as nausea, vomiting, and headaches. More serious cases, particularly organophosphate poisonings, may result in permanent damage to the nervous system and other body organs. Almost any pesticide can kill a human if the dose is large enough. Fortunately, pesticide application, packaging, and distribution regulations have led to decreased poisonings since the mid-1990s.

WHO estimates that, globally, pesticides poison more than 4 million people each year; of these, at least a million require hospitalization and about 300,000 die. The incidence of pesticide poisoning is highest in countries where pesticide regulations are minimal. Pesticide users in many countries often are not trained in the safe handling and storage of pesticides, and safety regulations are generally more lax there.

LONG-TERM EFFECTS OF PESTICIDES

Many studies of farmworkers and workers in pesticide factories, many of whom are exposed to low levels of pesticides over many years, show an association between cancer and long-term exposure to pesticides (**Figure 22.8**). A type of lymphoma (a cancer of the lymph system) is associated with the herbicide 2,4-D. Other pesticides are linked to a variety of cancers, such as leukemia and cancers of the brain, lungs, and testicles. Researchers have noted a correlation between breast cancer and a high level of one or more pesticides in the breast's fatty tissue.

Long-term exposure to at least one pesticide may have resulted in sterility in thousands of farmworkers on banana and pineapple plantations. More than 26,000 workers in 12 countries sued their employers or pesticide manufacturers, and most of the defendants have paid settlements but have not admitted liability. Evidence shows that pregnant women who live near pesticide applications have higher rates of miscarriages. Other studies have indicated that the children of agricultural workers are at greater risk for birth defects, particularly stunted limbs.

Figure 22.8 **Farmworker applying pesticide.** Most people who suffer from pesticide-related illnesses are farmworkers or others whose occupations involve daily contact with pesticides.

Long-term exposure to pesticides may increase the risk of Parkinson's disease, which afflicts about 1 million people in the United States. When researchers injected rats with repeated doses of rotenone, a plant-derived pesticide, the rats developed the symptoms of Parkinson's disease, including difficulty in walking, tremors, and abnormal protein deposits in their brains. Many pesticides have chemical structures similar to rotenone, which suggests that other pesticides may increase the risk of Parkinson's disease as well. This connection between pesticide exposure and Parkinson's disease is tentative and will require additional research to prove or disprove. To date, no study has connected Parkinson's disease in humans with exposure to a specific pesticide.

PESTICIDES AS ENDOCRINE DISRUPTERS

In the 1990s, many articles were published that linked certain pesticides and other persistent toxic chemicals with reproductive problems in animals. These chemicals, known as **endocrine disrupters**, interfere with or mimic hormones in humans or other animals (**Table 22.2**; see Chapter 7 discussion on endocrine disrupters). River otters exposed to synthetic chemical pollutants were found to have abnormally small penises. Female seagulls in southern California exhibited behavioral aberrations by pairing with one another rather than with males during the mating season. In Florida, alligators exposed to a common herbicide produce eggs inside their testes. In many cases scientists have produced the

Table 22.2 Some Pesticides That Are Known Endocrine Disrupters*

Pesticide	General Information
Atrazine	Herbicide; still used
Chlordane	Insecticide; banned in United States in 1988
DDT (dichlorodiphenyl-trichloroethane)	Insecticide; banned in United States in 1972
Endosulfan	Insecticide; still used
Kepone	Insecticide; banned in United States in 1977
Methoxychlor	Insecticide; still used

*Based on experimental research with laboratory animals.

same abnormal symptoms in the laboratory, providing support that certain persistent chemicals cause the defects.

The suggestion in the 1996 book *Our Stolen Future*, by Theo Colborn, Dianne Dumanoski, and John Myers, that persistent toxic chemicals in the environment are disrupting human hormone systems ignited a barrage of media attention. Theo Colborn, then a senior scientist at the World Wildlife Fund, hypothesized that ubiquitous chemicals in the environment are linked to disturbing trends in human health. These include increases in breast and testicular cancer, increases in male birth defects, and decreases in sperm counts.

Many studies suggest that certain chemicals in the environment may affect human health, but direct cause-and-effect relationships between these chemicals and adverse effects on the human population have not been widely established. Some scientists think the potential danger is so great that persistent chemicals such as DDT should be internationally banned immediately. Other scientists are more cautious in their assessment of the danger, but few think the problem should be disregarded. Meanwhile, an international effort is under way to ban the production and use of nine pesticides suspected of being endocrine disrupters (see section on the global ban of persistent organic pollutants later in this chapter).

REVIEW

1. What are some of the long-term effects of pesticides on human health?
2. Why are endocrine-disrupting pesticides of particular concern?

Alternatives to Pesticides

LEARNING OBJECTIVE

- Describe alternative ways to control pests, including cultivation methods, biological controls, pheromones and hormones, reproductive controls, genetic controls, quarantine, integrated pest management, and food irradiation.

Fortunately, pesticides are not the only weapons in our arsenal against pests. Alternative ways to control pests include cultivation methods, biological controls, pheromones and hormones, reproductive controls, genetic controls, quarantine, and irradiation. A combination of these methods, often including a limited use of pesticides as a last resort, is known as *integrated pest management (IPM)*. IPM is the most effective way to control pests. (Also see Chapter 18 for a discussion of organic agriculture, in which foods are grown in the absence of pesticides and commercial fertilizers.)

USING CULTIVATION METHODS TO CONTROL PESTS

Sometimes agricultural practices are altered in such a way that a pest is adversely affected or discouraged from causing damage. Although some techniques, such as using vacuums for removing insect pests, are relatively new, other cultivation methods that discourage pests have been practiced for centuries.

One way to reduce damage from crop pests is to *intercrop* mixtures of plants, for example, by alternating rows of different plants. When corn was interplanted with sorghum in an experiment in Kenya, only about 5% of the corn crop was damaged, as compared with about 39% damage in the monoculture field of corn. Sorghum repels some insects and attracts wasps that lay their eggs inside corn borers, insects that eat through and destroy maturing corn stalks. Sorghum–corn intercropping is now common in parts of Africa.

A technique used with success in alfalfa crops is *strip cutting*, in which only one segment of the crop is harvested at a time. The unharvested portion of the crop provides an undisturbed habitat for natural predators and parasites of the pest species. The same type of benefit is derived from keeping strips of unplowed plants (i.e., wild plants, including weeds) along the margins of fields. A German study found that pest mortality was about 50% higher in oilseed rape fields with margin strips of other plants, as compared to fields without margins. The margins provided a refuge for three parasites of the pollen beetle, which is a significant pest on oilseed rape plants.

The proper timing of planting, fertilizing, and irrigating promotes healthy, vigorous plants that are more resistant to pests because they are not stressed by other environmental factors. The rotation of crops controls pests, such as corn rootworm.

BIOLOGICAL CONTROLS

As an example of a **biological control**, suppose that an insect species is accidentally introduced into a country where it was not found previously and becomes a pest. It might be possible to control this pest by going to its place of origin and looking at how the pest fit into its native ecosystem. Usually, some other organism—often a wasp or disease organism—is an exclusive predator or parasite of the pest species. That predator or parasite, if successfully introduced, may lower the population of the pest species.

biological control A method of pest control that involves the use of naturally occurring disease organisms, parasites, or predators to control pests.

The pest species typically does not evolve genetic resistance to the biological control agent in the same way it does to a pesticide, since both pest and predator are living organisms that are responsive to *natural selection*. As the pest evolves a way to resist the biological control agent, the agent in turn may evolve some sort of countermeasure against the pest. (Recall the discussion of the predator–prey "arms race" in Chapter 5; a similar arms race occurs here between the pest and its biological control agent.)

Cottony-cushion scale provides an example of successfully using one organism to control another. The cottony-cushion scale is a small insect that sucks the sap from the branches and bark of many fruit trees, including citrus trees. It is native to Australia but was accidentally introduced into the United States in the 1880s. A U.S. entomologist went to Australia and returned with several possible biological control agents. One, the vedalia beetle (**Figure 22.9**), was effective in controlling scales, which it eats voraciously and exclusively. Within two years of its introduction, the vedalia beetle had significantly reduced the cottony-cushion scale in citrus orchards. Today, both the cottony-cushion scale and the vedalia beetle are present in low numbers, and the scale is not considered an economically important pest.

More than 300 species have been introduced as biological control agents to North America. The Agricultural Research Service of the U.S. Department of Agriculture (USDA) is constantly investigating possible biological controls for insect and weed pests. Although some examples of biological control are quite spectacular, finding an effective parasite or predator is usually difficult. And just because a parasite or predator is identified does not mean it will become successfully established in a new environment. Slight variations in environmental conditions, such as temperature and moisture, can alter the effectiveness of the biological control organism in its new habitat.

Care must be taken to ensure that the introduced control agent does not attack unintended hosts and become a pest itself. To guard against this possibility when the control agent is an insect, scientists put the insects in cages with samples of important crops, ornamentals, and native plants to determine if the control agents will eat the plants when they are starving. Despite such tests, organisms introduced as biological controls sometimes cause unintended problems in their new environment, and once they are introduced, they cannot be recalled. For example, a weevil was introduced in 1968 to control the Eurasian musk thistle, a noxious weed that arrived in North America in the mid-19th century. Since then, the weevil has expanded its host range to include a North American thistle listed as a threatened species. The weevils significantly reduce seed production in the native thistles.

Figure 22.9 Vedalia beetle. Imported from Australia to reduce cottony-cushion scale infestation, the vedalia beetle is a biological control success story.

National Geographic/Getty Images

Australia, in particular, has been home to many spectacularly effective and spectacularly disastrous biological control efforts. Rabbits were introduced for meat production, but were so devastating to native plant communities that a rabbit disease, myxomatosis, was introduced to control them. Cane toads were introduced as a biological control against the sugar cane grub, yet failed entirely at controlling the pest, and cane toads themselves are now a widespread pest throughout Queensland, Australia. All introductions of non-native species should be carefully considered, especially on islands; many problems could be averted by relying on native species for agriculture and landscaping.

Biological Control Agents Insects are not the only biological control agents (indeed, cats have long been kept to control rodents). When DDT was banned in the United States, alternative solutions to mosquito problems were explored. One alternative has been to spray nematodes (microscopic worms) that attack mosquito larvae. Other nematodes are often sprayed on corn to kill corn borers. Nematodes (and bacteria, discussed later) can be used to control beetle grubs in lawns. The nematodes can be sprayed much like a chemical pesticide, but with the advantage that the nematode is usually specific to a single pest and harmless to other animals. Some nematodes are also effective against a range of pests, including weevils, grasshoppers, and locusts.

Another noninsect biological control agent targets desert locusts: a fungal spore known as the Green Muscle. Locusts periodically increase in number and swarm across the African Sahel, threatening crops across 12 million hectares (30 million acres) (see Figure 4.22 for a map showing the Sahel). During a major swarm in 1988, African farmers spent $300 million to spray massive quantities of pesticides into the environment to bring the locust population under control. Many people were concerned about the adverse ecological and health effects of using such large quantities of pesticides. The use of Green Muscle in Tanzania averted much of the threat from a 2009 locust infestation that could have impacted crops that feed as many as 15 million people.

Bacteria and viruses that harm insect pests are also used successfully as biological controls. *Bacillus popilliae*, which causes milky spore disease in insects, is applied as a dust on the ground to control the larval (immature) stage of Japanese beetles. The common soil bacterium *Bacillus thuringiensis*, or *Bt*, produces a natural pesticide toxic to some insects, such as the cabbage looper, a green caterpillar that damages many vegetable crops, and the corn earworm. When eaten by insect larvae, *Bt* toxin damages the intestinal tract, killing the young insect. The toxin does not persist in the environment and is not known to harm mammals, birds, or other noninsect species.

PHEROMONES AND HORMONES

Pheromones are commonly called sexual attractants because they are often produced to attract members of the opposite sex for mating. Each insect species produces its own specific pheromone, so once the chemical structure is known, it is possible to make use of pheromones to control individual

pheromone A natural substance produced by animals to stimulate a response in other members of the same species.

pest species. Pheromones lure insects such as Japanese beetles to traps, where they are killed. Alternatively, pheromones are released into the atmosphere to confuse insects so that they cannot locate mates.

Insect hormones are natural chemicals produced by insects to regulate their own growth and metamorphosis, which is the process in which an insect's body changes through several stages to an adult. Specific hormones must be present at certain times in the life cycle of the insect; if they are present at the wrong time, the insect develops abnormally and dies. Many such insect hormones are known, and synthetic hormones with similar structures have been made. Entomologists are actively pursuing the possibility of using these substances to control insect pests.

A synthetic version of the insect hormone *ecdysone*, which causes molting, was the first hormone approved for use. Known as MIMIC, the hormone triggers abnormal molting in insect larvae (caterpillars) of moths and butterflies. Since MIMIC affects some beneficial insects, its use has risks.

REPRODUCTIVE CONTROLS

Like biological controls, reproductive controls of pests involve the use of organisms. Instead of using another species to reduce the pest population, reproductive control strategies suppress a pest population by sterilizing some of its members. In the **sterile male technique**, large numbers of males are sterilized in a laboratory, usually with radiation or chemicals. Males are sterilized rather than females because male insects of species selected for this type of control mate several to many times, but females of that species mate only once. Thus, releasing a single sterile male may prevent successful reproduction by several females, whereas releasing a single sterile female would prevent successful reproduction by only that female.

The sterilized males are released into the wild, where they reduce the reproductive potential of the pest population by mating with normal females, which then lay eggs that never hatch. As a result, the population of the next generation is much smaller.

One disadvantage of the sterile male technique is that to be effective it must be carried out continually. If sterilization is discontinued, the pest population rebounds to a high level in a few short generations. The procedure is expensive because a production facility must rear and sterilize large numbers of insects. During the 1990 Mediterranean fruit fly (medfly) outbreak in California, as many as 400 million sterile male medflies were released each week. The medfly, which lays eggs on 250 different fruits and vegetables, is a particularly destructive pest.

The standard way to sterilize male insects is to irradiate them. However, this approach does not work on at least one major human health threat: mosquitoes. An Oxford University researcher has developed a way to genetically modify mosquitoes such that their offspring die off in the larval stage. This is a promising technique, but it is too new to be certain how effective it is and how broadly it can be applied. With their short life cycles and widespread distribution in nonagricultural areas (where pest control is difficult), mosquitoes will continue to challenge our pest-control abilities.

GENETIC CONTROLS

Traditional selective breeding has been used to develop many varieties of crops that are genetically resistant to disease organisms or insects. Traditional breeding of crop plants typically involves identifying individual plants that are present in an area where the pest is common but that are not damaged by the pest. These individuals are then crossed with standard crop varieties in an effort to produce a pest-resistant version. It may take as long as 10 to 20 years to develop a resistant crop variety, but the benefits are usually worth the time and expense.

Although traditional selective breeding has resulted in many disease-resistant crops and has enabled decreased use of pesticides, there are potential problems. Fungi, bacteria, and other plant pathogens evolve rapidly and quickly adapt to the disease-resistant host plant—meaning that the new pathogen strains can cause disease in the formerly disease-resistant plant variety. Plant breeders are in a continual race to keep one step ahead of plant pathogens.

Genetic engineering offers promise in breeding pest-resistant plants more quickly (see Chapter 18). For example, a gene from the soil bacterium *Bt* (already discussed in the section on biological controls) has been introduced into several plants, such as corn and cotton. Caterpillars that eat cotton leaves from these **genetically modified (GM)** plants die or exhibit stunted growth. On the other hand, when *Bt* toxin is present in plants on a large scale, caterpillars are more likely to evolve resistance to the toxin.

 CASE IN POINT

Bt, ITS POTENTIAL AND PROBLEMS

Bt has been marketed since the 1950s, but it was not sold on a large scale until recent decades, mainly because many different varieties of the *Bt* bacterium occur and each variety produces a slightly different protein toxin. Each is toxic to only a small group of insects. The *Bt* variety that works against corn borers, for example, would not be effective against potato beetles. As a result, *Bt* is not economically competitive compared to chemical pesticides, each of which could kill many different kinds of pests on many different crops.

Genetic engineers have greatly increased the potential of *Bt*'s toxin as a natural pesticide by modifying the gene coding for the toxin so that it affects a wider range of insect pests. They then inserted the *Bt* gene that codes for the toxin into at least 18 crop species, including corn, potato, and cotton. *Bt* corn, one of the first GM crops, was engineered to produce a continuous supply of toxin, which provides a natural defense against insects such as the European corn borer. Similarly, *Bt* tomato and *Bt* cotton are more resistant to pests such as the tomato pinworm and the cotton bollworm. Early ecological risk assessment studies, performed before the EPA approved their use, indicated that GM crops are essentially like their more conventional counterparts that are produced by selective breeding. Genetically modified crops do not become invasive pests or persist in the environment longer than crops that are not genetically modified. Significantly fewer pesticide applications are needed for GM crops with the *Bt* gene.

The future of the *Bt* toxin as an effective substitute for chemical pesticides is not completely secure, however. Beginning in the late 1980s, farmers began to notice that chemically applied *Bt* was not working as well against the diamondback moth as it had in the past. All farmers who reported this reduction in effectiveness had used *Bt* frequently and in large amounts. In 1996, many farmers growing *Bt* cotton reported that their crop succumbed to the cotton bollworm, which *Bt* is supposed to kill. It appears that certain insects, such as the diamondback moth and possibly the cotton bollworm, may have evolved resistance to this natural toxin. If *Bt* is used in greater and greater amounts, it is likely that more insect pests will develop genetic resistance to it, greatly reducing *Bt*'s potential as a natural pesticide. Scientists are developing resistance management strategies to slow the evolution of pest resistance to genetically engineered crops that make *Bt*.

GM crops with the *Bt* gene create situations where the *Bt* toxin is present in more places. John Losey, a Cornell entomologist, demonstrated in 1998 that *Bt* corn pollen, blown by wind onto milkweed plants (a common corn weed), could make the milkweed leaves deadly to monarch butterfly larvae. This research demonstrated the potential for *Bt* toxin to become a kind of environmental pollutant through pollen spread from *Bt* crops. ■

QUARANTINE

Governments attempt to prevent the importation of foreign pests and diseases by practicing **quarantine**, or restriction of the importation of exotic plant and animal material that might harbor pests. If a foreign pest is accidentally introduced, quarantine of the area where it is detected helps prevent its spread. If a foreign pest is detected on a farm, the farmer may be required to destroy the entire crop.

Quarantine is an effective, though not foolproof, means of control. The USDA has blocked the accidental importation of medflies on more than 100 separate occasions. On the few occasions when quarantine failed and these insects successfully passed into the United States, millions of dollars of crop damage were incurred in addition to the millions spent to eradicate the pests. Eradication efforts include the use of helicopters to spray the insecticide malathion over hundreds of square kilometers and the rearing and releasing of millions of sterile males to breed the medfly out of existence.

Experts were once concerned that repeated finds of medflies in California indicated that, rather than being accidentally introduced each year, the medfly had become established in the state. Other countries could stop importing California produce or require expensive inspections and treatments of every shipment to prevent the importation of the medfly into their countries. However, by 2008, the medfly had been eliminated, so quarantine remains an important strategy to minimize future medfly infestations.

THE SYSTEMS APPROACH: INTEGRATED PEST MANAGEMENT

Many pests are not controlled effectively with a single technique; rather, a combination of control methods is often more effective. **Integrated pest management (IPM)** combines a variety of biological, cultivation, and pesticide controls tailored to the conditions and crops of an individual farm, school, campus, city, or greenhouse. **Figure 22.10** provides an illustrated example of some ways in which IPM can control corn pests.

Biological and cultural controls, including pest-resistant GM crops, are used as much as possible, and conventional pesticides are used sparingly and only when other methods fail (see You Can Make a Difference: Avoiding Pesticides at Home). When pesticides are

> **integrated pest management (IPM)** A combination of pest control methods that, if used in the proper order and at the proper times, keep the size of a pest population low enough that it does not cause substantial economic loss.

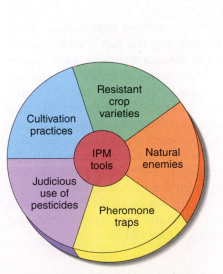

(a) IPM takes a systems approach to management of pests, providing interventions that impact various parts of pests' life cycles.

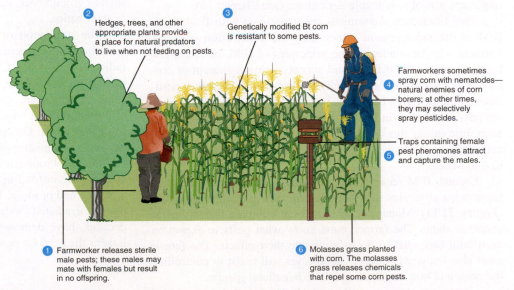

② Hedges, trees, and other appropriate plants provide a place for natural predators to live when not feeding on pests.

③ Genetically modified Bt corn is resistant to some pests.

④ Farmworkers sometimes spray corn with nematodes—natural enemies of corn borers; at other times, they may selectively spray pesticides.

⑤ Traps containing female pest pheromones attract and capture the males.

① Farmworker releases sterile male pests; these males may mate with females but result in no offspring.

⑥ Molasses grass planted with corn. The molasses grass releases chemicals that repel some corn pests.

(b) Examples of IPM strategies to control corn borers and other corn pests.

Figure 22.10   Tools of integrated pest management (IPM).

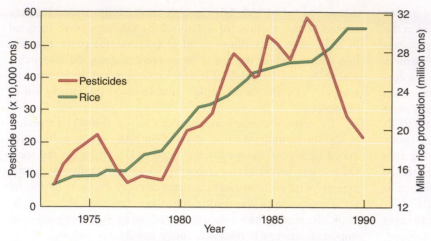

Figure 22.11   Rice production and pesticide use in Indonesia, 1972–1990. The decline in pesticide use in the late 1980s and early 1990s did not cause a decrease in rice yields. Instead, rice production increased during the four years following the new policy. Indonesia was the first Asian country to widely embrace integrated pest management. It phased out pesticide subsidies, banned the use of dozens of pesticides on rice, and trained more than 200,000 farmers in IPM techniques.

required, the least toxic pesticides are applied in the lowest possible quantities and at carefully scheduled times. Thus, IPM allows the farmer to control pests with a minimum of environmental disturbance and often at a minimal cost.

To be effective, IPM requires a thorough knowledge of the system, including life cycles, feeding habits, travel, and nesting habits of the pests as well as all their interactions with their hosts and other organisms. The timing of planting, cultivation, and treatments with biological controls is critical and is determined by carefully monitoring the concentration of pests. Integrated pest management optimizes natural controls by using agricultural techniques that discourage pests. Integrated pest management is an important part of sustainable agriculture (see Chapter 18).

Two fundamental premises are associated with IPM. First, IPM is the management rather than the eradication of pests. Farmers who have adopted the principles of IPM tolerate a low level of pests in their fields and accept a certain amount of economic damage from the pests. These farmers do not spray pesticides at the first sign of a pest. Instead, they periodically sample the pest population in the field to determine when it reaches an economic injury threshold, at which point the benefit of taking action (such as the judicious use of pesticides) exceeds the cost of that action.

Second, IPM requires that farmers be educated so that they know what strategies will work best in their particular situations (**Figure 22.11**). Managing pests is more complex than trying to eradicate them. The farmer must know what pests to expect on each crop and what to do to minimize their effects. The farmer must also know what beneficial species will assist in controlling the pests and how to encourage these beneficial species.

Cotton, which is attacked by many insect pests, has responded well to IPM. Cotton has the heaviest insecticide application of any crop: Although only about 1% of agricultural land in the United States is used for this crop, cotton accounts for almost 50% of all the insecticides used in agriculture! Applying simple techniques, such as planting a strip of alfalfa adjacent to the cotton field, lessens the need for chemical pesticides. Lygus bugs, a significant pest of cotton, move from the cotton field to the strip of alfalfa, which they prefer as a food. Thus, less damage is done to the cotton plants.

IPM is commonly used on school campuses. Pennsylvania, for example, mandates that state schools use IPM methods to control pests in and around school grounds. IPM helps reduce children's exposure to pesticides, and also helps reduce risk to school employees, especially cleaning staff.

U.S. farmers have increasingly adopted IPM since the 1960s, but the overall proportion of farmers using IPM is still small. IPM is not more widespread in part because the knowledge needed to use IPM is more sophisticated than that required to use pesticides. Also, marketing IPM crops is challenging, because no guarantees are given to consumers that the produce lacks pesticide residues.

IRRADIATING FOODS

It is possible to prevent insects and other pests from damaging harvested food without using pesticides. The food is harvested and then exposed to ionizing radiation, which kills many microorganisms, such as salmonella, a bacterium that causes food poisoning. This is called **food irradiation** or **cold pasteurization**. Numerous countries, such as Canada, much of Western Europe, Japan, Russia, and Israel, extend the shelf lives of foods with irradiation. The U.S. Food and Drug Administration (FDA) approved this process for fruits and vegetables as well as fresh poultry in 1986, and the first irradiated food was sold in the United States in 1992. In 2000, the USDA approved the irradiation of additional raw meats, such as ground beef, steaks, and pork chops, to eliminate bacterial contamination.

The irradiation of foods is somewhat controversial. Some consumers are concerned because they fear that irradiated food is radioactive; it is not. Critics of irradiation are concerned because irradiation forms traces of certain chemicals called *free radicals*, some of which are carcinogenic in laboratory animals. Critics also point out that we do not know the long-term effects of eating irradiated foods. Others are concerned about the potential security risk from misuse of the radiation source, usually cobalt-60 or cesium-137. Proponents of irradiation argue that free radicals normally occur in food and are produced by cooking methods such as frying and broiling. They assert that more than 1000 investigations of irradiated foods, conducted worldwide for more than three decades, have demonstrated it is safe. Furthermore, irradiation lessens the need for pesticides and food additives.

REVIEW

1. What is an example of using cultivation methods to control pests? Of using pheromones and hormones?
2. What is integrated pest management? Why is IPM considered a systems approach?

YOU CAN MAKE A DIFFERENCE

Avoiding Pesticides at Home

Animals, plants, and diseases are not just pests that affect agriculture and people in developing countries; they can be a constant challenge in most homes. Ants, moths, cockroaches, and rodents can damage property and spread disease. Weeds can overrun gardens, lawns, and houseplants. Mold and mildew can damage property and spoil food. While sprays and chemical traps can be quick and easy solutions, they are also household hazards—and they don't deal with the underlying causes of infestations.

Alternatives are available to homeowners to manage many pests. Consider mice, for example. Mice not only damage property by chewing through wood, furniture, and appliances; they can also carry disease. While mouse and cockroach droppings are an inhalation hazard even when they do not carry disease, the hantavirus, which is rare but often fatal, is transferred to humans through the urine and feces of deermice. One way to eliminate mice and rats is to put out poison—but rat poison causes tens of thousands of human and pet poisonings in the United States each year.

Consider the following set of control options (household integrated pest management) for mice:

- Remove the carrying capacity for mice to live in the home by keeping foods sealed and stored when not in use.
- Reduce potential entrances for mice by closing up holes in walls and carefully but thoroughly cleaning out any place where mice have built nests.

- Add a cat—a form of biological control that has been practiced for millennia.
- Set mousetraps in places where mice are known to travel.

Mildew thrives in warm, humid places and can be controlled by anti-mildew chemical sprays, but there are other options:

- The best solutions are preventative: Clean and dry mildew-prone areas regularly.
- Try to increase airflow to areas with mildew, and consider a dehumidifier or desiccant (a chemical that absorbs water from the air) for extreme cases.
- A mixture of lemon juice and salt can eliminate mildew from fabrics, but be sure to test for color-fastness. Other options include baking soda pastes and white vinegar.
- Bleach is a powerful mildew-control agent but must be used cautiously since it is also hazardous to humans and can damage materials.

Nonpesticide control measures are effective against a range of other household pests:

- Store food carefully, and clean up spills promptly. This includes pet food! This is the

best way to reduce cockroach, ant, and housefly infestations.

- Many plants that are considered lawn weeds are edible or beautiful (see photograph). Fostering a healthy lawn with proper mowing and reseeding can help reduce the worst weed problems.
- Mosquitoes need standing water to lay their eggs, so if you find they are a problem, check your yard and surrounding areas for water in old tires, flowerpots, unused pools, and other containers that might catch rainwater.
- The best way to control spiders is to eliminate their prey—typically flies and other small insects. Also, get to know spiders, since those that are not dangerous to humans, like the orb weaver spider, can help control pests.

Yellow wood sorrel (*Oxalis stricta*), an edible lawn weed.

Laws Controlling Pesticide Use

LEARNING OBJECTIVE

- Briefly summarize the three U.S. laws that regulate pesticides: the Food, Drug, and Cosmetics Act; the Federal Insecticide, Fungicide, and Rodenticide Act; and the Food Quality Protection Act.

The federal government has passed several laws to regulate pesticides in the interest of protecting human health and the environment. These include the Food, Drug, and Cosmetics Act; the Federal Insecticide, Fungicide, and Rodenticide Act; and the Food Quality Protection Act. The EPA and, to a lesser extent, the FDA and the USDA, oversee the implementation of these laws. In

addition, the EPA has some responsibility to regulate pesticides if they lead to violations of the Endangered Species Act.

FOOD, DRUG, AND COSMETICS ACT

The **Food, Drug, and Cosmetics Act (FDCA)**, passed in 1938, recognized the need to regulate pesticides found in food but did not provide a means of regulation. The FDCA was made more effective in 1954 with passage of the **Pesticide Chemicals Amendment**. This amendment, also called the *Miller Amendment*, required the establishment of acceptable and unacceptable levels of pesticides in food.

An amended FDCA, passed in 1958, contained an important section, the **Delaney Clause**, which stated that no substance capable of causing cancer in test animals or in humans would be permitted in processed food. Processed foods are prepared in some way—such as frozen, canned, dehydrated, or preserved—before being sold. The Delaney Clause recognized that pesticides tend to concentrate in condensed processed foods, such as tomato paste and applesauce.

The Delaney Clause, though commendable for its intent, contained two inconsistencies. First, it did not cover pesticides on raw foods such as fresh fruits and vegetables, milk, meats, fish, and poultry. As an example of this double standard, residues of a particular pesticide might be permitted on fresh tomatoes but not in tomato ketchup. Second, because the EPA lacked sufficient data on the cancer-causing risks of pesticides used for a long time, the Delaney Clause applied only to pesticides registered after strict tests were put into effect in 1978. This situation gave rise to one of the paradoxes of the Delaney Clause. There were cases in which a newer pesticide that posed minimal risk was banned because of the Delaney Clause, but an older pesticide, which the newer one was to have replaced, was still used even though it was many times more dangerous.

When the Delaney Clause was passed, the technologies for detecting pesticide residues could reveal only high levels of contamination. Modern scientific techniques are so sensitive that it is almost impossible for any processed food to meet the Delaney standard. As a result, the EPA found it difficult to enforce the strict standard required by the Delaney Clause. In 1988 the EPA began granting exceptions that permitted a "negligible risk" of one case of cancer in 70 years for every 1 million people. However, the EPA was taken to court because of its failure to follow the Delaney Clause as written, and the U.S. courts decided in 1994 that no exceptions could be granted unless Congress modified the Delaney Clause. (One of the key provisions of the 1996 Food Quality Protection Act, discussed shortly, did just that.)

Although the FDCA has not been amended since 2004, the EPA continues to update its regulations in response to new pesticides and newly recognized risks. For example, while many rules focus on cancer, birth defects, or acute effects, concern about *endocrine disrupters* is more recent. The EPA has identified 73 chemicals that it will review as possible endocrine disrupters.

FEDERAL INSECTICIDE, FUNGICIDE, AND RODENTICIDE ACT

The **Federal Insecticide, Fungicide, and Rodenticide Act (FIFRA)** was originally passed in 1947 to regulate the effectiveness of pesticides—that is, to prevent people from buying pesticides that did not work. FIFRA has been amended over the years to require testing and registration of the active ingredients of pesticides. Any pesticide that does not meet the tolerance standards established by the FDCA must be denied registration by FIFRA.

In 1972 the EPA was given the authority to regulate pesticide use under the terms of the FDCA and FIFRA. Since that time, the EPA has banned or restricted the use of many chlorinated hydrocarbons. In 1972 the EPA banned DDT for almost all uses. Aldrin and dieldrin were outlawed in 1974 after more than 80% of all dairy products, fish, meat, poultry, and fruits were found to contain residues of these insecticides. The banning of kepone occurred in 1977 and of chlordane and heptachlor in 1988.

FIFRA was amended in 1988 to require reregistration of older pesticides, which subjected them to the same toxicity tests that

new pesticides face. Although the 1988 amendment is stricter than previous legislation, it represented a compromise between agricultural interests, including pesticide manufacturers, and those opposed to all uses of pesticides. The new law did not address an important issue, the contamination of groundwater by pesticides. Nor did the law address the establishment of standards for pesticide residues on foods and the safety of farmworkers who are exposed to high levels of pesticides (the Occupational Safety and Health Administration regulates workplace pesticide exposures).

FIFRA also did not require pesticide companies to disclose the inert ingredients in their formulations. Many pesticide products contain as much as 99% inert ingredients, which are not supposed to have active properties. The National Coalition for Alternatives to Pesticides determined that 394 of the more than 2500 chemicals listed as "inert" were once listed as active ingredients. Many inert ingredients are generally recognized as safe (including pine oil, ethanol, silicone, and water). Others are known toxins (including asbestos, benzene, formaldehyde, lead, and cadmium). Inert ingredients in some herbicides are particularly toxic to amphibians. More than 200 inert ingredients are classified as hazardous air and water pollutants, and 21 are known or suspected to cause cancer.

FOOD QUALITY PROTECTION ACT

The **Food Quality Protection Act** of 1996 amended both the FDCA and FIFRA. It revised the Delaney Clause by establishing identical pesticide residue limits—those that pose a negligible risk—for both raw produce and processed foods. The law requires that the increased susceptibility of infants and children to pesticides be considered when establishing pesticide residue limits for some 9700 pesticide uses on specific crops. The pesticide limits are established for all health risks, not just cancer. For example, the EPA must develop a program to test pesticides for endocrine-disrupting properties. Another key provision of the Food Quality Protection Act is that it reduces the time it takes to ban a pesticide considered dangerous, from 10 years to 14 months.

REVIEW

1. What three laws regulate pesticides in the United States?

The Manufacture and Use of Banned Pesticides

LEARNING OBJECTIVE

- Define *persistent organic pollutant* and describe the purpose of the Stockholm Convention on Persistent Organic Pollutants.

Many pesticides have been identified as too hazardous to permit for any use and thus are banned in the United States and other highly developed countries. The UN Food and Agriculture Organization (FAO) is attempting to help developing countries become more aware of dangerous pesticides. It established a "red alert" list of more than 50 pesticides banned in five or more countries. The FAO further requires that the manufacturers of these pesticides inform importing countries about why such pesticides are banned. The United States supports these international guidelines and exports banned pesticides only with the informed consent of the importing country. However, this information often does not trickle down to the local level, and many farmers worldwide never receive any guidelines or training on the safe handling and application of pesticides.

Another concern is that unwanted stockpiles of leftover, deteriorated pesticides are accumulating, particularly in developing countries. A 2008 UN estimate found that more than 500,000 tons of these obsolete pesticides are stockpiled worldwide, with 120,000 tons in Africa alone. They are often stored in drums at waste sites in the countryside because developing countries have few or no hazardous waste disposal facilities. Over time, chemicals leach from such waste sites into the soil and groundwater (see Chapter 23).

IMPORTATION OF FOOD TAINTED WITH BANNED PESTICIDES

The fact that many dangerous pesticides are no longer being used in the United States is no guarantee that traces of those pesticides are not in our food. Although many pesticides are restricted or banned in the United States, they are widely used in other parts of the world. Much of our food—some 1.2 million shipments annually—is imported from other countries, particularly in Latin America. Some produce contains traces of banned pesticides such as DDT, dieldrin, chlordane, and heptachlor. The 2007 discovery of contaminated pet food and toothpaste from China highlighted the risk of imported foods.

The FDA monitors toxic residues on incoming fruits and vegetables, but it inspects only about 1% of the food shipments that enter the United States each year. In addition, the General Accounting Office reports that some food importers continue to sell tainted food after the FDA finds the company is in violation of the law. When caught, these companies face fines that are not severe enough to discourage such practice.

GLOBAL BAN ON PERSISTENT ORGANIC POLLUTANTS

The **Stockholm Convention on Persistent Organic Pollutants**, which went into effect in 2004, is an important international treaty. As last amended in 2014 (with some exemptions), it seeks to protect human health and the environment from the 23 most toxic chemicals, classified as **persistent organic pollutants (POPs)** (**Table 22.3**; also see the discussion of long-distance transport of air pollution in Chapter 19). Nine of these POPs are pesticides. Some POPs disrupt the endocrine system, others cause cancer, and still others adversely affect the developmental processes of organisms.

persistent organic pollutants (POPs)
A group of persistent, toxic chemicals that bioaccumulate in organisms and can travel thousands of kilometers through air and water to contaminate sites far removed from their source.

The Stockholm Convention requires that countries develop plans to eliminate the production and use of intentionally produced POPs. A notable exception to this requirement is that DDT can still be produced and used to control mosquitoes that carry the malaria pathogen in countries where no affordable alternatives exist. (DDT is inexpensive, and many of these countries cannot afford safer alternatives.)

Table 22.3	Persistent Organic Pollutants: The "Dirty Dozen"

Persistent Organic Pollutants Regulated Under the Stockholm Convention (2014)

Aldrin Pesticide	Hexabromobiphenyl* Industrial chemical
Chlordane Pesticide	Lindane Pesticide
Chlordecone Pesticide	Mirex Pesticide
DDT Pesticide	Pentachlorobenzene Pesticide, Industrial chemical, by-product
Dieldrin Pesticide	Perfluorooctane sulfonic acid Industrial chemical
Endrin Pesticide	Polychlorinated biphenyls Industrial chemical
Endosulfan Pesticide	Polychlorinated dibenzo-p-dioxins By-product
Heptachlor Pesticide	Polychlorinated dibenzofurans By-product
Hexachlorobenzene Pesticide	(Tetra- and penta-)bromodiphenyl ether Industrial chemical
Hexachlorocyclo-hexane* Pesticide, by-product	Toxaphene Pesticide

* multiple forms regulated

Although the Stockholm Convention goes a long way toward achieving the goal of a global ban on these chemicals, it applies only to the 179 countries that have ratified the treaty; as of 2014, the United States is not among them. To be more effective, the United States will have to ratify it and adhere to its rules, and countries that have ratified the treaty will have to abide by its conditions.

REVIEW

1. What is the Stockholm Convention on Persistent Organic Pollutants?

E N V I R O N E W S

Pest Management in Kitchen Gardens

Pesticide use in small-scale home gardens does not affect as many acres as agricultural pesticide use. However, suburban pesticide use often exceeds agricultural use in the amount of pesticides used per acre. Furthermore, people using pesticides at home have no required training in handling or proper use of these products. Pesticides available to the public in hardware stores and garden centers are just as toxic as those available to farmers and commercial applicators. Unfortunately, many people purchasing these products assume they are safer, the same way that a pain killer at the pharmacy is typically a milder or safer version of drugs available by prescription. In 2012, over 3000 poison control center calls addressed potential pesticide poisonings, mostly by home use pesticides.

Fortunately, many organic pesticides or biological controls are now available for gardeners and homeowners, and county agricultural extension services also offer advice about IPM techniques for pest control. Parasitoid wasps, which do not sting people, can be released to control tomato hornworms, one of the pests of greatest concern in many gardens. Beer attracts slugs, which die when they consume it. Mulches and newspaper, which are impractical to use in large-scale agriculture, can serve as excellent weed control in kitchen gardens.

REVIEW OF LEARNING OBJECTIVES WITH SELECTED KEY TERMS

- **Distinguish between narrow-spectrum and broad-spectrum pesticides, and describe the various types of pesticides, such as insecticides and herbicides.**

Pesticides are toxic chemicals used to kill **pests** such as insects (**insecticides**), weeds (**herbicides**), fungi (**fungicides**), and rodents (**rodenticides**). The ideal pesticide would be a **narrow-spectrum pesticide** that kills only the target organism. Most pesticides are **broad-spectrum pesticides**, which kill a variety of organisms, including beneficial organisms, in addition to the target pest.

- **Describe the benefits of pesticides in disease control and crop protection.**

Pesticides help prevent malaria and other diseases transmitted by insects. Pesticides reduce crop losses from pests, thereby increasing agricultural productivity. Pesticides reduce competition with weeds, crop consumption by insects, and diseases caused by plant **pathogens** such as certain fungi and bacteria.

- **Explain why monocultures are more susceptible to pest problems.**

A **monoculture** is the cultivation of only one type of plant over a large area. Thus, it represents a system that is out of balance compared to what would be found without human intervention. Agricultural fields are monocultures that provide abundant food for pest organisms. Many natural predators are absent from monocultures.

- **Summarize the problems associated with pesticide use, including development of genetic resistance; creation of imbalances in the ecosystem; persistence, bioaccumulation, and biological magnification; and mobility in the environment.**

Genetic resistance is any inherited characteristic that decreases the effect of a pesticide on a pest. Pesticides affect species other than those for which they are intended, causing imbalances in ecosystems; in some instances, the use of a pesticide has resulted in a pest problem that did not exist before. A pesticide that demonstrates **persistence** takes a long time to break down into less toxic forms. **Bioaccumulation** is the buildup of a persistent pesticide or other toxic substance in an organism's body. **Biological magnification** is the increased concentration of toxic chemicals, such as certain pesticides, in the tissues of organisms at higher trophic levels in food webs. Many pesticides move through the soil, water, and air, sometimes for long distances.

- **Describe pesticide resistance and resistance management.**

The **pesticide treadmill** is a predicament faced by pesticide users, in which the cost of applying pesticides increases (because they have to be applied more frequently or in larger doses) while their effectiveness decreases (as a result of increasing genetic resistance in the target pest). **Resistance management** consists of strategies for managing genetic resistance to maximize the period in which a pesticide is useful.

- **Discuss pesticide risks to human health, including short-term effects and long-term effects.**

Humans may be poisoned by short-term exposure to a large amount of pesticide. Lower levels of many pesticides may pose a long-term threat of cancer. Certain persistent pesticides may interfere with the actions of natural hormones.

- **Explain endocrine disruption.**

Endocrine disruption occurs when a chemical interferes with or mimics a hormone associated with growth and development in humans or other animals. Endocrine disrupters include pesticides such as atrazine and DDT, and problems include abnormal physical features—such as deformed reproductive organs—and aberrant behavior.

- **Describe alternative ways to control pests, including cultivation methods, biological controls, pheromones and hormones, reproductive controls, genetic controls, quarantine, integrated pest management, and food irradiation.**

Cultivation techniques such as strip cutting, interplanting, and crop rotation are effective in controlling pests. **Biological control** uses naturally occurring disease organisms, parasites, or predators to control pests. A **pheromone** is a natural substance produced by animals to stimulate a response in other members of the same species; pheromones can be used to lure insects to traps or to confuse insects so that they cannot locate mates. Insect hormones are natural substances produced by insects to regulate their own growth and metamorphosis; a hormone present at the wrong time in an insect's life cycle disrupts its normal development. Reproductive controls include reducing the pest population by sterilizing some of its members (the **sterile male technique**) or genetically modifying some of its members so that offspring are not viable. Genetic control of pests involves producing varieties of crops and livestock animals that are genetically resistant to pests; some **genetically modified (GM)** crops contain the *Bacillus thuringiensis* (*Bt*) gene that codes for a toxin used against insect pests. **Quarantine** involves restricting the importation of exotic plant and animal material that might harbor pests; if a foreign pest is accidentally introduced, quarantine of the area where it is detected helps prevent its spread. **Integrated pest management (IPM)** is a systems approach that combines several pest control methods that, if used in the proper order and at the proper times, keep the size of a pest population low enough that it does not cause substantial economic loss. **Food irradiation** controls pests after food is harvested.

- **Briefly summarize the three U.S. laws that regulate pesticides: the Food, Drug, and Cosmetics Act; the Federal Insecticide, Fungicide, and Rodenticide Act; and the Food Quality Protection Act.**

The **Food, Drug, and Cosmetics Act (FDCA)**, as originally passed, recognized the need to regulate pesticides in food but did not provide a means of regulation. The *Miller Amendment* required the establishment of acceptable and unacceptable levels of pesticides in food, and the **Delaney**

Clause stated that no substance capable of causing cancer in laboratory animals or in humans would be permitted in processed food. The **Federal Insecticide, Fungicide, and Rodenticide Act (FIFRA)** was amended over the years to require testing and registration of the active ingredients of pesticides; the 1988 version required the reregistration of older pesticides, which subjected them to the same toxicity tests that new pesticides face. The **Food Quality Protection Act** amended both the FDCA and FIFRA and revised the Delaney Clause by establishing identical pesticide residue limits—those that pose a negligible risk—for both raw and processed foods.

- **Define *persistent organic pollutant* and describe the purpose of the Stockholm Convention on Persistent Organic Pollutants**

Persistent organic pollutants (POPs) are a group of persistent, toxic chemicals that bioaccumulate in organisms and can travel thousands of kilometers through air and water to contaminate sites far removed from their source. The **Stockholm Convention on Persistent Organic Pollutants** seeks to protect human health and the environment from the 23 most toxic chemicals on Earth. As of 2014, the United States has not ratified this treaty.

Critical Thinking and Review Questions

1. Distinguish among insecticides, herbicides, fungicides, and rodenticides.

2. Describe the general characteristics of each of the following groups of insecticides: chlorinated hydrocarbons, organophosphates, and carbamates.

3. Overall, do you think the benefits of pesticide use outweigh its disadvantages? Give at least two reasons for your answer.

4. Sometimes pesticide use increases the damage done by pests. Explain.

5. The widely used herbicide Roundup is starting to lose its effectiveness in killing certain weeds. Explain why.

6. How is the buildup of insect resistance to insecticides similar to the increase in bacterial resistance to antibiotics?

7. How does genetic change in response to biological control agents differ from genetic resistance to pesticides?

8. Biological control is often much more successful on a small island than on a continent. Offer at least one reason why this might be the case.

9. It is more effective to use the sterile male technique when an insect population is small than when it is large. Explain.

10. Define *integrated pest management (IPM)*. List five tools of IPM, and give an example of each.

11. How is IPM related to ecological concepts such as food webs and energy flow?

12. Which of the following uses of pesticides do you think are most important? Which are least important? Explain your views.

    a. Keeping roadsides free of weeds

    b. Controlling malaria

    c. Controlling crop damage

    d. Producing blemish-free fruits and vegetables

13. Why is pesticide misuse increasingly viewed as a global environmental problem?

14. Propose an integrated pest management plan to control rabbits in a garden. Your approach may include a rodenticide.

15. Climate change may lead to a greater need for pesticides to control mosquitoes. Suggest some ways in which other pesticide uses could increase or decrease as the Earth warms and precipitation patterns shift over the next century.

16. Do you think this cartoon reasonably depicts the challenge of reducing pesticide use?

"Sure it costs more. We have to squash bugs by hand."

Food for Thought

Many home gardeners take pride in the fact that their produce is not only fresher than purchased produce, but that their use of organic pest control techniques can save money, reduce pesticide exposures, and finally, reduce greenhouse gas production inherent in food transport. Look into the pest management methods used where you live. Are they pesticide intensive? Are they effective? Explore some ways in which you could get equal or better pest control with fewer chemicals.

23

Solid and Hazardous Wastes

Obsolete computer equipment. The computers at this Texas electronic recycler are being disassembled. Functional tubes will be exported to Thailand, where they will be used to manufacture inexpensive televisions. Broken tubes will be recycled or disposed of in the United States.

In the United States, Canada, and other highly developed countries, the average computer is replaced every 18 to 24 months, not because it is broken but because rapid technological developments and new generations of software make it obsolete. Old computers may still be in working order, but they have no resale value and are even difficult to give away. As a result, they sit in warehouses, garages, and basements—or are frequently thrown away with the trash and end up in landfills or incinerators. U.S. residents generate an average of 30 kg per person annually of such electronic waste, or *e-waste*.

This disposal represents a huge waste of the high-quality plastics and metals (e.g., aluminum, copper, tin, nickel, palladium, silver, and gold) that make up computers. Computers also contain toxic heavy metals—lead, cadmium, mercury, and chromium—which could potentially leach from landfills into soil and groundwater. Twenty-five states have passed legislation requiring businesses and residents to *ecycle* consumer electronics—that is, recycle PCs, monitors, cell phones, and televisions.

E-waste has provided business opportunity for certain companies, which disassemble old computers, monitors, fax machines, printers, and other electronic equipment

that they obtain from businesses and individuals (see photograph). These ecycling businesses reclaim the glass from monitors, plastics from casings, and metal from wires and circuit boards and ship the components to manufacturers that *upcycle* the materials. (Upcycling is a type of recycling that takes discarded materials and makes a new product with a value higher than that of the original materials.)

Environmentally minded consumers are increasingly purchasing electronic equipment that is recyclable. Some electronics manufacturers, for example, take back their e-waste and recycle it into new products; other manufacturers take back obsolete electronic devices for ecycling by another company. Responsibility for disposition and reutilization of a product after it is obsolete is known as *cradle-to-cradle accounting*.

Although many U.S. companies handle obsolete computers, millions of old U.S. computers are shipped overseas to developing countries such as India, Pakistan, and China. There the computers are disassembled, often by methods potentially dangerous to the workers taking them apart. For example, circuit boards are often burned to obtain the small amount of gold in them, and burning releases hazardous fumes into the air.

Like it or not, humans produce solid waste, and it is our responsibility to deal with our waste in safe, cost-effective, and environmentally sensitive ways. Using a systems perspective, for example, we could design computers to allow more reuse, particularly of the cabinets that house the computer components. No single solution exists to the challenge of solid waste disposal today. A combination of waste reduction, reuse, recycling, composting, burning, and burying in sanitary landfills is currently the optimal way to manage solid waste. In this chapter you will examine the problems and the opportunities associated with the management of solid waste. You will then consider hazardous waste, which contains toxic materials and requires special disposal.

In Your Own Backyard…
What are the major sources of solid waste that you produce, and how could you produce fewer solid wastes?

Solid Waste

The United States generates more solid waste, per capita, than any other country. (Canada is a close second.) Each person in the United States produces an average of 1.99 kg (4.38 lb) of solid waste per day. This amount corresponded to a total of 251 million tons in 2012, a slight increase over previous years.

The solid waste problem was made abundantly clear by several highly publicized instances of garbage barges wandering from port to port and from country to country, trying to find someone willing to accept their cargo. One example is the tugboat *Break of Dawn*, which in 1987 towed a garbage barge from New York to North Carolina. When North Carolina refused to accept the solid waste, the *Break of Dawn* set off on a journey of many months. In total, six states and three countries rejected the waste, which was eventually returned to New York to be incinerated.

Waste generation is an unavoidable consequence of prosperous, high-technology, industrial economics. It is a problem not only in the United States but also in Canada and other highly developed countries. Many products that could be repaired, reused, or recycled are simply thrown away. Others, including paper napkins and disposable diapers, are intended to be used once and then discarded. Packaging—which not only makes a product more attractive and more likely to sell but also protects it, keeps it sanitary, and deters theft—contributes to waste. Nobody likes to think about solid waste, but the fact is that it is a concern of modern society—we keep producing it, and places to dispose of it safely are limited.

TYPES OF SOLID WASTE

Municipal solid waste (MSW) is a heterogeneous mixture composed primarily of paper and paperboard; yard waste; plastics; food waste; metals; materials such as rubber, leather, and textiles; wood; and glass (**Figure 23.1**). The proportions of the major types of solid waste in this mixture change over time. Today's solid waste contains more paper and plastics than in the past, whereas the amounts of glass and steel have declined.

municipal solid waste (MSW) Solid materials discarded by homes, office buildings, retail stores, restaurants, schools, hospitals, prisons, libraries, and other commercial and institutional facilities.

nonmunicipal solid waste Solid waste generated by industry, agriculture, and mining.

MSW is a relatively small portion of all the solid waste produced (about 2%). **Nonmunicipal solid waste**, which includes wastes from mining (mostly waste rock, about 75% of the total solid waste production), agriculture (about 13%), and industry (about 10%), is produced in substantially larger amounts than MSW. Thus, most solid waste generated in the United States is from nonmunicipal sources.

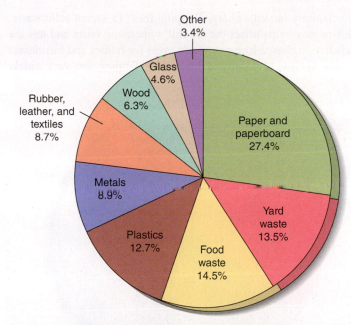

Figure 23.1 **Composition of municipal solid waste, 2012.** (EPA—Municipal Solid Waste Generation, Recycling, and Disposal in the United States: Facts and Figures for 2012)

OPEN DUMPS

Solid waste has traditionally been regarded as material that is no longer useful and that should be disposed of. The four ways to get rid of solid waste are to dump it, bury it, burn it, or compost it. The old method of solid waste disposal was dumping. Open dumps were unsanitary, malodorous places in which disease-carrying vermin such as rats and flies proliferated. Methane gas was released into the surrounding air as microorganisms decomposed the solid waste, and fires polluted the air with acrid smoke. Liquid that oozed and seeped through the solid waste heap ultimately found its way into the soil, surface water, and groundwater. Hazardous materials that were dissolved in this liquid often contaminated soil and water.

SANITARY LANDFILLS

Open dumps have been replaced by **sanitary landfills**, which receive about 54% of the solid waste generated in the United States today (**Figure 23.2**). Sanitary landfills differ from open dumps in that the solid waste is placed in a hole, compacted, and covered with a thin layer of soil every day (**Figure 23.3**). This process reduces the number of rats and other vermin usually associated with solid waste, lessens the danger of fires, and decreases the amount of odor. If a sanitary landfill is operated in accordance with approved guidelines for solid waste management, it does not pollute local surface water and groundwater. Safety is ensured by layers of compacted clay and plastic sheets at the bottom of the landfill, which prevent liquid waste from seeping into groundwater. Newer landfills possess a double-liner system (plastic, clay, plastic, clay) and use sophisticated systems to collect **leachate** (liquid that seeps through the solid waste) and gases that form during decomposition.

sanitary landfill The most common method of disposal of solid waste, by compacting it and burying it under a shallow layer of soil.

Sanitary landfills charge "tipping fees" to accept solid waste. This money helps offset the landfill's operating costs and lets the jurisdiction charge lower property taxes for homes and businesses located in close proximity to the landfill. Tipping fees vary widely

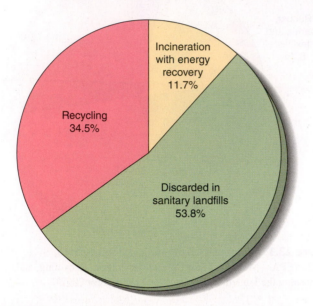

Figure 23.2   U.S. disposal of municipal solid waste in 2012. (EPA—Municipal Solid Waste Generation, Recycling, and Disposal in the United States: Facts and Figures for 2012)

from one state to another. Some jurisdictions cannot handle all their waste, so they export their solid waste to nearby states with lower tipping fees. For example, sanitary landfills in Ohio and Indiana accept solid waste from New York and New Jersey.

The location of an "ideal" sanitary landfill is based on a variety of factors, including the geology of the area, soil drainage properties, and the proximity of nearby bodies of water and wetlands. The landfill should be far enough away from centers of dense population that it is inoffensive but close enough that high transportation costs are not required. Landfill designs should take into account an area's climate, such as rainfall, snowmelt, and the likelihood of flooding.

Problems Associated with Sanitary Landfills   Although the operation of sanitary landfills has improved over the years with passage of stricter and stricter guidelines, few landfills are ideal. Most sanitary landfills in operation today do not meet current legal standards for new landfills.

One problem associated with sanitary landfills is the production of methane gas by microorganisms that decompose organic material anaerobically (in the absence of oxygen). This methane may seep through the solid waste and accumulate in underground pockets, creating the possibility of an explosion. It is even possible for methane to seep into basements of nearby homes, which is an extremely dangerous situation. Conventionally, landfills collected the methane and burned it off in flare systems. A growing number of landfills have begun to use the methane for gas-to-energy projects. About 620 U.S. landfills currently use methane gas to generate electricity.

Another problem associated with sanitary landfills is the potential contamination of surface water and groundwater by leachate that seeps from unlined landfills or through cracks in the lining of lined landfills. New York's Fresh Kills landfill, which was the largest in the United States before it closed in 2002, continues to produce an estimated 1 million gallons of leachate each day. Because household trash contains hazardous chemicals such as heavy metals, pesticides, and organic compounds that can seep into groundwater and surface water, the leachate must be collected and treated, even though the sanitary landfill has closed and is now being transformed into a park.

Landfills are not an indefinite remedy for waste disposal because they are filling up. From 1988 to 2009, the number of U.S. landfills in operation decreased from 7924 to 1908. Many of the landfills closed because they had reached their capacity. Other landfills closed because they did not meet state or federal environmental standards.

Fewer new sanitary landfills are being opened to replace the closed ones, although new landfills are generally much larger than in the past. The reasons for opening fewer new landfills are many and complex. Many

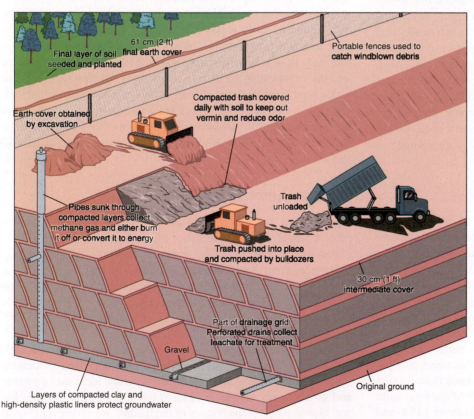

Figure 23.3   Sanitary landfill. Sanitary landfills constructed today have protective liners of compacted clay and high-density plastic and sophisticated leachate collection systems that minimize environmental problems such as contamination of groundwater. Solid waste is spread in a thin layer, compacted into small sections, and covered with soil.

desirable sites are already taken. Also, people living near a proposed site are usually adamantly opposed to a landfill near their homes. This attitude is partly the result of past problems with landfills, ranging from offensive odors to dangerous contamination of drinking water. It is also caused by the fear that a nearby facility will lower property values.

Once a sanitary landfill is full, closing it involves considerable expense. Because groundwater pollution and gas explosions remain possible for a long time, the Environmental Protection Agency (EPA) currently requires owners to continuously monitor a landfill for 30 years after the landfill is closed. In addition, no homes or other buildings can be built on a closed sanitary landfill for many years.

The Special Problem of Plastic

The amount of plastic in our solid waste is growing faster than any other component of MSW. More than half of this plastic is from packaging. Bottled water, for example, is the world's fastest-growing beverage. Plastics are chemical **polymers** composed of chains of repeating carbon compounds. The properties of the many types of plastics—polypropylene, polyethylene, and polystyrene, to name a few—differ on the basis of their chemical compositions.

Most plastics are chemically stable and do not readily break down, or decompose. This characteristic, which is essential in the packaging of certain products, such as food, causes long-term problems. Indeed, most plastic debris disposed of in sanitary landfills will probably last for centuries.

In response to concerns about the volume of plastic waste, some areas have actually banned the use of certain types of plastic, such as the polyvinyl chloride employed in packaging. Special plastics that have the ability to degrade or disintegrate have been developed. Some of these are **photodegradable**; that is, they break down only after being exposed to sunlight, which means they will not break down in a sanitary landfill. Other plastics are **biodegradable**; that is, they are decomposed by microorganisms such as bacteria. Whether biodegradable plastics actually break down under the conditions found in a sanitary landfill is not yet clear, although several studies indicate that they probably do not. Many factors, such as temperature, amount of oxygen present, and composition of the microbial community, affect biodegradation of plastics and other organic materials. (Other waste management options for plastic are discussed later in this chapter.)

The Special Problem of Tires

One of the most difficult materials to manage is rubber. Discarded tires—about 300 million each year in the United States—are made of vulcanized rubber, which cannot be melted and reused for tires. Millions of old tires have accumulated in tire dumps, as well as along roadsides and in vacant lots. Disposal of tires in sanitary landfills is a real problem because tires, being relatively large and light, have a tendency to move upward through the accumulated solid waste. After a period, they work their way to the surface of the landfill. These tires are a fire hazard, causing fires that are difficult to extinguish. Old tires also collect rainwater, providing a good breeding place for mosquitoes. Accordingly, most states either ban tires from sanitary landfills or require that they be shredded to save space and prevent water from pooling in them. (Other waste management options for tires are discussed later in the chapter.)

INCINERATION

When solid waste is incinerated, two positive things are accomplished. First, the volume of solid waste is reduced up to 90%. The ash that remains is, of course, much more compact than solid waste that has not been burned. Second, incineration produces heat that can make steam to warm buildings or generate electricity. In 2014, the EPA listed 86 U.S. waste-to-energy incinerators, which burned almost 12% of the nation's solid waste. In comparison, less than 1% of U.S. solid waste was incinerated in 1970. Waste-to-energy incinerators typically produce less carbon dioxide emissions than equivalent power plants that burn coal. (Recall from Chapter 20 that carbon dioxide is a potent greenhouse gas.)

Some materials are best removed from solid waste before incineration occurs. Glass does not burn, and when it melts, it is difficult to remove from the incinerator. Although food waste burns, its high moisture content often decreases the efficiency of incineration, so it is better to remove it before incineration. Removal of batteries, thermostats, and fluorescent lights is desirable because it eliminates most mercury emissions produced during combustion. The best materials for incineration are paper, plastics, and rubber.

Paper is a good candidate for incineration because it burns readily and produces a great amount of heat. Several studies have examined the economic and environmental costs and benefits of various waste paper management options. Many of these studies have concluded that waste-to-energy incineration is better than recycling, which in turn is better than disposal in a sanitary landfill. (The studies do not reach unanimous conclusions because economists do not agree about the cash value that should be applied for environmental benefits and costs. For example, estimates of the environmental cost of emitting 1 kg of carbon dioxide range from $1 to more than $50.) One potential environmental complication associated with burning paper is the presence of hazardous compounds in the ink and paper that might be emitted during incineration. Some types of paper release *dioxins* into the atmosphere when burned; dioxins are discussed later in the chapter.

Plastic produces a lot of heat when it is incinerated. In fact, one kilogram of plastic waste yields almost as much heat as a kilogram of fuel oil. As with paper, the pollutants that might be emitted during the incineration of plastic are of some concern. Polyvinyl chloride, a common component of many plastics, may release dioxins and other hazardous compounds when incinerated.

One of the best ways to dispose of old tires is incineration because burning rubber produces much heat. Some electric utilities in the United States and Canada burn tires instead of or in addition to coal (**Figure 23.4**). Tires produce as much heat as coal and often generate less pollution. About 45% of all tires discarded annually are incinerated.

Types of Incinerators

The three types of incinerators are mass burn, modular, and refuse-derived fuel. Most **mass burn incinerators** are large and are designed to recover the energy produced from combustion (**Figure 23.5**). **Modular incinerators** are smaller incinerators that burn all solid waste. They are assembled at factories and so are less expensive to build. In **refuse-derived fuel incinerators**, only the combustible

mass burn incinerator
A large furnace that burns all solid waste except for unburnable items such as refrigerators.

portion of solid waste is burned. First, noncombustible materials such as glass and metals are removed by machine or by hand. The remaining solid waste, including plastic and paper, is shredded or shaped into pellets and burned.

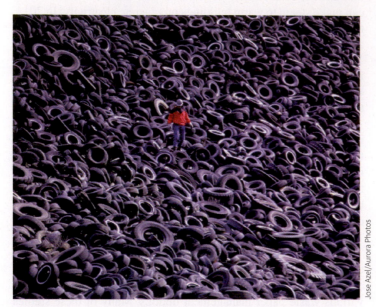

Figure 23.4   Tires that will be burned to generate electricity. This mountain of tires in Westley, California, contains 4 million to 6 million old tires. The power plant that burns them supplies electricity to 3500 homes. (The person wearing red gives a sense of scale.) How would burning more tires affect climate change?

Problems Associated with Incineration   The combustion of any fuel, whether it is coal or MSW, yields some air pollution. The possible production of hazardous air pollutants is the main reason people oppose incineration. Incinerators can pollute the air with carbon monoxide, particulates, heavy metals such as mercury, and other hazardous materials, unless air pollution–control devices are used. Such devices include **lime scrubbers**, towers in which a chemical spray neutralizes acidic gases, and **electrostatic precipitators**, which give ash a positive electrical charge so that it adheres to negatively charged plates rather than going out the chimney (see Figure 19.9 for diagrams of a scrubber and an electrostatic precipitator).

Incinerators produce large quantities of bottom ash and fly ash. **Bottom ash**, or slag, is the residual ash left at the bottom of the incinerator when combustion is completed. **Fly ash** is the ash from the flue (chimney) that is trapped by air pollution–control devices. Fly ash usually contains more hazardous materials, including heavy metals and possibly dioxins, than bottom ash.

Currently, both types of incinerator ash are best disposed of in specially licensed hazardous waste landfills (discussed later in this chapter). What happens to the hazardous materials in incinerator ash when it is placed in an average sanitary landfill is unknown; it is possible that the hazardous materials could contaminate groundwater.

As in the case of sanitary landfills, site selection for incinerators is controversial. People may recognize the need for an incinerator, but they do not want it near their homes. Another drawback of incinerators is their high cost. Prices have escalated because costly pollution-control devices are now required.

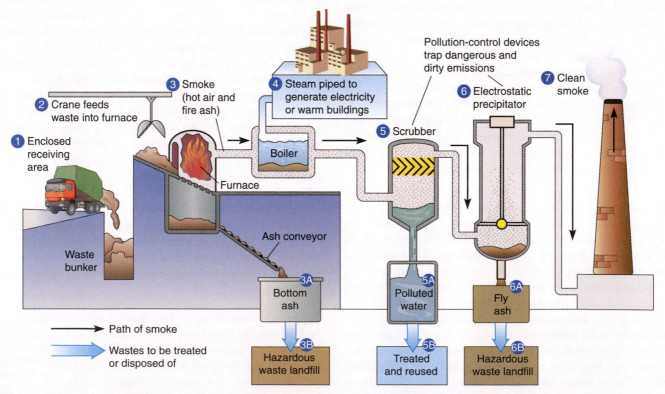

Figure 23.5   Mass burn, waste-to-energy incinerator. Modern incinerators have pollution-control devices such as lime scrubbers and electrostatic precipitators to trap dangerous and dirty emissions. (Begin with step 1 on the left side of the figure.)

COMPOSTING

Yard waste, such as grass clippings, branches, and leaves, is a substantial component of MSW (see Figure 23.1). As space in sanitary landfills becomes more limited, other ways to dispose of yard waste are being implemented. One of the best ways is to convert organic waste into soil conditioners such as compost or mulch (see You Can Make a Difference: Composting and Mulching, in Chapter 14). In 2012, about 58% of yard wastes in the United States were composed or otherwise recycled, per the EPA. Food scraps, sewage sludge, and agricultural manure are other forms of solid waste that can be used to make compost. Compost and mulch are used for landscaping in public parks and playgrounds or as part of the daily soil cover at sanitary landfills. Compost and mulch are also sold to gardeners.

Composting as a way to manage solid waste first became popular in Europe. In 2012, 3120 municipalities in the United States offered composting facilities as part of their comprehensive solid waste management plans, and many states have banned yard waste from sanitary landfills. This trend is likely to continue, making composting even more desirable.

Municipal solid waste composting is the large-scale composting of the entire organic portion of a community's garbage. Because approximately two-thirds of all household garbage is organic (paper, yard wastes, food wastes, and wood), MSW composting substantially reduces demand for sanitary landfills. Numerous city and county governments compost leaves and yard wastes in an effort to reduce the amount of solid waste sent to landfills. Although this endeavor is undeniably beneficial, MSW composting encompasses much more than yard wastes. It involves composting food wastes, paper, and anything else in the solid waste stream that is organic. In 2012, more than 2.4 million U.S. households could participate in food waste composting programs.

Initial composting occurs quickly—in three to four days—because conditions such as moisture and the carbon-nitrogen ratio are continually monitored and adjusted (e.g., by adding water or fertilizer) for maximum decomposition. The decay process is carried out by billions of bacteria and fungi, which convert the organic matter into carbon dioxide, water, and humus. So many decomposers eat, reproduce, and die in the compost heap that the mass heats up, killing off potentially dangerous organisms such as disease-causing bacteria (**Figure 23.6**). When the material emerges, it is placed outside for several months to cure, during which time additional decomposition occurs. Finally, it is sold as compost.

The potential market for compost is huge. Professional nurseries, landscapers, greenhouses, and golf courses use compost. Also, tons of compost could be used to reclaim the 167 million hectares (413 million acres) of badly eroded farmland in the United States. Compost could improve the fertility of badly eroded rangeland, forestland, and strip mines. Should certain technical problems be resolved, composting on a large scale could become economically feasible.

Technical problems include concerns over the presence of pesticide residues and heavy metals in the compost. Pesticides sprayed on urban and suburban landscapes would naturally find

Figure 23.6 **Municipal solid waste composting.** In the early stages of composting, a paddlewheel turns MSW at a Minnesota composting facility. Harmful bacteria are killed by the heat generated through decomposition.

their way into compost material on leaves, grass clippings, and other yard wastes. Some herbicides can be highly persistent in compost; many others are decomposed by microbes and high temperatures in the compost heap.

More troubling is the concern over heavy metals, such as lead and cadmium. Heavy metals can enter compost from sewage sludge, which may contain industrial wastewater, or from consumer products such as batteries. (Sewage sludge is often added to compost because it is a rich source of nitrogen for the decomposing microorganisms.) Two ways to reduce heavy metal contamination in municipal compost are sorting out heavy metal sources before everything is dumped into the composting drum, and requiring industries to pre-treat their industrial wastewater before it gets to the sewage treatment facility.

REVIEW

1. What is the difference between municipal solid waste and nonmunicipal solid waste?
2. What are three features of a sanitary landfill?
3. What are the main features of a mass burn incinerator?

Waste Prevention

LEARNING OBJECTIVES

- Summarize how source reduction, reuse, and recycling help reduce the volume of solid waste.
- Define *integrated waste management*.

Given the problems associated with sanitary landfills and incinerators, it makes sense to do whatever we can to lessen the need for these waste disposal methods. The three goals of waste prevention, in order of priority, are to reduce the amount of waste as much as possible, reuse products as much as possible, and recycle materials as much as possible.

Reducing the amount of waste includes purchasing products that last longer, are repairable, or have less packaging (**Figure 23.7**). Consumers can also decrease their consumption of products to reduce waste. Before deciding to purchase a product, a consumer should ask, "Do I really *need* this product, or do I merely *want* it?" Many U.S. consumers have participated actively for more than a decade in efforts to convert their throwaway economy into a waste prevention economy. Individual efforts have focused mainly on recycling, however, and much remains to be done in the areas of waste reduction and reuse (see Meeting the Challenge: Reusing and Recycling Old Automobiles).

We already discussed dematerialization, reuse, and recycling in Chapter 15, in the context of conservation of mineral resources. Now let's examine the impact of these practices on solid waste.

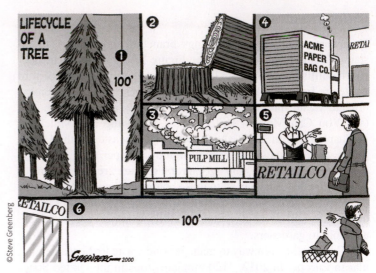

Figure 23.7   The six steps of wasteful packaging. What harmful environmental effects of wasteful packaging are depicted here?

source reduction An aspect of waste management in which products are designed and manufactured in ways that decrease the volume of solid waste and the amount of hazardous waste in the solid waste stream.

REDUCING THE AMOUNT OF WASTE: SOURCE REDUCTION

The most underused aspect of waste management is **source reduction**. Source reduction is accomplished in a variety of ways, such as substituting raw materials that introduce less waste during the manufacturing process and reusing and recycling wastes at the plants where they are generated. Innovations and product modifications can reduce the waste produced after a consumer has used a product. Dry-cell batteries, for example, contain much less mercury today than they did in the early 1980s. The 35% weight reduction in aluminum cans since the 1970s is another example of source reduction.

MEETING THE CHALLENGE

Reusing and Recycling Old Automobiles

In the United States, about 11 million cars and trucks are discarded each year. Although by weight about 75% of a discarded car is easily reused as secondhand parts or recycled as scrap metal, the remaining 25% is not easy to recycle and usually ends up in sanitary landfills. Because automobiles typically contain about 600 materials—glass, metals, plastics, fabrics, rubber, foam, leather, and so on—identifying ways to reuse or recycle old parts is complex. Economics is an important aspect of the problem, for reuse and recycling companies must make money as they sort and process auto components.

How does a car disassembly factory work? Workers typically begin disassembling a used car by draining all fluids—such as antifreeze, gasoline, transmission fluid, oil, and brake fluid—and recycling the fluids or processing them for disposal. Reusable parts and components, such as the engine, tires, and battery,

are then removed, cleaned, tested, and inventoried before being sold. Body shops, new and used car dealers, repair shops, and auto and truck fleets are the main buyers of used parts. Some parts are disassembled for their materials. For example, catalytic converters are disassembled because they contain valuable amounts of platinum and rhodium.

An automotive recycling facility then sends the remaining vehicle "shell" to a scrap processor. At the scrap-processing facility, a giant machine shreds the entire automobile into small pieces. Magnets and other machines sort the pieces into piles of steel, iron, copper, aluminum, and "fluff," which consists of the remaining materials, such as plastic, rubber, upholstery, and glass.

About 37% of the iron and steel scrap reprocessed in the United States comes from old automobiles. Recycling iron and steel saves energy and reduces pollution. According to the Environmental Protection Agency, recycling scrap iron and steel produces 86% less air pollution and 76% less water pollution than

mining and refining an equivalent amount of iron ore.

Recycling plastic is one of the biggest challenges in auto recycling. Plastic is lightweight, and, as a result, automakers use a lot of plastic to improve fuel efficiency. Because no industry standards for plastic parts currently exist, the kinds and amounts of plastic from which cars are made vary. As many as 15 plastics comprise some dashboards, and because many of these plastics are chemically incompatible, they cannot be melted together for recycling.

Auto manufacturers around the world have begun to address the challenge of reusing and recycling old cars. Japan and the European Union have mandated that by 2015, 95% of each discarded car must be recoverable. Toyota has developed a way to recover urethane foam and other shredded materials to make soundproofing material. Honda, Mercedes-Benz, Peugeot, Toyota, Volkswagen, Volvo, and other auto manufacturers have started to design cars so that each part of an old automobile can be reused or recycled.

The **Pollution Prevention Act** of 1990 was the first U.S. environmental law to focus on the reduced generation of pollutants at their point of origin rather than the reduction of pollutants or repair of damage caused by such substances. The act was written to increase the adoption of cost-effective source reduction measures. It requires the EPA to develop source reduction models, and it requires manufacturing facilities to report to the EPA annually on their source reduction and recycling activities.

Dematerialization, the progressive decrease in the size and weight of a product as a result of technological improvements, is an example of source reduction only if the new product is as durable as the one it replaced. If smaller, lighter products have shorter life spans and must be replaced more often, source reduction is not accomplished.

REUSING PRODUCTS

One example of reuse is refillable glass beverage bottles. Years ago, refillable beverage bottles were used a great deal in the United States. Today they are rarely used. For a glass bottle to be reused, it must be considerably thicker (and heavier) than one-use bottles. Because of the increased weight, transportation costs are higher. In the past, reuse of glass bottles made economic sense because many small bottlers were scattered across the United States, minimizing transportation costs. Today approximately one-tenth as many bottlers exist, making it economically difficult to go back to the days of refillable bottles. The price of beverages might have to increase to absorb increased transportation costs and the energy used to sterilize dirty bottles.

ENVIRONEWS

The U.S.–China Recycling Connection

More than half of recyclable wastes generated by Americans—from scrap metal to old cardboard boxes to used soda bottles—are redeveloped into U.S. products, but the rest are exported, primarily to China. During the early 2000s, China became the biggest importer of America's recyclable materials, collectively called scrap. When the scrap arrives in China, it becomes the raw materials for Chinese factories, paper mills, and steel mills. According to the Institute of Scrap Recycling Industries, scrap is now the third-largest export from the United States to China.

China does not have the natural resource base that countries such as the United States are lucky to possess. To fuel its economic growth, China relies on scrap—used paper for wood pulp and steel scrap to replace its dearth of iron ore. Some of the scrap shipped from the United States to China makes a round trip, returning to the United States as auto parts, polyester shirts, and toys. Because Chinese workers are paid much less than American workers, products made in China are generally less expensive for U.S. consumers than equivalent domestic products. The economic downside of the U.S. scrap–Chinese product cycle is that Chinese imports reduce the number of jobs available in the United States. Environmental disadvantages include degradation to Chinese ecosystems resulting from scrap processing.

Although the quantity of reusable glass bottles in the United States has declined, certain countries still reuse glass. In Japan, almost all beer and sake bottles are reused as many as 20 times. Bottles in Ecuador may remain in use for 10 years or longer. European countries such as Denmark, Finland, Germany, the Netherlands, Norway, Sweden, and Switzerland have passed legislation that promotes the refilling of beverage containers. Parts of Canada and 11 U.S. states also have deposit laws.

RECYCLING MATERIALS

It is possible to collect and reprocess many materials found in solid waste into new products of the same or a different type. Recycling is preferred over landfill disposal because it conserves our natural resources and is more environmentally benign. Every ton of recycled paper saves 17 trees, 7000 gallons of water, 4100 kilowatt-hours of energy, and 3 cubic yards of landfill space. Recycling also has a positive effect on the economy by generating jobs and revenues (from selling the recycled materials). However, recycling does have environmental costs. It uses energy (as does any human activity), and it generates pollution (as does any human activity). For example, the de-inking process in paper recycling requires energy and produces a toxic sludge that contains heavy metals.

The many materials in MSW must be separated from one another before recycling. It is easy to sort materials such as glass bottles and newspapers, but difficult to separate materials in items with complex compositions. Some food containers are composed of thin layers of metal foil, plastic, and paper, and trying to separate these layers is a daunting prospect.

The number of U.S. communities with recycling programs increased remarkably during the 1990s but leveled off somewhat in the early 2000s. Recycling programs, which are tracked by the EPA, include curbside collection, drop-off centers, buy-back programs, and deposit systems. The United States had about 9000 curbside recycling programs in 2009 (latest data available). The annual recycling rate in 2012 of aluminum and steel cans, plastic bottles, glass containers, newspapers, cardboard, and other materials was 189 kg (416 lb) per person.

Recyclables are usually sent to a **materials recovery facility**, where they are either hand-sorted or separated using a variety of technologies, including magnets, screens, and conveyor belts, and prepared for remanufacturing. Currently, the United States recycles 34.5% of its MSW, including composting of yard trimmings; this amount is greater than in other highly developed countries. (Recall, however, that the United States also generates more MSW than any other country.)

Most people think recycling involves merely separating certain materials from the solid waste stream, but that is only the first step. For recycling to work, a market must exist for the recycled goods, and the recycled products must be used in preference to virgin products. Prices paid by processors for old newspapers, used aluminum cans, used glass bottles, and the like vary significantly from one year to the next, depending largely on the demand for recycled products. In some places, recycling—particularly curbside collection—is not economically feasible.

Recycling Paper   The United States currently recycles more than 64% of its paper and paperboard. Many highly developed countries have greater paper-recycling rates. Denmark, for example, recycles about 97% of its paper. Part of the reason paper is not recycled more in the United States is that many older paper mills are not equipped to process waste paper. The number of mills that can process waste paper has increased in recent years, in part because of consumer demand. Most new mills in the United States are located near cities to take advantage of a local supply of scrap paper.

Demand is growing for U.S. waste paper in other countries. China, Mexico, Taiwan, and South Korea import large quantities of waste paper and cardboard from the United States.

Recycling Glass   Glass is another component of solid waste appropriate for recycling. The United States currently recycles about 34% of its glass containers. Recycled glass costs less than glass made from virgin materials, largely because of the energy savings (**Figure 23.8**). Glass food and beverage containers are crushed to form **cullet**, which glass manufacturers can melt and use to make new products. Cullet is more valuable when glass containers of different colors are separated before being crushed. Cullet made from a mixture of colors has some uses; for example, it is used to make glassphalt, a composite of glass and asphalt that makes an attractive roadway.

Recycling Aluminum and Other Metals   The recycling of aluminum is one of the best success stories in U.S. recycling, largely because of economic factors. Making a new aluminum can from a recycled one requires a fraction of the energy it would take to make a new can from raw metal (see Figure 23.8). Because energy costs for new cans are high, a strong economic incentive exists to recycle aluminum. According to the EPA, almost 55% of discarded aluminum beverage cans were recycled in 2012; about 36 barrels (1665 gallons) of gasoline are saved for each ton of recycled aluminum cans.

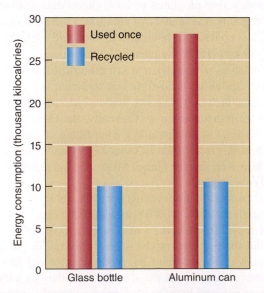

Figure 23.8   **Energy savings for recycled bottles and cans.** A comparison of the energy required to manufacture 350-mL (12-oz) glass bottles and aluminum cans that are either used once or recycled. (Data from Argonne National Laboratory)

Other recyclable metals include lead, gold, iron and steel, silver, and zinc. One obstacle to recycling metal products discarded in MSW is that their metallic compositions are often unknown. It is also difficult to extract metal from products, such as stoves, that contain other materials besides metal (e.g., plastic, rubber, or glass). In contrast, any waste metal produced at factories is recycled easily because its composition is known.

The economy has a large influence on whether metal is recycled or discarded. Greater recycling generally occurs when the price of metallic ores is more expensive than the price of recycled metals. Thus, although the supply of metal waste is fairly constant, the amount of recycling varies from year to year.

One exception to this generalization is steel. Before the 1970s, almost all steel was produced in large mills that processed raw ores. Starting in the 1970s and continuing to the present, "mini-mills" that produce steel products from up to 100% scrap have become increasingly important. These mills are located near many U.S. cities, so they can process local scrap more profitably (because there is no cost for transporting the scrap long distances). Mini-mills usually have electric arc furnaces that are energy-efficient and less polluting than the furnaces in old steel mills. According to the Institute of Scrap Recycling Industries, new steel products contain an average of 56% recycled scrap steel.

Recycling Plastic   In 2012, about 14% of plastic containers and packaging were recycled. Depending on the economic situation, it is sometimes less expensive to make plastic from raw materials (petroleum and natural gas) than to recycle it. In other words, plastic recycling—indeed, all recycling—is affected by economics. Some local and state governments support or require the recycling of plastic.

Polyethylene terphthalate (PET), the plastic used in bottled water and soda bottles, is recycled more than any other plastic. According to the EPA, 31% of the PET soft drink and water bottles sold annually are recycled to make such diverse products as carpeting, automobile parts, tennis ball felt, and polyester cloth (**Figure 23.9**); it takes about 25 plastic bottles to make one polyester sweater.

Polystyrene (one form of which is Styrofoam) is an example of a plastic that has great recycling potential but is currently not recycled appreciably. Cups, tableware, and packaging materials made of polystyrene can be recycled into a variety of products, such as coat hangers, flowerpots, foam insulation, and toys. Because approximately 2.3 billion kg (5 billion lb) of polystyrene are produced each year, large-scale recycling would make a major dent in the amount of polystyrene that ends up in landfills.

So many kinds of plastic present a challenge in recycling them. About 50 plastics are common in consumer products, and many products contain multiple kinds. A plastic ketchup bottle, for example, may have up to six layers of different plastics bonded together. To allow for effective recycling of high-quality plastic, the different types must be meticulously sorted or separated. If two or more resins are recycled together, the resultant plastic is of lower quality.

Low-quality plastic mixtures are used to make a construction material similar to wood. Because of its durability, this "plastic lumber" is particularly useful for outside products, such as fence posts, planters, highway retaining walls, decks, picnic tables, and park benches.

Recycling Tires   Almost 300 million tires are discarded in the United States each year, and 45% of them were recycled in 2012. Relatively few kinds of products are made from old tires: examples include retread tires, playground equipment, trashcans, garden hoses, and rubberized asphalt for pavement. More recently, rubber from old tires has been used to make carpets, roofing materials, and molded products. Research in product development continues, and almost all states now have tire-recycling programs.

INTEGRATED WASTE MANAGEMENT

The most effective way to deal with solid waste is through a combination of techniques. In **integrated waste management**, a variety of options that minimize waste, including the three R's of waste prevention (reduce, reuse, and recycle), are incorporated into an overall waste management plan (**Figure 23.10**). Even on a large scale, recycling and source reduction will not entirely eliminate the need for disposal facilities such as incinerators and landfills. However, recycling and source reduction will substantially reduce the amount of solid waste requiring disposal in incinerators and landfills.

> **integrated waste management** A combination of the best waste management techniques into a consolidated, systems-based program to deal effectively with solid waste.

REVIEW

1. What is source reduction?
2. How do source reduction, reuse, and recycling reduce the volume of solid waste?
3. What is integrated waste management?

Figure 23.9   Recycled plastic. A wide variety of products are made from recycled polyethylene terphthalate (PET), shown, and other plastics.

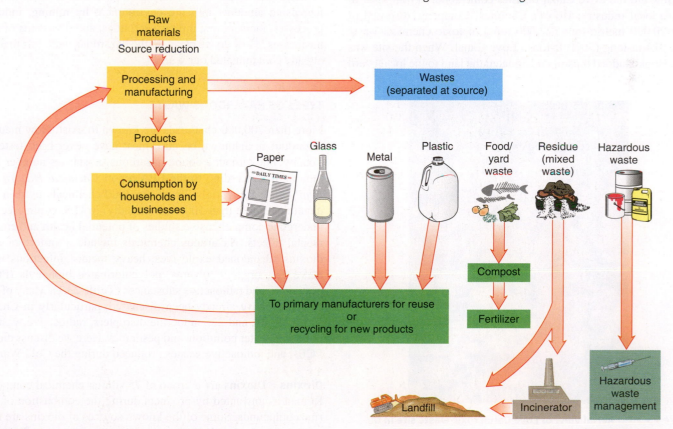

Figure 23.10   Integrated waste management. Source reduction, reuse, recycling, and composting are part of integrated, systems-based waste management, in addition to incineration and disposal in landfills.

Hazardous Waste

LEARNING OBJECTIVES

- Define *hazardous waste* and briefly characterize representative hazardous wastes (dioxins, PCBs, and radioactive wastes).
- Contrast the Resource Conservation and Recovery Act and the Comprehensive Environmental Response, Compensation, and Liability Act (the Superfund Act).
- Explain how green chemistry is related to source reduction.

hazardous waste Any discarded material that threatens human health or the environment.

Hazardous waste accounts for about 1% of the solid waste stream in the United States. Hazardous waste includes dangerously reactive, corrosive, ignitable, or toxic chemicals. The chemicals may be solids, liquids, or gases.

Hazardous waste has held national attention since 1977, when it was discovered that hazardous waste from an abandoned chemical dump had contaminated homes and possibly people in **Love Canal**, a small neighborhood on the edge of Niagara Falls, New York (**Figure 23.11**). **Lois Gibbs**, a housewife in Love Canal, led a successful crusade to evacuate the area after she discovered what seemed a high number of serious illnesses in the neighborhood, particularly among children. As a result of the publicity, Love Canal became synonymous with chemical pollution caused by negligent hazardous waste management. In 1978, it became the first location ever declared a national emergency disaster area because of hazardous waste; some 700 families were evacuated.

How did the Love Canal disaster come about? From 1942 to 1953, a local industry, Hooker Chemical Company, disposed of about 20,000 metric tons (22,000 tons) of toxic chemical waste in the 914-m-long (3000-ft-long) Love Canal. When the site was filled, Hooker added topsoil and donated the land to the local board of education and apparently provided documentation that wastes were present. A school and houses were built on the site, which began oozing hazardous waste several years later. Over 300 chemicals, many of them carcinogenic, have been identified in Love Canal's hazardous waste.

Tons of contaminated soil were removed during the cleanup that followed, but because the canal was so huge, the federal government decided to contain the waste and construct drainage trenches to prevent hazardous wastes from leaking from the site. In 1990, after almost 10 years of cleanup, the EPA and the New York Department of Health declared the nearby area safe for resettlement. Today, the canal is a 40-acre mound covered by clay and surrounded by a chain-link fence and warning signs.

Which health problems in residents were caused by the contamination remains unclear, because we know little about the effects of such complex mixtures of toxins. On average, Love Canal residents seem to have had more health problems than the average, from miscarriages and birth defects to psychological disorders.

The Love Canal episode resulted in passage of the federal Superfund law, which held polluters accountable for the cost of cleanups (discussed shortly). Love Canal also generated concern over hazardous waste that has been with us ever since. For example, a 1997 California study reported that women living within a quarter mile of untreated hazardous waste sites (Superfund sites) were at greater risk of having babies with serious birth defects, such as malformed neural tubes and defective hearts.

Other countries have the same problems with hazardous waste management. How should we deal with the bewildering array of hazardous waste continually generated and released in ever-increasing amounts into the environment by mining, industrial processes, incinerators, military activities, and thousands of small businesses? How do we clean up the hazardous materials that have already contaminated our world?

TYPES OF HAZARDOUS WASTE

More than 700,000 chemicals are known to exist. How many are hazardous is unknown because most have never been tested for toxicity, but without a doubt, hazardous substances number in the thousands. According to the EPA report *Chemical Hazard Data Availability Study*, only 7% of the 3000 chemicals used in large quantities (more than 500 tons annually) in U.S. commerce have undergone comprehensive studies of potential health and environmental effects. Hazardous chemicals include a variety of acids, dioxins, abandoned explosives, heavy metals, infectious waste, nerve gas, organic solvents, polychlorinated biphenyls (PCBs), pesticides, and radioactive substances (**Table 23.1**). Many of these chemicals have already been discussed, particularly in Chapters 7, 11, 19, 21, and 22: endocrine disrupters, radioactive waste, air pollution, water pollution, and pesticides. Here we discuss dioxins, PCBs, and radioactive wastes produced during the Cold War.

Dioxins **Dioxins** are a group of 75 similar chemical compounds formed as unwanted by-products during the combustion of chlorine compounds. Some of the known sources of dioxins are medical waste and municipal waste incinerators, iron ore mills, copper smelters, cement kilns, metal recycling, coal combustion, pulp

Figure 23.11 **Aerial view of Love Canal toxic waste site in the early 1980s.** All the homes shown in this photograph were evacuated and demolished.

Andy Levin/Science Source

Table 23.1   Examples of Hazardous Waste

Hazardous Material	Some Possible Sources
Acids	Ash from power plants and incinerators; petroleum products
CFCs (chlorofluorocarbons)	Coolant in air conditioners and refrigerators
Cyanide	Metal refining; fumigants in ships, railway cars, and warehouses
Dioxins	Emissions from incinerators and pulp and paper plants
Explosives	Old military installations
Heavy metals:	
Arsenic	Industrial processes, pesticides, additives to glass, paints, electronics discarded in landfills
Cadmium	Rechargeable batteries, incineration, paints, plastics, discarded electronics
Lead	Lead-acid storage batteries, stains and paints, TV picture tubes and electronics discarded in landfills
Mercury	Coal-burning power plants; paints, household cleaners (disinfectants), industrial processes, medicines, seed fungicides, discarded electronics
Infectious waste	Hospitals, research labs
Nerve gas	Old military installations
Organic solvents	Industrial processes; household cleaners, leather, plastics, pet maintenance (soaps), adhesives, cosmetics
PCBs (polychlorinated biphenyls)	Older appliances (built before 1980); electrical transformers and capacitors
Pesticides	Household products
Radioactive waste	Nuclear power plants, hospitals, and weapons production/dismantling facilities

and paper plants that use chlorine for bleaching, and chemical accidents. Incineration of medical and municipal wastes accounts for 70% to 95% of known human emissions of dioxins. In Japan, where nearly 75% of the solid waste is burned in incinerators, the air contains nearly 10 times the amount of dioxins found in other highly developed countries. The more than 6000 hospital waste incinerators in the United States are probably the largest dioxin polluters because they are so numerous, and they generally have unsophisticated pollution controls.

Dioxins are emitted in smoke and then settle on plants, the soil, and bodies of water; from there, they are incorporated into the food web. When humans and other animals ingest dioxins, they are stored and accumulate in fatty tissues (see the discussion of bioaccumulation and biological magnification in Chapter 7). Humans are primarily exposed when they eat contaminated meat, dairy products, and fish. Because dioxins are so widely distributed in the environment, virtually everyone has dioxins in his or her body fat.

Just how dangerous dioxins are to humans is somewhat controversial. Dioxins are known to cause several kinds of cancer in laboratory animals, but the data are conflicting on their cancer-causing ability in humans. A 1997 study of residents of Seveso, Italy, who were exposed to high levels of dioxin after a chemical accident in 1976, revealed a statistically significant increase in cancer deaths. Another study published in 2002 linked high levels of dioxin exposure to an increased incidence of breast cancer in Italian women living near the accident site. According to the EPA, dioxins probably cause several kinds of cancer in humans.

Other concerns center on the effects of dioxins on the human reproductive, immune, and nervous systems. Dioxins may delay fetal development and cause cognitive damage, lead to endometriosis in women (the growth of uterine tissue in abnormal locations in the body), and decreased sperm production in men. Dioxins are also linked to an increased risk of heart disease. Because human milk contains dioxins, nursing infants are considered particularly at risk.

PCBs   Polychlorinated biphenyls (PCBs) are a group of 209 industrial chemicals composed of carbon, hydrogen, and chlorine. These clear or light-yellow, oily liquids or waxy solids were manufactured in the United States between 1929 and 1979. PCBs were used as cooling fluids in electrical transformers, electrical capacitors, vacuum pumps, and gas-transmission turbines. They were also found in hydraulic fluids, fire retardants, adhesives, lubricants, pesticide extenders, inks, and other materials.

The first evidence that PCBs were dangerous emerged in 1968, when Japanese who ate rice bran oil contaminated with PCBs experienced liver and kidney damage. A similar mass poisoning, also attributed to PCB-contaminated rice oil, occurred in Taiwan in 1979. Since then, toxicity tests conducted on animals indicate that PCBs harm the skin, eyes, reproductive organs, and gastrointestinal system. PCBs are endocrine disrupters because they interfere with hormones released by the thyroid gland. Several studies suggest that children exposed to PCBs before birth have certain intellectual impairments, such as poor reading comprehension, memory problems, and difficulty paying attention. Several studies suggest that PCBs may be carcinogenic.

PCBs are chemically stable and resist chemical and biological degradation. Like dioxins, PCBs accumulate in fatty tissues and are subject to biological magnification in food webs. The general human population is mainly exposed to PCBs by eating food that became contaminated through biological magnification. One way that PCBs enter aquatic food webs is via benthic invertebrates that live in contaminated sediments. (PCBs tend to bind to organic particles in aquatic sediments.) Small fish eat these invertebrates, and as larger fish eat the smaller fish, the PCBs bioaccumulate. Human populations whose diets consist primarily of fish and marine mammals, such as the Inuit of northern Canada, are exposed to large amounts of PCBs.

Prior to the EPA ban in the 1970s, PCBs were dumped in large quantities into landfills, sewers, and fields. Such improper disposal is one of the reasons PCBs are still a threat today. Also, when sealed electrical transformers and capacitors leak or catch fire, PCB contamination of the environment occurs.

High-temperature incineration is one of the most effective ways to destroy PCBs. However, incineration is not practical for the removal of PCBs that have leached into the soil and water because, among other difficulties, the cost of incinerating large quantities of soil is prohibitively high.

Several bacteria can degrade PCBs. However, when PCB-eating bacteria are sprayed on the surface of the soil, they cannot decompose the PCBs that have already leached into the soil or groundwater systems. These microorganisms show promise in removing PCBs from the environment, but additional research is needed to make the biological degradation of PCBs practical. (Additional discussion about using bacteria to break down hazardous waste is found later in this chapter.)

CASE IN POINT

HANFORD NUCLEAR RESERVATION

U.S. nuclear weapons facilities are no longer actively manufacturing nuclear weapons, but they present us with a greater challenge—reducing and managing radioactive and toxic wastes that have accumulated at numerous sites around the United States since the 1940s. Every step in the production of nuclear warheads generated radioactive and chemical wastes. We focus our discussion on the Hanford Nuclear Reservation, a 1400-km² (560-mi²) area on the Columbia River in south-central Washington State (**Figure 23.12**). Hanford, the main production site for plutonium used in nuclear weapons, is the largest, most seriously contaminated site in the U.S. nuclear weapons infrastructure. Cleanup there began with a 1989 agreement between the EPA, the U.S. Department of Energy (DOE), and the State of Washington Department of Ecology.

The immensity of the cleanup task at Hanford has proved daunting. Tons of highly radioactive solid and liquid wastes were stored or dumped into trenches, pits, tanks, ponds, and underground cribs—a total of 1700 waste sites. (These methods of disposal were standard practice at the time.) Two concrete pools of water stored more than 100,000 spent fuel rods. As they corroded, the rods released highly radioactive uranium, plutonium, cesium, and strontium into the water, contaminating soil and groundwater and endangering the Columbia River. The fuel rods have been removed and placed in canister storage until a national or regional spent fuel repository is available (see Chapter 12).

The Columbia River is also threatened by millions of gallons of toxic chemical and radioactive liquid wastes stored in 177 large underground tanks. Liquids in some of these tanks are so reactive that they boiled for years from the heat of their own radioactivity or chemical activity, but most tanks are now covered by semisolid crusts that formed from chemical reactions within the mixtures. Some of the tanks are potentially explosive: Chemical reactions produce hydrogen and other hazardous gases. Vents allow the gas to escape and reduce the danger of ignition. Many of these tanks were designed to last only 10 to 20 years and are now leaking their poisons into the ground.

(a) Location of Hanford along the Columbia River in Washington State.

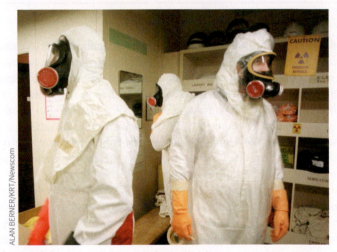

(b) Suited workers prepare to manage high-level radioactive waste at Hanford.

Figure 23.12   Hanford Nuclear Reservation.

Cleanup of this hazardous waste is complicated because the extent of radioactive pollution and the kinds of hazardous mixtures present are not well known (environmental records of hazardous waste production generally were not kept until the 1970s). Scientists and engineers were required to assess the damage, prioritize the cleanup process, and determine how best to proceed for each type of contamination. The cleanup, which is under the direction of DOE, has cost $30 billion so far and has fallen behind schedule, primarily because of budget restraints. Originally expected to be near completion, cleanup will likely continue for decades and cost tens of billions of dollars more.

The cleanup has sparked legal battles over environmental concerns and workers' health. In fact, cleanup may be a more dangerous occupation than working at Hanford when it was at full operation producing nuclear arms. Some of the cleanup workers have developed serious chronic illnesses because they come into close contact with toxic materials, such as beryllium, that were used to make bombs. Inhalation of beryllium particles can cause an incurable lung disease.

After the cleanup is finished, Hanford will remain hazardous for hundreds or even thousands of years, in part because we do not have the technologies to address the widespread soil contamination. Also, much of the buried radioactive waste may be left where it is because there is no other place to move it. Long-term monitoring will need to be maintained at the site to limit human exposure to remaining hazards. ■

MANAGEMENT OF HAZARDOUS WASTE

Humans have the technology to manage hazardous waste in an environmentally responsible way, but it is extremely expensive. Although great strides have been made in educating the public about the problems of hazardous waste, we have only begun to address many of the issues of hazardous waste disposal. No country currently has an effective hazardous waste management program, but several European countries have led the way by producing smaller amounts of hazardous waste and using fewer hazardous substances.

Chemical Accidents   When a chemical accident occurs in the United States, whether at a factory or during the transport of hazardous chemicals, the National Response Center is notified. Most chemical accidents reported to the National Response Center involve oil, gasoline, or other petroleum spills. The remaining accidents involve more than 1000 other hazardous chemicals, such as ammonia, sulfuric acid, and chlorine.

Chemical safety programs have traditionally stressed accident mitigation and the addition of safety systems to existing procedures. More recently, industry and government agencies have focused on accident prevention through the **principle of inherent safety**, in which industrial processes are redesigned to involve less hazardous materials so that dangerous accidents are prevented. The principle of inherent safety, which is an example of systems thinking, is an important aspect of source reduction.

Current Hazardous Waste Policies   Currently, two federal laws dictate how hazardous waste should be managed: (1) the Resource Conservation and Recovery Act, which is concerned with managing hazardous waste being produced now, and (2) the Superfund Act, which provides for the cleanup of abandoned and inactive hazardous waste sites.

The **Resource Conservation and Recovery Act (RCRA)** was passed in 1976 and amended in 1984. Among other things, RCRA instructs the EPA to identify which waste is hazardous and to provide guidelines and standards to states for hazardous waste management programs. RCRA bans hazardous waste from land disposal unless it is treated to meet the EPA's standards of reduced toxicity. In 1992, the EPA initiated a major reform of RCRA to expedite cleanups and streamline the permit system to encourage hazardous waste recycling.

In 1980, the **Comprehensive Environmental Response, Compensation, and Liability Act (CERCLA)**, commonly known as the **Superfund Act**, established a program to clean up abandoned and illegal hazardous waste sites across the United States. At many of these sites, hazardous chemicals have migrated deep into the soil and have polluted groundwater. The greatest threat to human health from hazardous waste sites comes from drinking water laced with such contaminants.

Cleaning Up Existing Hazardous Waste: The Superfund Program   The federal government estimates that the United States has more than 400,000 hazardous waste sites with leaking chemical storage tanks and drums (both above and below ground), pesticide dumps, and piles of mining waste (**Figure 23.13a**). This estimate does not include the hundreds or thousands of hazardous waste sites at military bases and nuclear weapons facilities.

CERCLA lists an inventory of more than 10,000 hazardous waste sites that the EPA has deemed as qualifying for cleanup. These sites are not identified according to any particular criteria. Some are dumps that local or state officials have known about for years, whereas concerned citizens identify others. The sites in the inventory are evaluated and ranked to identify those with extremely serious hazards, based on data from preliminary assessments, site inspections, and expanded site inspections, which include contamination tests of soil and groundwater and sampling of hazardous waste.

The sites that pose the greatest threat to public health and the environment are placed on the **Superfund National Priorities List**, which means that the federal government will assist in their cleanup (**Figure 23.13b**). As of 2014, a total of 1326 sites were on the National Priorities List. The five states with the greatest number of sites on the National Priorities List as of 2014 are New Jersey (114 sites), California (97 sites), Pennsylvania (95 sites), New York (86 sites), and Michigan (65 sites).

As of 2014, a total of 375 hazardous waste sites had been cleaned up enough to be deleted from the National Priorities List, and 60 other sites were partially corrected. The 1158 sites designated "construction completions" have had extensive cleanups but remain on the Superfund list because of continued problems with groundwater contamination. The average cost of cleaning up a site is $20 million.

One reason for the urgency in cleaning up sites on the National Priorities List is their locations. Most were originally in rural areas on the outskirts of cities. With the growth of cities and their suburbs, residential developments now surround many of the dumps. One in three Americans lives within 5 km (3 mi) of one or more Superfund sites.

Because the federal government cannot assume major responsibility for cleaning up every old dumping ground in the United States, CERCLA requires that the current landowner, prior owners, and anyone who has dumped waste or has transported waste to a particular site share cleanup costs. For some sites, many parties are considered liable for cleanup costs. The cleanup process has been mired in litigation—mostly by companies charged with polluting, which are suing each other. Despite the urgency of cleaning up sites on the National Priorities List, it will take many years to complete the job.

Although critics decry the slow pace and high cost of cleaning up Superfund sites, the very existence of CERCLA is a deterrent to further polluting. Companies that produce hazardous waste are

Courtesy USDA

(a) Hazardous waste in deteriorating drums at a site near Washington, DC. The metal drums in which much of the waste is stored have corroded and started to leak. Old hazardous waste dumps are commonplace around the United States.

©Greg Smith/Corbis

(b) Cleanup of a Superfund site near Houston, Texas. Removal and destruction of the wastes are complicated by the fact that often nobody knows what chemicals are present.

Figure 23.13   Cleaning up hazardous waste.

now fully aware of the costs of liability and cleanup and are more likely to take steps to dispose of their hazardous wastes in proper fashion.

The Biological Treatment of Hazardous Contaminants   A variety of methods are employed to clean up soil contaminated with hazardous waste. Because most of these processes are prohibitively expensive, innovative approaches such as bioremediation and phytoremediation are being developed to deal with hazardous waste. **Bioremediation** is the use of bacteria and other microorganisms to break down hazardous waste into relatively harmless components. **Phytoremediation** is the use of plants to absorb and accumulate hazardous materials from the soil. (The prefix *phyto-* comes from the Greek word meaning "plant.")

More than 1000 species of bacteria and fungi have been demonstrated to clean up various forms of organic pollution. Bioremediation takes a little longer to work than traditional hazardous waste disposal methods, but it accomplishes the cleanup at a fraction of the cost. In bioremediation, the contaminated site is exposed to an army of microorganisms, which gobble up poisons such as petroleum and other hydrocarbons and leave behind harmless substances such as carbon dioxide, water, and chlorides. Bioremediation encourages the natural processes in which bacteria consume organic molecules such as hydrocarbons. During bioremediation, conditions at the hazardous waste site are modified so that the desired bacteria will thrive in large enough numbers to be effective. Environmental engineers might pump air through the soil (to increase its oxygen level) and add a few soil nutrients such as phosphorus or nitrogen. They might install a drainage system to pipe any contaminated water that leaches through the soil back to the surface for another exposure to the bacteria.

In phytoremediation, plant species known to remove specific hazardous materials from the soil are grown at a contaminated site. Phytoremediation refers to several different ways that plants remove pollutants (**Figure 23.14**). In *phytoextraction*, some roots

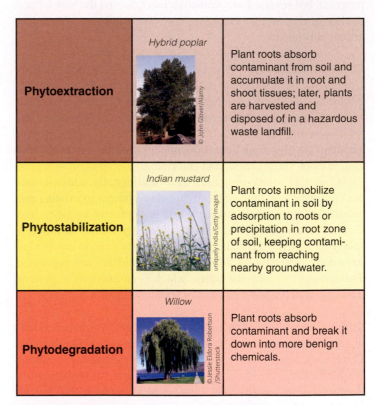

Phytoextraction	*Hybrid poplar* © John Glover/Alamy	Plant roots absorb contaminant from soil and accumulate it in root and shoot tissues; later, plants are harvested and disposed of in a hazardous waste landfill.
Phytostabilization	*Indian mustard* uniquely india/Getty Images	Plant roots immobilize contaminant in soil by adsorption to roots or precipitation in root zone of soil, keeping contaminant from reaching nearby groundwater.
Phytodegradation	*Willow* ©Jessie Eldora Robertson /Shutterstock	Plant roots absorb contaminant and break it down into more benign chemicals.

Figure 23.14   Examples of phytoremediation.

selectively absorb toxins from the soil and accumulate them in root and shoot tissues; later, the plants are harvested and disposed of in a hazardous waste landfill. Alternatively, some plants break down a hazardous chemical into more benign chemicals (*phytodegradation*). In *phytostabilization*, some plants immobilize toxins in soil, keeping them from reaching nearby groundwater.

Specific plants are known to remove such hazardous materials as trinitrotoluene (TNT), radioactive strontium and uranium, selenium, lead, and other heavy metals. Hybrid poplars and willows, for example, can remove chlorinated solvents from groundwater; Indian mustard removes heavy metals such as lead from the soil.

Phytoremediation costs less than conventional methods to clean up hazardous waste sites, but it does have limitations. Plants cannot remove contaminants present in the soil deeper than their roots normally grow. Also, insects and other animals might eat the plants, thereby introducing the toxins into the food web.

Managing the Hazardous Waste We Are Producing Now

Many people think, incorrectly, that establishment of the Superfund has eliminated the problem of hazardous waste. The Superfund deals only with hazardous waste produced in the past; it does nothing to eliminate the large amount of hazardous waste produced today. We manage hazardous waste by source reduction, conversion to less hazardous materials, and long-term storage.

As with MSW, the most effective of the three methods is *source reduction*—that is, reducing the amount of hazardous materials used in industrial processes and substituting less hazardous or nonhazardous materials for hazardous ones. Source reduction relies on an increasingly important subdiscipline of chemistry known as **green chemistry** (see Case in Point: Green Chemistry in Chapter 21).

For example, chlorinated solvents are widely used in electronics, dry cleaning, foam insulation, and industrial cleaning. It is sometimes possible to accomplish source reduction by substituting a less hazardous water-based solvent for a chlorinated solvent. Reducing solvent emissions also results in substantial source reduction of chlorinated solvents. Most chlorinated solvent pollution gets into the environment by evaporation during industrial processes. Installing solvent-saving devices benefits the environment and also provides economic gains, because smaller amounts of chlorinated solvents must be purchased. No matter how efficient source reduction becomes, however, it will never entirely eliminate hazardous waste.

The second-best way to deal with hazardous waste is to *reduce its toxicity* by chemical, physical, or biological means, depending on the nature of the hazardous waste. One method to detoxify organic compounds is high-temperature incineration. The high heat of combustion reduces these dangerous compounds, such as pesticides, PCBs, and organic solvents, into safe products such as water and carbon dioxide. The incineration ash, however, is hazardous and must be disposed of in a landfill designed specifically for hazardous materials. One method of reducing this toxic by-product is incineration using a *plasma torch*, which produces such high temperatures (up to 10,000°C, as much as five times hotter than conventional incinerators) that hazardous waste is almost completely converted to mostly harmless gases. (However, CO_2, which alters climate, is a harmful product of the plasma torch.) Hazardous waste that is produced in spite of source reduction and that is not completely detoxified must be placed in *long-term storage*.

Hazardous waste landfills are subject to strict environmental criteria and design features. They are located as far as possible from aquifers, streams, wetlands, and residences. These landfills have many special features, including several layers of compacted clay and high-density plastic liners at the bottom of the landfill to prevent leaching of hazardous substances into surface water and groundwater (**Figure 23.15**). *Leachate* is collected and treated to remove contaminants. The entire facility and nearby groundwater deposits are carefully monitored to make sure no leaking

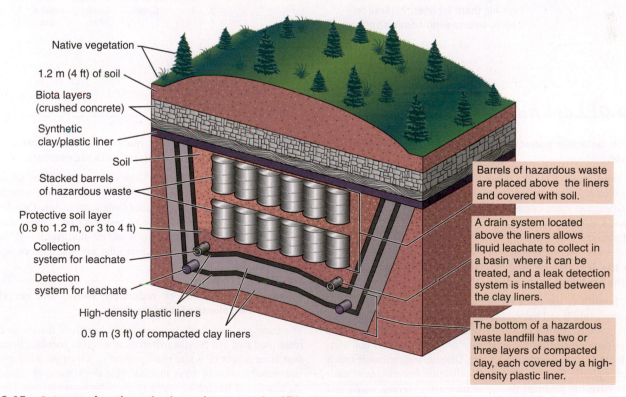

Native vegetation

1.2 m (4 ft) of soil

Biota layers (crushed concrete)

Synthetic clay/plastic liner

Soil

Stacked barrels of hazardous waste

Protective soil layer (0.9 to 1.2 m, or 3 to 4 ft)

Collection system for leachate

Detection system for leachate

High-density plastic liners

0.9 m (3 ft) of compacted clay liners

Barrels of hazardous waste are placed above the liners and covered with soil.

A drain system located above the liners allows liquid leachate to collect in a basin where it can be treated, and a leak detection system is installed between the clay liners.

The bottom of a hazardous waste landfill has two or three layers of compacted clay, each covered by a high-density plastic liner.

Figure 23.15   **Cutaway view through a hazardous waste landfill.** (Adapted from Rocky Mountain Arsenal Remediation Venture Office)

occurs. Only solid chemicals (not liquids) that have been treated to detoxify them as much as possible are accepted at hazardous waste landfills. These chemicals are placed in sealed barrels before being stored in the hazardous waste landfill.

Few landfills are certified to handle hazardous waste. Currently, only 23 commercial hazardous waste landfills are open in the United States, although many larger companies are licensed to treat their hazardous waste on-site. As a result, much of our hazardous waste is still placed in sanitary landfills, burned in incinerators that lack the required pollution-control devices, or discharged into sewers.

Some liquid hazardous wastes, such as certain organic compounds, fuels, explosives, and pesticides, are stored in the Earth's crust by **deep well injection**. These wells extend into impermeable rock layers several thousand feet below the surface. The Safe Drinking Water Act and the Underground Injection Control Program regulate the placement and number of such wells, which can be located only where minimal danger of groundwater contamination exists.

REVIEW

1. What is hazardous waste?

2. How are the Resource Conservation and Recovery Act and the Comprehensive Environmental Response, Compensation, and Liability Act alike? What is the focus of each?

3. How does green chemistry relate to source reduction?

ENERGY AND CLIMATE CHANGE

Storing, Reusing, or Incinerating Used Tires

As discussed in this chapter, the nearly 300 million used tires discarded each year in the United States present a variety of health and safety concerns. However, storing old tires can also be a form of carbon sequestration; that is, burying them can store the carbon they contain, rather than adding it to the atmosphere (see Chapter 20). Alternatively, old tires can be shredded and used for a variety of purposes, such as replacing some of the asphalt and gravel in roads. Finally, tires can be incinerated, and the energy they contain can be turned into electricity. Burning tires for energy releases about 78% as much carbon as does burning coal, but somewhat more than is produced when burning natural gas (see figure).

Carbon dioxide produced per unit energy produced. Burning used tires for energy produces less CO_2 than does producing the same amount of energy from burning coal, but slightly more than burning natural gas. Do you think we are better off burying used tires, turning them into other products, or burning them for energy? (Source: Energy Information Administration)

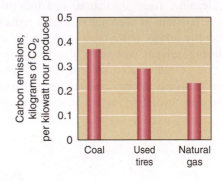

REVIEW OF LEARNING OBJECTIVES WITH SELECTED KEY TERMS

- **Distinguish between municipal solid waste and nonmunicipal solid waste.**

Municipal solid waste (MSW) consists of solid materials discarded by homes, office buildings, retail stores, restaurants, schools, hospitals, prisons, libraries, and other commercial and institutional facilities. **Nonmunicipal solid waste** consists of solid waste generated by industry, agriculture, and mining.

- **Describe the features of modern sanitary landfills and explain some of the problems associated with them.**

A **sanitary landfill** is the most common method of disposal of solid waste, by compacting it and burying it under a shallow layer of soil. The location of a sanitary landfill must take into account the geology of the area, soil drainage properties, the proximity of surface waters and wetlands, and distance from population centers. Despite design features such as high-density plastic liners and **leachate** collection systems, most sanitary landfills have the potential to contaminate soil, surface water, and groundwater.

- **Describe the features of a mass burn incinerator and explain some of the problems associated with incinerators.**

A **mass burn incinerator** is a large furnace that burns all solid waste except for unburnable items such as refrigerators. Most mass burn incinerators recover the energy produced from combustion. One drawback of incineration is the great expense of installing pollution-control devices on the incinerators. These controls reduce the toxicity of the gaseous emissions from incinerators but make the ash that remains behind more hazardous.

- **Summarize how source reduction, reuse, and recycling help reduce the volume of solid waste.**

The three goals of waste prevention are to reduce the amount of waste, reuse products, and recycle materials as much as possible. **Source reduction** is an aspect of waste management in which products are designed and manufactured in ways that decrease the volume of solid waste and the amount of hazardous waste in the solid waste stream. One example of reuse is refillable glass beverage bottles. Recycling involves collecting and reprocessing materials into new products. Many communities recycle paper, glass, metals, and plastic.

- **Define *integrated waste management*.**

Integrated waste management is a combination of the best waste management techniques into a consolidated, systems-based program to deal effectively with solid waste.

- **Define *hazardous waste* and briefly characterize representative hazardous wastes (dioxins, PCBs, and radioactive wastes).**

Hazardous waste consists of any discarded chemical that threatens human health or the environment. **Dioxins** are hazardous chemicals formed as unwanted by-products during the combustion of many chlorine compounds. **Polychlorinated biphenyls (PCBs)** are hazardous, oily, industrial chemicals composed of carbon, hydrogen, and chlorine. Radioactive and chemical wastes have accumulated at numerous U.S. nuclear weapons facilities around the United States; the largest, most seriously contaminated site is Hanford Nuclear Reservation in Washington State.

- **Contrast the Resource Conservation and Recovery Act and the Comprehensive Environmental Response, Compensation, and Liability Act (the Superfund Act).**

The **Resource Conservation and Recovery Act (RCRA)** instructs the EPA to identify which waste is hazardous and to provide guidelines and standards to states for hazardous waste management programs. The **Comprehensive Environmental Response, Compensation, and Liability Act (CERCLA)**, also known as the **Superfund Act**, addresses the challenge of cleaning up abandoned and illegal hazardous waste sites in the United States.

- **Explain how green chemistry is related to source reduction.**

Green chemistry is a subdiscipline of chemistry in which commercially important chemical processes are redesigned to significantly reduce environmental harm. Source reduction is the best way to reduce hazardous waste, so green chemists redesign processes with source reduction in mind.

Critical Thinking and Review Questions

1. What is solid waste?

2. Compare the advantages and disadvantages of disposing of waste in sanitary landfills and by incineration.

3. List what you think are the best ways to treat each of the following types of solid waste, and explain the benefits of the processes you recommend: paper, plastic, glass, metals, food waste, yard waste.

4. What initiatives can (or do) college campuses—or students on them—take to reduce the volume of various solid wastes they generate?

5. Why is creating a demand for recycled materials sometimes referred to as "closing the loop"?

6. How does recycling link the world's largest economy (the United States) to the world's fastest-growing economy (China)?

7. What are dioxins, and how are they produced? What harm do they cause?

8. Suppose hazardous chemicals were suspected to be leaking from an old dump near your home. Outline the steps you would take to (1) have the site evaluated to determine if it is dangerous and (2) mobilize the local community to get the site cleaned up.

9. What are the goals, strengths, and weaknesses of the Superfund program?

10. The Organization for African Unity has vigorously opposed the export of hazardous waste from industrialized nations to developing countries. It calls this practice "toxic terrorism." Explain.

11. What is integrated waste management? Why must a sanitary landfill always be included in any integrated waste management plan?

12. Compare integrated pest management, discussed in Chapter 22, to integrated waste management. How does each reduce potential damage to the environment?

13. How does the system of integrated waste management depicted in Figure 23.10 compare to a natural ecosystem?

14. How do the three R's of waste prevention reduce the release of climate-altering CO_2 into the environment?

15. In an effort to reduce municipal solid waste, some communities require customers to pay for garbage collection according to the amount of garbage they generate, known as unit pricing. The figure shows the effects of unit pricing in San Jose, California, on garbage sent to landfills and on wastes diverted through recycling and separation of yard wastes. How did unit pricing affect the amount of garbage sent to landfills? How did unit pricing affect the quantity of materials recycled? Of yard wastes collected?

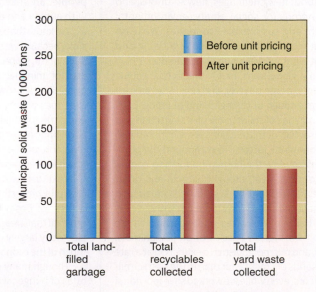

Food for Thought

Keep track of the amount of food waste you produce daily, both the food itself and any packaging. What percent of the waste do you throw away? Compost? Recycle? What changes in your purchasing habits could you make to reduce the amount of packaging you discard? What changes in your diet or meal preparation could you make to avoid discarding food? Could you reduce your food waste without diminishing your quality of life? How?

24

Tomorrow's World

Children planting trees. Such activities teach tomorrow's generation about environmental issues, such as the importance of forests.

We, the authors of this text, are—like our readers—students and citizens of environmental science ourselves. We learn continually about new challenges, new solutions, and new heroes for social equity, environmental health, and sustainable economies. We learn about what good work others are doing and about what work we ourselves can do.

On a global level, nearly one in four people lives in extreme poverty, and one in seven cannot get enough or the right kind of foods. Each year, millions of children die of malnutrition and disease; half of these deaths are attributable to environmental factors. To compound the problem of poverty, the human population continues to grow. Heat, drought, storms and other events destabilize the lives of individuals and of whole countries. Forests are cut without new ones being planted (see photograph). We are driving the world's species of plants, animals, fungi, and microorganisms to extinction at a rate thousands of times faster than in the past 65 million years, even while we work to save individual species and habitats. Fossil fuel extraction takes more energy and incurs more risk of pollution and accidents as the sources of these fuels become scarcer. Water quality and distribution is threatened by misuse and by climate change.

This chapter contains a five-point strategy for action to confront these critical global issues. Each of these strategies already has capable people working on them in a variety of places, in many states and countries. More people are needed for this work—people with energy and education and hope.

We know that each of these solutions is possible, because we see solutions at work in many parts of the globe. Some countries offer examples of minimal poverty and stable human populations. In Sweden, strong social supports exist for parents, from access to playgrounds and parks for children to parental leave policies and financial support for young families. Children and families are not hungry or homeless, and the average number of children per family is just under two. Costa Rica has shown leadership in protecting natural resources. Over a quarter of Costa Rica is protected land, with much of that area in national parks. Costa Rica's electricity largely comes from hydropower, and much of the economy is conservation-centered.

Though many countries have relatively low rates of hunger, we have learned that reducing gender inequality in agricultural land access is an excellent way to reduce hunger further. In this chapter, we discuss how novel methods of agricultural financing (Meeting the Challenge: Microcredit Programs) can help reduce hunger; these programs particularly help women, who do not qualify for financial assistance in many countries. We know that women farmers are more likely to use their labor to grow food, rather than cash crops.

Some of the solutions we need come from hard lessons. After Hurricane Sandy, residents of New Jersey saw that communities with intact sand dunes suffered far less storm damage. Protecting and re-creating healthy sand dunes will help mitigate storm damage, which will be a larger threat to many areas as climate change continues.

Finally, all of these solutions have the potential to make life more pleasant for all of us. Cities around the globe have joined the Sustainable Cities Network, not simply because sustainable cities are a moral imperative or a duty, but because sustainable cities are more pleasant: cleaner, with more trees and gardens, with less traffic, and less trash. Sustainable cities and sustainable countries are places where people, wildlife, and habitats can all thrive.

In Your Own Backyard...
What programs does your school have in place to encourage sustainability? Which of these programs can you continue with beyond school?

Living Sustainably

- Define *sustainability*.
- Discuss how the natural environment is linked to sustainable development.

Sustainability is a concept that people have discussed for many years. *Our Common Future*, the Brundtland Report of the World Commission on Environment and Development (1987), presented the closely related concept of **sustainable development**. The goals of sustainable development are to improve living conditions for all people while maintaining a healthy environmental system in which natural resources are not overused and excessive pollution is not generated. Sustainable development balances social justice, economic growth, and environmental conservation.

Our Common Future also observed that the environment's ability to meet present and future needs is directly related to the state of technology and social organization existing at a given time and in a given place. The number of people alive, their affluence (i.e., their level of **consumption**), and their choices of technology all interact to produce the total effect of a given society, or of society at large, on the sustainability of the environment.

Even with the use of the best technologies we could imagine, the productivity of Earth still has limits. Human population size and use of natural resources both fit within the Earth's limits of productivity. The world does not contain enough resources to sustain everyone at the level of consumption enjoyed in most developed nations (**Figure 24.1**).

Although we cannot accurately predict the consequences of particular kinds of economic development, all economic development takes place within the **carrying capacity (K)** of the ecosystems that support it. The carrying capacity of a given ecosystem is determined ultimately by its ability to absorb wastes and renew itself. In understanding Earth's carrying capacity for the human population and in designing appropriate economic development strategies, we, authors and students and other researchers, examine and question the living standards and expectations in different areas.

Sustainability is a term for behaviors and cultures that can continue for generations. We sometimes have a hard time predicting the results of our technologies even a single generation ahead, as we have seen when we study problems resulting from pesticides, plastics, and fossil fuel extraction. We rarely see the effects of a single action on a wide range of people, beyond our own neighborhoods. Therefore, sustainability is, as Kai N. Lee wrote in *Compass and Gyroscope: Integrating Science and Politics for the Environment*, "like freedom or justice, a direction in which we strive, along which we search for a life good enough to warrant

sustainability The ability to meet humanity's current needs without compromising the ability of future generations to meet their needs.

sustainable development Economic development that meets the needs of the present without compromising the ability of future generations to meet their own needs.

consumption The human use of materials and energy.

carrying capacity (K) The maximum number of individuals of a given species that a particular environment can support for an indefinite period, assuming the environment doesn't change.

Figure 24.1   A shopping mall in Kuala Lampur, Malaysia. While consumer purchasing in the U.S. still exceeds consumer purchasing in most other areas of the world, shopping malls demonstrate that marketing of consumer goods can be effective in almost any culture.

our comforts." Sustainability is not a reachable destination, but a direction for increasing well-being of people and the Earth.

REVIEW

1. What is sustainability?
2. How is the natural environment—Earth's living organisms and ecosystems—an essential part of sustainable development?

Sustainable Living: A Plan of Action

- Define *poverty* and briefly describe this global problem.
- Discuss problems relating to loss of forests and declining biological diversity, including the important ecosystem services that these resources provide.
- Describe the extent of food insecurity and identify at least two ways to increase food production sustainably.
- Define the *enhanced greenhouse effect* and explain how stabilizing climate is related to energy use.
- Describe at least two challenges for cities and explain how they can be addressed.

We have organized this section around the five recommendations for sustainable living presented in *Plan B 2.0: Rescuing a Planet Under Stress and a Civilization in Trouble*, by **Lester R. Brown.** We think that if people, both individuals and collectively as governments, focus their efforts and financial support on Brown's plan, the quality of human life will be much improved. Brown's five recommendations for sustainable living are as follows:

1. Eliminate poverty and stabilize the human population.
2. Protect and restore Earth's resources.
3. Provide adequate food for all people.
4. Mitigate climate change.
5. Design sustainable cities.

Seriously addressing these recommendations now offers us hope for the kind of future we all want for our children and grandchildren.

RECOMMENDATION 1: ELIMINATE POVERTY AND STABILIZE THE HUMAN POPULATION

The ultimate goal of economic development is to improve the quality of human life, making it possible for humans throughout the world to enjoy long, healthy, and fulfilling lives. A serious complication lies in the fact that the distribution of the world's resources is unequal. Those who live in highly developed countries, a rapidly shrinking 18% of the global population, control about 79% of the world's finances, as measured by summing gross domestic products. The *per capita GNI PPP* (gross national income in purchasing power parity) in countries with very high human development in 2012 (the latest data available) was about $35,390; the per capita GNI PPP of the other 82% of the people in the world was about $9480.

 poverty A condition in which people are unable to meet their basic needs for adequate food, clothing, shelter, education, or health.

For many of the world's women and children, life is an endless struggle for survival, focusing on the daily requirements for firewood, clean water, and food. Such **poverty** is a global problem. These problems cannot be solved without addressing the great inequalities between rich and poor, both within cities and between countries.

We who live in the United States, Canada, and other highly developed countries generally enjoy an abundance of what the world has to offer. We are collectively the wealthiest people who have ever existed, with the highest standard of living. Because the United States, with less than 5% of the world's people, controls approximately 25% of the world's economy, we depend on many other nations for our prosperity. In our actions we often underestimate our effects on the environment that supports us.

Failing to confront the problem of poverty around the world continues to make it impossible to attain global sustainability. For example, about 18,000 infants and children under the age of five die each day (2012 data from United Nations Children's Fund). Most of these deaths could have been prevented by access to adequate supplies of food, clean water, and basic medical care. As U.S. President **Franklin Delano Roosevelt** said so well in his second inaugural address in 1937, "The test of our progress is not whether we add more to the abundance of those who have much; it is whether we provide enough for those who have too little."

Improving the quality of life in lower-income countries will require increasing their economic growth so that issues of health, nutrition, and education can be addressed adequately (see Meeting the Challenge: Microcredit Programs). The role

MEETING THE CHALLENGE

Microcredit Programs

"The poor stay poor, not because they're lazy but because they have no access to capital." This statement by Peruvian economist **Hernando de Soto** provides the philosophical basis for **microcredit**, financial services used to empower women and attack poverty. Microcredit consists of small loans (*microloans* of US $50 to US $500) extended to very poor people to help them establish self-employment projects that generate income. These people lack the assets to qualify for loans from commercial banks.

The poor use these loans for a variety of projects. Some purchase used sewing machines to make clothing faster than sewing by hand. Others open small grocery stores after purchasing used refrigerators to store food so that it does not spoil. Still others use the money to bake bread, weave mats, or raise chickens (see photograph).

The Foundation for International Community Assistance (FINCA) is a nonprofit agency that administers a global network of microcredit banks. FINCA uses *village banking*, in which a group of very poor neighbors guarantee one another's loans, administer group lending and savings activities, and provide mutual support. Thus, village banks give autonomy to local people.

FINCA primarily targets women, and for good reason, because an estimated 70% of the world's poorest people are women. Also, the number of households in which women provide the sole support for their children is growing worldwide. FINCA believes that the best way to alleviate the effects of poverty and hunger on children is to provide their mothers with a means of self-employment. A woman's status in the community is raised as she begins earning income from her business.

The loans help people rise above poverty because they can generate more income and even save money for the future. Many women who borrow from FINCA spend their earnings to improve the nutrition and health of their children. For these women, the next priority after improved nutrition and health is education for their children.

Microcredit was first tried in Bangladesh in the 1970s, where it has helped thousands of urban and rural Bangladeshis start successful businesses. **Muhammad Yunus**, the Bangladeshi who founded the *Grameen Bank* and pioneered microcredit, was awarded a Nobel Peace Prize in 2006. Microcredit has helped about 92 million of the world's poor so far.

But the need remains great: An estimated 200 million more of the world's poor could benefit from microcredit programs.

RAFIQUR RAHMAN/Reuters/Landov

Microcredit. This Bangladeshi woman feeds chickens at her poultry farm. She received her first microcredit loan to buy a few chickens and has built the farm into a thriving business.

of women requires special attention, since the improvement of their status can make a significant contribution to the stability and prosperity of those communities. As **Nafis Sadik**, head of the UN Population Fund from 1987 to 2000, has pointed out, women hold a paradoxical place in many societies. As part of their traditional duties as mothers and wives, they are expected to bear the whole responsibility for childcare; at the same time, they are often expected to contribute significantly to the family income by direct labor (**Figure 24.2**). In many developing countries, women have few rights and little legal ability to protect their property, their rights to their children, their income, or anything else. Improving the status of women is a crucial aspect of sustainable development.

Raising the standard of living for poor countries will require that special attention be given to the poorest segments of all populations—that is, to those without much hope. In this context, the universal education of children and the reduction of illiteracy are of critical importance in raising and maintaining appropriate standards of living in every country.

Goods, services, and money flow throughout the world and greatly affect the nations among which they move. We have entered a new era of global trade, with a growing need for new guidelines for national, corporate, and individual behavior, even though our understanding of the dynamics of global systems is imperfect. For example, the flow of money from developing to highly developed countries has exceeded the flow in the other direction for many years. Former West German Chancellor **Willy Brandt** termed this phenomenon "a blood transfusion from the sick to the healthy."

Debt forgiveness for the poorest countries can help increase environmental sustainability and social justice. A well diversified, strong, locally centered economy is generally one of the most important paths that lead away from poverty. Wealthy nations can support healthy economic development through collaboration and development assistance, as well as through policies that support self-sufficiency in neighbor nations.

Stabilizing the Human Population   This overall situation is clearly unstable and must be corrected by the determination and adoption of acceptable levels of population and consumption for all regions.

The world population grew from 2.5 billion in 1950 to almost 7.2 billion in 2014. In 2012, the United Nations calculated that world population will reach approximately 9.6 billion people in 2050. Population growth rates are generally highest where poverty is most intense. Increasing literacy rates among women and increasing access to family planning services can both help reduce unwanted births (**Figure 24.3**). Family planning is important in

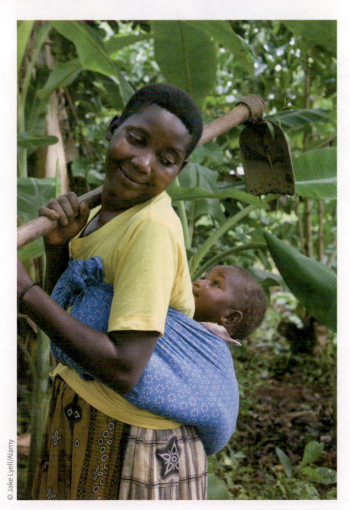

Figure 24.2   A young mother in Uganda, East Africa. In addition to caring for their children, women are the farmers in much of Africa.

Figure 24.3   Family planning in Agra, India. A health worker gives family planning and birth-control advice to two women.

both developing countries and in highly developed countries, which consume most of the world's resources even though they constitute a minority of the world's population.

The Links among Immigration, Poverty, and Population

Another important way in which we are linked to the developing world is the massive immigration of poor people from the tropics and subtropics to the industrialized nations of the temperate zone. U.S. Customs and Border Protection estimates that more than 1.1 million potential migrants are apprehended annually at or near U.S. borders, and many others successfully enter (see Chapter 8). The U.S. government and organizations such as the Pew Hispanic Center estimate that immigration now accounts for at least 50% of population growth in the United States.

The same pattern of migration to flee poverty can be seen worldwide. According to the Natural Resources and Environmental Institute in Cairo, in 2005 an estimated 30 million people were environmental refugees who left their homes in search of food, often crossing national borders. The number of such people seeking to enter the United States, Canada, and other highly developed countries is likely to increase greatly. This pattern is the direct result of mounting populations, political and ethnic tension, economic pressures, and environmental degradation in developing countries. Each of these immigrations represents a social or economic failure for an individual or family in another country.

RECOMMENDATION 2: PROTECT AND RESTORE EARTH'S RESOURCES

To build and maintain a sustainable society, renewable resources such as forests, biodiversity, soils, fresh water, and fisheries must be treated in ways that ensure their long-term productivity. Their capacity for renewal must be understood and respected, both because of the ecosystem services they offer and because of their direct role in human food supplies.

The World's Forests   The world's forests, which are being cut, burned, or seriously altered at a frightening rate, are being lost for two principal reasons. First, they are converted to cash. Like natural resources all over the world, forests are being exploited and sold. Until concepts of natural productivity and the central role of environmental sustainability become important aspects of economic calculations, susceptible natural resources will continue to be consumed unsustainably, driven by short-term economics. In many places, when the value of living forests become clear, reforestation and forest stewardship increase.

The second reason that the world's forests are being lost is the pressure created by rapid population growth and widespread poverty. In many developing countries, the forests have traditionally served as a "safety valve" for the poor, who, by consuming small tracts of forest on a one-time basis and moving on, find a source of food, shelter, and clothing for themselves and their families. When the global poor have support in reforestation and forest conservation efforts (**Figure 24.4**), the poverty cycle is mitigated, both through increasing economic resources and through increased availability of food, fuel, and other forest-based resources.

Mint Images - Frans Lanting/Getty Images

Figure 24.4   **Reforestation project in Montes Clares, Brazil.** Globally, many people and organizations are working to help reforestation efforts.

E N V I R O N E W S

Reforestation in Kenya

The East African country of Kenya is predominantly arid or semiarid. By the 1970s, most of its tropical dry forest had been destroyed, and the little that remained was disappearing rapidly. A young woman, **Wangari Maathai**, was concerned about the destruction of her country's natural resources. Dr. Maathai, who earned a PhD from the University of Nairobi, founded a grassroots, nongovernmental tree-planting organization called the Green Belt Movement (GBM). The GBM mobilized local groups of women to work for the continued improvement of Kenya's environment. Since 1977, these groups have planted about 30 million trees across eastern and southern Africa. The trees provide many ecosystem services, including cover that protects Kenya from encroaching desert. In 2002, Dr. Maathai was elected to Kenya's parliament and appointed an Assistant Minister for the Environment. In 2004, she was awarded the Nobel Peace Prize for her environmental work. The Nobel Committee chair selected Ms. Maathai because "peace depends on our ability to secure our environment."

Tropical rain forests—biologically the world's richest terrestrial areas—have been reduced to less than half their original area. We know little about replacing most tropical forests with productive agriculture and forestry. Short-term exploitation by logging or clearing (often by burning) frequently results in irreversible destruction of the tropical soil's potential productivity.

Clearing the forests and prairies of Eurasia and North America led to the establishment of productive farms, for the soil was rich. In contrast, clearing the forests of tropical Africa or Latin America often produces wastelands. The relative infertility of many tropical soils, their thin and easily disturbed surface layer of organic matter, and the high temperature and precipitation levels of tropical regions often combine to make the attainment of sustainable agriculture or forestry systems difficult.

Tropical forests are rapidly being cleared and destroyed not only because of the needs of the people who live in or near them but also because of the demands of the global economy. Many products—foods such as beef, bananas, coffee, and tea; medicines; and hardwoods—come to the industrialized world from the tropics. As timber is harvested or trees are cleared for other reasons, however, only a limited amount of replanting is taking place. Few nations have forestry plans, and there is almost no coordination of forestry policies.

Figure 24.5 **Endangered black rhinoceros in Africa.** This species has experienced a slight population increase but still remains critically endangered, largely because of the demand for rhino horns in traditional Asian medicine. Photographed in Zimbabwe.

Loss of Biodiversity

Over the next few decades, we can expect the rate of extinction to climb from dozens of species a day to hundreds of species a day, the great majority unknown to science. How big a loss is this? Unfortunately, we still have a limited knowledge about the world's **biological diversity**. An estimated

biological diversity The number and variety of Earth's organisms.

five-sixths of all kinds of organisms have not yet been recognized and described scientifically. Earth has but one living library; people have read few of its books and don't even have a complete catalog of the volumes it contains—but the library is being burned unread. The library analogy implies, though, that biodiversity is something to enjoy in our leisure time, rather than the fundamental life support system that it is.

Groups like the Nature Conservancy, which both protect vulnerable habitats and work with local peoples to help them understand the economic and health value of these intact habitats, are extremely important in protecting our remaining biodiversity (**Figure 24.5**). Pragmatically, we have a clear interest in protecting Earth's biological diversity and managing it sustainably. We obtain from living organisms all our food, most medicines, many building and clothing materials, biomass for energy, and numerous other products. In addition, communities of organisms and ecosystems provide an enormous array of **ecosystem services** without which we would not survive. These essential services include watersheds and soils, fertile agricultural lands, local clean air, climate, and global habitats for beneficial animals and plants. In view of these ecosystem services, the ultimate reason for caring for the community of life on Earth is a selfish one.

Economic development will succeed only if it is carried out in such a way as to maintain the sustainable productivity of the biosphere. The human population of 7.2 billion people uses an estimated 32% of annual land-based net primary productivity (see Chapter 3). We also use approximately 55% of accessible, renewable supplies of fresh water. This puts enormous and unprecedented pressures on the life-support systems that nature provides

and on the sustainability of Earth's renewable resources: forests, soils, fresh water, fisheries, and biological diversity. Preserving Earth's biological diversity makes the whole Earth system stronger, and the more variety that is preserved, the more interesting, productive, resilient, and beautiful the world is.

Biological diversity and human cultural diversity are intertwined: They are, in fact, two sides of the same coin. **Cultural diversity** is Earth's variety of human communities, each with its individual languages, traditions, and identities (**Figure 24.6**). Cultural diversity enriches the collective human experience. Unfortunately, cultural diversity sometimes promotes distrust—a feeling of "us" against "them." For that reason, the United Nations Educational, Scientific, and Cultural Organization (UNESCO) supports the protection of minorities, peace, and equity in the context of cultural diversity.

Sustainable Development and Protecting and Restoring Earth's Resources

Sustainable development can offer real improvements in the quality of human life. Earth's biological and physical systems are essential in achieving economic development. The economic development strategies most suitable for a given locality depend on the biological, physical, and human factors in operation there, and our management decisions will be more effective if we understand our ecosystems.

Many, if not most, environmental problems extend across boundaries. International alliances are necessary to preserve the organisms, lands, and waters on which our livelihoods depend. Throughout the world, communities are finding that sustainable activities preserve their local environments and improve their standards of living. Getting the best available information to rural communities is essential to helping the people who live there make the most appropriate choices for themselves. Once attitudes have shifted at the individual and community levels, then national strategies can be devised that safeguard the natural resources of each country for the future.

© Picture Contact BV/Alamy

Figure 24.6 **Human cultural diversity in the Amazon region of Brazil.** Humans in increasing numbers are encroaching on the traditional environments of native peoples such as this Xingu Indian boy, shown with his pet parrot.

RECOMMENDATION 3: PROVIDE ADEQUATE FOOD FOR ALL PEOPLE

A humanitarian crisis exists that rarely makes the evening news: Globally, 1 billion people lack access to the food needed for healthy, productive lives. This estimate, according to the UN Food and Agriculture Organization, includes a high percentage of children. Children are particularly susceptible to food deficiencies because their brains and bodies cannot develop properly without adequate nutrition. The World Health Organization (WHO) estimates that 182 million children under the age of five are seriously underweight for their age. Most malnourished people live in rural areas of the poorest developing countries. The link between poverty and food insecurity is inescapable.

food insecurity The condition in which people live with chronic hunger and malnutrition.

Reducing hunger also requires that we rethink where, how, and by whom food is produced. Farmers already use the land most appropriate for modern large-scale annual crops. Consequently, future agriculture will come from growing in different contexts. This includes planting crops or raising animals in ecosystems previously not tapped. Alternatively, farmers can plant on fertile lands in urban and suburban areas, which have historically been built up, paved over, or decoratively landscaped. These sort of shifts require new techniques, new policies, and different varieties of seeds and animals. Understanding how to cultivate and protect healthy soils in these new contexts will help us maintain the fertility of our agricultural lands. (**Figure 24.7**).

Replacing furrow irrigation, where farmers flow water down small trenches between crop rows, with *drip irrigation* can reduce water use by about 50%. Improving water efficiency is crucial, because in many places, lack of water is the overriding factor reducing agricultural productivity. Growth of more water-efficient crops, like pearl millet instead of rice, can also help increase productivity. These crops do not produce as high yields as rice or wheat, but they can be produced in places where rice and wheat could not thrive.

Supporting local agriculture so that crops are grown near where they are consumed conserves energy. By one estimate, most food eaten in the United States travels about 2400 km (1500 mi) from farm to plate, using massive quantities of fossil fuels in the process. Farmers markets, food cooperatives, and community supported agriculture (CSA) can all help connect people with locally produced food.

Sustainable agricultural systems can support improved dietary standards, such as the inclusion of high-quality protein in diets in developing countries. Several methods of animal protein production are particularly promising. India has dramatically increased its dairy production by feeding cows cornstalks, wheat straw, and even grass collected along roads. Using these plant materials, which would normally "go to waste," is far more efficient than feeding the cows high-quality grain. China's expanding use of *aquaculture* is another example of efficient protein production. The carp that are raised in Chinese aquaculture are very efficient at converting food into high-quality protein. In China, fish production by aquaculture now exceeds poultry production. Insect cultivation and cultivation of forest-dwelling animals such as iguanas can help increase protein production in areas where cattle are not ecologically appropriate.

Klaus Vedfelt/Getty Images

Figure 24.7 **Husband & wife harvesting carrots in their garden.**

CASE IN POINT

PENNSYLVANIA ASSOCIATION FOR SUSTAINABLE AGRICULTURE

Most organizations that support farmers do just that: help farmers get access to seeds, information, products, ideas, and techniques that will improve their ability to farm well. Rarely, though, do these organizations help connect farmers with their customers. Pennsylvania Association for Sustainable Agriculture (PASA) does just that, however. PASA simultaneously provides opportunities and information for farmers, while attracting other members, who eat but don't farm, through events like "Bike Fresh, Bike Local" in which rural routes are mapped and opened for bicyclists and farms opened to offer freshly prepared food for participants. Members of PASA are notified of events and farm produce available in their area.

Each winter, PASA holds a multi-day conference, with workshops open to producers and consumers alike. Consumers attend to learn more about food production, while producers attend to learn from each other and connect with consumers—their markets. Students attend to learn about agriculture.

PASA has helped expand an economic opportunity called *agritourism*, in which nonfarmers visit and participate in farm activities as part of a family outing or vacation. While this kind of opportunity has long existed in limited contexts, such as dude ranches, pumpkin patches, Christmas tree farms, and Amish communities, PASA has helped expand agritourism in Pennsylvania to a wider range of farmers and consumers both.

Farm holidays. Farmers and food consumers increasingly connect through agritourism. Simple housing and picturesque livestock help many city dwellers recharge while learning more about the sources of their food; their tourism dollars help support farmers. ◼

RECOMMENDATION 4: MITIGATE CLIMATE CHANGE

One of the most widely discussed consequences of the pressure we are exerting on the global environment is climate change caused by the **enhanced greenhouse effect**. Both highly developed and developing countries contribute to major increases in CO_2 in the atmosphere, as well as to the increasing amounts of methane, nitrous oxide, tropospheric ozone, and chlorofluorocarbons (CFCs). The most important greenhouse gas, CO_2, is produced primarily when we burn fossil fuels.

> **enhanced greenhouse effect** The additional warming produced by increased levels of gases that absorb infrared radiation.

Although Earth's climate has been relatively stable during the present interglacial period (the past 10,000 years), mounting evidence indicates that human activities are causing it to change. According to the Intergovernmental Panel on Climate Change, the average global temperature increased by almost 1°C during the 20th century; more than half of that warming occurred during the past 30 years. The scientific literature overwhelmingly indicates that Earth's climate will continue to change rapidly during the 21st century.

The outcomes of these likely changes, which were discussed in Chapter 20, are serious, because modern society has evolved and successfully adapted to conditions as they are. Keeping in mind that the change from the last ice age to the present was accompanied by an increase in global temperatures of just 5°C puts the consequences of the present change, the most rapid of the last 10,000 years, into perspective.

Many policy makers say that we should wait until scientific knowledge of climate change is complete. This reasoning is flawed because the Earth system is extremely complex, and there are limits to our ability to predict the consequences of climate change. We may never have a complete scientific understanding of Earth as a system.

For example, we often say that an increase in atmospheric CO_2 leads to climate warming. However, the increase in CO_2, like all human impacts, is not a simple cause-and-effect relationship but instead a cascade of interacting responses that ripple through the Earth system. Increasing atmospheric CO_2 also affects how plants grow, but different species react to increasing CO_2 in different ways; we can expect changes in plant community composition as the plants that are more competitive in the new conditions thrive and replace less competitive plants. We cannot begin to predict how these vegetation changes will affect humans or the rest of the biosphere.

Stabilizing the Climate Requires a Comprehensive Energy Plan   A comprehensive energy plan must be implemented for the entire world. This plan would involve phasing out fossil fuels in favor of renewable energy (such as solar, geothermal, and wind power), increasing energy conservation, and improving energy efficiency; there would also be limited roles for nuclear power and carbon sequestration (**Figure 24.8**). Many national and local governments as well as corporations and environmentally aware individuals are setting goals to cut carbon emissions from fossil fuels. Other nations, however, have not recognized the urgency of the global climate problem. We need a global consensus to address climate change, even though each country will have to implement

Anita Stizzoli/Getty Images

Energy sources

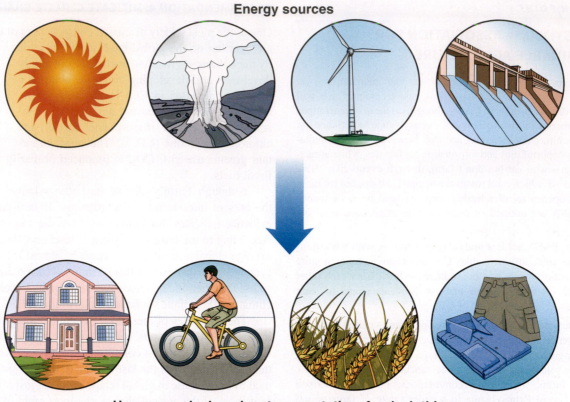

Human needs: housing, transportation, food, clothing

Figure 24.8   Comprehensive energy plan. In a comprehensive energy plan, locally appropriate energy sources such as solar, geothermal, wind, and hydropower are matched with basic human needs (housing, transportation, food, and clothing) in an inclusive process with adaptations made for local cultures.

a national energy policy that is custom-tailored to include its specific renewable energy resources.

According to the World Energy Council's 2013 survey, the 82% of the world's people who live in developing countries use about 40% of the world's commercial energy, the bulk of which is fossil fuel. The number of people in developing countries is growing rapidly, and even without any increase in their standard of living, they will use much larger quantities of energy in the future. Of special concern in this connection is China, an industrializing country of more than 1.3 billion people with the world's second-largest economy and huge deposits of coal. If we reach a global consensus on limiting CO_2 in the atmosphere, those of us who live in highly developed countries will find ourselves insisting that developing countries use clean, renewable energy sources as they industrialize. Such a strategy seems unlikely to prevail unless we all share in paying to reduce carbon emissions. Highly developed countries must realize that implementing a comprehensive energy plan for the developing world is a necessary ingredient for their own future security, as well as for global climate stability.

As countries around the world find their own ways to succeed economically, clean energy will be increasingly important. Clean energy economies are entirely possible, and energy production can be decentralized (using small solar panels or wind turbines, or locally grown biofuels) so that access to energy is not tied exclusively to access to a national electric grid.

RECOMMENDATION 5: DESIGN SUSTAINABLE CITIES

At the beginning of the Industrial Revolution, in approximately 1800, only 3% of the world's people lived in cities; 97% were rural, living on farms or in small towns. In the two centuries since then, population distribution has changed radically—toward the cities. More people live in Mexico City today than were living in all the cities of the world 200 years ago. This is a staggering difference in the way people live.

Just over half of the world's population now live in cities, and the percentage continues to grow. For example, although the population of Brazil has grown rapidly to its present 202 million people, the population in rural areas of Brazil is actually declining. In industrialized countries like the United States and Canada, almost 80% of the people live in cities.

CASE IN POINT

DURBAN, SOUTH AFRICA

As a member of the international group, Sustainable Cities Network, the South African city of Durban has made great steps and great plans toward greater sustainability (**Figure 24.9**). Community participation helps both recognize efforts and target where further effort is necessary. Residents of Durban have contributed to a map of green features, from transportation to parks, called eThekwini.

Steve Allan/Getty Images

Figure 24.9   Durban, South Africa. Planning, waste management, transportation, and food systems in Durban have made this city a model in moving toward sustainability. Note the green roof on the building in the foreground.

Open communication and planning also occur in a series of forums called Imagine Durban, which is modeled after a similar program in Calgary, Canada. Both of these efforts reflect how urban sustainability often involves similar types of solutions (transportation, access to nature, participatory urban planning) tailored to the culture and needs of an individual place, whether Canada or South Africa.

Food gardens have been planted in schools and throughout the community to increase food access. Two Durban schools, The Birches Pre-Primary, and Sithembile Junior Primary, have been recognized internationally for their success in improving nutrition for schoolchildren.

Waste is reduced through a cardboard recycling program. The program, Asiye eTafuleni's (AeT) Inner-City Cardboard Recycling Project, matches businesses generating used cardboard with small-scale recyclers who are able to use the material. This kind of matching between materials and need is a model for other resources and other places; often a material that is "waste" in one place is considered a resource for someone else.

Many homes in Durban have solar hot water, thanks to an initiative called Shisa Solar, which helped neighborhoods obtain reduced prices (a 10% discount) for bulk installation of solar hot water panels. This program was itself modeled after and adapted from a solar electric initiative in Portland, Oregon.

Although the particulars of sustainability in any city will suit the climate, terrain, ecosystems, and cultures of that city, themes and ideas can be shared among cities, allowing good ideas to spread globally.

How to Design a Sustainable City   City planners around the world are trying a variety of approaches to make cities more livable for their inhabitants. Many cities are developing urban transportation systems to reduce the use of cars and the problems associated with them, such as congested roads, large areas devoted to parking, and air pollution. Urban transportation can be as elaborate and expensive as mass transit subways and light rails or as inexpensive as bicycle and pedestrian pathways. Investing in urban transportation other than building more highways encourages commuters to use forms of transportation other than automobiles. To encourage

mass transit, some cities also tax people using highways to get into and out of cities during business hours.

When a city is built around people and bicycles instead of cars, urban residents have an improved quality of life. Parks and open spaces can be established instead of highways and parking lots. Also, commuters spend less time sitting in congested traffic. Air pollution, including climate-warming CO_2 emissions, is substantially reduced. Bicycle transportation also provides healthy exercise. To encourage bicycle use in the Netherlands, city streets have dedicated bike lanes, dedicated traffic signals, and plenty of parking for bicycles (**Figure 24.10**).

Some cities encourage their residents to tend small garden plots on vacant lots in their neighborhoods. Other residents grow small plots of vegetables in their yards or on rooftops. Small-scale farming in cities provides multiple benefits. Urban farms provide food for a city's inhabitants. Urban farmers in Singapore, for example, grow about one-fourth of the vegetables consumed by its residents. Gardening provides healthy exercise and, when multiple people garden on a vacant lot, promotes a sense of community with neighbors who might otherwise remain strangers. Also, there is something innately satisfying to the human spirit in tending plants and watching them grow and flourish.

Water scarcity is a major issue for many cities of the world. Some city planners think that innovative approaches must be adopted where water resources are scarce. These approaches would replace the traditional one-time water use that involves water purification before use, treatment of sewage and industrial wastes after use, and then discharge of the treated water. Some cities, such as Singapore, recycle some of their wastewater after it has been treated. Installing gray water systems, in which toilets are flushed using water from sinks or showers, can be an easy first step toward two-time water use.

In cities where growth has outstripped the ability to treat sewage, the *composting toilet* is a promising alternative to sewage treatment. The composting toilet converts human wastes and table scraps into compost without using water. It is relatively

© Ian Masterton/Alamy

Figure 24.10   Bicycles parked near a train station in central Amsterdam, the Netherlands. Each resident in the Netherlands rides a bicycle an average of 917 km (573 mi) per year. Currently, about 30% of trips in the Netherlands are made by bicycle.

inexpensive and has several advantages: Composting toilets reduce water use, reduce energy use (required to pump water and treat sewage), reduce the amount of sewage needing treatment (or reduce water pollution in cities without adequate water treatment), and produce compost, which can be added to soil as a conditioner.

Effectively dealing with the problems in urban squatter settlements is an urgent need. Evicting squatters does not address the underlying problem of abject poverty. Instead, cities should incorporate some sort of plan for the eventual improvement of squatter settlements. Providing basic services, such as clean water to drink, transportation (so people can leave their shelters to find gainful employment), and garbage pickup, would help improve the quality of life for the poorest of the poor.

REVIEW

1. What is the global extent of poverty?
2. What are two ecosystem services that natural resources such as forests and biological diversity provide?
3. What is the global extent of food insecurity?
4. How is stabilizing climate related to energy use?
5. What are two serious problems in cities in the developing world?

Changing Personal Attitudes and Practices

LEARNING OBJECTIVES

- Explain how unsustainable consumption threatens environmental sustainability.
- Define *sustainable consumption*.

sustainable consumption The use of goods and services that satisfy basic human needs and improve the quality of life but that minimize the use of resources so they are available for future generations.

Like sustainable development, **sustainable consumption** is a concept that forces us to address whether our actions undermine the environment's long-term ability to meet the needs of future generations. Factors that affect sustainable consumption include population, economic activities, technology choices, social values, and government policies.

In less developed countries, survival is often such an overwhelming concern that sustainable consumption may seem irrelevant. Sustainable consumption requires the eradication of poverty, which means that poor people must increase their consumption of certain essential resources. This increased consumption will not be sustainable unless the lifestyles and consumption patterns of prosperous people change. Some examples of promoting sustainable consumption include switching from motor vehicles to public transport and bicycles, and developing durable, repairable, recyclable products.

THE ROLE OF EDUCATION

If people are to adopt new lifestyles, they must be educated so that they understand the reasons for changing practices that may be highly ingrained or traditional. People are generally concerned about the environment, but their concerns do not naturally translate into action. Furthermore, most people believe that individual actions do not really make a difference, whereas, in fact, they are the only factors that do so, both individually and collectively. If people understand the way natural systems function, they can appreciate their own place in the natural world and can value sustainable actions. Formal and informal education are important both in bringing about change and in contributing to the sustainable management of resources (**Figure 24.11**). Accurate information must be made widely available; the media have an important part to play in such efforts. No national plan can be successful without a clear statement of its strategic goals, accompanied by an educational program that lays out the fundamental reasons why these goals are important and what individuals can do to help realize them.

In a democracy, many choices that require environmental knowledge are also made at local, national, and global levels. Should we ratify the Convention on Biological Diversity or remain one of a handful of countries that have not done so? Should we support international commitments to limit greenhouse gas emissions, like the Kyoto Treaty? Should we promote world trade—and, if so, what kind of trade? How important is it to reduce emissions from coal-fired plants? To answer these and similar questions, we can read and study. To help our fellow citizens understand these issues, we can do the following:

1. Establish environmental and sustainability curricula in primary and secondary schools and in colleges.
2. Encourage the activities of and collaborate with environmental organizations.
3. Support institutions such as natural history museums, zoos, aquaria, and botanical gardens, all of which promote conservation and sustainability.
4. Encourage the inclusion of relevant material in the programs of churches, social groups, and other institutions that logically should emphasize them.

© Frances Roberts/Alamy

Figure 24.11 **Environmental education program in the Bronx, New York City.** These second graders are learning about recycling and decomposition.

Most people are interested in the environment—their own local environment—in their own way. We must work to create a democratic society in which everyone will have the opportunity to learn and to contribute.

REVIEW

1. How does consumption affect environmental sustainability?
2. What is sustainable consumption?

What Kind of World Do We Want?

LEARNING OBJECTIVE

- Write an essay or blog post, build a web page, create a video, or otherwise describe the kind of world you want to leave for your children and your children's children.

Highly developed countries consume a disproportionate share of resources and must act forcefully to reduce their levels of consumption if we are all going to achieve sustainability. We in highly developed countries continue to strive to increase our high standard of living, from a level already considered utopian by most people on Earth. We have a responsibility to our children and our grandchildren and to children of other people worldwide, so these next generations will be able to live in anything like the comfort that we experience now.

In the last 50 years or so, we have come to recognize the nature of the impact of human activities on the biosphere. With that knowledge, we have found ways to address many environmental problems. We need to grow food, walk more, and use the sun and wind instead of fossil fuels to power our homes. We in the highly developed world need to learn to live with less energy, and to help people living with insufficient food, water, and housing to enjoy some of the benefits of our developed world without using the same amount of energy we do.

This change will require reconnecting with the natural environment. That means, at a personal level, taking opportunities to go outside (even if it is a city park or a backyard garden), to listen to the wind, and to look at the exquisite variety of plants, insects, and other life-forms with which we coexist. Humans evolved in nature: Our immensely multidimensional brains evolved as we interacted with growing things, weather patterns, and other animals. The world we have created now screens us from all that. The sophisticated devices we imagined and manufactured—things such as smart phones, touchscreens, laptop computers, and highly engineered automobiles—have come to define our world. Our challenge will be to use technology as a tool but not to let it define our interaction with the world.

The new environmental revolution will require that we revalue ourselves according to a different set of ideals. It will require reinventing economic constructs, such as building the cost of environmental impact and damage into our accounting systems, which will cause market forces to work in favor of environmental protection. Business activities must involve developing cooperative partnerships with people all over the world and making decisions based on long-term benefits to the environment.

This responsibility may seem daunting, but it is not overwhelming, and the rewards will be great. The choices we make now will have a greater impact on the future than those that any generation has made before. Even choosing to do nothing will have profound consequences for the future. At the same time, it is an incredible opportunity. **Margaret Mead** (1901–1978), the noted American anthropologist, once said, "Never doubt that a small group of thoughtful, committed people can change the world; indeed, it's the only thing that ever has!" This is a time in history when the best of human qualities—vision, courage, imagination, and concern—will play a critical role in establishing the nature of tomorrow's world.

REVIEW

1. How would you describe the world you want your children to live in? How does the future world you envision differ from the realities of today?

ENERGY AND CLIMATE CHANGE

Environmental Diplomacy

Political scientist **Richard N. L. Andrews** suggests that *environmental diplomacy* is an essential component of any effective climate policy. This idea represents a major change from policy decisions of the past, which have been made mainly at local or national levels. Diplomacy has played some role in managing ocean resources, combating the ozone hole, and limiting acid rain. But regarding climate change, the actions of all people affect all other people. And the stakes include who uses how much of which kinds of energy. The future of Earth's climate relies as much on effective international negotiation as it does on understanding climate science.

REVIEW OF LEARNING OBJECTIVES WITH SELECTED KEY TERMS

- **Define *sustainability*.**

Sustainability is the ability to meet humanity's current needs without compromising the ability of future generations to meet their needs. Currently, we are facing serious environmental problems, and we are not managing the world's resources sustainably.

- **Discuss how the natural environment is linked to sustainable development.**

Sustainable development is economic development that meets the needs of the present without compromising the ability of future generations to meet their own needs. Earth's productivity is limited. Sustainable development must allow for the maintenance of the life-support systems on which our lives and the lives of all other species depend.

- **Briefly describe the consumption habits of people in highly developed countries and relate consumption to carrying capacity.**

Consumption is the human use of materials and energy; generally speaking, people in highly developed countries are excessive consumers. **Carrying capacity (K)** is the maximum number of individuals of a given species that a particular environment can support for an indefinite period, assuming the environment doesn't change. The world does not contain enough resources to sustain more than 6 billion people at the level of consumption enjoyed in the United States.

- **Define *poverty* and briefly describe this global problem.**

Poverty is a condition in which people are unable to meet their basic needs for adequate food, clothing, shelter, education, or health. For many of the world's women and children, life is an endless struggle for survival to obtain the daily requirements of firewood, clean water, and food.

- **Discuss problems relating to loss of forests and declining biological diversity, including the important ecosystem services that these resources provide.**

The world's forests are being cut, burned, or seriously altered for timber and other products that the global economy requires. Also, rapid population growth and widespread poverty in many developing countries are putting pressure on their forests, as poor people try to eke out a living in them. **Biological diversity**, the number and variety of Earth's organisms, is declining at an alarming rate. Humans are part of Earth's web of life and are entirely dependent on that web, with all its interactions. Organisms and their ecosystems provide important **ecosystem services**.

- **Describe the extent of food insecurity and identify at least two ways to increase food production sustainably.**

Food insecurity is the condition in which people live with chronic hunger and malnutrition. Globally, about 1 billion people lack access to the food needed for healthy, productive lives. Replacing furrow irrigation, where farmers flow water down small trenches between crop rows, with drip irrigation can reduce water use; in many places, lack of water is the overriding factor reducing agricultural productivity. Urban agriculture on vacant lots and rooftops can help increase food access for people who have little access to farmland.

- **Define the *enhanced greenhouse effect* and explain how stabilizing climate is related to energy use.**

The **enhanced greenhouse effect** is the additional warming produced by increased levels of gases that absorb infrared radiation. Both highly developed and developing countries contribute to major increases in CO_2 in the atmosphere, as well as to the increasing amounts of the greenhouse gases methane, nitrous oxide, tropospheric ozone, and CFCs. An increase in atmospheric CO_2, produced mostly when fossil fuels are burned, leads to climate warming and associated climate changes. To stabilize climate, we must phase out fossil fuels and replace them with renewable energy, increased energy conservation, and improved energy efficiency.

- **Describe at least two challenges for cities and explain how they can be addressed.**

Many cities are developing urban transportation systems to reduce the use of cars and the problems associated with them. Illegal squatter settlements proliferate in cities; here the poorest inhabitants build dwellings on vacant land using whatever materials they can scavenge. Squatter settlements have the worst water, sewage, and solid waste problems. To improve the quality of life in squatter settlements, cities should develop plans to provide basic services, such as clean drinking water, transportation, and garbage pickup.

- **Explain how unsustainable consumption threatens environmental sustainability.**

Unsustainable consumption overuses natural resources, adds pollution, and accumulates waste. Also, unsustainable consumption tends to occur when many people's labor is undervalued and underpaid.

- **Define *sustainable consumption*.**

Sustainable consumption is the use of goods and services that satisfy basic human needs and improve the quality of life but that minimize the use of resources so that they are available for future generations.

Critical Thinking and Review Questions

1. What is environmental sustainability? How are people in highly developed countries not living sustainably? How are people in developing countries not living sustainably?

2. How are the natural environment and sustainable development linked?

3. What are the consumption habits of people in highly developed countries? How is consumption related to human carrying capacity?

4. What are the five recommendations for sustainable living discussed in this chapter?

5. How pervasive is poverty around the world?

6. What important ecosystem services does biological diversity provide?

7. How pervasive is food insecurity around the world?

8. Describe three serious problems associated with cities in the developing world.

9. Explain how unsustainable consumption threatens environmental sustainability.

10. What is sustainable consumption?

11. Discuss four specific environmental goals that you hope are achieved in your lifetime.

12. How is stabilizing climate related to fossil-fuel use?

13. At home, place an uncracked egg in a small glass bowl and cover it with vinegar. Allow it to sit in the vinegar for 24 hours.

Take the egg out and examine the shell. How does this simple experiment demonstrate a possible effect of increased atmospheric CO_2 on marine organisms with calcium carbonate shells?

14. Explain why the global increase in CO_2 is not a simple cause-and-effect relationship with climate warming but is instead a cascade of interacting responses that ripple through the Earth system.

Food for Thought

Research how your food options might be different if you lived in Durban, South Africa, or Nairobi, Kenya. How might your food choices be different, and what are some similarities between your food options and those elsewhere?

Glossary

abiotic Nonliving. Compare *biotic*. Compare hadal benthic zone.

acid A substance that releases hydrogen ions (H⁺) in water. Acids have a sour taste and turn blue litmus paper red. Compare *base*.

acid deposition Sulfur dioxide and nitrogen dioxide emissions that react with water vapor in the atmosphere to form acids that return to the surface as either dry or wet deposition. See *acid precipitation*.

acid mine drainage Pollution caused when sulfuric acid and toxic dissolved materials such as lead, arsenic, and cadmium wash from coal and metal mines into nearby lakes and streams.

acid precipitation Precipitation that is acidic as a result of both sulfur and nitrogen oxides forming acids when they react with water in the atmosphere; partially due to the combustion of coal; includes acid rain, acid snow, and acid fog.

acre-foot The amount of water needed to cover an acre of land 1 foot deep; equal to 1233 m³ (326,000 gallons), which is enough to supply eight people for one year.

active solar heating A system in which a series of collection devices mounted on a roof or in a field is used to absorb solar energy. Pumps or fans distribute the collected heat. Compare *passive solar heating*.

acute toxicity Adverse effects that occur within a short period after exposure to a toxicant. Compare *chronic toxicity*.

adaptation An evolutionary modification of an individual that improves that individual's chances of survival and reproductive success in its environment.

additivity A phenomenon in which two or more pollutants interact in such a way that their combined effects are what one would expect given the individual effects of each component of the mixture. Compare *antagonism* and *synergism*.

aerosols Tiny particles of natural and human-produced air pollution that are so small they remain suspended in the atmosphere for days or even weeks.

age structure The number and proportion of people at each age in a population.

agroecosystem An agricultural community composed of organisms that interact with one another and with their environment.

agroforestry Concurrent use of forestry and agricultural techniques to improve degraded soil and offer economic benefits.

A-horizon The topsoil; located just beneath the O-horizon of the soil. The A-horizon is rich in various kinds of decomposing organic matter.

air pollution Various chemicals (gases, liquids, or solids) present in high enough levels in the atmosphere to harm humans, other animals, plants, or materials. Excess noise and heat are also considered air pollution.

air toxic See hazardous air pollutant.

albedo The proportional reflectance of Earth's surface; glaciers and ice sheets have high albedos and reflect most of the sunlight hitting their surfaces, whereas the ocean and forests have low albedos.

alcohol fuel A liquid fuel such as methanol or ethanol that may be used in internal combustion engines; mixing gasoline with 10% ethanol produces a cleaner-burning mixture known as gasohol.

algae Unicellular or simple multicellular photosynthetic organisms; important producers in aquatic ecosystems.

alfisol Clay soil typical of hardword forest ecosystems.

alpine tundra A distinctive ecosystem located in the higher elevations of mountains, above the tree line; characteristic vegetation includes grasses, sedges, and small tufted plants. Compare *tundra*.

alternating current Electrical current that periodically changes the direction of electron flow. Contrast with *direct current*.

ammonification The conversion of nitrogencontaining organic compounds to ammonia (NH₃) and ammonium ions (NH₄⁺) by certain bacteria (ammonifying bacteria) in the soil; part of the nitrogen cycle.

antagonism A phenomenon in which two or more pollutants interact in such a way that their combined effects are less severe than the sum of their individual effects. Compare *additivity* and *synergism*.

antibiotics Medicines that prevent or combat bacterial, fungal, or viral diseases.

aquaculture The rearing of aquatic organisms (fishes, seaweeds, and shellfish), either freshwater or marine, for human consumption. See *mariculture*.

aquatic Pertaining to the water. Compare *terrestrial*.

aqueduct A large pipe or conduit constructed to carry water from a distant source.

aquifer The underground caverns and porous layers of underground rock in which groundwater is stored. See *confined aquifer* and *unconfined aquifer*.

aquifer depletion The removal by humans of more groundwater than can be recharged by precipitation or melting snow. Also called *groundwater depletion*.

archaea A kingdom of single-celled microorganisms lacking a cell nucleus, but sharing some enzymes with eukaryotes.

arctic tundra See *tundra*.

aridisol Calcium-rich soils typical of desert habitats' assisted colonization.

arid land A fragile ecosystem in which lack of precipitation limits plant growth. Arid lands are found in both temperate and tropical regions. Also called *desert*. Compare *semiarid land*.

artesian aquifer See *confined aquifer*.

artificial eutrophication Overnourishment of an aquatic ecosystem by nutrients such as nitrates and phosphates. In artificial eutrophication, the pace of eutrophication is rapidly accelerated due to human activities such as agriculture and discharge from sewage treatment plants. Also called *cultural eutrophication*. See *eutrophication*.

artificial insemination A technique in which sperm collected from a suitable male of a rare species is used to artificially impregnate a female (perhaps located in another zoo).

assimilation The incorporation of a substance into the cells of an organism.

asthenosphere The region of the mantle where rocks become hot and soft.

atmosphere The gaseous envelope surrounding Earth.

atoll A ring-shaped reef or island(s) formed of coral.

atom The smallest quantity of an element that retains the chemical properties of that element; composed of protons, neutrons, and electrons.

atomic mass A number that represents the sum of the number of protons and neutrons in the nucleus of an atom. The atomic mass represents the relative mass of an atom. Compare *atomic number*.

atomic number A number that represents the number of protons in the nucleus of an atom. Each element has its own characteristic atomic number. Compare *atomic mass*.

autotroph See *producer*.

baby boom The large wave of births that followed World War II, from 1945 to 1962.

background extinction The continuous, low-level extinction of species that has occurred throughout much of the history of life. Compare *mass extinction*.

bacteria Unicellular, prokaryotic microorganisms. Most bacteria are decomposers, but some are autotrophs and some are parasites.

bacterial source tracking Using molecular biological techniques to identify the source of dangerous bacteria in a stream or other body of water.

barrier reef A coral reef parallel to and separated from shore by deep water.

base A compound that releases hydroxide ions (OH⁻) when dissolved in water. A base turns red litmus paper blue. Compare *acid*.

benthos Bottom-dwelling marine organisms that fix themselves to one spot, burrow into the sand, or simply walk about on the ocean floor. Compare *plankton and nekton*.

B-horizon The light-colored, partially weathered soil layer underneath the A-horizon; subsoil. The B-horizon contains much less organic material than the A-horizon.

bioaccumulation The buildup of a persistent toxic substance, such as certain pesticides, in an organism's body, often in fatty tissues. Also called *bioconcentration.*

biocentric Focusing on all life-forms as equally important. Contrast *anthropocentric.*

biocentric preservationist A person who believes in protecting nature because all forms of life deserve respect and consideration.

biochemical oxygen demand (BOD) The amount of oxygen needed by microorganisms to decompose (by aerobic respiration) the organic material in a given volume of water. Also called *biological oxygen demand.*

bioconcentration See *bioaccumulation.*

biodegradable Referring to a chemical pollutant capable of being decomposed (broken down) by organisms or by other natural processes. Compare *nondegradable.*

biodiversity See *biological diversity.*

biodiversity hotspots Relatively small areas of land that contain an exceptional number of endemic species and are at high risk from human activities.

biofuels Any of a range of solid, liquid, and gaseous fuels that can be produced from biomass; includes biodiesel, ethyl alcohol, charcoal, and biogas. Compare *biomass (2).*

biogas A clean fuel, usually composed of a mixture of gases, whose combustion produces fewer pollutants than either coal or biomass. Biogas is produced from the anaerobic digestion of organic materials.

biogeochemical cycle Process by which matter cycles from the living world to the nonliving physical environment and back again. Examples of biogeochemical cycles include the carbon cycle, nitrogen cycle, phosphorus cycle, and sulfur cycle.

biological amplification See *biological magnification.*

biological control A method of pest control that involves the use of naturally occurring disease organisms, parasites, or predators to control pests.

biological diversity The number, variety, and variability of Earth's organisms; consists of three components: genetic diversity, species richness, and ecosystem diversity. Also called *biodiversity.*

biological magnification The increased concentration of toxic chemicals such as PCBs, heavy metals, and certain pesticides in the tissues of organisms at higher trophic levels in food webs. Also called *biological amplification.*

biological oxygen demand (BOD) See *biochemical oxygen demand.*

biomass (1) A quantitative estimate of the total mass, or amount, of living material. Often expressed as the dry weight of all the organic material that comprises organisms in a particular ecosystem. (2) Plant and animal materials used as fuel. Compare *biofuels.*

biome A large, relatively distinct terrestrial region characterized by similar climate, soil, plants, and animals, regardless of where it occurs on Earth; because it is so large in area, a biome encompasses many interacting ecosystems.

biomining A mining process in which microorganisms facilitate removal of minerals from ore.

bioprospecting Investigation of organisms in a forest, coral reef, or other natural area as potential sources of chemical products, including drugs, flavorings, fragrances, and natural pesticides.

bioremediation A method employed to clean up a hazardous waste site that uses microorganisms to break down the toxic pollutants. Compare *phytoremediation.*

biosphere The parts of Earth's atmosphere, ocean, land surface, and soil that contain all living organisms.

biotic Living. Compare *abiotic.*

biotic pollution Introduction of non-native species which can be disruptive to existing ecosystems. See *invasive species.*

birth rate The number of births per 1000 people per year. Also called *natality.*

bisphenol A (BPA) An organic compound that serves as a building block for many polymer products (including some plastic baby bottles and toys); thought to be an endocrine disruptor.

bitumen See *tar sand.*

bituminous coal The most common form of coal; produces a high amount of heat and is used extensively by electric power plants. Also called *soft coal.* Compare *anthracite, subbituminous coal,* and *lignite.*

bloom Large algal population growth caused by the sudden presence of large amounts of essential nutrients (such as nitrates and phosphates) in surface waters.

body mass index (BMI) A measure that relates an individual's body fat to his or her height as an estimate of physical fitness.

boreal forest A region of coniferous forest (such as pine, spruce, and fir) in the Northern Hemisphere; located just south of the tundra. Also called *taiga.*

botanical A plant-derived chemical used as a pesticide. See *synthetic botanical.*

bottom ash The residual ash left at the bottom of an incinerator when combustion is completed. Also called slag. Compare *fly ash.*

breeder nuclear fission A type of nuclear fission in which nonfissionable U-238 is converted to fissionable Pu-239.

broad-leaf herbicide An herbicide that kills plants with broad leaves but does not kill grasses (such as corn, wheat, and rice).

broad-spectrum pesticide A pesticide that kills a variety of organisms in addition to the pest against which it is used. Most pesticides are broad-spectrum pesticides. Also called *wide-spectrum pesticide.* Compare *narrow-spectrum pesticide.*

brownfield An urban area of abandoned, vacant factories, warehouses, and residential sites that may be contaminated from past uses.

bycatch The fishes, marine mammals, sea turtles, seabirds, and other animals caught unintentionally in a commercial fishing catch.

calorie A unit of heat (thermal energy); the amount of heat required to raise 1 g of water 1°C.

cancer A malignant tumor anywhere in the body. Cancers tend to spread throughout the body.

cancer potency An estimate of the expected increase in cancer associated with a unit increase in exposure to a chemical.

carbamates A class of broad-spectrum pesticides that are derived from carbamic acid.

carbohydrate An organic compound containing carbon, hydrogen, and oxygen in the ratio of 1C:2H:1O. Carbohydrates include sugars and starches, molecules metabolized readily by the body as a source of energy.

carbon capture and storage See *carbon sequestration.*

carbon cycle The worldwide circulation of carbon from the abiotic environment into organisms and back into the abiotic environment.

carbon footprint The amount of carbon (as CO_2) emitted by an activity or group of people; shows the demand placed on Earth's climate by burning fossil fuels.

carbon intensity In energy production, the amount of carbon dioxide released into the atmosphere per unit energy produced.

carbon management Ways to separate and capture the carbon dioxide produced during the combustion of fossil fuels and then sequester it away from the atmosphere.

carbon oxides Molecules consisting of a carbon atom and at least one oxygen atom.

carbon sequestration Placing carbon that has been produced when generating usable energy from fossil fuels into some sort of permanent storage. Also called *carbon capture and storage.*

carbon-silicate cycle Interaction between carbon and silicon cycles, on a geologic scale of millions of years.

carcinogen Any substance that causes cancer or accelerates its development.

carnivore An animal that feeds on other animals; flesh-eater. See *secondary consumer.* Compare *herbivore* and *omnivore.*

carrying capacity (K) The maximum number of individuals of a given species that a particular environment can support sustainably (long term), assuming there are no changes in that environment.

Cartagena Protocol on Biosafety The regulations resulting from the 1992 UN Convention on Biological Diversity that lessen the threat of gene transfer from genetically modified (GM) organisms to their wild relatives by providing appropriate procedures in the handling and use of GM organisms.

castings Bits of soil that have passed through the gut of an earthworm.

catalytic converter Emissions control device used immediately following combustion; oxidizes most unburned gases.

cell The basic structural and functional unit of life. Simple organisms are composed of single cells, whereas complex organisms are composed of many cells.

cellular respiration A process in which the energy of organic molecules is released within cells. Compare *aerobic respiration* and *anaerobic respiration.*

CFCs See *chlorofluorocarbons.*

chain reaction A reaction maintained because it forms as products the very materials that are used as reactants in the reaction. For example, a chain reaction occurs during nuclear fission when neutrons collide with other U-235 atoms and those atoms are split, releasing more neutrons, which then collide with additional U-235 atoms.

chaparral A biome with a mediterranean climate (mild, moist winters and hot, dry summers). Chaparral vegetation is characterized by small-leaved evergreen shrubs and small trees.

chemosynthesis The biological process by which certain bacteria take inorganic compounds from their environment and use them to obtain energy and make carbohydrate molecules; light is not required for this process. Compare *photosynthesis.*

chlorinated hydrocarbon A synthetic organic compound that contains chlorine and is used as a pesticide (for example, DDT) or an industrial compound (for example, PCBs).

chlorofluorocarbons Human-made organic compounds composed of carbon, chlorine, and fluorine that had many industrial and commercial applications but were banned because they attack the stratospheric ozone layer. Also called CFCs.

chlorophyll A green pigment that absorbs radiant energy for photosynthesis.

C-horizon The partly weathered layer in the soil located beneath the B-horizon. The C-horizon borders solid parent material.

chronic bronchitis A disease in which the air passages (bronchi) of the lungs become permanently inflamed, causing breathlessness and chronic coughing.

chronic toxicity Adverse effects that occur after a long period of exposure to a toxicant. Compare *acute toxicity.*

circumpolar vortex A mass of cold air that circulates around the southern polar region, in effect isolating it from the warmer air in the rest of the world.

citizen science Scientific research conducted by amateur or nonprofessional scientists (citizens), often through crowdsourcing.

clay The smallest inorganic soil particles. Compare *sand* and *silt.*

Clean Air Act (1970) Requires the U.S. EPA to establish national air quality standards for common pollutants, based on scientific data about health; implemented at state level.

Clean Water Act (1972) Requires U.S. EPA to implement pollution control for wastewater; strong legislation for point source pollution but less effective for nonpoint source pollution.

clear-cutting A forest management technique that involves the removal of all trees from an area at a single time. Compare *seed tree cutting, selective cutting,* and *shelterwood cutting.*

climate The typical patterns of weather that occur in a place over a period of years. Includes temperature and precipitation. Compare *weather.*

closed system A system that does not exchange energy with its surroundings. Compare *open system.*

coal A black, combustible solid composed mainly of carbon, water, and trace elements found in Earth's crust. Coal was formed from the remains of ancient plants that lived millions of years ago. See *fossil fuel.*

coal ash The residue left behind after coal is combusted.

coal gas A gas similar to natural gas that is produced by heating coal under anoxic (oxygen free) conditions.

coal gasification The technique of producing a synthetic gaseous fuel (such as methane) from solid coal.

coal liquefaction The process by which solid coal is used to produce a synthetic liquid fuel similar to oil.

coastal wetlands Marshes, bays, tidal flats, and swamps that are found along a coastline. See *mangrove forest, salt marsh,* and *wetlands.*

coevolution The interdependent evolution of two or more species that occurs as a result of their interactions over a long period. Flowering plants and their animal pollinators are an example of coevolution because each has profoundly affected the other's characteristics.

cogeneration An energy technology that involves recycling "waste" heat so that two useful forms of energy (electricity and either steam or hot water) are produced from the same fuel. Also called *combined heat and power.*

cohort A group of individuals who share a common experience, such as exposure to a chemical.

cold pasteurization A process in which foods are subjected to radiation to destroy pathogens.

Colorado River Compact (1922) Agreement among seven states bordering the Colorado River regarding water use and appropriation for agriculture and industry; allots water for upper and lower Colorado basins.

combined heat and power See *cogeneration.*

combined sewer overflow A problem that arises in a combined sewer system when too much water (from heavy rainfall or snowmelt) enters the system. The excess flows into nearby waterways without being treated.

combined sewer system A municipal sewage system in which human and industrial wastes are mixed with urban runoff from storm sewers before flowing into the sewage treatment plant.

combustion The process of burning by which organic molecules are rapidly oxidized, converting them into carbon dioxide and water with an accompanying release of heat and light.

command and control regulation Pollution-control laws that require specific technologies or limits. Compare *incentive-based regulation.*

commensalism A type of symbiosis in which one organism benefits and the other one is neither harmed nor helped. See *symbiosis.* Compare *mutualism* and *parasitism.*

commercial extinction Depletion of the population of a commercially important species to the point that it is unprofitable to harvest.

commercial harvest The collection of commercially important organisms from the wild. Examples include the commercial harvest of parrots (for the pet trade) and cacti (for houseplants).

commercial inorganic fertilizer See *fertilizer.*

common-pool resources Those resources of our environment that are available to everyone but for which no single individual has responsibility. Formerly called *global commons.*

community An association of different species living together at the same time in a defined habitat with some degree of mutual interdependence. Compare *ecosystem.*

community stability The ability of a community to withstand environmental disturbances.

compact development The design of cities so that tall, multiple-unit residential buildings are close to shopping and jobs, all of which are connected by public transportation.

comparative risk analysis Analysis of any of a range of federal, state, regional, and local projects designed to prioritize environmental regulation decisions by comparing or ranking risks along several dimensions, including impacts to human health and ecosystems.

competition The interaction among organisms that vie for the same resources in an ecosystem (such as food, living space, or other resources). See *interspecific competition* and *intraspecific competition.*

competitive exclusion The concept that no two species with identical living requirements can occupy the same ecological niche indefinitely. Eventually, one species will be excluded by the other as a result of interspecific competition for a resource in limited supply.

compost A natural soil and humus mixture that improves soil fertility and soil structure.

Comprehensive Environmental Response, Compensation, and Liability Act (CERCLA) (1980) U.S. law that created a tax on chemical and petroleum industries and authorized federal response to release of hazardous substances.

compressed air energy storage A technique for storing energy in which air is compressed, and can later be released through a turbine to produce electricity.

condenser Part of a steam turbine electricity generation system that condenses low-energy steam back to liquid water.

confined aquifer A groundwater storage area trapped between two impermeable layers of rock. Also called artesian aquifer. Compare *unconfined aquifer.*

conifer Any of a group of woody trees or shrubs (gymnosperms) that bear needle-like leaves and seeds in cones.

conservation The sensible and careful management of natural resources. Compare *preservation.*

conservation-based pricing Water supply pricing structure that rewards consumers for using less water; often comes in the form of low prices for water use up to some level, and involves stepped-up prices as use increases.

conservation biology A multidisciplinary science that focuses on the study of how humans impact organisms and on the development of ways to protect biological diversity; includes in situ and ex situ conservation.

conservation easement A legal agreement that protects privately owned forest or other property from development for a specified number of years.

Conservation Reserve Program U.S. program in which farmers keep environmentally sensitive areas out of cultivation in exchange for a yearly land rental payment.

conservation tillage A method of cultivation in which residues from previous crops are left in the soil, partially covering it and helping to hold it in place until the newly planted seeds are established. See *no-tillage.* Compare *conventional tillage.*

conservationist A person who supports the conservation of natural resources.

consumer An organism that cannot synthesize its own food from inorganic materials and therefore must use the bodies of other organisms as sources of energy and body-building materials. Also called *heterotroph*. Compare *producer*.

consumption The human use of materials and energy; generally speaking, people in highly developed countries are extravagant consumers, and their use of resources is greatly out of proportion to their numbers.

containment building A safety feature of nuclear power plants that provides an additional line of defense against any accidental leak of radiation.

continental shelf The submerged, relatively flat ocean bottom that surrounds continents. The continental shelf extends out into the ocean to the point where the ocean floor begins a steep descent.

contour plowing Plowing that matches the natural contour of the land, that is, the furrows run around a hill rather than in straight lines; lessens erosion. See *strip cropping*.

contraceptive A device or drug used to intentionally prevent pregnancy.

control group In a scientific experiment, the group of subjects for which the variable of interest is left unchanged. The control group provides a standard of comparison to verify the results of an experiment. Compare *experimental group*.

control rod A graphite rod used in the generation of nuclear power to absorb neutrons and thus reduce the rate of nuclear fission.

conventional tillage The traditional method of cultivation in which the soil is broken up by plowing before seeds are planted. Compare *conservation tillage*.

Convention on Biological Diversity (1992) A convention of the United Nations Environment Programme; provides a global legal framework for action on biodiversity.

Convention on International Trade in Endangered Species (CITES) (1973) International agreement to ensure that trade in wild flora and fauna does not threaten their survival.

convergent plate boundary Area where two lithospheric plates meet and one plate descends under the other in the process of subduction.

cooling tower Part of an electric power generating plant within which heated water is cooled.

coral reef Structure built from accumulated layers of calcium carbonate ($CaCO_3$); found in warm, shallow seawater. The living portion is composed principally of colonies of millions of tiny coral animals or of red coralline algae.

Coriolis effect The tendency of moving air or water to be deflected from its path to the right in the Northern Hemisphere and to the left in the Southern Hemisphere. Caused by the direction of Earth's rotation.

corporate average fuel economy (CAFE) A standard that determines that the average fuel economy (in miles per gallon) of passenger vehicles sold by each major automobile manufacturer in the United States.

corridor See *habitat corridor*.

cost-benefit analysis A mechanism that helps policymakers make decisions about environmental issues. Compares estimated costs of a particular action with potential benefits that would occur if that action was implemented.

cost-effectiveness analysis An economic tool that assesses and places a monetary value on the combined costs and benefits of a particular decision or activity.

Council on Environmental Quality U.S. agency that coordinates federal environmental efforts with the office of the President.

cover crop A crop planted with the purpose of protecting soil from erosion rather than for a harvestable product.

crop rotation The planting of different crops in the same field over a period of years. Crop rotation reduces mineral depletion of the soil because the mineral requirements of each crop vary.

crude oil See *petroleum*.

cryptic coloration Colors or markings that help certain organisms hide from predators by blending into their surroundings.

cullet Crushed glass food and beverage containers that glass manufacturers can melt and use to make new products.

cultural eutrophication See *artificial eutrophication*.

cultural diversity The range and number of human cultures and ethnic groups and practices in an area.

culture The ideas and customs of a group of people at a given period; culture, which is passed from generation to generation, evolves over time.

data (sing., *datum*) The information, or facts, with which science works and from which conclusions are inferred.

DDT Dichlorodiphenyltrichloroethane; a chlorine-containing organic compound that has insecticidal properties. Because it is slow to degrade and therefore persists in the environment and inside organisms, DDT has been banned in the United States and many other countries.

dead zone Low-oxygen water in a lake or ocean, caused by excessive nutrient pollution from human activities.

death rate The number of deaths per 1000 people per year. Also called *mortality*.

debt-for-nature swap The cancellation of part of a country's foreign debt in exchange for its agreement to protect certain lands (or other resources) from detrimental development.

decommission To dismantle an old nuclear power plant after it closes. Compare *entombment*.

decomposer A heterotroph that breaks down organic material and uses the decomposition products to supply itself with energy. Decomposers are organisms of decay. Also called *saprotroph*. Compare *detritivore*.

deductive reasoning Reasoning that operates from generalities to specifics and can make a relationship among data more apparent. Compare *inductive reasoning*.

deep ecology worldview An understanding of our place in the world based on harmony with nature, a spiritual respect for life, and the belief that humans and all other species have an equal worth.

deep well injection Disposal of liquid hazardous wastes in deep repositories located several thousand feet below Earth's surface.

deforestation The temporary or permanent clearance of large expanses of forests for agriculture or other uses.

degradation The natural or human-induced process that decreases the future ability of the land (or soil) to support crops or livestock. Also called *soil degradation*.

Delaney Clause 1958 law stating that no food can be sold with carcinogens, modified in recent years to reflect changing lab standards for carcinogen levels.

delta A deposit of sand or soil at the mouth of a river.

demand-side management A way that electric utilities can meet future power needs by helping consumers conserve energy and increase energy efficiency.

dematerialization The decrease in the size and weight of a product as a result of technological improvements that occur over time.

demographic transition The process whereby a country that is industrializing moves from relatively high birth and death rates to relatively low death rates, followed within a few generations by reduced birth rates.

demographics The application of demographic science that provides information on the populations of various countries or groups of people.

demography The science that deals with population structure and growth.

denitrification The conversion of nitrate (NO_3^-) to nitrogen gas (N_2) by certain bacteria (denitrifying bacteria) in the soil; part of the nitrogen cycle.

density The mass per unit volume of a substance.

density-dependent factor An environmental factor whose effects on a population change as population density changes; density-dependent factors tend to retard population growth as population density increases and enhance population growth as population density decreases.

density-independent factor An environmental factor that affects the size of a population but is not influenced by changes in population density.

derelict land Land area that was degraded by mining.

desalination See *desalinization*.

desalinization The removal of salt from ocean or brackish (somewhat salty) water. Also called *desalination*.

desert See *arid land*.

desertification Degradation of once-fertile rangeland, agricultural land, or tropical dry forest into nonproductive desert. Caused partly by soil erosion, deforestation, and overgrazing.

detritivore An organism (such as an earthworm or crab) that consumes fragments of dead organisms. Also called *detritus feeder*. Compare *decomposer*.

detritus Organic matter that includes dead organisms (such as animal carcasses and leaf litter) and wastes (such as feces).

detritus feeder See *detritivore*.

deuterium An isotope of hydrogen that contains one proton and one neutron per atom. Compare *tritium*.

developed country See *highly developed country*.

developing country A country that is not highly industrialized and is characterized by high fertility rates, high infant mortality rates, and low per capita incomes. Developing countries fall into two subcategories: moderately developed and less developed. Compare *highly developed country.*

dioxin Any of a family of mildly to extremely toxic chlorinated hydrocarbon compounds that are formed as byproducts in certain industrial processes.

direct current (DC) In electricity, a current in which electrons flow only in one direction. Contrast *alternating current.*

disease A departure from the body's normal healthy state as a result of infectious organisms, environmental stressors, or some inherent weakness.

disease-causing agents Viruses, bacteria, prions, or fungi which cause illness. See also *pathogens.*

dispersal The movement of individuals among populations, from one region or country to another. See *immigration* and *emigration.*

distillation A desalinization process in which water is evaporated and then recondensed to purify or separate it from other components of a complex mixture. Saltwater or brackish water may be distilled to remove the salt from the water. Compare *reverse osmosis.*

DNA Deoxyribonucleic acid. Present in a cell's chromosomes, DNA contains all of an organism's genetic information.

domestication The process of taming wild animals or adapting wild plants to serve humans; domestication markedly alters the characteristics of domesticated organisms.

dose In toxicology, the amount of a toxicant that enters the body of an exposed organism.

dose-response curve In toxicology, a graph or equation that shows the effect of different doses on a population of test organisms.

doubling time The time it takes for a population to double in size, assuming that its current rate of increase doesn't change.

drainage basin A land area that delivers water into a stream or river system. Also called *watershed.*

drift net A plastic net up to 64 km (40 mi) long that entangles thousands of fishes and other marine organisms.

drip irrigation See *microirrigation.*

Dust Bowl, American A semiarid region of the Great Plains that became desert-like as a result of human mismanagement, extended drought, and severe dust storms during the 1930s.

dust dome A dome of heated air that surrounds an urban area and contains a lot of air pollution. Compare *urban heat island.*

dynamic equilibrium The condition in which the rate of change in one direction is the same as the rate of change in the opposite direction.

ecological footprint The amount of land and ocean needed to supply an individual with food, energy, water, housing, transportation, and waste disposal.

ecological niche The relationship of a species to its ecosystem; typically describes food or sunlight needs, water, reproduction, and competitive ability.

ecological pyramid A graphic representation of the relative energy value at each trophic level. See *pyramid of biomass, pyramid of energy,* and *pyramid of numbers.*

ecological restoration See *restoration ecology.*

ecological risk assessment The process by which the ecological consequences of human activities are estimated. See *risk assessment.*

ecological succession See *succession.*

ecologically sustainable forest management Forest management that seeks not only to conserve forests for the commercial harvest of timber and nontimber forest products, but also to sustain biological diversity, prevent soil erosion, protect the soil, and preserve watersheds that produce clean water. Also called *sustainable forest management* or *sustainable forestry.*

ecology The study of systems that includes interrelationships among organisms and between organisms and their environment.

economic development An expansion in a government's economy, viewed by many as the best way to raise the standard of living.

economics The study of how people and entities (individuals, businesses, or countries) use their limited resources to fulfill their needs and wants. Economics encompasses the production, consumption, and distribution of goods.

ecosystem The interacting system that encompasses a community and its nonliving physical environment. In an ecosystem, all the biological, physical, and chemical components of an area form a complex interacting network of energy flow and materials cycling. Compare *community.*

ecosystem diversity Biological diversity that encompasses the variety among ecosystems, such as forests, grasslands, deserts, lakes, estuaries, and the ocean. Compare *genetic diversity* and *species richness.*

ecosystem management A conservation focus that emphasizes restoring and sustaining ecosystem quality rather than conserving of individual species.

ecosystem services Important environmental benefits that ecosystems provide to people, including clean air to breathe, clean water to drink, and fertile soil in which to grow crops.

ecotone The transitional zone where two ecosystems or biomes intergrade.

ecotourism A type of tourism in which tourists pay to observe wildlife in natural settings.

ecotoxicology The study of contaminants in the biosphere, including their harmful effects on ecosystems.

effective dose-50% (ED_{50}) In toxicology, the dose that causes 50% of a population to exhibit whatever biological response is under study.

efficiency An economics term used to describe getting the greatest amount of goods or services from a limited set of resources. Resources are allocated efficiently when two or more individuals or firms would be unwilling to further trade resources with each other.

E-horizon A heavily leached soil area that sometimes develops between the A- and B-horizons.

El Niño–Southern Oscillation (ENSO) A periodic, large-scale warming of surface waters of the tropical eastern Pacific Ocean that affects both ocean and atmospheric circulation patterns. Compare *La Niña.*

electricity The presence and flow of electrical charge, often as the flow of electrons through wires.

electrolysis The separation of water into hydrogen (H_2) and oxygen (O_2) through the application of electrical current.

electromagnetic spectrum The continuous range of wavelengths of electromagnetic energy; includes radio waves, microwaves, infrared waves, visible light, ultraviolet radiation, X-rays, and gamma rays.

electrostatic precipitator An air pollution–control device that gives ash a positive electrical charge so that it adheres to negatively charged plates.

embryo transfer A captive breeding technique in which eggs are produced in and removed from a female of a rare species, fertilized, and the embryos are transferred to a female from a related but less rare species, who later gives birth to offspring of the rare species.

Emergency Wetlands Resources Act (1986) Authorized the U.S. federal purchase of wetlands and the ability to establish entrance fees at National Wildlife Refuges.

emerging disease A disease that is new or rapidly increasing in incidence.

emigration A type of dispersal in which individuals leave a population and thus decrease its size. Compare *immigration.*

emission charge A government policy that controls pollution by charging the polluter for each given unit of emissions, that is, by establishing a tax on pollution.

emissions scenarios In climate modeling, assumptions about the amounts, rates, and mix of greenhouse gases in the future.

emphysema A disease in which the air sacs (alveoli) in the lungs become irreversibly distended, decreasing the efficiency of respiration and causing breathlessness and wheezy breathing.

endangered species A species whose numbers are so severely reduced that it is in imminent danger of becoming extinct in all or a significant part of its range. Compare *threatened species.*

Endangered Species Act (ESA) 1973 law that authorizes federal agencies to protect and support habitat for imperiled species; species can be listed as "endangered" (at risk of extinction throughout its range) or "threatened" (at risk of becoming endangered).

endemic disease A disease that is consistently present in a human population.

endemic species Localized, native species that are not found anywhere else in the world.

endocrine disrupter A chemical that interferes with the actions of the endocrine system (the body's hormones). Includes certain plastics such as polycarbonate; chlorine compounds such as PCBs and dioxin; the heavy metals lead and mercury; and some pesticides such as DDT, kepone, chlordane, and endosulfan.

energy The capacity or ability to do work.

energy conservation Saving energy by reducing energy use and waste. Carpooling to work or school is an example of energy conservation.

energy density The amount of energy contained within a fixed mass of an energy source. Gasoline has a higher energy density than does dry wood, which, in turn, has a higher energy density than wet wood.

energy efficiency A measure of the fraction of energy used relative to the total energy available in a given source. Efficiency ranges from 0 to 100%; use of natural gas for heating has an efficiency of close to 100%, while the efficiency of burning natural gas to generate electricity has a maximum efficiency of about 60% and is usually lower.

energy flow   The passage of energy in a one-way direction through an ecosystem.

energy intensity   A statistical estimate of energy efficiency, as, for example, a country's or region's total energy consumption divided by its gross national product.

energy services   The benefits we get from various sources of energy.

enhanced greenhouse effect   The increasing accumulation of heat in Earth's atmosphere as a consequence of anthropogenic greenhouse gas emissions.

enrichment   (1) The process by which uranium ore is refined after mining to increase the concentration of fissionable U-235. (2) The fertilization of a body of water, caused by the presence of high levels of plant and algal nutrients such as nitrogen and phosphorus; see eutrophication, which is a type of enrichment.

entombment   An option after the closing of an old nuclear power plant in which the entire power plant is permanently encased in concrete. Compare *decommission*.

entropy   A measure of the randomness or disorder of a system.

environment All the external conditions, both abiotic and biotic, that affect an organism or group of organisms.

environmental ethics   A field of applied ethics that considers the moral basis of environmental responsibility and how far this responsibility extends; environmental ethicists try to determine how we humans should relate to nature.

environmental impact statement (EIS)   A document that summarizes the potential and expected adverse impacts on the environment associated with a project, as well as alternatives to the proposed project. Typically mandated by law for public and/or private projects.

environmental justice   The right of every citizen, regardless of age, race, gender, social class, or other factor, to adequate protection from environmental hazards.

environmental movement   See *environmentalist*.

Environmental Performance Index   An index developed at Yale University as an indicator of a country's commitment to environmental and natural resource management. Compare to *national income account*.

Environmental Protection Agency   U.S. federal agency, created in 1970, to enforce environmental protection standards, conduct environmental research, provide support for combating pollution, and assist the Council on Environmental Quality in working with the President on environmental protection.

Environmental Protection Protocol to the Antarctic Treaty   (1998) International treaty prohibiting Antarctic mining.

environmental resistance   Limits set by the environment that prevent organisms from reproducing indefinitely at their intrinsic rate of increase; includes the limited availability exponential of food, water, shelter, and other essential resources, as well as limits imposed by disease and predation.

Environmental Risk Assessment (ERA)   A systematic procedure for predicting human or environmental health risks due to chemical exposure.

environmental science   The interdisciplinary study of how humanity interacts with other organisms and the nonliving physical environment.

environmental stressor   An environmental factor, whether natural or human-induced, that taxes an organism's ability to thrive.

environmental sustainability See *sustainability*.

environmental worldview   A worldview that helps us make sense of how the environment works, our place in the environment, and right and wrong environmental behaviors.

environmentalist   A person who works to solve environmental problems such as overpopulation; pollution of Earth's air, water, and soil; and depletion of natural resources. Environmentalists are collectively known as the *environmental movement*.

epicenter   The site at Earth's surface located directly above an earthquake's focus.

epidemiology   The study of the effects of toxic chemicals and diseases on human populations.

epiphyte   A small organism that grows on another organism but is not parasitic on it. Small plants that live attached to the bark of a tree's branches are epiphytes.

estuary   A coastal body of water that connects to the ocean, in which fresh water from a river mixes with salt water from the ocean.

ethanol   A colorless, flammable liquid, C_2H_5OH. Also called *ethyl alcohol*.

ethics   The branch of philosophy that deals with human values. See *environmental ethics*.

ethyl alcohol   See *ethanol*.

eukarya   Organisms whose cells contain a nucleus and other membrane-bound organelles.

euphotic zone   The upper reaches of the pelagic environment, from the surface to a maximum depth of 150 m in the clearest open-ocean water; sufficient light penetrates the euphotic zone to support photosynthesis.

eutrophic lake   A lake enriched with nutrients such as nitrates and phosphates and consequently overgrown with plants or algae; water in a eutrophic lake contains little dissolved oxygen. Compare oligotrophic lake.

eutrophication   The enrichment of a lake, estuary, or slow-flowing stream by nutrients that cause increased photosynthetic productivity. Eutrophication that occurs naturally is a slow process in which the body of water gradually fills in and converts to a marsh, eventually disappearing. See *enrichment (2)* and *artificial eutrophication*.

evaporation   The conversion of water from a liquid to a vapor. Also called *vaporization*.

evolution   The cumulative genetic changes in populations that occur during successive generations.

Evolution explains the origin of all the organisms that exist today or have ever existed.

ex situ conservation   Conservation efforts that involve conserving biological diversity in human-controlled settings. See *conservation biology*. Compare *in situ conservation*.

experimental group   In a scientific experiment, the group of subjects for which the variable of interest is altered in a known way. Compare *control group*.

exponential population growth   The accelerating population growth that occurs when optimal conditions allow a constant rate of increase over time. When the increase in population number versus time is plotted on a graph, growth produces a characteristic J-shaped curve.

externality   In economics, the effect (usually negative) of a firm that does not have to pay all the costs associated with its production.

extinction   The elimination of a species from Earth; occurs when the last individual member of a species dies.

extrapolation   In toxicology or epidemiology, estimating the expected effects at some dose of interest from the effects at known doses.

fall turnover   A mixing of the lake waters in temperate lakes, caused by falling temperatures in autumn. Compare spring turnover.

family planning   Providing the services, including information about birth-control methods, to help people have the number of children they want.

famine   Widespread starvation caused by a drastic shortage of food. Famine is caused by crop failures that are brought on by drought, war, flood, or some other catastrophic event.

Farm Bill   Agricultural and food policy bill, passed every five years. Funds a wide array of programs from crop subsidies to school lunches.

fault   A fracture in the crust along which rock moves forward and backward, up and down, or from side to side. Fault zones are often found at plate boundaries.

fecal coliform test   A water quality test for the presence of *E. coli* (fecal bacteria common in the intestinal tracts of people and animals). The presence of fecal bacteria in a water supply indicates a chance that pathogenic organisms may be present as well.

Federal Insecticide, Fungicide, and Rodenticide Act (FIFRA)   Controls pesticide distribution, sale, and use of U.S. pesticides, through licensing by the EPA.

Federal Land Policy and Management Act   (1976) U.S. law that governs the management of public lands by the Bureau of Land Management.

fertilizer   A material containing plant nutrients that is put on the soil to enhance plant growth. Fertilizers may be organic (animal manure, crop residues, bone meal, and compost) or inorganic (manufactured from inorganic chemical compounds).

first-generation pesticides   Pre-1940 pesticides, often with simple elemental ingredients such as copper, lead, and arsenic, but also including early organic compounds, generally highly toxic.

first law of thermodynamics   Energy cannot be created or destroyed, although it can be transformed from one form into another. See *thermodynamics.* Compare *second law of thermodynamics.*

fission A nuclear reaction in which certain elements are split into smaller atoms and subatomic particles, along with the release of a large amount of energy. Also called *nuclear fission.* Compare *fusion.*

flagship species In conservation biology, a high-profile species used to promote biodiversity conservation for a particular area or resource.

flexible fuel vehicles Vehicles that can be operated on more than one fuel, such gasoline and ethanol.

floodplain The area bordering a river that is subject to flooding.

flowing-water ecosystem A freshwater ecosystem such as a river or stream in which the water flows. Compare *standing-water ecosystem.*

fluidized-bed combustion A clean-coal technology in which crushed coal is mixed with particles of limestone in a strong air current during combustion; the limestone neutralizes the acidic sulfur compounds produced during combustion.

fly ash The ash from the flue (chimney) that is trapped by electrostatic precipitators. Compare *bottom ash.*

flyway An established route that ducks, geese, and shorebirds follow during their annual migrations.

focus The site, often far below Earth's surface, where an earthquake begins.

food access The level of economic, physical, and social access people have to healthy, fresh food options.

food chain The successive series of organisms through which energy flows in an ecosystem. Each organism in the series eats or decomposes the preceding organism in the chain. Compare *food web.*

food desert A neighborhood or area where opportunities to purchase fresh food are limited by transportation, safety, or availability.

Food, Drug, and Cosmetics Act (FDCA) (1938) U.S. law defines products identified as food and regulates ingredients, testing, and liability for food, drug and cosmetic products.

food insecurity The condition in which people live with chronic hunger and malnutrition.

food irradiation Exposing food products to ionizing radiation with the goal of increasing shelf life.

Food Quality Protection Act (FQPA) (1996) Revised FIFRA and FDCA to reduce pesticide exposure limits for infants and children.

food security The goal of all people having access to sufficient, safe, nutritious food at all times; the condition in which people do not live in hunger or fear of starvation.

Food Security Act See *Farm Bill.*

food sovereignty The right of people to access culturally appropriate, healthy food produced through sustainable, locally suitable agriculture.

food web A complex interconnection of all the food chains in an ecosystem. Compare *food chain.*

forest decline A gradual deterioration (and often death) of many trees in a forest; may be the result of a combination of environmental stressors, such as acid precipitation, toxic heavy metals in the soil, and surface-level ozone.

Forest Legacy Program (FLP) U.S. program that supports state efforts to protect forest lands. A

voluntary program, FLP facilitates conservation easements, restricting development and promoting sustainable forestry.

forest management See *ecologically sustainable forest management.*

fossil fuel Combustible deposits in Earth's crust, composed of the remnants (fossils) of prehistoric organisms that existed millions of years ago. Coal, oil (petroleum), and natural gas are the three types of fossil fuel.

fractional risk attribution A method for assigning the responsibility for harm to multiple causes, where each cause is assigned a fraction or percent of the responsibility based on its relative contribution to the harm.

fragmentation See *habitat fragmentation.*

freshwater wetland Land that is usually covered by shallow fresh water for at least part of the year and that has a characteristic soil and water-tolerant vegetation; includes marsh and swamp.

fringing reef A coral reef which grows directly from the shoreline; the most common reef found in the Caribbean and the Red Sea.

frontier attitude The attitude of most Americans during the 1700s and early 1800s that, because the natural resources of North America were seemingly inexhaustible, there was no reason not to conquer and exploit nature as much and as quickly as possible.

fuel assemblies Collections of fuel rods in a nuclear reactor.

fuel cell A device that directly converts chemical energy into electricity without the intermediate step of needing to produce steam and use a turbine and generator; the fuel cell requires hydrogen from a tank or other source and oxygen from the air.

fuel rods Closed tubes containing uranium dioxide pellets.

full-cost accounting The process of evaluating and presenting to decision makers the relative benefits and costs of various alternatives.

fundamental niche The potential ecological niche that an organism could have if there was no competition from other species. See *niche.* Compare *realized niche.*

fungicide A toxic chemical that kills fungi.

fusion A nuclear reaction in which two smaller atoms are combined to make one larger atom with the release of a large amount of energy. Also called *nuclear fusion.* Compare *fission.*

gas hydrates See *methane hydrates.*

gender inequality The social construct that results in women not having the same rights, opportunities, or privileges as men.

gender parity The level of equal access to education among males and females.

gene A segment of DNA that serves as a unit of hereditary information.

General Mining Law of 1872 Governs the transfer of rights to mine precious metals and other hardrock minerals from federal lands.

General Revision Act (1891) U.S. act authorized Congress to reserve remaining public forest lands, effectively creating federal forest reserves.

generator In electricity, a device that transforms mechanical energy into electricity.

genetic diversity Biological diversity that encompasses the genetic variety among

individuals within a single species. Compare *ecosystem diversity* and *species richness.*

genetic engineering The manipulation of genes, for example, by taking a specific gene from a cell of one species and placing it into a cell of an unrelated species, where it is expressed.

genetic resistance An inherited characteristic that decreases the effect of a pesticide on a pest. Over time, the repeated exposure of a pest population to a pesticide causes an increase in the number of individuals that can tolerate the pesticide.

genetically modified (GM) organism An organism that has had its genes intentionally manipulated.

gentrification A shift in an urban community toward more affluent residents and businesses; accompanied by increased property values and possibly increased property taxes.

geoengineering Controversial techniques proposed to mitigate climate change; examples include seeding the ocean with iron to allow more carbon to be taken up by algae and ejecting sulfur particles into the atmosphere to reflect sunlight.

geographic information system (GIS) Computer storage and analysis of maps and other geographic information; one use of GIS is low-energy precision application of irrigation water.

geothermal energy The natural heat within Earth that arises from ancient heat within Earth's core, from friction where continental plates slide over one another, and from decay of radioactive elements; geothermal energy can be used for space heating and to generate electricity.

geothermal heat pump (GHP) A device that heats and cools buildings by taking advantage of the difference in temperature between Earth's surface and subsurface at depths from about 1 to 100 m.

germplasm Any plant or animal material that may be used in breeding; includes seeds, plants, and plant tissues of traditional crop varieties and the sperm and eggs of traditional livestock breeds.

gigajoules A billion joules, equivalent to about 278 kilowatt hours.

global commons See *common-pool resources.*

global distillation effect The process whereby volatile chemicals evaporate from land as far away as the tropics and are carried by air currents to higher latitudes, where they condense and fall to the ground.

globalization The process by which people around the world are increasingly linked through economics, communication, transportation, governance, and culture.

grain stockpiles See *world grain stocks.*

grass herbicide an herbicide which is primarily effective in killing grass species.

greywater Water that has already been used for a relatively nonpolluting purpose, such as showers, dishwashing, and laundry; gray water is not potable but can be reused for toilets, watering plants, or washing cars.

green architecture The practice of designing and building homes with environmental considerations such as energy efficiency, recycling, and conservation of natural resources in mind.

green chemistry A subdiscipline of chemistry in which commercially important chemical processes are redesigned to significantly reduce environmental harm.

green power   Electricity produced from renewable sources such as solar power, wind, biomass, geothermal, and small hydroelectric plants.

green revolution   The period during the 20th century when plant scientists developed genetically uniform, high-yielding varieties of important food crops such as rice and wheat.

greenhouse effect   The increase of heat in a system where energy enters (often as light), is absorbed as heat, and released some time later; because the heat has a residence time within the system, the overall temperature of the system will be higher than its surroundings. See *enhanced greenhouse effect*.

greenhouse gas   A gas that absorbs infrared radiation; examples include carbon dioxide, methane, nitrous oxide, chlorofluorocarbons, and tropospheric ozone, all of which are accumulating in the atmosphere as a result of human activities, thereby increasing Earth's temperature.

gross primary productivity   The rate at which energy accumulates in an ecosystem (as biomass) during photosynthesis. Compare *net primary productivity*.

groundwater   The supply of fresh water under Earth's surface. Groundwater is stored in underground caverns and porous layers of underground rock called aquifers. Compare *surface water*.

groundwater depletion   See *aquifer depletion*.

growth rate (*r*)   The rate of change of a population's size, expressed in percent per year. In populations with little or no dispersal, it is calculated by subtracting the death rate from the birth rate. Also called *natural increase in human populations*.

gyre   A circular, prevailing wind that generates circular ocean currents.

habitat   The local environment in which an organism, population, or species lives.

habitat conservation plan (HCP)   Under the U.S. Endangered Species Act, a planning document required when human activity is likely to result in harm to the habitat of an endangered or threatened species.

habitat corridor   A protected zone that connects unlogged or undeveloped areas; wildlife corridors are thought to provide escape routes and allow animals to disperse so they can interbreed. Also called *wildlife corridor*.

habitat fragmentation   The division of habitats that formerly occupied large, unbroken areas into smaller, isolated areas by roads, fields, cities, and other land-transforming activities. Also called *fragmentation*.

hadal benthic zone   The ocean's benthic environment that extends from 6000 meters to the bottom of the deepest ocean trenches. Compare *abyssal benthic zone*.

half-life   See *radioactive half-life*.

hard coal   See *anthracite*.

hazard   A condition that has the potential to cause harm; differs from risk in not having an associated probability.

hazardous air pollutant   A potentially harmful air pollutant that may pose long-term health risks to people who live and work around chemical factories, incinerators, or other facilities that produce or use it. Also called *air toxic*.

hazardous waste   Any discarded material (solid, liquid, or gas) that threatens human health or the environment; may be flammable, chemically reactive, corrosive, and/or toxic.

heat   The kinetic energy of randomly moving atoms, ions, or molecules that spontaneously flows by virtue of a difference in temperature—that is, from a hot object to a cold object; measured in joules or calories.

herbicide   A toxic chemical that kills plants.

herbivore   An animal that feeds on plants or algae. See *primary consumer*. Compare *carnivore* and *omnivore*.

heterotroph   See *consumer*.

high-grade ore   An ore that contains relatively large amounts of a particular mineral. Compare *low-grade ore*.

high-input agriculture   See *industrialized agriculture*.

high-level radioactive waste   Any radioactive solid, liquid, or gas that initially gives off large amounts of ionizing radiation. Compare *low-level radioactive waste*.

highly developed country   An industrialized country that is characterized by a low fertility rate, low infant mortality rate, and high per capita income. Also called *developed country*. Compare *developing country*.

horizons, soil   See *soil horizons*.

hormesis   The favorable effects at low doses of physical, chemical, or biological agents that have adverse effects at higher dose levels.

hormone   A chemical messenger produced by a living organism in minute quantities to regulate its growth, reproduction, and other important biological functions. Also see *endocrine disrupter*.

hotspot (1)   A rising plume of magma that flows from deep within Earth's rocky mantle through an opening in the crust. (2) An area of great species diversity that is at risk of destruction by human activities.

Hubbard Brook Experimental Forest   An area of land in the White Mountains of New Hampshire, site of a Long Term Ecological Research project on the effects of acid rain on forest, stream, and soil health.

Hubbert's peak   See *Peak Oil*.

humus   Black or dark-brown decomposed organic material.

hydrocarbons   A diverse group of organic compounds that contain only hydrogen and carbon.

hydrogen bond   Bond between two water molecules, formed by the attraction between a negative (oxygen) end of one molecule and positive (hydrogen) end of another molecule.

hydrogen fuel cell vehicle (HFCV)   A vehicle that is impelled by a hydrogen fuel cell.

hydrologic cycle   The water cycle, which includes evaporation, precipitation, and flow to the seas. The hydrologic cycle supplies terrestrial organisms with a continual supply of fresh water. Also called water cycle.

hydrology   The science that deals with Earth's waters, including the availability and distribution of fresh water.

hydropower   A form of renewable energy that relies on flowing or falling water to generate mechanical energy or electricity.

hydrosphere   Earth's supply of water (both liquid and frozen, fresh and salty).

hydrothermal process   Any process involving water heated deep in Earth's crust.

hydrothermal reservoir   Large underground reservoir of hot water and possibly also steam; some of the hot water or steam may escape to the surface, creating hot springs or geysers.

hydrothermal vent   A hot spring on the seafloor where a solution of hot, mineral-rich water rises to the surface. Many hydrothermal vents support thriving communities.

hypertension   High blood pressure.

hypothesis   An educated guess that might be true and that is testable by observation and experimentation. Compare *theory*.

hypoxia   Low dissolved oxygen concentrations that occur in many bodies of water when nutrients stimulate the growth of algae that subsequently die and are decomposed by oxygen-using bacteria. Hypoxia often causes too little oxygen for other aquatic life.

igneous rocks   Rock formed by volcanic activity.

illuviation   The deposition of material in the lower layers of soil from the upper layers; caused by leaching.

immigration   A type of dispersal in which individuals enter a population and thus increase its size. Compare *emigration*.

in situ conservation   Conservation efforts that concentrate on preserving biological diversity in the wild. See *conservation biology*. Compare *ex situ conservation*.

incentive-based regulation   Pollution-control laws that work by establishing emission targets and providing industries with incentives to reduce emissions. Compare *command* and *control regulation*.

indicator species   An organism that provides an early warning of environmental damage. Examples include lichens, which are sensitive to air pollution, and amphibians, which are sensitive to pesticides and other environmental contaminants.

inductive reasoning   Reasoning that uses specific examples to draw a general conclusion or discover a general principle. Compare *deductive reasoning*.

industrial ecology   The study of energy and resource flows to, through, and from industrial systems.

industrial ecosystem   A complex web of interactions among various industries, in which "wastes" produced by one industrial process are sold to other companies as raw materials. Finding profitable uses for a company's wastes is an extension of sustainable manufacturing. See *sustainable manufacturing*.

industrial smog   The traditional, London-type smoke pollution, which consists principally of sulfur oxides and particulate matter. Compare *photochemical smog*.

industrial stage   Third stage in the demographic transition, characterized by a decline in birth rate; takes place during a country's or region's industrialization process.

industrialized agriculture   Modern agricultural methods that use large inputs of energy (from fossil fuels), mechanization, water, and agrochemicals (fertilizers and pesticides) to produce large crop and livestock yields. Also called *high-input agriculture.* Compare *subsistence agriculture.*

infant mortality rate   The number of infant deaths per 1000 live births. (An infant is a child in its first year of life.)

infectious disease   A disease caused by a microorganism (such as a bacterium or fungus) or infectious agent (such as a virus). Infectious diseases can be transmitted from one individual to another.

infrared radiation   Electromagnetic radiation with a wavelength longer than that of visible light but shorter than that of radio waves. Most of the energy absorbed by Earth is re-radiated as infrared radiation, which can be absorbed by greenhouse gases. Humans perceive infrared radiation as invisible waves of heat.

inherent safety   See *principle of inherent safety.*

inorganic chemical   A chemical that does not contain carbon and is not associated with life. Inorganic chemicals that are pollutants include mercury compounds, road salt, and acid drainage from mines.

inorganic plant nutrient   A nutrient such as phosphate or nitrate that stimulates plant or algal growth. Excessive amounts of inorganic plant nutrients (from animal wastes, plant residues, and fertilizer runoff) can cause soil and water pollution.

insecticide   A toxic chemical that kills insects.

integrated pest management (IPM)   A combination of pest control methods (biological, chemical, and cultivation) that, if used in the proper order and at the proper times, keep the size of a pest population low enough that it does not cause substantial economic loss.

integrated waste management   A combination of the best waste management techniques into a consolidated, systems-based program to deal effectively with solid waste.

intercropping   A form of intensive subsistence agriculture that involves growing several crops simultaneously on the same field. See *polyculture.*

interspecific competition   Competition between members of different species. See *competition.* Compare *intraspecific competition.*

intertidal zone   The shoreline area between the low tide mark and the high tide mark.

intraspecific competition   Competition among members of the same species. See *competition.* Compare *interspecific competition.*

intrinsic rate of increase   The exponential growth of a population that occurs under constant conditions. Sometimes called *biotic potential.*

invasive species Foreign species whose introduction causes economic or environmental harm.

ionizing radiation   Radiation that contains enough energy to eject electrons from atoms, forming positively charged ions. Ionizing radiation can damage living tissue.

IPAT **equation**   A model that shows the mathematical relationship between environmental impacts and the forces that drive them (number of people, affluence per person, and environmental effects of technologies used to obtain and consume resources).

isotope   An alternate form of the same element that has a different atomic mass; an isotope has a different number of neutrons but the same number of protons and electrons.

Jameel Poverty Action Lab (J-PAL) Kinetic energy storage any technique that stores energy by putting or keeping an object in motion, usually in order to later transform that energy into electricity.

K **selection**   A reproductive strategy in which a species typically has a large body size, slow development, long life span, and does not devote a large proportion of its metabolic energy to the production of offspring. Compare *r selection.*

kelps   Large brown algae that are common in relatively shallow, cooler temperate marine water along rocky coastlines.

kerogen   See *oil shales.*

keystone species   A species that is crucial in determining the nature and structure of the entire ecosystem in which it lives; other species of a community depend on or are greatly affected by the keystone species, whose influence is much greater than would be expected by its relative abundance.

kilocalorie   A unit of heat equivalent to the energy required to raise the temperature of 1 kilogram of water by 1°C.

kilojoule   A unit of energy; 1 kilocalorie equals 4.184 kilojoules.

kinetic energy   The energy of a body that results from its motion. Compare *potential energy.*

krill   Tiny shrimplike animals that are important in the Antarctic food web.

Kyoto Protocol   An international treaty stipulating that highly developed countries must cut their emissions of CO_2 and other gases that cause climate change by an average of 5.2% by 2012.

La Niña   A periodic occurrence in the eastern Pacific Ocean in which the surface water temperature becomes unusually cool, and westbound trade winds become unusually strong; often occurs after an El Niño event. Compare El Niño–Southern Oscillation.

land degradation   The natural or human-induced process that decreases the future ability of the land to support crops or livestock. See also *degradation.*

land use planning   The process of deciding the best uses for land in a given area.

land tenure   The explicit and functional property rights of those using land in a society.

landscape   A region that includes several interacting ecosystems.

landscape ecology   A subdiscipline in ecology that focuses on connections among ecosystems in a particular area.

landslide   A sliding mass of earth or rock from a steep slope.

latitude   The distance, measured in degrees north or south, from the equator.

lava   Magma (molten rock) that reaches Earth's surface.

leachate   The liquid that seeps through solid waste at a sanitary landfill or other waste disposal site.

leaching   The process by which dissolved materials (nutrients or contaminants) are washed away or filtered down through the various layers of the soil.

less-developed country   A developing country with a low level of industrialization, a high fertility rate, a high infant mortality rate, and a low per capita income (relative to highly developed countries). Compare moderately developed country and highly developed country.

lethal dose-50% (LD₅₀)   In toxicology, the dose lethal to 50% of a population of test animals.

life expectancy The amount of time that the typical person in a population or cohort is expected to live.

life table   A table of statistics on life expectancy and mortality for a population.

lignite   A low-grade brown or brown-black coal that has a soft, woody texture (softer than subbituminous coal). Compare *anthracite, bituminous coal,* and *subbituminous coal.*

lime scrubber   An air pollution–control device in which a chemical spray neutralizes acidic gases. See *scrubbers.*

limiting resource   Any environmental resource that, because it is scarce or at unfavorable levels, restricts the ecological niche of an organism.

limnetic zone   The open-water area away from the shore of a lake or pond that extends down as far as sunlight penetrates (and therefore photosynthesis occurs). Compare *littoral zone* and *profundal zone.*

liquefied petroleum gas   A mixture of liquefied propane and butane. Liquefied petroleum gas is stored in pressurized tanks.

lithosphere   Earth's outermost rigid rock layer that is composed of seven large plates, plus a few smaller ones.

littoral zone   The shallow-water area along the shore of a lake or pond. Compare *limnetic zone* and *profundal zone.*

loam   An ideal agricultural soil that has an optimum combination of soil particle sizes: approximately 40% each of sand and silt, and about 20% of clay.

logistic population growth   Population growth which is limited by density-dependent factors, ultimately approaching the carrying capacity (K), for that population.

longlines   Commercial fishing lines, up to 130 km (80 mi) long, with thousands of baited hooks.

Love Canal   A small neighborhood on the edge of Niagara Falls, New York, that became synonymous with chemical pollution caused by negligent management after a large number of serious illnesses, particularly among children, were associated with buried industrial wastes.

low-carbon fuel standard (LCFS)   A standard set that defines which fuels will lead to reduced amounts of carbon dioxide releases.

low-energy precision application (LEPA)   A water-conserving agricultural techique that makes use of geographic information systems.

low-grade ore   An ore that contains relatively small amounts of a particular mineral. Compare *high-grade ore.*

low-level radioactive waste   Any radioactive solid, liquid, or gas that gives off small amounts of ionizing radiation. Compare *high-level radioactive waste.*

Madrid Protocol   (1991) The Protocol on Environmental Protection to the Antarctic Treaty; provides for protection of the Antarctic environment and related ecosystems.

magma   Molten rock formed within Earth. Compare lava.

magmatic concentration   Layering of minerals in Earth's crust caused by cooling and solidification of magma.

Magnuson Fishery Conservation Act (Magnuson-Stevens Fishery Conservation and Management Act)   (1976, 2006) U.S. law to reduce bycatch, support enforcement of international fishing agreements, promote conservation of ocean resources, protect habitat, and administrate plans and councils to steward fishery resources.

malnutrition   Poor nutritional status; results from dietary intake either below or above required needs. Compare *undernutrition*.

manganese nodule   A small rock that contains manganese and other minerals; common on parts of the ocean floor.

mangrove forest   Swamps of mangrove trees that grow along many tropical coasts.

mantle   The thick layer of dense rock that surrounds Earth's core.

marasmus   Progressive emaciation caused by a diet low in both total calories and protein.

marginal cost of abatement   The cost to reduce a unit of pollution.

marginal cost of pollution   The cost in environmental damage of a unit of pollution emitted into the environment.

mariculture   The rearing of marine organisms (fishes, seaweeds, and shellfish) for human consumption; a subset of aquaculture.

marine snow   Organic debris that drifts into the darkened regions of the oceanic province from the upper, lighted regions.

marketable waste-discharge permit   See *tradable permit*.

marsh   A treeless wetland dominated by grasses; freshwater marshes are found inland along lakes and rivers, and salt marshes are found in bays and rivers near the ocean or in protected coastal areas. See *salt marsh*.

mass burn incinerator   A large furnace that burns all solid waste except for unburnable items such as refrigerators.

mass extinction   The extinction of numerous species during a relatively short period of geologic time. Compare *background extinction*.

materials recovery facility   A facility at which recyclable materials are either hand-sorted or separated using a variety of technologies, including magnets, screens, and conveyor belts, and prepared for remanufacturing.

maximum contaminant level   The maximum permissible amount (by law) of a water pollutant that might adversely affect human health.

Mediterranean climate   A climate with hot, dry summers and cool, wet winters, located in middle latitudes.

megacity   A city with more than 10 million inhabitants.

meltdown   The melting of a nuclear reactor vessel. A meltdown would cause the release of a substantial amount of radiation into the environment. See *reactor vessel*.

mesosphere   The layer of the atmosphere between the stratosphere and the thermosphere. It is characterized by the lowest atmospheric temperatures.

metal   A malleable, lustrous element that is a good conductor of heat and electricity. See *mineral*. Compare *nonmetal*.

metamorphic rock   Rock changed in form by heat and pressure.

methane   The simplest hydrocarbon, CH_4, which is an odorless, colorless, flammable gas.

methane hydrates   Reserves of ice-encrusted natural gas located in porous rock in the arctic tundra (under the permafrost) and in the deep-ocean sediments of the continental slope and ocean floor. Also called *gas hydrates*.

methanol   A colorless, flammable liquid, CH_3OH. Also called *methyl alcohol*.

methyl alcohol   See *methanol*.

microclimate   Local variation in climate produced by differences in elevation, in the steepness and direction of slopes, and in exposure to prevailing winds.

microcredit   Small loans available to borrowers who lack collateral or other standard qualifications for borrowing.

microirrigation   A type of water-conserving irrigation in which pipes with tiny holes bored into them convey water directly to individual plants. Also called *drip* or *trickle irrigation*.

Millennium Development Goals   collection of goals that make up a plan of action launched by the UN to try to meet the needs of the world's poor.

Millennium Ecosystem Assessment   assessment requested by the UN to gather scientific information on ecosystem changes and the effects of those changes on humans.

mineral   An inorganic solid, occurring naturally in or on Earth's crust, with characteristic chemical and physical properties. See *metal* and *nonmetal*.

mineral reserve   A mineral deposit that has been identified and is currently profitable to extract. See *total resources*. Compare *mineral resource*.

mineral resource   Any undiscovered mineral deposits or known deposits of low-grade ore that are currently unprofitable to extract. Mineral resources are potential resources that may be profitable to extract in the future. See *total resources*. Compare *mineral reserve*.

mitigation   In environmental management, methods or processes that reduce the effects of pollution that has been released.

mixed oxide fuel   A nuclear reactor fuel that contains a combination of uranium oxide and plutonium oxide; the plutonium can come from reprocessed spent fuel or from other plutonium stockpiles, including dismantled weapons.

mobile source   In air quality management, a source of pollution that is in motion, such as a bus or train. Contrast *stationary source*.

model   (1) A formal statement that describes a system and can be used to understand the present or predict the future course of events. (2) A simulation, using powerful computers, that represents the overall effect of competing factors to describe an environmental system in numerical terms.

moderately developed country   A developing country with a medium level of industrialization, a high fertility rate, a high infant mortality rate, and a low per capita income (all relative to highly developed countries). Compare less developed country and highly developed country.

modern synthesis   Unified explanation of evolution.

modular incinerator   A small, relatively inexpensive incinerator that burns solid waste.

mollisol   Grassland soils with a thick A-horizon.

monoculture   The cultivation of only one type of plant over a large area. Compare *polyculture*.

Montreal Protocol   International negotiations that resulted in a timetable to phase out CFC production.

mortality   See *death rate*.

mountaintop removal   A coal-mining technique in which trees, herbs, soil, and rock are relocated from a mountaintop to an adjacent valley to allow access to a coal seam.

mulch   Material placed on the surface of soil around the bases of plants; helps maintain soil moisture and reduce soil erosion. Organic mulches decompose over time, thereby enriching the soil.

municipal solid waste   (MSW) Solid materials discarded by homes, office buildings, retail stores, restaurants, schools, hospitals, prisons, libraries, and other commercial and institutional facilities. Compare *nonmunicipal solid waste*.

municipal solid waste composting   The large-scale composting of the organic portion of a community's solid waste.

mutation   A change in the DNA (that is, a gene) of an organism. A mutation in reproductive cells may be passed on to the next generation, where it may result in birth defects or genetic disease or, less commonly, an advantageous trait.

mutualism   A symbiotic relationship in which both partners benefit from the association. See *symbiosis*. Compare *parasitism* and *commensalism*.

mycelium   The nonreproductive part of a fungus, typically a network of fine threads, like roots, within a soil.

mycorrhiza   (pl., *mycorrhizae*) A mutualistic association between a fungus and the roots of a plant. Most plants form mycorrhizal associations with fungi, which enable plants to absorb adequate amounts of essential minerals from the soil.

nanomaterials   Any of a variety of materials that are designed to be functional at the scale of around a nanometer.

nanotechnology   Manufactured materials that are around one nanometer (10^{-9} meters) in diameter.

narrow-spectrum pesticide   An "ideal" pesticide that kills only the organism for which it is intended and does not harm any other species. Compare *broad-spectrum pesticide*.

natality   See *birth rate*.

national conservation strategy   An approach to that coordinates the reduction of energy demand across transportation, buildings, manufacturing, and other uses to most efficiently manage resources.

national emission limitations   The maximum permissible amount (by law) of a particular pollutant that can be discharged into the nation's rivers, lakes, and oceans from point sources.

National Environmental Policy Act   (NEPA) A U.S. federal law that is the cornerstone of U.S. environmental policy. It requires that the federal government consider the environmental impact of any construction project that it funds.

national income account A measure of the total income of a country for a given year; examples include the gross domestic product and net domestic product. Compare *Environmental Performance Index*.

national marine sanctuary A marine ecosystem set aside to minimize human impacts and protect unique natural resources and historical sites.

National Park System (1916) Federal bureau in U.S. Department of the Interior with responsibility of administering national parks and monuments.

National Wildlife Refuge System (1966) Act providing guidelines for management of areas for the protection of fish, birds, and other wildlife.

natural capital All of Earth's resources and processes that sustain living organisms, including humans; natural capital includes minerals, forests, soils, groundwater, clean air, wildlife, and fisheries.

natural gas A mixture of gaseous, energy-rich hydrocarbons (primarily methane) that occurs, often with oil deposits, in Earth's crust. See *fossil fuel*.

natural increase See *growth rate*.

natural resources See *resources (1)*.

natural selection The tendency of organisms that possess adaptations favorable to their environment to survive and become the parents of the next generation; evolution occurs when natural selection results in genetic changes in a population. Natural selection is the mechanism of evolution first proposed by Charles Darwin.

Nature Conservancy Nonprofit agency that brokers conservation easements, land purchases, and cooperative agreements to conserve ecologically important habitats.

negative externality See *externality*.

negative feedback system A system in which a change in some condition triggers a response that counteracts, or reverses, the changed condition. Compare *positive feedback system*.

nekton Relatively strong-swimming aquatic organisms such as fish and turtles. Compare *plankton* and *benthos*.

neritic province Open ocean from the shoreline to a depth of 200 meters. Compare *oceanic province*.

net primary productivity (**NPP**) Productivity after respiration losses are subtracted; that is, the amount of biomass found in excess of that broken down by a plant's cellular respiration. Compare *gross primary productivity*.

niche The totality of an organism's adaptations, its use of resources, and the lifestyle to which it is fitted. The niche describes how an organism uses materials in its environment as well as how it interacts with other organisms. Also called *ecological niche*. See *fundamental niche* and *realized niche*.

nitrification The conversion of ammonia (NH_3) and ammonium ions (NH_4^+) by certain bacteria (nitrifying bacteria) in the soil to nitrate (NO_3^-); part of the nitrogen cycle.

nitrogen cycle The worldwide circulation of nitrogen from the abiotic environment into organisms and back into the abiotic environment.

nitrogen fixation The conversion of atmospheric nitrogen (N_2) to ammonia (NH_3) by nitrogen-fixing bacteria and cyanobacteria; part of the nitrogen cycle.

nitrogen oxides Compounds containing a nitrogen molecule and one or more oxygen molecules.

noise pollution A loud or disagreeable sound, particularly when it results in physiological or psychological harm.

nomadic herding A traditional grazing method in which nomadic herdsmen wander freely over rangelands in search of good grazing for their livestock.

nondegradable Referring to a chemical pollutant (such as the toxic elements mercury and lead) that cannot be decomposed (broken down) by organisms or by other natural processes. Compare *biodegradable*.

nonmetal A nonmalleable, nonlustrous mineral that is a poor conductor of heat and electricity. See *mineral*. Compare *metal*.

nonmunicipal solid waste Solid waste generated by industry, agriculture, and mining. Compare *municipal solid waste*.

nonpoint source pollution Pollutants that enter bodies of water over large areas rather than being concentrated at a single point of entry. Examples include agricultural fertilizer runoff and sediments from construction. Also called *polluted runoff*. Compare *point source pollution*.

nonrenewable resources Natural resources that are present in limited supplies and are depleted by use; include minerals such as copper and tin and fossil fuels such as oil and natural gas. Compare *renewable resources*.

nonselective herbicide An herbicide that kills most or all plant species. **nonurban land** See *rural land*.

Northwest Forest Plan (1993) A federal plan to foster cooperation and resolve conflict among loggers, conservationists, and local communities regarding management of federally owned forest land in Oregon and Washington.

no-tillage A method of conservation tillage that does not involve tilling; leaves both the surface and subsurface soil undisturbed. Special machines punch holes in the soil for seeds. See *conservation tillage*. Compare *conventional tillage*.

nuclear energy The energy released from the nucleus of an atom in a nuclear reaction (fission or fusion) or during radioactive decay.

nuclear fission See *fission*.

nuclear fuel cycle The processes involved in producing the fuel used in nuclear reactors and in disposing of radioactive wastes (also called *nuclear wastes*).

nuclear fusion See *fusion*.

nuclear reactor A device that initiates and maintains a controlled nuclear fission chain reaction to produce energy, usually used as electricity.

Nuclear Waste Policy Act (1992) U.S. law that establishes a disposal program for highly radioactive wastes.

nutrient cycling The pathways of various nutrient minerals or elements from the environment through organisms and back to the environment; biogeochemical cycles are examples of nutrient cycling.

nutrition transition The emergence of overnutrition in countries that also have widespread undernutrition.

ocean temperature gradient The differences in temperature at various ocean depths.

ocean conveyor belt Global circulation of shallow and deep currents; moves cold, salty, deep-sea water from higher to lower latitudes.

oceanic province That part of the open ocean that is deeper than 200 meters and comprises most of the ocean. Compare *neritic province*.

ocean thermal energy conversion (OTEC) Indirect form of solar energy that would take advantage of ocean temperature gradients to produce electricity or cool buildings.

Ogallala Aquifer A massive groundwater deposit under eight midwestern states.

O-horizon The uppermost layer of certain soils, composed of dead leaves and other organic matter.

oil See *petroleum*.

Oil Pollution Act (1990) U.S. law, created in response to the spill from the Exxon *Valdez*, that enables resources and response to oil spills.

oil sand See *tar sand*.

oil shales Sedimentary "oily rocks" that contain a mixture of hydrocarbons known as kerogen; to yield oil, oil shales must be crushed, heated to high temperatures, and refined after they are mined.

oligotrophic lake A deep, clear lake that has minimal nutrients. Water in an oligotrophic lake contains a high level of dissolved oxygen. Compare *eutrophic lake*.

omnivore An animal that eats a variety of plant and animal material. Compare *herbivore* and *carnivore*.

open management A policy in which all fishing boats of a particular country are given unrestricted access to fish in their national waters.

open system A system that exchanges energy with its surroundings. Compare *closed system*.

open-pit mining A type of surface mining in which a giant hole (quarry) is dug to extract iron, copper, stone, or gravel. See *surface mining*. Compare *strip mining*.

optimal amount of pollution In economics, the amount of pollution that is most economically efficient. It is determined by plotting two curves, the marginal cost of pollution and the marginal cost of abatement; the point where the two curves meet is the optimum amount of pollution.

ore Rock that contains a large enough concentration of a particular mineral that it can be profitably mined and the mineral extracted.

organic agriculture Growing crops and livestock without the use of synthetic pesticides or commercial inorganic fertilizers. Organic agriculture makes use of natural organic fertilizers (such as manure and compost) and chemical-free methods of pest control.

organic compound A compound that contains the element carbon and is either naturally occurring (in organisms) or synthetic (manufactured by humans). Many synthetic organic compounds persist in the environment for an extended period, and some are toxic to organisms.

Organic Food Production Act   The 1990 U.S. law that regulates which foods may be certified as organic.

organophosphate   A synthetic organic compound that contains phosphorus and is toxic; used as an insecticide.

overburden   Overlying layers of soil and rock over mineral deposits. The overburden is removed during surface mining.

overgrazing   The destruction of an area's vegetation that occurs when too many animals graze on the vegetation, consuming so much of it that it does not recover.

oxide   A compound in which oxygen is chemically combined with some other element.

oxisols   Highly weathered, tropical soils with few nutrient minerals but typically abundant iron and aluminum.

ozone   A pale blue gas (O_3) that is both a pollutant in the lower atmosphere (troposphere) and an essential component that screens out UV radiation in the upper atmosphere (stratosphere).

pandemic   A disease that reaches nearly every part of the world and has the potential to infect almost every person.

paradigm   A generally accepted understanding of how some aspect of the world works; for example, evolution is the underlying paradigm of modern biology.

parasitism   A symbiotic relationship in which one member (the parasite) benefits and the other (the host) is adversely affected. See *symbiosis*. Compare *mutualism* and *commensalism*.

particulate matter   Solid particles and liquid droplets suspended in the atmosphere.

parts per billion   (ppb)   The number of molecules of a particular substance found in 1 billion molecules of air, water, or some other material.

parts per million   (ppm)   The number of molecules of a particular substance found in 1 million molecules of air, water, or some other material.

passive solar heating   A system that uses the sun's energy without requiring mechanical devices (pumps or fans) to distribute the collected heat. Compare *active solar heating*.

pathogen   An agent (usually a microorganism) that causes disease. See *disease-causing agent*.

payback time   The amount of time required to recover a capital expenditure through the savings from that initial expenditure.

PCBs   Polychlorinated biphenyls; chlorine-containing organic compounds that had a wide variety of industrial uses until their dangerous properties, including the fact that they are slow to degrade and therefore persist in the environment, were recognized.

peak demand   In electricity use, the time of day or year in which the most power is being drawn in an electrical grid.

Peak Oil   The point at which global oil production has reached a maximum rate; by some estimates, Peak Oil is already past, but best estimates have it occurring around 2020 or later. Also known as Hubbert's peak, after the U.S. geologist who came up with the idea.

pelagic environment   The aquatic environment which is not in close contact with either shoreline or seafloor.

permafrost   Permanently frozen subsoil characteristic of frigid areas such as the tundra.

persistence   A characteristic of certain chemicals that are extremely stable and may take many years to be broken down into simpler forms by natural processes.

persistent organic pollutants   (POPs) A group of persistent, toxic chemicals that bioaccumulate in organisms and can travel long distances through air and water to contaminate sites far removed from their source; some disrupt the endocrine system, cause cancer, or adversely affect the developmental processes of organisms.

pest   Any organism that interferes in some way with human welfare or activities.

pesticide   Any toxic chemical used to kill pests. See *fungicide, herbicide, insecticide,* and *rodenticide*.

Pesticide Chemicals Amendment   (1954) Modified the FDCA to prevent sale or shipment of any agricultural commodity with unsafe levels of pesticides.

pesticide treadmill   A predicament faced by pesticide users, in which the cost of applying pesticides increases because they have to be applied more frequently or in larger doses, while their effectiveness decreases as a result of increasing genetic resistance in the target pest.

petrochemicals   Chemicals, obtained from crude oil, that are used in the production of such diverse products as fertilizers, plastics, paints, pesticides, medicines, and synthetic fibers.

petroleum   A flammable liquid hydrocarbon mixture found in Earth's crust, formed from the remains of ancient microscopic organisms. When petroleum is refined, the mixture is separated into various hydrocarbon compounds, including gasoline, kerosene, fuel oil, lubricating oils, paraffin, and asphalt. Also called *crude oil*. See *fossil fuel*.

pH   A number from 0 to 14 that indicates the degree of acidity or basicity of a substance.

pH scale   A measure of acidity or basicity of an aqueous solution.

pheromone   A substance secreted into the environment by one organism; influences the development or behavior of other members of the same species.

phosphorus cycle   The worldwide circulation of phosphorus from the abiotic environment into organisms and back into the abiotic environment.

photochemical smog   A brownish-orange haze formed by complex chemical reactions involving sunlight, nitrogen oxides, and hydrocarbons. Some of the pollutants in photochemical smog include peroxyacetyl nitrates (PANs), ozone, and aldehydes. Compare *industrial smog*.

photodegradable   Breaking down upon exposure to sunlight.

photosynthesis   The biological process that captures light energy and transforms it into the chemical energy of organic molecules (such as glucose), which are manufactured from carbon dioxide and water. Photosynthesis is performed by plants, algae, and several kinds of bacteria. Compare *chemosynthesis*.

photovoltaic solar cell   A wafer or thin film of a solid-state material, such as silicon or gallium arsenide, that is treated with certain metals so that it generates electricity (a flow of electrons) when it absorbs solar energy.

phytoplankton   Microscopic floating algae that are the base of most aquatic food chains. See *plankton*. Compare *zooplankton*.

phytoremediation   A method employed to clean up a hazardous waste site that uses plants to absorb and accumulate toxic materials. Compare *bioremediation*.

pioneer community   The first organisms (such as lichens or mosses) to colonize an area and begin the first stage of ecological succession. See *succession*.

plankton   Small or microscopic aquatic organisms that are relatively feeble swimmers and thus, for the most part, are carried about by currents and waves. Composed of phytoplankton and zooplankton. Compare *nekton* and *benthos*.

plasma   An ionized gas formed at high temperatures when electrons are stripped from the gas atoms; plasma is formed during fusion reactions.

plate boundary   Any area where two tectonic plates meet. Plate boundaries are often sites of intense geologic activity. See *fault*.

plate tectonics   The study of the processes by which the lithospheric plates move over the asthenosphere.

plug-in hybrid electric vehicle (PHEV)   A vehicle that can use more than one source of energy where one of them is electricity.

point source pollution   Water pollution that can be traced to a specific spot (such as a factory or sewage treatment plant) because it is discharged into the environment through pipes, sewers, or ditches. Compare *nonpoint source pollution*.

polar   Referring to areas close to the North or South Poles, characterized by cold weather, long winter nights, and long summer days.

polar easterly   A prevailing wind that blows from the northeast near the North Pole or from the southeast near the South Pole.

pollutant   A physical, chemical, or biological substance, generated by human activity and emitted into the environment, that has the potential to have harmful effects on the health of organisms, including humans.

polluted runoff   See *nonpoint source pollution*.

pollution   Any undesirable alteration of air, water, or soil that harms the health, survival, or activities of humans and other living organisms.

Pollution Prevention Act   (1990) Encourages reduction of U.S. pollution sources through changes in production, operation, and raw materials use.

polychlorinated biphenyls   See *PCBs*.

polyculture   A type of intercropping in which several kinds of plants that mature at different times are planted together. See *intercropping*. Compare *monoculture*.

polymers   Chains of repeating carbon compounds used to create plastics.

population   A group of organisms of the same species that live in the same geographic area at the same time.

population crash   An abrupt decline from high to low population density.

population density   The number of individuals of a species per unit of area or volume at a given time.

population ecology   The branch of biology that deals with the numbers of a particular species that are found in an area and how and why those numbers change (or remain fixed) over time.

population growth momentum The continued growth of a population after fertility rates have declined, as a result of a population's young age structure; population growth momentum can be either positive or negative but is usually discussed in a positive context. Also called *population momentum*.

population momentum See *population growth momentum*.

positive feedback system A system in which a change in some condition triggers a response that intensifies the changing condition. Compare *negative feedback system*.

postindustrial stage Fourth stage in the demographic transition, characterized by low birth and death rates.

potential energy Stored energy that is the result of the relative position of matter instead of its motion. Compare *kinetic energy*.

poverty A condition in which people are unable to meet their basic needs for adequate food, clothing, or shelter.

precautionary principle The idea that no action should be taken or product introduced when the science is inconclusive but unknown risks may exist.

predation The consumption of one species (the prey) by another (the predator); includes both animals eating other animals and animals eating plants.

preindustrial CO_2 level The level of carbon dioxide in the atmosphere before large-scale use of fossil fuels began in 1800s.

preindustrial stage First stage in the demographic transition, characterized by high birth and death rates and modest population growth rates.

preservation Setting aside undisturbed areas, maintaining them in a pristine state, and protecting them from human activities that might alter the natural state. Compare *conservation*.

prevailing wind A major surface wind that blows more or less continually.

primary air pollutant A harmful chemical that enters directly into the atmosphere from either human activities or natural processes (such as volcanic eruptions). Compare *secondary air pollutant*.

primary consumer An organism that consumes producers. Also called herbivore. Compare *secondary consumer*.

primary productivity The total energy in an ecosystem captured through photosynthesis by plants and other autotrophs.

primary sludge A slimy mixture of bacteria-laden solids that settles out from sewage wastewater during primary treatment.

primary succession An ecological succession that occurs on land that has not previously been inhabited by plants; no soil is present initially. Compare *secondary succession*.

primary treatment Treatment of wastewater by removing suspended and floating particles (such as sand and silt) by mechanical processes (such as screens and physical settling). Compare *secondary treatment* and *tertiary treatment*.

prime farmland Land with the ideal combination of physical and chemical traits to produce crops.

principle of inherent safety Chemical safety programs that stress accident prevention by redesigning industrial processes to involve less toxic materials so that dangerous accidents are prevented.

producer An organism (such as a chlorophyll-containing plant) that manufactures complex organic molecules from simple inorganic substances. In most ecosystems, producers are photosynthetic organisms. Also called *autotroph*. Compare *consumer*.

profundal zone The deepest zone of a large lake. Compare limnetic zone and littoral zone.

Public Rangelands Improvement Act (1978) Establishes U.S. commitment to protect and inventory rangeland conditions, to charge equitable fees for public grazing use, and to protect and manage wild equines.

pumped hydroelectric storage Storage of energy by pumping water from a low elevation to a higher elevation. The water can then be run through a turbine to generate electricity.

pure electric vehicle (PEV) A vehicle that runs exclusively on electrical power.

purse seine net A net, as long as 2 km (more than 1 mi), set out by small powerboats to encircle large schools of tuna and other fishes.

pyramid of biomass An ecological pyramid that illustrates the total biomass (for example, the total dry weight of all organisms) at each successive trophic level in an ecosystem. See *ecological pyramid*. Compare *pyramid of energy* and *pyramid of numbers*.

pyramid of energy An ecological pyramid that shows the energy flow through each trophic level in an ecosystem. See ecological pyramid. Compare *pyramid of biomass* and *pyramid of numbers*.

pyramid of numbers An ecological pyramid that shows the number of organisms at each successive trophic level in a given ecosystem. See ecological pyramid. Compare *pyramid of biomass* and *pyramid of energy*.

quarantine Practice in which the importation of exotic plant and animal material that might be harboring pests is restricted.

quarry See *open-pit mining*.

r **selection** A reproductive strategy in which a species typically has a small body size, rapid development, and a short life span, and devotes a large proportion of its metabolic energy to the production of offspring. Compare *K selection*.

radiation The emission of fast-moving particles or rays of energy from the nuclei of radioactive atoms.

radiative forcing The capacity of a gas to affect the balance of energy that enters and leaves Earth's atmosphere; measured in units of power per unit area, usually watts per square meter (W/m^2).

radioactive atoms Atoms of unstable isotopes that spontaneously emit radiation.

radioactive decay The emission of energetic particles or rays from unstable atomic nuclei; includes positively charged alpha particles; negatively charged beta particles; and high-energy, electromagnetic gamma rays.

radioactive half-life The time required for one-half of a radioactive substance to change into a different material. Also called *half-life*.

radioisotope An unstable isotope that spontaneously emits radiation.

radon A colorless, tasteless, radioactive gas produced during the radioactive decay of uranium in Earth's crust.

rain shadow An area on the downwind side of a mountain range with little precipitation. Deserts often occur in rain shadows.

range The area of Earth in which a particular species occurs.

rangeland Land that is not intensively managed and is used for grazing livestock.

rare earth metals Elements such as dysprosium and terbium that are important in high-technology applications like hybrid-car batteries, wind turbines, and laser-guided missiles.

reactor core The part of a nuclear reactor that contains the fuel (uranium or plutonium).

reactor vessel A huge steel potlike structure encasing the uranium fuel in a nuclear reactor. The reactor vessel is a safety feature designed to prevent the accidental release of radiation into the environment.

realized niche The lifestyle that an organism actually pursues, including the resources that it actually uses. An organism's realized niche is narrower than its fundamental niche because of competition from other species. See *niche*. Compare *fundamental niche*.

reclaimed water Treated wastewater that is reused in some way, such as for irrigation, manufacturing processes that require water for cooling, wetlands restoration, or groundwater recharge.

recycling Conservation of the resources in used items by converting them into new products. For example, used aluminum cans are recycled by collecting, remelting, and reprocessing them into new cans. Compare *reuse*.

red tide A red, orange, or brown coloration of water caused by a bloom, or population explosion, of algae; many red tides cause serious environmental harm and threaten the health of humans and animals.

refuse-derived fuel incinerator An incinerator in which only the combustible portion of solid waste is burned, after noncombustible materials such as glass and metals has been removed.

renewable resources Resources that are replaced by natural processes and can be used forever, provided they are not overexploited in the short term. Examples include fresh water in lakes and rivers, fertile soil, and trees in forests. Compare *nonrenewable resources*.

replacement-level fertility The average number of children a couple must produce to "replace" themselves; the number is greater than two because some children die before reaching reproductive age.

reprocessing Reusing highly radioactive spent nuclear fuel by chemically separating the unused uranium and plutonium from other radioactive waste products.

reservoir An artificial lake produced by building a dam across a river or stream; allows water to be stored for use.

resistance management An approach to dealing with genetic resistance so that the period in which a pesticide is useful is maximized; involves efforts to delay the development of genetic resistance in pests.

Resource Conservation and Recovery Act (RCRA) (1976) Gives the U.S. EPA authority to control hazardous waste from "cradle to grave"; 1986 amendments allow EPA to address problems from underground chemical and petroleum storage.

resource partitioning The reduction in competition for environmental resources, such as food, that occurs among coexisting species as a result of each species' niche differing from those of the others in one or more ways.

resource recovery The process of removing any material—sulfur or metals, for example—from polluted emissions or solid waste and selling it as a marketable product.

resources (1) Any parts of the natural environment that are used to promote the welfare of people or other species; examples include clean air, fresh water, soil, forests, minerals, and organisms. Also called *natural resources.* (2) Anything from the environment that meets a particular species' needs.

respiration See *cellular respiration.*

response In toxicology, the type and amount of damage caused by exposure to a particular dose.

restoration ecology The field of science in which the principles of ecology are used to help return a degraded environment as close as possible to its former, undisturbed state. Also called *ecological restoration.*

reuse Conservation of the resources in used items by using them over and over again. For example, glass bottles can be collected, washed, and refilled again. Compare *recycling.*

reverse osmosis A desalinization process that involves forcing salt water through a membrane permeable to water but not to salt. Compare *distillation.*

riparian area The thin patch of vegetation along the bank of a stream or river that interfaces between terrestrial and aquatic habitats; protects the habitat of salmon, trout, and other aquatic species from sedimentation caused by soil erosion. Also called *riparian buffer.*

riparian buffer See *riparian area.*

risk The probability that a particular adverse effect will result from some exposure or condition.

risk assessment The use of statistical methods to quantify the harmful effects on human health or the environment of exposure to a particular danger. Risk assessments can provide a systems perspective when risks are compared with one another. See *ecological risk assessment.*

risk management Determining whether there is a need to reduce or eliminate a particular risk and, if so, what should be done; based on data from risk assessment as well as political, economic, and social considerations.

rock A naturally formed aggregate, or mixture, of minerals; rocks have varied chemical compositions.

rock cycle The cycle of rock transformation that involves all parts of Earth's crust.

rodenticide A toxic chemical that kills rodents.

runoff The movement of fresh water from precipitation and snowmelt to rivers, lakes, wetlands, and, ultimately, the ocean.

rural land Sparsely populated areas, such as forests, grasslands, deserts, and wetlands. Also called *nonurban land.*

salination See *salinization.*

salinity The concentration of dissolved salts (such as sodium chloride) in a body of water.

salinity gradient The difference in salt concentrations that occurs at different depths in the ocean and at different locations in estuaries.

salinization The gradual accumulation of salt in a soil, often as a result of improper irrigation methods. Most plants cannot grow in salinized soil. Also called *salination.*

salt marsh An estuarine wetland dominated by salt-tolerant grasses.

saltwater intrusion The movement of seawater into a freshwater aquifer located near the coast; caused by aquifer depletion. Saltwater intrusion also occurs in low-lying parts of the world due to sea-level rise.

sand Inorganic soil particles that are larger than clay or silt. Compare *clay* and *silt.*

sanitary landfill The most common method of disposal of solid waste, by compacting it and burying it under a shallow layer of soil.

saprotroph See *decomposer.*

savanna A tropical grassland with widely scattered trees or clumps of trees; found in areas of low rainfall with prolonged dry periods.

schistosomiasis A disease caused by parasitic blood flukes (schistosomes), acquired through contact with unfiltered tropical or subtropical fresh water.

science A human endeavor that seeks to reduce the apparent complexity of the natural world to general principles that can be used to make predictions, solve problems, or provide new insights.

scientific method The way a scientist approaches a problem (by formulating a hypothesis and then testing it by means of an experiment).

scrubbers Desulfurization systems that are used in smokestacks to decrease the amount of sulfur released in the air by 90% or more. One type is a lime scrubber.

scrub savanna A grassland interspersed with shrubs and small trees, typically tropical.

sea grasses Flowering plants that grow in quiet, shallow ocean water in temperate, subtropical, and tropical waters.

second green revolution (1) An effort to increase rice yields (primarily in Asia) and increase tolerance to drought, pests, floods, and disease; methods include genetic modification; (2) agricultural productivity increases through urban agriculture, smaller farm size, organic agriculture, and distribution through local produce markets.

second-generation pesticide Synthetic chemical compound created for use as a pesticide; may be more or less toxic than first-generation pesticides.

second law of thermodynamics When energy is converted from one form to another, some of it is degraded into a lower-quality, less useful form. Thus, with each successive energy transformation, less energy is available to do work. See *thermodynamics.* Compare *first law of thermodynamics.*

secondary air pollutant A harmful substance formed in the atmosphere when a primary air pollutant reacts with substances normally found in the atmosphere or with other air pollutants. Compare *primary air pollutant.*

secondary consumer An organism that consumes primary consumers. Also called *carnivore.* Compare *primary consumer.*

secondary sludge A slimy mixture of bacteria-laden solids that settles out from sewage wastewater during secondary treatment.

secondary succession An ecological succession that takes place after some disturbance destroys the existing vegetation; soil is already present. Compare *primary succession.*

secondary treatment Treatment of wastewater biologically to decompose suspended organic material; secondary treatment reduces the water's biochemical oxygen demand; occurs after primary treatment. Compare *primary treatment* and *tertiary treatment.*

sediment pollution Excessive amounts of soil particles that enter waterways as a result of erosion.

sedimentary rocks Rock formed from mineral and organic material accumulating on land or in water.

sedimentation (1) The process in which eroded particles are transported by water and deposited as sediment on river deltas and the seafloor. If exposed to sufficient heat and pressure, sediments can solidify into sedimentary rock. (2) Letting solids settle out of wastewater by gravity during primary treatment.

seed bank (1) the stock, in species and abundance, of plant seeds in soil; (2) a place where plant seeds are stored in case of future need.

seed tree cutting A forest management technique in which almost all trees are harvested from an area in a single cutting, but a few desirable trees are left behind to provide seeds for the regeneration of the forest. Compare *clear-cutting, selective cutting,* and *shelterwood cutting.*

seismic waves Vibrations that travel through rock as a result of an earthquake.

selective cutting A forest management technique in which mature trees are cut individually or in small clusters while the rest of the forest remains intact so that the forest can regenerate quickly (and naturally). Compare *clear-cutting, seed tree cutting,* and *shelterwood cutting.*

selective herbicide An herbicide which kills some plant species and leaves others unaffected.

semiarid land Land that receives more precipitation than a desert but is subject to frequent and prolonged droughts. Compare *arid land.*

sewage The wastewater released from drains or sewers (from toilets, washing machines, sinks, and showers); includes human wastes, soaps, and detergents.

sewage sludge See *primary sludge* and *secondary sludge.*

shaft mine In coal or mineral mining, a mine in which a hole is dug down into the ground.

shelterbelt A row of trees planted as a windbreak to reduce soil erosion of agricultural land.

shelterwood cutting A forest management technique in which all mature trees in an area are harvested in a series of partial cuttings over time; typically two or three harvests occur during a decade. Compare *clear-cutting, seed tree cutting,* and *selective cutting.*

shifting cultivation   A traditional form of subsistence agriculature in which short periods of cultivation are followed by longer periods of fallow (land left uncultivated), during which times the natural ecosystems may become reestablished. See *slash-and-burn agriculture.*

sick building syndrome   The condition inside a building in which eye irritations, nausea, headaches, respiratory infections, depression, and fatigue are caused by the presence of air pollution.

silviculture   The practice of forest management for regeneration, composition, health and quality; may include management of forest carbon resources.

silt   Medium-sized inorganic soil particles. Compare *clay* and *sand.*

sink   In environmental science, the part of the natural environment that receives an input of materials; economies depend on sinks for waste products. Compare *source.*

sinkhole   A large surface cavity or depression caused when an underground cave roof has collapsed; occurs most frequently when a drought or excessive pumping causes a lowering of the water table.

slag   See *bottom ash.*

slash-and-burn agriculture   A type of shifting cultivation in tropical forests in which a patch of vegetation is burned, leaving nutrient minerals in the ash. Crops are planted for a few years until the soil is depleted of nutrient minerals, after which the land must lie fallow for many years to recover. See *shifting cultivation.*

slope mine   In coal or mineral mining, a mine in which a hole is dug horizontally into a sloped surface.

Smart Grid   an electrical grid in which generators and consumers are coordinated and managed using a computer network.

smart growth   Urban planning and transportation strategy that mixes land uses.

smelting   A process in which ore is melted at high temperatures to help separate impurities from the molten metal.

smog   Air pollution caused by a variety of pollutants. See *industrial smog* and *photochemical smog.*

soft coal   See *bituminous coal.*

soil   The uppermost layer of Earth's crust, which supports terrestrial plants, animals, and microorganisms. Soil is a complex mixture of inorganic minerals (from the parent material), organic material, water, air, and organisms.

soil air   Air held among sand, silt, and clay particles.

Soil Conservation Act   (1935) Act establishing the Soil Conservation Service to promote soil management methods to control floods, reduce erosion, and maintain soil fertility.

soil degradation   See *degradation.*

soil erosion   The wearing away or removal of soil, especially top soil, from the land; caused by wind and flowing water. Although soil erosion occurs naturally from precipitation and runoff, human activities such as clearing land accelerate it.

soil horizons   The horizontal layers into which many soils are organized. May include the O-horizon (surface litter), A-horizon (topsoil), E-horizon, B-horizon (subsoil), and C-horizon (partly weathered parent material).

soil pollution   Any physical or chemical change in soil that adversely affects the health of plants and other organisms living in and on it.

soil profile   A vertical section through the soil, from the surface to the parent material, that reveals the soil horizons.

soil remediation   Use of one or more techniques to remove contaminants from the soil.

soil salinization   See *salinization.*

solar energy   Energy from the sun. Solar energy includes both direct solar radiation and indirect solar energy (such as wind, hydropower, and biomass).

soil taxonomy   The naming and classification of soils. See *state factors.*

solar thermal electric generation   A means of producing electricity in which the sun's energy is concentrated by mirrors or lenses to either heat a fluid-filled pipe or drive a Stirling engine.

solid waste   Any unwanted materials that are discarded as trash. See *municipal solid waste* and *nonmunicipal solid waste.*

soil water   Water held among sand, silt, and clay particles.

solar cell   A single photovoltaic cell used to generate electricity using direct solar energy.

source   In environmental science, the part of the environment from which materials move; economies depend on sources for raw materials. Compare *sink.*

source reduction   An aspect of waste management in which products are designed and manufactured in ways that decrease not only the volume of solid waste but also the amount of hazardous materials in the solid waste that remains.

species   A group of similar organisms whose members freely interbreed with one another in the wild to produce fertile offspring; members of one species generally do not interbreed with other species.

species richness   Biological diversity that encompasses the number of species in an area (or community). Compare *genetic diversity* and *ecosystem diversity.*

spent fuel   The used fuel elements that were irradiated in a nuclear reactor.

spodosol   Acidic, leached soils typical of coniferous or boreal forest.

spoil bank   A hill of loose rock created when the overburden from a new trench is put into the old (already excavated) trench during strip mining.

sprawl, suburban   See *suburban sprawl.*

spring turnover   A mixing of the lake waters in temperate lakes that occurs in spring as ice melts and the surface water reaches 4°C, its temperature of greatest density. Compare *fall turnover.*

stable runoff   The share of runoff from precipitation available throughout the year. Most geographic areas have a heavy runoff during a few months (the spring months, for example) when precipitation and snowmelt are highest. Stable runoff can be depended on every month.

standing-water ecosystem   A body of fresh water surrounded by land and whose water does not flow; a lake or pond. Compare *flowing-water ecosystem.*

state factor   A condition that affects soil formation; includes mineral type, climate, topography, time, and biotic potential.

stationary source   In air quality management, a source of pollution that does no move, such as a power plant or furnace. Contrast *mobile source.*

sterile male technique   A method of insect control that involves rearing, sterilizing, and releasing large numbers of males of the pest species.

stewardship   The concept that humans share responsibility for the sustainable care and management of our planet.

Stirling engine   An engine that produces kinetic energy through expansion and contraction of gases.

Stockholm Convention   International treaty (signed 2001, effective 2004) that aims to eliminate or restrict persistent organic pollutants from use and production.

Strategic Petroleum Reserve   An emergency supply of up to 1 billion barrels of oil stored in underground salt caverns along the coast of the Gulf of Mexico; mandated by the U.S. Energy Policy and Conservation Act.

stratosphere   The layer of the atmosphere between the troposphere and the mesosphere. It contains a thin ozone layer that protects life by filtering out much of the sun's ultraviolet radiation.

stratospheric ozone thinning   The accelerated destruction of ozone in the stratosphere by human-produced chlorine- and bromine-containing chemicals.

stressor   See environmental stressor.

strip cropping Different crops laid out in alternating, narrow strips along natural contours, as on a hillside. See *contour plowing.*

strip mining   A type of surface mining in which a trench is dug to extract the minerals, then a new trench is dug parallel to the old one; the overburden from the new trench is put into the old trench, creating a hill of loose rock known as a spoil bank. See *surface mining.* Compare *open-pit mining.*

structural trap   An underground geological structure that tends to trap oil or natural gas if it is present.

subbituminous coal   A grade of coal, intermediate between lignite and bituminous, that has a relatively low heat value and sulfur content. Compare *anthracite, bituminous coal,* and *lignite.*

subduction   The process in which one tectonic plate descends under an adjacent plate.

subsidence   The sinking or settling of land caused by aquifer depletion (as groundwater supplies are removed).

subsidy   A form of government support (such as monetary payments, public financing, tax benefits, or tax exemptions) given to a business or institution to promote the activity performed by that group.

subsistence agriculture   Traditional agricultural methods that are dependent on labor and a large amount of land to produce enough food to feed oneself and one's family, with little left over to sell or reserve for hard times. Subsistence agriculture uses humans and draft animals as its main source of energy. Compare *industrialized agriculture.*

subsurface mining   The extraction of mineral and energy resources from deep underground deposits. Compare *surface mining.*

suburban sprawl   A patchwork of low-density vacant and developed tracts around the edges of cities.

succession   The sequence of changes in a plant community over time. Also called *ecological succession*.

sulfide   A compound in which an element is combined chemically with sulfur.

sulfur cycle   The worldwide circulation of sulfur from the abiotic environment into organisms and back into the abiotic environment.

sulfur oxides   Compounds containing a sulfur molecule and one or more oxygen molecules.

surface mining   The extraction of mineral and energy resources near Earth's surface by first removing the soil, subsoil, and overlying rock strata (i.e., the overburden). See *strip mining* and *open-pit mining*. Compare *subsurface mining*.

surface water   Fresh water found on Earth's surface in streams and rivers, lakes, ponds, reservoirs, and wetlands. Compare *groundwater*.

superconducting magnetic energy storage   Storage of energy as magnetism in superconducting magnets so that it can later be converted to electricity.

Superfund National Priorities List   A list of the most contaminated superfund (hazardous waste) sites in the United States.

Surface Mining Control and Reclamation Act   (SMCRA) (1977) Requires mining operators to restore land to a condition capable of supporting the uses it could support prior to mining, or to support "higher and better uses" like prisons and big-box retailers.

survivorship   The probability that a given individual in a population will survive to a particular age; usually presented as a survivorship curve.

sustainability   The ability to meet the current human need for natural resources without compromising the ability of future generations to meet their needs; assumes the environment can function indefinitely without going into a decline from the stresses imposed by human society on natural systems such as fertile soil, water, and air. Also called *environmental sustainability*.

sustainable agriculture   Agricultural methods that maintain soil productivity and a healthy ecological balance while having minimal long-term impacts.

sustainable city   A city with a livable environment, a strong economy, and a social and cultural sense of community; sustainable cities enhance the well-being of current and future generations of urban dwellers.

sustainable consumption   The use of goods and services that satisfy basic human needs and improve the quality of life but that also minimize the use of nonrenewable and renewable resources so they are available for future generations.

sustainable development   Economic development that meets the needs of the present generation without compromising the ability of future generations to meet their own needs. Also called *sustainable economic development*.

sustainable economic development   See *sustainable development*.

sustainable forest management   See *ecologically sustainable forest management*.

sustainable forestry   See *ecologically sustainable forest management*.

sustainable manufacturing   A manufacturing system based on minimizing waste by industry. Sustainable manufacturing involves such practices as reuse, recycling, and source reduction. See *industrial ecosystem*.

sustainable soil use   The wise use of soil resources, without a reduction in soil fertility, so that the soil remains productive for future generations.

sustainable water use   The use of water resources in a fashion that does not harm the essential functions of the hydrologic cycle or the ecosystems on which present and future humans depend.

swamp   A wetland dominated by trees; freshwater swamps are found inland, and saltwater swamps occur along coastal areas. See *mangrove forest*.

symbionts   The partners of a symbiotic relationship.

symbiosis   An intimate relationship between two or more organisms of different species. See *commensalism, mutualism,* and *parasitism*.

synergism   A phenomenon in which two or more pollutants interact in such a way that their combined effects are more severe than the sum of their individual effects. Compare *additivity* and *antagonism*.

synfuel   A liquid or gaseous fuel synthesized from coal or other naturally occurring sources and used in place of oil or natural gas. Also called *synthetic fuel*.

synthetic botanical   Any of a group of human-made insecticides that are produced by chemically modifying the structure of natural botanicals. See *botanical*.

synthetic fuel   See *synfuel*.

Syr Darya Control and Northern Aral Sea Project   World Bank-sponsored project with goals of ecological restoration and commercial fishery recovery of the Aral Sea.

system   A set of components that interact and function as a whole.

taiga   See *boreal forest*.

tailings   Piles of loose rock produced when a mineral such as uranium is mined and processed (extracted and purified from the ore).

TAO/TRITON array   System of monitoring buoys with instruments tracking oceanic and weather data; allow prediction of ENSO events.

tar sand   An underground sand deposit permeated with a thick, asphalt-like oil known as bitumen. The bitumen can be separated from the sand by heating. Also called *bitumen or oil sand*.

Taylor Grazing Act   (1934) Established grazing districts and a permit system to manage livestock grazing on U.S. federally owned rangeland.

temperate deciduous forest   A forest biome that occurs in temperate areas where annual precipitation ranges from about 75 cm to 125 cm.

temperate grassland   A grassland characterized by hot summers, cold winters, and less rainfall than is found in a temperate deciduous forest biome.

temperate rain forest   A coniferous biome characterized by cool weather, dense fog, and high precipitation.

temperature inversion   A deviation from the normal temperature distribution in the atmosphere, resulting in a layer of cold air temporarily trapped near the ground by a warmer, upper layer. Also called *thermal inversion*.

terracing   A soil conservation method that involves building dikes on hilly terrain to produce level areas for agriculture.

terrestrial   Pertaining to the land. Compare *aquatic*.

tertiary consumers   A carnivore that feeds on other carnivores (secondary consumers).

tertiary treatment   An advanced wastewater treatment method that occurs after primary and secondary treatments and may include a variety of biological, chemical, and physical processes. Compare *primary treatment* and *secondary treatment*.

theory   An integrated explanation of numerous hypotheses, each of which is supported by a large body of observations and experiments. Compare *hypothesis*.

thermal expansion   The phenomenon in which water expands in volume as its temperature increases above 2°C (4°F).

thermal inversion   See *temperature inversion*.

thermal pollution   Water pollution that occurs when heated water produced during many industrial processes is released into waterways.

thermal stratification   The marked layering (separation into warm and cold layers) of temperate lakes during the summer. See *thermocline*.

thermocline   A marked and abrupt temperature transition in temperate lakes between warm surface water and cold deeper water. See thermal stratification.

thermodynamics   The branch of physics that deals with energy and its various forms and transformations. See *first law of thermodynamics* and *second law of thermodynamics*.

thermosphere   The outermost layer of the atmosphere. Temperatures are high due to the absorption of X-rays and shortwave ultraviolet radiation.

threatened species   A species in which the population is low enough for it to be at risk of becoming endangered in the foreseeable future throughout all or a significant portion of its range. Compare *endangered species*.

threshold   In toxicology, the maximum dose that has no measurable effect (or the minimum dose that produces a measurable effect).

tidal energy   A form of renewable energy that relies on the ebb and flow of the tides to generate electricity.

topography   A region's surface features, such as the presence or absence of mountains and valleys.

total fertility rate   The average number of children born per woman, given the population's current birth rate.

total resources   The combination of a mineral's reserves and resources. See mineral reserve and mineral resource. Also called *world reserve base*.

toxic waste   A type of hazardous waste harmful to the health of humans or other organisms.

toxicant   A chemical with adverse effects on health.

toxicology   The study of harmful chemicals (toxicants) that have adverse effects on health, as well as ways to counteract their toxicity.

tradable permit   A permit issued for an allowable amount of wastes; the permit allows the owner to either release the wastes or sell the right to emit to another party. Also called *marketable waste-discharge permit*.

trade wind A prevailing tropical wind that blows from the northeast (in the Northern Hemisphere) or from the southeast (in the Southern Hemisphere).

transform plate boundary A fault at which plates slide against one another in opposite directions.

transitional stage Second stage in the demographic transition, characterized by a lowered death rate and high birth rate, leading to rapid population growth.

transpiration The loss of water vapor from the aerial surfaces of plants.

trawl bag A weighted, funnel-shaped net pulled along the bottom of the ocean to catch bottom-feeding fishes and shrimp; as much as 27 metric tons (30 tons) of fishes, shrimp, and other seafood are caught in a single net.

trickle irrigation See *microirrigation*.

tritium An isotope of hydrogen that contains one proton and two neutrons per atom. Compare *deuterium*.

trophic level Each level in a food chain. All producers belong to the first trophic level, all herbivores belong to the second trophic level, and so on.

tropical cyclone Giant, rotating tropical storms with winds of at least 119 km per hour (74 mph); the most powerful tropical cyclones have wind velocities greater than 250 km per hour (155 mph). Called hurricanes in the Atlantic, typhoons in the Pacific, and cyclones in the Indian Ocean.

tropical dry forest A tropical forest where enough precipitation falls to support trees but not enough to support the lush vegetation of a tropical rain forest. Many tropical dry forests occur in areas with pronounced rainy and dry seasons.

tropical rain forest A lush, species-rich forest biome that occurs in tropical areas where the climate is moist throughout the year. Tropical rain forests tend to be characterized by old, infertile soils.

troposphere The atmosphere from Earth's surface to the stratosphere. It is characterized by the presence of clouds, turbulent winds, and decreasing temperature with increasing altitude.

tsunami A giant sea wave caused by an underwater earthquake, volcanic eruption, or landslide.

tundra The treeless biome in the far north that consists of boggy plains covered by lichens and small plants such as mosses. The tundra is characterized by harsh, cold winters and extremely short summers. Also called *arctic tundra*. Compare *alpine tundra*.

turbine A device in which spinning wires within a magnetic field or spinning magnets within a coil of wires generates electricity.

ultraviolet radiation That part of the electromagnetic spectrum with wavelengths just shorter than visible light; a high-energy form of radiation that can be lethal to organisms at higher levels of exposure. Also called *UV radiation*.

UN Convention on the Law of the Sea (UNCLOS) International treaty designating national territorial limits from coastline into the sea, safeguarding marine environments, and protecting freedom of travel on open seas.

unconfined aquifer A groundwater storage area located above a layer of impermeable rock. Water in an unconfined aquifer is replaced by surface water that drains from directly above it. Compare *confined aquifer*.

undernutrition Malnutrition caused when a person receives fewer calories or nutrients in the diet than are needed to maintain a healthy body.

unfunded mandate A federal requirement imposed on state and local governments without providing a way to pay for the cost of compliance.

upwelling A rising ocean current that transports colder, nutrient-laden water to the surface.

urban agglomeration An urbanized core region that consists of several adjacent cities or megacities and their surrounding developed suburbs; an example is the Tokyo-Yokohama-Osaka-Kobe agglomeration in Japan.

urban ecology The study of living organisms and their habitats within urban environments.

urban ecosystem The combination of buildings, lawns, parks, rooftops, culverts, streams, and woods in an urban area, along with the humans and other organisms who reside or rest (such as during migration) in these spaces.

urban growth The rate at which a city's population grows.

urban heat island Local heat buildup in an area of high population density. Compare *dust dome*.

urbanization The process in which people increasingly move from rural areas to densely populated cities; also involves the transformation of rural areas into urban areas.

use zone Urban areas restricted to specific land uses.

utilitarian conservationist A person who values natural resources because of their usefulness for practical purposes but uses them sensibly and carefully.

utility An economic term referring to the benefit that an individual gets from some good or service. Rational actors try to maximize utility.

UV radiation See *ultraviolet radiation*.

values The principles that an individual or society considers important or worthwhile.

vapor extraction A technique of soil remediation that involves injecting or pumping air into soil to remove organic compounds that are volatile (evaporate quickly).

vapor recovery The removal of unburned gasoline vapors from gasoline containers, including underground tanks at gas stations and automobile gas tanks; recovered vapors are either burned or condensed and recovered.

vaporization See *evaporation*.

variable A factor that influences a process. In scientific experiments, all variables are kept constant except for one. See *control group*.

vector (1) An organism that transmits a parasite from one host to another. (2) In genetic engineering, an agent that transfers genetic information from one cell to another.

vitrification A method of safely storing high-level radioactive liquid wastes in solid form, as enormous glass logs.

volatile organic compound (VOC) Carbon-based chemical that evaporates easily at room temperature.

voluntary simplicity A way of life that involves wanting and spending less.

warning coloration Bold, distinctive pattern of colors that tends to indicate toxicity or unpalatability to potential predators.

water cycle See *hydrologic cycle*.

water quality benchmark A chemical concentration in water or sediment, above which animals or humans in the ecosystem may be at health risk.

water pollution Any physical or chemical change in water that adversely affects the health of humans and other organisms.

water table The uppermost level of an unconfined aquifer, below which the ground is saturated with water.

watershed See *drainage basin*.

weather The general condition of the atmosphere (temperature, moisture, cloudiness) at a particular time and place. Compare *climate*.

weathering process A biological, chemical, or physical process that helps form soil from rock; during weathering, the rock is gradually broken down into smaller and smaller particles.

westerly A prevailing wind that blows in the midlatitudes from the southwest (in the Northern Hemisphere) or from the northwest (in the Southern Hemisphere).

Western worldview An understanding of our place in the world based on human superiority and dominance over nature, the unrestricted use of natural resources, and increased economic growth to manage an expanding industrial base.

wet deposition A form of acid deposition in which acid falls to Earth as precipitation. Compare *dry deposition*.

wetlands Lands that are transitional between aquatic and terrestrial ecosystems and are covered with water for at least part of the year.

wide-spectrum pesticide See *broad-spectrum pesticide*.

Wild Bird Conservation Act (1992) Restricts trade of protected birds in the United States; issues limited permits for trade for purposes of scientific research or some captive conservation or breeding.

wilderness Any area that has not been greatly disturbed by human activities and that humans may visit but do not permanently inhabit.

Wilderness Act (1964) Authorized U.S. government to set aside federal land without human habitation or development.

wildfire An uncontrolled fire in an area of combustible vegetation.

wildlife corridor See *habitat corridor*.

wildlife management An applied field of conservation biology that focuses on the continued productivity of plants and animals; includes the regulation of hunting and fishing and the management of food, water, and habitat for wildlife populations.

wildlife ranching An alternative use of land in which private landowners maintain herds of wild animals and earn money from tourists, photographers, and sport hunters as well as from animal hides, leather, and meat.

wind Surface air currents that are caused by the solar warming of air.

wind energy Electric or mechanical energy obtained from surface air currents caused by solar warming of air. See *wind farm*.

wind farm An array of wind turbines for utilizing wind energy by capturing it and converting it to electricity.

World Conservation Strategy Plan designed to conserve biological diversity worldwide, formulated in 1980 by the IUCN, the WWF, and the UN Environment Programme.

world grain carryover stocks See *world grain stocks*.

world grain stocks The amounts of rice, wheat, corn, and other grains stored by governments from previous harvests, as a cushion against poor harvests and rising prices. Also called *grain stockpiles* and *world grain carryover stocks*.

world reserve base See *total resources*.

worldview One of many perspectives based on a collection of our basic values; worldviews help us make sense of the world, understand our place in it, and determine right and wrong behaviors. See *environmental worldview*.

yield In agriculture, the amount of a food crop produced per unit of land.

Yucca Mountain A mountain in southern Nevada that has been proposed as a site for long-term storage of high-level nuclear waste.

zero-net-energy A building or device that, over the course of an extended period (typically a year), uses no net energy.

zero population growth When the birth rate equals the death rate. A population with zero population growth remains the same size.

zooplankton The nonphotosynthetic organisms— tiny shrimp, larvae, and other drifting animals— that are part of the plankton. See *plankton*. Compare *phytoplankton*.

zooxanthellae Algae that live inside coral animals and have a mutualistic relationship with them.

Index